TM

Selected papers on

GAUGE THEORY OF WEAK AND ELECTROMAGNETIC INTERACTIONS

Edited by C. H. Lai

D1447518

World Scientific Singapore
1981

World Scientific Publishing Co. Pte Ltd
P. O. Box 128
Farrer Road
Singapore 9128

This volume consists of selected papers on the Gauge Theory of Weak and Electromagnetic Interactions that have appeared in *Annals of Physics, IL Nuovo Cimento, Nuclear Physics B, Physics Letters B, Physical Review, Physical Review Letters and Theoretical and Mathematical Physics.* Included also are the 1979 Physics Nobel Lectures and an article by Prof. A Salam from the Proceedings of the Eighth Nobel Symposium on Elementary Particles Theory published by Almqvist & Wiksell Forlag AB. The editor and publisher are indebted to the original authors, journals, Nobel Foundation and other publishers for their assistance and permission to reproduce these papers.

ISBN 9971–83–023–x
ISBN 9971–83–022–1 pbk

CONTENTS

Page

PREFACE

THIS VOLUME of selected papers is not meant to serve as an introduction to the gauge theory of weak and electromagnetic interactions. Rather, the purpose of its publication is to provide physicists new to the field, particularly those in developing countries, with a compilation of some of the original literature, which are scattered in various journals and over many years. For the seasoned worker, this collection could perhaps still serve as a handy reference. The importance of the concept of gauge theories can hardly be exaggerated these days, and it is hoped that this volume will be consulted not only for the technical details but also to gain some flavor of the logical development of many of the ideas and techniques involved.

Many good summer school lectures and review articles exist and should be consulted for a coherent introduction. We find the following particularly useful:

C. Quigg, "Introduction to Gauge Theories of the Strong, Weak and Electromagnetic Interactions", Lectures given at the NATO Advanced Study Institute, "Techniques and Concepts of High Energy Physics", held at St. Croix, U.S. Virgin Islands, July 2–13, 1980. (Available as FERMILAB-Conf-80/64 THY)

J. C. Taylor, "Gauge Theories of Weak Interactions", Cambridge University Press, 1976.

E. S. Abers and B. W. Lee, "Gauge Theories", Physics Reports 9C, 1 (1973).

Preface

The Nobel lectures by the 1979 Laureates, Professors S. L. Glashow, A. Salam and S. Weinberg, are reprinted here: We feel that they provide interesting (if highly personal) accounts of the evolution of our present understanding of the electroweak interactions, as well as some ideas on the current outlook.

This selection of papers is not complete nor definitive by any standard and hence is far from giving proper credit to all contributions. Limitation of space, considerations as regarding the usefulness of the volume, and our limited expertise in the field have been deciding factors in the compilation. We apologize for any deserving papers that we have wittingly or unwittingly omitted. Two remarks about the sections on quantization and renormalization of gauge theories are perhaps appropriate here. These topics are generally considered to be the more formal aspects of gauge theories, and many of the papers are fairly difficult to understand. Thus forewarned, the readers should not feel discouraged if they find the papers in these sections rough going. Also, the papers on these topics that are included here are only representative fragments of the existing literature – many other relevant and influential papers have been dropped from our preliminary list at the suggestions of many physicists we consulted. It is likely that these additional papers, along with papers on the gauge theory of the strong interactions, grand unified theories and related topics, will appear in a sequel to this volume.

We would like to thank Professors C. N. Yang, A. Salam, M. Jacob and A. Zee for their interests and suggestions, and Professor R. N. Mohapatra, consulting editor of this volume, for his many contributions. Dr. K. K. Phua initiated this project and provided much encouragement and the necessary pressure during the final stages of the compilation. We are also indebted to the various journals and the Nobel Foundation for their permissions to reproduce the papers in this volume.

C. H. LAI

November 1980

1979 NOBEL LECTURES

Conceptual foundations of the unified theory of weak and electromagnetic interactions*†

Steven Weinberg

Lyman Laboratory of Physics, Harvard University and Harvard–Smithsonian Center for Astrophysics, Cambridge, Massachusetts 02138 U.S.A.

Our job in physics is to see things simply, to understand a great many complicated phenomena in a unified way, in terms of a few simple principles. At times, our efforts are illuminated by a brilliant experiment, such as the 1973 discovery of neutral current neutrino reactions. But even in the dark times between experimental breakthroughs, there always continues a steady evolution of theoretical ideas, leading almost imperceptibly to changes in previous beliefs. In this talk, I want to discuss the development of two lines of thought in theoretical physics. One of them is the slow growth in our understanding of symmetry, and in particular, broken or hidden symmetry. The other is the old struggle to come to terms with the infinities in quantum field theories. To a remarkable degree, our present detailed theories of elementary particle interactions can be understood deductively, as consequences of symmetry principles and of a principle of renormalizability which is invoked to deal with the infinities. I will also briefly describe how the convergence of these lines of thought led to my own work on the unification of weak and electromagnetic interactions. For the most part, my talk will center on my own gradual education in these matters, because that is one subject on which I can speak with some confidence. With rather less confidence, I will also try to look ahead, and suggest what role these lines of thought may play in the physics of the future.

Symmetry principles made their appearance in twentieth century physics in 1905 with Einstein's identification of the invariance group of space and time. With this as a precedent, symmetries took on a character in physicists' minds as *a priori* principles of universal validity, expressions of the simplicity of nature at its deepest level. So it was painfully difficult in the 1930s to realize that there are internal symmetries, such as isospin conservation,[1] having nothing to do with space and time, symmetries which are far from self-evident, and that only govern what are now called the strong interactions. The 1950s saw the discovery of another internal symmetry—the conservation of strangeness[2]—which is not obeyed by the weak interactions, and even one of the supposedly sacred symmetries of space-time—parity—was also found to be violated by weak interactions.[3] Instead of moving toward unity, physicists were learning that different interactions are apparently governed by quite different symmetries. Matters became yet more confusing with the recognition in the early 1960s of a symmetry group—the "eightfold way"—which is not even an exact symmetry of the strong interactions.[4]

These are all "global" symmetries, for which the symmetry transformations do not depend on position in space and time. It had been recognized[5] in the 1920s that quantum electrodynamics has another symmetry of a far more powerful kind, a "local" symmetry under transformations in which the electron field suffers a phase change that can vary freely from point to point in space-time, and the electromagnetic vector potential undergoes a corresponding gauge transformation. Today this would be called a U(1) gauge symmetry, because a simple phase change can be thought of as multiplication by a 1×1 unitary matrix. The extension to more complicated groups was made by Yang and Mills[6] in 1954 in a seminal paper in which they showed how to construct an SU(2) gauge theory of strong interactions. (The name "SU(2)" means that the group of symmetry transformations consists of 2×2 unitary matrices that are "special," in that they have determinant unity.) But here again it seemed that the symmetry, if real at all, would have to be approximate, because at least on a naive level gauge invariance requires that vector bosons like the photon would have to be massless, and it seemed obvious that the strong interactions are not mediated by massless particles. The old question remained: if symmetry principles are an expression of the simplicity of nature at its deepest level, then how can there be such a thing as an approximate symmetry? Is nature only approximately simple?

Sometime in 1960 or early 1961, I learned of an idea which had originated earlier in solid state physics and had been brought into particle physics by those like Heisenberg, Nambu, and Goldstone, who had worked in both areas. It was the idea of "broken symmetry," that the Hamiltonian and commutation relations of a quantum theory could possess an exact symmetry, and that the physical states might nevertheless not provide neat representations of the symmetry. In particular, a symmetry of the Hamiltonian might turn out to be not a symmetry of the vacuum.

As theorists sometimes do, I fell in love with this idea. But as often happens with love affairs, at first I was rather confused about its implications. I thought (as it turned out, wrongly) that the approximate symmetries—parity, isospin, strangeness, the eightfold way—might really be exact *a priori* symmetry principles, and that the observed violations of these symmetries might somehow be brought about by spontaneous symmetry breaking. It was therefore rather disturbing

*This lecture was delivered December 8, 1979, on the occasion of the presentation of the 1979 Nobel Prizes in Physics.

for me to hear of a result of Goldstone,[7] that in at least one simple case the spontaneous breakdown of a continuous symmetry like isospin would necessarily entail the existence of a massless spin zero particle—what would today be called a "Goldstone boson." It seemed obvious that there could not exist any new type of massless particle of this sort which would not already have been discovered.

I had long discussions of this problem with Goldstone at Madison in the summer of 1961, and then with Salam while I was his guest at Imperial College in 1961–62. The three of us soon were able to show that Goldstone bosons must in fact occur whenever a symmetry like isospin or strangeness is spontaneously broken, and that their masses then remain zero to all orders of perturbation theory. I remember being so discouraged by these zero masses that when we wrote our joint paper on the subject,[8] I added an epigraph to the paper to underscore the futility of supposing that anything could be explained in terms of a noninvariant vacuum state: it was Lear's retort to Cordelia, "Nothing will come of nothing: speak again." Of course, *The Physical Review* protected the purity of the physics literature, and removed the quote. Considering the future of the noninvariant vacuum in theoretical physics, it was just as well.

There was actually an exception to this proof, pointed out soon afterwards by Higgs, Kibble, and others.[9] They showed that if the broken symmetry is a local, gauge symmetry, like electromagnetic gauge invariance, then although the Goldstone bosons exist formally, and are in some sense real, they can be eliminated by a gauge transformation, so that they do not appear as physical particles. The missing Goldstone bosons appear instead as helicity zero states of the vector particles, which thereby acquire a mass.

I think that at the time physicists who heard about this exception generally regarded it as a technicality. This may have been because of a new development in theoretical physics, which suddenly seemed to change the role of Goldstone bosons from that of unwanted intruders to that of welcome friends.

In 1964 Adler and Weisberger[10] independently derived sum rules which gave the ratio g_A/g_V of axial-vector to vector coupling constants in beta decay in terms of pion–nucleon cross sections. One way of looking at their calculation (perhaps the most common way at the time) was as an analog to the old dipole sum rule in atomic physics: a complete set of hadronic states is inserted in the commutation relations of the axial vector currents. This is the approach memorialized in the name of "current algebra."[11] But there was another way of looking at the Adler–Weisberger sum rule. One could suppose that the strong interactions have an approximate symmetry, based on the group $SU(2)\times SU(2)$, and that this symmetry is spontaneously broken, giving rise among other things to the nucleon masses. The pion is then identified as (approximately) a Goldstone boson, with small but nonzero mass, an idea that goes back to Nambu.[12] Although the $SU(2)\times SU(2)$ symmetry is spontaneously broken, it still has a great deal of predictive power, but its predictions take the form of approximate formulas, which give the matrix elements for low energy pionic reactions. In this approach, the Adler–Weisberger sum rule is obtained by using the predicted pion nucleon scattering lengths in conjunction with a well-known sum rule,[13] which years earlier had been derived from the dispersion relations for pion-nucleon scattering.

In these calculations one is really using not only the fact that the strong interactions have a spontaneously broken approximate $SU(2)\times SU(2)$ symmetry, but also that the currents of this symmetry group are, up to an overall constant, to be identified with the vector and axial vector currents of beta decay. (With this assumption g_A/g_V gets into the picture through the Goldberger-Treiman relation,[14] which gives g_A/g_V in terms of the pion decay constant and the pion nucleon coupling.) Here, in this relation between the currents of the symmetries of the strong interactions and the physical currents of beta decay, there was a tantalizing hint of a deep connection between the weak interactions and the strong interactions. But this connection was not really understood for almost a decade.

I spent the years 1965–67 happily developing the implications of spontaneous symmetry breaking for the strong interactions.[15] It was this work that led to my 1967 paper on weak and electromagnetic unification. But before I come to that I have to go back in history and pick up one other line of thought, having to do with the problem of infinities in quantum field theory.

I believe that it was Oppenheimer and Waller in 1930[16] who independently first noted that quantum field theory when pushed beyond the lowest approximation yields ultraviolet divergent results for radiative self-energies. Professor Waller told me last night that when he described this result to Pauli, Pauli did not believe it. It must have seemed that these infinities would be a disaster for the quantum field theory that had just been developed by Heisenberg and Pauli in 1929–30. And indeed, these infinities did lead to a sense of discouragement about quantum field theory, and many attempts were made in the 1930s and early 1940s to find alternatives. The problem was solved (at least for quantum electrodynamics) after the war, by Feynman, Schwinger, and Tomonaga,[17] and Dyson.[19] It was found that all infinities disappear if one identifies the observed finite values of the electron mass and charge, not with the parameters m and e appearing in the Lagrangian, but with the electron mass and charge that are *calculated* from m and e, when one takes into account the fact that the electron and photon are always surrounded with clouds of virtual photons and electron–positron pairs.[18] Suddenly all sorts of calculations became possible, and gave results in spectacular agreement with experiment.

But even after this success, opinions differed as to the significance of the ultraviolet divergences in quantum field theory. Many thought—and some still do think—that what had been done was just to sweep the real problems under the rug. And it soon became clear that there was only a limited class of so-called "renormalizable" theories in which the infinities could be eliminated by absorbing them into a redefinition, or a "renormalization," of a finite number of physical parameters. (Roughly speaking, in renormalizable theories no coupling constants can have the dimensions of negative powers of mass. But every time we add a field or a space-time derivative to an interaction, we reduce the dimensionality of the associated coupling constant. So only a few simple types of interaction can be renormalizable.) In particular, the existing Fermi theory of weak interactions clearly was not renormalizable. (The Fermi coupling constant has the dimensions of $[\text{mass}]^{-2}$.) The sense of discouragement about quantum field theory persisted into the 1950s and 1960s.

I learned about renormalization theory as a graduate student, mostly by reading Dyson's papers.[19] From the beginning it seemed to me to be a wonderful thing that very few quantum field theories are renormalizable. Limitations of this sort are, after all, what we most *want;* not mathematical methods which can make sense out of an infinite variety of physically irrelevant theories, but methods which carry constraints, because these constraints may point the way toward the one true theory. In particular, I was very impressed by the fact that quantum electrodynamics could in a sense be *derived* from symmetry principles and the constraint of renormalizability; the only Lorentz invariant and gauge invariant renormalizable Lagrangian for photons and electrons is precisely the original Dirac Langrangian of QED. Of course, that is not the way Dirac came to his theory. He had the benefit of the information gleaned in centuries of experimentation on electromagnetism, and in order to fix the final form of his theory he relied on ideas of simplicity (specifically, on what is sometimes called minimal electromagnetic coupling). But we have to look ahead, to try to make theories of phenomena which have not been so well studied experimentally, and we may not be able to trust purely formal ideas of simplicity. I thought that renormalizability might be the key criterion, which also in a more general context would impose a precise kind of simplicity on our theories and help us to pick out the one true physical theory out of the infinite variety of conceivable quantum field theories. As I will explain later, I would say this a bit differently today, but I am more convinced than ever that the use of renormalizability as a constraint on our theories of the observed interactions is a good strategy. Filled with enthusiasm for renormalization theory, I wrote my Ph.D. thesis under Sam Treiman in 1957 on the use of a limited version of renormalizability to set constraints on the weak interactions,[20] and a little later I worked out a rather tough little theorem[21] which completed the proof by Dyson[19] and Salam[22] that ultraviolet divergences really do cancel out to all orders in nominally renormalizable theories. But none of this seemed to help with the important problem, of how to make a renormalizable theory of weak interactions.

Now, back to 1967. I had been considering the implications of the broken SU(2)× SU(2) symmetry of the strong interactions, and I thought of trying out the idea that perhaps the SU(2)× SU(2) symmetry was a "local," not merely a "global," symmetry. That is, the strong interactions might be described by something like a Yang–Mills theory, but in addition to the vector ρ mesons of the Yang–Mills theory, there would also be axial vector $A1$ mesons. To give the ρ meson a mass, it was necessary to insert a common ρ and $A1$ mass term in the Lagrangian, and the spontaneous breakdown of the SU(2)× SU(2) symmetry would then split the ρ and $A1$ by something like the Higgs mechanism, but since the theory would not be gauge invariant the pions would remain as physical Goldstone bosons. This theory gave an intriguing result, that the $A1/\rho$ mass ratio should be $\sqrt{2}$, and in trying to understand this result without relying on perturbation theory, I discovered certain sum rules, the "spectral function sum rules,"[23] which turned out to have a variety of other uses. But the SU(2)× SU(2) theory was not gauge invariant, and hence it could not be renormalizable,[24] so I was not too enthusiastic about it.[25] Of course, if I did not insert the ρ-$A1$ mass term in the Lagrangian, then the theory would be gauge invariant and renormalizable, and the $A1$ would be massive. But then there would be no pions and the ρ mesons would be massless, in obvious contradiction (to say the least) with observation.

At some point in the fall of 1967, I think while driving to my office at MIT, it occurred to me that I had been applying the right ideas to the wrong problem. It is not the ρ meson that is massless: it is the photon. And its partner is not the $A1$, but the massive intermediate bosons, which since the time of Yukawa had been suspected to be the mediators of the weak interactions. The weak and electromagnetic interactions could then be described[26] in a unified way in terms of an exact but spontaneously broken gauge symmetry. [Of course, not necessarily SU(2)× SU(2).] And this theory would be renormalizable like quantum electrodynamics because it is gauge invariant like quantum electrodynamics.

It was not difficult to develop a concrete model which embodied these ideas. I had little confidence then in my understanding of strong interactions, so I decided to concentrate on leptons. There are two left-handed electron-type leptons, the ν_{eL} and e_L, and one right-handed electron-type lepton, the e_R, so I started with the group U(2)× U(1): all unitary 2×2 matrices acting on the left-handed e-type leptons, together with all unitary 1×1 matrices acting on the right-handed e-type lepton. Breaking up U(2) into unimodular transformations and phase transformations, one could say that the group was SU(2)× U(1)× U(1). But then one of the U(1)'s could be identified with ordinary lepton number, and since lepton number appears to be conserved and there is no massless vector particle coupled to it, I decided to exclude it from the group. This left the four-parameter group SU(2)× U(1). The spontaneous breakdown of SU(2)× U(1) to the U(1) of ordinary electromagnetic gauge invariance would give masses to three of the four vector gauge bosons: the charged bosons $W^{\pm}$, and a neutral boson that I called the Z^0. The fourth boson would automatically remain massless, and could be identified as the photon. Knowing the strength of the ordinary charged current weak interactions like beta decay which are mediated by $W^{\pm}$, the mass of the $W^{\pm}$ was then determined as about 40 GeV/sinθ, where θ is the γ-Z^0 mixing angle.

To go further, one had to make some hypothesis about the mechanism for the breakdown of SU(2)× U(1). The only kind of field in a renormalizable SU(2)× U(1) theory whose vacuum expectation values could give the electron a mass is a spin zero SU(2) doublet (ϕ^+, ϕ^0), so for simplicity I assumed that these were the only scalar fields in the theory. The mass of the Z^0 was then determined as about 80 GeV/sin2θ. This fixed the strength of the neutral current weak interactions. Indeed, just as in QED, once one decides on the menu of fields in the theory, all details of the theory are completely determined by symmetry principles and renormalizability, with just a few free parameters: the lepton charge and masses, the Fermi coupling constant of beta decay, the mixing angle θ, and the mass of the scalar particle. (It was of crucial importance to impose the constraint of renormalizability; otherwise weak interactions would receive contributions from SU(2)× U(1)-invariant four-fermion couplings as well as from vector boson exchange, and the theory would lose most of its predictive power.) The naturalness of the whole theory is well demonstrated by the fact that much the same theory was independently developed[27] by Salam in 1968.

The next question now was renormalizability. The Feynman rules for Yang–Mills theories with unbroken gauge symmetries had been worked out[28] by deWitt, Faddeev, and Popov and others, and it was known that such theories are renormalizable. But in 1967 I did not know how to prove that this renormalizability was not spoiled by the spontaneous symmetry breaking. I worked on the problem on and off for several years, partly in collaboration with students,[29] but I made little progress. With hindsight, my main difficulty was that in quantizing the vector fields I adopted a gauge now known as the unitarity gauge[30]: this gauge has several wonderful advantages, it exhibits the true particle spectrum of the theory, but it has the disadvantage of making renormalizability totally obscure.

Finally, in 1971 't Hooft[31] showed in a beautiful paper how the problem could be solved. He invented a gauge, like the "Feynman gauge" in QED, in which the Feynman rules manifestly lead to only a finite number of types of ultraviolet divergence. It was also necessary to show that these infinities satisfied essentially the same constraints as the Lagrangian itself, so that they could be absorbed into a redefinition of the parameters of the theory. (This was plausible, but not easy to prove, because a gauge invariant theory can be quantized only after one has picked a specific gauge, so it is not obvious that the ultraviolet divergences satisfy the same gauge invariance constraints as the Lagrangian itself.) The proof was subsequently completed[32] by Lee and Zinn–Justin and by 't Hooft and Veltman. More recently, Becchi, Rouet, and Stora[33] have invented an ingenious method for carrying out this sort of proof, by using a global supersymmetry of gauge theories which is preserved even when we choose a specific gauge.

I have to admit that when I first saw 't Hooft's paper in 1971, I was not convinced that he had found the way to prove renormalizability. The trouble was not with 't Hooft, but with me: I was simply not familiar enough with the path integral formalism on which 't Hooft's work was based, and I wanted to see a derivation of the Feynman rules in 't Hooft's gauge from canonical quantization. That was soon supplied (for a limited class of gauge theories) by a paper of Ben Lee,[34] and after Lee's paper I was ready to regard the renormalizability of the unified theory as essentially proved.

By this time, many theoretical physicists were becoming convinced of the general approach that Salam and I had adopted: that is, the weak and electromagnetic interactions are governed by some group of exact local gauge symmetries; this group is spontaneously broken to U(1), giving mass to all the vector bosons except the photon; and the theory is renormalizable. What was not so clear was that our specific simple model was the one chosen by nature. That, of course, was a matter for experiment to decide.

It was obvious even back in 1967 that the best way to test the theory would be by searching for neutral current weak interactions, mediated by the neutral intermediate vector boson, the Z^0. Of course, the possibility of neutral currents was nothing new. There had been speculations[35] about possible neutral currents as far back as 1937 by Gamow and Teller, Kemmer, and Wentzel, and again in 1958 by Bludman and Leite–Lopes. Attempts at a unified weak and electromagnetic theory had been made[36] by Glashow and Salam and Ward in the early 1960s, and these had neutral currents with many of the features that Salam and I encountered in developing the 1967–68 theory. But since one of the predictions of our theory was a value for the mass of the Z^0, it made a definite prediction of the strength of the neutral currents. More important, now we had a comprehensive quantum field theory of the weak and electromagnetic interactions that was physically and mathematically satisfactory in the same sense as quantum electrodynamics—a theory that treated photons and intermediate vector bosons on the same footing, that was based on an exact symmetry principle, and that allowed one to carry calculations to any desired degree of accuracy. To test this theory, it had now become urgent to settle the question of the existence of the neutral currents.

Late in 1971, I carried out a study of the experimental possibilities.[37] The results were striking. Previous experiments had set upper bounds on the rates of neutral current processes which were rather low, and many people had received the impression that neutral currents were pretty well ruled out, but I found that in fact the 1967–68 theory *predicted* quite low rates, low enough in fact to have escaped clear detection up to that time. For instance, experiments[38] a few years earlier had found an upper bound of 0.12 ± 0.06 on the ratio of a neutral current process, the elastic scattering of muon neutrinos by protons, to the corresponding charged current process, in which a muon is produced. I found a predicted ratio of 0.15 to 0.25, depending on the value of the Z^0-γ mixing angle θ. So there was every reason to look a little harder.

As everyone knows, neutral currents were finally discovered[39] in 1973. There followed years of careful experimental study on the detailed properties of the neutral currents. It would take me too far from my subject to survey these experiments,[40] so I will just say that they have confirmed the 1967–68 theory with steadily improving precision for neutrino–nucleon and neutrino–electron neutral current reactions, and since the re-from the spontaneous breakdown of the grand unified gauge group. There is nothing impossible in this, but I have not been able to think of any reason why it should happen. (The problem may be related to the old mystery of why quantum corrections do not produce an enormous cosmological constant; in both cases, one is concerned with an anomalously small "super-renormalizable" term in the effective Lagrangian which has to be adjusted to be zero. In the case of the cosmological constant, the adjustment must be precise to some fifty decimal places.) With elementary scalars of small or zero mass, enormous ratios of symmetry breaking scales can arise quite naturally.[59] On the other hand, if there are no elementary scalars which escape getting superlarge masses from the breakdown of the grand unified gauge group, then as I have already mentioned, there must be extra strong forces to bind the composite Goldstone and Higgs bosons that are associated with the spontaneous breakdown of $SU(2)\times U(1)$. Such forces can occur rather naturally in grand unified theories. To take one example, suppose that the grand gauge group breaks, not into $SU(3) \times SU(2)\times U(1)$, but into $SU(4)\times SU(3)\times SU(2)\times U(1)$. Since $SU(4)$ is a bigger group than $SU(3)$, its coupling constant rises with decreasing energy more rapidly than the QCD coupling, so the $SU(4)$ force becomes strong at a much higher energy than the few hundred MeV at which the QCD force becomes strong. Ordinary quarks and leptons would be neutral under $SU(4)$, so they would not feel this force, but other fermions might carry $SU(4)$ quantum numbers, and so get rather large masses. One can even imagine a sequence of increasingly large subgroups of the grand gauge group, which would fill in the vast energy range up to 10^{15} or 10^{19} GeV with particle masses that are produced by these successively strong interactions.

If there are elementary scalars whose vacuum expectation values are responsible for the masses of ordinary quarks and leptons, then these masses can be affected in order α by radiative corrections involving the superheavy vector bosons of the grand gauge group, and it will probably be impossible to explain the value of quantities like m_e/m_μ without a complete grand unified theory. On the other hand, if there are no such elementary scalars, then almost all the details of the grand unified theory are forgotten by the effective field theory that describes physics at ordinary energies, and it ought to be possible to calculate quark and lepton masses purely in terms of processes at accessible energies. Unfortunately, no one so far has been able to see how in this way anything resembling the observed pattern of masses could arise.[60]

Putting aside all these uncertainties, suppose that there is a truly fundamental theory, characterized by an energy scale of order 10^{16} to 10^{19} GeV, at which strong, electroweak, and gravitational interactions are all united. It might be a conventional renormalizable quantum field theory, but at the moment, if we include gravity, we do not see how this is possible. (I leave the topic of supersymmetry and supergravity for Professor Salam's talk.) But if it is not renormalizable, what then determines the infinite set of coupling constants that are needed to absorb all the ultraviolet divergences of the theory?

I think the answer must lie in the fact that the quantum field theory, which was born just fifty years ago from the marriage of quantum mechanics with relativity, is a beautiful but not a very robust child. As Landau and Källén recognized long ago, quantum field theory at superhigh energies is susceptible to all sorts of diseases—tachyons, ghosts, etc.—and it needs special medicine to survive. One way that a quantum field theory can avoid these diseases is to be renormalizable and asymptotically free, but there are other possibilities. For instance, even an infinite set of coupling constants may approach a nonzero fixed point as the energy at which they are measured goes to infinity. However, to require this behavior generally imposes so many constraints on the couplings that there are only a finite number of free parameters left[61]—just as for theories that are renormalizable in the usual sense. Thus, one way or another, I think that quantum field theory is going to go on being very stubborn, refusing to allow us to describe all but a small number of possible worlds, among which, we hope, is ours.

I suppose that I tend to be optimistic about the future of physics. And nothing makes me more optimistic than the discovery of broken symmetries. In the seventh book of *The Republic*, Plato describes prisoners who are chained in a cave and can see only shadows that things outside cast on the cave wall. When released from the cave at first their eyes hurt, and for a while they think that the shadows they saw in the cave are more real than the objects they now see. But eventually their vision clears, and they can understand how beautiful the real world is. We are in such a cave, imprisoned by the limitations on the sorts of experiments we can do. In particular, we can study matter only at relatively low temperatures, where symmetries are likely to be spontaneously broken, so that nature does not appear very simple or unified. We have not been able to get out of this cave, but by looking long and hard at the shadows on the cave wall, we can at least make out the shapes of symmetries, which though broken, are exact principles governing all phenomena, expressions of the beauty of the world outside.

It has only been possible here to give references to a very small part of the literature on the subjects discussed in this talk. Additional references can be found in the following reviews: E. S. Abers and B. W. Lee, "Gauge Theories" (Phys. Rep. C **9**, No. 1, 1973); W. Marciano and H. Pagels, "Quantum Chromodynamics" (Phys. Rep. C **36**, No. 3, 1978); J. C. Taylor, *Gauge Theories of Weak Interactions* (Cambridge University, 1976).

REFERENCES

1. Tuve, M. A., Heydenberg, N. and Hafstad, L. R. Phys. Rev. *50,* 806 (1936); Breit, G., Condon, E. V. and Present, R. D. Phys. Rev. *50,* 825 (1936); Breit, G. and Feenberg, E. Phys. Rev. *50,* 850 (1936).
2. Gell-Mann, M. Phys. Rev. *92,* 833 (1953); Nakano. T. and Nishijima, K. Prog. Theor. Phys. *10,* 581 (1955).
3. Lee, T. D. and Yang, C. N. Phys. Rev. *104,* 254 (1956); Wu. C. S. et. al. Phys. Rev. *105,* 1413 (1957); Garwin, R., Lederman, L. and Weinrich, M. Phys. Rev. *105,* 1415 (1957); Friedman, J. I. and Telegdi V. L. Phys. Rev. *105,* 1681 (1957).
4. Gell-Mann, M. Cal. Tech. Synchotron Laboratory Report CTSL-20 (1961), unpublished; Ne'eman, Y. Nucl. Phys. *26,* 222 (1961).
5. Fock, V. Z. f. Physik *39,* 226 (1927); Weyl, H. Z. f. Physik *56,* 330 (1929). The name "gauge invariance" is based on an analogy with the earlier speculations of Weyl, H. in *Raum, Zeit, Materie,* 3rd edn, (Springer, 1920). Also see London, F. Z. f. Physik *42,* 375 (1927). (This history has been reviewed by Yang, C. N. in a talk at City College, (1977).)
6. Yang, C. N. and Mills, R. L. Phys. Rev. *96,* 191 (1954).
7. Goldstone, J. Nuovo Cimento *19,* 154 (1961).
8. Goldstone, J., Salam, A. and Weinberg, S. Phys. Rev. *127,* 965 (1962).
9. Higgs, P. W. Phys. Lett. *12,* 132 (1964); *13,* 508 (1964); Phys. Rev. *145,* 1156 (1966); Kibble, T. W. B. Phys. Rev. *155,* 1554 (1967); Guralnik, G. S., Hagen, C. R. and Kibble, T. W. B. Phys. Rev. Lett. *13,* 585 (1964); Englert, F. and Brout, R. Phys. Rev. Lett. *13,* 321 (1964); Also see Anderson, P. W. Phys. Rev. *130,* 439 (1963).
10. Adler, S. L. Phys. Rev. Lett. *14,* 1051 (1965); Phys. Rev. *140,* B736 (1965); Weisberger, W. I. Phys. Rev. Lett. *14,* 1047 (1965); Phys. Rev. *143,* 1302 (1966).
11. Gell-Mann, M. Physics *1,* 63 (1964).
12. Nambu, Y. and Jona-Lasinio, G. Phys. Rev. *122,* 345 (1961); *124,* 246 (1961); Nambu, Y. and Lurie, D. Phys. Rev. *125,* 1429 (1962); Nambu. Y. and Shrauner, E. Phys. Rev. *128,* 862 (1962); Also see Gell-Mann, M. and Levy, M., Nuovo Cimento *16,* 705 (1960).
13. Goldberger, M. L., Miyazawa, H. and Oehme, R. Phys. Rev. *99,* 986 (1955).
14. Goldberger, M. L., and Treiman, S. B. Phys. Rev. *111,* 354 (1958).
15. Weinberg, S. Phys. Rev. Lett. *16,* 879 (1966); *17,* 336 (1966); *17,* 616 (1966); *18,* 188 (1967); Phys. Rev. *166,* 1568 (1967).
16. Oppenheimer, J. R. Phys. Rev. *35,* 461 (1930); Waller, I. Z. Phys. *59,* 168 (1930); ibid., *62,* 673 (1930).
17. Feynman, R. P. Rev. Mod. Phys. *20,* 367 (1948); Phys. Rev. *74,* 939, 1430 (1948); *76,* 749, 769 (1949); *80,* 440 (1950); Schwinger, J. Phys. Rev. *73,* 146 (1948); *74,* 1439 (1948); *75,* 651 (1949); *76,* 790 (1949); *82,* 664, 914 (1951); *91,* 713 (1953); Proc. Nat. Acad. Sci. *37,* 452 (1951); Tomonaga, S. Progr. Theor. Phys. (Japan) *1,* 27 (1946); Koba, Z., Tati, T. and Tomonaga, S. ibid. *2,* 101 (1947); Kanazawa, S. and Tomonaga, S. ibid. *3,* 276 (1948); Koba, Z. and Tomonaga, S. ibid *3,* 290 (1948).
18. There had been earlier suggestions that infinities could be eliminated from quantum field theories in this way, by Weisskopf, V. F. Kong. Dansk. Vid. Sel. Mat.-Fys. Medd. *15* (6) 1936, especially p. 34 and pp. 5–6; Kramers, H. (unpublished).
19. Dyson, F. J. Phys. Rev. *75,* 486, 1736 (1949).
20. Weinberg, S. Phys. Rev. *106,* 1301 (1957).
21. Weinberg, S. Phys. Rev. *118,* 838 (1960).
22. Salam, A. Phys. Rev. *82,* 217 (1951); *84,* 426 (1951).
23. Weinberg, S. Phys. Rev. Lett. *18,* 507 (1967).
24. For the non-renormalizability of theories with intrinsically broken gauge symmetries, see Komar, A. and Salam, A. Nucl. Phys. *21,* 624 (1960); Umezawa, H. and Kamefuchi, S. Nucl. Phys. *23,* 399 (1961); Kamefuchi, S., O'Raifeartaigh, L. and Salam, A. Nucl. Phys. *28,* 529 (1961); Salam, A. Phys. Rev. *127,* 331 (1962); Veltman, M. Nucl. Phys. *B7,* 637 (1968); *B21,* 288 (1970); Boulware, D. Ann. Phys. (N. Y.) *56,* 140 (1970).
25. This work was briefly reported in reference 23, footnote 7.
26. Weinberg, S. Phys. Rev. Lett. *19,* 1264 (1967).
27. Salam, A. In *Elementary Particle Physics* (Nobel Symposium No. 8), ed. by Svartholm, N. (Almqvist and Wiksell, Stockholm, 1968), p. 367.
28. deWitt, B. Phys. Rev. Lett. *12,* 742 (1964); Phys. Rev. *162,* 1195 (1967); Faddeev L. D., and Popov, V. N. Phys. Lett. *B25,* 29 (1967); Also see Feynman, R. P. Acta. Phys. Pol. *24,* 697 (1963); Mandelstam, S. Phys. Rev. *175,* 1580, 1604 (1968).
29. See Stuller, L. M. I. T., Thesis, Ph. D. (1971), unpublished.
30. My work with the unitarity gauge was reported in Weinberg, S. Phys. Rev. Lett. *27,* 1688 (1971), and described in more detail in Weinberg, S. Phys. Rev. *D7,* 1068 (1973).
31. 't Hooft, G. Nucl. Phys. *B35,* 167 (1971).
32. Lee, B. W. and Zinn-Justin, J. Phys. Rev. *D5,* 3121, 3137, 3155 (1972); 't Hooft, G. and Veltman, M. Nucl. Phys. *B44,* 189 (1972); *B50,* 318 (1972). There still remained the problem of possible Adler–Bell–Jackiw anomalies, but these nicely cancelled; see D. J. Gross and R. Jackiw, Phys. Rev. *D6,* 477 (1972) and C. Bouchiat, J. Iliopoulos, and Ph. Meyer, Phys. Lett. *38B,* 519 (1972).
33. Becchi, C., Rouet, A. and Stora R. Comm. Math. Phys. *42,* 127 (1975).
34. Lee, B. W. Phys. Rev. *D5,* 823 (1972).
35. Gamow, G. and Teller, E. Phys. Rev. *51,* 288 (1937); Kemmer, N. Phys. Rev. *52,* 906 (1937); Wentzel, G. Helv. Phys. Acta. *10,* 108 (1937); Bludman, S. Nuovo Cimento *9,* 433 (1958); Leite-Lopes, J. Nucl. Phys. *8,* 234 (1958).
36. Glashow, S. L. Nucl. Phys. *22,* 519 (1961); Salam, A. and Ward, J. C. Phys. Lett. *13,* 168 (1964).
37. Weinberg, S. Phys. Rev. *5,* 1412 (1972).
38. Cundy, D. C. et. al., Phys. Lett. *31B,* 478 (1970).
39. The first published discovery of neutral currents was at the Gargamelle Bubble Chamber at CERN: Hasert, F. J. et. al., Phys. Lett. *46B,* 121, 138 (1973). Also see Musset, P. Jour, de Physique 11/12 T34 (1973). Muonless events were seen at about the same time by the HPWF group at Fermilab, but when publication of their paper was delayed, they took the opportunity to rebuild their detector, and then did not at first find the same neutral current signal. The HPWF group published evidence for neutral currents in Benvenuti, A. et. al., Phys. Rev. Lett. *32,* 800 (1974).
40. For a survey of the data see Baltay, C. *Proceedings of the 19th International Conference on High Energy Physics,* Tokyo, 1978. For theoretical analyses, see Abbott, L. F. and Barnett, R. M. Phys. Rev. *D19,* 3230 (1979); Langacker, P., Kim, J. E., Levine, M., Williams, H. H. and Sidhu, D. P. Neutrino Conference '79; and earlier references cited therein.
41. Prescott, C. Y. et. al., Phys. Lett. *77B,* 347 (1978).
42. Glashow, S. L. and Georgi, H. L. Phys. Rev. Lett. *28,* 1494 (1972). Also see Schwinger, J. Annals of Physics (N. Y.) *2,* 407 (1957).
43. Glashow, S. L., Iliopoulos, J. and Maiani, L. Phys. Rev. *D2,* 1285 (1970). This paper was cited in ref. 37 as providing a possible solution to the problem of strangeness changing neutral currents. However, at that time I was skeptical about the quark model, so in the calculations of ref. 37 baryons were incorporated in the theory by taking the protons and neutrons to form an SU(2) doublet, with strange particles simply ignored.
44. Politzer, H. D. Phys. Rev. Lett. *30,* 1346 (1973); Gross, D. J. and Wilczek, F. Phys. Rev. Lett. *30,* 1343 (1973).
45. Energy dependent effective couping constants were introduced by Gell-Mann, M. and Low, F. E. Phys. Rev. *95,* 1300 (1954).
46. Bloom, E. D. et. al., Phys. Rev. Lett. *23,* 930 (1969); Breidenbach, M. et. al., Phys. Rev. Lett. *23,* 935 (1969).
47. Weinberg, S. Phys. Rev. *D8,* 605 (1973).
48. Gross, D. J. and Wilczek, F. Phys. Rev. *D8,* 3633 (1973); Weinberg, S. Phys. Rev. Lett. *31,* 494 (1973). A similar idea had been proposed before the discovery of asymptotic freedom by Fritzsch, H., Gell-Mann, M. and Leutwyler, H. Phys. Lett. *47B,* 365 (1973).
49. Greenberg, O. W. Phys. Rev. Lett. *13,* 598 (1964); Han, M. Y. and Nambu, Y. Phys. Rev. *139,* B1006 (1965); Bardeen, W. A., Fritzsch, H. and Gell-Mann, M. in *Scale and Conformal Symmetry in Hadron Physics,* ed. by Gatto, R. (Wiley, 1973), p. 139; etc.
50. 't Hooft, G. Phys. Rev. Lett. *37,* 8 (1976).
51. Such "dynamical" mechanisms for spontaneous symmetry breaking were first discussed by Nambu, Y. and Jona-Lasinio, G. Phys. Rev. *122,* 345 (1961); Schwinger, J. Phys. Rev. *125,* 397 (1962); *128,* 2425 (1962); and in the context of modern gauge theories by Jackiw, R. and Johnson, K. Phys. Rev. *D8,* 2386 (1973); Cornwall, J. M. and Norton, R. E. Phys. Rev. *D8,* 3338 (1973). The implications of dynamical symmetry breaking have been considered by Weinberg, S. Phys. Rev. *D13,* 974 (1976); *D19,* 1277 (1979); Susskind, L., Phys. Rev. *D20,* 2619 (1979).
52. Weinberg, S. ref. 51. The possibility of pseudo-Goldstone bosons was originally noted in a different context by Weinberg, S. Phys. Rev. Lett. *29,* 1698 (1972).
53. Weinberg, S. ref. 51. Models involving such interactions have also been discussed by Susskind, L. ref. 51.
54. Weinberg, S. Phys. Rev. *135,* B1049 (1964).
55. Weinberg, S. Phys. Lett. *9,* 357 (1964); Phys. Rev. *B138,* 988 (1965); *Lectures in Particles and Field Theory,* ed. by Deser, S. and Ford, K. (Prentice-Hall, 1965), p. 988; and ref. 54. The program of deriving general relativity from quantum mechanics and special relativity was completed by Boulware, D. and Deser, S. Ann. Phys. *89,* 173 (1975). I understand that similar ideas were developed by Feynman, R. in unpublished lectures at Cal. Tech.
56. Georgi, H., Quinn, H. and Weinberg, S. Phys. Rev. Lett. *33,* 451 (1974).
57. An example of a simple gauge group for weak and electromagnetic interactions (for-which $\sin^2\theta = \frac{1}{4}$) was given by S. Weinberg, Phys. Rev. *D5,* 1962 (1972). There are a number of specific models of weak, electromagnetic, and strong interactions based on simple gauge groups, including those of Pati, J. C. and Salam, A. Phys. Rev. *D10,* 275 (1974); Georgi, H. and Glashow, S. L. Phys. Rev. Lett. *32,* 438 (1974); Georgi, H. in *Particles and Fields* (American Institute of Physics, 1975); Fritzsch, H. and Minkowski, P. Ann. Phys. *93,* 193 (1975); Georgi, H. and Nanopoulos, D. V. Phys. Lett. *82B,* 392 (1979); Gursey, F. Ramond, P. and Sikivie, P. Phys. Lett. *B60,* 177 (1975); Gursey, F. and Sikivie, P. Phys. Rev. Lett. *36,* 775 (1976); Ramond, P. Nucl. Phys. *B110,* 214 (1976); etc; all these violate baryon and lepton conservation, because they have quarks and leptons in the same multiplet; see Pati, J. C. and Salam, A. Phys. Rev. Lett. *31,* 661 (1973); Phys. Rev. *D8,* 1240 (1973).
58. Buras, A., Ellis, J., Gaillard, M. K. and Nanopoulos, D. V. Nucl. Phys. *B135,* 66 (1978); Ross, D. Nucl. Phys. *B140,* 1 (1978); Marciano, W. J. Phys. Rev. *D20,* 274 (1979); Goldman, T. and Ross, D. CALT 68-704, to be published; Jarlskog, C. and Yndurain, F. J. CERN preprint, to be published. Machacek, M. Harvard preprint HUTP-79/AO21, to be published in Nuclear Physics; Weinberg, S. paper in preparation. The phenomenology of nucleon decay has been discussed in general terms by Weinberg, S. Phys. Rev. Lett. *43,* 1566 (1979); Wilczek, F. and Zee, A. Phys. Rev. Lett. *43,* 1571 (1979).
59. Gildener, E. and Weinberg, S. Phys. Rev. *D13,* 3333 (1976); Weinberg, S. Phys. Letters *82B,* 387 (1979). In general there should exist at least one scalar particle with physical mass of order 10 GeV. The spontaneous symmetry breaking in models with zero bare scalar mass was first considered by Coleman, S. and Weinberg, E., Phys. Rev. *D7,* 1888 (1973).
60. This problem has been studied recently by Dimopoulos, S. and Susskind, L. Nucl. Phys. *B155,* 237 (1979); Eichten, E. and Lane, K. Physics Letters, to be published; Weinberg, S. unpublished.
61. Weinberg, S. in *General Relativity – An Einstein Centenary Survey,* ed. by Hawking, S. W. and Israel, W. (Cambridge Univ. Press, 1979), Chapter 16.

With Paul Matthews, we started on an exploration of renormalizability of meson theories. Finding that renormalizability held only for spin-zero mesons and that these were the only mesons that empirically existed then, (pseudoscalar pions, invented by Kemmer, following Yukawa) one felt thrillingly euphoric that with the triplet of pions (considered as the carriers of the strong nuclear force between the proton–neutron doublet) one might resolve the dilemma of the origin of this particular force. By the same token, the so-called weak nuclear force—the force responsible for β radioactivity (and described then by Fermi's nonrenormalizable theory) had to be mediated by some unknown spin-zero mesons if it was to be renormalizable. If massive charged spin-one mesons were to mediate this interaction, the theory would be nonrenormalizable, according to the ideas then.

Now this agreeably renormalizable spin-zero theory for the pion was a field theory, but not a gauge field theory. There was no conserved charge which determined the pionic interaction. As is well known, shortly after the theory was elaborated, it was found wanting. The $(\frac{3}{2}, \frac{3}{2})$ resonance Δ effectively killed it off as a fundamental theory; we were dealing with a complex dynamical system, not "structureless" in the field-theoretic sense.

For me, personally, the trek to gauge theories as candidates for fundamental physical theories started in earnest in September 1956—the year I heard at the Seattle Conference, Professor Yang expound his and Professor Lee's ideas (Lee and Yang, 1956) on the possibility of the hitherto sacred principle of left–right symmetry, being violated in the realm of the *weak nuclear force*. Lee and Yang had been led to consider abandoning left–right symmetry for weak nuclear interactions as a possible resolution of the (τ, θ) puzzle. I remember traveling back to London on an American Air Force (MATS) transport flight. Although I had been granted, for that night, the status of a Brigadier or a Field Marshal—I don't quite remember which—the plane was very uncomfortable, full of crying servicemen's children—that is, the children were crying, not the servicemen. I could not sleep. I kept reflecting on why Nature should violate left–right symmetry in weak interactions. Now the hallmark of most weak interactions was the involvement in radioactivity phenomena of Pauli's neutrino. While crossing over the Atlantic, came back to me a deeply perceptive questión about the neutrino which Professor Rudolf Peierls had asked when he was examining me for a Ph.D. a few years before. Peierls' question was: "The photon mass is zero because of Maxwell's principle of a gauge symmetry for electromagnetism; tell me, why is the neutrino mass zero?" I had then felt somewhat uncomfortable at Peierls, asking for a Ph.D. viva, a question of which he himself said he did not know the answer. But during that comfortless night the answer came. The analog for the neutrino of the gauge symmetry for the photon existed: it had to do with the masslessness of the neutrino, with symmetry under the γ_5 transformation (Salam, 1957a) (later christened "chiral symmetry"). The existence of this symmetry for the massless neutrino must imply a combination $(1+\gamma_5)$ or $(1-\gamma_5)$ for the neutrino interactions. Nature had the choice of an aesthetically satisfying but a left–right symmetry violating theory, with a neutrino which travels exactly with the velocity of light; or alternatively a theory where left–right symmetry is preserved, but the neutrino has a tiny mass—some ten thousand times smaller than the mass of the electron.

It appeared at that time clear to me what choice Nature must have made. Surely, left–right symmetry must be sacrificed in all neutrino interactions. I got off the plane the next morning, naturally very elated. I rushed to the Cavendish, worked out the Michel parameter and a few other consequences of γ_5 symmetry, rushed out again, got onto a train to Birmingham where Peierls lived. To Peierls I presented my idea: he had asked the original question; could he approve of the answer? Peierls' reply was kind but firm. He said "I do not believe left–right symmetry is violated in weak nuclear forces at all." Thus rebuffed in Birmingham, like Zuleika Dobson, I wondered where I could go next and the obvious place was CERN in Geneva, with Pauli—the father of the neutrino—nearby in Zurich. At that time CERN lived in a wooden hut just outside Geneva airport. Besides my friends, Prentki and d'Espagnat, the hut contained a gas ring on which was cooked the staple diet of CERN—Entrecôte à la creme. The hut also contained Professor Villars of MIT, who was visiting Pauli the same day in Zurich. I gave him my paper. He returned the next day with a message from the Oracle: "Give my regards to my friend Salam and tell him to think of something better." This was discouraging, but I was compensated by Pauli's excessive kindness a few months later, when Mrs. Wu's (Wu *et al.*, 1957), Lederman's (Garwin *et al.*, 1957) and Telegdi's (Friedman and Telegdi, 1957) experiments were announced showing that left–right symmetry was indeed violated and ideas similar to mine about chiral symmetry were expressed independently by Landau (1957) and Lee and Yang (1957). I received Pauli's first, somewhat apologetic letter on 24 January 1957. Thinking that Pauli's spirit should by now be suitably crushed, I sent him two short notes (Salam, 1957b)[1] I had written in the meantime. These contained suggestions to extend chiral symmetry to electrons and muons, assuming that their masses were a consequence of what has come to be known as dynamical spontaneous symmetry breaking. With chiral symmetry for electrons, muons, and neutrinos, the only mesons that could mediate weak decays of the muons would have to carry spin one. Reviving thus the notion of charged intermediate *spin-one* bosons, one could then postulate for these a type of gauge invariance which I called the "neutrino gauge." Pauli's reaction was swift and terrible. He wrote on 30th January 1957, then on 18 February and later on 11, 12, and 13 March: "I am reading (along the shores of Lake Zurich) in bright sunshine quietly your paper ..." "I am very much startled on the title of your paper 'Universal Fermi Interaction'... For quite a while I have for myself the rule if a theoretician says *universal* it just means pure nonsense. This holds particularly in connection with

[1]For reference, see Footnote 7, p. 89, of Marshak, Riazuddin, and Ryan (1969), and W. Pauli's letters (CERN Archives).

the Fermi interaction, but otherwise too, and now you too, Brutus, my son, come with this word...." Earlier, on 30 January, he had written "There is a similarity between this type of gauge invariance and that which was published by Yang and Mills... In the latter, of course, no γ_5 was used in the exponent." and he gave me the full reference of Yang and Mills' paper, [Phys. Rev. 96, 191 (1954)]. I quote from his letter: "However, there are dark points in your paper regarding the vector field B_μ. If the rest mass is infinite (or very large), how can this be compatible with the gauge transformation $B_\mu \rightarrow B_\mu - \partial_\mu \Lambda$?" and he concludes his letter with the remark: "Every reader will realize that you deliberately conceal here something and will ask you the same questions." Although he signed himself "With friendly regards," Pauli had forgotten his earlier penitence. He was clearly and rightly on the warpath.

Now the fact that I was using gauge ideas similar to the Yang–Mills [non-Abelian SU(2)-invariant] gauge theory was no news to me. This was because the Yang–Mills theory (Yang and Mills, 1954) (which married gauge ideas of Maxwell with the internal symmetry SU(2) of which the proton–neutron system constituted a doublet) had been independently invented by a Ph.D. pupil of mine, Ronald Shaw (1955), at Cambridge at the same time as Yang and Mills had written. Shaw's work is relatively unknown; it remains buried in his Cambridge thesis. I must admit I was taken aback by Pauli's fierce prejudice against universalism—against what we would today call unification of basic forces—but I did not take this too seriously. I felt this was a legacy of the exasperation which Pauli had always felt at Einstein's somewhat formalistic attempts at unifying gravity with electromagnetism—forces which in Pauli's phrase "cannot be joined—for God hath rent them asunder." But Pauli was absolutely right in accusing me of darkness about the problem of the masses of the Yang–Mills fields; one could not obtain a mass without wantonly destroying the gauge symmetry one had started with. And this was particularly serious in this context, because Yang and Mills had conjectured the desirable renormalizability of their theory with a proof which relied heavily and exceptionally on the masslessness of their spin-one intermediate mesons. The problem was to be solved only seven years later with the understanding of what is now known as the Higgs mechanism, but I will come back to this later.

Be that as it may, the point I wish to make from this exchange with Pauli is that already in early 1957, just after the first set of parity experiments, many ideas coming to fruition now, had started to become clear. These are:

(1) First was the idea of chiral symmetry leading to a $V-A$ theory. In those early days my humble suggestion (Salam, 1957a, b) of this was limited to neutrinos, electrons, and muons only, while shortly after, that year, Marshak and Sudarshan (Marshak and Sudarshan, 1957 and 1958)[2] Feynman and Gell-Mann (Feynman and Gell-Mann, 1958), and Sakurai (1958) had the courage to postulate γ_5 symmetry for baryons as well as leptons, making this into a universal principle of physics.[3]

Concomitant with the $(V-A)$ theory was the result that if weak interactions are mediated by intermediate mesons, these mesons must carry spin one.

(2) Second was the idea of spontaneous breaking of chiral symmetry to generate electron and muon masses, though the price which those latter-day Shylocks, Nambu and Jona-Lasinio (Nambu and Jona-Lasinio, 1961) and Goldstone [Nambu (1960) and Goldstone (1961)] exacted for this (i.e., the appearance of massless scalars), was not yet appreciated.

(3) And finally, though the use of a Yang–Mills–Shaw (non-Abelian) gauge theory for describing spin-one intermediate charged mesons was suggested already in 1957, the giving of masses to the intermediate bosons through spontaneous symmetry breaking, in such a manner as to preserve the renormalizability of the theory, was to be accomplished only during a long period of theoretical development between 1963 and 1971.

Once the Yang–Mills–Shaw ideas were accepted as relevant to the charged weak currents—to which the charged intermediate mesons were coupled in this theory—during 1957 and 1958 was raised the question of what was the third component of the SU(2) triplet, of which the charged weak currents were the two members. There were the two alternatives: the electroweak unification suggestion, where the electromagnetic current was assumed to be this third component; and the rival suggestion that the third component was a neutral current unconnected with electroweak unification. With hindsight, I shall call these the Klein (1938) (see Klein, 1939) and the Kemmer (1937) alternatives. The Klein suggestion, made in the context of a Kaluza-Klein five-dimensional space-time, was a real tour-de-force; it combined two hypothetical spin-one charged mesons with the photon in one multiplet, deducing from the compactification of the fifth dimension, a theory which looks like Yang–Mills–Shaw's. Klein intended his charged mesons for *strong* interactions, but if we read charged *weak* mesons for Klein's *strong* ones, one obtains the theory independently suggested by Schwinger (1957), though Schwinger, unlike Klein, did not build in any non-Abelian gauge aspects. With just these non-Abelian Yang–Mills gauge aspects very much to the fore, the idea of uniting weak interactions with electromagnetism was developed by Glashow (1959) and Ward and myself (Salam and Ward, 1959) in late 1958.
The rival Kemmer suggestion of a global SU(2)-invariant triplet of weak charged and neutral currents was independently suggested by Bludman (1958) in a gauge context and this is how matters stood till 1960.

To give you the flavor of, for example, the year 1960,

[2]The idea of a universal Fermi interaction for (P,N), (ν_e, e), and (ν_μ, μ) doublets goes back to Tiomno and Wheeler (1949a,b) and Yang and Tiomno (1950). Tiomno (1956) considered γ_5 transformations of Fermi fields linked with mass reversal.

[3]Today we believe protons and neutrons are composites of quarks, so that γ_5 symmetry is now postulated for the elementary entities of today—the quarks. If the neutrino also turns out to be massive, γ_5-symmetry is spontaneously broken for it, as it is for electrons, muons, and quarks.

there was a paper written that year by Ward and myself (Salam and Ward, 1961) with the statement "Our basic postulate is that it should be possible to generate strong, weak and electromagnetic interaction terms with all their correct symmetry properties (as well as with clues regarding their relative strengths) by making local gauge transformations on the kinetic energy terms in the free Lagrangian for all particles. This is the statement of an ideal which, in this paper at least, is only very partially realized." I am not laying a claim that we were the only ones who were saying this, but I just wish to convey to you the temper of the physics of twenty years ago—qualitatively no different today from then. But what a quantitative difference the next twenty years made, first with new and far-reaching developments in theory—and then, thanks to CERN, Fermilab, Brookhaven, Argonne, Serpukhov, and SLAC, in testing it!

So far as theory itself is concerned, it was the next seven years between 1961–67 which were the crucial years of quantitative comprehension of the phenomenon of spontaneous symmetry breaking and the emergence of the SU(2)× U(1) theory in a form capable of being tested. The story is well known and Steve Weinberg has already spoken about it. So I will give the barest outline. First there was the realization that the two alternatives mentioned above, a pure electromagnetic current versus a pure neutral current—Klein–Schwinger versus Kemmer–Bludman—were not alternatives; they were complementary. As was noted by Glashow (1961) and independently by Ward and myself (Salam and Ward, 1964), both types of currents and the corresponding gauge particles ($W^{\pm}$, Z^0, and γ) were needed in order to build a theory that could simultaneously accommodate parity violation for weak and parity conservation for the electromagnetic phenomena. Second, there was the influential paper of Goldstone in 1961 which, utilizing a nongauge self-interaction between scalar particles, showed that the price of spontaneous breaking of a continuous internal symmetry was the appearance of zero mass scalars—a result foreshadowed earlier by Nambu. In giving a proof of this theorem (Goldstone *et al.*, 1962) with Goldstone, I collaborated with Steve Weinberg, who spent a year at Imperial College in London. I would like to pay here a most sincerely felt tribute to him and to Sheldon Glashow for their warm and personal friendship.

I shall not dwell on the now well-known contributions of Anderson (1963), Higgs (1964a, 1964b, 1966), Brout and Englert (Englert and Brout, 1964; Englert *et al.*, 1966), Guralnik, Hagen, and Kibble (1964; Kibble, 1967) starting from 1963, which showed how spontaneous symmetry breaking using spin-zero fields could generate vector-meson masses, defeating Goldstone at the same time. This is the so-called Higgs mechanism.

The final steps towards the electroweak theory were taken by Weinberg (1967) and by myself (Salam, 1968) (with Kibble at Imperial College tutoring me about the Higgs phenomena). We were able to complete the present formulation of the spontaneously broken SU(2)× U(1) theory so far as leptonic weak interactions were concerned—with one parameter $\sin^2\theta$ describing all weak and electromagnetic phenomena and with one isodoublet Higgs multiplet. An account of this development was given during the contribution (Salam, 1968) to the Nobel Symposium (organized by Nils Svartholm and chaired by Lamek Hulthén held at Gothenburg after some postponements, in early 1968). As is well known, we did not have then, and still do not have, a prediction for the scalar Higgs mass.

Both Weinberg and I suspected that this theory was likely to be renormalizable.[4] Regarding spontaneously broken Yang–Mills–Shaw theories in general this had earlier been suggested by Englert, Brout, and Thiry (1966). But this subject was not pursued seriously except at Veltman's school at Utrecht, where the proof of renormalizability was given by 't Hooft (1971a, b) in 1971. This was elaborated further by that remarkable physicist, the late Benjamin Lee (Lee, 1972; Lee and Zinn-Justin, 1972, 1973), working with Zinn-Justin, and by 't Hooft and Veltman (1972a, 1972b).[5] This followed on the earlier basic advances in Yang–Mills calculational technology by Feynman (1963), DeWitt (1967a, b), Faddeev and Popov (1967), Mandelstam (1968a, b), Fradkin and Tyutin (1970), Boulware (1970), Taylor (1971), Slavnov (1972), Strathdee and Salam (Salam and Strathdee, 1970). In Coleman's eloquent phrase "'t Hooft's work turned the Weinberg–Salam frog into an enchanted prince." Just before had come the GIM (Glashow, Iliopoulos, and Maiani) mechanism (Glashow *et al.*, 1970), emphasizing that the existence of the fourth charmed quark (postulated earlier by several authors) was essential to the natural resolution of the dilemma posed by the absence of strangeness-violating currents. This tied in naturally with the understanding of the Steinberger–Schwinger–Rosenberg–Bell–Jackiw–Adler anomaly (see Jackiw, 1972) and its removal for SU(2) × U(1) by the parallelism of four quarks and four leptons, pointed out by Bouchiat, Iliopoulos, and Meyer (1972) and independently by Gross and Jackiw (1972).

If one has kept a count, I have so far mentioned around fifty theoreticians. As a failed experimenter, I have always felt envious of the ambience of large experimental teams and it gives me the greatest pleasure to acknowledge the direct or the indirect contributions of the "series of minds" to the spontaneously broken SU(2)× U(1) gauge theory. My profoundest personal appreciation goes to my collaborators at Imperial College, and Cambridge and the Trieste Centre, John Ward, Paul Matthews, Jogesh Pati, John Strathdee, Tom Kibble, and to Nicholas Kemmer.

In retrospect, what strikes me most about the early part of this story is how uninformed all of us were, not

[4]When I was discussing the final version of the SU(2)×U(1) theory and its possible renormalizability in Autumn 1967 during a postdoctoral course of lectures at Imperial College, Nino Zichichi from CERN happened to be present. I was delighted because Zichichi had been badgering me since 1958 with persistent questioning as to of what theoretical avail his precise measurements on (g-2) for the muon as well as those of the muon lifetime were, when not only the magnitude of the electromagnetic corrections to weak decays was uncertain, but also conversely the effect of nonrenormalizable weak interactions on "renormalized" electromagnetism was so unclear.

[5]An important development in this context was the invention of the dimensional regularization technique by Bollini and Giambiagi (1972), Ashmore (1972), and 't Hooft and Veltman.

only of each other's work, but also of work done earlier. For example, only in 1972 did I learn of Kemmer's paper written at Imperial College in 1937. Kemmer's argument essentially was that Fermi's weak theory was not globally SU(2) invariant and should be made so—though not for its own sake but as a prototype for strong interactions. Then this year I learnt that earlier, in 1936, Kemmer's Ph.D. supervisor, Gregor Wentzel (1937), had introduced (the yet undiscovered) analogs of lepto-quarks, whose mediation could give rise to neutral currents after a Fierz reshuffle. And only this summer, Cecilia Jarlskog at Bergen rescued Oscar Klein's paper from the anonymity of the Proceedings of the International Institute of Intellectual Cooperation of Paris, and we learnt of his anticipation of a theory similar to Yang–Mills–Shaw long before these authors. As I indicated before, the interesting point is that Klein was using his triplet, of two charged mesons plus the photon, not to describe weak interaction but for strong nuclear force unification with the electromagnetic—something our generation started on only in 1972—and not yet experimentally verified. Even in this recitation I am sure I have inadvertently left off some names of those who have in some way contributed to SU(2)× U(1). Perhaps the moral is that not unless there is the prospect of quantitative verification, does a qualitative idea make its impress in physics.

And this brings me to experiment, and the year of the Gargamelle (Hasert *et al.*, 1973). I still remember Paul Matthews and I getting off the train at Aix-en-Provence for the 1973 European Conference and foolishly deciding to walk with our rather heavy luggage to the student hostel where we were billeted. A car drove from behind us, stopped, and the driver leaned out. This was Musset whom I did not know well personally then. He peered out of the window and said: "Are you Salam?" I said "Yes." He said: "Get into the car. I have news for you. We have found neutral currents." I will not say whether I was more relieved for being given a lift because of our heavy luggage or for the discovery of neutral currents. At the Aix-en-Provence meeting that great and modest man, Lagarrigue, was also present and the atmosphere was that of a carnival—at least this is how it appeared to me. Steve Weinberg gave the rapporteur's talk with T. D. Lee as the chairman. T. D. was kind enough to ask me to comment after Weinberg finished. That summer Jogesh Pati and I had predicted proton decay within the context of what is now called grand unification, and in the flush of this excitement I am afraid I ignored weak neutral currents as a subject which had already come to a successful conclusion, and concentrated on speaking of the possible decays of the proton. I understand now that proton decay experiments are being planned in the United States by the Brookhaven, Irvine and Michigan and the Wisconsin–Harvard groups and also by a European collaboration to be mounted in the Mont Blanc Tunnel Garage No. 17. The later quantitative work on neutral currents at CERN, Fermilab, Brookhaven, Argonne and Serpukhov is, of course, history, but a special tribute is warranted to the beautiful SLAC–Yale–CERN experiment (Taylor, 1979) of 1978 which exhibited the effective Z^0-photon interference in accordance with the predictions of the theory. This was foreshadowed by Barkov *et al.*'s experiments (Barkov, 1979) at Novosibirsk in the USSR in their exploration of parity-violation in the atomic potential for bismuth. There is the apocryphal story about Einstein, who was asked what he would have thought if experiment had not confirmed the light deflection predicted by him. Einstein is supposed to have said, "Madam, I would have thought the Lord has missed a most marvelous opportunity." I believe, however, that the following quote from Einstein's Herbert Spencer lecture of 1933 expresses his, my colleagues', and my own views more accurately. "Pure logical thinking cannot yield us any knowledge of the empirical world; all knowledge of reality starts from experience and ends in it." This is exactly how I feel about the Gargamelle–SLAC experience.

III. THE PRESENT AND ITS PROBLEMS

Thus far we have reviewed the last twenty years and the emergence of SU(2)× U(1), with the twin developments of a gauge theory of basic interactions, linked with internal symmetries, and of the spontaneous breaking of these symmetries. I shall first summarize the situation as we believe it to exist now and the immediate problems. Then we turn to the future.

(1) To the level of energies explored, we believe that the following sets of particles are "structureless" (in a field-theoretic sense) and, at least to the level of energies explored hitherto, constitute the elementary entities of which all other objects are made.

$$\text{SU}_c(3)\ \text{triplets}$$

$$\text{Family I quarks}\begin{Bmatrix} u_R, u_Y, u_B \\ d_R, d_Y, d_B \end{Bmatrix}\quad \text{leptons}\begin{bmatrix} \nu_e \\ e \end{bmatrix}\quad \text{SU(2)doublets}$$

$$\text{Family II quarks}\begin{Bmatrix} c_R, c_Y, c_B \\ s_R, s_Y, s_B \end{Bmatrix}\quad \text{leptons}\begin{bmatrix} \nu_\mu \\ \mu \end{bmatrix}\quad \text{SU(2)doublets}$$

$$\text{Family III quarks}\begin{Bmatrix} t_R, t_Y, t_B \\ b_R, b_Y, b_B \end{Bmatrix}\quad \text{leptons}\begin{bmatrix} \nu_\tau \\ \tau \end{bmatrix}\quad \text{SU(2)doublets}$$

Together with their antiparticles each family consists of 15 or 16 two-component fermions (15 or 16 depending on whether the neutrino is four-component or not). The third family is still conjectural, since the top quark (t_R, t_Y, t_B) has not yet been discovered. Does this family really follow the pattern of the other two? Are there more families? Does the fact that the families are replicas of each other imply that Nature has discovered a dynamical stability about a system of 15 (or 16) objects, and that by this token there is a more basic layer of structure underneath? (See Pati and Salam, 1975a; Pati *et al.*, 1975a; Harari, 1979; Schupe, 1979; Curtwright and Fruend, 1979).

(2) Note that quarks come in three colors: Red (R), Yellow (Y), and Blue (B). Parallel with the electroweak SU(2)× U(1), a gauge field[6] theory ($\text{SU}_c(3)$) of

strong (quark) interactions (quantum chromodynamics, QCD[7] has emerged which gauges the three colors. The indirect discovery of the (eight) gauge bosons associated with QCD (gluons), has already been surmised by the groups at DESY.[8]

(3) All known baryons and mesons are singlets of color $SU_c(3)$. This has led to a hypothesis that color is always confined. One of the major unsolved problems of field theory is to determine if QCD—treated nonperturbatively—is capable of confining quarks and gluons.

(4) In respect of the electroweak SU(2)× U(1), all known experiments on weak and electromagnetic phenomena below 100 GeV carried out to date agree with the theory which contains one theoretically undetermined parameter $\sin^2\theta = 0.230 \pm 0.009$ (Winter, 1979). The predicted values of the associated gauge boson ($W^{\pm}$ and Z^0) masses are: $m_W \approx 77-84$ GeV, $m_Z \approx 89-95$ GeV, for $0.25 \geq \sin^2\theta \geq 0.21$.

(5) Perhaps the most remarkable measurement in electroweak physics is that of the parameter $\rho = (m_W/m_Z \cos\theta)^2$. Currently this has been determined from the ratio of neutral to charged current cross sections. The predicted value $\rho = 1$ for weak *iso-doublet Higgs* is to be compared with the experimental[9] $\rho = 1.00 \pm 0.02$.

(6) Why does Nature favor the simplest suggestion in SU(2)× U(1) theory of the Higgs scalars being iso-doublet?[10] Is there just one physical Higgs? Of what mass? At present the Higgs interactions with leptons and quarks as well as their self-interactions are non-gauge interactions. For a three-family (six-quark) model, 21 out of the 26 parameters needed are attributable to the Higgs interactions. Is there a basic principle, as compelling and as economical as the gauge principle, which embraces the Higgs sector? Alternatively, could the Higgs phenomenon itself be a manifestation of a dynamical breakdown of the gauge symmetry?[10]

(7) Finally there is the problem of the families; is there a distinct SU(2) for the first, another for the second, as well as a third SU(2), with spontaneous symmetry breaking such that the SU(2) apprehended by present experiment is a diagonal sum of these "family" SU(2)'s? To state this in another way, how far in energy does the $e-\mu$ universality (for example) extend? Are there more[11] Z^0's than just one, effectively differentially coupled to the e and the μ systems? (If there are, this will constitute minimodifications of the theory, but not a drastic revolution of its basic ideas.)

In the next section I turn to a direct extrapolation of the ideas which went into the electroweak unification, so as to include strong interactions as well. Later I shall consider the more drastic alternatives which may be needed for the unification of all forces (including gravity)—ideas which have the promise of providing a deeper understanding of the charge concept. Regretfully, by the same token, I must also become more technical and obscure for the nonspecialist. I apologize for this. The nonspecialist may sample the flavor of the arguments in the next section (Sec. IV), ignore the Appendices, and then go on to Sec. V, which is perhaps less technical.

IV. DIRECT EXTRAPOLATION FROM THE ELECTROWEAK TO THE ELECTRONUCLEAR

A. The three ideas

The three main ideas which have gone into the electronuclear—also called grand—unification of the electroweak with the *strong* nuclear force (and which date back to the period 1972–1974), are the following:

(1) First: the psychological break (for us) of grouping quarks and leptons in the *same* multiplet of a unifying group G, suggested by Pati and myself in 1972 (see Bjorken, 1972; Pati and Salam, 1973a). The group G must contain $SU(2)\times U(1)\times SU_c(3)$, and must be non-Abelian, if all quantum numbers (flavor, color, lepton, quark, and family numbers) are to be automatically quantized and the resulting gauge theory asymptotically free.

(2) Second: an extension, proposed by Georgi and Glashow (1974) which places not only (left-handed) quarks and leptons but also their antiparticles in the same multiplet of the unifying group.

Appendix I displays some examples of the unifying groups presently considered.

---

[6] "To my mind the most striking feature of theoretical physics in the last thirty-six years is the fact that not a single new theoretical idea of a fundamental nature has been successful. The notions of relativistic quantum theory....have in every instance proved stronger than the revolutionary ideas....of a great number of talented physicists. We live in a dilapidated house and we seem to be unable to move out. The difference between this house and a prison is hardly noticeable"—Res Jost (1963), "In Praise of Quantum Field Theory" (Siena European Conference).

[7] Pati and Salam. See the review by Bjorken (1972). See also Fritzsch and Gell-Mann (1972), Fritzsch, Gell-Mann, and Leutwyler (1973), Weinberg (1973a,b), and Gross and Wilczek (1973). For a review see Marciano and Pagels (1978).

[8] See the Tasso Collaboration (Brandelik *et al.*, 1979) and the Mark-J Collaboration (Barber *et al.*, 1979). See also the reports of the Jade, Mark-J, Pluto, and Tasso Collaborations to the International Symposium on Lepton and Photon Interactions at High Energies, Fermilab, August 1979.

[9] The one-loop radiative corrections to ρ suggest that the maximum mass of leptons contributing to ρ is less than 100 GeV (Ellis, 1979).

[10] To reduce the arbitrariness of the Higgs couplings and to motivate their iso-doublet character, one suggestion is to use supersymmetry. Supersymmetry is a Fermi-Bose symmetry, so that iso-doublet leptons like (ν_e, e) or (ν_μ, μ) in a supersymmetric theory must be accompanied in the same multiplet by iso-doublet Higgs. Alternatively, one may identify the Higgs as composite fields associated with bound states of a yet new level of elementary particles and new (so-called techni-color) forces (Dimopoulos and Susskind, 1979; Weinberg, 1979a; and 't Hooft) of which, at present low energy, we have no cognizance, and which may manifest themselves in the 1–100 TeV range. Unfortunately, both these ideas at first sight appear to introduce complexities, though in the context of a wider theory, which spans energy scales up to much higher masses, a satisfactory theory of the Higgs phenomena, incorporating these, may well emerge.

[11] See Pati and Salam (1974); Mohapatra and Pati (1975a,b); Elias, Pati, and Salam (1978a); and Pati and Rajpoot (1978).

Now a gauge theory based on a "simple" (or with discrete symmetries, a "semisimple") group G contains one basic gauge constant. This constant would manifest itself physically above the "grand unification mass" M, exceeding all particle masses in the theory—these themselves being generated (if possible) hierarchially through a suitable spontaneous symmetry-breaking mechanism.

(3) The third crucial development was by Georgi, Quinn, and Weinberg (1974) who showed how, using renormalization group ideas, one could relate the observed low-energy couplings $\alpha(\mu)$ and $\alpha_s(\mu)(\mu \sim 100$ GeV) to the magnitude of the grand unifying mass M and the observed value of $\sin^2\theta(\mu)$; ($\tan\theta$ is the ratio of the U(1) to the SU(2) couplings).

(4) If one extrapolates with Jowett,[12] that nothing essentially new can possibly be discovered—i.e., if one assumes that there are no new features, no new forces, or no new "types" of particles to be discovered, till we go beyond the grand unifying energy M—then the Georgi, Quinn, Weinberg method leads to a startling result: this featureless "plateau" with no "new physics" heights to be scaled stretches to fantastically high energies. More precisely, if $\sin^2\theta(\mu)$ is as large as 0.23, then the grand unifying mass M cannot be smaller than 1.3×10^{13} GeV (Marciano, 1979). (Compare with Planck mass $m_P \approx 1.2\times 10^{19}$ GeV related to Newton's constant where gravity must come in.)[13] The result follows from the formula (Marciano, 1979; Salam, 1979).

$$\frac{11\alpha}{3\pi}\ln\frac{M}{\mu}=\frac{\sin^2\theta(M)-\sin^2\theta(\mu)}{\cos^2\theta(M)}, \qquad (1)$$

if it is assumed that $\sin^2\theta(M)$—the magnitude of $\sin^2\theta$ for energies of the order of the unifying mass M—equals 3/8 (see Appendix B).

This startling result will be examined more closely in Appendix B. I show there that it is very much a consequence of the assumption that the SU(2)× U(1) symmetry survives intact from the low regime energies μ right upto the grand unifying mass M. I will also show that there already is some experimental indication that this assumption is too strong, and that there may be likely peaks of new physics at energies of 10 TeV upwards.

B. Tests of electronuclear grand unification

The most characteristic prediction from the existence of the electronuclear force is proton decay, first discussed in the context of grand unification at the Aix-en-Provence Conference (1973) (Pati and Salam, 1973b). For "semisimple" unifying groups with multiplets containing quarks and leptons only (but no antiquarks nor antileptons) the lepto-quark composites have masses (determined by renormalization group arguments) of the order of $\approx 10^5$–10^6 GeV (Elias *et al.*, 1978b; Rajpoot and Elias, 1978). For such theories the characteristic proton decays (proceeding through exchanges of *three* lepto-quarks) conserve quark number + lepton number, i.e., $P = qqq \to lll$, $\tau_P \sim 10^{29}$–10^{34} years. On the contrary, for the "simple" unifying family groups like SU(5) (Georgi and Glashow, 1974) or SO(10) (Fritzsch and Minkowski, 1975, 1976; Georgi, 1975; Georgi and Nanopoulos, 1979) (with multiplets containing antiquarks and antileptons) proton decay proceeds through an exchange of *one* lepto-quark into an antilepton (plus pions, etc.) ($P \to \bar{l}$).

An intriguing possibility in this context is that investigated recently for the maximal unifying group SU(16)—the largest group to contain a sixteenfold fermionic family $(q, l, \bar{q}, \bar{l})$. This can permit four types of decay modes: $P \to 3l$ *as well as* $P \to \bar{l}, P \to l$ (e.g., $P \to l^- + \pi^+ + \pi^+$), and $P \to 3\bar{l}$ (e.g., $N \to 3\bar{\nu} + \pi^0, P \to 2\bar{\nu} + e^+ + \pi^0$), the relative magnitudes of these alternative decays being model-dependent on how precisely SU(16) breaks down to SU(3)× SU(2)× U(1). Quite clearly, it is the central fact of the existence of the proton decay for which the present generation of experiments must be designed, rather than for any specific type of decay modes.

Finally, grand unifying theories predict mass relations like (Buras *et al.*, 1978):

$$\frac{m_d}{m_e}=\frac{m_s}{m_\mu}=\frac{m_b}{m_\tau}\approx 2.8$$

for six (or at most eight) flavors *below the unification mass*. The important remark for proton decay and for mass relations of the above type as well as for an understanding of baryon excess[14] in the universe,[15] is that for the present *these are essentially characteristic of the fact of grand unification—rather than of specific models*.

"Yet each man kills the thing he loves" sang Oscar Wilde anguishedly in his famous Ballad of the Reading

[12]The universal urge to extrapolate from what we know today and to believe that nothing new can possibly be discovered is well expressed in the following:

"I come first, My name is Jowett
I am the Master of this College,
Everything that is, I know it
If I don't, it isn't knowledge"

—The Balliol Masque.

[13]On account of the relative proximity of $M \approx 10^{13}$ GeV to m_P (and the hope of eventual unification with gravity), Planck mass m_P is now the accepted "natural" mass scale in particle physics. With this large mass as the input, the great unsolved problem of grand unification is the "natural" emergence of mass hierarchies (m_P, αm_P, $\alpha^2 m_P$, ...) or $m_P \exp(-c_n/\alpha)$, where c_n's are constants. [$m_e/m_P \sim 10^{-22}$.]

[14]See Yoshimura (1978), Dimopoulos and Susskind (1978), Toussiant *et al.* (1979), Ellis *et al.* (1979), Weinberg (1979b), and Nanopoulos and Weinberg (1979).

[15]The calculation of baryon excess in the universe—arising from a combination of CP and baryon number violations—has recently been claimed to provide teleological arguments for grand unification. For example, Nanopoulos (1979) has suggested that the "existence of human beings to measure the ratio n_B/n_γ (where n_B is the number of baryons and n_γ the number of photons in the universe) necessarily imposes severe bounds on this quantity: i.e., $10^{-11} \approx (m_e/m_P)^{1/2} \lesssim n_B/n_\gamma \lesssim 10^{-4}$ ($\approx O(\alpha^2)$)." Of importance in deriving these constraints are the upper (and lower) bounds on the numbers of flavors (≈ 6) deduced (1) from mass relations above, (2) from cosmological arguments which seek to limit the numbers of massless neutrinos, (3) from asymptotic freedom, and (4) from numerous (one-loop) radiative calculations. It is clear that lack of accelerators as we move up in energy scale will force particle physics to reliance on teleology and cosmology (which, in Landau's famous phrase, is "often wrong, but never in doubt").

Gaol. Like generations of physicists before us, some in our generation also (through a direct extrapolation of the electroweak gauge methodology to the electronuclear)—and with faith in the assumption of no "new physics," which leads to a grand unifying mass ~10^{13} GeV—are beginning to believe that the end of the problems of elementarity as well as of fundamental forces is nigh. They may be right, but before we are carried away by this prospect, it is worth stressing that even for the simplest grand unifying model [Georgi and Glashow's SU(5) with just two Higgs (a 5 and a 24)], the number of presently *ad hoc* parameters needed by the model is still unwholesomely large—22, compared with 26 for the six-quark model based on the humble $SU(2)\times U(1)\times SU_c(3)$. We cannot feel proud.

V. ELEMENTARITY: UNIFICATION WITH GRAVITY AND NATURE OF CHARGE

In some of the remaining parts of this lecture I shall be questioning two of the notions which have gone into the direct extrapolation of Sec. IV—first, do quarks and leptons represent the correct elementary[16] fields, which should appear in the matter Lagrangian, and which are structureless for renormalizability; second, could some of the presently considered gauge fields themselves be composite? This part of the lecture relies heavily on an address I was privileged to give at the European Physical Society meeting in Geneva in July this year (Salam, 1979).

A. The quest for elementarity, prequarks (preons and pre-preons)

If quarks and leptons are elementary, we are dealing with $3\times 15=45$ elementary entities. The "natural" group of which these constitute the fundamental representation is SU(45) with 2024 elementary gauge bosons. It is possible to reduce the size of this group to SU(11) for example (see Appendix A) with only 120 gauge bosons, but then the number of fermions increases to 501 (of which presumably $3\times 15=45$ objects are of low and the rest of Planckian mass). Is there any basic reason for one's instinctive revulsion when faced with these vast numbers of elementary fields?

The numbers by themselves would perhaps not matter so much. After all, Einstein, in his description of gravity (Einstein, 1916), chose to work with 10 fields $[g_{\mu\nu}(x)]$ rather than with just one (scalar field) as Nördstrom [(1912; 1913a, b; 1914a, b); see also Einstein (1912a, b)] before him. Einstein was not perturbed by the multiplicity he chose to introduce, since he relied on the sheet-anchor of a fundamental principle (the equivalence principle) which permitted him to relate the ten fields for gravity $g_{\mu\nu}$ with the ten components of the physically relevant quantity, the tensor $T_{\mu\nu}$ of energy and momentum. *Einstein knew that nature was not economical of structures:* only of principles of fundamental applicability. The question we must ask ourselves is this: Have we yet discovered such principles in our quest for elementarity, to justify having fields with such large numbers of components as elementary?

Recall that quarks carry at least three charges (color, flavor, and a family number). Should one not, by now, entertain the notions of quarks (and possibly of leptons) as being composites of some more basic entities[17] (prequarks or preons), which each carry but *one* basic charge? (Pati and Salam, 1975a; Pati *et al.*, 1975a; Harari, 1979; Schupe, 1979; Curtright and Freund, 1979) These ideas have been expressed before but they have become more compulsive now, with the growing multiplicity of quarks and leptons. Recall that it was similar ideas which led from the eightfold of baryons to a triplet of (Sakatons and) quarks in the first place.

The preon notion is now new. In 1975, among others, Pati, Salam, and Strathdee (1975a) introduced 4 chromons (the fourth color corresponding to the lepton number) and 4 flavons, the basic group being SU(8)—of which the family group $SU_F(4)\times SU_C(4)$ was but a subgroup. As an extension of these ideas, we now believe these preons carry magnetic charges and are bound together by very strong short-range forces, with quarks and leptons as their magnetically neutral composites (Pati and Salam, 1980). The important remark in this context is that in a theory containing *both* electric and magnetic generalized charges, the analogs of the well-known Dirac quantization condition (Dirac, 1931) give relations like $eg/4\pi=n/2$ for the strength of the two types of charges. Clearly, magnetic monopoles[18] of strength g and mass $mw/\alpha\approx 10^4-10^5$ GeV of opposite polarity, are likely to bind much more tightly than electric charges, yielding composites whose nonelementary nature will reveal itself only for very high energies. This appears to be the situation at least for leptons if they are composites.

In another form the preon idea has been revived this year by Curtright and Freund (1979), who, motivated by ideas of extended supergravity (to be discussed in the next subsection), reintroduce an SU(8) of 3 chromons (R, Y, B), 2 flavons, and 3 familons (horrible names). The family group SU(5) could be a subgroup of this SU(8). In the Curtright–Freund scheme, the $3\times 15=45$ fermions of SU(5) (Georgi and Glashow, 1974) can be found among the $8+28+56$ of SU(8) [or alternatively the $3\times 16=48$ of SO(10) among the vectorial 56 fermions of SU(8)]. (The next succession after the preon level may be the pre-preon level. It was suggested at the Geneva

[16] I would like to quote Feynman in a recent interview in *Omni* magazine: "As long as it looks like the way things are built [is] with wheels within wheels, then you are looking for the innermost wheel—but it might not be that way, in which case you are looking for whatever the hell it is you find!" In the same interview he remarks "a few years ago I was very skeptical about the gauge theories. . . . I was expecting mist, and now it looks like ridges and valleys after all."

[17] One must emphasize, however, that zero mass neutrinos are the hardest objects to conceive of as composites.

[18] According to 't Hooft's theorem, a monopole corresponding to the $SU_L(2)$ gauge symmetry is expected to possess a mass with the lower limit m_W/α ('t Hooft, 1974; Polyakov, 1974). Even if such monopoles are confined, their indirect effects must manifest themselves, if they exist. (Note that m_W/α is very much a lower limit for a grand unified theory like SU(5) for which the monopole mass is α^{-1} times the heavy leptoquark mass.) The monopole force may be the techni-force of Footnote 10.

TABLE I. Prognosis for the next decade.

Decade	1950–1960	1960–1970	1970–1980	1980→
Discovery in early part of the decade	The strange particles	The 8-fold way, Ω^-	Confirmation of neutral currents	W, Z, Proton decay
Expectation for the rest of the decade		SU(3) resonances		Grand Unification, Tribal Groups
Actual discovery		Hit the next level of elementarity with quarks		May hit the preon level, and composite structure of quarks

Conference (see Salam, 1979) that with certain developments in field theory of composite fields it could be that just two pre-preons may suffice. But at this stage this is pure speculation.)

Before I conclude this section, I would like to make a prediction regarding the course of physics in the next decade, extrapolating from our past experience of the decades gone by (See Table I).

B. Post-Planck physics, supergravity, and Einstein's dreams

I now turn to the problem of a deeper comprehension of the charge concept (the basis of gauging)—*which, in my humble view, is the real quest of particle physics.* Einstein, in the last 35 years of his life, lived with two dreams: one was to unite gravity with matter (the photon)—he wished to see the "base wood" (as he put it) which makes up the stress tensor $T_{\mu\nu}$ on the right-hand side of his equation $R_{\mu\nu}-\frac{1}{2}g_{\mu\nu}R=-T_{\mu\nu}$ transmuted through this union, into the "marble" of gravity on the left-hand side. The second (and the complementary) dream was to use this unification to comprehend the nature of electric charge in terms of space-time geometry in the same manner as he had successfully comprehended the nature of gravitational charge in terms of space-time curvature.

In case someone imagines[19] that such deeper comprehension is irrelevant to quantitative physics, let me adduce the tests of Einstein's theory versus the proposed modifications to it [Brans–Dicke (Brans and Dicke, 1961) for example]. Recently (1976), the *strong* equivalence principle (i.e., the proposition that gravitational forces contribute equally to the inertial and the gravitational masses) was tested[20] to one part in 10^{12} [i.e., to the same accuracy as achieved in particle physics for $(g-2)_e$] through lunar-laser ranging measurements [Williams *et al.* (1976), Shapiro *et al.* (1976). For a discussion see Salam (1977)]. These measurements determined departures from Kepler equilibrium distances of the moon, the earth, and the sun to better than ±30 cm and triumphantly vindicated Einstein.

There have been four major developments in realizing Einstein's dreams:

(1) The Kaluza-Klein (Kaluza, 1921; Klein, 1926) miracle: An Einstein Lagrangian (scalar curvature) in five-dimensional space-time (where the fifth dimension is compactified in the sense of all fields being explicitly independent of the fifth coordinate) precisely reproduces the *Einstein-Maxwell* theory in four dimensions, the $g_{\mu 5}(\mu=0,1,2,3)$ components of the metric in five dimensions being identified with the Maxwell field A_μ. From this point of view, Maxwell's field is associated with the extra components of curvature implied by the (conceptual) existence of the fifth dimension.

(2) The second development is the recent realization by Cremmer, Scherk, Englert, Brout, Minkowski, and others that the compactification of the extra dimensions (Cremmer and Scherk, 1976a, b, c; Minkowski, 1977)—(their curling up to sizes perhaps smaller than Planck length $\lesssim 10^{-33}$ cm and the very high curvature associated with them)—might arise through a spontaneous symmetry breaking (in the first 10^{-43} sec) which reduced the higher-dimensional space-time effectively to the four-dimensional that we apprehend directly.

(3) So far we have considered Einstein's second dream, i.e., the unification of electromagnetism (and presumably of other gauge forces) with gravity, giving a space-time significance to gauge charges as corresponding to extended curvature in extra bosonic dimensions. A full realization of the first dream (unification of spinor matter with gravity and with other gauge fields) had to await the development of supergravity[21]—and an extension to extra fermionic dimensions of superspace (Salam and Strathdee, 1974a) (with extended torsion being brought into play in addition to curvature). I discuss this development later.

(4) And finally there was the alternative suggestion by Wheeler (Fuller and Wheeler, 1962; Wheeler, 1964) and Schemberg that electric charge may be associated with space-time topology--with worm-holes, with space-time Gruyère-cheesiness. This idea has re-

[19]The following quotation from Einstein is relevant here. "We now realize, with special clarity, how much in error are those theorists who believe theory comes inductively from experience. Even the great Newton could not free himself from this error (*Hypotheses non fingo*)." This quote is complementary to the quotation from Einstein at the end of Sec. II.

[20]The *weak* equivalence principle (the proposition that all but the gravitational force contribute equally to the inertial and the gravitational masses) was verified by Eötvös to 1:10^8 and by Dicke and Braginsky and Panov to 1:10^{12}.

[21]See Freedman, van Nieuwenhuizen, and Ferrara (1976) and Deser and Zumino (1976). For a review and comprehensive list of references, see Freedman (1979). See also Arnowitt, Nath, and Zumino (1975), Zumino (1975), Wess and Zumino (1977), Akulov, Volkov, and Soroka (1975), and Brink *et al.* (1978).

cently been developed by Hawking[22] and his collaborators (Hawking, 1978, 1979a, b; Gibbons *et al.*, 1978).

C. Extended supergravity, SU(8) preons, and composite gauge fields

Now so far I have reviewed the developments in respect of Einstein's dreams as reported at the Stockholm Conference held in 1978 in this hall and organized by the Swedish Academy of Sciences.

A remarkable new development was reported during 1979 by Julia and Cremmer (Cremmer *et al.* (1978); Cremmer and Julia (1978, 1979); see also Julia (1979)] which started with an attempt to use the ideas of Kaluza and Klein to formulate extended supergravity theory in a higher (compactified) space-time—more precisely, in eleven dimensions. This development links up, as we shall see, with preons and composite Fermi fields—and even more important—possibly with the notion of composite gauge fields.

Recall that simple supergravity[23] is the gauge theory of supersymmetry[24]—the gauge particles being the (helicity ± 2) gravitons and (helicity $\pm\frac{3}{2}$) gravitinos.[25] *Extended supergravity* gauges supersymmetry combined with SO(N) internal symmetry. For $N=8$, the (tribal) supergravity multiplet consists of the following SO(8) families.[26]

Helicity	± 2	1
	$\pm\frac{3}{2}$	8
	± 1	28
	$\pm\frac{1}{2}$	56
	0	70

As is well known, SO(8) is too small to contain SU(2) $\times$ U(1) $\times$ SU$_c$(3). Thus this tribe has no place for $W^{\pm}$ (though Z^0 and γ are contained) and no places for μ or τ or the t quark.

This was the situation last year. This year, Cremmer and Julia (see Footnote 26) attempted to write down the $N=8$ supergravity Lagrangian explicitly, using an extension of the Kaluza–Klein ansatz which states that *extended supergravity* [with SO(8) internal symmetry] has the same Lagrangian in four space-time dimensions as *simple supergravity* in (compactified) eleven dimensions. This formal—and rather formidable—ansatz, when carried through, yielded a most agreeable bonus. *The supergravity Lagrangian possesses an unsuspected SU(8) "local" internal symmetry* although one started with an internal SO(8) only.

The tantalizing questions which now arise are the following.

(1) Could this internal SU(8) be the symmetry group of the 8 preons (3 chromons, 2 flavons, 3 familons) introduced earlier?

(2) When SU(8) is gauged, there should be 63 spin-one fields. The supergravity tribe contains only 28 spin-one fundamental objects which are not minimally coupled. Are the 63 fields of SU(8) to be identified with composite gauge fields made up of the 70 spin-zero objects of the form $V^{-1}\partial_\mu V$? Do these composites propagate, in analogy with the well-known recent result in CP^{n-1} theories (D'Adda *et al.*, 1978), where a composite gauge field of this form propagates as a consequence of quantum effects (quantum completion)?

The entire development I have described—the unsuspected extension of SO(8) to SU(8) when extra compactified space-time dimensions are used, and the possible existence and quantum propagation of composite gauge fields—is of such crucial importance for the future prospects of gauge theories that one begins to wonder how much of the extrapolation which took SU(2) $\times$ U(1) $\times$ SU$_c$(3) into the electronuclear grand unified theories is likely to remain unaffected by these new ideas now unfolding.

But where in all this is the possibility to appeal directly to experiment? For grand unified theories, it was the proton decay. What is the analog for supergravity? Perhaps the spin-$\frac{3}{2}$ massive gravitino, picking its mass from a super-Higgs effect [Cremmer *et al.* (1979); see also Ferrara (1979) and references therein] provides the answer. Fayet (1977, 1979) has shown that for a spontaneously broken globally supersymmetric weak theory the introduction of a local gravitational interaction leads to a super-Higgs effect. Assuming that supersymmetry breakdown is at mass scale m_W, the gravitino acquires a mass and an effective interaction, but of conventional weak rather than of the gravitational strength—an enhancement by a factor of 10^{34}. One may thus search for the gravitino among the neutral decay modes of J/ψ—the predicted rate being 10^{-3}–10^{-5} times smaller than the observed rate for $J/\psi \to e^+e^-$. This will surely tax all the ingenuity of the great men (and women) at SLAC and DESY. Another effect suggested by Scherk (1979) is antigravity—a cancellation of the attractive gravitational force with the force produced by spin-one gravi-photons which exist in all extended supergravity theories. Scherk shows that the Compton wavelength of the gravi-photon is either smaller than 5 cm or is between 10 and 850 meters in order that there will be no conflict with what

---

[22]The Einstein Lagrangian allows large fluctuations of metric and topology on Planck-length scale. Hawking has surmised that the dominant contributions to the path integral of quantum gravity come from metrics which carry one unit of topology per Planck volume. On account of the intimate connection (de Rham, Atiyah-Singer) (Atiyah and Singer, 1963) of curvature with the measures of space-time topology (Euler number, Pontryagin number) the extended Kaluza-Klein and Wheeler-Hawking points of view may find consonance after all.

[23]See Freedman, van Nieuwenhuizen, and Ferrara (1976), and Deser and Zumino (1976). For a review and comprehensive list of references, see Freedman (1979).

[24]See Gol'fand and Likhtman (1971), Volkov and Akulov (1972), Wess and Zumino (1974), Salam and Strathdee (1974a,b,c). For a review, see Salam and Strathdee (1978).

[25]Supersymmetry algebra extends Poincaré group algebra by adjoining to it supersymmetric charges Q_α which transform bosons to fermions. $\{Q_\alpha, Q_\beta\} = (\gamma_\mu P_\mu)_{\alpha\beta}$. The currents which correspond to these charges (Q_α and P_μ) are $J_{\mu\alpha}$ and $T_{\mu\nu}$—these are essentially the currents which in gauged supersymmetry (i.e., supergravity) couple to the gravitino and the graviton, respectively.

[26]See Footnote 23 and Cremmer, Julia, and Scherk (1978) and Cremmer and Julia (1978, 1979). See also Julia (1979).

TABLE A.I. Examples of grand unifying groups.

Semisimple groups[27]	Multiplet	Exotic gauge particles	Proton decay
(with left-right symmetry)	$G_L \to \begin{pmatrix} q \\ l \end{pmatrix}_L$, $G_R \to \begin{pmatrix} q \\ l \end{pmatrix}_R$	Lepto-quarks $\to (\bar{q}l)$	Lepto-quarks $\to W$ + (Higgs) or
Example $[SU(6)_F \times SU(6)_c]_{L\to R}$	$G = G_L \times G_R$	Unifying mass $\approx 10^6$ GeV	Proton $= qqq \to lll$
Examples Family groups $\to$ SU(5) or SO(10) ↓ ↓ Tribal groups $\to$ SU(11) SO(22)	$G \to \begin{pmatrix} q \\ l \\ \bar{q} \\ \bar{l} \end{pmatrix}_L$	Diquarks $\to (qq)$ Dileptons $\to (ll)$ Lepto-quarks $\to (\bar{q}l), (ql)$ Unifying mass $\approx 10^{13} - 10^{15}$ GeV	$qq \to \bar{q}l$ i.e. Proton $P = qqq \to \bar{l}$ Also possible, $P \to l, P \to 3\bar{l}, P \to 3l$

is presently known about the strength of the gravitational force.

Let me summarize: it is conceivable of course, that there is indeed a grand plateau—extending even to Planck energies. If so, the only eventual laboratory for particle physics will be the early universe, where we shall have to seek for the answers to the questions on the nature of charge. There may, however, be indications of a next level of structure around 10 TeV; there are also beautiful ideas (like, for example, those of electric and magnetic monopole duality) which may manifest at energies of the order of $\alpha^{-1} m_W (= 10$ TeV). Whether even this level of structure will give us the final clues to the nature of charge, one cannot predict. All I can say is that I am forever and continually being amazed at the depth revealed at each successive level we explore. I would like to conclude, as I did at the 1978 Stockholm Conference, with a prediction which J. R. Oppenheimer made more than twenty-five years ago and which has been fulfilled today in a manner he did not live to see. More than anything else, it expresses the faith for the future with which this greatest of decades in particle physics ends: "Physics will change even more. . . . If it is radical and unfamiliar. . . we think that the future will be only more radical and not less, only more strange and not more familiar, and that it will have its own new insights for the inquiring human spirit" (J. R. Oppenheimer, Reith Lectures, BBC, 1953).

APPENDIX A: EXAMPLES OF GRAND UNIFYING GROUPS

Appendix A is contained in Table A.1: Examples of grand unifying groups.

APPENDIX B: DOES THE GRAND PLATEAU REALLY EXIST

The following assumptions went into the derivation of the formula (*I*) in the text.

(a) $SU_L(2) \times U_{L,R}(1)$ survives intact as the electroweak symmetry group from energies $\approx \mu$ right up to M. This intact survival implies that one eschews, for example, all suggestions that (i) low-energy $SU_L(2)$ may be the diagonal sum of $SU_L^{I}(2), SU_L^{II}(2), SU_L^{III}(2)$, where I, II, III, refer to the (three?) known families; (ii) or that the $U_{L,R}(1)$ is a sum of pieces, where $U_R(1)$ may have differentially descended from a $(V+A)$-symmetric $SU_R(2)$ contained in G, or (iii) that U(1) contains a piece from a four-color symmetry $SU_c(4)$ (with lepton number as the fourth color) and with $SU_c(4)$ breaking at an intermediate mass scale to $SU_c(3) \times U_c(1)$.

(b) The second assumption which goes into the derivation of the formula above is that there are no unexpected heavy fundamental fermions, which might make $\sin^2\theta(M)$ differ from $\frac{3}{8}$—its value for the low mass fermions presently known to exist.[28]

(c) If these assumptions are relaxed, for example, for the three family group $G = [SU_F(6) \times SU_c(6)]_{L\to R}$, where $\sin^2\theta(M) = \frac{9}{28}$, we find the grand unifying mass M tumbles down to 10^6 GeV.

(d) The introduction of intermediate mass scales [for example, those connoting the breakdown of family universality, or of left-right symmetry, or of a breakdown of 4-color $SU_c(4)$ down to $SU_c(3) \times U_c(1)$] will as a rule push the magnitude of the grand unifying mass M upwards [see Salam (1979) and Shafi and Wetterich (1979)]. In order to secure a proton decay life, consonant with present empirical lower limits ($\sim 10^{30}$ years) (Learned *et al.*, 1979) this is desirable anyway. (τ_{proton} for $M \sim 10^{13}$ GeV is unacceptably low $\sim 6 \times 10^{23}$

[27]Grouping quarks (q) and leptons (l) together implies treating lepton number as the fourth color, i.e., $SU_c(3)$ extends to $SU_c(4)$ (Pati and Salam, 1974). A tribal group, by definition, contains all known families in its basic representation. Favored representations of tribal SU(11) (Georgi, 1979) and tribal SO(22) [Gell-Mann (1979) *et al.*] contain **561** and **2048** fermions!

[28]If one does not know G, one way to infer the parameter $\sin^2\theta(M)$ is from the formula:

$$\sin^2\theta(M) = \frac{\Sigma T_{3L}^2}{\Sigma Q^2} = \left[\frac{9N_q + 3N_l}{20N_q + 12N_l}\right].$$

Here N_q and N_l are the numbers of the fundamental quark and lepton SU(2) doublets (assuming these are the only multiplets that exist). If we make the further *assumption* that $N_q = N_l$ (from the requirement of anomaly cancellation between quarks and leptons) we obtain $\sin^2\theta(M) = \frac{3}{8}$. This assumption, however, is not compulsive; for example, anomalies cancel also if (heavy) mirror fermions exist (Pati *et al.*, 1975b; Pati and Salam, 1975b,c; Pati, 1975). This is the case for $[SU(6)]^4$ for which $\sin^2\theta(M) = 9/28$.

years unless there are 15 Higgs.) There is, from this point of view, an indication of there being in particle physics one or several intermediate mass scales which possibly start from around 10^4 GeV upwards. *This is the end result which I wished this Appendix to lead up to.*

REFERENCES

Akulov, V. P., D. V. Volkov, and V. A. Soroka, 1975, JETP Lett. **22**, 187.

Anderson, P. W., 1963, Phys. Rev. **130**, 439.

Arnowitt, R., P. Nath, and B. Zumino, 1975, Phys. Lett. B **56**, 81.

Ashmore, J., 1972, Nuovo Cimento Lett. **4**, 289.

Atiyah, M. F., and I. M. Singer, 1963, Bull. Am. Math. Soc. **69**, 422.

Barber, D. P., *et al.*, 1979, Phys. Rev. Lett. **43**, 830.

Barkov, L. M., 1979, in *Proceedings of the 19th International Conference on High Energy Physics* (Physical Society of Japan, Tokyo), p. 425, edited by S. Homma, M. Kawaguchi, and H. Miyazawa.

Bjorken, J. D., 1972, in the *Proceedings of the XVI International Conference on High Energy Physics* at Chicago-Batavia, edited by J. D. Jackson and A. Roberts (NAL, Batavia, Illinois), Vol. 2, p. 304.

Bludman, S., 1958, Nuovo Cimento **9**, 433.

Bollini, C., and J. Giambiagi, 1972, Nuovo Cimento B **12**, 20.

Bouchiat, C., J. Iliopoulos, and P. Meyer, 1972, Phys. Lett. B **38**, 519.

Boulware, D. G., 1970, Ann. Phys. (NY) **56**, 140.

Brandelik, R., *et al.*, 1979, Phys. Lett. B **86**, 243.

Brans, C. H., and R. H. Dicke, 1961, Phys. Rev. **124**, 925.

Brink, L., M. Gell-Mann, P. Ramond, and J. H. Schwarz, 1978, Phys. Lett. B **74**, 336.

Buras, A., J. Ellis, M. K. Gaillard, and D. V. Nanopoulos, 1978, Nucl. Phys. B **135**, 66.

Chamseddine, A. H., and P. C. West, 1977, Nucl. Phys. B **129**, 39.

Cremmer, E., and B. Julia, 1978, Phys. Lett. B **80**, 48.

Cremmer, E., and B. Julia, 1979, Ecole Normale Supérieure preprint, LPTENS 79/6, March.

Cremmer, E., B. Julia, and J. Scherk, 1978, Phys. Lett. B **76**, 409.

Cremmer, E., and J. Scherk, 1976a, Nucl. Phys. B **103**, 399.

Cremmer, E., and J. Scherk, 1976b, Nucl. Phys. B **108**, 409.

Cremmer, E., and J. Scherk, 1976c, Nucl. Phys. B **118**, 61.

Cremmer, E., *et al.*, 1979, Nucl. Phys. B **147**, 105.

Curtwright, T. L., and P. G. O. Freund, 1979, Enrico Fermi Institute preprint EFI 79/25, April.

D'Adda, A., M. Lüscher, and P. Di Vecchia, 1978, Nucl. Phys. B **146**, 63.

Deser, S., and B. Zumino, 1976, Phys. Lett. B **62**, 335.

DeWitt, B. S., 1967a, Phys. Rev. **162**, 1195.

DeWitt, B. S., 1967b, Phys. Rev. **162**, 1239.

Dimopoulos, S., and L. Susskind, 1978, Phys. Rev. D **18**, 4500.

Dimopoulos, S., and L. Susskind, 1979, Nucl. Phys. B **155**, 237.

Dirac, P. A. M., 1931, Proc. R. Soc. (London) A **133**, 60.

Einstein, A., 1912a, Ann. Phys. (Leipzig) **38**, 355.

Einstein, A., 1912b, Ann. Phys. (Leipzig) **38**, 433.

Einstein, A., 1916, Ann. Phys. (Leipzig) **49**, 769 [English translation in *The Principle of Relativity* (Methuen, 1923), reprinted by Dover (New York), p. 35].

Elias, V., J. C. Pati, and A. Salam, 1978a, Phys. Lett. B **73**, 451.

Elias, V., J. C. Pati, and A. Salam, 1978b, Phys. Rev. Lett. **40**, 920.

Ellis, J., 1979, in *Neutrino-79*, Proceedings of the International Conference on Neutrinos, Weak Interactions, and Cosmology, June 1979, Bergen, edited by A. Haatuft and C. Jarlskog, p. 451.

Ellis, J., M. K. Gaillard, and D. V. Nanopoulos, 1979, Phys. Lett. B **80**, 360 [**82**, 464(E) (1979)].

Englert, F., and R. Brout, 1964, Phys. Rev. Lett. **13**, 321.

Englert, F., R. Brout, and M. F. Thiry, 1966, Nuovo Cimento **48**, 244.

Faddeev, L. D., and V. N. Popov, 1967, Phys. Lett. B **25**, 29.

Fayet, P., 1977, Phys. Lett. B **70**, 461.

Fayet, P., 1979, Phys. Lett. B **84**, 421.

Ferrara, S., 1979, in Proceedings of the Second Marcel Grossmann Meeting, Trieste (in preparation), and references therein.

Feynman, R. P., 1963, Acta Phys. Pol. **24**, 297.

Feynman, R. P., and M. Gell-Mann, 1958, Phys. Rev. **109**, 193.

Fradkin, E. S., and I. V. Tyutin, 1970, Phys. Rev. D **2**, 2841.

Freedman, D. Z., 1979, in *Proceedings of the 19th International Conference on High Energy Physics* (Physical Society of Japan, Tokyo), edited by S. Homma, M. Kawaguchi, and H. Miyazawa.

Freedman, D. Z., P. van Nieuwenhuizen, and S. Ferrara, 1976, Phys. Rev. D **13**, 3214.

Friedman, J. I., and V. L. Telegdi, 1957, Phys. Rev. **105**, 1681.

Fritzsch, H., and M. Gell-Mann, 1972, in *Proceedings of the XVI International Conference on High Energy Physics*, Chicago-Batavia, edited by J. D. Jackson and A. Roberts (NAL, Batavia, Illinois), Vol. 2, p. 135.

Fritzsch, H., M. Gell-Mann, and H. Leutwyler, 1973, Phys. Lett. B **47**, 365.

Fritzsch, H., and P. Minkowski, 1975, Ann. Phys. (NY) **93**, 193.

Fritzsch, H., and P. Minkowski, 1976, Nucl. Phys. B **103**, 61.

Fuller, R. W., and J. A. Wheeler, 1962, Phys. Rev. **128**, 919.

Garwin, R., L. Lederman, and M. Weinrich, 1957, Phys. Rev. **105**, 1415.

Gell-Mann, M., 1979 (unpublished).

Georgi, H., 1975, in *Particles and Fields-1974* (APS/DPF, Williamsburg), edited by C. E. Carlson (AIP, New York), p. 575.

Georgi, H., 1979, Harvard University Report No. HUTP-29/A013.

Georgi, H., and S. L. Glashow, 1974, Phys. Rev. Lett. **32**, 438.

Georgi, H., and D. V. Nanopoulos, 1979, Phys. Lett. B **82**, 392.

Georgi, H., H. R. Quinn, and S. Weinberg, 1974, Phys. Rev. Lett. **33**, 451.

Gibbons, G. W., S. W. Hawking, and M. J. Perry, 1978, Nucl. Phys. B **138**, 141.

Glashow, S. L., 1959, Nucl. Phys. **10**, 107.

Glashow, S. L., 1961, Nucl. Phys. **22**, 579.

Glashow, S., J. Iliopoulos, and L. Maiani, 1970, Phys. Rev. D **2**, 1285.

Goldstone, J., 1961, Nuovo Cimento **19**, 154.

Goldstone, J., A. Salam, and S. Weinberg, 1962, Phys. Rev. **127**, 965.

Gol'fand, Yu. A., and E. P. Likhtman, 1971, JETP Lett. **13**, 323.

Gross, D. J., and R. Jackiw, 1972, Phys. Rev. D **6**, 477.

Gross, D. J., and J. Wilczek, 1973, Phys. Rev. D **8**, 3633.

Guralnik, G. S., C. R. Hagen, and T. W. B. Kibble, 1964, Phys. Rev. Lett. **13**, 585.

Harari, H., 1979, Phys. Lett. B **86**, 83.

Hasert, F. J., *et al.*, 1973, Phys. Lett. B **46**, 138.

Hawking, S. W., 1978, Phys. Rev. D **18**, 1747.

Hawking, S. W., 1979a, in *General Relativity: An Einstein Centenary Survey* (Cambridge University, Cambridge, England).

Hawking, S. W., 1979b, "Euclidean quantum gravity," DAMTP, Univ. of Cambridge preprint.

Higgs, P. W., 1964a, Phys. Lett. **12**, 132.

Higgs, P. W., 1964b, Phys. Rev. Lett. **13**, 508.

Higgs, P. W., 1966, Phys. Rev. **145**, 1156.
Jackiw, R., 1972, in *Lectures on Current Algebra and Its Applications*, by S. B. Treiman, R. Jackiw, and D. J. Gross (Princeton University, Princeton, N. J.).
Julia, B., 1979, in Proceedings of the Second Marcel Grossmann Meeting, Trieste (in preparation).
Kaluza, T., 1921, Sitzungsber. Preuss. Akad. Wiss. p. 966.
Kemmer, N., 1937, Phys. Rev. **52**, 906.
Kibble, T. W. K., 1967, Phys. Rev. **155**, 1554.
Klein, O., 1926, Z. Phys. **37**, 895.
Klein, O., 1939, "On the theory of charged fields," in *Le Magnétisme*, Proceedings of the Conference organized at the University of Strasbourg by the International Institute of Intellectual Cooperation, Paris.
Landau, L., 1957, Nucl. Phys. **3**, 127.
Learned, J., F. Reines, and A. Soni, 1979, Phys. Lett. **43**, 907.
Lee, B. W., 1972, Phys. Rev. D **5**, 823.
Lee, T. D., and C. N. Yang, 1956, Phys. Rev. **104**, 254.
Lee, T. D., and C. N. Yang, 1957, Phys. Rev. **105**, 1671.
Lee, B. W., and J. Zinn-Justin, 1972, Phys. Rev. D **5**, 3137.
Lee, B. W., and J. Zinn-Justin, 1973, Phys. Rev. D **7**, 1049.
MacDowell, S. W., and F. Mansouri, 1977, Phys. Rev. Lett. **38**, 739.
Mandelstam, S., 1968a, Phys. Rev. **175**, 1588.
Mandelstam, S., 1968b, Phys. Rev. **175**, 1604.
Marciano, W. J., 1979, Phys. Rev. D **20**, 274.
Marciano, W., and H. Pagels, 1978, Phys. Rep. C **36**, 137.
Marshak, R. E., Riazuddin, and C. P. Ryan, 1969, *Theory of Weak Interactions in Particle Physics* (Wiley-Interscience, New York).
Marshak, R. E., and E. C. G. Sudarshan, 1957, in *Proceedings of the Padua-Venice Conference on Mesons and Recently Discovered Particles* (Societa Italiana di Fisica).
Marshak, R. E., and E. C. G. Sudarshan, 1958, Phys. Rev. **109**, 1860.
Minkowski, P., 1977, University of Berne preprint, October.
Mohapatra, R. N., and J. C. Pati, 1975a, Phys. Rev. D **11**, 566.
Mohapatra, R. N., and J. C. Pati, 1975b, Phys. Rev. D **11**, 2558.
Nambu, Y., 1960, Phys. Rev. Lett. **4**, 380.
Nambu, Y., and G. Jona-Lasinio, 1961, Phys. Rev. **122**, 345.
Nanopoulos, D. V., 1979, CERN preprint TH2737, September.
Nanopoulos, D. V., and S. Weinberg, 1979, Harvard University preprint HUTP-79/A023.
Nördstrom, G., 1912, Phys. Z. **13**, 1126.
Nördstrom, G., 1913a, Ann. Phys. (Leipzig) **40**, 856.
Nördstrom, G., 1913b, Ann. Phys. (Leipzig) **42**, 533.
Nördstrom, G., 1914a, Ann. Phys. (Leipzig) **43**, 1101.
Nördstrom, G., 1914b, Phys. Z. **15**, 375.
Ogievetsky, V., and E. Sokatchev, 1978, Phys. Lett. B **79**, 222.
Pati, J. C., 1975, in *Theories and Experiments in High-Energy Physics*, edited by A. Perlmutter and S. Widmayer, Orbis Scientiae II, Coral Gables, Florida (Plenum, New York), p. 253.
Pati, J. C., and S. Rajpoot, 1978, Phys. Lett. B **79**, 65.
Pati, J. C., and A. Salam, 1973a, Phys. Rev. D **8**, 1240.
Pati, J. C., and A. Salam, 1973b, Phys. Rev. Lett. **31**, 661.
Pati, J. C., and A. Salam, 1974, Phys. Rev. D **10**, 275.
Pati, J. C., and A. Salam, 1975a, ICTP, Trieste, IC/75/106, Palermo Conference, June.
Pati, J. C., and A. Salam, 1975b, Phys. Rev. D **11**, 1137.
Pati, J. C., and A. Salam, 1975c, Phys. Rev. D **11**, 1149.
Pati, J. C., and A. Salam, 1980 (in preparation).
Pati, J. C., A. Salam, and J. Strathdee, 1975a, Phys. Lett. B **59**, 265.
Pati, J. C., A. Salam, and J. Strathdee, 1975b, Nuovo Cimento A **26**, 72.
Polyakov, A. M., 1974, JETP Lett. **20**, 194.
Rajpoot, S., and V. Elias, 1978, ICTP, Trieste, preprint IC/78/159.
Sakurai, J. J., 1958, Nuovo Cimento **7**, 1306.
Salam, A., 1957a, Nuovo Cimento **5**, 299.
Salam, A., 1957b, preprint, Imperial College, London.
Salam, A., 1968, in *Elementary Particle Theory*, Proceedings of the 8th Nobel Symposium, edited by N. Svartholm (Almqvist and Wiksell, Stockholm).
Salam, A., 1977, in *Physics and Contemporary Needs*, edited by Riazuddin (Plenum, New York), Vol. 1, p. 301.
Salam, A., 1979, in the Proceedings of the International Conference on High-Energy Physics, Geneva (European Physical Society, Cern), p. 853.
Salam, A., and J. Strathdee, 1970, Phys. Rev. D **2**, 2869.
Salam, A., and J. Strathdee, 1974a, Nucl. Phys. B **79**, 477.
Salam, A., and J. Strathdee, 1974b, Nucl. Phys. B **80**, 499.
Salam, A., and J. Strathdee, 1974c, Phys. Lett. B **51**, 353.
Salam, A., and J. Strathdee, 1978, Fortschr. Phys. **26**, 57.
Salam, A., and J. C. Ward, 1959, Nuovo Cimento **11**, 568.
Salam, A., and J. C. Ward, 1961, Nuovo Cimento **19**, 165.
Salam, A., and J. C. Ward, 1964, Phys. Lett. **13**, 168.
Scherk, J., 1979, Ecole Normale Supérieure preprint, LPTENS 79/17, September.
Schupe, M., 1979, Phys. Lett. B **86**, 87.
Schwinger, J., 1957, Ann. Phys. (NY) **2**, 407.
Shafi, Q., and C. Wetterich, 1979, Phys. Lett. B **85**, 52.
Shapiro, I. I., *et al.*, 1976, Phys. Rev. Lett. **36**, 555.
Shaw, R., 1955, "The problem of particle types and other contributions to the theory of elementary particles," Ph.D. thesis, Cambridge University (unpublished).
Siegel, W., 1977, Harvard University preprint HUTP-77/A068 (unpublished).
Slavnov, A., 1972, Theor. Math. Phys. **10**, 99.
Taylor, J. C., 1971, Nucl. Phys. B **33**, 436.
Taylor, J. G., 1977, King's College, London, preprint (unpublished).
Taylor, R. E., 1979, in *Proceedings of the 19th International Conference on High Energy Physics*, edited by S. Homma, M. Kawaguchi, and H. Miyazawa (Physical Society of Japan, Tokyo), p. 422.
't Hooft, G., 1971a, Nucl. Phys. B **33**, 173.
't Hooft, G., 1971b, Nucl. Phys. B **35**, 167.
't Hooft, G., 1974, Nucl. Phys. B **79**, 276.
't Hooft, G., and M. Veltman, 1972a, Nucl. Phys. B **44**, 189.
't Hooft, G., and M. Veltman, 1972b, Nucl. Phys. B **50**, 318.
Tiomno, J., 1956, Nuovo Cimento **1**, 226.
Tiomno, J., and J. A. Wheeler, 1949a, Rev. Mod. Phys. **21**, 144.
Tiomno, J., and J. A. Wheeler, 1949b, Rev. Mod. Phys. **21**, 153.
Toussiant, B., S. B. Treiman, F. Wilczek, and A. Zee, 1979, Phys. Rev. D **19**, 1036.
Volkov, D. V., and V. P. Akulov, 1972, JETP Lett. **16**, 438.
Weinberg, S., 1967, Phys. Rev. Lett. **27**, 1264.
Weinberg, S., 1973a, Phys. Rev. Lett. **31**, 494.
Weinberg, S., 1973b, Phys. Rev. D **8**, 4482.
Weinberg, S., 1979a, Phys. Rev. D **19**, 1277.
Weinberg, S., 1979b, Phys. Rev. Lett. **42**, 850.
Wentzel, G., 1937, Helv. Phys. Acta **10**, 108.
Wess, J., and B. Zumino, 1974, Nucl. Phys. B **70**, 39.
Wess, J., and B. Zumino, 1977, Phys. Lett. B **66**, 361.
Wheeler, J. A., 1964, in *Relativity Groups and Topology*, Proceedings of the Les Houches Summer School, 1963, edited by B. S. DeWitt and C. M. DeWitt (Gordon and Breach, New York).
Williams, J. G., *et al.*, 1976, Phys. Rev. Lett. **36**, 551.
Winter, K., 1979, in Proceedings of the International Symposium on Lepton and Photon Interactions at High Energies, Fermilab, August.
Wu, C. S., *et al.*, 1957, Phys. Rev. **105**, 1413.
Yang, C. N., and R. L. Mills, 1954, Phys. Rev. **96**, 191.
Yang, C. N., and J. Tiomno, 1950, Phys. Rev. **75**, 495.
Yoshimura, M., 1978, Phys. Rev. Lett. **41**, 381.
Zumino, B., 1975, in *Gauge Theories and Modern Field Theory*, Proceedings of a Conference held at Northeastern University, Boston, edited by R. Arnowitt and P. Nath (MIT, Cambridge, Mass.).

Towards a unified theory: Threads in a tapestry*

Sheldon Lee Glashow

Lyman Laboratory of Physics, Harvard University, Cambridge, Massachusetts, 02138

INTRODUCTION

In 1956, when I began doing theoretical physics, the study of elementary particles was like a patchwork quilt. Electrodynamics, weak interactions, and strong interactions were clearly separate disciplines, separately taught and separately studied. There was no coherent theory that described them all. Developments such as the observation of parity-violation, the successes of quantum electrodynamics, the discovery of hadron resonances and the appearance of strangeness were well-defined parts of the picture, but they could not be easily fitted together.

Things have changed. Today we have what has been called a "standard theory" of elementary particle physics in which strong, weak, and electromagnetic interactions all arise from a local symmetry principle. It is, in a sense, a complete and apparently correct theory, offering a qualitative description of all particle phenomena and precise quantitative predictions in many instances. There are no experimental data that contradict the theory. In principle, if not yet in practice, all experimental data can be expressed in terms of a small number of "fundamental" masses and coupling constants. The theory we now have is an integral work of art: the patchwork quilt has become a tapestry.

Tapestries are made by many artisans working together. The contributions of separate workers cannot be discerned in the completed work, and the loose and false threads have been covered over. So it is in our picture of particle physics. Part of the picture is the unification of weak and electromagnetic interactions and the prediction of neutral currents, now being celebrated by the award of the Nobel Prize. Another part concerns the reasoned evolution of the quark hypothesis from mere whimsy to established dogma. Yet another is the development of quantum chromodynamics into a plausible, powerful, and predictive theory of strong interactions. All is woven together in the tapestry; one part makes little sense without the other. Even the development of the electroweak theory was not as simple and straightforward as it might have been. It did not arise full blown in the mind of one physicist, nor even of three. It, too, is the result of the collective endeavor of many scientists, both experimenters and theorists.

Let me stress that I do not believe that the standard theory will long survive as a correct and complete picture of physics. All interactions may be gauge interactions, but surely they must lie within a unifying group. This would imply the existence of a new and very weak interaction which mediates the decay of protons. All matter is thus inherently unstable, and can be observed to decay. Such a synthesis of weak, strong, and electromagnetic interactions has been called a "grand unified theory", but a theory is neither grand nor unified unless it includes a description of gravitational phenomena. We are still far from Einstein's truly grand design.

Physics of the past century has been characterized by frequent great but unanticipated experimental discoveries. If the standard theory is correct, this age has come to an end. Only a few important particles remain to be discovered, and many of their properties are alleged to be known in advance. Surely this is not the way things will be, for Nature must still have some surprises in store for us.

Nevertheless, the standard theory will prove useful for years to come. The confusion of the past is now replaced by a simple and elegant synthesis. The standard theory may survive as a part of the ultimate theory, or it may turn out to be fundamentally wrong. In either case, it will have been an important way-station, and the next theory will have to be better.

In this talk, I shall not attempt to describe the tapestry as a whole, nor even that portion which is the electroweak synthesis and its empirical triumph. Rather, I shall describe several old threads, mostly overwoven, which are closely related to my own researches. My purpose is not so much to explain who did what when, but to approach the more difficult question of why things went as they did. I shall also follow several new threads which may suggest the future development of the tapestry.

EARLY MODELS

In the 1920's, it was still believed that there were only two fundamental forces: gravity and electromagnetism. In attempting to unify them, Einstein might have hoped to formulate a universal theory of physics. However, the study of the atomic nucleus soon revealed the need for two additional forces: the strong force to hold the nucleus together and the weak force to enable it to decay. Yukawa asked whether there might be a deep analogy between these new forces and electromagnetism. All forces, he said, were to result from the exchange of mesons. His conjectured mesons were originally intended to mediate both the strong and weak interactions: they were strongly coupled to nucleons and weakly coupled to leptons. This first attempt to unify strong and weak interactions was fully forty years premature. Not only this, but Yukawa could have predicted the existence of neutral currents. His neutral meson, essentially to provide the charge independence of nuclear forces, was also weakly coupled to pairs of leptons.

Not only is electromagnetism mediated by photons, but it arises from the requirement of local gauge invariance. This concept was generalized in 1954 to apply to non-Abelian local symmetry groups (Yang and Mills, 1954; Shaw, 1954). It soon became clear that a more far-reaching analogy might exist between elec-

*This lecture was delivered December 8, 1979, on the occasion of the presentation of the 1979 Nobel Prizes in Physics.

tromagnetism and the other forces. They, too, might emerge from a gauge principle.

A bit of a problem arises at this point. All gauge mesons must be massless, yet the photon is the only massless meson. How do the other gauge bosons get their masses? There was no good answer to this question until the work of Weinberg (1967) and Salam (1968) as proven by 't Hooft (for spontaneously broken gauge theories) ('t Hooft, 1971a, 1971b; Lee and Zinn-Justin, 1972; 't Hooft and Veltman, 1972) and of Gross, Wilczek, and Politzer (for unbroken gauge theories) (Gross and Wilczek, 1973; Politzer, 1973). Until this work was done, gauge meson masses had simply to be put in *ad hoc*.

Sakurai suggested in 1960 that strong interactions should arise from a gauge principle (Sakurai, 1960). Applying the Yang–Mills construct to the isospin-hypercharge symmetry group, he predicted the existence of the vector mesons ρ and ω. This was the first phenomenological $SU(2) \times U(1)$ gauge theory. It was extended to local SU(3) by Gell-Mann and Ne'eman in 1961 (see Gell-Mann and Ne'eman, 1964). Yet, these early attempts to formulate a gauge theory of strong interactions were doomed to fail. In today's jargon, they used "flavor" as the relevant dynamical variable, rather than the hidden and then unknown variable "color." Nevertheless, this work prepared the way for the emergence of quantum chromodynamics a decade later.

Early work in nuclear beta decay seemed to show that the relevant interaction was a mixture of S, T, and P. Only after the discovery of parity violation, and the undoing of several wrong experiments, did it become clear that the weak interactions were in reality $V-A$. The synthesis of Feynman and Gell-Mann and of Marshak and Sudarshan was a necessary precursor to the notion of a gauge theory of weak interactions (Feynmann and Gell-Mann, 1958; Marshak and Sudarshan, 1958). Bludman formulated the first SU(2) gauge theory of weak interactions in 1958 (Bludman, 1958). No attempt was made to include electromagnetism. The model included the conventional charged current interactions, and in addition, a set of neutral current couplings. These are of the same strength and form as those of today's theory in the limit in which the weak mixing angle vanishes. Of course, a gauge theory of weak interactions alone cannot be made renormalizable. For this, the weak and electromagnetic interactions must be unified.

Schwinger, as early as 1956, believed that the weak and electromagnetic interactions should be combined into a gauge theory (Schwinger, 1958). The charged massive vector intermediary and the massless photon were to be the gauge mesons. As his student, I accepted this faith. In my 1958 Harvard thesis, I wrote: "It is of little value to have a potentially renormalizable theory of beta processes without the possibility of a renormalizable electrodynamics. We should care to suggest that a fully acceptable theory of these interactions may only be achieved if they are treated together..." (Glashow, 1958). We used the original SU(2) gauge interaction of Yang and Mills. Things had to be arranged so that the charged current, but not the neutral (electromagnetic) current, would violate parity and strangeness. Such a theory is technically possible to construct, but it is both ugly and experimentally false (Georgi and Glashow, 1972). We know now that neutral currents do exist and that the electroweak gauge group must be larger than SU(2).

Another electroweak synthesis without neutral currents was put forward by Salam and Ward in 1959 (Salam and Ward, 1959). Again, they failed to see how to incorporate the experimental fact of parity violation. Incidentally, in a continuation of their work in 1961, they suggested a gauge theory of strong, weak, and electromagnetic interactions based on the local symmetry group $SU(2) \times SU(2)$ (Salam and Ward, 1961). This was a remarkable portent of the $SU(3) \times SU(2) \times U(1)$ model which is accepted today.

We come to my own work (Glashow, 1961), done in Copenhagen in 1960, and done independently by Salam and Ward (Salam and Ward, 1964). We finally saw that a gauge group larger than SU(2) was necessary to describe the electroweak interactions. Salam and Ward were motivated by the compelling beauty of gauge theory. I thought I saw a way to a renormalizable scheme. I was led to the group $SU(2) \times U(1)$ by analogy with the approximate isospin-hypercharge group which characterizes strong interactions. In this model there were two electrically neutral intermediaries: the massless photon and a massive neutral vector meson which I called B but which is now known as Z. The weak mixing angle determined to what linear combination of $SU(2) \times U(1)$ generators B would correspond. The precise form of the predicted neutral current interaction has been verified by recent experimental data. However, the strength of the neutral current was not prescribed, and the model was not in fact renormalizable. These glaring omissions were to be rectified by the work of Salam and Weinberg and the subsequent proof of renormalizability. Furthermore, the model was a model of leptons—it could not evidently be extended to deal with hadrons.

RENORMALIZABILITY

In the late 50's, quantum electrodynamics and pseudoscalar meson theory were known to be renormalizable, thanks in part to work of Salam. Neither of the customary models of weak interactions—charged intermediate vector bosons or direct four-fermion couplings—satisfied this essential criterion. My thesis at Harvard, under the direction of Julian Schwinger, was to pursue my teacher's belief in a unified electroweak gauge theory. I had found some reason to believe that such a theory was less singular than its alternatives. Feinberg, working with charged intermediate vector mesons, discovered that a certain type of divergence would cancel for a special value of the meson anomalous magnetic moment (Feinberg, 1958). It did not correspond to a "minimal electromagnetic coupling," but to the magnetic properties demanded by a gauge theory. Tzou Kuo-Hsien examined the zero-mass limit of charged vector meson electrodynamics (Kuo-Hsien, 1957). Again, a sensible result is obtained only for a very special choice of the magnetic dipole moment and electric quadrupole moment, just the values assumed in a gauge theory. Was it just coincidence that the electromagnetism of a charged vector meson was least pathological in a gauge theory?

Inspired by these special properties, I wrote a notorious paper (Glashow, 1959). I alleged that a softly broken gauge theory, with symmetry breaking provided by explicit mass terms, was renormalizable. It was quickly shown that this is false.

Again, in 1970, Iliopoulos and I showed that a wide class of divergences that might be expected would cancel in such a gauge theory (Glashow and Iliopoulos, 1971). We showed that the naive divergences of order $(\alpha\Lambda^4)^n$ were reduced to "merely" $(\alpha\Lambda^2)^n$, where Λ is a cut-off momentum. This is probably the most difficult theorem that Iliopoulos or I had ever proven. Yet, our labors were in vain. In the spring of 1971, Veltman informed us that his student Gerhart 't Hooft had established the renormalizability of spontaneously broken gauge theory.

In pursuit of renormalizability, I had worked diligently but I completely missed the boat. The gauge symmetry is an exact symmetry, but it is hidden. One must not put in mass terms by hand. The key to the problem is the idea of spontaneous symmetry breakdown: the work of Goldstone as extended to gauge theories by Higgs and Kibble in 1964.[1] These workers never thought to apply their work on formal field theory to a phenomenologically relevant model. I had had many conversations with Goldstone and Higgs in 1960. Did I neglect to tell them about my $SU(2)\times U(1)$ model, or did they simply forget?

Both Salam and Weinberg had had considerable experience in formal field theory, and they had both collaborated with Goldstone on spontaneous symmetry breaking. In retrospect, it is not so surprising that it was they who first used the key. Their $SU(2)\times U(1)$ gauge symmetry was spontaneously broken. The masses of the W and Z and the nature of neutral current effects depend on a single measurable parameter, not two as in my unrenormalizable model. The strength of the neutral currents was correctly predicted. The daring Weinberg-Salam conjecture of renormalizability was proven in 1971. Neutral currents were discovered in 1973 (Hasert *et al.*, 1973; Hasert *et al.*, 1974; Benvenuti *et al.*, 1974), but not until 1978 was it clear that they had just the predicted properties (Prescott *et al.*, 1978).

THE STRANGENESS-CHANGING NEUTRAL CURRENT

I had more or less abandoned the idea of an electroweak gauge theory during the period 1961–1970. Of the several reasons for this, one was the failure of my naive foray into renormalizability. Another was the emergence of an empirically successful description of strong interactions—the SU(3) unitary symmetry scheme of Gell-Mann and Ne'eman. This theory was originally phrased as a gauge theory, with ρ, ω, and κ^* as gauge mesons. It was completely impossible to imagine how both strong and weak interactions could be gauge theories: there simply wasn't room enough for commuting structures of weak and strong currents. Who could foresee the success of the quark model, and the displacement of SU(3) from the arena of flavor to that of color? The predictions of unitary symmetry were being borne out—the predicted Ω^- was discovered in 1964. Current algebra was being successfully exploited. Strong interactions dominated the scene.

When I came upon the $SU(2)\times U(1)$ model in 1960, I had speculated on a possible extension to include hadrons. To construct a model of leptons alone seemed senseless: nuclear beta decay, after all, was the first and foremost problem. One thing seemed clear. The fact that the charged current violated strangeness would force the neutral current to violate strangeness as well. It was already well known that strangeness-changing neutral currents were either strongly suppressed or absent. I concluded that the Z^0 had to be made very much heavier than the $W^\pm$. This was an arbitrary but permissible act in those days: the symmetry breaking mechanism was unknown. I had "solved" the problem of strangeness-changing neutral currents by suppressing all neutral currents: The baby was lost with the bath water.

I returned briefly to the question of gauge theories of weak interactions in a collaboration with Gell-Mann in 1961 (Gell-Mann and Glashow, 1961). From the recently developing ideas of current algebra we showed that a gauge theory of weak interactions would inevitably run into the problem of strangeness-changing neutral currents. We concluded that something essential was missing. Indeed it was. Only after quarks were invented could the idea of the fourth quark and the GIM (Glashow-Iliopoulos-Maini) mechanism arise.

From 1961 to 1964, Sidney Coleman and I devoted ourselves to the exploitation of the unitary symmetry scheme. In the spring of 1964, I spent a short leave of absence in Copenhagen. There, Bjorken and I suggested that the Gell-Mann-Zweig system of three quarks should be extended to four (Bjorken and Glashow, 1964). [Other workers had the same idea at the same time (Amati *et al.*, 1964; Hara, 1964; Okun, 1964; Maki and Ohnuki, 1964; Nauenberg, 1964; Teplitz and Tarjanne, 1963).] We called the fourth quark the charmed quark. Part of our motivation for introducing a fourth quark was based on our mistaken notions of hadron spectroscopy. But we also wished to enforce an analogy between the weak leptonic current and the weak hadronic current. Because there were two weak doublets of leptons, we believed there had to be two weak doublets of quarks as well.

The weak current Bjorken and I introduced in 1964 was precisely the GIM current. The associated neutral current, as we noted, conserved strangeness. Had we inserted these currents into the earlier electroweak theory, we would have solved the problem of strangeness-changing neutral currents. We did not. I had apparently quite forgotton my earlier ideas of electroweak synthesis. The problem which was explicitly posed in 1961 was solved, in principle, in 1964. No one, least of all me, knew it. Perhaps we were all befuddled by the chimera of relativistic SU(6), which arose at about this time to cloud the minds of theorists.

Five years later, John Iliopoulos, Luciano Maiani and I returned to the question of strangeness-changing neutral currents (Glashow, Iliopoulos, and Maiani,

[1]Many authors are involved with this work: R. Brout, F. Englest, J. Goldstone, G. Guralnik, C. Hagen, P. Higgs, G. Jona-Lasinio, T. Kibble, and Y. Nambu.

1970). It seems incredible that the problem was totally ignored for so long. We argued that unobserved effects (a large K_1K_2 mass difference; decays like $K \rightarrow \pi\nu\bar{\nu}$; etc.) would be expected to arise in any of the known weak interaction models: four fermion couplings; charged vector meson models; or the electroweak gauge theory. We worked in terms of cut-offs, since no renormalizable theory was known at the time. We showed how the unwanted effects would be eliminated with the conjectured existence of a fourth quark. After languishing for a decade, the problem of the selection rules of the neutral current was finally solved. Of course, not everyone believed in the predicted existence of charmed hadrons.

This work was done fully three years after the epochal work of Weinberg and Salam and was presented in seminars at Harvard and at M.I.T. Neither I, nor my co-workers, nor Weinberg, sensed the connection between the two endeavors. We did not refer, nor were we asked to refer, to the Weinberg–Salam work in our paper.

The relevance became evident only a year later. Due to the work of 't Hooft, Veltman, Benjamin Lee, and Zinn–Justin, it became clear that the Weinberg–Salam *ansatz* was in fact a renormalizable theory. With GIM, it was trivially extended from a model of leptons to a theory of weak interactions. The ball was now squarely in the hands of the experimenters. Within a few years, charmed hadrons and neutral currents were discovered, and both had just the properties they were predicted to have.

FROM ACCELERATORS TO MINES

Pions and strange particles were discovered by passive experiments which made use of the natural flux of cosmic rays. However, in the last three decades, most discoveries in particle physics were made in the active mode, with the artificial aid of particle accelerators. Passive experimentation stagnates from a lack of funding and lack of interest. Recent developments in theoretical particle physics and in astrophysics may mark an imminent rebirth of passive experimentation. The concentration of virtually all high-energy physics endeavors at a small number of major accelerator laboratories may be a thing of the past.

This is not to say that the large accelerator is becoming extinct; it will remain an essential if not exclusive tool of high-energy physics. Do not forget that the existence of Z^0 at ~100 GeV is an essential but quite untested prediction of the electroweak theory. There will be additional dramatic discoveries at accelerators, and these will not always have been predicted in advance by theorists.

Consider the successes of the electroweak synthesis, and the fact that the only plausible theory of strong interactions is also a gauge theory. We must believe in the ultimate synthesis of strong, weak, and electromagnetic interactions. It has been shown how the strong and electroweak gauge groups may be put into a larger but simple gauge group (Georgi and Glashow, 1974). Grand unification—perhaps along the lines of the original SU(5) theory of Georgi and me—must be essentially correct. This implies that the proton, and indeed all nuclear matter, must be inherently unstable. Sensitive searches for proton decay are now being launched. If the proton lifetime is shorter than 10^{32} years, as theoretical estimates indicate, it will not be long before it is seen to decay.

Once the effect is discovered (and I am sure it will be), further experiments will have to be done to establish the precise modes of decay of nucleons. The selection rules, mixing angles, and space-time structure of a new class of effective four-fermion couplings must be established. The heroic days of the discovery of the nature of beta decay will be repeated.

The first generation of proton decay experiments is cheap, but subsequent generations will not be. Active and passive experiments will compete for the same dwindling resources.

Other new physics may show up in elaborate passive experiments. Today's theories suggest modes of proton decay which violate both baryon number and lepton number by unity. Perhaps this $\Delta B = \Delta L = 1$ law will be satisfied. Perhaps $\Delta B = -\Delta L$ transitions will be seen. Perhaps, as Pati and Salam suggest, the proton will decay into three leptons. Perhaps two nucleons will annihilate in $\Delta B = 2$ transitions. The effects of neutrino oscillations resulting from neutrino masses of a fraction of an electron volt may be detectable. "Superheavy isotopes" which may be present in the Earth's crust in small concentrations could reveal themselves through their multi-GeV decays. Neutrino bursts arising from distant astronomical catastrophes may be seen. The list may be endless or empty. Large passive experiments of the sort now envisioned have never been done before. Who can say what results they may yield?

PREMATURE ORTHODOXY

The discovery of the J/ψ in 1974 made it possible to believe in a system involving just four quarks and four leptons. Very quickly after this a third charged lepton (the tau) was discovered, and evidence appeared for a third $Q = -\frac{1}{3}$ quark (the b quark). Both discoveries were classic surprises. It became immediately fashionable to put the known fermions into families or generations:

$$\begin{pmatrix} u & \nu_e \\ d & e \end{pmatrix} \quad \begin{pmatrix} c & \nu_\mu \\ s & \mu \end{pmatrix} \quad \begin{pmatrix} t & \nu_\tau \\ b & \tau \end{pmatrix} .$$

The existence of a third $Q = \frac{2}{3}$ quark (the t quark) is predicted. The Cabibbo–GIM scheme is extended to a system of six quarks. The three-family system is the basis to a vast and daring theoretical endeavor. For example, a variety of papers have been written putting experimental constraints on the four parameters which replace the Cabibbo angle in a six quark system. The detailed manner of decay of particles containing a single b quark has been worked out. All that is wanting is experimental confirmation. A new orthodoxy has emerged, one for which there is little evidence, and one in which I have little faith.

The predicted t quark has not been found. While the upsilon mass is less than 10 GeV, the analogous $\bar{t}t$ particle, if it exists at all, must be heavier than 30 GeV. Perhaps it doesn't exist.

Howard Georgi and I, and others before us, have been working on models with no t quark (Georgi and Glashow, 1979). We believe this unorthodox view is as attractive as its alternative. And, it suggests a number of exciting experimental possibilities.

We assume that b and τ share a quantum number, like baryon number, that is essentially exactly conserved. (Of course, it may be violated to the same extent that baryon number is expected to be violated.) Thus, the b, τ system is assumed to be distinct from the lighter four quarks and four leptons. There is, in particular, no mixing between b and d or s. The original GIM structure is left intact. An additional mechanism must be invoked to mediate b decay, which is not present in the $SU(3) \times SU(2) \times U(1)$ gauge theory.

One possibility is that there is an additional SU(2) gauge interaction whose effects we have not yet encountered. It could mediate such decays of b as these

$$b \to \tau^+ + (e^- \text{ or } \mu^-) + (d \text{ or } s).$$

All decays of b would result in the production of a pair of leptons, including a τ^+ or its neutral partner. There are other possibilities as well, which predict equally bizarre decay schemes for b matter. How the b quark decays is not yet known, but it soon will be.

The new SU(2) gauge theory is called upon to explain CP violation as well as b decay. In order to fit experiment, three additional massive neutral vector bosons must exist, and they cannot be too heavy. One of them can be produced in e^+e^- annihilation, in addition to the expected Z^0. Our model is rife with experimental predictions—for example, a second Z^0, a heavier version of b and of τ, the production of τb in ep collisions, and the existence of heavy neutral unstable leptons which may be produced and detected in e^+e^- or in νp collisions.

This is not the place to describe our views in detail. Our models are highly speculative and hence they are likely to be wrong. The point I wish to make is simply that it is too early to convince ourselves that we know the future of particle physics. There are too many points at which the conventional picture may be wrong or incomplete. The $SU(3) \times SU(2) \times U(1)$ gauge theory with three families is certainly a good beginning, not to accept but to attack, extend, and exploit. We are far from the end.

ACKNOWLEDGMENTS

I wish to thank the Nobel Foundation for granting me the greatest honor to which a scientist may aspire. There are many without whom my work would never have been. Let me thank my scientific collaborators, especially James Bjorken, Sidney Coleman, Alvaro De Rújula, Howard Georgi, John Iliopoulos, and Luciano Maiani; the Niels Bohr Institute and Harvard University for their hospitality while my research on the electroweak interaction was done; Julian Schwinger for teaching me how to do scientific research in the first place; the Public School System of New York City, Cornell University, and Harvard University for my formal education; my high-school friends Gary Feinberg and Steven Weinberg for making me learn too much too soon of what I might otherwise have never learned at all; my parents and my two brothers for always encouraging a child's dream to be a scientist. Finally, I wish to thank my wife and my children for the warmth of their love.

REFERENCES

Amati, D., H. Bacry, J. Nuyts, and J. Prentki, 1964, Nuovo Cimento **34**, 1932.
Benvenuti, A., *et al.*, 1974, Phys. Rev. Lett. **32**, 800.
Bjorken, J., and S. L. Glashow, 1964, Phys. Lett. **11**, 84.
Bludman, S., 1958, Nuovo Cimento **9**, 433.
Feinberg, G., 1958, Phys. Rev. **110**, 1482.
Feynman, R., and M. Gell-Mann, 1958, Phys. Rev. **109**, 193.
Gell-Mann, M., and S. L. Glashow, 1961, Ann. Phys. **15**, 437.
Gell-Mann, M., and Y. Ne'eman, 1964, *The Eightfold Way* (Benjamin, New York).
Georgi, H., and S. L. Glashow, 1972, Phys. Rev. Lett. **28**, 1494.
Georgi, H., and S. L. Glashow, 1974, Phys. Rev. Lett. **33**, 438.
Georgi, H., and S. L. Glashow, 1979, Harvard preprint HUTP-79/A053.
Glashow, S. L., 1958, Ph.D. thesis (Harvard University).
Glashow, S. L., 1959, Nucl. Phys. **10**, 107.
Glashow, S. L., 1961, Nucl. Phys. **22**, 579.
Glashow, S. L., and J. Iliopoulos, 1971, Phys. Rev. D **3**, 1043.
Glashow, S. L., J. Iliopoulos, and L. Maiani, 1970, Phys. Rev. D **2**, 1285.
Gross, D. J., and F. Wilczek, 1973, Phys. Rev. Lett. **30**, 1343.
Hara, Y., 1964, Phys. Rev. **134**, B701.
Hasert, F. J., *et al.*, 1973, Phys. Lett. B **46**, 138.
Hasert, F. J., *et al.*, 1974, Nucl. Phys. B **73**, 1.
Kuo-Hsien, Tzou, 1957, C. R. Acad. Sci. (Paris) **245**, 289.
Lee, B. W., and J. Zinn-Justin, 1972, Phys. Rev. D **5**, 3121-3160.
Maki, Z., and Y. Ohnuki, 1964, Prog. Theor. Phys. **32**, 144.
Marshak, R., and E. C. G. Sudarshan, 1958, Phys. Rev. **109**, 1860.
Nauenberg, M., 1964 (unpublished).
Okun, L. B., 1964, Phys. Lett. **12**, 250.
Politzer, H. D., 1973, Phys. Rev. Lett. **30**, 1346.
Prescott, C. Y., *et al.*, 1978, Phys. Lett. B **77**, 347.
Sakurai, J. J., 1960, Ann. Phys. **11**, 1.
Salam, A., 1968, in *Elementary Particle Theory*, edited by N. Svartholm (Almqvist & Wiksells, Stockholm), p. 367.
Salam, A., and J. Ward, 1959, Nuovo Cimento **11**, 568.
Salam, A., and J. Ward, 1961, Nuovo Cimento **19**, 165.
Salam, A., and J. Ward, 1964, Phys. Lett. **13**, 168.
Schwinger, J., 1958, Ann. Phys. **2**, 407.
Shaw, R., 1954 (unpublished).
Teplitz, V., and P. Tarjanne, 1963, Phys. Rev. Lett. **11**, 447.
't Hooft, G., 1971a, Nucl. Phys. B **33**, 173.
't Hooft, G., 1971b, Nucl. Phys. B **35**, 167.
't Hooft, G., and M. Veltman, 1972, Nucl. Phys. B **44**, 189.
Weinberg, S., 1967, Phys. Rev. Lett. **19**, 1264.
Yang, C. N., and R. Mills, 1954, Phys. Rev. **96**, 191.

I. YANG-MILLS FIELD AND EARLY ATTEMPTS TO UNIFY WEAK AND ELECTROMAGNETIC INTERACTIONS

Conservation of Isotopic Spin and Isotopic Gauge Invariance*

C. N. Yang † and R. L. Mills
Brookhaven National Laboratory, Upton, New York
(Received June 28, 1954)

It is pointed out that the usual principle of invariance under isotopic spin rotation is not consistant with the concept of localized fields. The possibility is explored of having invariance under local isotopic spin rotations. This leads to formulating a principle of isotopic gauge invariance and the existence of a **b** field which has the same relation to the isotopic spin that the electromagnetic field has to the electric charge. The **b** field satisfies nonlinear differential equations. The quanta of the **b** field are particles with spin unity, isotopic spin unity, and electric charge $\pm e$ or zero.

INTRODUCTION

THE conservation of isotopic spin is a much discussed concept in recent years. Historically an isotopic spin parameter was first introduced by Heisenberg[1] in 1932 to describe the two charge states (namely neutron and proton) of a nucleon. The idea that the neutron and proton correspond to two states of the same particle was suggested at that time by the fact that their masses are nearly equal, and that the light stable even nuclei contain equal numbers of them. Then in 1937 Breit, Condon, and Present pointed out the approximate equality of $p-p$ and $n-p$ interactions in the ^1S state.[2] It seemed natural to assume that this equality holds also in the other states available to both the $n-p$ and $p-p$ systems. Under such an assumption one arrives at the concept of a total isotopic spin[3] which is conserved in nucleon-nucleon interactions. Experi-

* Work performed under the auspices of the U. S. Atomic Energy Commission.

† On leave of absence from the Institute for Advanced Study, Princeton, New Jersey.

[1] W. Heisenberg, Z. Physik **77**, 1 (1932).

[2] Breit, Condon, and Present, Phys. Rev. **50**, 825 (1936). J. Schwinger pointed out that the small difference may be attributed to magnetic interactions [Phys. Rev. **78**, 135 (1950)].

[3] The total isotopic spin **T** was first introduced by E. Wigner, Phys. Rev. **51**, 106 (1937); B. Cassen and E. U. Condon, Phys. Rev. **50**, 846 (1936).

ments in recent years[4] on the energy levels of light nuclei strongly suggest that this assumption is indeed correct, An implication of this is that all strong interactions such as the pion-nucleon interaction, must also satisfy the same conservation law. This and the knowledge that there are three charge states of the pion, and that pions can be coupled to the nucleon field *singly*, lead to the conclusion that pions have isotopic spin unity. A direct verification of this conclusion was found in the experiment of Hildebrand[5] which compares the differential cross section of the process $n+p\rightarrow\pi^0+d$ with that of the previously measured process $p+p\rightarrow\pi^++d$.

The conservation of isotopic spin is identical with the requirement of invariance of all interactions under isotopic spin rotation. This means that when electromagnetic interactions can be neglected, as we shall hereafter assume to be the case, the orientation of the isotopic spin is of no physical significance. The differentiation between a neutron and a proton is then a purely arbitrary process. As usually conceived, however, this arbitrariness is subject to the following limitation: once one chooses what to call a proton, what a neutron, at one space-time point, one is then not free to make any choices at other space-time points.

It seems that this is not consistent with the localized field concept that underlies the usual physical theories. In the present paper we wish to explore the possibility of requiring all interactions to be invariant under *independent* rotations of the isotopic spin at all space-time points, so that the relative orientation of the isotopic spin at two space-time points becomes a physically meaningless quantity (the electromagnetic field being neglected).

We wish to point out that an entirely similar situation arises with respect to the ordinary gauge invariance of a charged field which is described by a complex wave function ψ. A change of gauge[6] means a change of phase factor $\psi\rightarrow\psi'$, $\psi'=(\exp i\alpha)\psi$, a change that is devoid of any physical consequences. Since ψ may depend on x, y, z, and t, the relative phase factor of ψ at two different space-time points is therefore completely arbitrary. In other words, the arbitrariness in choosing the phase factor is local in character.

We define *isotopic gauge* as an arbitrary way of choosing the orientation of the isotopic spin axes at all space-time points, in analogy with the electromagnetic gauge which represents an arbitrary way of choosing the complex phase factor of a charged field at all space-time points. We then propose that all physical processes (not involving the electromagnetic field) be invariant under an isotopic gauge transformation, $\psi\rightarrow\psi'$, $\psi'=S^{-1}\psi$, where S represents a space-time dependent isotopic spin rotation.

To preserve invariance one notices that in electrodynamics it is necessary to counteract the variation of α with x, y, z, and t by introducing the electromagnetic field A_μ which changes under a gauge transformation as

$$A_\mu'=A_\mu+\frac{1}{e}\frac{\partial\alpha}{\partial x_\mu}.$$

In an entirely similar manner we introduce a B field in the case of the isotopic gauge transformation to counteract the dependence of S on x, y, z, and t. It will be seen that this natural generalization allows for very little arbitrariness. The field equations satisfied by the twelve independent components of the B field, which we shall call the **b** field, and their interaction with any field having an isotopic spin are essentially fixed, in much the same way that the free electromagnetic field and its interaction with charged fields are essentially determined by the requirement of gauge invariance.

In the following two sections we put down the mathematical formulation of the idea of isotopic gauge invariance discussed above. We then proceed to the quantization of the field equations for the **b** field. In the last section the properties of the quanta of the **b** field are discussed.

ISOTOPIC GAUGE TRANSFORMATION

Let ψ be a two-component wave function describing a field with isotopic spin $\frac{1}{2}$. Under an isotopic gauge transformation it transforms by

$$\psi=S\psi', \tag{1}$$

where S is a 2×2 unitary matrix with determinant unity. In accordance with the discussion in the previous section, we require, in analogy with the electromagnetic case, that all derivatives of ψ appear in the following combination:

$$(\partial_\mu-i\epsilon B_\mu)\psi.$$

B_μ are 2×2 matrices such that[7] for $\mu=1$, 2, and 3, B_μ is Hermitian and B_4 is anti-Hermitian. Invariance requires that

$$S(\partial_\mu-i\epsilon B_\mu')\psi'=(\partial_\mu-i\epsilon B_\mu)\psi. \tag{2}$$

Combining (1) and (2), we obtain the isotopic gauge transformation on B_μ:

$$B_\mu'=S^{-1}B_\mu S+\frac{i}{\epsilon}S^{-1}\frac{\partial S}{\partial x_\mu}. \tag{3}$$

The last term is similar to the gradiant term in the gauge transformation of electromagnetic potentials. In analogy to the procedure of obtaining gauge invariant field strengths in the electromagnetic case, we

[4] T. Lauritsen, Ann. Rev. Nuclear Sci. 1, 67 (1952); D. R. Inglis, Revs. Modern Phys. 25, 390 (1953).
[5] R. H. Hildebrand, Phys. Rev. 89, 1090 (1953).
[6] W. Pauli, Revs. Modern Phys. 13, 203 (1941).

[7] We use the conventions $\hbar=c=1$, and $x_4=it$. Bold-face type refers to vectors in isotopic space, not in space-time.

define now

$$F_{\mu\nu}=\frac{\partial B_\mu}{\partial x_\nu}-\frac{\partial B_\nu}{\partial x_\mu}+i\epsilon(B_\mu B_\nu-B_\nu B_\mu). \tag{4}$$

One easily shows from (3) that

$$F_{\mu\nu}'=S^{-1}F_{\mu\nu}S \tag{5}$$

under an isotopic gauge transformation.‡ Other simple functions of B than (4) do not lead to such a simple transformation property.

The above lines of thought can be applied to any field ψ with arbitrary isotopic spin. One need only use other representations S of rotations in three-dimensional space. It is reasonable to assume that different fields with the same total isotopic spin, hence belonging to the same representation S, interact with the same matrix field B_μ. (This is analogous to the fact that the electromagnetic field interacts in the same way with any charged particle, regardless of the nature of the particle. If different fields interact with different and independent B fields, there would be more conservation laws than simply the conservation of total isotopic spin.) To find a more explicit form for the B fields and to relate the B_μ's corresponding to different representations S, we proceed as follows.

Equation (3) is valid for any S and its corresponding B_μ. Now the matrix $S^{-1}\partial S/\partial x_\mu$ appearing in (3) is a linear combination of the isotopic spin "angular momentum" matrices T^i ($i=1, 2, 3$) corresponding to the isotopic spin of the ψ field we are considering. So B_μ itself must also contain a linear combination of the matrices T^i. But any part of B_μ in addition to this, $\bar{B}_\mu$, say, is a scalar or tensor combination of the T's, and must transform by the homogeneous part of (3), $\bar{B}_\mu'=S^{-1}\bar{B}_\mu S$. Such a field is extraneous; it was allowed by the very general form we assumed for the B field, but is irrelevant to the question of isotopic gauge. Thus the relevant part of the B field is of the form

$$B_\mu=2\mathbf{b}_\mu\cdot\mathbf{T}. \tag{6}$$

(Bold-face letters denote three-component vectors in isotopic space.) To relate the $\mathbf{b}_\mu$'s corresponding to different representations S we now consider the product representation $S=S^{(a)}S^{(b)}$. The B field for the combination transforms, according to (3), by

$$B_\mu'=[S^{(b)}]^{-1}[S^{(a)}]^{-1}BS^{(a)}S^{(b)}$$

$$+\frac{i}{\epsilon}[S^{(a)}]^{-1}\frac{\partial S^{(a)}}{\partial x_\mu}+\frac{i}{\epsilon}[S^{(b)}]^{-1}\frac{\partial S^{(b)}}{\partial x_\mu}.$$

But the sum of $B_\mu^{(a)}$ and $B_\mu^{(b)}$, the B fields corresponding to $S^{(a)}$ and $S^{(b)}$, transforms in exactly the same way, so that

$$B_\mu=B_\mu^{(a)}+B_\mu^{(b)}$$

(plus possible terms which transform homogeneously, and hence are irrelevant and will not be included). Decomposing $S^{(a)}S^{(b)}$ into irreducible representations, we see that the twelve-component field $\mathbf{b}_\mu$ in Eq. (6) is the same for all representations.

To obtain the interaction between any field ψ of arbitrary isotopic spin with the $\mathbf{b}$ field one therefore simply replaces the gradiant of ψ by

$$(\partial_\mu-2i\epsilon\mathbf{b}_\mu\cdot\mathbf{T})\psi, \tag{7}$$

where T^i ($i=1, 2, 3$), as defined above, are the isotopic spin "angular momentum" matrices for the field ψ.

We remark that the nine components of $\mathbf{b}_\mu$, $\mu=1, 2, 3$ are real and the three of $\mathbf{b}_4$ are pure imaginary. The isotopic-gauge covariant field quantities $F_{\mu\nu}$ are expressible in terms of $\mathbf{b}_\mu$:

$$F_{\mu\nu}=2\mathbf{f}_{\mu\nu}\cdot\mathbf{T}, \tag{8}$$

where

$$\mathbf{f}_{\mu\nu}=\frac{\partial\mathbf{b}_\mu}{\partial x_\nu}-\frac{\partial\mathbf{b}_\nu}{\partial x_\mu}-2\epsilon\mathbf{b}_\mu\times\mathbf{b}_\nu. \tag{9}$$

$\mathbf{f}_{\mu\nu}$ transforms like a vector under an isotopic gauge transformation. Obviously the same $\mathbf{f}_{\mu\nu}$ interact with all fields ψ irrespective of the representation S that ψ belongs to.

The corresponding transformation of $\mathbf{b}_\mu$ is cumbersome. One need, however, study only the infinitesimal isotopic gauge transformations,

$$S=1-2i\mathbf{T}\cdot\delta\boldsymbol{\omega}.$$

Then

$$\mathbf{b}_\mu'=\mathbf{b}_\mu+2\mathbf{b}_\mu\times\delta\boldsymbol{\omega}+\frac{1}{\epsilon}\frac{\partial}{\partial x_\mu}\delta\boldsymbol{\omega}. \tag{10}$$

FIELD EQUATIONS

To write down the field equations for the $\mathbf{b}$ field we clearly only want to use isotopic gauge invariant quantities. In analogy with the electromagnetic case we therefore write down the following Lagrangian density:[8]

$$-\tfrac{1}{4}\mathbf{f}_{\mu\nu}\cdot\mathbf{f}_{\mu\nu}.$$

Since the inclusion of a field with isotopic spin $\frac{1}{2}$ is illustrative, and does not complicate matters very much, we shall use the following total Lagrangian density:

$$\mathfrak{L}=-\tfrac{1}{4}\mathbf{f}_{\mu\nu}\cdot\mathbf{f}_{\mu\nu}-\bar{\psi}\gamma_\mu(\partial_\mu-i\epsilon\boldsymbol{\tau}\cdot\mathbf{b}_\mu)\psi-m\bar{\psi}\psi. \tag{11}$$

One obtains from this the following equations of motion:

$$\begin{aligned}\partial\mathbf{f}_{\mu\nu}/\partial x_\nu+2\epsilon(\mathbf{b}_\nu\times\mathbf{f}_{\mu\nu})+\mathbf{J}_\mu&=0,\\ \gamma_\mu(\partial_\mu-i\epsilon\boldsymbol{\tau}\cdot\mathbf{b}_\mu)\psi+m\psi&=0,\end{aligned} \tag{12}$$

‡ *Note added in proof.*—It may appear that B_μ could be introduced as an auxiliary quantity to accomplish invariance, but need not be regarded as a field variable by itself. It is to be emphasized that such a procedure violates the principle of invariance. Every quantity that is not a pure numeral (like 2, or M, or any definite representation of the γ matrices) should be regarded as a dynamical variable, and should be varied in the Lagrangian to yield an equation of motion. Thus the quantities B_μ must be regarded as independent fields.

[8] Repeated indices are summed over, except where explicitly stated otherwise. Latin indices are summed from 1 to 3, Greek ones from 1 to 4.

where

$$\mathbf{J}_\mu = i\epsilon\bar{\psi}\gamma_\mu\boldsymbol{\tau}\psi. \tag{13}$$

The divergence of $\mathbf{J}_\mu$ does not vanish. Instead it can easily be shown from (13) that

$$\partial\mathbf{J}_\mu/\partial x_\mu = -2\epsilon\mathbf{b}_\mu\times\mathbf{J}_\mu. \tag{14}$$

If we define, however,

$$\mathfrak{J}_\mu = \mathbf{J}_\mu + 2\epsilon\mathbf{b}_\nu\times\mathbf{f}_{\mu\nu}, \tag{15}$$

then (12) leads to the equation of continuity,

$$\partial\mathfrak{J}_\mu/\partial x_\mu = 0. \tag{16}$$

$\mathfrak{J}_{1,2,3}$ and $\mathfrak{J}_4$ are respectively the isotopic spin current density and isotopic spin density of the system. The equation of continuity guarantees that the total isotopic spin

$$\mathbf{T} = \int \mathfrak{J}_4 d^3x$$

is independent of time and independent of a Lorentz transformation. It is important to notice that $\mathfrak{J}_\mu$, like $\mathbf{b}_\mu$, does not transform exactly like vectors under isotopic space rotations. But the total isotopic spin,

$$\mathbf{T} = -\int \frac{\partial\mathbf{f}_{4i}}{\partial x_i} d^3x,$$

is the integral of the divergence of $\mathbf{f}_{4i}$, which transforms like a true vector under isotopic spin space rotations. Hence, under a general isotopic gauge transformation, if $S\rightarrow S_0$ on an infinitely large sphere, $\mathbf{T}$ would transform like an isotopic spin vector.

Equation (15) shows that the isotopic spin arises both from the spin-$\frac{1}{2}$ field ($\mathbf{J}_\mu$) and from the $\mathbf{b}_\mu$ field itself. Inasmuch as the isotopic spin is the source of the $\mathbf{b}$ field, this fact makes the field equations for the $\mathbf{b}$ field nonlinear, even in the absence of the spin-$\frac{1}{2}$ field. This is different from the case of the electromagnetic field, which is itself chargeless, and consequently satisfies linear equations in the absence of a charged field.

The Hamiltonian derived from (11) is easily demonstrated to be positive definite in the absence of the field of isotopic spin $\frac{1}{2}$. The demonstration is completely identical with the similar one in electrodynamics.

We must complete the set of equations of motion (12) and (13) by the supplementary condition,

$$\partial\mathbf{b}_\mu/\partial x_\mu = 0, \tag{17}$$

which serves to eliminate the scalar part of the field in $\mathbf{b}_\mu$. This clearly imposes a condition on the possible isotopic gauge transformations. That is, the infinitesimal isotopic gauge transformation $S=1-i\boldsymbol{\tau}\cdot\delta\boldsymbol{\omega}$ must satisfy the following condition:

$$2\mathbf{b}_\mu\times\frac{\partial}{\partial x_\mu}\delta\boldsymbol{\omega}+\frac{1}{\epsilon}\frac{\partial^2}{\partial x_\mu{}^2}\delta\boldsymbol{\omega}=0. \tag{18}$$

This is the analog of the equation $\partial^2\alpha/\partial x_\mu{}^2=0$ that must be satisfied by the gauge transformation $A_\mu' = A_\mu + e^{-1}(\partial\alpha/\partial x_\mu)$ of the electromagnetic field.

QUANTIZATION

To quantize, it is not convenient to use the isotopic gauge invariant Lagrangian density (11). This is quite similar to the corresponding situation in electrodynamics and we adopt the customary procedure of using a Lagrangian density which is not obviously gauge invariant:

$$\mathfrak{L} = -\frac{1}{2}\frac{\partial\mathbf{b}_\mu}{\partial x_\nu}\cdot\frac{\partial\mathbf{b}_\mu}{\partial x_\nu}+2\epsilon(\mathbf{b}_\mu\times\mathbf{b}_\nu)\frac{\partial\mathbf{b}_\mu}{\partial x_\nu} - \epsilon^2(\mathbf{b}_\mu\times\mathbf{b}_\nu)^2+\mathbf{J}_\mu\cdot\mathbf{b}_\mu-\bar{\psi}(\gamma_\mu\partial_\mu+m)\psi. \tag{19}$$

The equations of motion that result from this Lagrangian density can be easily shown to imply that

$$\frac{\partial^2}{\partial x_\nu{}^2}\mathbf{a}+2\epsilon\mathbf{b}_\nu\times\frac{\partial}{\partial x_\nu}\mathbf{a}=0,$$

where

$$\mathbf{a}=\partial\mathbf{b}_\mu/\partial x_\mu.$$

Thus if, consistent with (17), we put on one space-like surface $\mathbf{a}=0$ together with $\partial\mathbf{a}/\partial t=0$, it follows that $\mathbf{a}=0$ at all times. Using this supplementary condition one can easily prove that the field equations resulting from the Lagrangian densities (19) and (11) are identical.

One can follow the canonical method of quantization with the Lagrangian density (19). Defining

$$\boldsymbol{\Pi}_\mu = -\partial\mathbf{b}_\mu/\partial x_4 + 2\epsilon(\mathbf{b}_\mu\times\mathbf{b}_4),$$

one obtains the equal-time commutation rule

$$[b_\mu{}^i(x), \Pi_\nu{}^j(x')]_{t=t'} = -\delta_{ij}\delta_{\mu\nu}\delta^3(x-x'), \tag{20}$$

where $b_\mu{}^i$, $i=1,2,3$, are the three components of $\mathbf{b}_\mu$. The relativistic invariance of these commutation rules follows from the general proof for canonical methods of quantization given by Heisenberg and Pauli.[9]

The Hamiltonian derived from (19) is identical with the one from (11), in virtue of the supplementary condition. Its density is

$$H = H_0 + H_{\text{int}},$$

$$H_0 = -\tfrac{1}{2}\boldsymbol{\Pi}_\mu\cdot\boldsymbol{\Pi}_\mu+\frac{1}{2}\frac{\partial\mathbf{b}_\mu}{\partial x_j}\cdot\frac{\partial\mathbf{b}_\mu}{\partial x_j}+\bar{\psi}(\gamma_j\partial_j+m)\psi,$$

$$H_{\text{int}} = 2\epsilon(\mathbf{b}_i\times\mathbf{b}_4)\cdot\boldsymbol{\Pi}_i - 2\epsilon(\mathbf{b}_\mu\times\mathbf{b}_j)\cdot(\partial\mathbf{b}_\mu/\partial x_j) + \epsilon^2(\mathbf{b}_i\times\mathbf{b}_j)^2 - \mathbf{J}_\mu\cdot\mathbf{b}_\mu. \tag{21}$$

The quantized form of the supplementary condition is the same as in quantum electrodynamics.

[9] W. Heisenberg and W. Pauli, Z. Physik 56, 1 (1929).

PROPERTIES OF THE *b* QUANTA

The quanta of the **b** field clearly have spin unity and isotopic spin unity. We know their electric charge too because all the interactions that we proposed must satisfy the law of conservation of electric charge, which is exact. The two states of the nucleon, namely proton and neutron, differ by charge unity. Since they can transform into each other through the emission or absorption of a **b** quantum, the latter must have three charge states with charges $\pm e$ and 0. Any measurement of electric charges of course involves the electromagnetic field, which necessarily introduces a preferential direction in isotopic space at all space-time points. Choosing the isotopic gauge such that this preferential direction is along the z axis in isotopic space, one sees that for the nucleons

$$Q=\text{electric charge}=e(\tfrac{1}{2}+\epsilon^{-1}T^z),$$

and for the **b** quanta

$$Q=(e/\epsilon)T^z.$$

The interaction (7) then fixes the electric charge up to an additive constant for all fields with any isotopic spin:

$$Q=e(\epsilon^{-1}T^z+R). \tag{22}$$

The constants R for two charge conjugate fields must be equal but have opposite signs.[10]

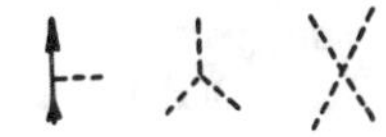

FIG. 1. Elementary vertices for **b** fields and nucleon fields. Dotted lines refer to **b** field, solid lines with arrow refer to nucleon field.

We next come to the question of the mass of the **b** quantum, to which we do not have a satisfactory answer. One may argue that without a nucleon field the Lagrangian would contain no quantity of the dimension of a mass, and that therefore the mass of the **b** quantum in such a case is zero. This argument is however subject to the criticism that, like all field theories, the **b** field is beset with divergences, and dimensional arguments are not satisfactory.

One may of course try to apply to the **b** field the methods for handling infinities developed for quantum electrodynamics. Dyson's approach[11] is best suited for the present case. One first transforms into the interaction representation in which the state vector Ψ

[10] See M. Gell-Mann, Phys. Rev. **92**, 833 (1953).

[11] F. J. Dyson, Phys. Rev. **75**, 486, 1736 (1949).

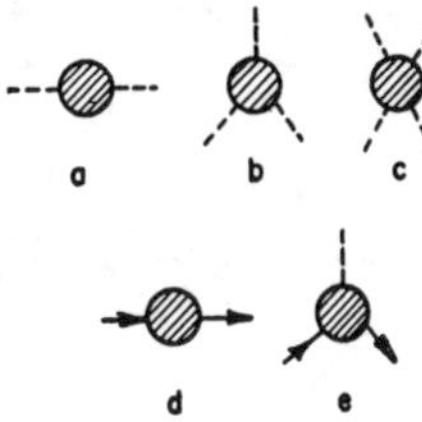

FIG. 2. Primitive divergences.

satisfies

$$i\partial\Psi/\partial t=H_{\text{int}}\Psi,$$

where H_{int} was defined in Eq. (21). The matrix elements of the scattering matrix are then formulated in terms of contributions from Feynman diagrams. These diagrams have three elementary types of vertices illustrated in Fig. 1, instead of only one type as in quantum electrodynamics. The "primitive divergences" are still finite in number and are listed in Fig. 2. Of these, the one labeled a is the one that effects the propagation function of the **b** quantum, and whose singularity determines the mass of the **b** quantum. In electrodynamics, by the requirement of electric charge conservation,[12] it is argued that the mass of the photon vanishes. Corresponding arguments in the **b** field case do not exist[13] even though the conservation of isotopic spin still holds. We have therefore not been able to conclude anything about the mass of the **b** quantum.

A conclusion about the mass of the **b** quantum is of course very important in deciding whether the proposal of the existence of the **b** field is consistent with experimental information. For example, it is inconsistent with present experiments to have their mass less than that of the pions, because among other reasons they would then be created abundantly at high energies and the charged ones should live long enough to be seen. If they have a mass greater than that of the pions, on the other hand, they would have a short lifetime (say, less than 10^{-20} sec) for decay into pions and photons and would so far have escaped detection.

[12] J. Schwinger, Phys. Rev. **76**, 790 (1949).

[13] In electrodynamics one can formally prove that $G_{\mu\nu}k_\nu=0$, where $G_{\mu\nu}$ is defined by Schwinger's Eq. (A12). ($G_{\mu\nu}A_\nu$ is the current generated through virtual processes by the arbitrary external field A_ν.) No corresponding proof has been found for the present case. This is due to the fact that in electrodynamics the conservation of charge is a consequence of the equation of motion of the electron field alone, quite independently of the electromagnetic field itself. In the present case the **b** field carries an isotopic spin and destroys such general conservation laws.

A Theory of the Fundamental Interactions

JULIAN SCHWINGER

Harvard University, Cambridge, Massachusetts

"The axiomatic basis of theoretical physics cannot be extracted from experience but must be freely invented."

A. Einstein.

This note is an account of some developments in an effort to find a description of the present stock of elementary particles within the framework of the theory of quantized fields[1].

The theory of fields suggests that the spin values of ½ for F(ermi)-D(irac) fields, and of 0 and 1 for B(ose)-E(instein) fields are not only exceptional in their simplicity but are likely to be unique in the possibility of constructing a consistent formalism for particles with mass and electric charge. We shall attempt to describe the massive, strongly interacting particles by means of fields with the smallest spin appropriate to the statistics, 0(B.E.) and ½(F.D.). Spin 1 remains a possibility for B.E. fields but will be assumed to refer to a different family of particles, of which the electromagnetic field may be a special example. If the spin values are thus limited, the origin of the diversity of known particles must be sought in internal degrees of freedom. We suppose that the various intrinsic degrees of freedom are dynamically exhibited by specific interactions, each with its characteristic symmetry properties, and that the final effect of interactions with successively lower symmetry[2] is to produce a spectrum of physically distinct particles from initially degenerate states. Thus we attempt to relate the observed masses to the same couplings responsible for the production and interaction of these particles

The general multicomponent Hermitian field χ separates into the F.D. field ψ and the B.E. field ϕ. The representation of spin ½ requires 4 components, while spin 0 demands 5 components, decomposable into a scalar and a vector. The existence of internal degrees of freedom is expressed by an additional multiplicity

[1] The initial stages of this work are described in a previous paper (*1*). The further considerations recorded here were first presented in a series of lectures delivered at Harvard and Massachusetts Institute of Technology, Oct.–Dec., 1956.

[2] The concept of a hierarchy of interactions has been discussed before by Pais (*2*), but with reference to particle stability rather than the mass spectrum.

or degeneracy of each set of fields, with the corresponding freedom of transformations in the internal symmetry space. In virtue of physical positive-definite requirements, referring to the commutation relations for the ψ field, and to the energy for the ϕ field, the metric of the symmetry space is necessarily Euclidean. This is indicated in the structure of the general Lagrange function

$$\mathcal{L} = \tfrac{1}{4}(\chi A^{\mu}\partial_{\mu}\chi - \partial_{\mu}\chi A^{\mu}\chi) - \mathcal{H}(\chi)$$

by stating that the matrices A^{μ} refer only to the space-time properties of the various fields. A classification of internal degrees of freedom follows from the consideration of continuous transformation groups with various numbers of independent infinitesimal operations, as represented by antisymmetrical, imaginary matrices T_a. Associated with each symmetry transformation T is a physical quantity T, described by a flux vector

$$j_T^{\mu} = -\frac{i}{2}\chi A^{\mu} T\chi,$$

which is conserved if the transformation is a dynamical invariance property. Then the total content of property T, symbolized by the Hermitian operator

$$\mathbf{T} = \int d\sigma_{\mu} j_T^{\mu},$$

is a constant of the motion. From an evident analogy we speak of an individual property of this type as a charge. The requirement that a number of symmetry operations form a continuous group is expressed by the commutation relations

$$[T_a, T_b] = \textstyle\sum_c t_{abc} T_c,$$

which must also be obeyed by the Hermitian operators $\mathbf{T}_a$, as the generators of an isomorphic group of unitary transformations. The connection between the two groups is provided by the field commutation relations

$$[\chi, \mathbf{T}_a] = T_a\chi.$$

A study of the group requirement shows that the internal symmetry group can be factored into a completely commutative group, and an essentially noncommutative one with a structure characterized by the property that the matrices

$$t_b = (t_{abc})$$

constitute a T matrix representation of dimensionality equal to n, the number of transformation parameters.

The first examples of the latter type are encountered for $n = 3$ and $n = 6$. For $n = 3$, the group structure is uniquely that of the three-dimensional Euclidean rotation group. We can conclude that the group with $n = 6$ contains the three-dimensional rotation group as a subgroup, if we make explicit use of the physical requirement that a symmetry group possess the groups of lower order as subgroups in order to effect a systematic reduction of internal symmetry through the addition of interactions with lower symmetry. It then follows that the six parameter group factors into the product of two three-dimensional rotation groups. This group can also be described as the four-dimensional Euclidean rotation group, for which the six independent matrices

$$T_{ab} = -T_{ba} \qquad a, b = 1 \cdots 4$$

obey the commutation relations

$$\frac{1}{i}[T_{ab}, T_{cd}] = \delta_{ac}T_{bd} - \delta_{bc}T_{ad} + \delta_{bd}T_{ac} - \delta_{ad}T_{bc}.$$

The factorization is accomplished by the definitions

$$T_3 = \tfrac{1}{2}(T_{12} + T_{34}), \qquad Z_3 = \tfrac{1}{2}(T_{12} - T_{34}) \tag{1}$$

and their cyclic permutations.

This investigation of the simplest group structures is supplemented by an examination of the symmetries that can be represented by matrices of small dimensionality, ν. The number of independent, antisymmetrical matrices, $\frac{1}{2}\nu(\nu - 1)$, coincides with the number of rotation planes in a ν-dimensional space and we thus encounter elementary representations of the various rotation groups, as provided by the matrices.

$$(T_{ab})_{cd} = \frac{1}{i}(\delta_{ac}\delta_{bd} - \delta_{ad}\delta_{bc}).$$

Each matrix has the eigenvalues 0, ± 1, with the exception of $\nu = 2$ where the 0 eigenvalue does not occur. Hence, beginning with $\nu = 2$, there is one independent antisymmetrical matrix and the symmetry space is two-dimensional; three dimensional symmetries cannot be described. For $\nu = 3$, with three antisymmetrical matrices, we obtain the $T = 1$ representation of the three-dimensional rotation group, appropriate to the rotations of a vector in that space. With $\nu = 4$, and six antisymmetrical matrices, we can describe the rotations of a vector in the four-dimensional space, which combine the $T = 0$ and $T = 1$ representations of the three-dimensional group. Alternatively, we can use (1) to construct the matrices

$$T_k = \tfrac{1}{2}\tau_k, \qquad Z_k = \tfrac{1}{2}\zeta_k, \qquad k = 1 \cdots 3,$$

which give two $T = \frac{1}{2}$ representations of the three-dimensional rotation group. Thus the same four-dimensional matrices can be viewed as referring to the three-dimensional representations $T = 0, 1$, or to a two-fold $T = \frac{1}{2}$ representation.

The distinction between integral and half-integral T would only appear when four-dimensional symmetries are reduced to three-dimensional ones, since we have the choice of selecting the three-dimensional transformations $T_1 = \frac{1}{2}\tau_1, \cdots$ ($T = \frac{1}{2}$) or the transformations $T_1 = T_{23}, \cdots$ ($T = 0, 1$). It may be noted here that, in the further reduction to two-dimensional symmetry that physically distinguishes T_{12}, the relation of this matrix to its three-dimensional origins differs in the two possibilities. Thus, for $T = 0, 1$, we have simply $T_{12} = T_3$ while, for $T = \frac{1}{2}$, there appears $T_{12} = T_3 + Z_3$, where $Z_3 = \frac{1}{2}\zeta_3$. The two forms can be united by writing

$$T_{12} = T_3 + \tfrac{1}{2}Y, \tag{2}$$

in which Y, the hypercharge, is defined to be zero for integral T and is represented by the matrix ζ_3, with eigenvalues ± 1, for $T = \frac{1}{2}$.

Since the fundamental $T = \frac{1}{2}$ representation of the three-dimensional rotation group (familiar as isotopic spin) is first encountered within the framework of four-dimensional symmetries, it is natural to suppose that the latter is the underlying symmetry, and that fields exist realizing the two-fold $T = \frac{1}{2}$ representation and the equivalent $T = 0, 1$ representation. The reduction to three-dimensional symmetries should distinguish physically the two $T = \frac{1}{2}$ representations (labelled by $Y = \pm 1$) as well as the $T = 0$ and $T = 1$ representations. Now it is well known that the electromagnetic field is the physical agency that destroys the three-dimensional isotopic space symmetry and leaves two dimensional symmetries. Thus the physical property defined by the two-dimensional rotations is explicitly electrical charge, and (2) automatically provides the identification

$$Q = T_3 + \tfrac{1}{2}Y. \tag{3}$$

This general connection between isotopic spin, electrical charge, and hypercharge enables us to recognize that the simple representations to which we have been led are realized[3] completely for the heavy F.D. particles, where one encounters two charge doublets, $N(Y = +1)$ and $\Xi(Y = -1)$ [although Ξ^0 is unknown experimentally], a charge triplet, Σ, and a charge singlet, Λ. For the heavy B.E. particles, there exist two charge doublets, $K(Y = +1)$, $\tilde{K}(Y = -1)$, and a charge triplet, π, but no charge singlet (σ) is known. Of course, the absence of such a stable or metastable particle does not necessarily mean that a field of the corresponding type does not exist. Thus, if the σ-field were scalar, as contrasted with the known pseudoscalar character of the π-field, and the corresponding particles had masses greater than $2\mu_\pi$, these σ-particles would be highly unstable against rapid disintegration into two π-mesons. It is also possible that the B.E. field differs from the F.D. field in realizing only $T = \frac{1}{2}$, and not $T = 0, 1$,

[3] Indeed, the first successful classification of the heavy unstable particles (by Gell-Mann and Nishijima) employed an empirical relation having the form of (3).

despite the four-dimensional kinematical equivalence between integral and half-integral T. The origin of a $T = 1$ representation can then be found in the matrices defined by the structure of the commutation relations for the four-dimensional rotations, which are themselves representations of the symmetry operations. These matrices are six-dimensional and describe the transformations of an antisymmetrical tensor. Through the analog of the definitions (1), this tensor can be decomposed into two, each of which is self-dual (to within a sign change). A single self-dual tensor possesses three independent components and provides a basis for the $T = 1$ representation in three dimensions, without an accompanying $T = 0$ representation.

Current field theory describes the dynamics of fields by adding a coupling term in the Lagrange function to those of the uncoupled fields, which are, for spin 0 and spin ½ fields,

$$\mathfrak{L}^{(0)} = -\tfrac{1}{2}[\phi^{\lambda}.\partial_{\lambda}\phi - \phi.\partial_{\lambda}\phi^{\lambda} - \phi^{\lambda}\phi_{\lambda} + \mu_0^2\phi^2]$$

and

$$\mathfrak{L}^{(\frac{1}{2})} = -\tfrac{1}{2}\left[\psi.\beta\gamma^{\mu}\frac{1}{i}\partial_{\mu}\psi + m_0\psi.\beta\psi\right].$$

The dot indicates symmetrical and antisymmetrical multiplication, respectively, as is appropriate to the statistics of these fields. Also $x_k = x^k$, $k = 1 \cdots 3$, $x_0 = -x^0$, and the matrix $\beta = \gamma^0$ is antisymmetrical and imaginary, while the γ^k are symmetrical and imaginary. The constants μ_0 and m_0 are identified as the masses of the field quanta in the absence of interaction. It should also be noted that, in the natural units we are employing ($\hbar = c = 1$), the B.E. field ϕ is an inverse length $[L^{-1}]$ while the F.D. field ψ has the dimensions $[L^{-3/2}]$, since the Lagrange function is in the nature of an inverse four-dimensional volume. In seeking possible forms for the interaction term in $\mathfrak{L}$ we shall be guided by the heuristic principle that the coupling between fields is described by simple algebraic functions of the field operators in which only dimensionless constants appear. This principle expresses the attitude that present theories, which are based on the infinite divisibility of the space-time manifold, contain no intrinsic standard of length. The mass constants of the individual fields are regarded as phenomenological manifestations of the unknown physical agency that produces the failure in the conventional space-time description and establishes the absolute scale of length and of mass. On this view, the coupling terms employed within the present formalism should not embody a unit of length that finds its dynamical origin outside of the domain of physical experience to which the theory of fields is applicable. For interacting spin 0 and spin ½ fields only two types of coupling terms are admitted by this principle, $\phi\psi\psi$ and $\phi\phi\phi\phi$.

We are interested in fields that possess internal degrees of freedom corresponding to the possible T values of 0, ½, and 1. Hence the Yukawa-type coupling term $\phi\psi\psi$, will be composed additively of structures of the type $\phi_{(0,1)}\psi_{(1/2)}\,\psi_{(1/2)}$, $\phi_{(0,1)}\psi_{(0,1)}\psi_{(0,1)}$, and $\phi_{(1/2)}\psi_{(1/2)}\psi_{(0,1)}$, according to the evident requirements stemming from the symmetry of the three-dimensional internal space. We must now examine the possibility of introducing dynamically the four-dimensional symmetry that provides the kinematical foundation for the three-dimensional isotopic space representations. The internal symmetry is superimposed upon the space-time structure which, according to the space reflection properties of the spin 0 B.E. field, can be $\phi\frac{1}{2}\psi\beta\psi$, or $\phi\frac{1}{2}\psi\beta\gamma_5\psi$, $\gamma_5 = \overset{0}{\gamma}\overset{1}{\gamma}\overset{2}{\gamma}\overset{3}{\gamma}$. It is important to remark that, since both matrices β and $\beta\gamma_5$ are antisymmetrical and imaginary, which are just the necessary symmetry and reality properties, any additional matrix referring to internal degrees of freedom must be symmetrical and real.

In considering the coupling term $\phi_{(0,1)}\psi_{(1/2)}\psi_{(1/2)}$, let us assume first that the integral T B.E. field is a self-dual antisymmetrical tensor, containing only three independent components ϕ_k, $k = 1 \cdots 3$, which must be combined with the similar tensor formed from the three matrices τ_k. The latter are antisymmetrical, imaginary matrices, however, and thus unacceptable according to the preceeding comments. Hence, to form an interaction of this type the $T = \frac{1}{2}$ F.D. fields must possess yet another internal degree of freedom in the nature of a charge, which will be identified with nucleonic charge N and which is represented by the antisymmetrical imaginary matrix

$$\nu = \begin{pmatrix} 0 & -i \\ i & 0 \end{pmatrix},$$

defined in an independent two-dimensional nucleonic charge space. The ensuing interaction term appears as

$$g_\pi \phi_{(1)} \frac{1}{2} \psi_{(1/2)}\beta\gamma_5\nu\tau\psi_{(1/2)}$$

where g_π occurs as a pure number measuring the strength of the interaction. A definite choice of space-time structure has been made, in accordance with the known pseudoscalar nature of the π-field, $\phi_{(1)}$.

The field $\psi_{(1/2)}$ now comprises, in its internal multiplicity, 2×4 Hermitian components which can be acted on by the two-dimensional matrix $\nu = \nu_3$, together with ν_1, ν_2, and the four-dimensional matrices τ_k, ζ_k. But, as we have observed, $T = \frac{1}{2}$ is only one possible interpretation of the transformations in the four-dimensional space to which the latter matrices refer. The matrix representation illustrated by

$$\tau_3 = \begin{pmatrix} 0 & -i & 0 & 0 \\ i & 0 & 0 & 0 \\ 0 & 0 & 0 & -i \\ 0 & 0 & i & 0 \end{pmatrix}$$

can equally be regarded as acting upon a four-dimensional vector, with the first three rows and columns referring to $T = 1$ and the last row and column to $T = 0$. In this way all statements pertaining to the field $\psi_{(1/2)}$ of internal multiplicity 2×4, can be translated into properties of the 2×4 fold degenerate field $\psi_{(0,1)}$. We are thus led to the concept of a universal π-heavy fermion interaction in which both integral and half-integral isotopic spin F.D. fields take part, in a manner that maintains the four-dimensional equivalence between the two distinct three-dimensional interpretations of the common kinematical structure. This universality also implies that nucleonic charge is a general property of the heavy F.D. particle field, without regard to isotopic spin. Thus the π-field emerges as the dynamical agency that defines nucleonic charge[4], the absolute conservation of which is the quantitative expression of the stability of nuclear matter. The complete π-coupling term that embodies these considerations is

$$\mathfrak{L}_\pi = g_\pi \phi_{(1)} \frac{1}{2} \Big[\psi_{(1/2)} \beta \gamma_5 \nu \tau \psi_{(1/2)} + \psi_{(0)} \beta \gamma_5 i \nu \psi_{(1)} - \psi_{(1)} \beta \gamma_5 i \nu \psi_{(0)} + \psi_{(1)} \beta \gamma_5 \nu \frac{1}{i} \times \psi_{(1)} \Big], \tag{4}$$

where the vector operation $(1/i)\times$ can also be expressed by the three-dimensional matrices t_k that provide the $T = 1$ representation. The general concept of the universal π-interaction is also compatible with the opposite choice of relative sign between the integral T and half-integral T terms.

The internal symmetry properties possessed by this coupling term are described by the following infinitesimal rotations:

$$\begin{aligned}
&N \text{ rotation:} && \psi \to (1 + i\delta\alpha\nu)\psi \\
& && \phi_{(1)} \to \phi_{(1)}; \\
&\tau \text{ rotation:} && \psi_{(1/2)} \to \left(1 + \frac{i}{2}\delta\omega\tau\right)\psi_{(1/2)}, \\
& && \psi_{(0)} \to \psi_{(0)} - \frac{1}{2}\delta\omega\psi_{(1)}, \\
& && \psi_{(1)} \to \left(1 + \frac{i}{2}\delta\omega t\right)\psi_{(1)} + \frac{1}{2}\delta\omega\psi_{(0)}, \\
& && \phi_{(1)} \to (1 + i\delta\omega t)\phi_{(1)};
\end{aligned} \tag{5}$$

[4] This theory can be regarded as the precise realization of some general remarks of Wigner (*3*).

ζ rotation:

$$\psi_{(1/2)} \to \left(1 + \frac{i}{2}\,\delta\epsilon\zeta\right)\psi_{(1/2)},$$

$$\psi_{(0)} \to \psi_{(0)} + \frac{1}{2}\,\delta\epsilon'\psi_{(1)}, \tag{6}$$

$$\psi_{(1)} \to \left(1 + \frac{i}{2}\,\delta\epsilon' t\right)\psi_{(1)} - \frac{1}{2}\,\delta\epsilon'\psi_{(0)},$$

$$\phi_{(1)} \to \phi_{(1)}.$$

The ζ rotations of the $\psi_{(1/2)}$ and $\psi_{(0,1)}$ fields are quite independent, which enables us to define the three-dimensional T rotation by superimposing the ζ rotation with $\delta\epsilon' = \delta\omega$ on the four-dimensional τ rotation:

T rotation:

$$\psi_{(1/2)} \to \left(1 + \frac{i}{2}\,\delta\omega\tau\right)\psi_{(1/2)},$$

$$\psi_{(0)} \to \psi_{(0)},$$

$$\psi_{(1)} \to (1 + i\delta\omega t)\psi_{(1)},$$

$$\phi_{(1)} \to (1 + i\delta\omega t)\phi_{(1)}.$$

Associated with each infinitesimal rotation is a conserved physical property. This is illustrated by the N rotation, which defines the current vector of nucleonic charge

$$j_N{}^\mu = \tfrac{1}{2}\psi\beta\gamma^\mu\nu\psi,$$

and thereby the total nucleonic charge,

$$N = \int d\sigma_\mu \frac{1}{2}\,\psi\beta\gamma^\mu\nu\psi.$$

The latter operator obeys commutation relations that express the nucleonic charge properties of the various fields,

$$[\psi, N] = \nu\psi, \qquad [\phi, N] = 0.$$

In addition to the symmetry properties described by rotations, there may also exist discrete symmetry operations in the nature of reflections. The introduction of the self-dual tensor $\phi_{(1)}$ excludes any such reflection invariance in the four-dimensional space. But, for the nucleonic charge, as represented by the matrix ν_3 with eigenvalues ± 1, there exists a transformation interchanging these eigenvalues—a nucleonic charge reflection. This is given by

$$\psi \to \nu_1\psi, \qquad \nu_1 = \begin{pmatrix} 1 & 0 \\ 0 & -1 \end{pmatrix},$$

$$\phi_{(1)} \to -\phi_{(1)},$$

the π-field reversal being required to maintain the structure of $\mathcal{L}_\pi$. Associated with this transformation is a unitary nucleonic charge reflection operator R_N, which is such that

$$R_N^{-1}\psi R_N = \nu_1\psi, \qquad R_N^{-1}\phi_{(1)}R_N = -\phi_{(1)}$$

$$R_N^{-1}NR_N = -N.$$

States of zero nucleonic charge can be characterized by the eigenvalues of R_N, which are consistently chosen as ± 1, and the π-field possesses only matrix elements connecting states with opposite values of this nucleonic charge parity.

The kinematical discussion of internal symmetry spaces applies equally to B.E. and F.D. fields. But the manner in which the two types of fields appear dynamically is quite distinct, as emphasized by the structure of the Yukawa coupling, $\phi\psi\psi$. The quadratic dependence upon ψ permits the application of four-dimensional invariance requirements including equivalence between integral and half-integral T representations, whereas the latter have quite different meanings for the field ϕ. Hence there is no fundamental objection to the four-dimensional representations being realized in the B.E. field only as $\phi_{(1/2)}$. Nevertheless we must ask whether there is some possibility of using the integral T interpretation of the four-dimensional representation and thereby introducing the B.E. field $\phi_{(0,1)}$. If the σ-field $\phi_{(0)}$ is to be incorporated with the universal π-coupling, it can only appear as $\phi_{(0)}\frac{1}{2}\psi\beta\psi$ or $\phi_{(0)}\frac{1}{2}\psi\beta\gamma_5\psi$, for, with the exception of the unit matrix there is no matrix of the internal symmetry spaces that is acceptable in its action upon both $\psi_{(1/2)}$ and $\psi_{(0,1)}$. Furthermore, if the complete field $\phi_{(0,1)}$ is pseudoscalar, there is no possibility of exhibiting transformations under which $\phi_{(0,1)}$ behaves as a four-dimensional vector, since $\psi\beta\gamma_5\psi$ is invariant under all transformations referring solely to the internal symmetry space. But such four-vector transformations do exist if the σ-field is scalar, as distinguished from the pseudoscalar π-field, and if due account is taken of the dynamical origin of mass. The σ-field coupling term to be added to $\mathcal{L}$ is then

$$\mathcal{L}_\sigma = g_\pi\phi_{(0)}\tfrac{1}{2}[\psi_{(1/2)}\beta\psi_{(1/2)} + \psi_{(0)}\beta\psi_{(0)} + \psi_{(1)}\beta\psi_{(1)}].$$

First let us note the special features that accompany the assumptions $m_0 = 0$ and $\mu_0 = 0$. The absence of the Dirac field mass term implies invariance of $\mathcal{L}^{(1/2)}$ under the rotation $\psi \to (1 + \delta\varphi\gamma_5)\psi$, while the Lagrange term $\mathcal{L}^{(0)}$ will be invariant to within an added divergence under the spin 0 gauge transformation: $\phi \to \phi + \lambda$, $\phi^\mu \to \phi^\mu$, with λ an arbitrary constant, if $\mu_0 = 0$. Now, if both mass constants are zero, the complete Lagrange function is invariant under the following infinitesimal transformation, which represents a partial union of space-time with the Euclidean internal space,

$$\psi_{(1/2)} \to (1 + \tfrac{1}{2}\, \gamma_5 \nu \delta\omega \tau)\, \psi_{(1/2)}\,,$$

$$\psi_{(0)} \to \psi_{(0)} + \frac{i}{2}\, \gamma_5 \nu \delta\omega \psi_{(1)}\,,$$

$$\psi_{(1)} \to \left(1 + \frac{1}{2}\, \gamma_5 \nu \delta\omega t\right) \psi_{(1)} - \frac{i}{2}\, \gamma_5 \nu \delta\omega \psi_{(0)}\,,$$

$$\phi_{(0)} \to \phi_{(0)} + \delta\omega \phi_{(1)}$$

$$\phi_{(1)} \to \phi_{(1)} - \delta\omega \phi_{(0)}\,.$$

If the integral T terms appear with signs opposite from those given in (4), the infinitesimal changes in $\psi_{(0,1)}$ also require the reversed signs. The invariance property we have described for $m_0 = \mu_0 = 0$ persists if only the μ_0 term is included. As to the Dirac mass term, it has the same structure as $\mathcal{L}_\sigma$ and forms the combination $m_0 - g_\pi \phi_{(0)}$. Hence if $m_0 \neq 0$, it is $\phi_{(0)} - (m_0/g_\pi)$ that transforms with $\phi_{(1)}$ as a four-vector and the addition of the constant to $\phi_{(0)}$ does not upset the invariance of $\mathcal{L}$ if $\mu_0 = 0$, according to the gauge invariance of $\mathcal{L}^{(0)}$ appropriate to that circumstance. Of the various possibilities, it is the latter one, $m_0 \neq 0$, $\mu_0 = 0$, that is to be preferred, for, within our dynamical scheme there is no effective source of a Dirac field mass term, whereas the $\phi\phi\phi\phi$ coupling can serve this purpose for the B.E. fields. The former statement is justified by the existence of an invariance transformation for the complete Lagrange function, namely,

$$\psi \to e^{1/2\pi\gamma_5}\psi = \gamma_5\psi, \qquad \phi \to -\phi,$$

which a term of the form $m_0 \psi\beta\psi$ would violate. In other words, the Dirac field with $m_0 = 0$ exhibits a space parity symmetry which is not removed by linear coupling with B.E. fields, and thus a mass term must be superimposed to eliminate this nonphysical degeneracy. As to the B.E. fields, a coupling of the form illustrated by

$$\mathcal{L}_\phi = -g_\phi^{2} \tfrac{1}{2}(\phi_{(0)}\phi_{(0)} + \phi_{(1)}\phi_{(1)}) \tfrac{1}{2}\phi_{(1/2)}\phi_{(1/2)}$$

will produce effective mass terms for each field through the action of the vacuum fluctuations of the other fields. This assertion does not contradict our discussion of the scale-independence of the present formalism, for the theory must be modified artificially (cut off) to convert the infinite quantity $\langle \phi^2 \rangle_0$ into a finite inverse square of a length. We conclude that a $\phi_{(0)}$ field can be introduced, in a four-dimensional Euclidean sense, provided it is a scalar field, and that the uncoupled B. E. field is massless. The $\phi^2\,\phi^2$ term that generates masses also disturbs the four-dimensional symmetry and leads to different observed masses for the σ-meson and the π-meson. If the former becomes sufficiently heavy ($\mu_\sigma > 2\mu_\pi$), the requirement for practical unobservability of this scalar particle will be satisfied.

The Yukawa coupling referring to the K-field, $\phi_{(1/2)}\psi_{(1/2)}\psi_{(0,1)}$, is indicated more precisely by the terms

$$\phi_{(1/2)}\psi_{(0)}\beta(1, \gamma_5)\psi_{(1/2)}, \qquad \phi_{(1/2)}i\tau\psi_{(1)}\beta(1, \gamma_5)\psi_{(1/2)},$$

in which the requirements of three-dimensional isotopic space symmetry are explicitly satisfied. The space reflection characteristics remain to be specified, as does the structure pertaining to other internal degrees of freedom. The physical information needed to determine the form of $\mathfrak{L}_K$ is supplied, in part, by the mass spectrum of the heavy F.D. particles. The coupling with the K-field must be the physical agency that differentiates N from Ξ, or, effectively decomposes the $\psi_{(1/2)}$ field into two parts labelled by the values of $\zeta_3\nu$, and it must also be the agency that produces different masses for Λ and Σ or, destroys the four-dimensional symmetry exhibited by $\psi_{(0,1)}$.

In seeking the physical property that distinguishes N from Ξ, it is well to notice that, while $\psi_{(0,1)}$ acts as a unit with respect to spatial reflection, a dependence of the reflection properties of $\psi_{(1/2)}$ upon ζ_3 is consistent with the universal π-coupling. Thus, it is conceivable that the two fields, $\frac{1}{2}(1 + \zeta_3\nu)\psi_{(1/2)}$ and $\frac{1}{2}(1 - \zeta_3\nu)\psi_{(1/2)}$, behave oppositely under spatial reflection, in accordance with the two possibilities available for Hermitian Dirac fields:

$$R_s\!:\psi(\mathbf{x}, x^0) \to \pm i\gamma^0\psi(-\mathbf{x}, x_0).$$

If we accept this tentative interpretation of the physical situation, the choice between the matrices 1 and γ_5 multiplying $\psi_{(1/2)}$ is resolved by the combination

$$\left(\frac{1 + \zeta_3\nu}{2}\gamma_5 + \frac{1 - \zeta_3\nu}{2}\right)\psi_{(1/2)},$$

in which γ_5 has been (arbitrarily) associated with the nucleon field $\frac{1}{2}(1 + \zeta_3\nu)\psi_{(1/2)}$ and the coefficients have been chosen to permit the invariance operation

$$\psi_{(1/2)} \to -\zeta_1\nu\gamma_5\psi_{(1/2)}, \qquad \phi_{(1/2)} \to i\zeta_2\phi_{(1/2)}$$

for $\mathfrak{L}_K$. Of course the mass term in $\mathfrak{L}^{(1/2)}$ is not invariant under this transformation. If the reflection properties of $\psi_{(0,1)}$ are standardized, the complete list of space reflection transformations (omitting the spatial coordinates) is given by

$$\begin{aligned} R_s\colon \psi_{(1/2)} &\to (\pm)\zeta_3 \nu i\gamma^0 \psi_{(1/2)}\,, \\ \psi_{(0,1)} &\to i\gamma^0 \psi_{(0,1)}\,, \\ \phi_{(1/2)} &\to -(\pm)\phi_{(1/2)}, \\ \phi_{(0)} &\to \phi_{(0)}\,, \qquad \phi_{(1)} \to -\phi_{(1)}\,, \end{aligned} \tag{7}$$

and the intrinsic space parity of the K-field can be specified further only by convention.

The second stage in determining the K-coupling term $\mathfrak{L}_K$ is facilitated by considering the behavior of the combinations $\psi_{(0)} \pm i\tau\psi_{(1)}$ under left and right multiplication by the matrices representing infinitesimal transformations. Thus

$$\psi_{(0)} + i\tau\psi_{(1)} \to \left(1 + \frac{i}{2}\,\delta\omega\tau\right)(\psi_{(0)} + i\tau\psi_{(1)})$$

implies

$$\begin{aligned} \psi_{(0)} &\to \psi_{(0)} - \tfrac{1}{2}\delta\omega\psi_{(1)}\,, \\ \psi_{(1)} &\to \psi_{(1)} - \tfrac{1}{2}\delta\omega \times \psi_{(1)} + \tfrac{1}{2}\delta\omega\psi_{(0)}\,, \end{aligned} \tag{8}$$

which is the τ transformation of (5), whereas

$$\psi_{(0)} + i\tau\psi_{(1)} \to (\psi_{(0)} + i\tau\psi_{(1)})\left(1 - \frac{i}{2}\,\delta\epsilon'\tau\right)$$

yields

$$\begin{aligned} \psi_{(0)} &\to \psi_{(0)} + \tfrac{1}{2}\delta\epsilon'\psi_{(1)}\,, \\ \psi_{(1)} &\to \psi_{(1)} - \tfrac{1}{2}\delta\epsilon' \times \psi_{(1)} - \tfrac{1}{2}\delta\epsilon'\psi_{(0)}\,, \end{aligned} \tag{9}$$

the ζ transformation of (6). The similar results for $\psi_{(0)} - i\tau\psi_{(1)}$ can be obtained by transposition. Thus the τ-transformation (8) is generated by

$$\psi_{(0)} - i\tau\psi_{(1)} \to (\psi_{(0)} - i\tau\psi_{(1)})\left(1 - \frac{i}{2}\,\delta\omega\tau\right),$$

while the ζ transformation (9) emerges from

$$\psi_{(0)} - i\tau\psi_{(1)} \to \left(1 + \frac{i}{2}\,\delta\epsilon'\tau\right)(\psi_{(0)} - i\tau\psi_{(1)}).$$

It may be noted here that the operator structure of the integral T field π-coupling term is given by

$$\psi_{(1)}\psi_{(0)} - \psi_{(0)}\psi_{(1)} + \psi_{(1)} \times \psi_{(1)} = \tfrac{1}{4} tr[(\psi_{(0)} - i\tau\psi_{(1)})i\tau(\psi_{(0)} + i\tau\psi_{(1)})],$$

where the trace applies to the four-dimensional matrices. The trace is properly invariant under the ζ transformation, and the τ-transformation induces the correct three-dimensional rotation.

We now observe that $\phi_{(1/2)}(\psi_{(0)} \pm i\tau\psi_{(1)})\psi_{(1/2)}$ possesses four-dimensional invariance properties with either sign since left or right hand factors multiplying $\psi_{(0)} \pm i\tau\psi_{(1)}$ can be compensated by appropriate transformations of $\phi_{(1/2)}$ or $\psi_{(1/2)}$. But one combination, namely $\psi_{(0)} - i\tau\psi_{(1)}$, is distinguished by leading to the same four-dimensional symmetries as the π-coupling term. Thus, under a τ-transformation $(\psi_{(0)} - i\tau\psi_{(1)})\psi_{(1/2)}$ is invariant, while a ζ-transformation of the integral T fields implies a corresponding transformation of $\phi_{(1/2)}$ without affecting $\psi_{(1/2)}$. Hence if $\mathfrak{L}_K$ contains only the combination $\psi_{(0)} - i\tau\psi_{(1)}$, the mass degeneracy of Λ and Σ would persist. We are led thereby, as one simple possibility, to suppose that $\psi_{(0,1)}$ enters $\mathfrak{L}_K$ only in the form $\psi_{(0)} + i\tau\psi_{(1)}$. The mass spectrum of the integral T F.D. particles is then viewed as the result of a clash between the different four-dimensional symmetries possessed by $\mathfrak{L}_\pi$ and $\mathfrak{L}_K$. When the latter is combined with the parity interpretation of the $N - \Xi$ mass splitting, we emerge with the following K-field interaction term

$$\mathfrak{L}_K = g_K\phi_{(1/2)}(\psi_{(0)} + i\tau\psi_{(1)})\beta\left(\gamma_5\frac{1+\zeta_3\nu}{2} + \frac{1-\zeta_3\nu}{2}\right)\psi_{(1/2)}. \tag{10}$$

This interaction is qualitatively satisfactory since it produces a complex mass spectrum for the fermions, and also implies a nondegenerate boson mass spectrum in virtue of the difference between the interactions $\mathfrak{L}_K$ and $\mathfrak{L}_\pi$, together with the breakdown of four dimensional symmetry implied by $\mathfrak{L}_\phi$, if the $\phi_{(0.1)}$ description is employed We need hardly emphasize that (10) is but an example of a general class of interaction term, the members of which can be distinguished only by their quantitative implications.

The invariance properties of the complete Lagrange function

$$\mathfrak{L} = \mathfrak{L}^{(0)} + \mathfrak{L}^{(1/2)} + \mathfrak{L}_\pi + \mathfrak{L}_K + \mathfrak{L}_\phi$$

characterize the class of possible interactions. These properties include invariance under the following rotations:

N rotation:
$$\psi \to (1 + i\delta\alpha\nu)\psi, \qquad \phi \to \phi,$$
$$j_N{}^\mu = \tfrac{1}{2}\psi\beta\gamma^\mu\nu\psi:$$

T rotation:
$$\psi_{(1/2)}, \phi_{(1/2)} \to \left(1 + \frac{i}{2}\,\delta\omega\tau\right)\psi_{(1/2)}, \phi_{(1/2)},$$
$$\psi_{(1)}, \phi_{(1)} \to (1 + i\delta\omega t)\psi_{(1)}, \phi_{(1)},$$
$$\psi_{(0)}, \phi_{(0)} \to \psi_{(0)}, \phi_{(0)},$$
$$j_T{}^\mu = \frac{1}{2}\psi_{(1/2)}\beta\gamma^\mu\frac{\tau}{2}\psi_{(1/2)} + \frac{1}{2}\psi_{(1)}\beta\gamma^\mu t\psi_{(1)} + i\phi_{(1/2)}{}^\mu\cdot\frac{\tau}{2}\phi_{(1/2)} + i\phi_{(1)}{}^\mu\cdot t\phi_{(1)};$$

Y rotation: $\psi_{(1/2)}, \phi_{(1/2)} \to (1 + i\delta\eta\zeta_3)\psi_{(1/2)}, \phi_{(1/2)}$

$$\psi_{(0,1)}, \phi_{(0,1)} \to \psi_{(0,1)}, \phi_{(0,1)}$$

$$j_Y^\mu = \tfrac{1}{2}\psi_{(1/2)}\beta\gamma^\mu\zeta_3\psi_{(1/2)} + i\phi_{(1/2)}{}^\mu . \zeta_3\phi_{(1/2)} .$$

The Y rotation, which defines the hypercharge, is equivalent to the ζ_3 rotation of the $T = \frac{1}{2}$ fields, $Z_3 = \frac{1}{2}Y$. There is no invariance under other ζ rotations, nor with respect to any τ rotation. Furthermore, in view of the physical coupling between nucleonic charge and hypercharge, nucleonic charge reflection is not an invariance operation, but must be supplemented by the reflection of hypercharge. The latter can be produced by the rotation $e^{\pi i Z_1}$ applied to the $T = \frac{1}{2}$ fields. In this way, we obtain the discrete symmetry operation described by

$$R_T = R_N e^{\pi i Z_1} ,$$

which we designate as the reflection operator for the three-dimensional isotopic space. Its effect as a unitary operator is indicated by

$$R_T^{-1}\psi_{(1/2)}R_T = i\zeta_1\nu_1\psi_{(1/2)} , \qquad R_T^{-1}\phi_{(1/2)}R_T = i\zeta_1\phi_{(1/2)} ,$$

$$R_T^{-1}\psi_{(0,1)}R_T = \nu_1\psi_{(0,1)} , \qquad R_T^{-1}\phi_{(1)}R_T = -\phi_{(1)} ,$$

$$R_T^{-1}\phi_{(0)}R_T = \phi_{(0)} ,$$

and

$$R_T^{-1}NR_T = -N, \qquad R_T^{-1}YR_T = -Y.$$

The states of zero nucleonic charge and zero hypercharge can be classified by the eigenvalues of R_T. It might be noted here that while

$$R_N^2 = +1,$$

we have

$$R_T^2 = e^{2\pi i Z_1} = e^{2\pi i Z_3} = (-1)^Y,$$

which is equivalent to the observation that the unitary operator R_T^2 reverses the sign of all fields that carry hypercharge ($T = \frac{1}{2}$), but leaves nonhypercharged fields unaffected. In a similar way, the repetition of the spatial reflection reverses the sign of all F.D. fields ($s = \frac{1}{2}$) and leaves B.E. fields unaltered. Since the former also carry nucleonic charge, this property can be expressed by

$$R_s^2 = (-1)^N. \tag{11}$$

The electromagnetic field fits easily into this theory of the heavy fermions and bosons. In addition to the Lagrange function of the pure Maxwell field,

$$\mathfrak{L}^{(M)} = -\tfrac{1}{2}[\tfrac{1}{2}F^{\mu\nu} . (\partial_\mu A_\nu - \partial_\nu A_\mu) - A_\nu . \partial_\mu F^{\mu\nu} - \tfrac{1}{2}F^{\mu\nu}F_{\mu\nu}],$$

we add the classic example of the coupling that contains no dimensional constant,

$$\mathfrak{L}_A = e j_Q^{\mu} \,.\, A^{\mu},$$

where the current vector of electrical charge follows from the identification (3),

$$j_Q^{\mu} = j_{T_3}^{\mu} + \frac{1}{2} j_Y^{\mu} = \frac{1}{2} \psi_{(1/2)} \beta \gamma^{\mu} \frac{\tau_3 + \zeta_3}{2} \psi_{(1/2)} + \frac{1}{2} \psi_{(1)} \beta \gamma^{\mu} t_3 \psi_{(1)}$$

$$+ i\phi_{(1/2)}{}^{\mu} \cdot \frac{\tau_3 + \zeta_3}{2} \phi_{(1/2)} + i\phi_{(1)}{}^{\mu} \cdot t_3 \phi_{(1)} \,.$$

This interaction term makes explicit the dynamical role of the electromagnetic field in reducing the three-dimensional T symmetries to the two-dimensional one described by T_3 rotations. At the same time, R_T is no longer a suitable reflection operation since it does not maintain the structure $\tau_3 + \zeta_3$. But if one superimposes the rotation $e^{\pi i T_1}$, the entire operator j_Q^{μ} reverses sign and an invariance operation is provided by R_Q, the electric charge reflection operator, with the following characteristics:

$$\begin{aligned}
R_Q^{-1} \psi_{(1/2)} R_Q &= -\, \zeta_1 \tau_1 \nu_1 \psi_{(1/2)} \,, & R_Q^{-1} \phi_{(1/2)} R_Q &= -\, \zeta_1 \tau_1 \phi_{(1/2)} \,, \\
R_Q^{-1} \psi_{(1)} R_Q &= (1 - 2t_1^2) \nu_1 \psi_{(1)} \,, & R_Q^{-1} \phi_{(1)} R_Q &= (2t_1^2 - 1) \phi_{(1)} \,, \\
R_Q^{-1} \psi_{(0)} R_Q &= \nu_1 \psi_{(0)} \,, & R_Q^{-1} \phi_{(0)} R_Q &= \phi_{(0)} \,, \\
R_Q^{-1} A_{\mu} R_Q &= -A_{\mu} \,, & R_Q^{-1} F_{\mu\nu} R_Q &= -F_{\mu\nu} \,,
\end{aligned}$$

and

$$R_Q^{-1} N R_Q = -N, \qquad R_Q^{-1} Y R_Q = -Y, \qquad R_Q^{-1} Q R_Q = -Q.$$

We also have

$$R_Q^2 = +1,$$

since

$$R_T^2 e^{2\pi i T_1} = e^{2\pi i (T_3 + 1/2 Y)} = 1.$$

Corresponding to the general characterization of the electric charge operator by rotations in a plane of the four-dimensional symmetry space,

$$Q = T_{12} \,,$$

we can express the electric charge reflection operator as

$$R_Q = R_N e^{\pi i T_{23}} \,, \tag{12}$$

with the interpretation of the reflection of the first axis in that two-dimensional charge space. This description includes the electromagnetic field if the latter is viewed as the [12] = 3 component of an axial vector in the three-dimensional isotopic space. It will be noted that the π-field acts as a polar vector; the first component reverses sign under the operation R_Q while the other components are unaltered. We observe, in this connection, that states of zero electric charge, zero hypercharge, and zero nucleonic charge can be classified by the eigenvalues of R_Q, the charge parity. A familiar example is a single π^0 meson which, according to the reflection properties of $\phi_{(1)3}$, defines a charge symmetric state. This neutral particle should decay into an even number of photons through the combined action of the $\mathcal{L}_\pi$ and $\mathcal{L}_A$ couplings. But, it is characteristic of our theory that the π-interaction involves the nucleonic charge while the electromagnetic interaction makes no reference to that property. Hence it is specifically the linking of nucleonic charge and hypercharge produced by the K-coupling that enables this transmutation to occur.

The theory thus far devised refers to heavy fermions and heavy bosons, together with the photon, and gives an account of their strong and electromagnetic interactions. Omitted are the light fermions (leptons) and the various physical processes that exhibit a very long time scale. The interactions responsible for these processes are certainly of lower symmetry than those already discussed, the total effect of the latter being described by various two-dimensional rotational symmetries, or charges, and a single charge reflection operation. Since the leptons carry electrical charge, they at least realize a two-dimensional internal symmetry space, which invites an attempt to correlate their properties with the aid of an internal space of higher dimensionality, but one which is presumably of lesser dimensionality than that employed for the heavy particles. Now, in our discussion of matrices with increasing dimensionality that culminated in the four-dimensional space necessary for the characterization of the heavy particles, we encountered the $T = 1$ representation of the three-dimensional rotation group which is thus naturally indicated for the description of the lepton family. According to the physical identification that accompanies the reduction to two-dimensional symmetry, we have

$$Q = T_3 ,$$

and the three-dimensional matrix t_3 has the distinct eigenvalue 1, 0, −1. It is tempting to extend to all fermions the property that the sign reversal of the field produced by repetition of a spatial reflection could also be generated by an internal rotation, in the sense of (11). This leads us to assign an analog of nucleonic charge to the leptons, which is called the leptonic charge L, and is represented by the matrix

$$\lambda = \begin{pmatrix} 0 & -i \\ i & 0 \end{pmatrix} .$$

Particles labelled by leptonic charge and electrical charge permit a complete identification with the known leptons; $L = +1 : \mu^+, \nu^0, e^-$, and $L = -1 : e^+, \bar{\nu}^0, \mu^-$, in which the leptonic charge serves to distinguish particles with the same electrical charge. This is analogous to the distinction afforded by the nucleonic charge between $N^+ (N = +1)$ and $\Xi^+ (N = -1)$, for example.

The mass spectrum of the leptons is characterized by the zero mass of its electrically neutral members, and the very striking mass asymmetry between the electrically charged particles with opposite values of $t_3\lambda$. Our general viewpoint encourages us to interpret the large mass of μ mesons ($t_3\lambda = +1$) by means of a comparatively strong interaction that serves to define dynamically the leptonic charge, and to remove the three-dimensional symmetries of the isotopic space. Of course, any proposed interaction must be reconciled with the apparent absence in known μ-meson phenomena of substantial nonelectromagnetic forces. It is at least interesting that an interaction which seems to possess the requisite properties can be exhibited with the aid of the hypothetical σ-field, $\phi_{(0)}$. To begin with, this scalar field enables us to establish a dynamical relation between the leptons and the strongly interacting particles, without upsetting thereby the higher internal symmetry characteristic of the massive particles. If the coupling of $\phi_{(0)}$ with the lepton field ψ_l is to reduce the lepton internal symmetry, the interaction must contain symmetrical real matrices other than unity, for which the only possibilities are t_3^2 and $t_3\lambda$. We choose the interaction as

$$\mathfrak{L}_\mu = g_\mu \phi_{(0)} \tfrac{1}{2} \psi_l \beta t_3 \tfrac{1}{2}(t_3 + \lambda) \psi_l ,$$

in which the particular matrix combination serves to select the μ-meson part of the lepton field. The unique properties of the σ-field can now be called upon again. As a field which is a scalar under all operations in the three-dimensional isotopic space and in space-time, $\phi_{(0)}$ has a nonvanishing expectation value in the vacuum. Although unable to affix the value implied by the strong interactions with heavy fermions, one could at least anticipate that $\langle \phi_{(0)} \rangle$ would have the magnitude of nucleon masses, and thus a suitable μ-meson mass constant might emerge from $g_\mu \langle \phi_{(0)} \rangle$ without requiring a particularly large coupling constant g_μ. It is the latter feature, combined with the supposedly large mass of the σ-particle, that may enable one to avoid conflict between the dynamical implications of $\mathfrak{L}_\mu$ and the present lack of evidence for significant nonelectromagnetic μ interactions. Naturally, one must eventually find such evidence if our hypothesis is to attain a measure of credibility.

Whatever the interpretation of the μ-meson mass may be, the symmetry properties of the lepton field can be generally stated at the dynamical level that incorporates electromagnetic interactions. These include rotations,

L rotation:
$$\psi_l \to (1 + i\delta\alpha\lambda)\psi_l ,$$
$$j_L^\mu = \tfrac{1}{2} \psi_l \beta \gamma^\mu \lambda \psi_l ,$$

T_3 rotation:
$$\psi_l \to (1 + i\delta\omega t_3)\psi_l\,,$$
$$(j_Q{}^\mu)_l = \tfrac{1}{2}\psi_l\beta\gamma^\mu t_3\psi_l\,,$$

and charge reflection,

$$R_Q:\psi_l \to (1 - 2t_1{}^2)\lambda_1\psi_l\,.$$

The latter transformation is implied by the extension of the general formula (12),

$$R_Q = R_N R_L e^{\pi i T_{23}}\,,$$

where

$$R_L{}^{-1}\psi_l R_L = \lambda_1\psi_l\,,$$

and thus

$$R_Q{}^{-1}LR_Q = -L.$$

There is also the special feature of the zero neutrino mass (although we have yet to produce a mechanism that accounts for the electron mass) which is described by invariance under the infinitesimal transformation

$$\psi_l \to [1 + i\delta\varphi(1 - t_3{}^2)i\gamma_5]\psi_l\,. \tag{13}$$

But there will be more of this shortly.

The symmetry that exists between the heavy bosons and fermions in their isotopic space properties prompts us to ask: Is there also a family of *bosons* that realizes the $T = 1$ representation of the three-dimensional rotation group? The exceptional position of the electromagnetic field in our scheme, and the formal suggestion that this field is the third component of a three-dimensional isotopic vector, encourage an affirmative answer. We are thus led to the concept of a spin one family of bosons, comprising the massless, neutral, photon and a pair of electrically charged particles that presumably carry mass, in analogy with the leptons (abstracting from the additional complexity that accompanies the fermion property of leptonic charge). These considerations are expressed by the Lagrange function term

$$\mathfrak{L}^{(1)} = -\tfrac{1}{2}[\tfrac{1}{2}Z^{\mu\nu}\,.\,(\partial_\mu Z_\nu - \partial_\nu Z_\mu) - Z_\nu\,.\,\partial_\mu Z^{\mu\nu} - \tfrac{1}{2}Z^{\mu\nu}Z_{\mu\nu}]$$

and the identification

$$Z_3{}^\mu = A^\mu, \qquad Z_3{}^{\mu\nu} = F^{\mu\nu},$$

together with the boson interaction

$$\mathfrak{L}_{Z\phi} = -g_{Z\phi}{}^2\tfrac{1}{2}\phi_{(0)}{}^2\tfrac{1}{2}Z^\mu t_3{}^2 Z_\mu\,,$$

in which we again use the σ-field to remove three-dimensional internal symmetries and produce masses for charged particles. At the same time, the coupling augments the mass of the σ-particle. This part of the Lagrange function referring to the Z field, and including the boson interaction that is the electromagnetic coupling with the charged Z particles, possesses the following symmetry properties,

T_3 rotation:
$$Z \to (1 + i\delta\omega t_3)Z,$$
$$(j_Q^{\mu})_Z = iZ^{\mu\nu} \,.\, t_3 Z_\nu\,,$$

and

$$R_Q : Z \to (1 - 2t_1^2)Z,$$

where the latter contains the known charge reflection property of the electromagnetic field, together with

$$R_Q^{-1} Z_1 R_Q = Z_1\,, \qquad R_Q^{-1} Z_2 R_Q = -Z_2\,.$$

There is also the expression of the null photon mass-invariance of $\mathfrak{L}$ (to within an added divergence) under the gauge transformation

$$Z_\mu \to Z_\mu + \frac{1}{e}(1 - t_2^2)\partial_\mu \lambda(x). \qquad Z_{\mu\nu} \to Z_{\mu\nu}\,,$$

when combined with the general transformation

$$\chi \to e^{i\lambda(x) T_{12}}\chi$$

of all charged fields, including the Z field.

Now we must face the problem of discovering the specific Yukawa interactions of the massive, charged Z particles. From its role as a partner of the electromagnetic field, we might expect that the charged Z field interacts universally with electric charge, or rather, changes of charge, without particular regard to other internal attributes. If this be so, the coupling with the Z field (henceforth understood to be the charged Z field) will produce further reductions of internal symmetry, which raises the hope that this general mechanism may be the underlying cause of the whole group of physical processes that are characterized by a long time scale. Indeed, that time scale becomes more comprehensible, without invoking inordinately weak interactions, if every observable process requires the virtual creation of a heavy particle. Our general viewpoint regarding the systematic reduction of internal symmetry also impels us to seek some internal symmetry aspect of the Z field that is destroyed by the coupling with various combinations of electrically charged and neutral fields. There is no question, presumably, of a breakdown of invariance under the two-dimensional rotations that define electrical charge, which leaves no choice other than a failure of the charge reflection symmetry property for the Z field interactions. (It must be admitted that, despite its natural place in our scheme, this conclusion required the stimulus of certain recent experiments, and was not drawn in the lectures upon which this article is based.)

To investigate the possibility of charge reflection invariance failure, let us consider the Z field interaction with the lepton field, as illustrated by

$$Z^{\mu}\tfrac{1}{2}\psi_l\beta\gamma_\mu t\psi_l$$

$$(Zt = Z_1t_1 + Z_2t_2),$$

which is invariant under rotations and reflections in the two-dimensional charge space. Another such coupling is obtained on replacing t_1 with $t_2 = (1/i)[t_3, t_1]$, and t_2 with $-t_1 = (1/i)[t_3, t_2]$, provided the charge reflection properties of Z are reversed to compensate the effect of t_3. Although it might appear that reflection invariance would be destroyed if both couplings were operative, that impression is misleading since a suitable relative rotation of the charge axes for the two fields suffices to remove either term. To obtain something fundamentally different, the charge matrix t must multiply t symmetrically, rather than entering in a commutator or alternatively, t could be multiplied by the leptonic charge matrix, although our previous comments would suggest that only electrical charge is relevant. But both of these matrices are symmetrical and are excluded in conjunction with the symmetrical matrices $\beta\gamma_\mu$. A breakdown of charge reflection invariance cannot be produced, therefore, unless one also replaces the symmetrical real matrices $\beta\gamma_\mu$ by the antisymmetrical imaginary matrices $\beta\gamma_\mu i\gamma_5$, which possess the opposite space reflection characteristics. We thus recognize that a failure of invariance under charge reflection must be accompanied by a failure of invariance under space reflection. The common origin of both breakdowns indicates, however, that the combination of the two reflections is still an invariance operation,

$$R = R_sR_Q,$$

which now appears as the ultimate expression of the complete equivalence of oppositely oriented spatial coordinate systems[5]. The eigenvalues of this reflection operator, the union of space parity and charge parity, will be designated as the *parity*.

The vector coupling between the Z field and the lepton field that emerges from these considerations is

$$\mathfrak{L}_{Zl} = g_Z Z^{\mu}\tfrac{1}{2}\psi_l\beta\gamma_\mu(t - i\gamma_5\{t_3, t\})\psi_l, \tag{14}$$

in which the relative coefficients of the two terms have been chosen to conform

[5] That the indiscernibility of left and right can be reconciled with "nonconservation of parity", in the specific sense of R_s, has been emphasized particularly by Landau (4).

with an invariance requirement that expresses the null neutrino mass. The relevant operation is just (13), extended by a suitable transformation for the charged leptons[6],

$$\psi_l \to [1 + i\delta\varphi((1 - t_3^2)i\gamma_5 - t_3)]\psi_l \,.$$

One could also employ the opposite sign in all γ_5 terms. This invariance property implies a conservation law for a lepton field quantity that we shall call neutrinic charge, n. It is described by the vector

$$j_n^\mu = \tfrac{1}{2}\psi_l\beta\gamma^\mu((1 - t_3^2)i\gamma_5 - t_3)\psi_l \,.$$

As thus defined, the neutrinic charge of the particles μ and e is the negative of their electrical charge. The neutrinic charge of the neutrino is represented by the matrix $i\gamma_5$, the eigenvalues of which also have the significance of the spin projection along the direction of motion of the massless particle. Thus a neutrino with $n = +1$ or -1 can be designated as a right or left polarized neutrino. In a process involving the creation of a pair of leptons through the intervention of the Z field, the conservation of neutrinic charge states that a lepton of positive electrical charge appears with a right polarized neutrino, and a negatively charged lepton with a left polarized neutrino.

The neutrinic charge conservation law is quite independent of that for leptonic charge[7] which asserts, for example, that a neutrino $\nu(\lambda = +1)$, as distinguished from an antineutrino $\bar{\nu}(\lambda = -1)$, accompanies a positive electron and a negative muon. Thus the detailed charge correlations are

$$Z^+ \leftrightarrow \mu^+ + \bar{\nu}_R \,, \qquad e^+ + \nu_R \,,$$

$$Z^- \leftrightarrow \mu^- + \nu_L \,, \qquad e^- + \bar{\nu}_L \,,$$

where R and L are polarization labels. One consequence of these assignments refers to the self-coupling of the lepton field through the intermediary of the Z field, which implies the physical process $\mu \to e + 2\nu$. The leptonic and neutrinic charge attributes of μ and e require that the electrically neutral particles possess the same leptonic charge but the opposite neutrinic charge,

$$\mu^+ \to e^+ + \nu_R + \nu_L \,,$$

$$\mu^- \to e^- + \bar{\nu}_L + \bar{\nu}_R \,.$$

[6] A similar invariance requirement has been discussed by Touschek (*5*), but lacks the restriction of the transformation to the lepton field, and the specific form of the transformation properties assigned to the non-neutrino fields.

[7] This shows that the neutrino theory developed here is not to be identified with the so-called two-component theory (*6*). What is called lepton (ic charge) conservation in that formalism is specifically the conservation of neutrinic charge. In this connection, see the paper on the conservation of the lepton charge by Pauli (*7*).

Thus the two neutrinos are oppositely polarized. This is consistent with the finite intensity observed at the high energy end of the electron spectrum, where the neutrinos travel in the same direction. Let us also observe that, when sufficient energy is available, the mass of the electron can be neglected and that particle is produced with a definite polarization. In virtue of the commutativity of $\beta\gamma_\mu$ with γ_5 and the antisymmetry of the latter matrix, a pair of essentially massless leptons generated by the coupling (14) will be oppositely polarized. The implication for the successive decays $\pi \to \mu + \nu$, $\mu \to e + 2\nu$, arises from the zero spin of the π meson according to which μ^-, say, in $\pi^- \to \mu^- + \nu_L$ must have its spin directed oppositely to its direction of motion. At the high energy end of the electron spectrum, it is the electron that carried the angular momentum about the axis of disintegration and, since e^- accompanies $\bar{\nu}_L$, the electron is right polarized. Hence the electron must be emitted predominantly in the direction of the μ spin, or oppositely to the initial direction of the μ meson. This is precisely the asymmetry that was so strikingly revealed in recent experiments (*8*).

The extension of the Z coupling to the heavy F.D. fields must be guided by the concept of universality since there is no analog of the neutrinic charge for these massive particles. The charged Z particles interact equally with all pairs of electrically charged and neutral leptons, thereby destroying charge reflection symmetry. An illustration of the attempt to transfer these properties to the heavy fermions is given by the following coupling term

$$\mathfrak{L}_{ZN} = 2^{-1/2} g_Z Z^\mu \frac{1}{2} \psi\beta\gamma_\mu \left(\tau - i\gamma_5 \left\{\frac{\tau_3 + \zeta_3}{2}, \tau\right\}\right)\psi, \tag{15}$$

where ψ stands for the complete field $\psi_{(1/2)} + \psi_{(0,1)}$. That is, with the physical emphasis placed on electrical charge, the three-dimensional isotopic classification is no longer meaningful and the two sets of fields $\psi_{(1/2)}$ and $\psi_{(0,1)}$ appear on the same footing. The possibility of writing the single term (15) makes very explicit use of the underlying four-dimensional symmetry and the accompanying ability to apply the matrices τ and ζ either to $\psi_{(1/2)}$ or to $\psi_{(0,1)}$. The implications of the compact symbolism are made more explicit by observing that we have extended the coupling of Z^+ with the fermion pairs $\mu^+\bar{\nu}$ and $e^+\nu$ to the fermion pairs $N^+\bar{N}^0$, $\Xi^0\bar{\Xi}^+$; $\Sigma^+\bar{\Sigma}^0$, $\Sigma^+\bar{\Lambda}$, $\Sigma^0\bar{\Sigma}^+$, $\Lambda\bar{\Sigma}^+$; $N^+\bar{\Sigma}^0$, $N^+\bar{\Lambda}$, $\Sigma^+\bar{N}^0$, $N^0\bar{\Sigma}^+$, $\Sigma^0\bar{\Xi}^+$, $\Lambda\bar{\Xi}^+$, $\Sigma^+\bar{\Xi}^0$, $\Xi^0\bar{\Sigma}^+$.

One formal point still demands attention. The general formula (15) assumes that the complete field ψ behaves as a unit under the reflection operation R,

$$R: \qquad \begin{aligned} &\psi \to i\gamma^0\nu_1 \quad && e^{1/2\pi i(\zeta_1+\tau_1)}\psi, \\ &Z_1^0 \to Z_1^0, && Z_2^0 \to -Z_2^0, \\ &Z_1^k \to -Z_1^k, && Z_2^k \to Z_2^k, \end{aligned} \tag{16}$$

(spatial coordinates omitted) and some revision is necessary if we accept the space parity distinction between the N field and the Ξ field. One procedure is to separate the $\psi_{(0,1)}\psi_{(1,2)}$ term from (15) and modify it appropriately. But we can also retain (15) by redefining the $\psi_{(1/2)}$ field.

$$\psi_{(1/2)} \to e^{\pm\frac{1}{4}\pi i(\zeta_3+\nu)}\psi_{(1/2)},$$

which converts the K coupling term into

$$\mathfrak{L}_K = g_K\phi_{(1/2)}(\psi_{(0)} + i\tau\psi_{(1)})\beta\left(\pm i\gamma_5\nu\,\frac{1+\zeta_3\nu}{2} + \frac{1-\zeta_3\nu}{2}\right)\psi_{(1/2)}\,.$$

The choice of sign is conventional if only strong and electromagnetic interactions are considered but becomes physically significant when the Z field coupling is included. With the new meaning of $\psi_{(1/2)}$, the behavior of this field under the operation R_T reads

$$R_T^{-1}\psi_{(1/2)}R_T = i\zeta_1\nu_1(-\zeta_3\nu_3)\psi_{(1/2)} = i\zeta_2\nu_2\psi_{(1/2)},$$

and the electric charge reflection transformation becomes

$$R_Q^{-1}\psi_{(1/2)}R_Q = -\zeta_1\tau_1\nu_1(-\zeta_3\nu_3)\psi_{(1/2)} = -\zeta_2\tau_1\nu_2\psi_{(1/2)}.$$

On forming the reflection operation R, we now find a common behavior for the fields contained in ψ if the lower sign in (7) is chosen. Thus the universal Z coupling determines the *parity* assignments of the heavy particle fields,

$$R:\qquad \begin{aligned} \psi &\to i\gamma^0\nu_1 e^{1/2\pi i(\zeta_1+\tau_1)}\psi, \\ \phi_{(1/2)} &\to -\zeta_1\tau_1\phi_{(1/2)}, \\ \phi_{(1)} &\to (1 - 2t_1^2)\phi_{(1)}\,, \\ \phi_{(0)} &\to \phi_{(0)}\,, \end{aligned}$$

to which is added the Z field transformation of (16), and

$$R:\qquad \begin{aligned} &A^0 \to -A^0, \qquad A^k \to A^k, \\ &\psi_l \to i\gamma^0\lambda_1(1 - 2t_1^2)\psi_l\,. \end{aligned}$$

All these transformations are summarized by

$$R = R_{(s)}R_{(N)}R_{(L)}e^{\pi i T_{23}},$$

where the individual reflection operators are simpler than those previously considered, for $R_{(N)}$ and $R_{(L)}$ induce transformations only on fermion fields, while $R_{(s)}$ generates a standard space reflection transformation that multiplies all fermion fields by $i\gamma^0$ and leaves scalar boson fields unaltered.

Through the intervention of the Z field, physical processes involving heavy particles take place that conserve nucleonic charge and electrical charge only of the list of internal attributes, to which *parity* should be added as a joint internal and space-time property. These processes include known particle decays: $\Sigma, \Lambda \to N + \pi$; $\Xi \to \Lambda + \pi$; $K \to 2\pi, 3\pi$, and the theory must meet various quantitative tests, including its effectiveness in suppressing the decay $\Xi \to N + \pi$. In the latter connection, it is interesting to observe an invariance property that the Lagrange function would possess if the four-dimensional symmetry of Λ and Σ were not destroyed, which is to say, if only $\psi_{(0)} - i\tau\psi_{(1)}$ occurred in $\mathfrak{L}_K$. The transformation is

$$\psi \to e^{i\varphi\zeta_3}\psi, \qquad \phi_{(1/2)} \to e^{i\varphi(\tau_3+\zeta_3)}\phi_{(1/2)}, \tag{17}$$

with all other fields unchanged. The importance of this comment stems from the possibility that such an idealization of processes in which Λ and Σ do not appear explicitly, and in which the K coupling is not directly responsible for the transition, may be justified by the relatively small $\Lambda - \Sigma$ mass splitting. Accepting this, we conclude that a reaction limited to $T = \frac{1}{2}$ fermions and π-mesons, with no K-particles in evidence, conserves hypercharge. Hence $\Xi \to N + \pi$ is (approximately) forbidden. We can also draw from (17) the conclusion that, in processes where no heavy fermion appears, the electrical charge carried by the K-particles is individually conserved. Were we to ignore the direct relevance of the K-coupling, we would infer the forbiddenness of charged K-particle decay, which may bear on the empirical observation that the shortest K-particle lifetime is that of a neutral K-particle, with *parity* $+1$.

The Z field also couples the heavy fermions with the light fermions, thereby implying such processes as $\mu + N \to \nu + N$, $\pi \to \mu + \nu$, $K \to \mu + \nu$, $K \to \pi + \mu + \nu$, as well as $N \to N + e + \nu$, $K \to \pi + e + \nu$, and $\pi \to e + \nu$, $K \to e + \nu$. It is the electron phenomena that present an immediate challenge. Thus, the latter decays, $\pi \to e + \nu$, $K \to e + \nu$ have not been observed although their μ-meson counterparts exist. Now it is encouraging that these electron processes would not occur if the electron mass were zero, for then electron and neutrino are oppositely polarized, which produces a net angular momentum about the axis of disintegration and contradicts the zero spin of π and K. It will be noted that the vector nature of the Z coupling is decisive in this argument. It can be concluded that the theory discriminates against the electron decay of the spinless bosons, but the precise ratio of decay probabilities may depend upon the specific dynamical origins of the electron and muon masses.

We then come to the comparatively well-known β-decay processes, where there is evidence that the lepton field appears in a tensor form combined with either a vector or a scalar coupling, the latter being currently favored by angular correlation measurements. The empirical tensor interaction fits naturally into the Z-particle picture, although two viewpoints are possible. It is familiar that effective tensor interactions are byproducts of fundamental vector couplings and one could imagine this to be the situation, which carries the implication that the weakness of Z couplings is illusory and confined to low-energy phenomena. This interpretation also requires that the electron possess a mass, and the smallness of the latter enhances the quantitative difficulties of the approach. Alternatively, one can invert matters by supposing that with the charged Z field Yukawa interactions we have reached the limits of the principle of scale independence and that a length appears explicitly, in a form analagous to an intrinsic magnetic moment coupling. The combined dynamical effects of the vector and tensor Z couplings then imply a mass difference between the charged and neutral leptons, and the general order of magnitude thus anticipated for the electron mass is not unreasonable[8]. But whether it be a phenomenological description or a fundamental interaction, the form of the tensor coupling of the Z field with leptons is determined by the neutrinic charge invariance property as

$$f\,\tfrac{1}{2}Z^{\mu\nu}\tfrac{1}{2}\psi_l\beta\sigma_{\mu\nu}(t - i\gamma_5[t_3, t])\psi_l\,,$$

where f has the dimensions of a length. The previous comments referring to the electron mass can now be understood in the following way: If the electron mass is zero, the vector coupling (15) is invariant under the electron field transformation:

$$\psi_l \rightarrow [1 + i\delta\varphi t_3\lambda(1 - t_3\lambda)\,(i\gamma_5 + t_3)]\psi_l\,,$$

which is equivalent to the possibility of specifying the electron polarization under these circumstances (e^+ is left polarized, e^- is right polarized). On the other hand, for a massless electron the tensor coupling term is invariant under the different transformation

$$\psi_l \rightarrow [1 + i\delta\varphi t_3\lambda(1 - t_3\lambda)\,(-i\gamma_5 + t_3)]\psi_l\,,$$

which also implies the production of electrons with definite polarization (e_R^+, e_L^-) Hence, the tensor coupling cannot be a dynamical byproduct of the vector coupling if the electron mass is zero, and, if both interactions are fundamental there is no γ_5 invariance property, which indicates that the interference between vector and tensor couplings generates an electron mass.

The tensor coupling term can be presented in a different way by employing the matrix property

$$\sigma_{\mu\nu}\gamma_5 = \tfrac{1}{2}\epsilon_{\mu\nu\lambda\kappa}\sigma^{\lambda\kappa}, \tag{18}$$

where ϵ is the alternating symbol specified by $\epsilon^{0123} = +1$. If we also observe

[8] Which is only to say that the value of the cutoff momentum that must be chosen is not exceptionally large.

that electrical charge conservation is expressed by

$$\psi[t_3, t]\psi Z = (\psi t\psi)t_3 Z,$$

(omitting irrelevant matrices) the following form is obtained,

$$f\tfrac{1}{2}\psi_l\beta\sigma_{\mu\nu}t\psi_l\tfrac{1}{2}[Z^{\mu\nu} - it_3(\epsilon Z)^{\mu\nu}],$$

in which

$$(\epsilon Z)^{\mu\nu} = \tfrac{1}{2}\epsilon^{\mu\nu\lambda\kappa}Z_{\lambda\kappa} \tag{19}$$

is the tensor dual to $Z^{\mu\nu}$. Thus the tensor coupling with leptons involves the Z field in an essentially self-dual combination,

$$\epsilon(Z - it_3\epsilon Z) = it_3(Z - it_3\epsilon Z),$$

where the matrix multiplication is in the sense of (19). The basic property used here,

$$\epsilon^2 = -1,$$

can be understood from the matrix version of (18),

$$\sigma\gamma_5 = \epsilon\sigma,$$

which indicates that ϵ is a six-dimensional matrix representation of γ_5.

The tensor combination $(1 - it_3\epsilon)Z$ makes very explicit the role of the Z field in destroying charge reflection and space reflection invariance, while maintaining the reflection property R. But the natural extension of the coupling to the heavy fermion field encounters difficulties. A universal coupling of the form $m(1 - it_3\epsilon)Z$ implies an effective self-interaction of the tensor sources of the Z field which, for large Z-particle mass, is dominated by

$$\tfrac{1}{2}m(1 - it_3\epsilon)(1 + it_3\epsilon)m = \tfrac{1}{2}m(1 - t_3^2)m$$
$$= 0.$$

Hence we do not obtain direct tensor coupling between heavy and light fermions. While this conclusion is modified by dynamical effects of the Z-particles, it is perhaps simpler to say that the tensor coupling is not entirely universal but distinguishes between heavy and light fermions through the appearance of the alternative tensor combinations $(1 \pm it_3\epsilon)Z$. Thus, a possible tensor coupling between the heavy fermions and the charged Z-particles is

$$2^{-1/2}f\,\frac{1}{2}\,Z^{\mu\nu}\,\frac{1}{2}\,\psi\beta\sigma_{\mu\nu}\left(\tau + i\gamma_5\left[\frac{\tau_3 + \zeta_3}{2}, \tau\right]\right)\psi.$$

If both vector and tensor couplings of the heavy F. D. particles are operative, a mass difference between electrically charged and neutral particles is implied beyond that produced by the electromagnetic field. Thus the role initially assigned to the electromagnetic field may be modified by the more complete theory for which it served as a model. We have not discussed linear interactions of the Z field with the heavy bosons. Perhaps it suffices to say that one cannot construct a bilinear vector combination of the spin 0 fields that realized the dynamical function of the Z field to destroy charge and space reflection invariance.

From the general suggestions of a family of bosons that is the isotopic analog of the leptons, and the identification of its neutral member as the photon, we have been led to a dynamics of a charged, unit spin Z-particle field that is interpreted as the invisible instrument of the whole class of weak interactions. The direct identification of this hypothetical particle will not be easy. Its linear couplings are neither so strong that it would be produced copiously, nor are they so weak that an appreciable lifetime would be anticipated. And as to the detailed implications of this model for the effective weak interactions that it seeks to comprehend, although the theory is definite enough about the fundamental predominance of vector and tensor coupling, the rest of the structure is hardly unique, and the profound effects of the various strong interactions obscure the actual predictions of the formalism. The definitive results of the group of experiments that exploit the newly discovered lepton polarization properties of the weak interactions will be particularly relevant in judging this hypothesis.

The heavy fermions and bosons, the leptons and the photon-Z particle family, have been viewed as physical realizations of four and three dimensional internal symmetry spaces, respectively. Do particles similarly attached to a two-dimensional symmetry space exist, or does every particle family contain an electrical neutral member? In the latter circumstance, we must descend to a one-dimensional internal space, to a neutral field that presumably possesses no internal properties and responds dynamically only to the space-time attributes of other systems. According to the generalization that introduced leptonic charge, this noninternally degenerate system is a B.E. field. If it follows the example of the electrically neutral fields associated with the other odd-dimensional space, it is massless, and the spin restrictions commented on at the beginning of this discussion need not apply to a single neutral massless field. It appears that in the hierarchy of fields there is a natural place for the gravitational field.

What has been presented here is an attempt to elaborate a complete dynamical theory of the elementary particles from a few general concepts. Such a connected series of speculations can be of value if it provides a convenient frame of reference in seeking a more coherent account of natural phenomena.

RECEIVED: July 31, 1957

REFERENCES

1. J. Schwinger, *Phys. Rev.* **104,** 1164 (1956).
2. A. Pais, *Proc. Nat. Acad. Sci. U. S.* **40,** 484 (1954).
3. E. P. Wigner, *Proc. Nat. Acad. Sci. U. S.* **38,** 449 (1952). See also M. Gell-Mann, *Phys. Rev.* **106,** 1296 (1957).
4. L. Landau, *Nuclear Phys.* **3,** 127 (1957).
5. B. Touschek, *Nuovo cimento* **5,** 1281 (1957).
6. T. Lee and C. Yang, *Phys. Rev.* **105,** 1671 (1957); A. Salam, *Nuovo cimento* **5,** 299 (1957); L. Landau, *Nuclear Phys.* **3,** 127 (1957).
7. W. Pauli, *Nuovo cimento* **6,** 204 (1957).
8. R. Garwin, L. Lederman, and M. Weinrich, *Phys. Rev.* **105,** 1415 (1957).

On a Gauge Theory of Elementary Interactions.

A. SALAM

Imperial College - London

J. C. WARD

Carnegie Insitute of Technology - Pittsburgh

(ricevuto il 15 Settembre 1960)

Summary. — A theory of strong as well as weak interactions is proposed using the idea of having only such interactions which arise from generalized gauge transformations.

1. – Introduction.

One of the problems engaging current interest in field theory is the problem of determining which fields are « elementary » in some fundamental sense, and which are not. An equally, if not more, important, problem is that of finding a guiding principe for writing fundamental interactions of fields. The only such principle which exists at the present time seems to be the gauge-principle. Whenever a symmetry property exists, the associated gauge transformation leads in a definite manner to the postulation of an interaction through the mediation of a number of intermediate particles. There exist, at present, numerous attempts to understand all known elementary interactions in this manner. In an earlier paper (1), the authors considered a gauge-transformation in [3] « charge-space » to generate weak and electro-magnetic interactions. Recently J. J. SAKURAI (2) has used similar ideas to postulate five intermediate vector mesons which may be responsible for mediating strong interactions.

All these attempts suffer from certain weaknesses. Our earlier attempt (1) to understand weak and electro-magnetic interactions produced only parity-

---

(1) A. SALAM and J. C. WARD: *Nuovo Cimento*, **11**, 568 (1959).

(2) J. SAKURAI: *Ann. of Phys.*, **11**, 1 (1960).

conserving interactions in a natural manner. Besides, the weak interactions did not obey the $\Delta I = \frac{1}{2}$ rule. Sakurai's strong Lagrangian suffers from the defect that it contains no Yukawa-like terms permitting single emission of pions or K-mesons by baryons.

In this note we wish to reconsider the problem. Our basic postulate is that it should be possible to generate strong, weak and electro-magnetic interaction terms (with all their correct symmetry properties and also with clues regarding their relative strengths), by making local gauge transformations on the kinetic-energy terms in the free Lagrangian for all particles. This is the statement of an ideal which, in this paper at least, is only very partially realized.

It may however be of interest to set down the procedure which has been followed.

2. – A simple model.

Consider the π-nucleon system. Following SCHWINGER we assume the existence of a scalar iso-scalar particle σ. Write the free Lagrangian kinetic energy terms in the form

$$N_L^+ \gamma_4 \gamma_\mu \partial_\mu N_L + N_R^+ \gamma_4 \gamma_\mu \partial_\mu N_R + \tfrac{1}{2}(\partial_\mu \boldsymbol{\pi})\cdot(\partial_\mu \boldsymbol{\pi}) + \tfrac{1}{2}(\partial_\mu \sigma)^2 \,, \tag{1}$$

where

$$N_L = \tfrac{1}{2}(1+\gamma_5)N \,,$$

$$N_R = \tfrac{1}{2}(1-\gamma_5)N \,.$$

Following a suggestion made by SCHWINGER [3], GELL-MANN and LEVY [4] one may consider $\begin{pmatrix} N_L \\ N_R \end{pmatrix}$ as forming a spinor and $\begin{pmatrix} \sigma \\ \boldsymbol{\pi} \end{pmatrix}$ a 4-vector in a [4] Euclidean space. Since there are 6 rotations in such a space, the gauge-principle will give rise to six fields X, Y with the interactions

$$\begin{aligned} L_{\text{int}} = \frac{1}{2}\left(\sigma \frac{\partial \boldsymbol{\pi}}{\partial x_\mu} - \frac{\partial \sigma}{\partial x_\mu}\boldsymbol{\pi} + \boldsymbol{\pi}\wedge \frac{\partial \boldsymbol{\pi}}{\partial x_\mu} - \frac{\partial \boldsymbol{\pi}}{\partial x_\mu}\wedge \boldsymbol{\pi}\right)\cdot \mathbf{X}_\mu + \\ + iN_L^+\gamma_4\gamma_\mu \boldsymbol{\tau}\cdot \mathbf{X}_\mu N_L + \frac{1}{8}(\mathbf{X}\cdot\boldsymbol{\pi})^2 + \frac{1}{8}(\boldsymbol{\pi}\wedge\mathbf{X})^2 + \frac{1}{8}\mathbf{X}^2\sigma^2 + \\ + \frac{1}{2}\left[\left(\sigma\frac{\partial \boldsymbol{\pi}}{\partial x_\mu} - \frac{\partial \sigma}{\partial x_\mu}\boldsymbol{\pi}\right) + \left(\frac{\partial \boldsymbol{\pi}}{\partial x_\mu}\wedge\boldsymbol{\pi} - \frac{\partial \boldsymbol{\pi}}{\partial x_\mu}\wedge\boldsymbol{\pi}\right)\right]\cdot \mathbf{Y}_\mu + \\ + iN_R^+\gamma_4\gamma_\mu\boldsymbol{\tau}\cdot\mathbf{Y}_\mu N_R + \frac{1}{8}(\mathbf{Y}\cdot\boldsymbol{\pi})^2 + \frac{1}{8}(\boldsymbol{\pi}\wedge\mathbf{Y})^2 + \frac{1}{8}\mathbf{Y}^2\sigma^2 \,. \end{aligned} \tag{2}$$

(3) J. SCHWINGER: *Ann. of Phys.*, **2**, 407 (1957).
(4) M. GELL-MANN and M. LEVY: *Nuovo Cimento*, **14**, 705 (1960).

The « free » Lagrangians for the $\boldsymbol{X}$ field is as follows

$$\left[\left(\frac{\partial}{\partial x_\mu}\boldsymbol{X}_\nu - \boldsymbol{X}_\mu\wedge\boldsymbol{X}_\nu\right) - \left(\frac{\partial}{\partial x_\nu}\boldsymbol{X}_\mu - \boldsymbol{X}_\nu\wedge\boldsymbol{X}_\mu\right)\right]^2,$$

with a similar expression for the $\boldsymbol{Y}$ field.

One can rewrite the above interaction slightly differently, introducing fields

$$\tfrac{1}{2}(\boldsymbol{X}+\boldsymbol{Y}) = \boldsymbol{u}\,,$$
$$\tfrac{1}{2}(\boldsymbol{X}-\boldsymbol{Y}) = \boldsymbol{v}\,.$$

Thus

$$(3)\qquad L_{\text{int}} = \left[\left(\boldsymbol{\pi}\wedge\frac{\partial\boldsymbol{\pi}}{\partial x_\mu} - \frac{\partial\boldsymbol{\pi}}{\partial x_\mu}\wedge\boldsymbol{\pi}\right) + N^+\gamma_4\gamma_\mu\boldsymbol{\tau}N\right]\cdot\boldsymbol{u} +$$
$$+\left[\sigma\frac{\partial\boldsymbol{\pi}}{\partial x_\mu} - \frac{\partial\sigma}{\partial x_\mu}\boldsymbol{\pi} + N^+\gamma_4\gamma_\mu\gamma_5\boldsymbol{\tau}N\right]\cdot\boldsymbol{v} + \text{terms quadratic in } \boldsymbol{u} \text{ and } \boldsymbol{v}\,.$$

We wish to identify the σ-containing part of the above Lagrangian as representing strong interactions and the remaining Lagrangian as representing weak interactions. The leptons (e^+, ν, e^-) or (μ^+, ν', μ^-) form a 3-vector in the space we are considering and a gauge-transformation will only link them with the u-field. The strength of the strong coupling comes about if we assume that the vacuum expectation value of σ ($\langle\sigma\rangle_0$) does not equal zero but equals $(g_s/2\boldsymbol{M}_N)(1/g_\omega)$ (*).

Thus terms in the Lagrangian with $\sigma(\partial\boldsymbol{\pi}/\partial x_\mu)\cdot\boldsymbol{v}$ and $M^+\gamma_4\gamma_\mu\gamma_5\boldsymbol{\tau}\cdot\boldsymbol{v}N$ together give the conventional pseudo-vector strong Yukawa interaction with pions emitted singly. This is not to say that we are considering σ as an alternative expression for the strong coupling constant. There is every possibility that σ-particles are emitted (and absorbed) as physical particles.

Notice that the weak interactions in this model conserve parity. This unfortunate situation seems to persist in subsequent work also (**).

3. – Extension to strange particles.

In the above model σ, $\boldsymbol{\pi}$ form a 4-vector in a [4]-Euclidean manifold while N_L and N_R form a 4-spinor. A direct extension of this to include K-mesons is possible, provided we consider σ, π and K-particles to form a vector in an [8]-space while the sixteen baryons (eight-baryons each decomposed into their left and right components) form a 16-component spinor in such a space.

(*) Notice the terms $\sigma^2(X^2+Y^2)$ in (2) could give the mass-terms for u and v particles.

(**) The theory of weak interactions recently proposed by GELL-MANN and LÉVY [4] effectively proceeds by identifying the terms containing $\boldsymbol{X}_\mu$ only in (2) with the strangeness-conserving weak Lagrangian. The gauge-transformations giving rise to $\boldsymbol{Y}_\mu$ fields are not considered.

The formalism we use was essentially developed by TIOMNO ([5]). Let us first recapitulate this. Write

$$\frac{1}{\sqrt{2}}(\Lambda^0 - \boldsymbol{\tau}\cdot\boldsymbol{\Sigma}) = \begin{pmatrix} Z^0 & \Sigma^+ \\ \Sigma^- & Y^0 \end{pmatrix} = (\Sigma_2, \Sigma_1)\,,$$

$$K^+ = K_1 - iK_2\,, \qquad K^0 = K_3 - iK_4\,.$$

If all K-coupling constants are equal, the conventional p.s. (or p.v.) strong K-Lagrangian can be written as

$$L_K = (N^+\Xi^+)(i\gamma_4\gamma_5)\begin{pmatrix} K^0 & K^+ \\ K^- & -\bar{K}^0 \end{pmatrix}\begin{pmatrix} \Sigma_1 \\ \Sigma_2 \end{pmatrix} + \text{h.c.}\,. \tag{4}$$

Write $\psi = \begin{pmatrix} N \\ \Xi \\ \Sigma_1 \\ \Sigma_2 \end{pmatrix}$; then L_K equals

$$\sum_{\alpha=1}^{4} \psi^+ i\gamma_4\gamma_5 \Gamma_\alpha K_\alpha \psi\,, \tag{5}$$

where

$$\Gamma = \begin{pmatrix} & \boldsymbol{\tau}\times 1 \\ \boldsymbol{\tau}\times 1 & \end{pmatrix}, \qquad \Gamma_4 = \begin{pmatrix} & i\times 1 \\ -\,i\times 1 & \end{pmatrix}. \tag{6}$$

These Γ matrices are 8×8 matrices pertaining to a (six or) seven dimensional manifold. The spinor ψ is an 8×1 column.

It is possible to write the conventional p.s. (or p.v.) π-interactions in term of ψ.

Define three additional matrices,

$$\Gamma_{5,6,7} = \begin{pmatrix} 1\times\boldsymbol{\tau} & & & \\ & 1\times\boldsymbol{\tau} & & \\ & & -1\times\boldsymbol{\tau} & \\ & & & -1\times\boldsymbol{\tau} \end{pmatrix} = \tau_3\times 1\times\boldsymbol{\tau}\,. \tag{7}$$

Then the matrices $\Gamma_1, \Gamma_2, \ldots, \Gamma_7$ anti-commute. If all π-couplings are equal (and in particular if $g_{\mu\mathcal{N}} = -g_{\pi\Sigma\Sigma}$), one can write the π-Lagrangian

$$L_\pi = \sum_{\alpha=5,6,7} \psi^+ i\gamma_4\gamma_5(\Gamma_\alpha \pi_\alpha)\psi\,. \tag{8}$$

It is clear that π's and K's form a vector in a [7]-space.

So much for Tiomno's formalism. We can now follow a procedure analogous to Section **2** and obtain an expression which would contain terms like

$$\sigma\frac{\partial\boldsymbol{\pi}}{\partial x_\mu}\cdot\boldsymbol{V}_\mu\,, \qquad \sigma\frac{\partial K_\alpha}{\partial x_\mu}Z_\alpha\,, \qquad \sum_{\alpha=1}^{4}\psi^+(i\gamma_4\gamma_\mu\gamma_5)\Gamma_\alpha Z_\alpha\psi\,,$$

([5]) J. TIOMNO: *Nuovo Cimento*, **6**, 69 (1957).

to give an effective strong p.v. Lagrangian of the Tiomno type. Write

(9) $$L_f = \psi_L^+ \gamma_4 \gamma_\mu \partial_\mu \psi_L + \psi_R^+ \gamma_4 \gamma_\mu \partial_\mu \psi_R + \tfrac{1}{2}[\partial_\mu \sigma + (\partial_\mu \boldsymbol{\pi})^2 + (\partial_\mu K_\alpha)^2] .$$

The sixteen-component entity $\begin{pmatrix}\psi_L\\ \psi_R\end{pmatrix}$ forms a spinor in [8]-space. One anti-commuting Dirac set for such a space is

(10) $$\begin{cases} \Gamma^{(8)}_{1,2,3} = \tau_1 \times (\tau_1 \times \boldsymbol{\tau}) \times 1 , \\ \Gamma^{(8)}_4 \;\; = \tau_1 \times (\tau_2 \times 1) \times 1 , \\ \Gamma^{(8)}_{5,6,7} = \tau_1 \times (\tau_3 \times 1) \times \boldsymbol{\tau} , \\ \Gamma^8_8 \;\;\; = \tau_2 \times 1 \times 1 \times 1 . \end{cases}$$

In [8] space there are 28 rotations. Seven of these rotations ($\sigma \to \sigma + \boldsymbol{\epsilon}\cdot\boldsymbol{\pi} + \varepsilon_\alpha K_\alpha$, $\boldsymbol{\pi} \to \boldsymbol{\pi} - \boldsymbol{\epsilon}\sigma$, etc.) with the corresponding spinor rotation matrices

$$\frac{1}{2i}(\Gamma^{(8)}_8 \Gamma^8_\alpha - \Gamma^{(8)}_\alpha \Gamma^{(8)}_8) ,$$

give

(11) $$L_{\text{int}} = \frac{1}{2}\left(\sigma \frac{\partial \boldsymbol{\pi}}{\partial x_\mu}\cdot v_\mu + \sigma \frac{\partial K_\alpha}{\partial x} Z_\alpha - \frac{\partial \sigma}{\partial x_\mu} \boldsymbol{\pi}\cdot \boldsymbol{v} - \frac{\partial \sigma}{\partial x_\mu} K_\alpha Z_\alpha\right)$$

i.e. Tiomno Lagrangian with $(i\gamma_4\gamma_5\boldsymbol{\tau}\cdot\boldsymbol{\pi})$ replaced by $(i\gamma_4\gamma_\mu\gamma_5\boldsymbol{\tau}\cdot\boldsymbol{v}_\mu)$ and $(i\gamma_4\gamma_5\Gamma'_\alpha K_\alpha)$ replaced by $i\gamma_4\gamma_\mu\gamma_5\Gamma_\alpha Z_{\alpha,\mu}$. The fields $\boldsymbol{V}_\mu$ and Z_μ behave, so far as isotopic spin, etc., is concerned just like $\boldsymbol{\pi}$ and K mesons.

In so far as these seven rotation matrices do not form a Lie-Algebra, the interaction Lagrangian must contain 21 other fields corresponding to the remaining 21 rotations. From the point of view adopted in Section **2**, these give weak interactions only. A general analysis of these terms has been given by GÜRSEY ([6]) in a recent paper which also adopts the Tiomno formalism to give an analogue of the Gell-Mann–Lévy theory of weak interactions.

It may be more profitable from our point of view to consider two fields σ and σ' in such a way that $(\sigma, \boldsymbol{\pi})$ form a 4-vector and (σ', K_α) a 5-vector. The resulting strong Lagrangian would then contain two coupling parameters $\langle\sigma\rangle_0$ and $\langle\sigma'\rangle_0$. $(\sigma, \boldsymbol{\pi})$ and (σ', K_α) spaces in a sense form two (disconnected) pieces of a [9] space. Even for the Tiomno Lagrangian it was possible to consider $\boldsymbol{\pi}$ and K-mesons as particles corresponding to disjunct pieces of the 7-dimensional space. Thus if we replace $\Gamma_\alpha \pi_\alpha$ in eq. (8) by $\Gamma'_\alpha \pi_\alpha$ where $\Gamma'_\alpha = 1 \times 1 \times \boldsymbol{\tau}$ (*i.e.* $g_{\pi NN} = + g_{\pi\Sigma\Sigma}$) we see that $\boldsymbol{\pi}$ and K no longer form a 7-vector. The per-

([6]) F. GÜRSEY: preprint.

mitted rotations have to be limited in such a way that π-mesons are not transformed into K-mesons.

Returning to the [9] space, if (σ, $\boldsymbol{\pi}$) form a [4] subspace and (σ', K) a [5] sub-space, it is clear that the total number of intermediate bosons will be $6+10=16$. Of these seven will mediate strong (p.v.) interaction, and 9 will mediate weak (v.) interactions. All these interactions conserve parity and isotopic-spin.

4. – $\Delta|\boldsymbol{I}|=\frac{1}{2}$ rule.

One simple way to introduce strangeness violation consistent with the $\Delta|\boldsymbol{I}|=\frac{1}{2}$ rule is to remark that the other field besides σ (or σ') which can have a non-zero expectation-value is the field corresponding to the θ^0 particle ($CP=+I$). In the notation above $\langle K_4\rangle \neq 0$ and would in fact be proportional to g_w even in the conventional theory. Thus all strangeness conserving terms like $\bar{\Lambda}N\,\boldsymbol{\tau}\cdot\boldsymbol{\pi}K$ describe also the matrix-elements for (parity-conserving) weak decay of $\Lambda\to\mathcal{N}+\pi$ consistent with $\Delta T=\frac{1}{2}$ provided we take the vacuum expectation value of the K-meson.

The non-zero expectation value of θ^0 is the perfect realization of the spurion idea of Wentzel so that it may not be necessary to introduce any additional fields to violate strangeness.

From what has been said above, it is clear that seven fields (three with transformation character of π-mesons and four with that of K-mesons) are necessary to mediate strong-interactions. The number of those necessary for weak interactions depends on the model used. However in all this work parity-violation for weak interactions remains a complete mystery.

* * *

We are deeply indebted to Professor R. G. SACHS for an invitation to the University of Wisconsin Summer Institute where part of the above work was completed.

Gauge Theories of Vector Particles*

Sheldon L. Glashow and Murray Gell-Mann

California Institute of Technology, Pasadena, California

The possibility of generalizing the Yang–Mills trick is examined. Thus we seek theories of vector bosons invariant under continuous groups of coordinate-dependent linear transformations. All such theories may be expressed as superpositions of certain "simple" theories; we show that each "simple" theory is associated with a simple Lie algebra. We may introduce mass terms for the vector bosons at the price of destroying the gauge-invariance for coordinate-dependent gauge functions.

The theories corresponding to three particular simple Lie algebras—those which admit precisely two commuting quantum numbers—are examined in some detail as examples. One of them might play a role in the physics of the strong interactions if there is an underlying super-symmetry, transcending charge independence, that is badly broken.

The intermediate vector boson theory of weak interactions is discussed also. The so-called "schizon" model cannot be made to conform to the requirements of partial gauge-invariance. It is possible, however, to find a formal theory of four intermediate bosons that is partially gauge-invariant and gives an approximate $|\Delta I| = \frac{1}{2}$ rule.

I. INTRODUCTION

The electromagnetic interaction of elementary particles is remarkably simple. It is of universal strength and form and is associated with a principle of gauge invariance. In fact, starting with the idea of invariance under gauge transformations with coordinate-dependent gauge functions, one can deduce the existence of a massless vector field coupled to a conserved current. If all charged fields are subjected to the same gauge transformation, then the electric charges of all particles are the same.

The fact that the weak interactions are vectorial in character (apart from nonconservation of parity) and nearly universal in strength has suggested to many physicists that they may be mediated by vector fields (*1*, *2*) and that there may be a useful parallel between them and electromagnetism, perhaps even extending to the notion of gauge invariance (*3*–*6*).

* Research supported in part by the U.S. Atomic Energy Commission and the Alfred P. Sloan Foundation. The research was begun while the authors were National Science Foundation Fellows.

The strong interactions, too, seem to exhibit some degree of universality. Moreover, the approximate conservation laws of isotopic spin and of strangeness, as well as the exact law of conservation of baryons, present an analogy with the conservation of charge and suggest that some principles of gauge invariance may be at work. Until recently, it seemed that the strong couplings were not vectorial, but there is mounting evidence that there are objects (like the $I = 1$, $J = 1$, $\pi\pi$ resonance) that can be interpreted as vector mesons and that may play a very significant role in the strong interactions (*7*, *8*).

There are two great difficulties in the way of constructing theories of weak and strong interactions by analogy with electrodynamics. One is that some of the relevant currents are not conserved. The isotopic spin and strangeness currents that may enter into a vectorial theory of the strong couplings fail to be conserved on account of electromagnetic and weak interactions, while the conservation of the weak current is broken not only by electromagnetism but, in the case of the axial vector and strangeness-changing parts, by masses and perhaps by strong interactions as well.

The other difficulty is that whereas photons are massless (as the quanta must be in a theory that is fully gauge invariant with a coordinate-dependent gauge function) the vector particles that mediate the strong and weak interactions must be massive if they exist at all.

Thus the notion has arisen (*3–8*) of a theory that is partially gauge-invariant. In each case we have a Lagrangian like the electromagnetic one, fully invariant under coordinate-dependent gauge transformations, plus other terms. The remaining terms are of two kinds:

(a) those which break the full gauge invariance, while leaving intact the conservation law and the invariance under constant gauge transformations;

(b) those which destroy the gauge invariance altogether, along with the conservation law.

In the case where the conservation law is exact (conservation of baryons) the terms of Type (b) are, of course, absent.

Now the idea of partial gauge-invariance poses a number of questions, to which we shall return briefly in Section VII. For the moment, let us concentrate on the straightforward part of the problem, the construction of the fully gauge-invariant part of the theory.

The coupling of a vector meson field to a single quantity like baryon number follows exactly the pattern of electromagnetic coupling to the charge, as long as the complete gauge-invariance is maintained. But when we go over to the case of three non-commuting quantities like the components of the isotopic spin current, the situation becomes different and a more sophisticated theory becomes necessary. The intermediary vector meson field now carries isotopic spin 1 and its own isotopic spin current contributes a source term. Thus the theory of the vector meson field becomes non-linear. The problem of constructing the theory in question has been solved by Yang and Mills (*9*) and by Shaw (*10*).

In the next two sections, we review the simple case of charge or baryon number and the more complicated case of isotopic spin. Then, in Section IV, we go on to the main point of this article—the description of all possible straightforward generalizations of the Yang–Mills trick. We are interested in such generalizations because we do not know, for either the strong or the weak interactions, exactly how many intermediate vector fields may be involved (if any). To give just one example, it has been suggested (*11–13*) that there may be four such (Hermitian) fields for the weak interactions—the so-called schizon model, set up to give $|\Delta \mathbf{I}| = \frac{1}{2}$ and $\Delta S = 0, \pm 1$ for the nonleptonic weak interactions of baryons and mesons. We shall show in Section VII that the ideas of partial gauge-invariance lead to severe restrictions on four-field models; in fact, the restrictions are so strong as to make it impossible to construct the schizon model according to the gauge principles of this article.

The classification of generalized Yang–Mills theories discussed in Section IV is described further in Section V; some examples are given in Section VI; and some possible physical applications are touched on briefly in Section VII.

II. THE ONE-PARAMETER GAUGE THEORY

The gauge formalism of electromagnetism is, of course, well known. The generalization from charge to baryon number was discussed by Yang and Lee (*14*); it is clear from their work that the generalization contradicts experiment unless either the coupling constant is ridiculously small or the gauge invariance is broken, say by a mass term for the vector field. Let us review the method.

We start with an additive quantity like charge or baryon number; call it Q. Let the fields $\psi_a(x)$ destroy particles of charge Q_a and create their antiparticles. We then discuss invariance under the infinitesimal gauge transformations

$$\psi_a(x) \rightarrow \psi_a(x) - iQ_a\Lambda(x)\psi_a(x). \tag{2.1}$$

Whenever the coordinate derivative ∂_α acts on ψ_a, it undergoes the transformation

$$\partial_\alpha \rightarrow \partial_\alpha - iQ_a\partial_\alpha\Lambda \quad (\text{on } \psi_a). \tag{2.2}$$

In order to cancel this change, we introduce a vector field $A_\alpha(x)$ that suffers the gauge transformation

$$A_\alpha(x) \rightarrow A_\alpha(x) - \partial_\alpha\Lambda(x) \tag{2.3}$$

and a field Lagrangian density L_A invariant under this transformation, say

$$L_A = -\tfrac{1}{4}(\partial_\alpha A_\beta - \partial_\beta A_\alpha)^2. \tag{2.4}$$

In the absence of the field A_α and its couplings, let the Lagrangian be $L_0(\psi_a)$ and let it conserve Q. Then the "minimal" gauge-invariant Lagrangian including A_α is

$$L = \tilde{L}_0(\psi_a) + L_A , \tag{2.5}$$

where $\tilde{L}_0$ is obtained from L_0 by the replacement

$$\partial_\alpha \rightarrow \partial_\alpha - iQ_a A_\alpha(x) \quad (\text{on } \psi_a). \tag{2.6}$$

It is evident that (2.5) gives us a gauge-invariant Lagrangian and certainly the procedure described by (2.5) is the usual one. But what do we mean by "minimal"? The point is that we could add to the Lagrangian (2.5) further gauge-invariant terms involving the field strength $\partial_\alpha A_\beta - \partial_\beta A_\alpha$. However, nature, in the case of electromagnetism, does not seem to make use of such terms.

Consider, for example, a Dirac particle of charge e, for which ψ is a spinor and the free Lagrangian density is

$$L_0 = -\bar{\psi}(\gamma_\alpha \partial_\alpha + m_0)\psi. \tag{2.7}$$

The substitution (2.6) gives the usual coupling,

$$ie\bar{\psi}\gamma_\alpha A_\alpha \psi, \tag{2.8}$$

but no Pauli moment. We generally suppose that the effective Pauli moments of nucleons arise from the ordinary electrical interaction of the meson cloud around the nucleon and not from a basic Pauli moment term in the Lagrangian:

$$i\mu\bar{\psi}\sigma_{\alpha\beta}\psi(\partial_\alpha A_\beta - \partial_\beta A_\alpha). \tag{2.9}$$

Hence the attempt (*15*) to state a principle of minimal electromagnetic interaction, that the electromagnetic field interacts only with electric charges in the normal way [as in (2.6)] and not through special field-dependent terms like (2.9) in the basic Lagrangian.

The difficulty (*16*, *17*) with any attempt to put the idea of minimal electromagnetic interaction in definite mathematical form is the following. Various Lagrangian densities (differing by divergences of four-vectors) can lead to the same equations of motion. But if we choose in this way a new L_0, the resulting electromagnetic coupling (and the equations of motion including electromagnetism) may become radically different. Thus we can obtain the Pauli moment term (2.9) by the "minimal" procedure (2.6) if we just add to the usual L_0 in (2.7) the term

$$(2\mu/e)\partial_\alpha(\bar{\psi}\sigma_{\alpha\beta}\partial_\beta\psi). \tag{2.10}$$

We see that the procedure (2.6) defines the "minimal" interaction only if the original Lagrangian density L_0 is chosen in a "minimal" way. We must assign a physical meaning to L_0 and say that (2.7) describes a Dirac particle properly, while if the term (2.10) is added we obtain the wrong Lagrangian density for a Dirac particle, even though the equation of motion without electromagnetism is just the Dirac equation in both cases.

Of course we have still not specified in a clear-cut way how to find the "minimal" L_0 in all cases. But that difficulty is not restricted to the problem of electromagnetic couplings. Even without electromagnetic interactions and without strong and weak interactions, we must still assign a physical significance to L_0 because it determines the gravitational coupling. If we add a term like (2.10) to L_0 and follow the usual procedure for constructing the stress-energy-momentum tensor, we will get a different answer. In fact, the gravitational interactions are constructed from L_0 in a way that is closely analogous to the method given in (2.6) for electromagnetism.

Now let us return to the theory described by the Lagrangian density (2.5). The equation of motion for the field A_α is

$$\Box^2 A_\alpha - \partial_\alpha \partial_\beta A_\beta = -D\tilde{L}_0(\psi_a)/DA_\alpha = -j_\alpha\,, \tag{2.11}$$

where D/DA_α is the Lagrangian derivative $\partial/\partial A_\alpha - \partial_\beta \partial/\partial(\partial_\beta A_\alpha)$.[1] The formula for the current can be re-expressed as follows. Consider a gauge transformation in which the ψ fields are affected as in (2.1) but A_α is *not* transformed. Denote partial derivatives with respect to Λ and $\partial_\alpha \Lambda$ under this condition by

$$[\partial/\partial\Lambda]_{\delta A=0} \qquad \text{and} \qquad [\partial/\partial(\partial_\alpha\Lambda)]_{\delta A=0}\,.$$

Then we remark that since $\tilde{L}_0$ is totally gauge-invariant, the derivative

$$[\partial/\partial(\partial_\alpha\Lambda)]_{\delta A=0}$$

has the effect of the negative of a derivative with $\delta\psi_a = 0$ and *only* A_α affected by the gauge transformation. But such a negative derivative is exactly $-D/DA_\alpha$. Thus we have the result

$$j_\alpha = [\partial \tilde{L}_0/\partial(\partial_\alpha\Lambda)]_{\delta A=0}\,. \tag{2.12}$$

The current is calculated from the Lagrangian (either $\tilde{L}_0$ or L) by a gauge transformation involving *only* the ψ_a fields and not A_α.

Next we note (*6*) that in any local gauge transformation, the Euler–Lagrange equation applies to the gauge function, even though it is not a field variable, as a consequence of the Euler–Lagrange equations for the field variables them-

[1] We are using a classical, not a quantum, action principle. The variational derivatives are thus with respect to c-number quantities. Although we neglect the difficulties encountered due to lack of commutativity of the quantized fields, our results are presumably independent of this omission in a properly formulated quantum theory.

selves. Thus we have

$$\begin{aligned} \partial_\alpha j_\alpha = \partial_\alpha[\partial \tilde{L}_0/\partial(\partial_\alpha \Lambda)]_{\delta A=0} &= [\partial \tilde{L}_0/\partial \Lambda]_{\delta A=0} \\ &= \partial \tilde{L}_0/\partial \Lambda. \end{aligned} \tag{2.13}$$

But the Lagrangian is invariant under gauge transformations with constant gauge function. Therefore the current is conserved:

$$\partial_\alpha j_\alpha = \partial \tilde{L}_0/\partial \Lambda = 0. \tag{2.14}$$

Looking back at the equation of motion (2.11), we see that the supplementary condition $\partial_\alpha A_\alpha = 0$ may be imposed.

Finally, we may identify the constant of the motion $-i\int j_4\, d^3x$ with the charge Q. So far we have looked at the equations classically; but in quantum mechanics, of course, Q is an operator and has the commutation relations

$$[\psi_a, Q] = Q_a \psi_a . \tag{2.15}$$

Now that we have sketched the fully gauge-invariant theory, we may discuss what happens when a term is added to L that breaks the full gauge invariance but leaves the invariance under gauge transformations of the first kind, that is, with constant Λ. We shall take the simple case of a mass term for the vector meson

$$-\mu_0^2 A_\alpha A_\alpha/2.$$

Evidently all that happens is that the equation of motion (2.11) becomes

$$(\Box^2 - \mu_0^2)A_\alpha - \partial_\alpha \partial_\beta A_\beta = -j_\alpha , \tag{2.16}$$

while the expression (2.12) for the current and the conservation law (2.14) remain unchanged. We have a vector meson coupled to a conserved current in a "partially gauge-invariant" theory.

III. THE THREE-PARAMETER GAUGE THEORY OF YANG AND MILLS

We now turn from the simple case of charge or baryon number to the case of the isotopic spin **I**, obeying the commutation relations

$$[I_i, I_j] = ie_{ijk} I_k . \tag{3.1}$$

This time our fields ψ carry isotopic spin; let us consider for simplicity a field N of isotopic spin ½ (the nucleon) and a field π of isotopic spin 1 (the pion) (*9*). The relations analogous to (2.15) are

$$\begin{aligned} [N, I_i] &= \tau_i N/2, \\ [\pi_j, I_i] &= -ie_{ijk}\pi_k . \end{aligned} \tag{3.2}$$

The infinitesimal gauge transformations analogous to (2.1) are then

$$\begin{aligned} N(x) &\to N(x) - i\gamma_0\boldsymbol{\tau}\cdot\boldsymbol{\Lambda}(x)N(x), \\ \boldsymbol{\pi}(x) &\to \boldsymbol{\pi}(x) + 2\gamma_0\boldsymbol{\Lambda}(x)\times\boldsymbol{\pi}(x). \end{aligned} \tag{3.3}$$

(We have denoted by γ_0 the bare coupling parameter.) Thus the coordinate derivative ∂_α acting on the fields N and $\boldsymbol{\pi}$ suffers the change

$$\begin{aligned} \partial_\alpha &\to \partial_\alpha - i\gamma_0\boldsymbol{\tau}\cdot\partial_\alpha\boldsymbol{\Lambda} \quad (\text{on } N), \\ \partial_\alpha &\to \partial_\alpha + 2\gamma_0\partial_\alpha\boldsymbol{\Lambda}x \quad (\text{on } \boldsymbol{\pi}). \end{aligned} \tag{3.4}$$

corresponding to (2.2).

To form a gauge-invariant theory, we must introduce a vector field $\mathbf{A}_\alpha(x)$ with isotopic spin one; its gauge transformation is essentially different from (2.3) because $\mathbf{A}_\alpha$ carries isotopic spin, whereas the photon carries no charge. Thus the field is not only displaced by the gradient of Λ but also undergoes isotopic rotation as $\boldsymbol{\pi}(x)$ does in (3.3). The gauge transformation is thus

$$\mathbf{A}_\alpha(x) \to \mathbf{A}_\alpha(x) - \partial_\alpha\boldsymbol{\Lambda}(x) + 2\gamma_0\boldsymbol{\Lambda}(x)\times\mathbf{A}_\alpha(x). \tag{3.5}$$

For the field Lagrangian density, we must choose an expression invariant under this gauge transformation. We note that

$$\mathbf{G}_{\alpha\beta} \equiv \partial_\alpha\mathbf{A}_\beta - \partial_\beta\mathbf{A}_\alpha + 2\gamma_0\mathbf{A}_\alpha\times\mathbf{A}_\beta \tag{3.6}$$

transforms according to the rule

$$\mathbf{G}_{\alpha\beta} \to \mathbf{G}_{\alpha\beta} + 2\gamma_0\boldsymbol{\Lambda}\times\mathbf{G}_{\alpha\beta}\,. \tag{3.7}$$

The simplest gauge-invariant Lagrangian is thus

$$L_A = -\tfrac{1}{4}\mathbf{G}_{\alpha\beta}\cdot\mathbf{G}_{\alpha\beta}\,, \tag{3.8}$$

which is, of course, nonlinear, unlike (2.4). In the equation of motion deducible from (3.8), the source of the A field is its own isotopic spin current.

Now, given a Lagrangian $L_0(N, \boldsymbol{\pi})$ not involving the A field but conserving isotopic spin, we can introduce the "minimal" gauge-invraiant Lagrangian density including A:

$$L = \tilde{L}_0(N, \boldsymbol{\pi}) + L_A\,, \tag{3.9}$$

with $\tilde{L}_0$ obtained from L_0 by the substitutions

$$\begin{aligned} \partial_\alpha &\to \partial_\alpha - i\gamma_0\boldsymbol{\tau}\cdot\mathbf{A}_\alpha \quad (\text{on } N), \\ \partial_\alpha &\to \partial_\alpha + 2\gamma_0\mathbf{A}_\alpha x \quad (\text{on } \boldsymbol{\pi}) \end{aligned} \tag{3.10}$$

analogous to (2.6).

The current, source of the field $\mathbf{A}_\alpha$, is given, not exactly as in (2.12), but by the formula

$$2\gamma_0\mathbf{I}_\alpha = [\partial\tilde{L}_0/\partial(\partial_\alpha\mathbf{\Lambda})]_{2\gamma_0\mathbf{\Lambda}\times\mathbf{A}}, \tag{3.11}$$

and is conserved, while the analog of the charge is just

$$-i2\gamma_0\int \mathbf{I}_4\, d^3x = 2\gamma_0\mathbf{I}. \tag{3.12}$$

If we now add a common mass term for the three kinds of vector meson,

$$-(\mu_0^2/2)\mathbf{A}_\alpha\cdot\mathbf{A}_\alpha,$$

the gauge invariance is broken (except for constant $\mathbf{\Lambda}$) but the isotopic spin current is still conserved. Unfortunately the renormalizability of the theory, at least in the conventional sense, is lost when a mass is added (*18, 19*).

IV. GENERALIZATIONS

We now come to grips with our problem, that of classifying the straightforward generalizations of the Yang–Mills trick. We imagine sets of N fields, like the two kinds of nucleon or the three kinds of pion in Section III, on which a gauge operation performs a linear transformation as in (3.3). We may write

$$\psi_i(x)\psi_i(x) - 2i\gamma_0\sum_{j=1}^{n}\sum_{k=1}^{N} M_j^{ik}\Lambda_j(x)\psi_k(x) \tag{4.1}$$

for our generalization. The n independent gauge functions $\Lambda_j(x)$ may be taken real, while the M_j are, for the moment, arbitrary complex $N \times N$ matrices.

The Lagrangian density $L_0(\psi_i)$ is presumed invariant under (4.1) for constant gauge functions Λ_j. Then (4.1) must be an infinitesimal unitary operation; the matrices M_j must be Hermitian. The coordinate derivative acting on ψ changes according to the rule

$$\partial_\alpha \to \partial_\alpha - 2i\gamma_0\sum_{j=1}^{n} M_j(\partial_\alpha\Lambda_j). \tag{4.2}$$

To cancel this change, we introduce n Hermitian fields $A_{\alpha i}$ to take up the gauges Λ_i. In place of (3.5) we have

$$A_{\alpha i}(x) \to A_{\alpha i}(x) - \partial_\alpha\Lambda_i(x) + 2\gamma_0\sum_{j,k=1}^{n} c_{ijk}\Lambda_j(x)A_{\alpha k}(x), \tag{4.3}$$

where all the indices in c_{ijk} run from 1 to n. The c's must be real to preserve the Hermiticity of the A fields. We must determine the properties of c_{ijk} that will permit the Yang–Mills trick to go through.

First of all, we must be able to find a gauge-invariant field Lagrangian for the $A_{\alpha i}$. We seek a field strength that transforms simply, like $\mathbf{G}_{\alpha\beta}$ in (3.7):

$$G_{\alpha\beta i} \equiv \partial_\alpha A_{\beta i} - \partial_\beta A_{\alpha i} + 2\gamma_0 \sum_{j,k=1}^{n} b_{ijk} A_{\alpha j} A_{\beta k} . \tag{4.4}$$

The transformation is, in general, very complicated. Let us use the summation convention. We obtain

$$\begin{aligned} G_{\alpha\beta i} \to G_{\alpha\beta i} &+ 2\gamma_0 (c_{ijk} - b_{ijk}) \partial_\alpha \Lambda_j A_{\beta k} - 2\gamma_0 (c_{ijk} + b_{ikj}) \partial_\beta \Lambda_j A_{\alpha k} \\ &+ 2\gamma_0 c_{ijk} \Lambda_j (\partial_\alpha A_{\beta k} - \partial_\beta A_{\alpha k}) \\ &+ 4\gamma_0^2 (b_{ijm} c_{jkl} + b_{ilj} c_{jkm}) \Lambda_k A_{\alpha l} A_{\beta m} . \end{aligned} \tag{4.5}$$

In order to obtain a law analogous to (3.7), we must put

$$b_{ijk} = c_{ijk} , \qquad b_{ikj} = -c_{ijk} , \qquad b_{ijm} c_{jkl} + b_{ilj} c_{jkm} = c_{ikj} b_{jlm} .$$

Then we have

$$G_{\alpha\beta i} \to G_{\alpha\beta i} + 2\gamma_0 c_{ijk} \Lambda_j G_{\alpha\beta k} , \tag{4.6}$$

with

$$G_{\alpha\beta i} = \partial_\alpha A_{\beta i} - \partial_\beta A_{\alpha i} + 2\gamma_0 c_{ijk} A_{\alpha j} A_{\beta k} , \tag{4.7}$$

$$c_{ijk} = -c_{ikj} , \tag{4.8}$$

$$c_{ijm} c_{jkl} + c_{ilj} c_{jkm} + c_{ikj} c_{jml} = 0. \tag{4.9}$$

Now the Lagrangian density

$$L_A = -\tfrac{1}{4} G_{\alpha\beta i} G_{\alpha\beta i} \tag{4.10}$$

will indeed be gauge-invariant provided we have

$$c_{ijk} = -c_{kji} . \tag{4.11}$$

The necessary and sufficient conditions for the construction of the generalized Yang–Mills field are thus:

$$c_{ijk} \text{ totally antisymmetric and real,} \tag{4.12a}$$

$$c_{ijm} c_{klj} + c_{kjm} c_{lij} + c_{ljm} c_{ikj} = 0. \tag{4.12b}$$

The Yang–Mills theory itself is the special case in which $n = 3$ and $c_{ijk} = e_{ijk}$, which obviously satisfies (4.12).

Now we must couple the field $A_{\alpha i}$ to the current generated by the gauge transformation (4.1) of the ψ_i. We have to construct from $L_0(\psi_i)$ a completely gauge-invariant quantity $\tilde{L}_0(\psi_i)$. In order to make the gauge transformations

(4.2) and (4.3) compensate each other, we use the prescription analogous to (3.10),

$$\partial_\alpha \to \partial_\alpha - 2i\gamma_0 \sum_{i=1}^{n} M_i A_{\alpha i} \quad (\text{on } \psi) \tag{4.13}$$

to construct $\tilde{L}_0$ from L. Under the unitary transformation (4.1), which we may rewrite in the form

$$\psi \to \psi - 2i\gamma_0 (\sum_j M_j \Lambda_j) \psi,$$

the M's transform according to the rule

$$M_i \to M_i - 2i\gamma_0 \sum_j [M_i, M_j] \Lambda_j, \tag{4.14}$$

while ∂_α and $A_{\alpha i}$ transform as in (4.2) and (4.3), respectively. Thus the prescription (4.13) yields a gauge-invariant Lagrangian density $\tilde{L}_0$ if and only if we have

$$[M_i, M_j] = i c_{ijk} M_k, \tag{4.15}$$

with the summation convention understood. Evidently this is the generalization of the commutation rule (3.1) for the isotopic spin.

In order that c_{ijk} define the commutation relation (4.15), it must obey just two conditions. First, the rule

$$[M_i, M_j] = -[M_j, M_i] \tag{4.16}$$

tells us that c_{ijk} must be antisymmetric in i and j; but we already know that from (4.12a). Second, the Jacobi identity

$$[M_i, [M_j, M_k]] + [M_j, [M_k, M_i]] + [M_k, [M_i, M_j]] = 0 \tag{4.17}$$

gives us just (4.12b).

There remains the condition that c_{ijk} be antisymmetric not only in i and j but in the other pairs of indices as well. We shall return to the consequences of this further condition shortly.

Suppose we can divide the indices k into two sets such that $c_{ijk} = 0$ whenever i belongs to one set and j to the other. Then the fields $A_{i\alpha}$ of one set and those of the other set are completely unconnected to each other by any of the gauge transformations we have discussed. Likewise, the operators M_i belonging to one set of indices commute with those belonging to the other set. We are then dealing with a linear superposition of two completely independent Yang–Mills theories, which may have vastly different coupling strengths and no direct physical connection. We might as well restrict our attention to one of these.

We can go further. Suppose our theory is not simplifiable as just discussed.

We may apply any real rotation in the n-dimensional space of the $A_{i\alpha}$, rotating at the same time the M_i and the gauges Λ_i. (The properties (4.12) of c_{ijk} are unaffected by such a rotation.) It may turn out then that our theory is simplifiable. In that case, let us restrict our attention to one of the parts. We continue this process until we reach a nonsimplifiable Yang–Mills theory, one for which we cannot, no matter how we rotate in the n-dimensional space, find two sets of indices that are unconnected by the c_{ijk}. From now on, we shall deal with these unsimplifiable or "simple" theories, from which the most general theory can be built up by ordinary superposition and rotation.

Simple theories with more than one vector meson have an important property —they are characterized by a single *universal* coupling constant. To see this, suppose there are two distinct multiplets of fermions $\psi^{(1)}$ and $\psi^{(2)}$, both coupled to the $A_{\alpha i}$ by means of the prescription (4.13) but possibly with different coupling strengths, $\gamma_{(1)}$ and $\gamma_{(2)}$. The scale of the matrices, $M_{(1)}$ and $M_{(2)}$, acting upon $\psi^{(1)}$ and $\psi^{(2)}$ is fixed by (4.15). For the interaction to be invariant under simultaneous rotations of the $\psi^{(\lambda)}$, (4.1), and the $A_{\alpha i}$, (4.3), it is clear that either: (a) the $A_{\alpha i}$ may be separated into two sets, one of which interacts only with $\psi^{(1)}$, the other with $\psi^{(2)}$—such a theory cannot be simple; or (b) $c_{ijk} \equiv 0$—this is possible only if the theory is a superposition of one or more trivial one-parameter theories; or (c) $\gamma_{(1)} \equiv \gamma_{(2)}$—the nontrivial simple theory must also be a universal theory.[2]

Now the condition that c_{ijk} be totally antisymmetric is easily shown to be equivalent to the condition,[3]

$$\mathrm{Tr}\, M_i M_j = (\text{const.})\delta_{ij}\,, \tag{4.18}$$

for a "simple" theory. For the original Yang–Mills theory, in which the M_i ($i = 1, 2, 3$) are isotopic spin matrices, Eq. (4.18) is evidently fulfilled.

We may now summarize the necessary and sufficient conditions for a simple generalized gauge theory. We must find an algebraic system, say of quantities

[2] The remarkable universality of the electric charge would be better understood were the photon not merely a singlet, but a member of a family of vector mesons comprising a simple partially gauge-invariant theory. One of the authors (S.L.G.) acknowledges a conversation with G. Feinberg in this connection.

[3] Define $d_{ijk} = -i\,\mathrm{Tr}\, M_i\,[M_j\,, M_k\,]$; define $g_{kl} = \mathrm{Tr}\, M_k\, M_l$. Clearly d_{ijk} is real and totally antisymmetric, g_{kl} is real and symmetric, and $d_{ijk} = \sum_l g_{il}\, c_{ljk}$. Now if $g_{il} = A\delta_{il}$, evidently $d_{ijk} = Ac_{ijk}$ and c is totally antisymmetric. To prove the converse, diagonalize g_{il} by an orthogonal transformation of the fields $A_{i\alpha}$. Then $g_{il} = F_i\, \delta_{il}$ and $d_{ijk} = F_i\, c_{ijk}$. But both d and c are totally antisymmetric and therefore $d_{jik} = F_j\, c_{jik} = F_i\, c_{ijk}$ and $F_j = F_i$ whenever i and j are connected by a nonzero coefficient c_{ijk}. For a *simple* theory, however, we can ultimately convert all elements to one another in this way and prove all the F_i are equal. Thus $g_{il} = A\delta_{il}$ (unless, of course, c_{ijk} vanishes; that gives the one-parameter theory).

S_i $(i = 1, \cdots, n)$, defined by a commutator $[S_i, S_j]$ obeying the antisymmetry and Jacobi laws (4.16) and 4.17) as well as the relation

$$[S_i, S_j] = ic_{ijk}S_k$$

with real, totally antisymmetric c_{ijk}. Furthermore, no real rotation of the S^i may result in a system that can be split into two commuting parts.

Such an algebraic system can always be represented by various sets of Hermitian matrices M_i obeying the same rules as well as the condition (4.18). The construction of a Yang–Mills theory then follows the pattern we have outlined.

Now the algebraic systems under discussion are well known to the mathematicians. One is the trivial one with $n = 1$, $c_{ijk} = 0$ that was discussed in Section II. All the others, including the Yang–Mills case with $n = 3$, $c_{ijk} = e_{ijk}$, are called simple Lie algebras (strictly speaking, simple Lie algebras in a special kind of real form). As such, they have been completely classified. All possible ones are known, and their representations by Hermitian matrices M_i have been studied. In the next section, we shall discuss the classification and some of the simpler cases.

Utiyama (*20*) has treated the connection of the Yang–Mills trick with Lie algebras, but he did not mention the severe restrictions of the Lie algebra that are necessary to obtain a vector meson theory with positive probabilities.[4]

V. ON SIMPLE LIE ALGEBRAS

Let us mention first the listing of all the simple Lie algebras by Cartan (*21*). Each one, of course, may be regarded as the algebra of the infinitesimal generators of a continuous group, which is called a simple Lie group.

(a) First of all, there is the infinite sequence of unitary unimodular groups $SU(\nu)(\nu = 2, 3, 4, \cdots)$. The group $SU(\nu)$ is made up of all unitary transformations with unit determinant in an ν-dimensional complex space. The infinitesimal generators are then isomorphic to the traceless Hermitian $\nu \times \nu$ matrices; evidently there are $\nu^2 - 1$ independent matrices of that kind and therefore the algebra has $\nu^2 - 1$ elements S_i. We have $n = \nu^2 - 1$.

We have, incidentally, constructed the smallest representation of the S_i by matrices M_i; we simply use the $\nu^2 - 1$ traceless Hermitian $\nu \times \nu$ matrices. They can, of course, be chosen to obey (4.18). Moreover, this representation is

[4] Note that if we set up the Einstein theory of gravity by gauge methods then the conclusions are slightly different. Instead of an isotopic rotation, we perform a 4-dimensional translation at each point of space. Thus we have, in place of the isotopic index i, another Lorentz index β, giving us a tensor field $A_{\alpha\beta}$ or $h_{\alpha\beta}$. But whereas the metric in isotopic space must be positive to give positive probabilities, the metric in Minkowski space is both positive and negative, and this causes no trouble with positive probabilities. Such a situation, although it occurs in the theory of gravity, cannot be permitted in a gauge theory of vector fields.

irreducible. (To avoid confusion, let us remark that the *algebra* of the S_i is already *simple*; no real rotation in the n-dimensional space of the $A_{i\alpha}$ can divide it into two parts that are unconnected by the c_{ijk}. However, the *representation* of the algebra by the $N \times N$ matrices M_i may be *reducible*. In other words, there may be a unitary transformation in the N-dimensional space of the ψ's that reduces all the M_i simultaneously to block form and allows us to pick out a smaller representation of the algebra. If such a reduction is impossible, the representation is irreducible.)

The isotopic spin algebra is that of $SU(2)$; we have $n = 2^2 - 1 = 3$ and $S_i = I_i$ $(i = 1, 2, 3)$. The irreducible representation by traceless Hermitian 2×2 matrices $\tau_i/2$ satisfying (4.18) is just the familiar spin ½ representation. We know, too, all the other irreducible representations, classified according to the value $I(I + 1)$ of the matrix $\sum_{i=1}^{3} M_i^2$, which commutes with all the I_i. The Ith representation ($I = 0, 1, 2, \cdots$) is said to correspond to isotopic spin I and has dimension $2I + 1$.

(b) Next, we have the infinite sequence of rotation groups $O(\nu)$ in real ν-dimensional spaces ($\nu = 7, 8, 9, \cdots$). We have omitted $O(2)$ because it is just the one-parameter group of electromagnetism and that degenerate case is not included among the simple Lie groups by the mathematicians. $O(3)$ is just the 3-dimensional rotation group and we know that is essentially the same as the isotopic spin group SU_2. The four-dimensional rotation group $O(4)$ is not simple; it is equivalent to the direct product $O(3) \times O(3)$. The groups $O(5)$ and $O(6)$ are omitted because they are essentially the same as $Sp(2)$ (see below) and $SU(4)$, respectively. Thus we begin with $O(7)$. The dimension n of the algebra of $O(\nu)$ is just the number of infinitesimal rotations $\nu(\nu - 1)/2$. In fact, the infinitesimal $\nu \times \nu$ rotation matrices (imaginary and antisymmetric) form an irreducible matrix representation of the algebra of the group.

(c) The third infinite sequence of simple Lie groups is that of the symplectic groups Sp($\nu/2$) with $\nu = 4, 6, 8, 10, 12, \cdots$. The algebra of the infinitesimal elements of $Sp(\nu/2)$ is just the algebra of the $\nu \times \nu$ skew-symplectic matrices.[5] Again we have a natural $\nu \times \nu$ irreducible matrix representation of the algebra. We note that $Sp(1)$ is omitted because it is the same as $SU(2)$.

(d) Finally, there are five more simple Lie groups and the corresponding Lie algebras. These are called exceptional Lie algebras and their names and dimensions are as follows: G_2–14, F_4–52, E_6–78, E_7–133, E_8–240.

In our listing, we have really defined each of the simple Lie algebras (except the

[5] A $\nu \times \nu$ matrix M is skew-symplectic if it is unitary and $M^T A M = A$, where $\| A_{ij} \| = \pm\delta_{\nu,i+j}$ depending upon whether or not $i > j$. $Sp(\nu/2)$ may alternatively be defined as the group of unitary transformations on $\nu/2$ quaternions. The generators of $Sp(\nu/2)$ are then the skew-Hermitian $\nu/2 \times \nu/2$ matrices over quaternions; there are evidently $\nu(\nu + 1)/2$ such matrices.

exceptional ones, for which the same can be done) by exhibiting one of its matrix representations. In each case, we understand that the $n\nu \times \nu$ matrices of the defining representation are to be taken Hermitian and satisfying (4.18). We then have the simple Lie algebra in "real form" with real, totally antisymmetric c_{ijk}. In each case, we fix the value of the constant in (4.18) for the defining representation; that fixes the scale of the M's and of the c's.

For any given Lie algebra the matrix $\sum_i M_i^2$, which commutes with all the M_i [as we can see from (4.15)], equals some number for each irreducible representation. (This situation is familiar for the isotopic spin algebra, as mentioned above.) Let the value of $\sum_i M_i^2$ for representation R be V_R. Then for that representation the constant in (4.18) is $V_R\, d_R/n$, where d_R is the dimension of the representation.

In our generalized Yang–Mills theory, the various fields ψ that are coupled to the $A_{i\alpha}$ fall into multiplets, with each multiplet corresponding to an irreducible representation of the algebra. As long as the symmetry is maintined under gauge transformations *with constant gauge function*, the members of a multiplet are degenerate. The number of particles in the multiplet is, of course, the dimension of the representation.

Now the n vector fields $A_{i\alpha}$ represent n vector particles that also form a degenerate multiplet. They too correspond to an irreducible representation of the algebra, called the adjoint representation, with the same dimension as the algebra itself. (For example, in the case of isotopic spin, for which $n = 3$, the adjoint representation is that with isotopic spin one, and the vector mesons form an isotopic triplet.) The matrices of the adjoint representation are easy to construct. They are simply

$$M_i^{jk} = -ic_{ijk} \quad \text{for adjoint representation.} \tag{5.1}$$

That this is so is obvious from the transformation properties of the $A_{i\alpha}$ in (4.3), with the gauge function taken constant, compared to those of the ψ_i in (4.1). In the adjoint representation, let

$$\operatorname{Tr} M_i M_j = A\delta_{ij} \quad \text{(adjoint representation).} \tag{5.2}$$

Then A defines the scale of the algebra. It is an arbitrary positive constant; from the above discussion it is clearly equal to the value of $\sum_i M_i^2$ for the adjoint representation. For any linear combinations S and T of the S_i, we can define a scalar product

$$(S, T) = \operatorname{Tr} M(S)M(T) \quad \text{(adjoint representation).} \tag{5.3}$$

Then we have $(S_i, S_j) = A\delta_{ij}$, $iAc_{ijk} = (S_k, [S_i, S_j])$.

We might now characterize each simple Lie algebra by the constants c_{ijk}, but they are subject to arbitrary orthogonal transformations on the n-dimensional space of the $A_{i\alpha}$. An invariant and physically useful characterization is constructed as follows. (We quote without proof the usual mathematical results (*22*).)

Each simple algebra has a certain maximum number f of elements that all commute with one another; let us call f the rank of the algebra. We may then enumerate the elements S_i of the algebra in this way:

$$C_1, C_2, \cdots C_f; \qquad \frac{E_1 + E_{-1}}{\sqrt{2}}; \quad \frac{E_2 + E_{-2}}{\sqrt{2}}, \cdots \frac{E_g + E_{-g}}{\sqrt{2}};$$
$$\frac{E_1 - E_{-1}}{\sqrt{2}i}, \cdots \frac{E_g - E_{-g}}{\sqrt{2}i}.$$

Here $g = (n - f)/2$. The C's are a maximal set of commuting elements. The E's are not real and are represented by non-Hermitian matrices, but E_α and $E_{-\alpha}$ are represented by Hermitian conjugate matrices. The corresponding vector fields are complex. The E's may be chosen to have these properties[6]:

$$(E_\alpha, E_\beta) = A\delta_{\alpha,-\beta}, \tag{5.4}$$

$$[C_i, E_\alpha] = \lambda_i^\alpha E_\alpha, \tag{5.5}$$

$$[E_\alpha, E_{-\alpha}] = \textstyle\sum_i \lambda_i^\alpha C_i, \tag{5.6}$$

$$\lambda_i^\alpha = -\lambda_i^{-\alpha}. \tag{5.7}$$

The C_i are analogous to I_z in the isotopic spin algebra, while the $E_{\pm\alpha}$ are analogous to the raising and lowering operators $I_\pm$. The λ_i^α are the possible eigenvalue differences of the operators C_i in any representation. They are real and can be regarded as $n - f$ distinct nonzero vectors in a real f-dimensional space.

The $(n - f)/2$ complex vector fields corresponding to the E_α give $n - f$ vector particles carrying specific values of the quantities C_i, namely, λ_i^α. Thus in the Yang–Mills theory the two charged mesons carry $I_z = \pm 1$.

The vectors λ_i^α in the real f-dimensional space of the C_i are called roots. Their lengths and relative angles are invariant properties of the algebra (except for the overall scale of length, proportional to $\sqrt{A}$). We may define a scalar product for the roots:

$$\langle \alpha, \beta \rangle = \textstyle\sum_i \lambda_i^\alpha \lambda_i^\beta. \tag{5.8}$$

When we add to one root λ_i^β integral multiples $k\lambda_i^\alpha$ of another, we may find further roots. When this occurs, it always happens only for a sequence of successive integers $k = p_{\beta\alpha}, \cdots q_{\beta\alpha}$. Evidently $p \leqq 0$ and $q \geqq 0$. When $q \geqq 1$,

[6] The C_i and E_α are known as a Weyl basis to the Lie algebra.

then $\lambda_i^\beta + \lambda_i^\alpha$ is a root. This situation is important for the commutation properties of the E_α :

$$[E_\alpha , E_\beta] = 0 \text{ unless } \beta = -\alpha \text{ or } \lambda_i^\alpha + \lambda_i^\beta \text{ is a root;} \tag{5.9}$$

$$[E_\alpha , E_\beta] = iN_{\alpha\beta}E_\gamma \text{ when } \lambda_i^\alpha + \lambda_i^\beta = \lambda_i^\gamma; \tag{5.10}$$

$$N_{\alpha\beta}^2 = \tfrac{1}{2} \langle\alpha,\alpha\rangle\, q_{\beta\alpha} (1 - p_{\beta\alpha}); \tag{5.11}$$

$$N_{\alpha\beta} = N_{-\alpha,-\beta} . \tag{5.12}$$

Even with the sign condition (5.12), the various relative signs of the E's must still be adjusted and a sign convention established for the $N_{\alpha\beta}$. But apart from that the algebra is now completely and invariantly described by its rank and the scalar products of its roots with one another. The commutation rules of all the C's and E's can then be constructed.

In the generalized Yang–Mills theory associated with the simple Lie algebra, we have gone over to a new particle representation. Instead of the n real fields $A_{i\alpha}$, we have f real fields coupled to the currents of commuting quantities and then $(n - f)/2$ complex fields coupled to the currents of raising and lowering operators for these commuting quantities. Instead of the c_{ijk} , we have the quantities λ_i^α and $N_{\alpha\beta}$ to describe the commutation rules and the amplitudes of the trilinear couplings among the vector mesons. (By going back to the real and imaginary parts of the complex fields, we can immediately recover the c_{ijk} in a particular form.) The particles of the complex fields carry the values λ_i^α of the quantities C_i and, since the C_i are conserved, the emission of the vector particle changes the value of C_i for the rest of the system by λ_i^α; the λ_i^α are indeed the possible eigenvalue differences of the C_i , whatever the representation.

In the next section, we shall give some examples of simple Lie algebras analyzed by the method of roots.

VI. EXAMPLES OF SIMPLE LIE ALGEBRAS

The simple Lie algebras of smallest dimension are those of the groups $SU(2)$, with $n = 3$; $SU(3)$, with $n = 8$; $Sp(2)$, with $n = 10$; G_2 , with $n = 14$; $SU(4)$, with $n = 15$; $Sp(3)$, with $n = 21$; $O(7)$, with $n = 21$; $SU(5)$, with $n = 24$; $O(8)$, with $n = 28$; $SU(6)$, with $n = 35$; $Sp(4)$, with $n = 36$; and $O(9)$, with $n = 36$. It is hard to imagine that any higher Lie algebras will be of physical interest.

The algebra of $SU(2)$ or $O(3)$ or $Sp(1)$ is just the isotopic spin algebra and has, of course, rank one. The next three algebras are the only ones of rank two, and we shall use them as examples. (We might mention, however, that the next algebra after these, that of $SU(4)$ or $O(6)$, with rank three, is familiar to physicists. It is the algebra of the traceless Dirac matrices and is also the algebra of

Wigner's old theory of nuclear supermultiplets (*23*).) For the three algebras of rank two, the roots are two-dimensional vectors, which are plotted in Figs. 1–3. The orientation and overall scale of length are arbitrary, as has been mentioned.

For any algebra, it is convenient to take one of the roots lying along the first axis and normalize its length to unity by proper choice of the constant A. For $SU(3)$, it doesn't matter which root is chosen. For each of the other two cases, there are two inequivalent choices; we can take either a long or a short vector.

With the "first" root taken along the first axis with length one, the elements E_1, E_{-1}, and C_1 form the components $J_+/\sqrt{2}$, $J_-/\sqrt{2}$, and J_z of an angular momentum, as we can see from the commutation rules (5.5 and 5.6). Moreover, the second commuting quantity C_2 commutes with all three components of $\mathbf{J}$.

Let us consider the algebra of $SU(3)$. We may, using our convention, read off the values of the six roots in Fig. 1:

$$(1, 0),\ (-1, 0),\ (\tfrac{1}{2}, \sqrt{\tfrac{3}{2}}),\ (-\tfrac{1}{2}, \sqrt{\tfrac{3}{2}}),\ (\tfrac{1}{2}, -\sqrt{\tfrac{3}{2}}),\ (-\tfrac{1}{2}, -\sqrt{\tfrac{3}{2}}).$$

The commuting elements C_1, C_2, can be thought of as belonging to a "root" $(0, 0)$. With respect to $\mathbf{J}$ spin and C_2, then, we have for the eight vector mesons the following: a triplet with $C_2 = 0$, a singlet with $C_2 = 0$, a doublet with $C_2 = \sqrt{\tfrac{3}{2}}$, and a doublet with $C_2 = -\sqrt{\tfrac{3}{2}}$. The triplet is coupled to the $\mathbf{J}$-spin current, the singlet to the C_2 current, and the two doublets to the currents of raising and lowering operators that change $\mathbf{J}$ by $\tfrac{1}{2}$ and C_2 by $\pm\sqrt{\tfrac{3}{2}}$.

Any representation of the algebra may be analyzed in terms of $\mathbf{J}$ and C_2. For example, consider the defining representation, of dimension 3. In order to accommodate all the operators enumerated above, it must contain a singlet and a doublet, with values of C_2 differing by $\sqrt{\tfrac{3}{2}}$.

To obtain the quantities $N^2_{\alpha\beta}$ characteristic of the commutators in (5.10), we must ask what roots, in Fig. 1, can be added to make other roots. Evidently, the only case is that of two roots at 120° to each other; when added, they give the root in between. We see by inspection that the numbers p and q of Eq. (5.11) are zero and one, respectively, in this case. Thus, $N^2_{\alpha\beta} = \tfrac{1}{2}\langle\alpha, \alpha\rangle$.

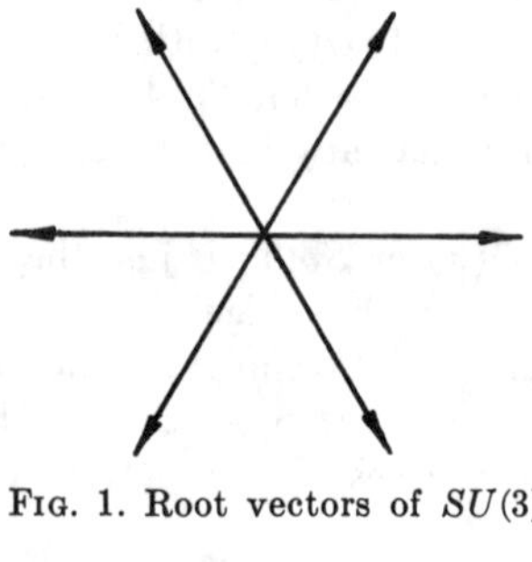

FIG. 1. Root vectors of $SU(3)$

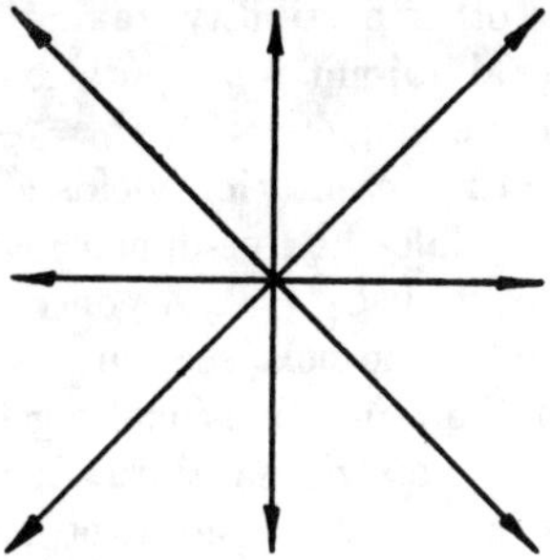

FIG. 2. Root vectors of $Sp(2) = O(5)$

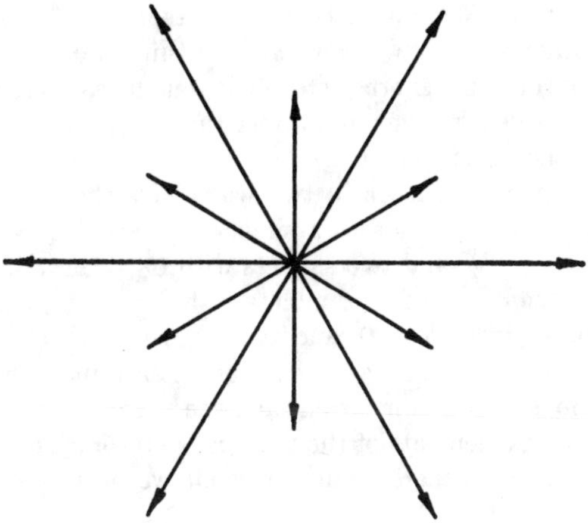

FIG. 3. Root vectors of G_2

Next, let us look at the 10-dimensional algebra, with roots as in Fig. 2. If we take one of the short vectors to be $(1, 0)$, then the root system is

$$(\pm 1, 1), \quad (0, 1), \quad (\pm 1, -1), \quad (0, -1), \quad (\pm 1, 0).$$

Including the two vector mesons coupled to C_1 and C_2, both treated as $(0, 0)$, we have a triplet with $C_2 = 0$, a triplet with $C_2 = +1$, a triplet with $C_2 = -1$, and a singlet with $C_2 = 0$.

If we treat the algebra as belonging to the group $Sp(2)$, we get a four-dimensional defining representation which corresponds, using the above analysis, to a doublet with $C_2 = +\frac{1}{2}$ and another with $C_2 = -\frac{1}{2}$. If we consider the algebra in connection with $O(5)$, then the defining representation is the 5-dimensional one, which consists, in our present language, of two singlets with $C_2 = \pm 1$ and a triplet with $C_2 = 0$.

Now we may consider the other possibility, taking one of the long vectors to be (1, 0). The ten-dimensional adjoint representation then corresponds to two doublets with $C_2 = \pm\frac{1}{2}$, three singlets with $C_2 = \pm 1$, 0, and a triplet with $C_2 = 0$. The four-dimensional representation yields a doublet with $C_2 = 0$ and two singlets with $C_2 = \pm\frac{1}{2}$, while the five-dimensional one gives two doublets with $C_2 = \pm\frac{1}{2}$ and a singlet with $C_2 = 0$. A conceivable physical application of this situation is mentioned in the next section.

The evaluation of $N^2_{\alpha\beta}$ for the ten-dimensional algebra involves two different situations in which adding two roots gives a third. As we see from Fig. 2, we can add a long vector to a short one at 135° from it, obtaining the short one at 45°. Or we can add two short vectors at right angles, obtaining the long one in between. In each case, N^2 comes out to equal the norm of the short vector.

Finally, we look at Fig. 3, showing the root system of G_2.[7] Here there are four different cases in which adding two roots gives a third. For three of these cases, N^2 is $\frac{3}{2}$ times the norm of the short vector. The fourth case is that of adding two long vectors at 120° to each other, obtaining the long one in the middle; N^2 is twice the norm of the short vector.

Again, the **J**-spin may be chosen in two ways. If a short vector is used, the adjoint representation corresponds to a singlet and a triplet with $C_2 = 0$, two quartets with $C_2 = \pm\sqrt{\frac{3}{2}}$, and two singlets with $C_2 = \pm\sqrt{3}$. If a long vector is taken to be (1, 0), then we get four doublets with $C_2 = \pm(1/2\sqrt{3}), \pm(\sqrt{3}/2)$, a singlet and a triplet with $C_2 = 0$, and two singlets with $C_2 = \pm\sqrt{3}$.

For each of the three algebras we have taken as examples, one may work out all the representations of low dimension, analyze them according to **J** and C_2, and calculate the matrix elements of the various operators. The whole procedure is a fairly straightforward generalization of what we do in the case of isotopic spin.

VII. POSSIBLE APPLICATIONS TO PHYSICS

Sakurai (*7*) has discussed a vector meson picture of the strong interactions in which three simple gauge theories are superposed. We have a one-parameter theory of a meson ω^0 coupled to the hypercharge current, a three-parameter theory of a meson ϱ coupled to the isotopic spin current, and another one-parameter theory of a meson B^0 coupled to the baryon current. In all three cases, gauge invariance with variable gauge function is broken by some kind of meson mass term. In the first two cases, the conservation of the current itself, corresponding to gauge invariance with constant gauge function, is broken by weak and electromagnetic interactions, respectively.

[7] A theory of the strong interactions of baryons and mesons whose invariance group is G_2 has been suggested by Behrends and Sirlin (*24*). They do not discuss the possibility of introducing vector gauge fields coupled to each of the fourteen conserved currents in order to secure invariance under coordinate-dependent transformations.

If we are willing to let some large effect, such as the $N - \Lambda$ mass difference or whatever causes it, break the gauge invariance with constant gauge function, then we may consider theories in which higher symmetries than isotopic spin play a role, and strangeness-changing currents are conserved to begin with. There are then strange vector mesons in the gauge theory, and we may be dealing with generalized Yang–Mills theories such as we have discussed.

It has been suggested (*8*) that the eight-dimensional algebra of $SU(3)$ may be used for such a theory. The **J** spin of the last section is taken to be the isotopic spin **I** and C_2 is taken to be $\sqrt{3/2}$ times the hypercharge. Then if the baryons N, Ξ, Λ, and Σ all have the same spin and parity they can form an irreducible representation of the algebra. So can the pseudoscalar mesons K, $\bar{K}$, π, and χ, where χ is a hypothetical isotopic singlet of zero strangeness. The vector mesons of the gauge theory then follow the same pattern, and consist of Sakurai's ρ and ω^0 and a pair of strange doublets M and $\bar{M}$. The sources of the strange mesons are then strangeness-changing currents, the conservation of which is broken by such things as the baryon mass differences.

Alternatively, we may imagine that the baryon supermultiplet does not consist of N, Ξ, Λ, and Σ. Using the algebra of $SU(3)$, we could take the three-dimensional irreducible representation and have it correspond to N and Λ (*25*).

We might even use the ten-dimensional algebra of $Sp(2)$, taking one of the long vectors to correspond to the root $(1, 0)$, as discussed in the previous section. If we interpret **J** as **I** and C_2 as half the hypercharge, then the baryons N, Λ, and Ξ could correspond to the five-dimensional irreducible representatiòn.

Besides the strong interactions, we may consider the possible application of vector gauge theories to the weak couplings.

Since there is no sign of charge-retention weak couplings among the leptons, one might try to describe all weak couplings by a $J_\alpha{}^+ J_\alpha$ model in which just two intermediate vector particles $X^\pm$ are involved. Since the operators to which X^+ and X^- are coupled cannot commute, the algebraic system involved is not closed and the theory cannot be of our type.

The possibility has been discussed (*4*, *5*) of correcting the situation by introducing the electromagnetic field as the third member of a Yang–Mills triplet including $X^\pm$. The introduction of a huge mass for the $X^\pm$ would totally ruin the symmetry and account for the short range and feeble strength of the weak interactions. A major difficulty in this approach is that the generating operators of X^+ and X^- violate parity conservation and it is hard to make their commutator equal the electric charge operator which does conserve parity. For the leptons, the problem can be solved (*26*) only with the introduction of a fourth neutral gauge field. In the resulting "nonsimple" $3 \oplus 1$ theory, the photon must be identified as a linear combination of the singlet gauge field and one member of the triplet.

A further difficulty is that with just $X^{\pm}$ for the weak interactions, one cannot justify the $|\Delta\mathbf{I}| = \frac{1}{2}$ rule for nonleptonic strangeness-violating weak couplings of baryons and mesons. One may, however, try forgetting the leptons and introducing, for baryons and mesons, charge-retention weak interactions mediated by neutral X's.

If just one X^0 is used, then the strangeness-changing part of the operator to which it is coupled must carry both $\Delta S = +1$ and $\Delta S = -1$ in order to be Hermitian. In the resulting weak interaction, one cannot then avoid having $|\Delta S| = 2$, which brings trouble with the $K_1^0 - K_2^0$ mass difference.

The "schizon" model (*11–13*) avoids $|\Delta S| = 2$ by using two neutral X's, X^0 and $\bar{X}^0$, along with $X^{\pm}$. Say X^+ is coupled to an operator B, X^- to B^+, X^0 to A, and $\bar{X}^0$ to A^+. Then we take $B = B_0 + B_1$, where B_0 conserves strangeness and changes isotopic spin by one unit, while B_1 lowers strangeness by one and changes isotopic spin by one-half unit. Similarly, $A = A_0 + A_1$, where A_0 conserves S and gives $|\Delta\mathbf{I}| = 0, 1$, while A_1 gives $\Delta S = -1$, $|\Delta\mathbf{I}| = \frac{1}{2}$. The operators B_1 and A_1 are chosen to be isotopic spin "partners"; the same is true of B_0 and the $|\Delta\mathbf{I}| = 1$ part of A_0. It is then easy to adjust the relative values of B_1 and A_1, B_0 and A_0 to give $|\Delta\mathbf{I}| = \frac{1}{2}$, $|\Delta S| = 1$ for the nonleptonic strangeness-violating weak interaction.

Unfortunately, it is impossible to apply the gauge principles we have discussed in this article to the "schizon" model, as we shall now see.

The only four-dimensional theories of our type are those made up of four one-dimensional theories superposed or else of one three-dimensional and one one-dimensional theory superposed. In the first case, the four operators A, B, A^+, and B^+ must all commute, which is clearly impossible. In the second case, if we take account of charge conservation, we must have (with suitable normalization and suitable choice of the arbitrary phase of A) the commutation rules:

$$\begin{aligned} &[B, \tfrac{1}{2}(A + A^+)] = B, \\ &[B, A - A^+] = 0, \\ &[B^+, B] = \tfrac{1}{2}(A + A^+), \\ &[A + A^+, A - A^+] = 0, \end{aligned} \tag{7.1}$$

characteristic of a "3 ⊕ 1" theory. Writing the last equation as $[A, A^+] = 0$ and taking the $\Delta S = 0$ part of the equation, we have

$$[A_0{}^+, A_0] + [A_1{}^+, A_1] = 0. \tag{7.2}$$

Now, A_1 is a strangeness-lowering operator. Let us suppose that a finite number of particles participate in the interaction. Consider those particles of lowest strangeness that are coupled to any of higher strangeness. For all these particles (treated as states Ψ_i), we have

$$A_1\Psi_i = 0. \tag{7.3}$$

On the other hand, $A_1^+\Psi_i$ cannot be zero for all the Ψ_i, since some of these particles are coupled to states of higher strangeness. Thus

$$\sum_i (\Psi_i, [A_1, A_1^+]\Psi_i) > 0. \tag{7.4}$$

But by (7.2) we then have

$$\sum_i (\Psi_i, [A_0^+, A_0]\Psi_i) > 0, \tag{7.5}$$

which is impossible, since the matrices A_0 and A_0^+ conserve strangeness and connect states Ψ_i only to states of the same set; the expression in (7.5) is thus the trace of the commutator of two finite matrices and vanishes.

The algebraic system of the "schizon" model, then, has no interesting representations.

It is interesting that if we insist on trying to reconcile our gauge notions with the idea of a weak interaction with four X's that explains the $|\Delta \mathbf{I}| = \frac{1}{2}$ rule and the $K_1^0 - K_2^0$ mass difference, we can construct a model that does all of those things. As usual, we must gloss over the leptons (with their apparent lack of neutral currents) and treat just baryons and mesons.

To start with, we forget the $K_1^0 - K_2^0$ mass difference and allow $|\Delta S| = 2$. Then we can use just three vector fields: X^+, coupled to B, X^-, coupled to B^+, and X^0, coupled to A. Later, we will add in a field Y^0 coupled to an operator C. The X's are described by the original 3-parameter Yang–Mills theory. As before, we have $B = B_0 + B_1$, but this time $A = A_0 + A_1 + A_1^+$.

The commutation rules of the Yang–Mills theory

$$\begin{aligned} [B, A] &= B, \\ [B^+, B] &= A, \end{aligned} \tag{7.6}$$

yield, for the whole system of A_0, A_1, B_0, B_1, strangeness S, and charge Q, an algebra which (if we introduce a simplification by making A_0 a linear function of S and Q) is equivalent to that of $SU(3)$. (Note that for the moment we do *not* introduce any isotopic spin rules; that would make the algebra still more complicated.) We may describe the theory, then, by using the smallest representation of $SU(3)$, involving three particles, which we will take to be n, p, and Λ.

Let us introduce rotated "particles" (*6*, *27*)

$$\begin{aligned} n' &= n \cos\theta + \Lambda \sin\theta, \\ \Lambda' &= \Lambda \cos\theta - n \sin\theta, \end{aligned} \tag{7.7}$$

and work only with the left-handed parts $(1 + \gamma_5)/2n' = n_L'$, etc., of these fields. Then we couple p_L and n_L' only, with $B = \tau_-'$, $B^+ = \tau_+'$, $A = \tau_z'$, where the τ' matrices are just like ordinary τ's, but with p_L, n_L' as a basis. We get the currents

$$\begin{aligned}
\bar{p}_L\gamma_\alpha n_L' &= \bar{p}_L\gamma_\alpha n_L \cos\theta + \bar{p}_L\gamma_\alpha \Lambda_L \sin\theta,\\
\bar{n}_L'\gamma_\alpha p_L &= \bar{n}_L\gamma_\alpha p_L \cos\theta + \bar{\Lambda}_L\gamma_\alpha p_L \sin\theta,\\
\frac{1}{\sqrt{2}}(\bar{p}\gamma_\alpha p_L - \bar{n}_L'\gamma_\alpha n_L) &= \frac{1}{\sqrt{2}}\{\bar{p}_L\gamma_\alpha p_L - \bar{n}_L\gamma_\alpha n_L \cos^2\theta\\
&\quad - \bar{\Lambda}_L\gamma_\alpha\Lambda_L \sin^2\theta - \cos\theta \sin\theta(\bar{\Lambda}_L\gamma_\alpha n_L + \bar{n}_L\gamma_\alpha\Lambda_L)\},
\end{aligned} \tag{7.8}$$

which lead to $\Delta S = 0$, $|\Delta S| = 1$, and $|\Delta S| = 2$ interactions.

The $|\Delta S| = 1$ interaction contains both $|\Delta \mathbf{I}| = \frac{1}{2}$ and $|\Delta \mathbf{I}| = \frac{3}{2}$. However, if θ is small, then the $|\Delta \mathbf{I}| = \frac{1}{2}$ interaction is of order θ while the $|\Delta \mathbf{I}| = \frac{3}{2}$ interaction is of order θ^3. Then the strangeness-changing interaction is weaker than the strangeness-conserving one and the $|\Delta \mathbf{I}| = \frac{3}{2}$ interaction is weaker than $|\Delta \mathbf{I}| = \frac{1}{2}$, but not zero.[8]

We must still cancel out the $|\Delta S| = 2$ contribution to the $K_1^0 - K_2^0$ mass difference. That can be done (*28*) by cancelling just the scalar part of the $|\Delta S| = 2$ interaction. If we couple the fourth boson Y^0 with *appropriate strength* to the current

$$i\bar{n}_R\gamma_\alpha\Lambda_R - i\bar{\Lambda}_R\gamma_\alpha n_R, \tag{7.9}$$

where we now work with the right-handed fields only, then the total $|\Delta S| = 2$ interaction can be made purely pseudoscalar and of order θ^2. It will give no $K_1^0 - K_2^0$ mass difference, but it will give a very small probability for $\Xi \to N + \pi$.

The model we have discussed is not seriously put forward as a physical theory, but it is a good illustration of the ideas involved in the gauge method. The fact that we are led to such an ugly model suggests either that partial gauge invariance does not apply to the weak interactions (for example, the simple $J_\alpha^+J_\alpha$ theory could be correct) or else that we are missing some important ingredient of the theory.

We have discussed several ways in which the strong interactions may constitute a partially gauge-invariant theory, and have sketched a formal partially gauge-invariant model of the weak interactions. In general, the "weak" and

[8] Observe that the sum of the squares of the coupling strengths to strangeness-saving charged currents and to strangeness-changing charged currents is just the square of the universal coupling strength. Should the gauge principle be extended to leptons—at least, for the charged currents—the equality between G_V and G_μ is no longer the proper statement of universality, for in this theory $G_V^2 + G_\Lambda^2 = G_\mu^2$ (G_Λ is the *unrenormalized* coupling strength for β-decay of Λ).

"strong" gauge symmetries will not be mutually compatible. There will also be conflicts with the electromagnetic gauge symmetry, conflicts that must be resolved in favor of electromagnetism, since its gauge invariance is exact. We have not attempted here to describe the three kinds of interaction together, but only to speculate about what the symmetry of each one might look like in an ideal limit where symmetry-breaking effects disappear.

Received: June 12, 1961

REFERENCES

1. R. P. Feynman and M. Gell-Mann, *Phys. Rev.* **109,** 193 (1958).
2. J. Schwinger, *Ann. Phys.* **2,** 407 (1957).
3. S. Bludman, *Nuovo cimento* [10], **9,** 433 (1958).
4. S. L. Glashow, *Nuclear Phys.* **10,** 107 (1959).
5. A. Salam and J. C. Ward, *Nuovo cimento* [10], **11,** 568 (1959).
6. M. Gell-Mann and M. Lévy, *Nuovo cimento* [10], **16,** 705 (1960).
7. J. Sakurai, *Ann. Phys.* **11,** 1 (1960).
8. M. Gell-Mann, "The Eightfold Way: A Theory of Strong Interaction Symmetry," California Institute of Technology Synchrotron Laboratory Report No. CTSL-20 (1961). See also Y. Ne'eman, *Nuclear Phys.* (to be published).
9. C. N. Yang and R. Mills, *Phys. Rev.* **96,** 191 (1954).
10. R. Shaw (unpublished).
11. T. D. Lee and C. N. Yang, *Phys. Rev.* **119,** 1410 (1960).
12. S. B. Treiman, *Nuovo cimento* [10], **15,** 916 (1959).
13. M. Gell-Mann, *Bull. Am. Phys. Soc.* **4,** 256(T) (1959).
14. C. N. Yang and T. D. Lee, *Phys. Rev.* **98,** 1501 (1955).
15. M. Gell-Mann, *Nuovo Cimento Suppl.* [10], **4** 848, (1956).
16. G. Wentzel, as reported by M. Goldberger, *in* "Proceedings of the 1960 Conference on High Energy Physics at Rochester." Interscience, New York, 1960.
17. C. G. Bollini, *Nuovo cimento* [10], **14,** 560 (1959).
18. H. Umezawa and S. Kamefuchi, *Nuclear Phys.* **23,** 399 (1961).
19. A. Komar and A. Salam, *Nuclear Phys.* **21,** 624 (1960).
20. R. Utiyama, *Phys. Rev.* **101,** 1597 (1956).
21. E. Cartan, "Sur la Structure des Groupes de Transformations Finis et Continus", thèse. Paris, 1894; 2nd ed., 1933.
22. "The Sophus Lie Seminars, 1954–1955." École Normale Supérieure, Paris, 1955.
23. E. Wigner, *Phys. Rev.* **51,** 106 (1937).
24. R. E. Behrends and A. Sirlin, *Phys. Rev.* **121,** 324 (1961).
25. A. Salam (to be published).
26. S. L. Glashow, *Nuclear Phys.* **22,** 579 (1961).
27. R. P. Feynman and M. Gell-Mann, *Proc. 2nd Intern. Conf. Peaceful Uses Atomic Energy* (1958).
28. S. L. Glashow, *Phys. Rev. Letters* **6,** 531 (1961).

II. SPONTANEOUS SYMMETRY BREAKING AND GOLDSTONE'S THEOREM

Field Theories with « Superconductor » Solutions.

J. GOLDSTONE

CERN - Geneva

(ricevuto l'8 Settembre 1960)

Summary. — The conditions for the existence of non-perturbative type « superconductor » solutions of field theories are examined. A non-covariant canonical transformation method is used to find such solutions for a theory of a fermion interacting with a pseudoscalar boson. A covariant renormalisable method using Feynman integrals is then given. A « superconductor » solution is found whenever in the normal perturbative-type solution the boson mass squared is negative and the coupling constants satisfy certain inequalities. The symmetry properties of such solutions are examined with the aid of a simple model of self-interacting boson fields. The solutions have lower symmetry than the Lagrangian, and contain mass zero bosons.

1. – Introduction.

This paper reports some work on the possible existence of field theories with solutions analogous to the Bardeen model of a superconductor. This possibility has been discussed by NAMBU (1) in a report which presents the general ideas of the theory which will not be repeated here. The present work merely considers models and has no direct physical applications but the nature of these theories seems worthwhile exploring.

The models considered here all have a boson field in them from the beginning. It would be more desirable to construct bosons out of fermions and this type of theory does contain that possibility (1). The theories of this paper have the dubious advantage of being renormalisable, which at least allows one to find simple conditions in finite terms for the existence of « supercon-

(1) Y. NAMBU: Enrico Fermi Institute for Nuclear Studies, Chicago, Report 60-21.

ducting » solutions. It also appears that in fact many features of these solutions can be found in very simple models with only boson fields, in which the analogy to the Bardeen theory has almost disappeared. In all these theories the relation between the boson field and the actual particles is more indirect than in the usual perturbation type solutions of field theory.

2. – Non-covariant theory.

The first model has a single fermion interacting with a single pseudoscalar boson field with the Lagrangian

$$L=\overline{\psi}\left(i\gamma^{\mu}\frac{\partial\psi}{\partial x^{\mu}}-m\psi\right)+\frac{1}{2}\left(\frac{\partial\varphi}{\partial x^{\mu}}\frac{\partial\varphi}{\partial x_{\mu}}-\mu_0^2\varphi^2\right)-g_0\overline{\psi}\gamma^5\psi\varphi-\frac{1}{24}\lambda_0\varphi^4\,.$$

(The last term is necessary to obtain finite results, as in perturbation theory.) The new solutions can be found by a non-covariant calculation which perhaps may show more clearly what is happening than the covariant theory which follows.

Let $\varphi(\boldsymbol{x})=(1/\sqrt{V})\sum q_{k}\exp[i\boldsymbol{k}\cdot\boldsymbol{x}]$ and let p_{k} be the conjugate momentum to q_{k}. Let $a^{\dagger}_{k\sigma}$, $b^{\dagger}_{k\sigma}$ be the creation operators for Fermi particles of momentum $\boldsymbol{k}$ spin σ and antiparticles of momentum $-\boldsymbol{k}$, spin $-\sigma$ respectively. Retain only the mode $\boldsymbol{k}=0$ of the boson field in the Hamiltonian. (The significance of this approximation appears below.) Then

$$H=H_F+H_B+H_I\,, \tag{1}$$

$$H_F=\sum_i E_i(a_i^{\dagger}a_i+b_i^{\dagger}b_i)\,,$$

(k, σ is replaced by a single symbol i)

$$H_B=\tfrac{1}{2}(p_0^2+\mu_0^2q_0^2)\,,$$

$$H_I=\frac{g_0}{\sqrt{V}}\,q_0\sum_i(a_i^{\dagger}b_i^{\dagger}+b_ia_i)+\frac{\lambda_0}{24V}\,q_0^4\,.$$

When H_I is treated as a perturbation, its only finite effects are to alter the boson mass and to scatter fermion pairs of zero total momentum. These effects can be calculated exactly (when $V\to\infty$) by writing $a_i^{\dagger}b_i^{\dagger}=c_i^{\dagger}$ and treating $c_i^{\dagger}$ as a boson creation operator ([2]). The Hamiltonian becomes

$$H'=\sum 2E_ic_i^{\dagger}c_i+\frac{1}{2}(p_0^2+\mu_0^2q_0^2)+\frac{g_0}{\sqrt{V}}\,q_0\sum(c_i^{\dagger}+c_i)\,.$$

([2]) N. N. Bogoliubov: *Žurn. Ėksp. Teor. Fiz.*, **34**, 73 (1958); *Sov. Phys. J.E.T.P.*, **7**, 51 (1958).

The $(1/V)q_0^4$ terms has no finite effects. Let

$$\frac{c_i^\dagger + c_i}{\sqrt{4E_i}} = q_i ; \qquad i\sqrt{E}(c_i^\dagger - c_i) = p_i ,$$

then

$$H' = \frac{1}{2}(p_0^2 + \mu_0^2 q_0^2) + \sum_i \frac{1}{2}(p_i^2 + 4E_i^2 q_i^2) + \frac{g_0}{\sqrt{V}} q_0 \sum_i \sqrt{4E_i} q_i - \sum_i E_i .$$

H' represents a set of coupled oscillators and is easily diagonalized. The frequencies of the normal modes are ω_0, ω_i, given by the roots of the equation

$$\mu_0^2 - \omega^2 = \frac{g_0^2}{V} \sum_i \frac{4E_i}{4E_i^2 - \omega^2} .$$

In the limit as $V \to \infty$, this becomes

$$(2) \qquad \mu_0^2 - \omega^2 = \frac{2g_0^2}{(2\pi)^3} \int \frac{4E_k}{4E_k^2 - \omega^2} \mathrm{d}^3\boldsymbol{k} = \frac{2g_0^2}{(2\pi)^3} \int \mathrm{d}^3\boldsymbol{k} \left\{ \frac{1}{E_k} + \frac{\omega^2}{4E_k^3} + \frac{\omega^4}{4E_k^3(4E_k^2 - \omega^2)} \right\} .$$

The first two terms in the integral diverge. This procedure corresponds exactly to the covariant procedure of calculating the poles of the boson propagator

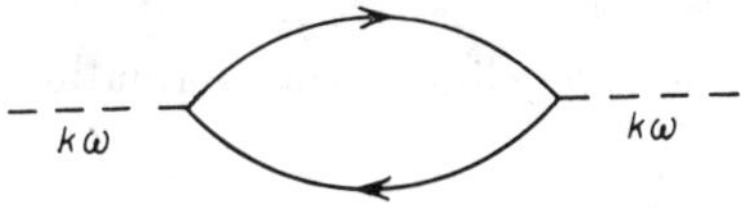

Fig. 1.

$$D(k, \omega) = \frac{1}{\omega^2 - k^2 - \mu_0^2 - \Pi(k, \omega)} ,$$

and including in Π only the lowest polarization part shown in Fig. 1.

This comparison shows that the two divergent terms can be absorbed into the renormalized mass and coupling constant. The renormalization is carried out at the point $k_\mu = 0$ instead of as usual at the one-boson pole of D. Thus if

$$\Pi(0, \omega) = A + B\omega^2 + g_0^2 \Pi_1(\omega^2) ,$$

then

$$D(0, \omega) = \frac{Z}{\omega^2 - \mu_1^2 - g_1^2 \Pi_1(\omega^2)} ,$$

$$Z = \frac{1}{1 - B} , \qquad \frac{\mu_1^2}{Z} = \mu_0^2 - A , \qquad g_1^2 = \frac{g_0^2}{Z} .$$

Equation (2) becomes

$$\mu_1^2 - \omega^2 = \frac{2g_1^2}{(2\pi)^3} \omega^4 \int \frac{\mathrm{d}^3\boldsymbol{k}}{4E_k^3 (4E_k^2 - \omega^2)} = F(\omega^2) .$$

The isolated root of this equation, ω_0^2, is found as shown in Fig. 2. When $\mu_1^2 > 0$, ω_0^2 is the square of the physical boson mass. When $\mu_1^2 \sim 4m^2$, this root disappears. This is the case when the boson can decay into a fermion pair. There is always another root for ω^2 large and negative. This corresponds to the well-known « ghost » difficulties and will be ignored here. When $\mu_1^2 < 0$ (but not too large), there is a negative root for ω^2. This is usually taken to mean simply that the theory with $\mu_1^2 < 0$ does not exist. Here it is taken to indicate that the approximation used is wrong and that the Hamiltonian (1) must be investigated further.

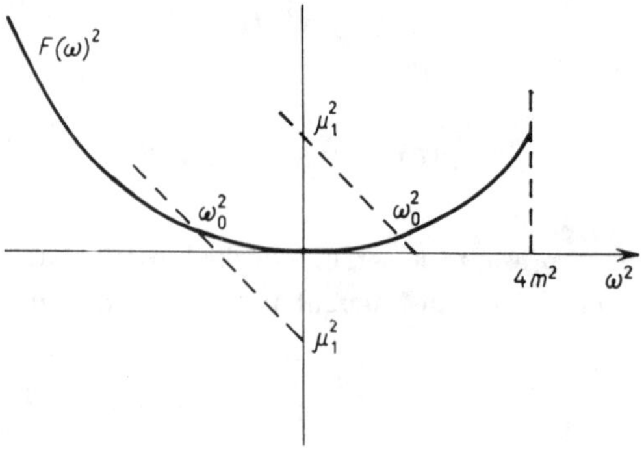

Fig. 2.

This is done by a series of canonical transformations based on the idea that in the vacuum state the expectation value of the boson field is not zero. Let

$$q_0 = q_0' + \frac{\Delta}{g_0}\sqrt{V}, \qquad p_0 = p_0',$$

Δ is a parameter to be fixed later. Also make a Bogoliubov transformation on the fermion field

$$a_i = \cos\frac{\theta_i}{2}\,\alpha_i - \sin\frac{\theta_i}{2}\,\beta_i^\dagger,$$

$$b_i = \sin\frac{\theta_i}{2}\,\alpha_i^\dagger + \cos\frac{\theta_i}{2}\,\beta_i,$$

$$\operatorname{tg}\theta_i = \frac{\Delta}{E_i}.$$

Then

$$H = H_0 + H_1 + H_2 + H_3,$$

$$H_0 = -\sum_i \left(\sqrt{E_i^2 + \Delta^2} - E_i\right) + \frac{\mu_0^2}{2}\frac{\Delta^2}{g_0^2}V + \frac{\lambda_0}{24}\frac{\Delta^4}{g_0^4}V,$$

$$H_1 = q_0'\sqrt{V}\left\{\mu_0^2\frac{\Delta}{g_0} + \frac{\lambda_0}{6}\frac{\Delta^3}{g_0^3} - \frac{g_0}{V}\sum_i \frac{\Delta}{\sqrt{E_i^2 + \Delta^2}}\right\},$$

$$H_2 = \frac{1}{2} p_0'^2 + \frac{1}{2} q_0'^2 \left(\mu_0^2 + \frac{\lambda_0}{2}\frac{\Delta^2}{g_0^2}\right) + \sum_i \sqrt{E_i^2 + \Delta^2}\,(\alpha_i^\dagger \alpha_i + \beta_i^\dagger \beta_i) + \frac{g_0}{\sqrt{V}} q_0' \sum_i \frac{E_i}{\sqrt{E_i^2 + \Delta^2}} (\alpha_i^\dagger \beta_i^\dagger + \beta_i \alpha_i)\,,$$

$$H_3 = \frac{\lambda_0}{24V} q_0'^4 + \frac{\lambda_0}{6}\frac{\Delta}{g_0}\frac{1}{\sqrt{V}} q_0'^3 + \frac{g_0}{\sqrt{V}} q_0' \sum_i \frac{\Delta}{\sqrt{E_i^2 + \Delta^2}} (\alpha_i^\dagger \alpha_i + \beta_i^\dagger \beta_i)\,.$$

Now choose Δ so that $H_1 = 0$. H_0 is just a constant proportional to V and H_3 has no finite effects. This leaves H_2, which can be diagonalised by the method used above for the original Hamiltonian (1).

One solution of $H_1 = 0$ is $\Delta = 0$. This gives the original approximation, which is no use when $\mu_1^2 < 0$. Other solutions are given by

$$(3) \qquad \mu_0^2 + \frac{1}{6}\lambda_0 \frac{\Delta^2}{g_0^2} = \frac{g_0^2}{V}\sum_i \frac{1}{\sqrt{E_i^2 + \Delta^2}} = \frac{2g_0^2}{(2\pi)^3}\int \mathrm{d}^3\boldsymbol{k} \left\{\frac{1}{E_k} - \frac{\Delta^2}{2E_k^3}\right\} + g_0^2 G(\Delta^2)\,,$$

where G is a finite function. Let

$$\mu_0^2 - \frac{2g_0^2}{(2\pi)^3}\int \frac{\mathrm{d}^3\boldsymbol{k}}{E_k} = \frac{\mu_1^2}{Z}\,,$$

$$\lambda_0 + \frac{6g_0^4}{(2\pi)^3}\int \frac{\mathrm{d}^3\boldsymbol{k}}{E_k^3} = \frac{\lambda_1}{Z^2}\,, \qquad g_0^2 = \frac{g_1^2}{Z}\,,$$

μ_1^2, λ_1, Z can be identified as the lowest order perturbation theoretic values of the renormalized boson mass, four-boson interaction constant and wave function renormalization, arising from the graphs in Fig. 1 and 3.

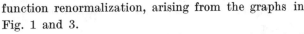

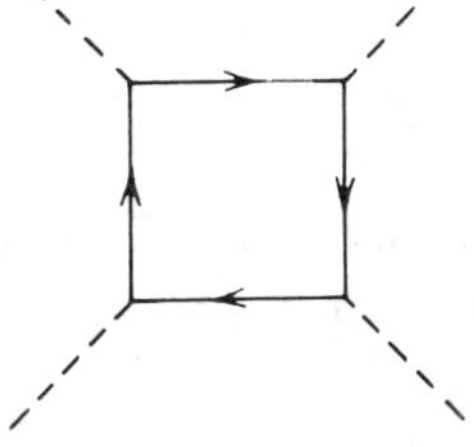

Fig. 3.

As before, the renormalizations are carried out at $k_\mu = 0$. Equation (3) becomes

$$(4) \qquad \mu_1^2 + \frac{1}{6}\lambda_1 \frac{\Delta^2}{g_1^2} = g_1^2 G(\Delta^2)\,,$$

Thus an equation for Δ^2 is obtained which is finite in terms of constants which would be the renormalized parameters of the theory in the ordinary solution. It will be shown below that when $\mu_1^2 < 0$, equation (4) has solutions for a certain range of λ_1, and that then there does exist a real boson.

3. – Covariant theory.

A first approach to a covariant theory can be made by calculating the fermion Green's function in a self-consistent field approximation. In perturbation theory the term represented by Fig. 4 vanishes by reflection invariance.

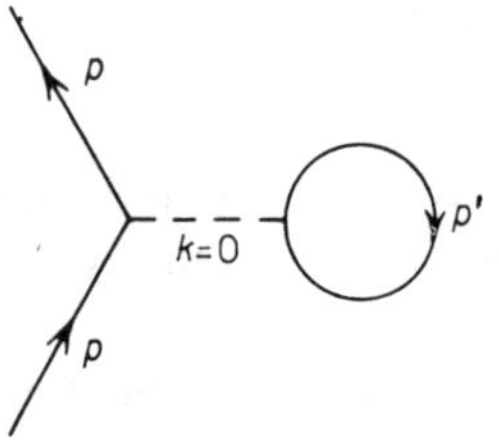

Fig. 4.

However, suppose it gives a contribution $\gamma_5 \Delta$ to the fermion self-energy. Then

$$S(p) = \frac{1}{\gamma_\mu p^\mu - m - \gamma_5 \Delta},$$

and evaluating Fig. 4 with this value of S gives

$$\Delta = -\frac{1}{\mu_0^2}\frac{g_0^2}{(2\pi)^4 i}\int \mathrm{d}^4 p \,\mathrm{Tr}\,\gamma_5 \frac{1}{\gamma_\mu p^\mu - m - \gamma_5 \Delta} = -\frac{4g_0^2}{\mu_0^2}\frac{1}{(2\pi)^4 i}\int \mathrm{d}^4 p \frac{\Delta}{p^2 - m^2 - \Delta^2},$$

which is the same as equation (3) without the λ term, and again has the perturbation solution $\Delta = 0$ and possibly other solutions.

A general covariant theory can be found by using the Feynman integral technique. This gives an explicit formula for the fermion Green's function

$$S(x', x) = \frac{\int S(x', x; \varphi) \exp[-iW(\varphi)] \exp[i\int L_M(\varphi)\,\mathrm{d}^4 x]\,\delta\varphi}{\int \exp[-iW(\varphi)] \exp[i\int L_M(\varphi)\,\mathrm{d}^4 x]\,\delta\varphi}.$$

Here $S(x'x\varphi)$ is the fermion Green's function calculated in an external boson field φ (and without interacting bosons). Exp $[-iW(\varphi)]$ is the vacuum-vacuum S-matrix amplitude in an external field and $L_M(\varphi)$ is the boson part of the Lagrangian. The integration are carried out over all fields $\varphi(\boldsymbol{x}, t)$.

Let

$$\varphi(\boldsymbol{x}t) = \frac{1}{\Omega}\sum_k \varphi_k \exp[ikx],$$

where k is now a 4-vector and Ω a large space-time volume. To obtain the Bardeen-type solutions, first do all the integrations except that over φ_0 (this has $\boldsymbol{k} = \omega = 0$) and put $\varphi_0 = \Omega\chi$ (χ finite). Then

$$S(x', x) = \frac{\int S(x', x; \chi) \exp[-i\Omega F(\chi)]\,\mathrm{d}\chi}{\int \exp[-i\Omega F(\chi)]\,\mathrm{d}\chi},$$

$$F(\chi) = w(\chi) + \frac{\mu_0^2}{2}\chi^2 + \frac{\lambda_0}{24}\chi^4,$$

S is the one particle Green's function calculated in a constant external field $\varphi(x) = \chi$ and including all the interacting boson degrees of freedom except $\boldsymbol{k} = \omega = 0$. $\text{Exp}\,[-i\Omega w]$ is the vacuum-vacuum amplitude calculated in the same way. The idea is to look for stationary points of $F(\chi)$ other than $\chi = 0$. If $F'(\chi_1) = 0$, then in the limit $\Omega \to \infty$, $S(x'x) = S(x'x\chi_1)$.

$-i\Omega w(\chi)$ is given by the sum of all connected vacuum diagrams. It is easy to see that

$$w(\chi) = \sum_{n=0}^{\infty} V_{2n} \frac{\chi^{2n}}{2n\,!},$$

where V_{2n} is the perturbation theory value of the $2n$-boson vertex with all external momenta zero. For example $V_2 = \Pi(0)$. In perturbation theory, $V_{2n} = V_{2n}^{(r)}/Z^n$ where $\sqrt{Z}$ is the boson wave function renormalization, and $V_{2n}^{(r)}$ is finite provided the terms μ_0^2 and λ_0 are absorbed into V_2 and V_4. As before, the renormalizations are carried out at $k = 0$. Thus if

$$\chi = \sqrt{Z}\,\chi^{(r)}, \qquad \text{then} \qquad F(\chi^{(r)}) = \sum \frac{V_{2n}^{(r)}}{2n\,!}\,\chi^{(r)2n},$$

an expression from which all the divergences have been removed. It also follows that if $F'(\chi_1^{(r)}) = 0$, the new values for the vertex parameters V_n are given by

$$V'_n = \frac{\partial^n}{\partial \chi_1^{(r)n}}\, F(\chi_1^{(r)})\,.$$

In particular the new boson mass is given by

$$\mu^2 = F''(\chi_1^{(r)})\,.$$

(This is really the mass operator at $k = 0$, not the mass.) Thus one condition for the existence of these abnormal solutions is that $F(\chi)$ should have a minimum at some non-zero value of χ.

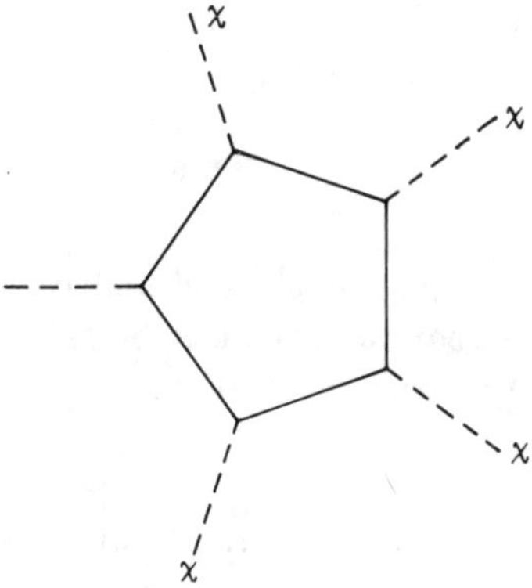

Fig. 5.

$F'(\chi)$ is in fact easier to evaluate than F. It is given by the sum of all diagrams with one external boson line. The previous approximations are recovered by putting $g_0\chi = \Delta$ and including only the diagrams in Fig. 5.

This gives

$$F'(\chi) = \mu_0^2\chi + \frac{\lambda_0}{6}\,\chi^3 + \frac{4g_0^2}{(2\pi)^4 i}\int \frac{\chi}{p^2 - m^2 - g_0^2\chi^2}\,\mathrm{d}^4p\,.$$

Hence

$$F'(\chi^{(r)}) = \mu_1^2 \chi^{(r)} + \frac{\lambda_1}{6} \chi^{(r)3} + \frac{4g_1^2}{(2\pi)^4 i} \int \frac{\chi^{(r)} g_1^4 \chi^{(r)4}}{(p^2 - m^2)^2 (p^2 - m^2 - \Delta^2)} \, d^4 p \, .$$

Thus Δ is given by

$$0 = \frac{\mu_1^2}{g_1^2} + \frac{\lambda_1}{6g_1^4} \Delta^2 - \frac{m^2}{4\pi^2} \left\{ \left(1 + \frac{\Delta^2}{m^2}\right) \log \left(1 + \frac{\Delta^2}{m^2}\right) - \frac{\Delta^2}{m^2} \right\} .$$

Let

$$\frac{4\pi^2 \mu_1^2}{m^2 g_1^2} = A \, , \qquad \frac{2\pi^2}{3} \frac{\lambda_1}{g_1^4} = B \, , \qquad \frac{\Delta^2}{m^2} = x \, .$$

Then

$$A + Bx = (1 + x) \log (1 + x) - x = h(x) \, .$$

The new boson mass is given by

$$\mu^2 = F''(\chi) = \frac{m^2 g_1^2}{2\pi^2} x \{ B - h'(x) \} \, .$$

It can be seen from Fig. 6 that there will be roots with $\mu^2 > 0$ only when $\mu_1^2 < 0$ and $B > B_{\text{crit}}$, where B_{crit} is given by $h'(x) = B$. Thus the abnormal solutions exist when

$$B > 0 \, , \qquad 0 > A > -(e^B - 1 - B) \, .$$

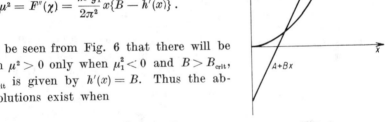

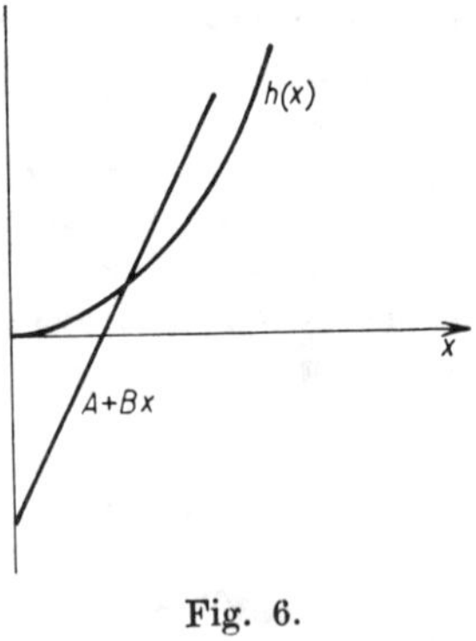

Fig. 6.

This calculation is exact in the limit $g_1 \to 0$ keeping μ_1^2/g_1^2, λ_1/g_1^4 and $g_1 \chi^{(r)}$ finite. Thus in at least one case a solution of the required type can be established as plausibly as the more usual perturbation theory solutions.

4. – Symmetry properties and a simple model.

It is now necessary to discuss the principal peculiar feature of this type of solution. The original Lagrangian had a reflection symmetry. From this it follows that $F(\chi)$ must be an even function. Thus is $\chi = \chi_1$ is one solution of $F'(\chi_1) = 0$, $\chi_1 = -\chi_1$ is another. By choosing one solution, the reflection symmetry is effectively destroyed. It is possible to make a very simple model which shows this kind of behaviour, and also demonstrates that so long as there is a boson field in the theory to start with, the essential features of the abnormal solutions have very little to do with fermion pairs.

Consider the theory of a single neutral pseudoscalar boson interacting with itself,

$$L=\frac{1}{2}\left(\frac{\partial\varphi}{\partial x_\mu}\frac{\partial\varphi}{\partial x^\mu}-\mu_0^2\varphi^2\right)-\frac{\lambda_0}{24}\varphi^4 .$$

Normally this theory is quantized by letting each mode of oscillation of the classical field correspond to a quantum oscillator whose quantum number gives the number of particles. When $\mu_0^2<0$, this approach will not work. However, if $\lambda_0>0$, the function

$$\frac{\mu_0^2}{2}\varphi^2+\frac{\lambda_0}{24}\varphi^4 ,$$

is as shown in Fig. 7.

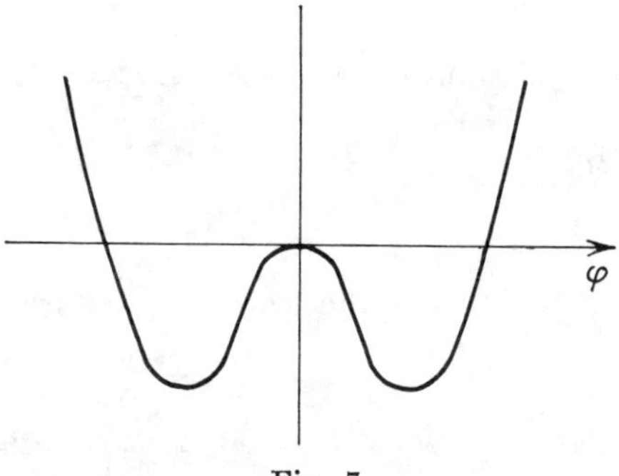

Fig. 7.

The classical equations

$$(\Box^2+\mu_0^2)\varphi+\frac{\lambda_0}{6}\varphi^3=0 ,$$

now have solutions $\varphi=\pm\sqrt{-6\mu_0^2/\lambda_0}$ corresponding to the minima of this curve. Infinitesimal oscillations round one of these minima obey the equation

$$(\Box^2-2\mu_0^2)\,\delta\varphi=0 .$$

These can now be quantized to represent particles of mass $\sqrt{-2\mu_0^2}$. This is simply done by making the transformation $\varphi=\varphi'+\chi$

$$\chi^2=-\frac{6\mu_0^2}{\lambda_0} .$$

Then

$$L=\frac{1}{2}\left(\frac{\partial\varphi'}{\partial x_\mu}\frac{\partial\varphi'}{\partial x^\mu}+2\mu_0^2\varphi'^2\right)-\frac{\lambda_0}{24}\varphi'^4-\frac{\lambda_0\chi}{6}\varphi'^3+\frac{3}{2}\frac{\mu_0^2}{\lambda_0} .$$

This new Lagrangian can be treated by the canonical methods.

In any state with a finite number of particles, the expectation value of φ is infinitesimally different from the vacuum expectation value. Thus the eigenstates corresponding to oscillations round $\varphi=\chi$ are all orthogonal to the usual states corresponding to oscillations round $\varphi=0$, and also to the eigenstates round $\varphi=-\chi$. This means that the theory has two vacuum states, with a complete set of particle states built on each vacuum, but that there is a superselection rule between these two sets so that it is only necessary to consider one of them. The symmetry $\varphi \to -\varphi$ has disappeared. Of course it can be restored by introducing linear combinations of states in the two sets but because of the superselection rule this is a highly artificial procedure.

Now consider the case when the symmetry group of the Lagrangian is continuous instead of discrete. A simple example is that of a complex boson field, $\varphi = (\varphi_1 + i\varphi_2)/\sqrt{2}$

$$L = \frac{\partial \varphi^*}{\partial x_\mu} \frac{\partial \varphi}{\partial x^\mu} - \mu_0^2 \varphi^* \varphi - \frac{\lambda_0}{6} (\varphi^* \varphi)^2 .$$

The symmetry is $\varphi \to \exp[i\alpha]\varphi$. The canonical transformation is $\varphi = \varphi' + \chi$

$$|\chi|^2 = -\frac{3\mu_0^2}{\lambda_0} .$$

The phase of χ is not determined. Fixing it destroys the symmetry. With χ real the new Lagrangian is

$$L = \frac{1}{2}\left(\frac{\partial \varphi_1'}{\partial x_\mu} \frac{\partial \varphi_1'}{\partial x^\mu} + 2\mu_0^2 \varphi_1'^2\right) + \frac{1}{2} \frac{\partial \varphi_2'}{\partial x_\mu} \frac{\partial \varphi_2'}{\partial x^\mu} - \frac{\lambda_0 \chi}{6} \varphi_1'(\varphi_1'^2 + \varphi_2'^2) - \frac{\lambda_0}{24} (\varphi_1'^2 + \varphi_2'^2)^2 .$$

The particle corresponding to the φ_2' field has zero mass. This is true even when the interaction is included, and is the new way the original symmetry expresses itself.

A simple picture can be made for this theory by thinking of the two dimensional vector φ at each point of space. In the vacuum state the vectors have magnitude χ and are all lined up (apart from the quantum fluctuations). The massive particles φ_1' correspond to oscillations in the direction of χ. The massless particles φ_2' correspond to « spin-wave » excitations in which only the direction of φ makes infinitesimal oscillations. The mass must be zero, because when all the $\varphi(x)$ rotate in phase there is no gain in energy because of the symmetry.

This time there are infinitely many vacuum states. A state can be specified by giving the phase of χ and then the numbers of particles in the two different oscillation modes. There is now a superselection rule on the phase of χ. States with a definite charge can only be constructed artificially by superposing states with different phases.

5. – Conclusion.

This result is completely general. Whenever the original Lagrangian has a continuous symmetry group, the new solutions have a reduced symmetry and contain massless bosons. One consequence is that this kind of theory cannot be applied to a vector particle without losing Lorentz invariance. A method of losing symmetry is of course highly desirable in elementary particle theory but these theories will not do this without introducing non-existent massless bosons (unless discrete symmetry groups can be used). SKYRME [3] has hoped that one set of fields could have excitations both of the usual type and of the « spin-wave » type, thus for example obtaining the π-mesons as collective oscillations of the four K-meson fields, but this does not seem possible in this type of theory. Thus if any use is to be made of these solutions something more complicated than the simple models considered in this paper will be necessary.

---

(3) T. H. R. SKYRME: *Proc. Roy. Soc.*, A **252**, 236 (1959).

RIASSUNTO (*)

Si esaminano le condizioni per l'esistenza di soluzioni « superconduttrici » del tipo non perturbativo delle teorie di campo. Si usa un metodo non covariante di trasformazioni canoniche per trovare queste soluzioni per la teoria di un fermione interagente con un bosone pseudoscalare. Si espone poi un metodo covariante rinormalizzabile usando integrali di Feynman. Si trova un soluzione « superconduttrice » ogniqualvolta nella soluzione normale del tipo perturbativo la massa bosonica elevata al quadrato è negativa e le costanti di accoppiamento soddisfano ad alcune ineguaglianze. Si esaminano le proprietà di simmetria di tali soluzioni con l'aiuto di un semplice modello di campi di bosoni auto-interagenti. Le soluzioni hanno simmetria inferiore al lagrangiano, e contengono bosoni di massa zero.

(*) *Traduzione a cura della Redazione.*

Dynamical Model of Elementary Particles Based on an Analogy with Superconductivity. I*

Y. Nambu and G. Jona-Lasinio†
The Enrico Fermi Institute for Nuclear Studies and the Department of Physics, The University of Chicago, Chicago, Illinois
(Received October 27, 1960)

It is suggested that the nucleon mass arises largely as a self-energy of some primary fermion field through the same mechanism as the appearance of energy gap in the theory of superconductivity. The idea can be put into a mathematical formulation utilizing a generalized Hartree-Fock approximation which regards real nucleons as quasi-particle excitations. We consider a simplified model of nonlinear four-fermion interaction which allows a γ_5-gauge group. An interesting consequence of the symmetry is that there arise automatically pseudoscalar zero-mass bound states of nucleon-antinucleon pair which may be regarded as an idealized pion. In addition, massive bound states of nucleon number zero and two are predicted in a simple approximation.
The theory contains two parameters which can be explicitly related to observed nucleon mass and the pion-nucleon coupling constant. Some paradoxical aspects of the theory in connection with the γ_5 transformation are discussed in detail.

I. INTRODUCTION

In this paper we are going to develop a dynamical theory of elementary particles in which nucleons and mesons are derived in a unified way from a fundamental spinor field.[1] In basic physical ideas, it has thus the characteristic features of a compound-particle model, but unlike most of the existing theories, dynamical treatment of the interaction makes up an essential part of the theory. Strange particles are not yet considered.

The scheme is motivated by the observation of an interesting analogy between the properties of Dirac particles and the quasi-particle excitations that appear in the theory of superconductivity, which was originated with great success by Bardeen, Cooper, and Schrieffer,[2] and subsequently given an elegant mathematical formulation by Bogoliubov.[3] The characteristic feature of the BCS theory is that it produces an energy gap between the ground state and the excited states of a superconductor, a fact which has been confirmed experimentally. The gap is caused due to the fact that the attractive phonon-mediated interaction between electrons produces correlated pairs of electrons with opposite momenta and spin near the Fermi surface, and it takes a finite amount of energy to break this correlation.

Elementary excitations in a superconductor can be conveniently described by means of a coherent mixture of electrons and holes, which obeys the following equations[3,4]:

$$E\psi_{p+}=\epsilon_p\psi_{p+}+\phi\psi_{-p-}^*,\qquad E\psi_{-p-}^*=-\epsilon_p\psi_{-p-}^*+\phi\psi_{p+}, \tag{1.1}$$

near the Fermi surface. ψ_{p+} is the component of the excitation corresponding to an electron state of momentum p and spin $+$(up), and ψ_{-p-}^* corresponding to a hole state of momentum p and spin $+$, which means an absence of an electron of momentum $-p$ and spin $-$(down). ϵ_p is the kinetic energy measured from the Fermi surface; ϕ is a constant. There will also be an equation complex conjugate to Eq. (1), describing another type of excitation.

Equation (1) gives the eigenvalues

$$E_p=\pm(\epsilon_p^2+\phi^2)^{\frac{1}{2}}. \tag{1.2}$$

The two states of this quasi-particle are separated in energy by $2|E_p|$. In the ground state of the system all the quasi-particles should be in the lower (negative) energy states of Eq. (2), and it would take a finite energy $2|E_p|\geqslant 2|\phi|$ to excite a particle to the upper state. The situation bears a remarkable resemblance to the case of a Dirac particle. The four-component Dirac equation can be split into two sets to read

$$E\psi_1=\sigma\cdot p\psi_1+m\psi_2,\qquad E\psi_2=-\sigma\cdot p\psi_2+m\psi_1,\qquad E_p=\pm(p^2+m^2)^{\frac{1}{2}}, \tag{1.3}$$

where ψ_1 and ψ_2 are the two eigenstates of the chirality operator $\gamma_5=\gamma_1\gamma_2\gamma_3\gamma_4$.

According to Dirac's original interpretation, the ground state (vacuum) of the world has all the electrons in the negative energy states, and to create excited states (with zero particle number) we have to supply an energy $\geqslant 2m$.

In the BCS-Bogoliubov theory, the gap parameter ϕ, which is absent for free electrons, is determined essentially as a self-consistent (Hartree-Fock) representation of the electron-electron interaction effect.

* Supported by the U. S. Atomic Energy Commission.

† Fulbright Fellow, on leave of absence from Instituto di Fisica dell' Universita, Roma, Italy and Istituto Nazionale di Fisica Nucleare, Sezione di Roma, Italy.

[1] A preliminary version of the work was presented at the Midwestern Conference on Theoretical Physics, April, 1960 (unpublished). See also Y. Nambu, Phys. Rev. Letters 4, 380 (1960); and Proceedings of the Tenth Annual Rochester Conference on High-Energy Nuclear Physics, 1960 (to be published).

[2] J. Bardeen, L. N. Cooper, and J. R. Schrieffer, Phys. Rev. 106, 162 (1957).

[3] N. N. Bogoliubov, J. Exptl. Theoret. Phys. (U.S.S.R.) 34, 58, 73 (1958) [translation: Soviet Phys.-JETP 34, 41, 51 (1958)]; N. N. Bogoliubov, V. V. Tolmachev, and D. V. Shirkov, *A New Method in the Theory of Superconductivity* (Academy of Sciences of U.S.S.R., Moscow, 1958).

[4] J. G. Valatin, Nuovo cimento 7, 843 (1958).

One finds that

$$\phi \approx \omega \exp[-1/\rho], \tag{1.4}$$

where ω is the energy bandwidth ($\approx$ the Debye frequency) around the Fermi surface within which the interaction is important; ρ is the average interaction energy of an electron interacting with unit energy shell of electrons on the Fermi surface. It is significant that ϕ depends on the strength of the interaction (coupling constant) in a nonanalytic way.

We would like to pursue this analogy mathematically. As the energy gap ϕ in a superconductor is created by the interaction, let us assume that the mass of a Dirac particle is also due to some interaction between massless bare fermions. A quasi-particle in a superconductor is a mixture of bare electrons with opposite electric charges (a particle and a hole) but with the same spin; correspondingly a massive Dirac particle is a mixture of bare fermions with opposite chiralities, but with the same charge or fermion number. Without the gap ϕ or the mass m, the respective particle would become an eigenstate of electric charge or chirality.

Once we make this analogy, we immediately notice further consequences of special interest. It has been pointed out by several people[3,5–8] that in a refined theory of superconductivity there emerge, in addition to the individual quasi-particle excitations, collective excitations of quasi-particle pairs. (These can alternatively be interpreted as moving states of bare electron pairs which are originally precipitated into the ground state of the system.) In the absence of Coulomb interaction, these excitations are phonon-like, filling the gap of the quasi-particle spectrum.

In general, they are excited when a quasi-particle is accelerated in the medium, and play the role of a backflow around the particle, compensating the change of charge localized on the quasi-particle wave packet. Thus these excitations are necessary consequences of the fact that individual quasi-particles are not eigenstates of electric charge, and hence their equations are not gauge invariant; whereas a complete description of the system must be gauge invariant. The logical connection between gauge invariance and the existence of collective states has been particularly emphasized by one of the authors.[8]

This observation leads to the conclusion that if a Dirac particle is actually a quasi-particle, which is only an approximate description of an entire system where chirality is conserved, then there must also exist collective excitations of bound quasi-particle pairs. The chirality conservation implies the invariance of the theory under the so-called γ_5 gauge group, and from its nature one can show that the collective state must be a pseudoscalar quantity.

[5] D. Pines and J. R. Schrieffer, Nuovo cimento **10**, 496 (1958).
[6] P. W. Anderson, Phys. Rev. **110**, 827, 1900 (1958); **114**, 1002 (1959).
[7] G. Rickayzen, Phys. Rev. **115**, 795 (1959).
[8] Y. Nambu, Phys. Rev. **117**, 648 (1960).

It is perhaps not a coincidence that there exists such an entity in the form of the pion. For this reason, we would like to regard our theory as dealing with nucleons and mesons. The implication would be that the nucleon mass is a manifestation of some unknown primary interaction between originally massless fermions, the same interaction also being responsible for the binding of nucleon pairs into pions.

An additional support of the idea can be found in the weak decay processes of nucleons and pions which indicate that the γ_5 invariance is at least approximately conserved, as will be discussed in Part II. There are some difficulties, however, that naturally arise on further examination.

Comparison between a relativistic theory and a nonrelativistic, intuitive picture is often dangerous, because the former is severely restricted by the requirement of relativistic invariance. In our case, the energy-gap equation (4) depends on the energy density on the Fermi surface; for zero Fermi radius, the gap vanishes. The Fermi sphere, however, is not a relativistically invariant object, so that in the theory of nucleons it is not clear whether a formula like Eq. (4) could be obtained for the mass. This is not surprising, since there is a well known counterpart in classical electron theory that a finite electron radius is incompatible with relativistic invariance.

We avoid this difficulty by simply introducing a relativistic cutoff which takes the place of the Fermi sphere. Our framework does not yet resolve the divergence difficulty of self-energy, and the origin of such an effective cutoff has to be left as an open question.

The second difficulty concerns the mass of the pion. If pion is to be identified with the phonon-like excitations associated with a gauge group, its mass must necessarily be zero. It is true that in real superconductors the collective charge fluctuation is screened by Coulomb interaction to turn into the plasma mode, which has a finite "rest mass." A similar mechanism may be operating in the meson case too. It is possible, however, that the finite meson mass means that chirality conservation is only approximate in a real theory. From the evidence in weak interactions, we are inclined toward the second view.

The observation made so far does not yet give us a clue as to the exact mechanism of the primary interaction. Neither do we have a fundamental understanding of the isospin and strangeness quantum numbers, although it is easy to incorporate at least the isospin degree of freedom into the theory from the beginning. The best we can do here is to examine the various existing models for their logical simplicity and experimental support, if any. We will do this in Sec. 2, and settle for the moment on a nonlinear four-fermion interaction of the Heisenberg type. For reasons of simplicity in presentation, we adopt a model without isospin and strangeness degrees of freedom, and possessing complete γ_5 invariance. Once the choice is made, we can explore the whole idea mathematically, using essentially the formulation developed in reference 8. It is gratifying that the various field-theoretical techniques can be fully utilized. Section 3 will be devoted to introduction of the Hartree-Fock equation for nucleon self-energy, which will make the starting point of the theory. Then we go on to discuss in Sec. 4 the collective modes. In addition to the expected pseudoscalar "pion" states, we find other massive mesons of scalar and vector variety, as well as a scalar "deuteron." The coupling constants of these mesons can be easily determined. The relation of the pion to the γ_5 gauge group will be discussed in Secs. 5 and 6.

The theory promises many practical consequences. For this purpose, however, it is necessary to make our model more realistic by incorporating the isospin, and allowing for a violation of γ_5 invariance. But in doing so, there arise at the same time new problems concerning the mass splitting and instability. This refined model will be elaborated in Part II of this work, where we shall also find predictions about strong and weak interactions. Thus the general structure of the weak interaction currents modified by strong interactions can be treated to some degree, enabling one to derive the decay processes of various particles under simple assumptions. The calculation of the pion decay rate gives perhaps one of the most interesting supports of the theory. Results about strong interactions themselves are equally interesting. We shall find specific predictions about heavier mesons, which are in line with the recent theoretical expectations.

II. THE PRIMARY INTERACTION

We briefly discuss the possible nature of the primary interaction between fermions. Lacking any radically new concepts, the interaction could be either mediated by some fundamental Bose field or due to an inherent nonlinearity in the fermion field. According to our postulate, these interactions must allow chirality conservation in addition to the conservation of nucleon number. The chirality X here is defined as the eigenvalue of γ_5, or in terms of quantized fields,

$$X=\int \bar{\psi}\gamma_4\gamma_5\psi d^3x. \tag{2.1}$$

The nucleon number is, on the other hand

$$N=\int \bar{\psi}\gamma_4\psi d^3x. \tag{2.2}$$

These are, respectively, generators of the γ_5- and ordinary-gauge groups

$$\psi\rightarrow \exp[i\alpha\gamma_5]\psi,\quad \bar{\psi}\rightarrow\bar{\psi}\exp[i\alpha\gamma_5], \tag{2.3}$$

$$\psi\rightarrow \exp[i\alpha]\psi,\quad \bar{\psi}\rightarrow\bar{\psi}\exp[-i\alpha], \tag{2.4}$$

where α is an arbitrary constant phase.

Furthermore, the dynamics of our theory would require that the interaction be attractive between particle and antiparticle in order to make bound-state formation possible. Under the transformation (2.3), various tensors transform as follows:

$$\begin{aligned}
&\text{Vector:} && i\bar{\psi}\gamma_\mu\psi\rightarrow i\bar{\psi}\gamma_\mu\psi,\\
&\text{Axial vector:} && i\bar{\psi}\gamma_\mu\gamma_5\psi\rightarrow i\bar{\psi}\gamma_\mu\gamma_5\psi,\\
&\text{Scalar:} && \bar{\psi}\psi\rightarrow\bar{\psi}\psi\cos 2\alpha+i\bar{\psi}\gamma_5\psi\sin 2\alpha,\\
&\text{Pseudoscalar:} && i\bar{\psi}\gamma_5\psi\rightarrow i\bar{\psi}\gamma_5\psi\cos 2\alpha-\bar{\psi}\psi\sin 2\alpha,\\
&\text{Tensor:} && \bar{\psi}\sigma_{\mu\nu}\psi\rightarrow\bar{\psi}\sigma_{\mu\nu}\psi\cos 2\alpha+i\bar{\psi}\gamma_5\sigma_{\mu\nu}\psi\sin 2\alpha.
\end{aligned} \tag{2.5}$$

It is obvious that a vector or pseudovector Bose field coupled to the fermion field satisfies the invariance. The vector case would also satisfy the dynamical requirement since, as in the electromagnetic interaction, the forces would be attractive between opposite nucleon charges. The pseudovector field, on the other hand, does not meet the requirement as can be seen by studying the self-consistent mass equation discussed later.

The vector field looks particularly attractive since it can be associated with the nucleon number gauge group. This idea has been explored by Lee and Yang,[9] and recently by Sakurai.[10] But since we are dealing with strong interactions, such a field would have to have a finite observed mass in a realistic theory. Whether this is compatible with the invariance requirement is not yet clear. (Besides, if the bare mass of both spinor and vector field were zero, the theory would not contain any parameter with the dimensions of mass.)

The nonlinear fermion interaction seems to offer another possibility. Heisenberg and his co-workers[11] have been developing a comprehensive theory of elementary particles along this line. It is not easy, however, to gain a clear physical insight into their results obtained by means of highly complicated mathematical machinery.

We would like to choose the nonlinear interaction in this paper. Although this looks similar to Heisenberg's theory, the dynamical treatment will be quite different and more amenable to qualitative understanding.

The following Lagrangian density will be assumed ($\hbar=c=1$):

$$L=-\bar{\psi}\gamma_\mu\partial_\mu\psi+g_0[(\bar{\psi}\psi)^2-(\bar{\psi}\gamma_5\psi)^2]. \tag{2.6}$$

The coupling parameter g_0 is positive, and has dimensions $[\text{mass}]^{-2}$. The γ_5 invariance property of the interaction is evident from Eq. (2.5). According to the Fierz theorem, it is also equivalent to

$$-\tfrac{1}{2}g_0[(\bar{\psi}\gamma_\mu\psi)^2-(\bar{\psi}\gamma_\mu\gamma_5\psi)^2]. \tag{2.7}$$

This particular choice of γ_5-invariant form was taken without a compelling reason, but has the advantage

[9] T. D. Lee and C. N. Yang, Phys. Rev. **98**, 1501 (1955).
[10] J. J. Sakurai, Ann. Phys. **11**, 1 (1960).
[11] W. Heisenberg, Z. Naturforsch. **14**, 441 (1959). Earlier papers are quoted there.

that it can be naturally extended to incorporate isotopic spin.[12]

Unlike Heisenberg's case, we do not have any theory about the handling of the highly divergent singularities inherent in nonlinear interactions. So we will introduce, as an additional and independent assumption, an *ad hoc* relativistic cutoff or form factor in actual calculations. Thus the theory may also be regarded as an approximate treatment of the intermediate-boson model with a large effective mass.

As will be seen in subsequent sections, the nonlinear model makes mathematics particularly easy, at least in the lowest approximation, enabling one to derive many interesting quantitative results.

III. THE SELF-CONSISTENT EQUATION FOR NUCLEON MASS

We will assume that all quantities we calculate here are somewhow convergent, without asking the reason behind it. This will be done actually by introducing a suitable phenomenological cutoff.

Without specifying the interaction, let Σ be the unrenormalized proper self-energy part of the fermion, expressed in terms of observed mass m, coupling constant g, and cutoff Λ. A real Dirac particle will satisfy the equation

$$i\gamma\cdot p+m_0+\Sigma(p,m,g,\Lambda)=0 \tag{3.1}$$

for $i\gamma\cdot p+m=0$. Namely

$$m-m_0=\Sigma(p,m,g,\Lambda)|_{i\gamma\cdot p+m=0}. \tag{3.2}$$

The g will also be related to the bare coupling g_0 by an equation of the type

$$g/g_0=\Gamma(m,g,\Lambda). \tag{3.3}$$

Equations (3.1) and (3.2) may be solved by successive approximation starting from m_0 and g_0. It is possible, however, that there are also solutions which cannot thus be obtained. In fact, there can be a solution $m\neq 0$ even in the case where $m_0=0$, and moreover the symmetry seems to forbid a finite m.

This kind of situation can be most easily examined by means of the generalized Hartree-Fock procedure[8,13] which was developed before in connection with the theory of superconductivity. The basic idea is not new in field theory, and in fact in its simplest form the method is identical with the renormalization procedure of Dyson, considered only in a somewhat different context.

Suppose a Lagrangian is composed of the free and interaction part: $L=L_0+L_i$. Instead of diagonalizing L_0 and treating L_i as perturbation, we introduce the self-energy Lagrangian L_s, and split L thus

$$\begin{aligned} L&=(L_0+L_s)+(L_i-L_s)\\ &=L_0'+L_i'. \end{aligned}$$

For L_s we assume quite general form (quadratic or bilinear in the fields) such that L_0' leads to linear field equations. This will enable one to define a vacuum and a complete set of "quasi-particle" states, each particle being an eigenmode of L_0'. Now we treat L_i' as perturbation, and determine L_s from the requirement that L_i' shall not yield additional self-energy effects. This procedure then leads to Eq. (3.2). The self-consistent nature of such a procedure is evident since the self-energy is calculated by perturbation theory with fields which are already subject to the self-energy effect.

In order to apply the method to our problem, let us assume that $L_s=-m\bar{\psi}\psi$, and introduce the propagator $S_F^{(m)}(x)$ for the corresponding Dirac particle with mass m. In the lowest order, and using the two alternative forms Eqs. (2.6) and (2.7), we get for Eq. (3.2)

$$\begin{aligned}\Sigma=2g_0[&\mathrm{Tr}S_F^{(m)}(0)-\gamma_5\,\mathrm{Tr}S_F^{(m)}(0)\gamma_5\\ &-\tfrac{1}{2}\gamma_\mu\,\mathrm{Tr}\gamma_\mu S_F^{(m)}(0)+\tfrac{1}{2}\gamma_\mu\gamma_5\,\mathrm{Tr}\gamma_\mu\gamma_5 S_F^{(m)}(0)]\end{aligned} \tag{3.4}$$

in coordinate space.

This is quadratically divergent, but with a cutoff can be made finite. In momentum space we have

$$\Sigma=-\frac{8g_0 i}{(2\pi)^4}\int\frac{m}{p^2+m^2-i\epsilon}d^4p\,F(p,\Lambda), \tag{3.5}$$

where $F(p,\Lambda)$ is a cutoff factor. In this case the self-energy operator is a constant. Substituting Σ from Eq. (3.5), Eq. (3.2) gives $(m_0=0)$

$$m=-\frac{g_0 m i}{2\pi^4}\int\frac{d^4p}{p^2+m^2-i\epsilon}F(p,\Lambda). \tag{3.6}$$

This has two solutions: either $m=0$, or

$$1=-\frac{g_0 i}{2\pi^4}\int\frac{d^4p}{p^2+m^2-i\epsilon}F(p,\Lambda). \tag{3.7}$$

The first trivial one corresponds to the ordinary perturbative result. The second, nontrivial solution will determine m in terms of g_0 and Λ.

If we evaluate Eq. (3.7) with a straight noninvariant cutoff at $|\mathbf{p}|=\Lambda$, we get

$$\frac{\pi^2}{g_0\Lambda^2}=\left(\frac{m^2}{\Lambda^2}+1\right)^{\frac{1}{2}}-\frac{m^2}{\Lambda^2}\ln\left[\left(\frac{\Lambda^2}{m^2}+1\right)^{\frac{1}{2}}+\frac{\Lambda}{m}\right]. \tag{3.8}$$

If we use Eq. (3.5) with an invariant cutoff at $p^2=\Lambda^2$ after the change of path: $p_0\rightarrow ip_0$, we get

$$\frac{2\pi^2}{g_0\Lambda^2}=1-\frac{m^2}{\Lambda^2}\ln\left(\frac{\Lambda^2}{m^2}+1\right) \tag{3.9}$$

[12] This will be done in Part II.

[13] N. N. Bogoliubov, Uspekhi Fiz. Nauk **67**, 549 (1959) [translation: Soviet Phys.-Uspekhi **67**, 236 (1959)].

Since the right-hand side of Eq. (3.8) or (3.9) is positive and $\leqslant 1$ for real Λ/m, the nontrivial solution exists only if

$$0<2\pi^2/g_0\Lambda^2<1. \tag{3.10}$$

Equation (3.9) is plotted in Fig. 1 as a function of m^2/Λ^2. As $g_0\Lambda^2$ increases over the critical value $2\pi^2$, m starts rising from 0. The nonanalytic nature of the solution is evident as m cannot be expanded in powers of g_0.

In the following we will assume that Eq. (3.10) is satisfied, so that the nontrivial solution exists. As we shall see later, physically this means that the nucleon-antinucleon interaction must be attractive ($g_0>0$) and strong enough to cause a bound pair of zero total mass. In the BCS theory, the nontrivial solution corresponds to a superconductive state, whereas the trivial one corresponds to a normal state, which is not the true ground state of the superconductor. We may expect a similar situation to hold in the present case.

In this connection, it must be kept in mind that our solutions are only approximate ones. We are operating under the assumption that the corrections to them are not catastrophic, and can be appropriately calculated when necessary. If this does not turn out to be so for some solution, such a solution must be discarded. Later we shall indeed find this possibility for the trivial solution, but for the moment we will ignore such considerations.

Let us define then the vacuum corresponding to the two solutions. Let $\psi^{(0)}$ and $\psi^{(m)}$ be quantized fields satisfying the equations

$$\gamma_\mu\partial_\mu\psi^{(0)}(x)=0, \tag{3.11a}$$

$$(\gamma_\mu\partial_\mu+m)\psi^{(m)}(x)=0, \tag{3.11b}$$

$$\psi^{(0)}(x)=\psi^{(m)}(x) \quad \text{for} \quad x_0=0. \tag{3.11c}$$

According to the standard procedure, we decompose the ψ's into Fourier components:

$$\begin{aligned}\psi_\alpha^{(i)}(x)&=\frac{1}{V^{\frac{1}{2}}}\sum_{\substack{\mathbf{p},s\\ p_0=(p^2+m^2)^{\frac{1}{2}}}}[u_\alpha^{(i)}(\mathbf{p},s)a^{(i)}(\mathbf{p},s)e^{ip\cdot x}\\ &\qquad+v_\alpha^{*(i)}(\mathbf{p},s)b^{(i)\dagger}(\mathbf{p},s)e^{-ip\cdot x}],\\ \psi_\alpha^{\dagger(i)}(x)&=\frac{1}{V^{\frac{1}{2}}}\sum_{\substack{\mathbf{p},s\\ p_0=(p^2+m^2)^{\frac{1}{2}}}}[u_\alpha^{(i)*}(\mathbf{p}\cdot s)a^{(i)\dagger}(\mathbf{p},s)\\ &\qquad\times e^{-ip\cdot x}+v_\alpha^{(i)}(\mathbf{p},s)b^{(i)}(\mathbf{p},s)e^{ip\cdot x}],\end{aligned} \tag{3.12}$$

$$i=0 \text{ or } m,$$

where $u_\alpha^{(i)}(\mathbf{p},s)$, $v_\alpha^{(i)}(\mathbf{p},s)$ are the normalized spinor eigenfunctions for particles and antiparticles, with momentum p and helicity $s=\pm1$, and

$$\begin{aligned}&\{a^{(i)}(\mathbf{p},s),a^{(i)\dagger}(\mathbf{p}',s')\}\\ &\quad=\{b^{(i)}(\mathbf{p},s),b^{(i)\dagger}(\mathbf{p}',s')\}=\delta_{\mathbf{pp}'}\delta_{ss'}, \text{ etc.}\end{aligned} \tag{3.13}$$

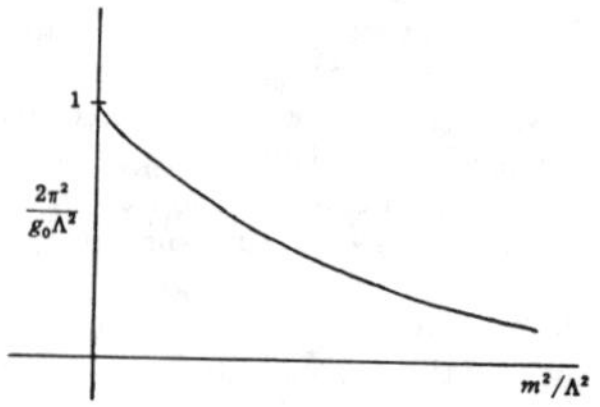

FIG. 1. Plot of the self-consistent mass equation (3.9).

The operator sets $(a^{(0)},b^{(0)})$ and $(a^{(m)},b^{(m)})$ are related by a canonical transformation because of Eq. (3.11c):

$$\begin{aligned}a^{(m)}(\mathbf{p},s)&=\sum_{\alpha,s'}[u_\alpha^{(m)*}(\mathbf{p},s)u_\alpha^{(0)}(\mathbf{p},s')a^{(0)}(\mathbf{p},s')\\ &\qquad+u_\alpha^{(m)*}(\mathbf{p},s)v_\alpha^{(0)*}(-\mathbf{p},s')b^{(0)\dagger}(-\mathbf{p},s')],\\ b^{(m)}(\mathbf{p},s)&=\sum_{\alpha,s'}[v_\alpha^{(m)*}(\mathbf{p},s)v_\alpha^{(0)}(\mathbf{p},s')b^{(0)}(\mathbf{p},s')\\ &\qquad+v_\alpha^{(m)*}(\mathbf{p},s)u_\alpha^{(0)*}(-\mathbf{p},s')a^{(0)\dagger}(-\mathbf{p},s')].\end{aligned} \tag{3.14}$$

Using Eq. (1.3), this is evaluated to give

$$\begin{aligned}a^{(m)}(\mathbf{p},s)&=[\tfrac{1}{2}(1+\beta_p)]^{\frac{1}{2}}a^{(0)}(\mathbf{p},s)\\ &\qquad+[\tfrac{1}{2}(1-\beta_p)]^{\frac{1}{2}}b^{(0)\dagger}(-\mathbf{p},s),\\ b^{(m)}(\mathbf{p},s)&=[\tfrac{1}{2}(1+\beta_p)]^{\frac{1}{2}}b^{(0)}(\mathbf{p},s)\\ &\qquad-[\tfrac{1}{2}(1-\beta_p)]^{\frac{1}{2}}a^{(0)\dagger}(-\mathbf{p},s),\\ \beta_p&=|\mathbf{p}|/(\mathbf{p}^2+m^2)^{\frac{1}{2}}.\end{aligned} \tag{3.15}$$

The vacuum $\Omega^{(0)}$ or $\Omega^{(m)}$ with respect to the field $\psi^{(0)}$ or $\psi^{(m)}$ is now defined as

$$a^{(0)}(\mathbf{p},s)\Omega^{(0)}=b^{(0)}(\mathbf{p},s)\Omega^{(0)}=0, \tag{3.16}$$

$$a^{(m)}(\mathbf{p},s)\Omega^{(m)}=b^{(m)}(\mathbf{p},s)\Omega^{(m)}=0. \tag{3.16'}$$

Both $\psi^{(0)}$, $\bar\psi^{(0)}$ and $\psi^{(m)}$, $\bar\psi^{(m)}$ applied to $\Omega^{(0)}$ always create particles of mass zero, whereas the same applied to $\Omega^{(m)}$ create particles of mass m.

From Eqs. (3.15) and (3.16) we obtain

$$\begin{aligned}\Omega^{(m)}&=\prod_{\mathbf{p},s}\{[\tfrac{1}{2}(1+\beta_p)]^{\frac{1}{2}}\\ &\qquad-[\tfrac{1}{2}(1-\beta_p)]^{\frac{1}{2}}a^{(0)\dagger}(\mathbf{p},s)b^{(0)\dagger}(-\mathbf{p},s)\}\Omega^{(0)}.\end{aligned} \tag{3.17}$$

Thus $\Omega^{(m)}$ is, in terms of zero-mass particles, a superposition of pair states. Each pair has zero momentum, spin and nucleon number, and carries ±2 units of chirality, since chirality equals minus the helicity s for massless particles.

Let us calculate the scalar product $(\Omega^{(0)},\Omega^{(m)})$ from Eq. (3.15):

$$\begin{aligned}(\Omega^{(0)},\Omega^{(m)})&=\prod_{\mathbf{p},s}[\tfrac{1}{2}(1+\beta_p)]^{\frac{1}{2}}\\ &=\exp\{\sum_{\mathbf{p},s}\tfrac{1}{2}\ln[\tfrac{1}{2}(1+\beta_p)]\}.\end{aligned} \tag{3.18}$$

For large p, $\beta_p\sim1-m^2/2\mathbf{p}^2$, so that the exponent diverges as $V\pi m^2\int d\mathbf{p}/(2\pi)^3$ (V=normalization volume). Hence

$$(\Omega^{(0)},\Omega^{(m)})=0. \tag{3.19}$$

It is easy to see that any two states $\Psi^{(0)}$ and $\Psi^{(m)}$, obtained by applying a finite number of creation operators on $\Omega^{(0)}$ and $\Omega^{(m)}$ respectively, are also orthogonal.

Thus the two "worlds" based on $\Omega^{(0)}$ and $\Omega^{(m)}$ are physically distinct and outside of each other. No interaction or measurement, in the usual sense, can bridge them in finite steps.

What is the energy difference of the two vacua? Since both are Lorentz invariant states, the difference can only be either zero or infinity. Using the expression

$$\begin{aligned} H^{(m)}&=\sum_{\mathbf{p},s}(\mathbf{p}^2+m^2)^{\frac{1}{2}}\{a^{(m)\dagger}(\mathbf{p},s)a^{(m)}(\mathbf{p},s) \\ &\qquad -b^{(m)}(\mathbf{p},s)b^{(m)\dagger}(\mathbf{p},s)\}, \\ H^{(0)}&=\sum_{\mathbf{p},s}|\mathbf{p}|\{a^{(0)\dagger}(\mathbf{p},s)a^{(0)}(\mathbf{p},s) \\ &\qquad -b^{(0)}(\mathbf{p},s)b^{(0)\dagger}(\mathbf{p},s)\}, \end{aligned} \tag{3.20}$$

we get for the respective energies

$$E^{(m)}-E^{(0)}=-2\sum_p[(\mathbf{p}^2+m^2)^{\frac{1}{2}}-|\mathbf{p}|], \tag{3.21}$$

which is negative and quadratically divergent. So $\Omega^{(m)}$ may be called the "true" ground state, as was expected.

There remains finally the question of γ_5 invariance. The original Hamiltonian allowed two conservations X and N, Eqs. (2.1) and (2.2). Both $\Omega^{(0)}$ and $\Omega^{(m)}$ belong to $N=0$, and their elementary excitations carry $N=\pm1$. In the case of X, the same is true for the space $\Omega^{(0)}$, but $\Omega^{(m)}$ as well as its elementary excitations are not eigenstates of X, as is clear from the foregoing results. If the latter solution is to be a possibility, there must be an infinite degeneracy with respect to the quantum number X. A ground state will be in general a linear combination of degenerate states with different $X=0$, $\pm2,\cdots$:

$$\Omega^{(m)}=\sum_{n=-\infty}^{\infty}C_{2n}\Omega_{2n}^{(m)}. \tag{3.22}$$

Equation (3.17) is in fact a particular case of this. The γ_5-gauge transformation Eq. (2.3) induces the change

$$\begin{aligned} a^{(0)}(\mathbf{p},\pm1)&\to e^{\mp i\alpha}a^{(0)}(\mathbf{p},\pm1), \\ b^{(0)}(\mathbf{p},\pm1)&\to e^{\mp i\alpha}b^{(0)}(\mathbf{p},\pm1), \\ a^{(0)\dagger}(\mathbf{p},\pm1)&\to e^{\pm i\alpha}a^{(0)\dagger}(\mathbf{p},\pm1), \\ b^{(0)\dagger}(\mathbf{p},\pm1)&\to e^{\pm i\alpha}b^{(0)\dagger}(\mathbf{p},\pm1), \end{aligned} \tag{3.23}$$

and the coefficients of Eq. (3.22) become

$$C_{2n}\to e^{-2ni\alpha}C_{2n}. \tag{3.24}$$

In particular

$$\begin{aligned} \Omega^{(m)}&\to\Omega_\alpha^{(m)} \\ &=\exp[-i\alpha X]\Omega^{(m)} \\ &=\prod_{\mathbf{p},\pm}\{[\tfrac{1}{2}(1+\beta_p)]^{\frac{1}{2}}-[\tfrac{1}{2}(1-\beta_p)]^{\frac{1}{2}} \\ &\qquad\times e^{\pm2i\alpha}a^{(0)\dagger}(p,\pm)b^{(0)\dagger}(-p,\pm)\}\Omega^{(0)}. \end{aligned} \tag{3.25}$$

The Dirac equation (3.11b), at the same time, is transformed into

$$[\gamma_\mu\partial_\mu+m\cos2\alpha+im\gamma_5\sin2\alpha]\psi=0. \tag{3.26}$$

The moral of this is that the self-consistent self-energy Σ is determined only up to a γ_5 transformation. This can be easily verified from Eq. (3.4), in which the second term on the right-hand side is nonvanishing when a propagator corresponding to Eq. (3.26) is used. Although Eq. (3.26) seems to violate parity conservation, it is only superficially so since $\Omega_\alpha^{(m)}$ is now not an eigenstate of parity. We could alternatively say that the parity operator undergoes transformation together with the mass operator. Despite the odd form of the equation (3.26), there is no change in the physical predictions of the theory. We shall see more of this later.

Let us calculate, as before, the scalar product of $\Omega_\alpha^{(m)}$ and $\Omega_{\alpha'}^{(m)}$. From Eqs. (3.17) and (3.25) we get

$$\begin{aligned} (\Omega_\alpha^{(m)},\Omega_{\alpha'}^{(m)}) &=\prod_{p,\pm}[\tfrac{1}{2}(1+\beta_p)-e^{\pm2i(\alpha'-\alpha)}\tfrac{1}{2}(1-\beta_p)] \\ &=\prod_{p,\pm}[1+(e^{\pm2i(\alpha'-\alpha)}-1)\tfrac{1}{2}(1-\beta_p)] \\ &=\exp\{\sum_{p,\pm}\ln[1+(e^{\pm2i(\alpha'-\alpha)}-1)\tfrac{1}{2}(1-\beta_p)]\}. \end{aligned} \tag{3.27}$$

For large $|\mathbf{p}|$, the exponent goes like

$$\frac{V}{(2\pi)^3}\sum_\pm(e^{\pm2i(\alpha'-\alpha)}-1)\int\frac{m^2}{4\mathbf{p}^2}d^3p.$$

The integral is again divergent. Hence

$$\begin{aligned} (\Omega_\alpha^{(m)},\Omega_{\alpha'}^{(m)})&=(\Omega^{(m)},\exp[-i(\alpha'-\alpha)X]\Omega^{(m)}) \\ &=0,\quad \alpha'\neq\alpha(\text{mod}2\pi), \end{aligned} \tag{3.28}$$

and, of course

$$(\Omega^{(0)},\Omega_\alpha^{(m)})=0. \tag{3.28'}$$

We can evaluate $(\Omega_\alpha^{(m)},\Omega_{\alpha'}^{(m)})$ alternatively from Eqs. (3.22) and (3.24). Then

$$\sum_{m=-\infty}^{\infty}|C_{2n}|^2e^{2ni(\alpha-\alpha')}=0,\quad \alpha\neq\alpha'(\text{mod}2\pi), \tag{3.29}$$

implying that

$$|C_0|=|C_{\pm2}|=|C_{\pm4}|=\cdots=C. \tag{3.30}$$

Thus there is an infinity of equivalent worlds described by $\Omega_\alpha^{(m)}$, $0\leqslant\alpha<2\pi$. The states Ω_{2n} of Eq. (3.22) are then expressed in terms of $\Omega_\alpha^{(m)}$ as

$$C_{2n}\Omega_{2n}^{(m)}=\frac{1}{2\pi}\int_0^{2\pi}e^{2ni\alpha}\Omega_\alpha^{(m)}d\alpha, \tag{3.31}$$

which form another orthogonal set. Since the original total H commutes with X, it will have no matrix elements connecting different "worlds." Moreover, as was the case with $\Omega^{(m)}$ and $\Omega^{(0)}$, no finite measurement can induce similar transitions. This is a kind of superselection rule, which effectively avoids the apparent degeneracy to show up as physical effects.[14] The usual description of the world by means of $\Omega^{(m)}$ and ordinary Dirac particles must be regarded as only the most convenient one.

We still are left with some paradoxes. The X conservation implies the existence of a conserved X current:

$$j_{\mu 5}=i\bar{\psi}\gamma_\mu\gamma_5\psi, \tag{3.32}$$

$$\partial_\mu j_{\mu 5}=0, \tag{3.32'}$$

which can readily be verified from Eq. (2.6). On the other hand, for a massive Dirac particle the continuity equation is not satisfied:

$$\partial_\mu\bar{\psi}^{(m)}\gamma_\mu\gamma_5\psi^{(m)}=2m\bar{\psi}^{(m)}\gamma_5\psi^{(m)}. \tag{3.33}$$

If a massive Dirac particle has to be a real eigenstate of the system, how can this be reconciled? The answer would be that the X-current operator taken between real one-nucleon states should not be given simply by $i\gamma_\mu\gamma_5$ because of the "radiative corrections." We expect instead

$$\langle p'|j_{\mu 5}|p\rangle=\bar{u}(p')X_\mu(p',p)u(p), \tag{3.34}$$

where the renormalized quantity $X_{\mu 5}$ should be, from relativistic invariance grounds, of the form

$$\begin{aligned} X_\mu(p',p)&=F_1(q^2)i\gamma_\mu\gamma_5+F_2(q^2)\gamma_5 q_\mu,\\ q&=p'-p,\quad p^2=p'^2=-m^2. \end{aligned} \tag{3.35}$$

The continuity equation (3.32'), together with Eq. (3.33), further reduces this to

$$\begin{aligned} F_1&=F_2q^2/2m\equiv F,\\ X_\mu(p',p)&=F(q^2)\left(i\gamma_\mu\gamma_5+\frac{2m\gamma_5 q_\mu}{q^2}\right). \end{aligned} \tag{3.36}$$

The real nucleon is not a point particle. Its X-current (3.36) is provided with the dramatic "anomalous" term.

To understand the physical meaning of the anomalous term, we have to make use of the dispersion relations. The form factors F_1 and F_2 will, in general, satisfy dispersion relations of the form

$$F_i(q^2)=F_i(0)-\frac{q^2}{\pi}\int\frac{\operatorname{Im}F_i(-\kappa^2)}{(q^2+\kappa^2-i\epsilon)\kappa^2}d\kappa^2, \tag{3.37}$$

assuming one subtraction. Each singularity at κ^2 corresponds to some physical intermediate state. Thus if $F(0)\neq 0$, Eq. (3.36) indicates that there is a pole at $q^2=0$ for F_2 (and no subtraction), which means in turn that there is an isolated intermediate state of zero mass.

[14] This was discussed by R. Haag, Kgl. Danske Videnskab. Selskab, Mat.-fys. Medd. 29, No. 12 (1955). See also L. van Hove, Physica 18, 145 (1952).

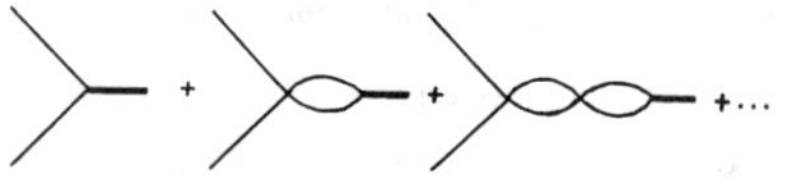

FIG. 2. Graphs corresponding to the Bethe-Salpeter equation in "ladder" approximation. The thick line is a bound state.

To see its nature, we take a time-like q in its own rest frame and go to the limit $q^2\to 0$. The anomalous term has then only the time component, and is proportional to the amplitude for creation of a nucleon pair in a $J=0^-$ state. Hence the zero mass state must have the same property as this pair. It belongs to nucleon number zero, so that we may call it a zero-mass pseudoscalar meson. *In order for a γ_5-invariant Hamiltonian such as Eq. (2.6) to allow massive nucleon states and a nonvanishing X current for $q=0$, it is therefore necessary to have at the same time pseudoscalar zero-mass mesons coupled with the nucleons.* Since we did not have such mesons in the theory, they must be regarded as secondary products, i.e., bound states of nucleon pairs. This conclusion would not hold if in Eq. (3.36) $F(q^2)=O(q^2)$ near $q^2=0$. A nucleon then would have always $X=0$. Such a possibility cannot be excluded. We will show, however, that the pseudoscalar zero-mass bound states do follow explicitly, once we assume the nontrivial solution of the self-energy equation.

IV. THE COLLECTIVE STATES

From the general discussion of Secs. 2 and 3, we may expect the existence of collective states of the fundamental field which would manifest themselves as stable or unstable particles. In particular we have argued that, as a consequence of the γ_5 invariance, a pseudoscalar zero-mass state must exist. We want now to discuss the problem in detail, trying to determine the mass spectrum of the collective excitations (at least its general features) and the strength of their coupling with the nucleons. These states must be considered as a direct effect of the same primary interaction which produces the mass of the nucleon, which itself is a collective effect. We will study the bound-state problem through the use of the Bethe-Salpeter equation, taking into account explicitly the self-consistency conditions. We first verify in the following the existence of the zero-mass pseudoscalar state.

The Bethe-Salpeter equation for a bound pair B deals with the amplitude

$$\Phi(x,y)=\langle 0|T(\psi(x)\bar{\psi}(y))|B\rangle. \tag{4.1}$$

As is well known, the equation is relatively easy to handle in the ladder approximation. In our case we have a four-spinor point interaction and the analog of the "ladder" approximation would be the iteration of the simplest closed loop (see Fig. 2) in which all lines represent dressed particles. We introduce the vertex function

Γ related to Φ by

$$\Phi(p)=S_F^{(m)}(p+\tfrac{1}{2}q)\Gamma(p+\tfrac{1}{2}q,\,p-\tfrac{1}{2}q)S_F^{(m)}(p-\tfrac{1}{2}q). \quad (4.2)$$

All we have to do then is to set up the integral equation generated by the chain of diagrams, looking for solutions having the symmetry properties of a pseudoscalar state. This means that our solutions must be proportional to γ_5. This requirement makes only the pseudoscalar and axial vector part of the interaction contribute to the integral equation. We have

$$\begin{aligned}\Gamma(p+\tfrac{1}{2}q,\,p-\tfrac{1}{2}q)&=\frac{2ig_0}{(2\pi)^4}\gamma_5\int \mathrm{Tr}[\gamma_5 S_F^{(m)}(p'+\tfrac{1}{2}q)\\&\quad\times\Gamma(p'+\tfrac{1}{2}q,\,p'-\tfrac{1}{2}q)S_F^{(m)}(p'-\tfrac{1}{2}q)]d^4p'\\&\quad-\frac{ig_0}{(2\pi)^4}\gamma_5\gamma_\mu\int \mathrm{Tr}[\gamma_5\gamma_\mu S^{(m)}(p'+\tfrac{1}{2}q)\\&\quad\times\Gamma(p'+\tfrac{1}{2}q,\,p'-\tfrac{1}{2}q)S_F^{(m)}(p'-\tfrac{1}{2}q)]d^4p'. \quad (4.3)\end{aligned}$$

For the moment let us ignore the pseudovector term on the right-hand side. It then follows that the equation has a constant solution $\Gamma=C\gamma_5$ if $q^2=0$. To see this, first observe that for the special case $q=0$, Eq. (4.3) reduces to

$$1=-\frac{8ig_0}{(2\pi)^4}\int\frac{d^4p}{p^2+m^2-i\epsilon}, \quad (4.4)$$

which is nothing but the self-consistency condition (3.7), provided that the same cutoff is applied. Since the pseudoscalar term of Eq. (4.3) gives a function of q^2 only, the same condition remains true as long as $q^2=0$.

When the pseudovector term is included, we have still the same eigenvalue $q^2=0$ with a solution of the form $\Gamma=C\gamma_5+iD\gamma_5\gamma\cdot q$, which is not difficult to verify (see Appendix).

We now add some remarks. First, the bound state amplitude for this solution spreads in space over a region of the order of the fermion Compton wavelength $1/m$ because of Eq. (4.2), making the zero-mass particle only partially localizable. We want also to stress the role played by the γ_5 invariance in the argument. We had in fact already inferred the existence of the pseudoscalar particle from relativistic and γ_5 invariance alone, and at first sight the same result seems to follow now essentially from the self-consistency equation. However, we must notice that only the scalar term of the Lagrangian appears in this equation while only the pseudoscalar part contributes in the Bethe-Salpeter equation. It is because of the γ_5-invariant Lagrangian that the Bethe-Salpeter equation can be reduced to the self-consistency condition.

Along the same line we could try to see whether other bound states exist in the "ladder" approximation. However, besides calculating the spectrum, it is also important to determine the interaction properties of these collective states with the fermions. For this purpose the study of the two-"nucleon" scattering amplitude appears much more suitable, as we shall realize after the following remark. Once we have recognized that in the ladder approximation the collective states would appear as real stable particles, we must expect to the same degree of approximation poles in the scattering matrix of two nucleons corresponding to the possibility of the virtual exchange of these particles. For definiteness we shall refer again as an example to the pseudoscalar zero-mass particle. Let us indicate by $J_p(q)$ the analytical expression corresponding to the graph whose iteration produces the bound state [Fig. 3(a)]. We construct next the scattering matrix generated by the exchange of all possible simple chains built with this element. This means that we consider the set of diagrams in Fig. 3(b). The series is easily evaluated and we obtain

$$2g_0 i\gamma_5\frac{1}{1-J_p(q)}i\gamma_5, \quad (4.5)$$

where the γ_5's refer to the pairs (1,1′) and (2,2′), respectively. The meaning of this result is clear: because of the self-consistent equation $J_p(0)=1$, Eq. (4.5) is equivalent to a phenomenological exchange term where the intermediate particle is our pseudoscalar massless boson (Fig. 4). The coupling constant G can now be evaluated by straightforward comparison. Before doing this calculation we need the explicit expression of $J_P(q)$. Using the ordinary rules for diagrams, we have

$$J_P(q)=-\frac{2ig_0}{(2\pi)^4}\times\int\frac{4(m^2+p^2)-q^2}{[(p+\tfrac{1}{2}q)^2+m^2][(p-\tfrac{1}{2}q)^2+m^2]}d^4p. \quad (4.6)$$

It is however more convenient to rewrite J_P in the form of a dispersive integral, and if we forget for a moment that it is a divergent expression, a simple manipulation gives

$$J_P(q)=\frac{g_0}{4\pi^2}\int_{4m^2}^{\Lambda^2}\frac{\kappa^2(1-4m^2/\kappa^2)^{\frac{1}{2}}}{q^2+\kappa^2}d\kappa^2. \quad (4.6')$$

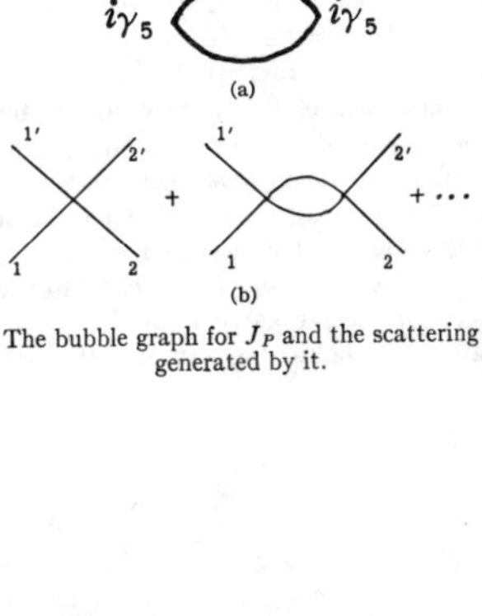

FIG. 3. The bubble graph for J_P and the scattering matrix generated by it.

In order for this expression to be meaningful, a new cutoff Λ must be introduced. There is no simple relation between this and the previous cutoffs. The dispersive form is more comfortable to handle and accordingly we shall reformulate the self-consistent condition $J_P(0)=1$, or

$$1=\frac{g_0}{4\pi^2}\int_{4m^2}^{\Lambda^2}(1-4m^2/\kappa^2)^{\frac{1}{2}}d\kappa^2. \tag{4.7}$$

It may be of interest to remark at this point that Eq. (4.7) can be obtained also if we think of our theory as a theory with intermediate pseudoscalar boson in the limit of infinite boson mass. We are now in a position to evaluate the phenomenological coupling constant G. From Eqs. (4.6′) and (4.7) we have

$$J_P(q^2)=1-q^2\frac{g_0}{4\pi^2}\int_{4m^2}^{\Lambda^2}\frac{(1-4m^2/\kappa^2)^{\frac{1}{2}}}{q^2+\kappa^2}d\kappa^2, \tag{4.8}$$

which leads immediately to the result

$$\frac{G_P^2}{4\pi}=2\pi\left[\int_{4m^2}^{\Lambda^2}\frac{(1-4m^2/\kappa^2)^{\frac{1}{2}}}{\kappa^2}d\kappa^2\right]^{-1}. \tag{4.9}$$

This equation is interesting since it establishes a connection between the phenomenological constant G_P and the cutoff independently of the value of the fundamental coupling g_0. This fact exhibits the purely dynamical origin of the phenomenological coupling G_P. Actually g_0 is buried in the value of the mass m.

So far we have exploited only the γ_5 vertex. What happens then if the scalar part is iterated to form chains of bubbles similar to those we have already discussed? The procedure just explained can be followed again, and a quantity $J_S(q)$ can be defined similarly with the result

$$J_S(q)=\frac{g_0}{4\pi^2}\int_{4m^2}^{\Lambda^2}\frac{(\kappa^2-4m^2)(1-4m^2/\kappa^2)^{\frac{1}{2}}}{q^2+\kappa^2}d\kappa^2. \tag{4.10}$$

It is immediately seen that because of Eq. (4.7)

$$J_S(-4m^2)=1, \tag{4.11}$$

which causes a new pole to appear in the S matrix for $q^2=-4m^2$. This means that we have another collective state of mass $2m$, parity + and spin 0! We observe that it is necessary to assume the same cutoff as in the pseudoscalar case in order that this result may be obtained. The choice of the same cutoff in both cases seems to be suggested by the γ_5 invariance as will be seen later. We also notice the peculiar symmetry existing between the pseudoscalar and the scalar state: the first has zero mass and binding energy $2m$, while the opposite is true for the scalar particle. So in the bound-state picture the scalar particle would not be a true bound state and should be, rather, interpreted as a

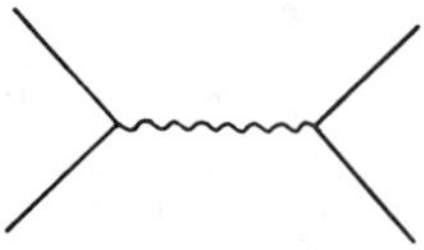

FIG. 4. The equivalent phenomenological one-meson exchange graph.

correlated exchange of pairs in the scattering process.[15] The "nucleon-nucleon" forces induced by the exchange of the scalar particle are, of course, of rather short range. The general physical implications of these results will be discussed more thoroughly later.

The phenomenological coupling constant G_S for the scalar meson is given by

$$\frac{G_S^2}{4\pi}=2\pi\left[\int_{4m^2}^{\Lambda^2}\frac{(1-4m^2/\kappa^2)^{\frac{1}{2}}}{(\kappa^2-4m^2)}d\kappa^2\right]^{-1}. \tag{4.12}$$

Let us next turn to the vector state generated by iteration of the vector interaction. In this case we obtain for each "bubble" a tensor

$$J_{V\mu\nu}=(\delta_{\mu\nu}-q_\mu q_\nu/q^2)J_V,$$
$$J_V=-\frac{g_0}{4\pi^2}\frac{q^2}{3}\int_{4m^2}^{\Lambda^2}\frac{d\kappa^2}{q^2+\kappa^2}\times\left(1+\frac{2m^2}{\kappa^2}\right)(1-4m^2/\kappa^2)^{\frac{1}{2}}. \tag{4.13}$$

Perhaps a remark is in order here regarding the evaluation of J_V. It suffers from an ambiguity of subtraction well known in connection with the photon self-energy problem. The above result is of the conventional gauge invariant form, which we take to be the proper choice.

Equation (4.13) leads to the scattering matrix

$$g_0\left[\gamma_\mu\frac{1}{1-J_V}\gamma_\mu-\gamma\cdot q\frac{J_V}{(1-J_V)q^2}\gamma\cdot q\right], \tag{4.14}$$

where the second term is, of course, effectively zero. It can be easily seen that the denominator can produce a pole below $4m^2$ for sufficiently small Λ^2. In fact, from Eqs. (4.7) and (4.13), we find

$$(8/3)m^2<\mu_V^2. \tag{4.15}$$

The coupling constant is given by

$$\frac{G_V^2}{4\pi}=3\pi\left[\int_{4m^2}^{\Lambda^2}d\kappa^2\frac{\kappa^2+2m^2}{(\kappa^2-\mu_V^2)^2}(1-4m^2/\kappa^2)^{\frac{1}{2}}\right]^{-1}. \tag{4.16}$$

It must be noted that the mass of the vector meson now depends on the cutoff, unlike the previous two cases.

Finally we are left with the pseudovector state. We

[15] Of course this and other heavy mesons will in general become unstable in higher order approximation, which is beyond the scope of the present paper.

find for the bubble[16]

$$J_{A\mu\nu}=-J_{V\mu\nu}+J_A'\delta_{\mu\nu},$$

$$J_A'=\frac{g_0m^2}{2\pi^2}\int_{4m^2}^{\Lambda^2}\frac{d\kappa^2}{q^2+\kappa^2}(1-4m^2/\kappa^2)^{\frac{1}{2}}. \tag{4.17}$$

In view of the self-consistency condition (4.7), it can be seen that this does not produce a pole of the scattering matrix for $-q^2<4m^2$, corresponding to a pseudovector meson.

So far we have considered only iterations of the same kind of interactions. In the ladder approximation there is actually a coupling between pseudoscalar and pseudovector interactions as was explicitly considered in Eq. (4.3). However, the coupling between scalar and vector interactions vanish because of the Furry's theorem.

This coupling of pseudoscalar and pseudovector interactions does not change the pion pole of the scattering matrix, but it affects the coupling of the pion to the nucleon since a chain of the pseudoscalar can join the external nucleon with an axial vector interaction. In other words, the pion-nucleon coupling is in general a mixture of pseudoscalar and derivative pseudovector types (Appendix).

We would like to inject here a remark concerning the trivial solution of the self-energy equation, against which we had no decisive argument. So let us also try to apply our scattering formula to this solution. For the pseudoscalar state we now find $J_P(q=0)>1$, provided that the cutoff Λ is kept fixed and m is set equal to zero in Eq. (4.6′). (The pseudovector interference vanishes.) In other words, there will be a pole for some $q^2>0$ ($\mu^2<0$). This is again a supporting evidence that the trivial solution could be unstable, capable of decaying by emitting such mesons. The final answer, however, depends on the exact nature of the cutoff.

Finally we would like to discuss the nucleon-nucleon scattering in the same spirit and approximation as for the nucleon-antinucleon scattering. In order to make a correspondence with the previous cases, it is convenient to rewrite the Hamiltonian in the following way:

$$\begin{aligned}H_1&=-g_0[\bar\psi\psi\bar\psi^c\psi^c-\bar\psi\gamma_5\psi\bar\psi^c\gamma_5\psi^c]\\&=\tfrac{1}{2}g_0[\bar\psi\gamma_\mu\psi^c\bar\psi^c\gamma_\mu\psi-\bar\psi\gamma_\mu\gamma_5\psi^c\bar\psi^c\gamma_\mu\gamma_5\psi]\\&=-\tfrac{1}{2}g_0[\bar\psi\gamma_\mu C\bar\psi\psi C^{-1}\gamma_\mu\psi-\bar\psi\gamma_\mu\gamma_5C\bar\psi\psi C^{-1}\gamma_\mu\gamma_5\psi],\end{aligned} \tag{4.18}$$

where ψ^c, $\bar\psi^c$ are the charge-conjugate fields.

The last form of Eq. (4.18) is suitable for our purpose. We note first that the vector part of the interaction is identically zero because of the anticommutativity of ψ. Thus only the pseudovector part survives. A "bubble" made of this interaction then is seen to give rise to the same integral J_A, Eq. (4.17). Since the interfering pseudoscalar interaction is missing in the present case, we get the complete scattering matrix by iterating J_A:

$$\begin{aligned}-\gamma_\mu\gamma_5C&\left[\frac{\delta_{\mu\nu}-q_\mu q_\nu/q^2}{1-J_A}+\frac{q_\mu q_\nu/q^2}{1-J_A'}\right]C^{-1}\gamma\gamma_5\\&=\gamma_\mu\gamma_5C\frac{1}{1-J_A}C^{-1}\gamma_\mu\gamma_5\\&\quad+\gamma\cdot q\gamma_5C\frac{J_V/q^2}{(1-J_A')(1-J_A)}C^{-1}\gamma\cdot q\gamma_5,\end{aligned} \tag{4.19}$$

$$J_A\equiv J_A'-J_V.$$

The first term, corresponding to a scattering in the $J=1^-$ state, does not have a pole. The second term can have one below $4m^2$ for $1=J_A'$. With Eqs. (4.7) and (4.17), this determines the mass μ_D:

$$2m^2<\mu_D^2. \tag{4.20}$$

In this second term of the scattering matrix, the wave function is proportional to $C\gamma\cdot q\gamma_5$, so that the bound state behaves like a scalar "deuteron" (a singlet S state). The residue of the pole determines the nucleon-"deuteron" coupling constant (derivative) G_D^2, which is positive as it should be.

TABLE I. Mass spectrum.

Nucleon number	Mass μ	Spin-parity	Spectroscopic notation
0	0	0^-	$^1S_0$
0	$2m$	0^+	$^3P_0$
0	$(8/3)m^2<\mu^2$	1^-	$^3P_1$
±2	$2m^2<\mu^2$	0^+	$^1S_0$

Table I summarizes the main results of this section. Although our approximation is a very crude one, we believe that it reflects the real situation at least qualitatively, because all the results are understandable in simple physical terms. Thus in the nonrelativistic sense, our Hamiltonian contains spin-independent attractive scalar and vector interactions plus a spin-dependent axial vector interaction between a particle and an antiparticle. Between particles, the vector part turns into a repulsion. Table I is just what we expect for the level ordering from this consideration.

V. PHENOMENOLOGICAL THEORY AND γ_5 INVARIANCE

In the previous section special subsets of diagrams were taken into account, and the existence of various boson states was established, together with their couplings with the nucleons. As was discussed there, we can reasonably expect that these results are essentially correct in spite of the very simple approximations. Because the bosons have in general small masses (compared to the unbound nucleon states), they will play important roles in the dynamics of strong interactions at least at energies comparable to these masses.

[16] We meet here again the problem of subtraction. Our choice follows naturally from comparison with the vector case, and is consistent with Eq. (3.33).

Thus if we are willing to accept the conclusions of our lowest order approximation, what we should do then is to study the dynamics of systems consisting of nucleons and the different kinds of bosons which all together represent the primary manifestation of the fundamental interaction. These particles will be now assumed to interact via their phenomenological couplings. So we may describe our purpose as an attempt to construct a theory in the conventional sense in which a separate field is introduced for each kind of particle. However, this is not a simple and unambiguous problem because our fundamental theory is completely γ_5 invariant and we must make sure that this invariance is preserved at any stage of our calculations in order that the results be meaningful. For a better understanding of the problem, let us consider our Lagrangian in the lowest self-consistent approximation. We have

$$L'=L_0'+L_I',$$

where

$$\begin{aligned} L_0'&=-(\bar{\psi}\gamma_\mu\partial_\mu\psi+m\bar{\psi}\psi),\\ L_I'&=g_0[(\bar{\psi}\psi)^2-(\bar{\psi}\gamma_5\psi)^2]+m\bar{\psi}\psi. \end{aligned} \tag{5.1}$$

L' is obviously γ_5 invariant. In order to preserve this invariance we must study the S matrix generated by L_I'. Some subsets of diagrams have been considered in the previous section and it will be shown now how those calculations comply with γ_5 invariance. This point must be understood clearly so that we shall discuss it in a rather systematic way. Let us recall first how we constructed the scattering matrix in the "ladder" approximation. The lowest-order contribution is certainly invariant as no internal massive line appears. But what will happen to the next-order terms [Fig. 3(b)]? To these diagrams corresponds the expression

$$J_S(q^2)-\gamma_5 J_P(q^2)\gamma_5+iJ_{SP}(q^2)\gamma_5+i\gamma_5 J_{PS}(q^2). \tag{5.2}$$

In the gauge in which our calculations were performed, the last two terms happened to be zero. We write down next the transformation properties of the quantities appearing above. By straightforward calculation we find

$$\begin{aligned} \gamma_5&\to\gamma_5\cos2\alpha+i\sin2\alpha,\\ 1&\to\cos2\alpha+i\gamma_5\sin2\alpha,\\ J_P&\to J_P\cos^2 2\alpha+J_S\sin^2 2\alpha,\\ J_S&\to J_S\cos^2 2\alpha+J_P\sin^2 2\alpha,\\ J_{SP}&\to(J_P-J_S)\sin2\alpha\cos2\alpha,\\ J_{PS}&\to(J_P-J_S)\sin2\alpha\cos2\alpha. \end{aligned} \tag{5.3}$$

By simple substitution the invariance follows easily. The argument can now be extended to all orders, provided at each order all the possible combinations of S and P are included. The invariance of the scattering in the "ladder" approximation is thus established. It may look surprising that the SP and PS contributions do not vanish identically. This can be understood by considering the fact that the γ_5 transformation changes the parity of the vacuum which will be in general a superposition of states of opposite parities. In this way products of fields of different parities (as the SP propagator) may have a nonvanishing average value in the vacuum state.

We may now attempt the construction of the phenomenological coupling by introducing two local fields Φ_P and Φ_S describing the pseudoscalar and the scalar particles, respectively. We start by observing that, in the same gauge in which the previous calculations were made, we can write the meson-nucleon interaction as

$$L_I=G_P i\bar{\psi}\gamma_5\psi\Phi_P+G_S\bar{\psi}\psi\Phi_S. \tag{5.4}$$

In order to find the general expression valid in any gauge, it is convenient to introduce the following two-dimensional notation

$$\varphi\equiv\begin{pmatrix} i\bar{\psi}\gamma_5\psi \\ \bar{\psi}\psi\end{pmatrix},\quad \Phi\equiv\begin{pmatrix}\Phi_P\\ \Phi_S\end{pmatrix},\quad G\equiv\begin{pmatrix} G_P & 0\\ 0 & G_S\end{pmatrix}. \tag{5.5}$$

The interaction Lagrangian Eq. (5.4) can be written in this notation in a compact form,

$$L_I=\varphi G\Phi. \tag{5.6}$$

The effect of the γ_5 transformation on φ is described with the aid of the matrix

$$U\equiv\begin{pmatrix}\cos2\alpha & -\sin2\alpha\\ \sin2\alpha & \cos2\alpha\end{pmatrix}, \tag{5.7}$$

which satisfies $UU^+=UU^{-1}=UU^T=1$. In other words, the γ_5 transformation induces a unitary transformation in the two-dimensional space, and Eq. (5.6) remains invariant if

$$G\to UGU^{-1},\quad \Phi\to U\Phi. \tag{5.8}$$

To complete the construction of the theory, the free Lagrangian for the fields Φ_P and Φ_S must be added. If we work again in the special gauge $\alpha=0$, we may write

$$L_0=-\tfrac{1}{2}\partial_\mu\Phi_P\partial_\mu\Phi_P-\tfrac{1}{2}\partial_\mu\Phi_S\partial_\mu\Phi_S-\tfrac{1}{2}\mu^2\Phi_S^2, \tag{5.9}$$

where $\mu^2=4m^2$. We use again the two-dimensional notation, and defining the mass operator

$$M^2\equiv\begin{pmatrix}0 & 0\\ 0 & \mu^2\end{pmatrix}, \tag{5.10}$$

we write Eq. (5.9) in the invariant form

$$L_0=-\tfrac{1}{2}\partial_\mu\Phi\partial_\mu\Phi-\tfrac{1}{2}\Phi M^2\Phi. \tag{5.11}$$

In this way we have given a formal prescription for the γ_5 transformation in the phenomenological treatment. We have to emphasize here that the Lagrangians (5.9) and (5.11) are *not* γ_5 invariant in the ordinary sense of the word. In our theory, where the mesons are only phenomenological substitutes which partially represent the dynamical contents of the theory, they may be, however, called γ_5 covariant. In other words, *the masses and the coupling constants are not fixed parameters, but rather dynamical quantities which are subject to transformations when the representation is changed.* It will be legitimate to ask whether this situation corresponds to the one obtained in the framework of the fundamental theory and discussed in the "ladder" approximation in the previous section. We shall examine the transformation rule for the mass operator M^2, since this illustrates the case in point. Let us calculate explicitly M^2 in an arbitrary gauge α. We have

$$M^2 \to UM^2U^{-1} = \mu^2\begin{pmatrix} \sin^2 2\alpha & -\sin 2\alpha \cos 2\alpha \\ -\sin 2\alpha \cos 2\alpha & \cos^2 2\alpha \end{pmatrix}. \quad (5.12)$$

The meaning of this equation is that the pseudoscalar and the scalar particle will have generally different masses in different gauges. In particular we see that the pseudoscalar particle has in the gauge α a mass $\sin 2\alpha\mu$. If this is the case we must expect that after the transformation the pole in the corresponding propagator will move from $q^2=0$ to $q^2=-(\sin^2 2\alpha)\mu^2$. This actually may be verified directly in the "ladder" approximation which shows that the pion propagator changes according to

$$iG_P{}^2\Delta_{FP} = \frac{2g_0}{1-J_P} \to \frac{2g_0}{1-J_P\cos^2 2\alpha - J_S \sin^2 2\alpha}. \quad (5.13)$$

Using the results of the previous section, it is seen that the denominator of the right-hand side vanishes for $q^2=-(\sin^2 2\alpha)4m^2$. In this way we have seen how our γ_5-invariant theory can be approximated by a phenomenological description in terms of pseudoscalar and scalar mesons. Of course one may add the vector meson as well. Such a description does not look γ_5 invariant. It is only γ_5 covariant, and the masses and coupling constants must be understood to be matrices which, however, can be simultaneously diagonalized.

The reason for this situation is the degeneracy of the vacuum and the world built upon it. Only after combining all the equivalent but nonintersecting worlds labeled with different α do we recover complete γ_5 invariance. Nevertheless, even in a particular world we can find manifestations of the invariance, such as the zero-mass pseudoscalar meson and the conserved γ_5 current.

VI. THE CONSERVATION OF AXIAL VECTOR CURRENT

In this section we will discuss another paradoxical aspect of the theory regarding the γ_5 invariance. In Sec. 3 we argued that the X current should really be conserved, and that this is possible if a nucleon X current possesses a peculiar anomalous term. We now verify the statement explicitly in our approximation.

First we have to realize that the problem is again how to keep the γ_5-invariant nature of the theory at every stage of approximation. It is well known in quantum electrodynamics that, in order to observe the ordinary gauge invariance, a certain set of graphs have to be combined together in a given approximation. The necessity for this is based on a general proof which makes use of the so-called Ward identity. In our present case there also exists an analog of the Ward identity. In order to derive it, let us first consider the proper self-energy part of our fermion in the presence of an external axial vector field B_μ with the interaction $L_B=-j_{\mu5}B_\mu$. The self-energy operator is now a matrix $\Sigma^{(B)}(p',p)$ depending on initial and final momenta. Expanding Σ in powers of B, we have

$$\Sigma^{(B)}(p',p)=\Sigma(p)+\Lambda_{\mu5}(p',p)B_\mu(p'-p)+\cdots. \quad (6.1)$$

We readily realize that the coefficient of the second term gives the desired X-current vertex correction.

On the other hand, the entire Lagrangian remains invariant under a *local* γ_5 transformation if Eq. (2.3) is accompanied by

$$B_\mu \to B_\mu - \partial_\mu \alpha, \quad (6.2)$$

where α is now an arbitrary function. In other words,

$$e^{i\alpha\gamma_5}\Sigma^{(B-\partial\alpha)}e^{i\alpha\gamma_5}=\Sigma^{(B)} \quad (6.3)$$

in a symbolical way of writing.[17]

Expanding (6.3) after putting $B=0$, we get

$$i\alpha(p'-p)[\gamma_5\Sigma(p)+\Sigma(p')\gamma_5] = i\alpha(p'-p)(p'-p)_\mu\Lambda_{\mu5}(p',p),$$

or

$$\gamma_5\Sigma(p)+\Sigma(p')\gamma_5=(p'-p)_\mu\Lambda_{\mu5}(p',p). \quad (6.4)$$

The entire vertex $\Gamma_{\mu5}=i\gamma_\mu\gamma_5+\Lambda_{\mu5}$ then satisfies

$$\gamma_5 L'(p)+L'(p')\gamma_5=-(p'-p)_\mu\Gamma_{\mu5}(p',p), \quad L'(p)\equiv -i\gamma\cdot p-\Sigma(p), \quad (6.5)$$

which is the desired generalized Ward identity.[18] The right-hand side of Eq. (6.5) is the divergence of the X current, while the left-hand side vanishes when p and p' are on the mass shell of the actual particle. The X-current conservation is thus established. Moreover, the way the anomalous term arises is now clear. For if we assume $\Sigma(p)=m$, Eq. (6.4) gives

$$2m\gamma_5=(p'-p)_\mu\Lambda_{\mu5}(p'-p), \quad (6.6)$$

so that we may write the longitudinal part of Λ as

$$\Lambda_{\mu5}{}^{(l)}(p',p)=2m\gamma_5 q_\mu/q^2, \quad q=p'-p, \quad (6.7)$$

which is of the desired form.

Next we have to determine what types of graphs

[17] We assume here that $\alpha(x)$ is different from zero only over a finite space-time region, so that the gauge of the nontrivial vacuum, which we may fix at remote past, is not affected by the transformation. The limiting process of going over to constant α is then ill-defined as we can see from the fact that the anomalous term in $\Gamma_{\mu5}$ has no limit as $q\to 0$.

[18] See also J. Bernstein, M. Gell-Mann, and L. Michel, Nuovo cimento **16**, 560 (1960).

should be considered for Γ_μ in our particular approximation of the self-energy. Examining the way in which the relation (6.3) is maintained in a perturbation expansion, we are led to the conclusion that our self-energy represented by Fig. 5(a) gives rise to the series of vertex graphs [Fig. 5(b)]. The summation of the graphs is easily carried out to give

$$\Lambda_{\mu 5}=i\gamma_5\frac{1}{1-J_P}J_{PA}, \tag{6.8}$$

where J_P was obtained before [Eq. (4.8)], and

$$\begin{aligned} J_{PA}&=\frac{2ig_0}{(2\pi)^4}\int \mathrm{Tr}\gamma_5 S(p+q/2)\gamma_\mu\gamma_5 S(p-q/2)d^4p \\ &=-\frac{g_0}{2\pi^2}imq_\mu\int_{4m^2}^{\Lambda^2}\frac{d\kappa^2}{q^2+\kappa^2}(1-4m^2/\kappa^2)^{\frac{1}{2}}. \end{aligned} \tag{6.9}$$

Thus

$$\begin{aligned} \Gamma_{\mu 5}&=i\gamma_\mu\gamma_5+\Lambda_{\mu 5} \\ &=i\gamma_\mu\gamma_5+2m\gamma^5 q_\mu/q^2, \end{aligned} \tag{6.10}$$

in agreement with the general formula. We see also that there is no form factor in this approximation.

This example will suffice to show the general procedure necessary for keeping γ_5 invariance. When we consider further corrections, the procedure becomes more involved, but we can always find a set of graphs which are sufficient to maintain the X-current conservation. We shall come across this problem in connection with the axial vector weak interactions.

VII. SUMMARY AND DISCUSSION

We briefly summarize the results so far obtained. Our model Hamiltonian, though very simple, has been found to produce results which strongly simulate the general characteristics of real nucleons and mesons. It is quite appealing that both the nucleon mass and the pseudoscalar "pion" are of the same dynamical origin, and the reason behind this can be easily understood in terms of (1) classical concepts such as attraction or repulsion between particles, and (2) the γ_5 symmetry.

According to our model, the pion is not the primary agent of strong interactions, but only a secondary effect. The primary interaction is unknown. At the present stage of the model the latter is only required to have appropriate dynamical and symmetry properties, although the nonlinear four-fermion interaction, which we actually adopted, has certain practical advantages.

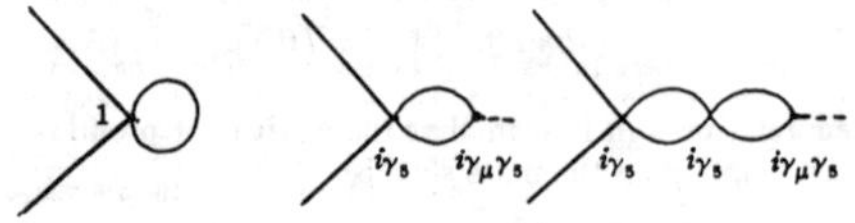

FIG. 5. Graphs for self-energy and matching radiative corrections to an axial vector vertex.

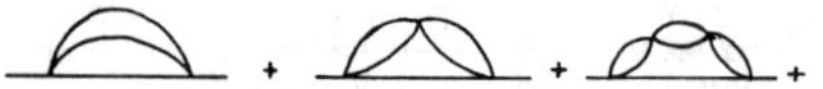

FIG. 6. A class of higher order self-energy graphs.

In our model the idealized "pion" occupies a special position in connection with the γ_5-gauge transformation. But there are also other massive bound states which may be called heavy mesons and deuterons. The conventional meson field theory must be regarded, from our point of view, as only a phenomenological description of events which are actually dynamic processes on a higher level of understanding, in the same sense that the phonon field is a phenomenological description of interatomic dynamics.

Our theory contains two parameters, the primary coupling constant and the cutoff, which can be translated into observed quantities: nucleon mass and the pion-nucleon coupling constant. It is interesting that the pion coupling depends only on the cutoff in our approximation. In order to make the pion coupling as big as the observed one (≈ 15) the cutoff has to be rather small, being of the same order as the nucleon mass.

We would like to make some remarks about the higher order approximations. If the higher order corrections are small, the usual perturbation calculation will be sufficient. If they are large compared to the lowest order estimation, the self-consistent procedure must be set up, including these effects from the beginning. This is complicated by the fact that the pions and other mesons have to be properly taken into account.

To get an idea about the importance of the corrections, let us take the next order self-energy graph (Fig. 6). This is only the first term of a class of corrections shown in Fig. 6, the sum of which we know already to give rise to an important collective effect, i.e., the mesons. It would be proper, therefore, to consider the entire class put together. The correction is then equivalent to the ordinary second order self-energy due to mesons, plus modifications arising at high momenta. Thus strict perturbation with respect to the bare coupling g_0 will not be an adequate procedure. Evaluating, for example, the pion contribution in a phenomenological way, we get

$$\frac{\delta m}{m}=\frac{G_P^2}{32\pi^2}\int_{m^2}^{\Lambda'^2}\frac{d\kappa^2}{\kappa^2}\left(1-\frac{m^2}{\kappa^2}\right), \tag{7.1}$$

where Λ' is an effective cutoff. Substituting G_P^2 from Eq. (4.9), this becomes

$$\frac{\delta m}{m}=\frac{1}{4}\int_{4m^2}^{4\Lambda'^2}\frac{d\kappa^2}{\kappa^2}\left(1-\frac{4m^2}{\kappa^2}\right)\bigg/\int_{4m^2}^{\Lambda^2}\frac{d\kappa^2}{\kappa^2}\left(1-\frac{4m^2}{\kappa^2}\right)^{\frac{1}{2}}. \tag{7.2}$$

As Λ and Λ' should be of the same order of magnitude, the higher order corrections are in general not negligible. We may point out, on the other hand, that there is a tendency for partial cancellation between contributions from different mesons or nucleon pairs.

We already remarked before that the model treated here is not realistic enough to be compared with the actual nucleon problem. Our purpose was to show that a new possibility exists for field theory to be richer and more complex than has been hitherto envisaged, even though the mathematics is marred by the unresolved divergence problem.

In the subsequent paper we will attempt to generalize the model to allow for isospin and finite pion mass, and draw various consequences regarding strong as well as weak interactions.

APPENDIX

We treat here, for completeness, the problem created by the coupling of pseudoscalar and pseudovector terms encountered in the text. As we have seen, such an effect is not essential for the discussion of γ_5 invariance, but rather adds to complication, which however naturally appears in the ladder approximation.

First let us write down the integral equation for a vertex part Γ:

$$\Gamma(p+\tfrac{1}{2}q,\,p-\tfrac{1}{2}q)$$
$$=\gamma(p+\tfrac{1}{2}q,\,p-\tfrac{1}{2}q)+\frac{2ig_0}{(2\pi)^4}\gamma_5\int\mathrm{Tr}[\gamma_5 S(p'+\tfrac{1}{2}q)$$
$$\times\Gamma(p'+\tfrac{1}{2}q,\,p-\tfrac{1}{2}q)S_F(p-\tfrac{1}{2}q)]d^4p'$$
$$-\frac{ig_0}{(2\pi)^4}\gamma_5\gamma_\mu\int\mathrm{Tr}[\gamma_5\gamma_\mu S(p'+\tfrac{1}{2}q)$$
$$\times\Gamma(p'+\tfrac{1}{2}q,\,p-\tfrac{1}{2}q)S_F(p-\tfrac{1}{2}q)]d^4p'. \quad \text{(A.1)}$$

This embraces three special cases depending on the inhomogeneous term γ:

(a) $\gamma=0$ for the Bethe-Salpeter equation for the pseudoscalar meson;
(b) $\gamma=i\gamma_\mu\gamma_5$ for the pseudovector vertex function $\Gamma_{\mu 5}$;
(c) $\gamma=2g_0(\gamma_5)_f(\gamma_5)_i-g_0(\gamma_\mu\gamma_5)_f(\gamma_\mu\gamma_5)_i$ for the nucleon-antinucleon scattering through these interactions.

Here i and f refer to initial and final states, and the integral kernel of Eq. (A.1) operates on the f part.

We will consider them successively.

(a) We make the ansatz $\Gamma=C\gamma_5+iD\gamma_5\gamma\cdot q$. The integrals in Eq. (A.1) then reduce to the standard forms considered in the text. Making use of Eqs. (4.9), (4.17), and (6.9), we get[16]

$$C=C-(C+2mD)q^2I,$$
$$D=(C+2mD)mI, \quad \text{(A.2)}$$
$$I(q^2)=\frac{g_0}{4\pi^2}\int\frac{d\kappa^2}{q^2+\kappa^2}\left(1-\frac{4m^2}{\kappa^2}\right)^{\frac{1}{2}},$$

which lead to $q^2=0$, and $C:D=1-2m^2I(0):mI(0)$. From Eq. (4.8), we have $0<2m^2I(0)<\frac{1}{2}$.

(b) Put
$$\Gamma_{\mu5}=(i\gamma_\mu\gamma_5+2m\gamma_5q_\mu/q^2)F_1(q^2)+(i\gamma_\mu\gamma_5-i\gamma\cdot q\gamma_5q_\mu/q^2)F_2(q^2). \quad \text{(A.3)}$$

This is seen to satisfy the integral equation if

$$F_1=1,$$
$$F_2=J_A(q^2)/[1-J_A(q^2)], \quad \text{(A.4)}$$
$$J_A(q^2)=2m^2I(q^2)-J_V(q^2),$$

where $J(q^2)$ was defined in Eq. (4.13).

On the mass shell, $\Gamma_{\mu5}$ reduces to

$$(i\gamma_\mu\gamma_5+2m\gamma_5q_\mu/q^2)F(q^2),$$
$$F(q^2)=1+F_2(q^2)=1/[1-J_A(q^2)]. \quad \text{(A.5)}$$

For $q^2=0$, we have $J(q^2)=0$ so that $1<F(0)=1/[1-2m^2I(0)]<2$.

(c) From the structure of the inhomogeneous term, it is clear that the scattering matrix is given by

$$M=2g_0(\Gamma_5)_f(\gamma_5)_i+g_0(\Gamma_{\mu5})_f(i\gamma_\mu\gamma_5)_i,$$

where Γ_5 is the pseudoscalar vertex function.

Again, from Eq. (A.1), Γ_5 is determined as

$$\Gamma_5=\gamma_5[1-2m^2I(q^2)]/q^2I(q^2)-mi\gamma\cdot q\gamma_5/q^2, \quad \text{(A.6)}$$

which has an entirely different behavior from the bare γ_5 for small q^2. The scattering matrix is then

$$M=(\gamma_5)_f(\gamma_5)_i 2g_0[1-2m^2I(q^2)]/q^2I(q^2)$$
$$-[(i\gamma\cdot q\gamma_5)_f(\gamma_5)_i-(\gamma_5)_f(i\gamma\cdot q\gamma_5)_i]2mg_0/q^2$$
$$-(i\gamma\cdot q\gamma_5)_f(i\gamma\cdot q\gamma_5)_i g_0J_A(q^2)/q^2[1-J_A(q^2)]$$
$$+(i\gamma_\mu\gamma_5)_f(i\gamma_\mu\gamma_5)_i g_0/[1-J_A(q^2)]. \quad \text{(A.7)}$$

The first three terms have a pole at $q^2=0$. The coupling constants of the pseudoscalar meson are then

pseudoscalar coupling:

$$G_p{}^2=2g_0[1-2m^2I(0)]/I(0),$$

pseudovector coupling:

$$G_{pv}{}^2=g_0J_A(0)/[1-J_A(0)]$$
$$=g_0 2m^2I(0)/[1-2m^2I(0)]. \quad \text{(A.8)}$$

Their relative sign is such that the equivalent pseudoscalar coupling on the mass shell is

$$G_p'^2=4m^2g_0\left\{\left[\frac{1-2m^2I(0)}{2m^2I(0)}\right]^{\frac{1}{2}}+\left[\frac{2m^2I(0)}{1-2m^2I(0)}\right]^{\frac{1}{2}}\right\}^2. \quad \text{(A.9)}$$

Broken Symmetries*

Jeffrey Goldstone
Trinity College, Cambridge University, Cambridge, England

AND

Abdus Salam and Steven Weinberg†
Imperial College of Science and Technology, London, England

(Received March 16, 1962)

Some proofs are presented of Goldstone's conjecture, that if there is continuous symmetry transformation under which the Lagrangian is invariant, then either the vacuum state is also invariant under the transformation, or there must exist spinless particles of zero mass.

I. INTRODUCTION

IN the past few years several authors have developed an idea which might offer hope of understanding the broken symmetrics that seem to be characteristic of elementary particle physics. Perhaps the fundamental Lagrangian is invariant under all symmetries, but the vacuum state[1] is not. It would then be impossible to prove the usual sort of symmetry relations among S-matrix elements, but enough symmetry might remain (perhaps at high energy) to be interesting.

But whenever this idea has been applied to specific models, there has appeared an intractable difficulty. For example, Nambu suggested that the Lagrangian might be invariant under a continuous chirality transformation $\psi \to \exp(i\boldsymbol{\theta}\cdot\boldsymbol{\tau}\gamma_5)\psi$ even if the fermion physical mass M were nonzero. But then there would be a conserved current J_λ, with matrix element

$$\langle p'|J_\lambda|p\rangle = f(q^2)\bar{u}'\gamma_5[i\gamma_\lambda - (2M/q^2)q_\lambda]u,$$

where $q=p-p'$. The pole at $q^2=0$ can only arise from a spinless particle of mass zero, which almost certainly does not exist. Of course, the pole would not occur if $f(0)=0$, which might be the case if we do not insist on identifying J_λ with the axial vector current of β decay. But Nambu showed that this unwanted massless "pion" also appears as a solution of the approximate Bethe-Salpeter equation.[1]

Goldstone[2] has examined another model, in which the manifestation of "broken" symmetry was the nonzero vacuum expectation value of a boson field. (This was suggested as an explanation of the $\Delta I=\frac{1}{2}$ rule by Salam and Ward.)[3] Here again there appeared a spinless particle of zero mass. Goldstone was led to conjecture that this will always happen whenever a continuous symmetry group leaves the Lagrangian but not the vacuum invariant.

* This research was supported in part by the U. S. Air Force under a contract monitored by the Air Force Office of Scientific Research of the Air Development Command and the Office of Naval Research.

† Alfred P. Sloan Foundation Fellow; Permanent address: University of California, Berkeley, California.

[1] Y. Nambu and G. Jona-Lasinio, Phys. Rev. 122, 345 (1961); W. Heisenberg, Z. Naturforsch. 14, 441 (1959).

[2] J. Goldstone, Nuovo cimento 19, 154 (1961).

[3] A. Salam and J. C. Ward, Phys. Rev. Letters 5, 512 (1960).

We will present here three proofs of this result. The first uses perturbation theory; the other two are much more general.

II. PERTURBATION THEORY

We will consider a multiplet of n spinless fields ϕ_i which interact among themselves and perhaps also with other fields. The Lagrangian is assumed to be invariant under a set of infinitesimal transformations:

$$\delta^\alpha\phi_i=\epsilon T_{ij}{}^\alpha\phi_j. \tag{1}$$

If the vacuum state were also invariant under these transformations, the vacuum expectation values of the ϕ_i would be subject to a set of linear relations,

$$T_{ij}{}^\alpha\langle\phi_j\rangle_0=0. \tag{2}$$

(Usually, there would be enough such relations to imply that all ϕ_i have zero vacuum expectation value. This is the case in the example to be discussed at the end of this section, where the ϕ_i transform as the defining representation of the orthogonal group, so that the T span the space of all antisymmetric matrices.)

We are going to examine the possibility that the vacuum state is not invariant under these transformations; in particular we will consider the consequences that ensue if

$$T_{ij}{}^\alpha\langle\phi_j\rangle_0\neq 0 \tag{3}$$

for some α and some i. It is inconvenient to work with fields with nonzero vacuum expectation value, we we will define

$$\phi_i=\chi_i+\eta_i, \tag{4}$$

where

$$\eta_i\equiv\langle\phi_i\rangle_0,$$

so that χ_i is a quantum field with

$$\langle\chi_i\rangle_0\equiv 0. \tag{5}$$

In perturbation theory this means that we should ignore all "tadpole" diagrams with a single external χ line.

The Lagrangian $L(\phi)$, although invariant under (1), will, in general, not be invariant under the "naive" transformations

$$\delta^\alpha\chi_i=\epsilon T_{ij}{}^\alpha\chi_j. \tag{6}$$

Hence, the vanishing of $\langle\chi_i\rangle_0$ provides a nontrivial self-consistency condition which allows us to calculate η_i up to some unavoidable ambiguities. We will now show that the value of η is such that propagators of some of the χ_i have a pole at zero mass.

We begin by defining a function $F(\eta)$ as the sum of all proper connected graphs with no external lines, and with the over-all energy momentum conservation factor $i(2\pi)^4\delta^4(0)$ omitted. Every factor λ_i in each term of $F(\lambda)$ represents a place where we might instead have an external line of type i. This can be seen in general by noting that the interaction Lagrangian density used in calculating these graphs is

$$L'(\chi,\eta)=L(\chi+\eta)-L_0(\chi),$$
$$L_0(\chi)=-\tfrac{1}{2}(\partial_\mu\chi_i)(\partial^\mu\chi_i)-\tfrac{1}{2}m^2\chi_i\chi_i. \tag{7}$$

It follows then that the sum $F^{(N)}$ of all connected proper diagrams with N external lines i, j, $\cdots$ carrying zero energy and momentum is (for $N\neq 2$)

$$F_{ij\ldots}{}^{(N)}=(\partial^N/\partial\eta_i\partial\eta_j\cdots)F(\eta). \tag{8}$$

For $N=2$ the mass term in $-L_0$ gives an additional contribution, so

$$F_{ij}{}^{(2)}=(\partial^2/\partial\eta_i\partial\eta_j)F(\eta)+m^2\delta_{ij}. \tag{9}$$

As defined here, $F^{(N)}$ does not include the propagators for its external lines or the over-all factor $i(2\pi)^4\delta(0)$.

It is clear from the definition of $F^{(1)}$ that

$$F_i{}^{(1)}=(\Delta'^{-1}(0))_{ij}\langle\chi_j\rangle_0. \tag{10}$$

Here $\Delta'(p)$ is the complete propagator given by

$$\Delta'_{ij}(p)=\int d^4x\, e^{-ip\cdot x}\langle T\{\chi_i(x),\chi_j(0)\}\rangle_0. \tag{11}$$

Because $\langle\chi_i\rangle_0$ vanishes, the most general improper diagram for $\Delta'(p)$ can be constructed by stringing together proper self-energy parts $\Pi^*(p)$ into a linear chain; hence the inverse of the propagator is given as usual by

$$(\Delta'^{-1}(p))_{ij}=(p^2+m^2)\delta_{ij}-\Pi^*{}_{ij}(p). \tag{12}$$

For zero momentum,

$$\Pi^*{}_{ij}(0)=F_{ij}{}^{(2)}, \tag{13}$$

and so, using (9), we have

$$(\Delta'^{-1})_{ij}=-(\partial^2/\partial\eta_i\partial\eta_j)F(\eta). \tag{14}$$

We are going to prove that (14) has no inverse, so it should be kept in mind that it is $\Delta'^{-1}(0)$ and not $\Delta'(0)$ that is well defined.

To complete the proof, we now must make use of the invariance of $L(\phi)$ under the transformation (1). This has the consequence that it is only the presence of the η terms that breaks invariance under (6), and hence that $F(\eta)$ is invariant under the corresponding transformations

$$\delta^\alpha\eta_i=\epsilon T_{ij}{}^\alpha\eta_j. \tag{15}$$

Thus

$$(\partial F/\partial\eta_i)T_{ij}{}^\alpha\eta_j=0. \tag{16}$$

Differentiating with respect to η, this gives

$$(\partial^2F/\partial\eta_i\partial\eta_j)T_{ik}{}^\alpha\eta_k+(\partial F/\partial\eta_i)T_{ij}{}^\alpha=0. \tag{17}$$

For physically allowed values of η this relation together with (10) and (14), yields at zero momentum

$$(\Delta'^{-1})_{ij}T_{ik}{}^\alpha\eta_k=0. \tag{18}$$

We see that for zero momentum the inverse of the propagator becomes singular and so some elements of the propagator become infinite. This does not prove that there is a pole at zero mass, but we certainly expect the propagator to be infinite at $P^2=0$ only if the theory involves particles of zero mass. The fields with nonvanishing matrix element between the vacuum and states of zero mass are

$$\chi^\alpha \equiv T_{ik}{}^\alpha \eta_k \chi_i. \tag{19}$$

Clearly, none of this trouble would occur if it were not for our assumption (3) that $T_{ik}{}^\alpha\eta_k \neq 0$. It is the broken symmetry, and not merely the nonzero vacuum expectation value η, that necessitates massless bosons.[4]

To see how this all works in a specific example, let us take as the Lagrangian

$$L(\phi) = -\bar\psi(\gamma\partial+M)\psi - \tfrac{1}{2}(\partial_\mu\phi_i)(\partial^\mu\phi_i) - \tfrac{1}{2}m^2\phi_i\phi_i - g\bar\psi O_i\psi\phi_i - \tfrac{1}{4}\lambda(\phi_i\phi_i)^2. \tag{20}$$

Defining $\phi_i=\chi_i+\eta_i$, this becomes

$$\begin{aligned} L(\phi) = L(\chi) &- m^2(\chi_i\eta_i) - \tfrac{1}{2}m^2(\eta_i\eta_i) - g\bar\psi O_i\psi\eta_i \\ &- \lambda(\chi_i\chi_i)(\chi_j\eta_j) - \lambda(\chi_i\eta_i)^2 - \tfrac{1}{2}\lambda(\chi_i\chi_i)(\eta_j\eta_j) \\ &- \lambda(\eta_i\eta_i)(\chi_j\eta_j) - \tfrac{1}{4}\lambda(\eta_i\eta_i)^2. \end{aligned} \tag{21}$$

The low-order contributions to the sum of all proper connected vacuum graphs are

$$\begin{aligned} F(\eta) = F(0) &- \tfrac{1}{2}m^2(\eta_i\eta_i) - \tfrac{1}{4}\lambda(\eta_i\eta_i)^2 + i(2\pi)^{-4}g\eta_i \\ &\times \int d^4p\, \mathrm{Tr}\{O_iS(p)\} + i(2\pi)^{-4}\lambda(\eta_i\eta_i + \tfrac{1}{2}\delta_{ij}\eta_k\eta_k) \\ &\times \int d^4p\, \Delta_{ij}(p) + \frac{i}{2}(2\pi)^{-4}g^2\eta_i\eta_j \int d^4p \\ &\times \mathrm{Tr}\{O_iS(p)O_jS(p)\} - i(2\pi)^{-4}\lambda^2(\eta_i\eta_j + \tfrac{1}{2}\delta_{ij}\eta_k\eta_k) \\ &\times (\eta_a\eta_b + \tfrac{1}{2}\delta_{ab}\eta_c\eta_c) \int d^4p\, \Delta_{ia}(p)\Delta_{jb}(p) \\ &- (2\pi)^{-8}\lambda^2\delta_{ij}\eta_k\delta_{ab}\eta_c \int d^4p\, d^4p' \\ &\times \{\Delta_{ia}(p)\Delta_{jb}(p')\Delta_{kc}(p+p') + \Delta_{ic}(p)\Delta_{jb}(p') \\ &\qquad \times \Delta_{ka}(p+p') + \Delta_{ib}(p)\Delta_{jc}(p')\Delta_{ka}(p+p')\}, \end{aligned} \tag{22}$$

where

$$S(p) = [-ip_\mu\gamma^\mu + M]^{-1},$$
$$\Delta_{ij}(p) = \delta_{ij}(p^2+m^2)^{-1}.$$

It can be readily seen that the derivative of $F(\eta)$ with respect to η_i is the sum $F_i^{(1)}$ of the proper connected diagrams with one external line, and that the second derivative with respect to η_i and η_j is the sum of the proper connected diagrams with two external lines, except that the $m^2\eta^2$ term in $F(\eta)$ does not contribute to $F^{(2)}$.

We will now assume that $L(\phi)$ is invariant under the group $SO(n)$ of orthogonal transformations on ϕ. This implies that the operators O_i are such that

$$\int d^4p\, \mathrm{Tr}\{O_iS(p)\} = 0, \tag{23}$$

$$\int d^4p\, \mathrm{Tr}\{O_iS(p)O_jS(p)\} = \delta_{ij}I. \tag{24}$$

Thus $F(\eta)$ is a function of $\eta^2=(\eta_i\eta_i)$,

$$\begin{aligned} F(\eta^2) = F(0) &- \tfrac{1}{2}m^2\eta^2 - \tfrac{1}{4}\lambda\eta^4 \\ &+ i(2\pi)^{-4}\eta^2(1+\tfrac{1}{2}n)\int d^4p(p^2+m^2)^{-1} \\ &+ \tfrac{1}{2}i(2\pi)^{-4}g^2\eta^2 I - i(2\pi)^{-4}\lambda^2\eta^4(2+\tfrac{1}{4}n) \\ &\times \int d^4p(p^2+m^2)^{-1}(p^2+m^2)^{-1} - (2\pi)^{-8}\lambda^2\eta^2(2+n) \\ &\times \int d^4p \int d^4p'(p^2+m^2)^{-1}(p'^2+m^2)^{-1} \\ &\qquad \times [(p+p')^2+m^2]^{-1}. \end{aligned} \tag{25}$$

The dependence on η^2 alone implies then that

$$F_i^{(1)} = 2\eta_iF', \tag{26}$$

$$F_{ij}^{(2)} = (m^2+2F')\delta_{ij} + 4\eta_i\eta_jF''. \tag{27}$$

(A prime denotes differentiation with respect to η^2.) We see that the contribution of the term $-\tfrac{1}{2}m^2\eta^2$ in $F^{(2)}$ is canceled by the term $m^2\delta_{ij}$ in (27) arising from $-L_0$.

We have shown that $F_i^{(1)}$ is proportional to $\langle\chi_i\rangle_0$ and so must vanish. This implies that either η is zero or it must satisfy the consistency condition:

$$F'(\eta^2) = 0. \tag{28}$$

Thus if η is not zero, it is determined up to an orthogonal transformation; we certainly could not expect a more unambiguous determination.[5]

---

[4] It is clear from (19) that the maximum number of zero-mass fields is L, the number of Lie generators. There may in special cases be fewer than L zero-mass fields if not all fields χ^α given by (19) are linearly independent. This happens for example when $T_{ik}{}^\alpha$ correspond to the "tensor" representations of simple Lie groups. For this case $T_{ik}{}^\alpha$ are antisymmetric for all three indices. Therefore $\eta_\alpha\chi^\alpha = \eta_\alpha T_{ik}{}^\alpha\eta_k\chi_i = 0$, and only $(L-1)$ of the fields χ^α are linearly independent. These results are unaltered even if we allow in the theory more than one set of scalar fields ϕ_i with nonzero vacuum expectation values. To take a concrete case, the spurion theory proposed by Salam and Ward (reference 3) to explain the $\Delta I=\tfrac{1}{2}$ rule rests on assuming $\langle K_1^0\rangle \neq 0$. This would mean that the three companion fields to K_1^0, i.e., K^-, K^+, and K_2^0, must possess zero masses.

[5] Basic to the entire self-consistency procedure is, of course, the conjecture that $F'(\eta^2)=0$ does possess a root for real η. By considering classical field theories, Goldstone states that a real root would exist provided the bare mass for the ϕ_i fields is pure imaginary. It is interesting to note that if the Lagrangian (20) contains no $-\tfrac{1}{4}\lambda(\phi_i\phi_i)^2$ term, an application of Lehmann's mass theorem [H. Lehmann, Nuovo cimento 11, 342 (1945)] shows that Goldstone's condition [(bare mass)$^2<0$] can never be satisfied. If, however, the $-\tfrac{1}{4}\lambda(\phi_i\phi_i)^2$ term is present in the Lagrangian, Lehmann's theorem gives no indication of the sign of (bare mass)^2.

We have also shown that

$$\Delta_{ij}'^{-1}(0)=m^2\delta_{ij}-F_{ij}^{(2)}$$
$$=-2F'\delta_{ij}-4\eta_i\eta_jF'' \quad (29)$$

If $\eta=0$, there is no reason to expect that $F'=0$, so that $(\Delta'^{-1})_{ij}$ is the nonsingular matrix $-2F'\delta_{ij}$. But if $\eta\neq 0$, then for physically allowed values of η we must have $F'=0$ so,

$$(\Delta'^{-1}(0))_{ij}=-4\eta_i\eta_jF'' \quad (30)$$

This is certainly a singular matrix. In fact

$$(\Delta'^{-1}(0))_{ij}u_j=0 \quad (31)$$

for any u orthogonal to η. All such u can be expressed as

$$u_i=T_{ij}\eta_j, \quad (32)$$

by choosing T_{ij} as an appropriate antisymmetric matrix. We see then that the space of u's is precisely the space indicated by the general considerations above.

III. GENERAL PROOFS

If the Lagrangian is invariant under an n-dimensional set of infinitesimal transformations which transform a general field ϕ_a according to

$$\delta\phi_a=\epsilon T_{ab}{}^{\alpha}\phi_b, \quad (33)$$

then there will exist a set of conserved currents

$$J^{\mu\alpha}=i\frac{\partial L}{\partial(\partial_\mu\phi_a)}T_{ab}{}^{\alpha}\phi_b, \quad (34)$$

$$\partial_\mu J^{\mu\alpha}=0. \quad (35)$$

The usual proof of the conservation equations (35) makes use only of the invariance of the Lagrangian, and hence should not be affected by the noninvariance of the vacuum. Also, from the canonical commutation relations we always expect that

$$[Q^\alpha,\phi_a]=T_{ab}{}^{\alpha}\phi_b, \quad (36)$$

where

$$Q^\alpha=\int d^3x\, J^{0\alpha}(x). \quad (37)$$

We will begin by assuming again that there exists a set of spinless fields ϕ_i transforming according to Eq. (1), i.e.,

$$[Q^\alpha,\phi_i]=T_{ij}{}^{\alpha}\phi_j. \quad (38)$$

The fields ϕ_i need not be "fundamental" here; all our remarks will apply equally well if the ϕ_i are synthetic objects like $\bar{\psi}O_i\psi$.

We shall show that if the vacuum is not annihilated by Q^α, so that

$$T_{ij}{}^{\alpha}\langle\phi_j\rangle\neq 0, \quad (39)$$

then the theory must involve massless particles.

The place we will look for zero-mass singularities is in the vacuum expectation value of the commutator of $J^{\mu\alpha}$ with ϕ_i. Using the usual Lehmann-Källén arguments, this can be written

$$\langle[J^{\mu\alpha}(x),\phi_i(y)]\rangle_0=\partial^\mu\int dm^2\,\Delta(x-y,m^2)\rho_i{}^{\alpha}(m^2), \quad (40)$$

where Δ is the usual causal Green's function for mass m and

$$(\Box^2-m^2)\Delta(x-y,m^2)=0, \quad (41)$$

$$(2\pi)^{-3}p^\mu\theta(p^0)\rho_i{}^{\alpha}(-p^2)=-\sum\delta(p-p^n)(0|J^{\mu\alpha}{}_{(0)}|n)\times\langle n|\phi_i(0)|0\rangle. \quad (42)$$

The current conservation condition (35) together with (40) and (41) implies that

$$m^2\rho_i{}^{\alpha}(m^2)=0, \quad (43)$$

and hence

$$\rho_i{}^{\alpha}(m^2)=N_i{}^{\alpha}\delta(m^2), \quad (44)$$

$$\langle[J^{\mu\alpha}(x),\phi_i(y)]\rangle_0=N_i{}^{\alpha}\partial^\mu D(x-y), \quad (45)$$

where D is Δ for $m=0$. We would normally expect no singularity in $\rho(m^2)$, or in other words, $N_i{}^{\alpha}=0$. (It is well known, for example, that the pion-decay matrix element would vanish if the axial vector current were conserved.) But because of (39) we can show that $N_i{}^{\alpha}\neq 0$. For

$$\begin{aligned}N_i{}^{\alpha}&=\int d^3x\, N_i{}^{\alpha}\partial^0 D(x)\\&=\langle[Q^\alpha,\phi_i(0)]\rangle_0\\&=T_{ij}{}^{\alpha}\langle\phi_j(0)\rangle_0\neq 0.\end{aligned} \quad (46)$$

Thus the sum in (42) must include states of zero mass.

It perhaps does not necessarily follow from the noninvariance of the vacuum that there exists (or can be constructed) a set of spinless fields ϕ_i with $T_{ij}{}^{\alpha}\langle\phi_j\rangle_0\neq 0$. We therefore wish to offer a simple nonrigorous argument that if there were no massless particles in the theory, then we would have to conclude that

$$0=|\alpha\rangle\equiv Q^\alpha|0\rangle; \quad (47)$$

for the conservation of current implies that Q^α and hence $|\alpha\rangle$ is invariant under the inhomogeneous Lorentz group. But then

$$\langle\alpha|J^{\alpha\mu}(x)|0\rangle=0, \quad (48)$$

so that

$$\langle\alpha|\alpha\rangle=\int d^3x\,\langle\alpha|J^{\alpha0}(x)|0\rangle=0. \quad (49)$$

Equation (47) follows from (49) and the positive-definiteness assumption.

This "proof" is probably unobjectionable in ordinary theories with no massless particles. But if there are massless particles, the integrals in (37) and (49) become somewhat poorly defined, because then there are states that are not Lorentz invariant but arbitrarily close to Lorentz invariance. If $|\alpha\rangle$ is such a state, the matrix element (48) will be small, but may give a large value to the integral (49).

As an example, let $|\mathbf{p}\rangle$ be a state containing a particle of mass zero, and construct the wave packet

$$|f\rangle=\int d^3p\ |\mathbf{p}|^{-\frac{1}{2}}f(|\mathbf{p}|)|\mathbf{p}\rangle. \tag{50}$$

The normalization condition is

$$1=4\pi\int|f(E)|^2EdE, \tag{51}$$

so that if we wish we can choose $f(0)\neq0$. Lorentz invariance requires that

$$\langle\mathbf{p}|J^{\mu\alpha}(x)|0\rangle=N^\alpha|\mathbf{p}|^{-\frac{1}{2}}p^\mu e^{-ip\cdot x}, \tag{52}$$

where $p^0=|\mathbf{p}|$.

For particles with mass, current conservation would require that $N^\alpha=0$, but no such conclusion can be drawn for massless particles. For the wave packet (50), we now have

$$\langle f|J^{\mu\alpha}(x)|0\rangle=N^\alpha\int d^3p(p^\mu/|p|)f^*(|\mathbf{p}|)e^{-ip\cdot x},$$

so that

$$\langle f|\alpha\rangle=\int d^3x\,\langle f|J^{0\alpha}(x)|0\rangle$$
$$=(2\pi)^3N^\alpha f^*(0).$$

If we choose $f(0)$ to be nonzero, then the factor $|\mathbf{p}|^{-\frac{1}{2}}$ in (50) gives $|f\rangle$ a more-or-less Lorentz-invariant component, which has nonzero matrix element with $|\alpha\rangle$.

This becomes a bit more understandable if we ask ourselves what is the meaning of the state

$$(\exp i\epsilon Q^\alpha)\ |0\rangle=|0\rangle+i\epsilon\ |\alpha\rangle.$$

Clearly, this state is degenerate with $|0\rangle$, and is in fact another possible vacuum. It involves an infinitesimal component containing massless bosons of preponderantly low momentum. The role of the massless particles is apparently just to give meaning to the various possible vacua.

IV. PROSPECTS FOR THE UNSYMMETRIC VACUUM

The general proofs of the last section rest entirely on the assumption that there exists a conserved current, and that the integral of its time-component satisfies (38). This follows formally from the invariance of the Lagrangian, but in a *quantum* field theory the noncommutativity of the factors in the current, and the possible nonconvergence of the integral of its time-component, make our arguments essentially nonrigorous.

Therefore, it seems reasonable to defer belief in the necessity of massless bosons in a theory with unsymmetric vacua until such a *bête noire* is found in an actual calculation based on such a theory. We have already shown in Sec. II that the massless bosons do appear when we perform calculations using perturbation theory, provided that the symmetry of the theory is broken only by the choice of the vacuum expectation value η of the boson field.

But this is not the most general possibility. The original work of Nambu[1] indicates that the choice of a fermion mass can also break a symmetry. In this theory the fermion mass is

$$-ip^\mu\gamma_\mu=m_1+i\gamma_5m_2,$$

where (m_1,m_2) transform under chirality transformations like the components of a 2-vector. If m_1 and m_2 are not zero they must satisfy a condition of form

$$F(m_1^2+m_2^2)=0.$$

Any particular choice of direction for the vector (m_1,m_2) breaks the chirality invariance. (It should be noted that Nambu's choice $m_2=0$ is purely arbitrary and not dictated by parity conservation. For a general mass we must simply define the matrix associated with parity transformations to be

$$[(m_1+im_2\gamma_5)/(m_1^2+m_2^2)^{\frac{1}{2}}]\beta,$$

rather then just β.)

In Nambu's theory there is no "bare" spinless boson, but it is possible to construct a two-vector

$$\phi_1=\bar\psi\psi,\quad \phi_2=i\bar\psi\gamma_5\psi.$$

With Nambu's definition of parity (i.e., $m_2=0$) the vacuum expectation value of ϕ_2 but not of ϕ_1 vanishes, so the vector $\langle\phi\rangle_0$ points in the 1-direction. An infinitesimal chirality transformation would rotate $\langle\phi\rangle_0$ towards the 2-axis, so we are led to conjecture that the propagator of ϕ_2 has a zero-mass pole. In fact, just such a pole was found by Nambu in an approximate treatment of the bound-state problem. However, to show that the pole remains at zero mass when more complicated diagrams are considered would require a more thorough understanding of the treatment of bound states in perturbation theory. We are attempting this at present.

In a more complicated situation we could have an invariance broken both by the choice of a vacuum expectation value of a "bare" field and also simultaneously by the choice of a mass. For example, if we specialize the model discussed in Sec. II to the case of chirality invariance, we must take

$$M=0,\quad O_1=1,\quad O_2=i\gamma_5,$$

so that

$$L=-\bar\psi(\partial\cdot\gamma)\psi-\tfrac{1}{2}(\partial_\mu\phi_i)(\partial^\mu\phi_i)-\tfrac{1}{2}m^2(\phi_i\phi_i)$$
$$-g\bar\psi\psi\phi_1-ig\bar\psi\gamma_5\psi\phi_2-\tfrac{1}{4}\lambda(\phi_i\phi_i)^2.$$

In this case our conjecture would be that:

(1) If part of the loss of symmetry is due to the choice of a two-vector $\langle\phi\rangle_0$, then there must appear a zero-mass pole in the part of ϕ perpendicular to this vacuum expectation value.

(2) If part of the loss of symmetry is due to the choice of a nonzero Fermion mass $m_1+im_2\gamma_5$ then there must appear a zero-mass pole in the propagator of

$$\phi' = -m_2\bar{\psi}\psi + m_1\bar{\psi}\gamma_5\psi.$$

[Presumably this is the same pole as for (1). Parity conservation would require (m_1,m_2) to be in the direction of $\langle\phi\rangle_0$.]

(3) If part of the loss of symmetry is due to the choice of a noninvariant boson mass (i.e., if the residue of the pole at mass m in the propagator of ϕ_i and ϕ_j is a matrix which is not just a constant times δ_{ij}), then there must appear a *two-boson* pole at zero mass in the propagator of ϕ^2.

These "conjectures" can be taken as proved if we accept the arguments of Sec. III. We believe that we will also soon be able to prove these conjectures, in general, within the framework of perturbation theory.

If this is so, then there seem only three roads open to an understanding of broken symmetries based on the noninvariance of the vacuum:

(A) The particle interpretation of such theories might be revised (as in the Gupta-Bleuler method) so that the massless particles are not physically present in final states if they are absent in initial states. However, all our attempts in this direction have failed.

(B) The massless particles might really exist. The argument against this based on the Eötvös experiment might not apply if the particles carry quantum numbers, since then the scattering cross section of two macroscopic bodies due to exchange of the massless bosons would be proportional only to the numbers of atoms in each body and not (as for Coulomb forces or gravitation) to the squares of the numbers of atoms. But the couplings of these massless particles would presumably be quite strong, and would have shown up in exotic decay modes.

(C) Goldstone has already remarked that nothing seems to go wrong if it is just discrete symmetries that fail to leave the vacuum invariant. A more appealing possibility is that the "ur symmetry" broken by the vacuum involves an inextricable combination of gauge and space-time transformations.

Note added in proof. Recently, one of us (S. W., Proceedings of the 1962 Geneva Conference on High Energy Nuclear Physics) has developed a method of rewriting any Lagrangian in order to introduce fields for bound as well as "elementary" particles. This allows the proof of Sec. II to be extended to the case where the field with nonvanishing vacuum expectation value is any scalar function of the elementary particle fields, hence completing our argument.

ACKNOWLEDGMENTS

We would like to thank Dr. S. Kamefuchi for some helpful discussions. One of us (S. W.) would like to thank (A. S.) for his hospitality at Imperial College, and we all wish to gratefully acknowledge the hospitality of Professor R. G. Sachs at the 1961 Summer Institute of the University of Wisconsin.

III. SPONTANEOUS BREAKING OF LOCAL GAUGE SYMMETRIES AND HIGGS MECHANISM

* * * * *

BROKEN SYMMETRIES, MASSLESS PARTICLES AND GAUGE FIELDS

P. W. HIGGS
Tait Institute of Mathematical Physics, University of Edinburgh, Scotland

Received 27 July 1964

Recently a number of people have discussed the Goldstone theorem [1,2]: that any solution of a Lorentz-invariant theory which violates an internal symmetry operation of that theory must contain a massless scalar particle. Klein and Lee [3] showed that this theorem does not necessarily apply in non-relativistic theories and implied that their considerations would apply equally well to Lorentz-invariant field theories. Gilbert [4], however, gave a proof that the failure of the Goldstone theorem in the nonrelativistic case is of a type which cannot exist when Lorentz invariance is imposed on a theory. The purpose of this note is to show that Gilbert's argument fails for an important class of field theories, that in which the conserved currents are coupled to gauge fields.

Following the procedure used by Gilbert [4], let us consider a theory of two hermitian scalar fields

$\varphi_1(x)$, $\varphi_2(x)$ which is invariant under the phase transformation

$$\varphi_1 \to \varphi_1 \cos\alpha + \varphi_2 \sin\alpha \ , \qquad \varphi_2 \to -\varphi_1 \sin\alpha + \varphi_2 \cos\alpha \ . \qquad (1)$$

Then there is a conserved current j_μ such that

$$i\left[\int d^3x\, j_0(x),\ \varphi_1(y)\right] = \varphi_2(y). \qquad (2)$$

We assume that the Lagrangian is such that symmetry is broken by the nonvanishing of the vacuum expectation value of φ_2. Goldstone's theorem is proved by showing that the Fourier transform of $i\langle[j_\mu(x),\ \varphi_1(y)]\rangle$ contains a term $2\pi\langle\varphi_2\rangle\epsilon(k_0)k_\mu\delta(k^2)$, where k_μ is the momentum, as a consequence of Lorentz-covariance, the conservation law and eq. (2).

Klein and Lee [3] avoided this result in the nonrelativistic case by showing that the most general form of this Fourier transform is now, in Gilbert's notation,

$$\text{F.T.} = k_\mu \rho_1(k^2, nk) + n_\mu \rho_2(k^2, nk) + C_3 n_\mu \delta^4(k) \ , \qquad (3)$$

where n_μ, which may be taken as (1, 0, 0, 0), picks out a special Lorentz frame. The conservation law then reduces eq. (3) to the less general form

$$\text{F.T.} = k_\mu \delta(k^2)\rho_4(nk) + [k^2 n_\mu - k_\mu(nk)]\rho_5(k^2, nk) + C_3 n_\mu \delta^4(k) \ . \qquad (4)$$

It turns out, on applying eq. (2), that all three terms in eq. (4) can contribute to $\langle\varphi_2\rangle$. Thus the Goldstone theorem fails if $\rho_4 = 0$, which is possible only if the other terms exist. Gilbert's remark that no special timelike vector n_μ is available in a Lorentz-covariant theory appears to rule out this possibility in such a theory.

There is however a class of relativistic field theories in which a vector n_μ does indeed play a part. This is the class of gauge theories, where an auxiliary unit timelike vector n_μ must be introduced in order to define a radiation gauge in which the vector gauge fields are well defined operators. Such theories are nevertheless Lorentz-covariant, as has been shown by Schwinger [5]. (This has, of course, long been known of the simplest such theory, quantum electrodynamics.) There seems to be no reason why the vector n_μ should not appear in the Fourier transform under consideration.

It is characteristic of gauge theories that the conservation laws hold in the strong sense, as a consequence of field equations of the form

$$j^\mu = \partial_\nu F'^{\mu\nu} \ ,$$

$$F_{\mu\nu}' = \partial_\mu A_\nu' - \partial_\nu A_\mu' \ . \qquad (5)$$

Except in the case of abelian gauge theories, the fields A_μ', $F_{\mu\nu}'$ are not simply the gauge field variables A_μ, $F_{\mu\nu}$, but contain additional terms with combinations of the structure constants of the group as coefficients. Now the structure of the Fourier transform of $i\langle[A_\mu'(x),\ \varphi_1(y)]\rangle$ must be given by eq. (3). Applying eq. (5) to this commutator gives us as the Fourier transform of $i\langle[j_\mu(x),\ \varphi_1(y)]\rangle$ the single term $[k^2 n_\mu - k_\mu(nk)]\rho(k^2, nk)$. We have thus exorcised both Goldstone's zero-mass bosons and the "spurion" state (at $k_\mu = 0$) proposed by Klein and Lee.

In a subsequent note it will be shown, by considering some classical field theories which display broken symmetries, that the introduction of gauge fields may be expected to produce qualitative changes in the nature of the particles described by such theories after quantization.

References

1) J. Goldstone, Nuovo Cimento 19 (1961) 154.
2) J. Goldstone, A. Salam and S. Weinberg, Phys. Rev. 127 (1962) 965.
3) A. Klein and B. W. Lee, Phys. Rev. Letters 12 (1964) 266.
4) W. Gilbert, Phys. Rev. Letters 12 (1964) 713.
5) J. Schwinger, Phys. Rev. 127 (1962) 324.

* * * * *

BROKEN SYMMETRY AND THE MASS OF GAUGE VECTOR MESONS*

F. Englert and R. Brout
Faculté des Sciences, Université Libre de Bruxelles, Bruxelles, Belgium
(Received 26 June 1964)

It is of interest to inquire whether gauge vector mesons acquire mass through interaction[1]; by a gauge vector meson we mean a Yang-Mills field[2] associated with the extension of a Lie group from global to local symmetry. The importance of this problem resides in the possibility that strong-interaction physics originates from massive gauge fields related to a system of conserved currents.[3] In this note, we shall show that in certain cases vector mesons do indeed acquire mass when the vacuum is degenerate with respect to a compact Lie group.

Theories with degenerate vacuum (broken symmetry) have been the subject of intensive study since their inception by Nambu.[4-6] A characteristic feature of such theories is the possible existence of zero-mass bosons which tend to restore the symmetry.[7,8] We shall show that it is precisely these singularities which maintain the gauge invariance of the theory, despite the fact that the vector meson acquires mass.

We shall first treat the case where the original fields are a set of bosons φ_A which transform as a basis for a representation of a compact Lie group. This example should be considered as a rather general phenomenological model. As such, we shall not study the particular mechanism by which the symmetry is broken but simply assume that such a mechanism exists. A calculation performed in lowest order perturbation theory indicates that those vector mesons which are coupled to currents that "rotate" the original vacuum are the ones which acquire mass [see Eq. (6)].

We shall then examine a particular model based on chirality invariance which may have a more fundamental significance. Here we begin with a chirality-invariant Lagrangian and introduce both vector and pseudovector gauge fields, thereby guaranteeing invariance under both local phase and local γ_5-phase transformations. In this model the gauge fields themselves may break the γ_5 invariance leading to a mass for the original Fermi field. We shall show in this case that the pseudovector field acquires mass.

In the last paragraph we sketch a simple argument which renders these results reasonable.

(1) Lest the simplicity of the argument be shrouded in a cloud of indices, we first consider a one-parameter Abelian group, representing, for example, the phase transformation of a charged boson; we then present the generalization to an arbitrary compact Lie group.

The interaction between the φ and the A_μ fields is

$$H_{\text{int}} = ieA_\mu \varphi^* \overleftrightarrow{\partial}_\mu \varphi - e^2 \varphi^* \varphi A_\mu A_\mu, \qquad (1)$$

where $\varphi = (\varphi_1 + i\varphi_2)/\sqrt{2}$. We shall break the symmetry by fixing $\langle\varphi\rangle \neq 0$ in the vacuum, with the phase chosen for convenience such that $\langle\varphi\rangle = \langle\varphi^*\rangle = \langle\varphi_1\rangle/\sqrt{2}$.

We shall assume that the application of the

theorem of Goldstone, Salam, and Weinberg[7] is straightforward and thus that the propagator of the field φ_2, which is "orthogonal" to φ_1, has a pole at $q = 0$ which is not isolated.

We calculate the vacuum polarization loop $\Pi_{\mu\nu}$ for the field A_μ in lowest order perturbation theory about the self-consistent vacuum. We take into consideration only the broken-symmetry diagrams (Fig. 1). The conventional terms do not lead to a mass in this approximation if gauge invariance is carefully maintained. One evaluates directly

$$\Pi_{\mu\nu}(q) = (2\pi)^4 ie^2[g_{\mu\nu}\langle\varphi_1\rangle^2 - (q_\mu q_\nu/q^2)\langle\varphi_1\rangle^2]. \quad (2)$$

Here we have used for the propagator of φ_2 the value $[i/(2\pi)^4]/q^2$; the fact that the renormalization constant is 1 is consistent with our approximation.[9] We then note that Eq. (2) both maintains gauge invariance ($\Pi_{\mu\nu}q_\nu = 0$) and causes the A_μ field to acquire a mass

$$\mu^2 = e^2\langle\varphi_1\rangle^2. \quad (3)$$

We have not yet constructed a proof in arbitrary order; however, the similar appearance of higher order graphs leads one to surmise the general truth of the theorem.

Consider now, in general, a set of boson-field operators φ_A (which we may always choose to be Hermitian) and the associated Yang-Mills field $A_{a,\mu}$. The Lagrangian is invariant under the transformation[10]

$$\delta\varphi_A = \sum_{a,A}\epsilon_a(x)T_{a,AB}\varphi_B,$$

$$\delta A_{a,\mu} = \sum_{c,b}\epsilon_c(x)c_{acb}A_{b,\mu} + \partial_\mu\epsilon_a(x), \quad (4)$$

where c_{abc} are the structure constants of a compact Lie group and $T_{a,AB}$ the antisymmetric generators of the group in the representation defined by the φ_B.

Suppose that in the vacuum $\langle\varphi_{B'}\rangle \neq 0$ for some B'. Then the propagator of $\sum_{A,B'}T_{a,AB'}\varphi_A$

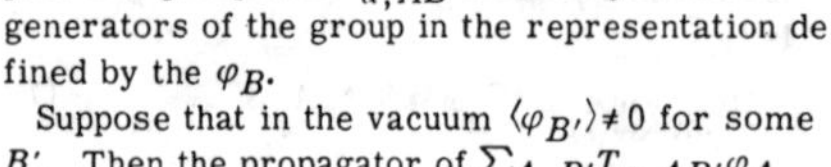

(a) (b)

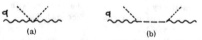

FIG. 1. Broken-symmetry diagram leading to a mass for the gauge field. Short-dashed line, $\langle\varphi_1\rangle$; long-dashed line, φ_2 propagator; wavy line, A_μ propagator. (a) $\to (2\pi)^4 ie^2 g_{\mu\nu}\langle\varphi_1\rangle^2$, (b) $\to -(2\pi)^4 ie^2(q_\mu q_\nu/q^2)\times\langle\varphi_1\rangle^2$.

$\times\langle\varphi_{B'}\rangle$ is, in the lowest order,

$$\left[\frac{i}{(2\pi)^4}\right]\sum_{A,B',C'}\frac{T_{a,AB'}\langle\varphi_{B'}\rangle T_{a,AC'}\langle\varphi_{C'}\rangle}{q^2}$$

$$\equiv\left[\frac{-i}{(2\pi)^4}\right]\frac{(\langle\varphi\rangle T_a T_a\langle\varphi\rangle)}{q^2}$$

With λ the coupling constant of the Yang-Mills field, the same calculation as before yields

$$\Pi_{\mu\nu}{}^a(q) = -i(2\pi)^4\lambda^2(\langle\varphi\rangle T_a T_a\langle\varphi\rangle)$$

$$\times[g_{\mu\nu} - q_\mu q_\nu/q^2],$$

giving a value for the mass

$$\mu_a{}^2 = -(\langle\varphi\rangle T_a T_a\langle\varphi\rangle). \quad (6)$$

(2) Consider the interaction Hamiltonian

$$H_{\text{int}} = -\eta\bar\psi\gamma_\mu\gamma_5\psi B_\mu - \epsilon\bar\psi\gamma_\mu\psi A_\mu, \quad (7)$$

where A_μ and B_μ are vector and pseudovector gauge fields. The vector field causes attraction whereas the pseudovector leads to repulsion between particle and antiparticle. For a suitable choice of ϵ and η there exists, as in Johnson's model,[11] a broken-symmetry solution corresponding to an arbitrary mass m for the ψ field fixing the scale of the problem. Thus the fermion propagator $S(p)$ is

$$S^{-1}(p) = \gamma p - \Sigma(p) = \gamma p[1-\Sigma_2(p^2)] - \Sigma_1(p^2), \quad (8)$$

with

$$\Sigma_1(p^2) \neq 0$$

and

$$m[1-\Sigma_2(m^2)] - \Sigma_1(m^2) = 0.$$

We define the gauage-invariant current $J_\mu{}^5$ by using Johnson's method[12]:

$$J_\mu{}^5 = -\eta\lim_{\xi\to 0}\bar\psi'(x+\xi)\gamma_\mu\gamma_5\psi'(x),$$

$$\psi'(x) = \exp[-i\textstyle\int_{-\infty}^x \eta B_\mu(y)dy^\mu\gamma_5]\psi(x). \quad (9)$$

This gives for the polarization tensor of the

pseudovector field

$$\Pi_{\mu\nu}{}^5(q) = \eta^2 \frac{i}{(2\pi)^4} \int \mathrm{Tr}\{S(p-\tfrac{1}{2}q)\Gamma_{\nu 5}(p-\tfrac{1}{2}q; p+\tfrac{1}{2}q) \times S(p+\tfrac{1}{2}q)\gamma_\mu \gamma_5 - S(p)[\partial S^{-1}(p)/\partial p_\nu] S(p)\gamma_\mu\} d^4p, \quad (10)$$

where the vertex function $\Gamma_{\nu 5} = \gamma_\nu \gamma_5 + \Lambda_{\nu 5}$ satisfies the Ward identity[5]

$$q_\nu \Lambda_{\nu 5}(p-\tfrac{1}{2}q; p+\tfrac{1}{2}q) = \Sigma(p-\tfrac{1}{2}q)\gamma_5 + \gamma_5 \Sigma(p+\tfrac{1}{2}q), \quad (11)$$

which for low q reads

$$q_\nu \Gamma_{\nu 5} = q_\nu \gamma_\nu \gamma_5 [1-\Sigma_2] + 2\Sigma_1 \gamma_5 - 2(q_\nu p_\nu)(\gamma_\lambda p_\lambda)(\partial \Sigma_2/\partial p^2)\gamma_5. \quad (12)$$

The singularity in the longitudinal $\Gamma_{\nu 5}$ vertex due to the broken-symmetry term $2\Sigma_1\gamma_5$ in the Ward identity leads to a nonvanishing gauge-invariant $\Pi_{\mu\nu}{}^5(q)$ in the limit $q \to 0$, while the usual spurious "photon mass" drops because of the second term in (10). The mass of the pseudovector field is roughly $\eta^2 m^2$ as can be checked by inserting into (10) the lowest approximation for $\Gamma_{\nu 5}$ consistant with the Ward identity.

Thus, in this case the general feature of the phenomenological boson system survives. We would like to emphasize that here the symmetry is broken through the gauge fields themselves. One might hope that such a feature is quite general and is possibly instrumental in the realization of Sakurai's program.[3]

(3) We present below a simple argument which indicates why the gauge vector field need not have zero mass in the presence of broken symmetry. Let us recall that these fields were introduced in the first place in order to extend the symmetry group to transformations which were different at various space-time points. Thus one expects that when the group transformations become homogeneous in space-time, that is $q \to 0$, no dynamical manifestation of these fields should appear. This means that it should cost no energy to create a Yang-Mills quantum at $q = 0$ and thus the mass is zero. However, if we break gauge invariance of the first kind and still maintain gauge invariance of the second kind this reasoning is obviously incorrect. Indeed, in Fig. 1, one sees that the A_μ propagator connects to intermediate states, which are "rotated" vacua. This is seen most clearly by writing $\langle\varphi_1\rangle = \langle[Q\varphi_2]\rangle$ where Q is the group generator. This effect cannot vanish in the limit $q \to 0$.

---

*This work has been supported in part by the U. S. Air Force under grant No. AFEOAR 63-51 and monitored by the European Office of Aerospace Research.

[1]J. Schwinger, Phys. Rev. 125, 397 (1962).

[2]C. N. Yang and R. L. Mills, Phys. Rev. 96, 191 (1954).

[3]J. J. Sakurai, Ann. Phys. (N.Y.) 11, 1 (1960).

[4]Y. Nambu, Phys. Rev. Letters 4, 380 (1960).

[5]Y. Nambu and G. Jona-Lasinio, Phys. Rev. 122, 345 (1961).

[6]"Broken symmetry" has been extensively discussed by various authors in the Proceedings of the Seminar on Unified Theories of Elementary Particles, University of Rochester, Rochester, New York, 1963 (unpublished).

[7]J. Goldstone, A. Salam, and S. Weinberg, Phys. Rev. 127, 965 (1962).

[8]S. A. Bludman and A. Klein, Phys. Rev. 131, 2364 (1963).

[9]A. Klein, reference 6.

[10]R. Utiyama, Phys. Rev. 101, 1597 (1956).

[11]K. A. Johnson, reference 6.

[12]K. A. Johnson, reference 6.

GLOBAL CONSERVATION LAWS AND MASSLESS PARTICLES*

G. S. Guralnik,† C. R. Hagen,‡ and T. W. B. Kibble
Department of Physics, Imperial College, London, England
(Received 12 October 1964)

In all of the fairly numerous attempts to date to formulate a consistent field theory possessing a broken symmetry, Goldstone's remarkable theorem[1] has played an important role. This theorem, briefly stated, asserts that if there exists a conserved operator Q_i such that

$$[Q_i, A_j(x)] = \sum_k t_{ijk} A_k(x),$$

and if it is possible consistently to take $\sum_k t_{ijk} \times \langle 0|A_k|0\rangle \neq 0$, then $A_j(x)$ has a zero-mass particle in its spectrum. It has more recently been observed that the assumed Lorentz invariance essential to the proof[2] may allow one the hope of avoiding such massless particles through the introduction of vector gauge fields and the consequent breakdown of manifest covariance.[3] This, of course, represents a departure from the assumptions of the theorem, and a limitation on its applicability which in no way reflects on the general validity of the proof.

In this note we shall show, within the framework of a simple soluble field theory, that it is possible consistently to break a symmetry (in the sense that $\sum_k t_{ijk}\langle 0|A_k|0\rangle \neq 0$) without requiring that $A(x)$ excite a zero-mass particle. While this result might suggest a general procedure for the elimination of unwanted massless bosons, it will be seen that this has been accomplished by giving up the global conservation law usually implied by invariance under a local gauge group. The consequent time dependence of the generators Q_i destroys the usual global operator rules of quantum field theory (while leaving the local algebra unchanged), in such a way as to preclude the possibility of applying the Goldstone theorem. It is clear that such a modification of the basic operator relations is a far more drastic step than that taken in the usual broken-symmetry theories in which a degenerate vacuum is the sole symmetry-breaking agent, and the operator algebra possesses the full symmetry. However, since superconductivity appears to display a similar behavior, the possibility of breaking such global conservation laws must not be lightly discarded.

Normally, the time independence of

$$Q_i = \int d^3x\, j_i^{\,0}(\vec{x},t)$$

is asserted to be a consequence of the local conservation law $\partial_\mu j^\mu = 0$. However, the relation

$$\partial_\mu \langle 0|[j_i^{\,\mu}(x), A_j(x')]|0\rangle = 0$$

implies that

$$\int d^3x \langle 0|[j_i^{\,0}(x), A_j(x')]|0\rangle = \text{const}$$

only if the contributions from spatial infinity vanish. This, of course, is always the case in a fully causal theory whose commutators vanish outside the light cone. If, however, the theory is not manifestly covariant (e.g., radiation-gauge electrodynamics), causality is a requirement which must be imposed with caution. Since Q_i consequently may not be time independent, it will not necessarily generate local gauge transformations upon $A_j(x')$ for $x^0 \neq x'^0$ despite the existence of the differential conservation laws $\partial_\mu j^\mu = 0$.

The phenomenon described here has previously been observed by Zumino[4] in the radiation-gauge formulation of two-dimensional electrodynamics where the usual electric charge cannot be conserved. The same effect is not present in the Lorentz gauge where zero-mass excitations which preserve charge conservation are found to occur. (These correspond to gauge parts rather than physical particles.) We shall, however, allow the possibility of the breakdown of such global conservation laws, and seek solutions of our model consistent only with the differential conservation laws.

We consider, as our example, a theory which was partially solved by Englert and Brout,[5] and bears some resemblance to the classical theory of Higgs.[6] Our starting point is the ordinary electrodynamics of massless spin-zero particles, characterized by the Lagrangian

$$\mathcal{L} = -\tfrac{1}{2}F^{\mu\nu}(\partial_\mu A_\nu - \partial_\nu A_\mu) + \tfrac{1}{4}F^{\mu\nu}F_{\mu\nu}$$
$$+ \varphi^\mu \partial_\mu \varphi + \tfrac{1}{2}\varphi^\mu \varphi_\mu + ie_0 \varphi^\mu q \varphi A_\mu,$$

where φ is a two-component Hermitian field, and q is the Pauli matrix σ_2. The broken-symmetry condition

$$ie_0 q\langle 0|\varphi|0\rangle = \eta \equiv \begin{pmatrix}\eta_1\\ \eta_2\end{pmatrix}$$

will be imposed by approximating $ie_0\varphi^\mu q\varphi A_\mu$ in the Lagrangian by $\varphi^\mu \eta A_\mu$. The resulting equations of motion,

$$F^{\mu\nu} = \partial^\mu A^\nu - \partial^\nu A^\mu,$$
$$\partial_\nu F^{\mu\nu} = \varphi^\mu \eta,$$
$$\varphi^\mu = -\partial^\mu \varphi - \eta A^\mu,$$
$$\partial_\mu \varphi^\mu = 0,$$

are essentially those of the Brout-Englert model, and can be solved in either the radiation[7] or Lorentz gauge. The Lorentz-gauge formulation, however, suffers from the fact that the usual canonical quantization is inconsistent with the field equations. (The quantization of A_μ leads to an indefinite metric for one component of φ.) Since we choose to view the theory as being imbedded as a linear approximation in the full theory of electrodynamics, these equations will have significance only in the radiation gauge.

With no loss of generality, we can take $\eta_2 = 0$, and find

$$(-\partial^2 + \eta_1^{\,2})\varphi_1 = 0,$$
$$-\partial^2 \varphi_2 = 0,$$
$$(-\partial^2 + \eta_1^{\,2})A_k^{\,T} = 0,$$

where the superscript T denotes the transverse part. The two degrees of freedom of $A_k^{\,T}$ combine with φ_1 to form the three components of a massive vector field. While one sees by inspection that there is a massless particle in the theory, it is easily seen that it is completely decoupled from the other (massive) excitations, and has nothing to do with the Goldstone theorem.

It is now straightforward to demonstrate the failure of the conservation law of electric charge. If there exists a conserved charge Q, then the relation expressing Q as the generator of rotations in charge space is

$$[Q, \varphi(x)] = e_0 q \varphi(x).$$

Our broken symmetry requirement is then

$$\langle 0 | [Q, \varphi_1(x)] | 0 \rangle = -i\eta$$

or, in terms of the soluble model considered here,

$$\int d^3x'\, \eta_1 \langle 0 | [\varphi_1^{\,0}(x'), \varphi_1(x)] | 0 \rangle = -i\eta_1.$$

From the result

$$\langle 0 | \varphi_1^{\,0}(x') \varphi_1(x) | 0 \rangle = \partial_0 \Delta^{(+)}(x'-x; \eta_1^{\,2}),$$

one is led to the consistency condition

$$\eta_1 \exp[-i\eta_1(x_0'-x_0)] = \eta_1,$$

which is clearly incompatible with a nontrivial η_1. Thus we have a direct demonstration of the failure of Q to perform its usual function as a conserved generator of rotations in charge space. It is well to mention here that this result not only does not contradict, but is actually required by, the field equations, which imply

$$(\partial_0^{\,2} + \eta_1^{\,2}) Q = 0.$$

It is also remarkable that if A_μ is given any bare mass, the entire theory becomes manifestly covariant, and Q is consequently conserved. Goldstone's theorem can therefore assert the existence of a massless particle. One indeed finds that in that case φ_1 has only zero-mass excitations.

In summary then, we have established that it may be possible consistently to break a symmetry by requiring that the vacuum expectation value of a field operator be nonvanishing without generating zero-mass particles. If the theory lacks manifest covariance it may happen that what should be the generators of the theory fail to be time-independent, despite the existence of a local conservation law. Thus the absence of massless bosons is a consequence of the inapplicability of Goldstone's theorem rather than a contradiction of it. Preliminary investigations indicate that superconductivity displays an analogous behavior.

The first named author wishes to thank Dr. W. Gilbert for an enlightening conversation, and two of us (G.S.G. and C.R.H.) thank Professor A. Salam for his hospitality.

---

*The research reported in this document has been sponsored in whole, or in part, by the Air Force Office of Scientific Research under Grant No. AF EOAR 64-46 through the European Office of Aerospace Research (OAR), U. S. Air Force.

†National Science Foundation Postdoctoral Fellow.

‡On leave of absence from the University of Rochester, Rochester, New York.

[1]J. Goldstone, Nuovo Cimento 19, 154 (1961); J. Goldstone, A. Salam, and S. Weinberg, Phys. Rev. 127, 965 (1962); S. A. Bludman and A. Klein, Phys. Rev. 131, 2364 (1963).

[2]W. Gilbert, Phys. Rev. Letters 12, 713 (1964).

[3]P. W. Higgs, Phys. Letters 12, 132 (1964).

[4]B. Zumino, Phys. Letters 10, 224 (1964).

[5]F. Englert and R. Brout, Phys. Rev. Letters 13, 321 (1964).

[6]P. W. Higgs, to be published.

[7]This is an extension of a model considered in more detail in another context by D. G. Boulware and W. Gilbert, Phys. Rev. 126, 1563 (1962).

Spontaneous Symmetry Breakdown without Massless Bosons*

PETER W. HIGGS†
Department of Physics, University of North Carolina, Chapel Hill, North Carolina
(Received 27 December 1965)

We examine a simple relativistic theory of two scalar fields, first discussed by Goldstone, in which as a result of spontaneous breakdown of $U(1)$ symmetry one of the scalar bosons is massless, in conformity with the Goldstone theorem. When the symmetry group of the Lagrangian is extended from global to local $U(1)$ transformations by the introduction of coupling with a vector gauge field, the Goldstone boson becomes the longitudinal state of a massive vector boson whose transverse states are the quanta of the transverse gauge field. A perturbative treatment of the model is developed in which the major features of these phenomena are present in zero order. Transition amplitudes for decay and scattering processes are evaluated in lowest order, and it is shown that they may be obtained more directly from an equivalent Lagrangian in which the original symmetry is no longer manifest. When the system is coupled to other systems in a $U(1)$ invariant Lagrangian, the other systems display an induced symmetry breakdown, associated with a partially conserved current which interacts with itself via the massive vector boson.

I. INTRODUCTION

THE idea that the apparently approximate nature of the internal symmetries of elementary-particle physics is the result of asymmetries in the stable solutions of exactly symmetric dynamical equations, rather than an indication of asymmetry in the dynamical equations themselves, is an attractive one. Within the framework of quantum field theory such a "spontaneous" breakdown of symmetry occurs if a Lagrangian, fully invariant under the internal symmetry group, has such a structure that the physical vacuum is a member of a set of (physically equivalent) states which transform according to a nontrivial representation of the group. This degeneracy of the vacuum permits nontrivial multiplets of scalar fields (which may be either fundamental dynamic variables or polynomials constructed from them) to have nonzero vacuum expectation values, whose appearance in Feynman diagrams leads to symmetry-breaking terms in propagators and vertices. That vacuum expectation values of scalar fields, or "vacuons," might play such a role in the breaking of symmetries was first noted by Schwinger[1] and by Salam and Ward.[2] Under the alternative name, "tadpole" diagrams, the graphs in which vacuons appear have been used by Coleman and Glashow[3] to account for the observed pattern of deviations from $SU(3)$ symmetry.

The study of field theoretical models which display spontaneous breakdown of symmetry under an internal Lie group was initiated by Nambu,[4] who had noticed[5] that the BCS theory of superconductivity[6] is of this type, and was continued by Glashow[7] and others.[8] All these authors encountered the difficulty that their theories predicted, *inter alia*, the existence of a number of massless scalar or pseudoscalar bosons, named "zerons" by Freund and Nambu.[9] Since the models which they discussed, being inspired by the BCS theory, used an attractive interaction between massless fermions and antifermions as the mechanism of symmetry breakdown, it was at first unclear whether zerons occurred as a result of the approximations (including the usual cutoff for divergent integrals) involved in handling the models or whether they would still be there in an exact solution. Some authors,

* This work was partially supported by the U. S. Air Force Office of Scientific Research under grant No. AF-AFOSR-153-64.

† On leave from the Tait Institute of Mathematical Physics, University of Edinburgh, Scotland.

[1] J. Schwinger, Phys. Rev. **104**, 1164 (1954); Ann. Phys. (N. Y.) **2**, 407 (1957).

[2] A. Salam and J. C. Ward, Phys. Rev. Letters **5**, 390 (1960); Nuovo Cimento **19**, 167 (1961).

[3] S. Coleman and S. L. Glashow, Phys. Rev. **134**, B671 (1964).

[4] Y. Nambu and G. Jona-Lasinio, Phys. Rev. **122**, 345 (1961); **124**, 246 (1961); Y. Nambu and P. Pascual, Nuovo Cimento **30**, 354 (1963).

[5] Y. Nambu, Phys. Rev. **117**, 648 (1960).

[6] J. Bardeen, L. N. Cooper, and J. R. Schrieffer, Phys. Rev. **106**, 162 (1957).

[7] M. Baker and S. L. Glashow, Phys. Rev. **128**, 2462 (1962); S. L. Glashow, *ibid.* 130, 2132 (1962).

[8] M. Suzuki, Progr. Theoret. Phys. (Kyoto) **30**, 138 (1963); **30**, 627 (1963); N. Byrne, C. Iddings, and E. Shrauner, Phys. Rev. **139**, B918 (1965); **139**, B933 (1965).

[9] P. G. O. Freund and Y. Nambu, Phys. Rev. Letters **13**, 221 (1964).

wishing to identify their zerons with known massive scalar or pseudoscalar mesons, were prepared to spoil the elegance of their theories by adding symmetry-breaking terms to the Lagrangian in order to generate masses.

That zerons must indeed be present in Lorentz invariant field theories in which an internal symmetry breaks down spontaneously was first shown by Goldstone.[10] He clarified the nature of the phenomenon considerably by exhibiting it in a model of a self-interacting scalar field, where the vacuon is the vacuum expectation value of the field itself, rather than that of a bilinear combination of fermion operators which occurs in the BCS model and its progeny. In a theory of this type the breakdown of symmetry occurs already at the level of classical field theory, where vacuons are just nontrivial translationally invariant solutions of the field equations, and zerons, whose existence is readily demonstrated, are small-amplitude waves (superimposed on a "vacuon" solution) whose frequency tends to zero as their wavelength tends to infinity. In a later paper[11] the proof of the Goldstone theorem, as it is now known, was generalized to allow for the possibility that vacuons might be formed from polynomials of any degree in the fundamental field variables of a dynamical system.

During the last few years the problem of how to avoid massless Goldstone bosons has received much attention. Attempts in this direction have been encouraged by the observation that the BCS model does not contain such excitations, provided that Coulomb interactions are taken into account.[12] Klein and Lee[13] showed that in a nonrelativistic theory the spectral representations upon which the more sophisticated proofs of Goldstone's theorem are based are not so restricted in form as to allow the proof to go through, and they conjectured that this might remain true in some relativistic theories. But Gilbert[14] pointed out that their extra terms are ruled out in relativistic theories by the requirement of manifest Lorentz covariance. The present writer restored the status quo to a limited extent by remarking[15] that radiation gauge formulations of gauge field theories, of which electrodynamics is the simplest and best known example, can describe Lorentz-invariant dynamical systems despite the lack of manifest covariance of some of the equations. The freedom which Klein and Lee hoped for in the spectral representations is thereby restored sufficiently to invalidate the Goldstone theorem. From a more physical standpoint one may regard this as an effect of Coulomb interactions, treated now as part of a relativistic field theory.

[10] J. Goldstone, Nuovo Cimento **19**, 154 (1961).
[11] J. Goldstone, A. Salam, and S. Weinberg, Phys. Rev. **127**, 965 (1962).
[12] P. W. Anderson, Phys. Rev. **112**, 1900 (1958).
[13] A. Klein and B. W. Lee, Phys. Rev. Letters **12**, 266 (1964).
[14] W. Gilbert, Phys. Rev. Letters **12**, 713 (1964).
[15] P. W. Higgs, Phys. Letters **12**, 132 (1964).

More recently Guralnik, Hagen, and Kibble[16] and Lange[17] have studied how the failure of global (as distinct from local) conservation laws in spontaneous breakdown theories is related to the existence of Goldstone bosons. Meanwhile, proofs of the theorem have reached new levels of sophistication within the language (or languages) of axiomatic field theory.[18] It has been pointed out that theories of the type proposed in Ref. 15 do not contradict the Goldstone theorem, but rather represent a departure from the assumptions, such as *manifest* covariance and *manifest* causality, upon which it is based. Such considerations do not seem relevant to the possible usefulness of such theories in generating zeron-free models of spontaneous symmetry breakdown, a point which seems to have been overlooked by those[19] who proclaim the failure of the Nambu-Glashow program.

In parallel with the development of "superconductor" models a program has emerged for describing the weak,[20] and possibly also the strong,[21] interactions by an extension of the gauge principle[22] which operates in electrodynamics. According to this principle the symmetry group of the Lagrangian is to be enlarged from global to local (i.e., coordinate-dependent) transformations: To maintain the invariance of derivative terms it is necessary to couple the currents of the group generators to a multiplet of vector fields belonging to the adjoint representation.[23] Like the "superconductor" theories, these gauge theories have suffered from a zero-mass difficulty: The gauge principle appears to guarantee that the associated vector field quanta are massless, in

[16] G. S. Guralnik, C. R. Hagen, and T. W. B. Kibble, Phys. Rev. Letters **13**, 585 (1964). These authors appear to attribute the failure of a local conservation law to yield a global conservation law to the lack of manifest covariance of a theory. In fact this happens even in manifestly covariant models of spontaneous breakdown.
[17] R. V. Lange, Phys. Rev. Letters **14**, 3 (1965).
[18] R. F. Streater, Proc. Roy. Soc. (London) **A287**, 510 (1965). The proof of the Goldstone theorem given here is based on axioms which include manifest causality. Radiation gauge theories escape *this* version of the theorem by virtue of their (quite innocuous) acausality. The question of the extent to which one may give up manifest covariance and causality in a theory without losing covariance and causality of the physics which it describes deserves further study. If there are contexts in which it is possible other than the gauge theories which we are discussing, then there are probably other escape routes from the Goldstone theorem. See also D. Kastler, D. W. Robinson, and A. Swieca, Commun. Math. Phys. (to be published).
[19] R. F. Streater, Phys. Rev. Letters **15**, 475 (1965). The generalized Goldstone theorem proved by this author and extended by N. Fuchs, Phys. Rev. Letters **15**, 911 (1965) also relies on manifest causality and is therefore inapplicable to gauge theories.
[20] S. A. Bludman, Phys. Rev. **100**, 372 (1955); Nuovo Cimento **9**, 433 (1958); S. L. Glashow, Nucl. Phys. **10**, 107 (1959); **22**, 579 (1961).
[21] A. Salam and J. C. Ward, Nuovo Cimento **11**, 568 (1959); Phys. Rev. **136**, B763 (1964); J. J. Sakurai, Ann. Phys. (N. Y.) **11**, 1 (1960).
[22] C. N. Yang and R. L. Mills, Phys. Rev. **96**, 191 (1954); R. Shaw, dissertation, Cambridge University, 1954 (unpublished); R. Utiyama, Phys. Rev. **101**, 1597 (1956).
[23] M. Gell-Mann and S. L. Glashow, Ann. Phys. (N. Y.) **15**, 437 (1961).

perturbation theory at least. But the only known massless vector boson is the photon; the existing evidence suggests[24] that all other vector bosons must be massive. In particular, the hypothetical intermediate vector bosons of weak interactions, which in a gauge theory belong to the gauge field multiplet, must be much heavier than the known hadrons. For the most part, advocates of gauge theories have met this difficulty either by spoiling the gauge invariance of their theories with explicit mass terms or by taking comfort from the argument of Schwinger[25] that a sufficiently strong gauge-field coupling might generate mass. Recently, however, it was shown by Englert and Brout[26] and by the present writer[27] that gauge vector mesons acquire mass if the symmetry to whose generators they are coupled breaks down spontaneously, however weak their coupling may be. In Ref. 27 this phenomenon was exhibited in a classical field theory, and it was pointed out that in such a theory the longitudinal polarization of the massive vector excitation replaces the massless scalar excitation which would occur in the absence of coupling to the gauge field. Thus it now appears that the spontaneous breakdown program of Nambu *et al.* and the gauge field program of Salam *et al.* stand or fall together. Each saves the other from its zero-mass difficulty.

The purpose of the present paper is to amplify and substantiate the assertions made in Refs. 15 and 27 by displaying the behavior of the simplest possible relativistic field theory which combines spontaneous breakdown of symmetry under a compact Lie group with the gauge principle. That is to say, we take the symmetry group to be the trivial Abelian group $U(1)$, we take as the fundamental dynamic variables a pair of Hermitian scalar fields $\Phi_1(x)$, $\Phi_2(x)$ together with the Hermitian vector gauge field $A_\mu(x)$, and we induce spontaneous breakdown by means of the simplest $U(1)$-invariant self-interaction of $\Phi_a(x)$ which will do the trick, namely, a combination of quadratic and quartic terms. In the absence of the gauge field coupling, the model is just one which Goldstone[10] first discussed.[28]

In Sec. II the behavior of the small-amplitude wave solutions of the classical field equations is used as a guide in formulating a perturbation theory in which the major effects of spontaneous breakdown are already taken into account in zero order. The radiation gauge commutators of the zero-order approximation are obtained and used to provide an explicit realization of the spectral representation which was proposed in Ref. 15. In Sec. III decay and scattering amplitudes are calculated in lowest order and it is verified that they are gauge-invariant, Lorentz-invariant, and causal despite the lack of manifest covariance and causality of the radiation gauge. In Sec. IV it is shown that the same amplitudes may be derived by a manifestly covariant and causal method from an equivalent Lagrangian which lacks the original symmetry. Finally, our conclusions are summarized in Sec. V, and the way in which coupling between a system of the kind described here and other symmetric dynamical systems may lead to partially conserved currents is sketched.

In subsequent papers we propose to elaborate these considerations, both with regard to symmetry groups and with regard to mechanisms of symmetry breakdown, so as to make closer contact with particle physics.

II. THE MODEL

The Lagrangian density from which we shall work is given by[29]

$$\mathcal{L}=-\tfrac{1}{4}g^{\kappa\mu}g^{\lambda\nu}F_{\kappa\lambda}F_{\mu\nu}-\tfrac{1}{2}g^{\mu\nu}\nabla_\mu\Phi_a\nabla_\nu\Phi_a +\tfrac{1}{4}m_0^2\Phi_a\Phi_a-\tfrac{1}{8}f^2(\Phi_a\Phi_a)^2. \quad (1)$$

In Eq. (1) the metric tensor $g^{\mu\nu}=-1$ $(\mu=\nu=0)$, $+1$ $(\mu=\nu\neq 0)$ or 0 $(\mu\neq\nu)$, Greek indices run from 0 to 3 and Latin indices from 1 to 2. The $U(1)$-covariant derivatives $F_{\mu\nu}$ and $\nabla_\mu\Phi_a$ are given by

$$F_{\mu\nu}=\partial_\mu A_\nu-\partial_\nu A_\mu,$$
$$\nabla_\mu\Phi_1=\partial_\mu\Phi_1-eA_\mu\Phi_2,$$
$$\nabla_\mu\Phi_2=\partial_\mu\Phi_2+eA_\mu\Phi_1.$$

At first sight this theory appears to be scalar electrodynamics augmented by a quartic self-interaction. However, what appears to be the bare-mass term has the wrong sign. In conjunction with the quartic term this feature has the consequence that the field equations

$$\partial_\nu F^{\mu\nu}=ej^\mu,\quad j_\mu=\Phi_2\nabla_\mu\Phi_1-\Phi_1\nabla_\mu\Phi_2, \qquad \nabla_\mu\nabla^\mu\Phi_a+\tfrac{1}{2}(m_0^2-f^2\Phi_b\Phi_b)\Phi_a=0, \quad (2)$$

in the classical theory possess a coordinate-independent solution

$$A_\mu=0,\quad \Phi_b\Phi_b=m_0^2/f^2. \quad (3)$$

The invariance of the Lagrangian (1) under the local $U(1)$ transformations

$$A_\mu(x)\to A_\mu(x)+e^{-1}\partial_\mu\Lambda(x),$$
$$\Phi_1(x)\to\Phi_1(x)\cos\Lambda(x)+\Phi_2(x)\sin\Lambda(x), \quad (4)$$
$$\Phi_2(x)\to-\Phi_1(x)\sin\Lambda(x)+\Phi_2(x)\cos\Lambda(x),$$

is reflected in the existence of a one-parameter family of static solutions defined by Eq. (3).[30] In the classical

[24] See, for example, S. Weinberg, Phys. Rev. Letters 13, 495 (1964).
[25] J. Schwinger, Phys. Rev. 125, 397 (1962); 128, 2425 (1962).
[26] F. Englert and R. Brout, Phys. Rev. Letters 13, 321 (1964).
[27] P. W. Higgs, Phys. Rev. Letters 13, 508 (1964).
[28] I understand from Dr. Goldstone (private communication) that he and W. Gilbert at one time considered adding a gauge field to the model.

[29] We do not explicitly perform the symmetrizations which a correct quantum-mechanical treatment would demand. They are not necessary for the purposes of the present paper and would, in any case, be dealt with more satisfactorily in a first-order formalism.
[30] Strictly speaking, global $U(1)$ invariance suffices to guarantee this result.

theory, any one of these solutions,

$$\Phi_1=\eta\cos\alpha,\quad \Phi_2=\eta\sin\alpha,$$

where $\eta=m_0/f$, defines a possible asymmetric configuration of stable equilibrium, stability being ensured by the sign of the quartic term in Eq. (1). Quantum mechanically each solution, regarded as the "bare" value of the vacuon $\langle\Phi_a\rangle$, corresponds to a different possible vacuum state.[31]

Let us choose $\alpha=\pi/2$ and linearize the classical field equations (2) by treating A_μ, Φ_1, and $\Phi_2-\eta$ as small quantities. We obtain

$$\partial_\nu F^{\mu\nu}=-e^2\eta^2B^\mu,\quad \partial_\mu B^\mu=0,$$
$$(\Box-m_0{}^2)\chi=0,$$

in which we have introduced the notation

$$B_\mu=A_\mu-(e\eta)^{-1}\partial_\mu\Phi,$$
$$\Phi=\Phi_1,\quad \chi=\Phi_2-\eta. \tag{5}$$

As we remarked in Ref. 27, these are the linear field equations which, after quantization, describe free vector bosons of mass $e\eta$ and free scalar bosons of mass m_0. The longitudinal vector excitation becomes the Goldstone scalar excitation in the limit $e\rightarrow 0$. The Lagrangian to which these field equations belong is given by

$$\mathcal{L}_0=-\tfrac{1}{4}F_{\mu\nu}F^{\mu\nu}-\tfrac{1}{2}g^{\mu\nu}(\partial_\mu\Phi-m_1A_\mu)(\partial_\nu\Phi-m_1A_\nu)$$
$$-\tfrac{1}{2}g^{\mu\nu}\partial_\mu\chi\partial_\nu\chi-\tfrac{1}{2}m_0{}^2\chi^2, \tag{6}$$

where we have written m_1 for the vector boson mass $e\eta$.

The foregoing analysis of classical small-amplitude wave propagation suggests the following perturbation theoretic treatment of the quantized theory. We rewrite Eq. (1) in the form $\mathcal{L}=\mathcal{L}_0+\mathcal{L}_{\rm int}$, where $\mathcal{L}_0$ is given by Eq. (6) apart from a trivial additive constant and

$$\mathcal{L}_{\rm int}=eA^\mu(\chi\partial_\mu\Phi-\Phi\partial_\mu\chi)-em_1\chi A_\mu A^\mu-\tfrac{1}{2}fm_0\chi(\Phi^2+\chi^2)$$
$$-\tfrac{1}{2}e^2A_\mu A^\mu(\Phi^2+\chi^2)-\tfrac{1}{8}f^2(\Phi^2+\chi^2)^2. \tag{7}$$

Note that the ancestry of (6) plus (7) in the symmetric Lagrangian (1) is embodied in the relation $m_1/m_0=e/f$ between the bare masses and the bare coupling constants. Our perturbation theory now consists in developing transition amplitudes in powers of e and f (or, equivalently, in inverse powers of η), the masses m_0 and m_1 being treated as of order zero.[32] Thus in Eq. (7) all five cubic vertices are of the first degree and all five quartic vertices are of the second in the expansion parameter. It will be found that, with few exceptions, gauge-invariant results are obtained only when all Feynman graphs of the same degree are summed.

We first write down the commutators and propagators corresponding to the bare Lagrangian $\mathcal{L}_0$. Apart from the terms in χ, this is just the second-order Lagrangian of a model proposed by Boulware and Gilbert[33] as an illustration of the possibility of a gauge-invariant theory describing a massive vector boson. We shall study it in a radiation gauge; the Lorentz gauge formulation, which even in quantum electrodynamics leads to unnecessary complications such as redundant states, is here inconsistent with the canonical com, mutation rules, as was pointed out by Guralnik, Hagen-and Kibble.[16] In a radiation gauge defined by the condition

$$(\partial A)+(n\partial)(nA)=0,$$

where n^μ is a constant timelike unit vector and (ab) denotes $a_\mu b^\mu$, the variables A_μ and Φ may be expressed in terms of the massive vector field B_μ which was introduced in Eq. (5):

$$A_\mu=B_\mu+m_1{}^{-1}\partial_\mu\Phi,$$
$$\Phi=-m_1[(\partial^2)+(n\partial)^2]^{-1}[(\partial B)+(n\partial)(nB)]. \tag{8}$$

Since $\mathcal{L}_0$, when expressed in terms of B_μ and χ, is just the usual second order Lagrangian for free vector and scalar bosons, we may immediately write down the covariant commutators:

$$[B_\mu(x),B_\nu(y)]=-i(g_{\mu\nu}-m_1{}^{-2}\partial_\mu\partial_\nu)\Delta(x-y,\,m_1{}^2),$$
$$[\chi(x),\chi(y)]=-i\Delta(x-y,\,m_0{}^2), \tag{9}$$

where $\Delta(x,m^2)=i(2\pi)^{-3}\int d^4ke^{i(kx)}\epsilon(k^0)\delta(k^2+m^2)$. Then Eq. (8) enables us to deduce the nonvanishing commutators of A_μ, Φ, and χ:

$$[A_\mu(x),A_\nu(y)]=-i\{g_{\mu\nu}-[(n_\mu\partial_\nu+n_\nu\partial_\mu)(n\partial)+\partial_\mu\partial_\nu]$$
$$\times[(\partial^2)+(n\partial)^2]^{-1}\}\Delta(x-y,\,m_1{}^2),$$
$$[A_\mu(x),\Phi(y)]=-im_1n_\mu(n\partial)$$
$$\times[(\partial^2)+(n\partial)^2]^{-1}\Delta(x-y,\,m_1{}^2), \tag{10}$$
$$[\Phi(x),\Phi(y)]=-i(n\partial)^2[(\partial^2)+(n\partial)^2]^{-1}\Delta(x-y,\,m_1{}^2),$$
$$[\chi(x),\chi(y)]=-i\Delta(x-y,\,m_0{}^2).$$

We also note the commutator relation

$$[B_\mu(x),\Phi(y)]=-im_1{}^{-1}[(\partial^2)n_\mu-(n\partial)\partial_\mu](n\partial)$$
$$\times[(\partial^2)+(n\partial)^2]^{-1}\Delta(x-y,\,m_1{}^2). \tag{11}$$

[31] The orthogonality of the worlds built upon different vacua may be understood as a consequence of the impossibility in the classical theory of a displacement of the system from one static configuration to another, on account of the infinite inertia associated with such a motion. To see this, imagine a one-dimensional model consisting of an infinite uniform elastic string subjected to a force field of cylindrical symmetry about the axis from which transverse displacements Φ_1, Φ_2 are measured. (We omit the gauge field from this do-it-yourself model.) If the force is such that stable equilibrium occurs when the whole string is at a distance η from the axis in any direction, then the system exhibits broken rotational symmetry. Displacement of the string from equilibrium at one orientation to equilibrium at another is impossible, since the moment of inertia about the axis is infinite. Waves on the string do not conserve angular momentum about the axis, since the string as a whole can emit or absorb angular momentum without recoiling.

[32] When one comes to consider radiative corrections, it becomes necessary to make these statements about the renormalized rather than the bare masses and coupling constants.

[33] D. C. Boulware and W. Gilbert, Phys. Rev. 126, 1563 (1962).

The generator $Q(t)$ of infinitesimal *global* $U(1)$ transformations (that is, transformations (4) with Λ constant and infinitesimal) on the hypersurface $(nx)+t=0$ is $\int d\sigma_\mu j^\mu$, where $d\sigma_\mu$ is the volume element of the hypersurface and $j_\mu(x)$ is given by Eq. (2). The invariance of the Lagrangian (1) under these transformations leads to the local conservation law, $\partial_\mu j^\mu=0$. However, even in the absence of the gauge field coupling, the four-dimensional integral of this equation fails to yield the usual global conservation law, $Q(t)=\text{constant}$, since the flux of j_μ across the surface of a large sphere bounding the hypersurface does not tend to zero as the radius tends to infinity. That this is so can be seen by noting that the lowest order approximation to j_μ is $-e\eta^2B_\mu$ (or $\eta\partial_\mu\Phi$ in the absence of the gauge field): Matrix elements of this operator do not decrease sufficiently rapidly at large spatial distances for the flux term to vanish. (In normal theories the lowest order term in j_μ is quadratic, giving a better asymptotic behavior of the matrix elements.) Strictly speaking, the "operator" $Q(t)$ is now not merely time-dependent but nonexistent, since $\int d\sigma_\mu j^\mu$ diverges as a result of the same bad asymptotic behavior of j^μ. However, certain commutators, such as $[Q(t),\Phi_a(y)]$, do still exist.[34]

The zero-order approximation to the commutator vacuum expectation value $\langle i[j_\mu(x),\Phi_1(y)]\rangle$, upon which so much of the discussion of the Goldstone theorem has centered,[35] is found by replacing j_μ by $-e\eta^2B_\mu$ and using Eq. (11): It is

$$\eta[(n\partial)\partial_\mu-(\partial^2)n_\mu](n\partial)[(\partial^2)+(n\partial)^2]^{-1}\Delta(x-y,m_1^2).$$

Its Fourier transform,

$$-2\pi\eta[(nk)k_\mu-(k^2)n_\mu](nk)\times[(k^2)+(nk)^2]^{-1}\epsilon(k^0)\delta(k^2+m_1^2), \quad (12)$$

provides an explicit realization of a spectral representation of the form obtained in Ref. 15, the Lorentz invariance of the spectrum of intermediate states now being made clear. We are led to conjecture that the vacuum expectation value of the exact commutator may be of the form

$$\langle\Phi_2\rangle[(n\partial)\partial_\mu-(\partial^2)n_\mu](n\partial)[(\partial^2)+(n\partial)^2]^{-1}\times\int_0^\infty dm^2\rho(m^2)\Delta(x-y,m^2), \quad (13)$$

where $\rho(m^2)$ is a nonnegative spectral function satisfying the sum rule

$$\int_0^\infty dm^2\rho(m^2)=1.$$

It may be noted that when $m_1=0$ in Eq. (12), corresponding to $e=0$, we recover the manifestly covariant spectral representation $-2\pi\eta k_\mu\epsilon(k^0)\delta(k^2)$ and with it the Goldstone theorem.

We define the propagators of the system described by $\mathcal{L}_0$ to be the quantities $\langle T^*A_\mu(x)A_\nu(y)\rangle$, etc., obtained from the corresponding commutators in Eq. (10) by substituting for Δ the scalar propagator Δ_F given by

$$\Delta_F(x,m^2)=(2\pi)^{-4}\int d^4k\, e^{i(kx)}(k^2+m^2-i\epsilon)^{-1}.$$

Then we may calculate S-matrix elements by using the simple Feynman rules based on the Nishijima-Wick expansion of the expression $T^*\exp(i\int d^4x\,\mathcal{L}_{\text{int}})$ for the S-operator in the interaction picture.[36] We thereby avoid the n_μ-dependent terms, in addition to those already introduced by the radiation gauge, which the use of simple chronological ordering and the Dyson-Wick expansion would entail.

III. DECAY AND SCATTERING AMPLITUDES

As an illustration of the physical content of the model we now calculate in lowest order the matrix elements for the simplest processes which it describes. We shall verify that, despite the unpromising appearance of the radiation gauge propagators, these matrix elements are gauge invariant and Lorentz invariant.

In applying the Feynman rules we shall need, in addition to the propagators, the wave functions $a_\mu(k,\sigma)$ and $\phi(k,\sigma)$ which correspond to the annihilation by the operators A_μ and Φ, respectively, of a vector meson from an incoming state of momentum k and spin component σ. They are related by the Fourier transform of Eq. (8) to the usual vector meson wave function $b_\mu(k,\sigma)$, which has the explicit form

$$b^\mu(k,0)=(\omega/m_1)(|\mathbf{k}|/\omega,\mathbf{k}/|\mathbf{k}|),\qquad b^\mu(k,\pm1)=2^{-1/2}(0,\mathbf{e}_1\pm i\mathbf{e}_2), \quad (14)$$

where $\omega=(|\mathbf{k}|^2+m_1^2)^{1/2}$ and $\mathbf{e}_1$, $\mathbf{e}_2$, $\mathbf{k}/|\mathbf{k}|$ form a right-handed orthonormal triad. Actually, all that we shall need is the relation

$$a_\mu=b_\mu+(ik_\mu/m_1)\phi, \quad (15)$$

by which matrix elements may be expressed in terms of wave functions b_μ and ϕ, the desired invariance being achieved by the cancellation of all terms containing factors ϕ. Similar considerations apply to out-

[34] In Ref. 18 it is proved that in a manifestly causal theory this commutator (or at least certain of its matrix elements) is independent of t, despite the breakdown of the global conservation law. The gauge field coupling destroys manifest causality and induces a time dependence in this commutator: In the zero-order approximation it oscillates at a frequency m_1.

[35] See Refs. 11, 13, 14, and 15. In Refs. 14 and 15 it is implied erroneously that the commutator $[Q(t),\Phi_a(y)]$ is independent of t. Fortunately, the discussion of the Goldstone theorem in these papers does not depend on this assumption.

[36] P. T. Matthews, Phys. Rev. 76, 684 (1949); K. Nishijima, Progr. Theoret. Phys. (Kyoto) 5, 405 (1950). The most general conditions for the validity of this expression have been stated by C. S. Lam, Nuovo Cimento 38, 1755 (1965).

going states and associated complex conjugate wave functions.

i. Decay of a Scalar Boson into Two Vector Bosons

The process occurs in first order (four of the five cubic vertices contribute), provided that $m_0 > 2m_1$. Let p be the incoming and k_1, k_2 the outgoing momenta. Then

$$M = i\{e[a^{*\mu}(k_1)(-ik_{2\mu})\phi^*(k_2) + a^{*\mu}(k_2)(-ik_{1\mu})\phi^*(k_1)] - e(ip_\mu)[a^{*\mu}(k_1)\phi^*(k_2) + a^{*\mu}(k_2)\phi^*(k_1)] - 2em_1 a_\mu^*(k_1)a^{*\mu}(k_2) - fm_0\phi^*(k_1)\phi^*(k_2)\}.$$

By using Eq. (15), conservation of momentum, and the transversality ($k_\mu b^\mu(k) = 0$) of the vector wave functions we reduce this to the form

$$M = -2iem_1 b^{*\mu}(k_1)b_\mu^*(k_2) - iem_1^{-1}(p^2 + m_0^2)\phi^*(k_1)\phi^*(k_2). \quad (16)$$

We have retained the last term, which we shall need in calculating scattering amplitudes; when the incident particle is on the mass shell it vanishes and we are left with the invariant expression

$$M = -2iem_1 b^{*\mu}(k_1)b_\mu^*(k_2). \quad (17)$$

Conservation of angular momentum allows three possibilities for the spin states of the decay products: They may be both right-handed, both left-handed, or both longitudinal ($\sigma_1 = \sigma_2 = +1, -1$, or 0). With the help of the explicit vectors (14), we find

$$M(+1, +1) = M(-1, -1) = 2iem_1,$$
$$M(0,0) = ifm_0(1 - 2e^2/f^2).$$

We note that as $e \to 0$ the amplitudes for decay to transverse states tend to zero, but the amplitude $M(0,0)$ tends to the value ifm_0 which we would calculate from the vertex $-\frac{1}{2}fm_0\Phi^2 X$ for the decay of one massive into two massless scalar bosons in the original Goldstone model. (The sign change arises from the factor i which is associated with the term ϕ in each b_μ.)

ii. Vector Boson-Vector Boson Scattering

Let k_1, k_2 be the incoming and k_1', k_2' the outgoing momenta. The process occurs as a second-order effect of the cubic vertices, by exchange of a scalar boson in the s, t, or u channel, where $s = -(p_1 + p_2)^2$, $t = -(p_1 - p_1')^2$, $u = -(p_1 - p_2')^2$. It also occurs as a direct effect of two of the quartic vertices. Equation (16) enables us to write down

$$M_s = i^2\{-2em_1 b_\mu^*(k_1')b^{*\mu}(k_2') + em_1^{-1}(s - m_0^2)\phi^*(k_1')\phi^*(k_2')\} \times i(s - m_0^2)^{-1}\{-2em_1 b_\nu(k_1)b^\nu(k_2) + em_1^{-1}(s - m_0^2)\phi(k_1)\phi(k_2)\}$$

and similar expressions for M_t and M_u. The quartic vertices yield a contribution given by

$$\begin{aligned} M_{\text{direct}} &= i(-2e^2)\{a_\mu^*(k_1')a^{*\mu}(k_2')\phi(k_1)\phi(k_2) + 5 \text{ similar terms}\} + i(-3f^2)\phi^*(k_1')\phi^*(k_2')\phi(k_1)\phi(k_2) \\ &= -2ie^2\{b_\mu^*(k_1')b^{*\mu}(k_2')\phi(k_1)\phi(k_2) + 5 \text{ similar terms}\} + i(4e^2 - 3f^2)\phi^*(k_1')\phi^*(k_2')\phi(k_1)\phi(k_2). \end{aligned}$$

It is only when we combine these four contributions that we obtain (after some algebra) the invariant expression

$$\begin{aligned} M_{\text{total}} &= M_s + M_t + M_u + M_{\text{direct}} \\ &= -4ie^2 m_1^2\{(s - m_0^2)^{-1}b^{*\mu}(k_1')b^{*\mu}(k_2')b_\nu(k_1)b^\nu(k_2) \\ &\quad + (t - m_0^2)^{-1}b_\mu^*(k_1')b^\mu(k_1)b_\nu^*(k_2')b^\nu(k_2) \\ &\quad + (u - m_0^2)^{-1}b_\mu^*(k_1')b^\mu(k_2)b_\nu^*(k_2')b^\nu(k_1)\}. \end{aligned} \quad (18)$$

iii. Vector Boson-Scalar Boson Scattering

Let k, p be the momenta of the incoming vector and scalar boson, respectively, and k', p' be their outgoing momenta. Again there are four contributions, M_s, M_t, M_u, and M_{direct}. In the s and u channels a vector boson is exchanged and it turns out that the various propagators, $\langle T^*A_\mu A_\nu\rangle$, $\langle T^*A_\mu\Phi\rangle$, and $\langle T^*\Phi\Phi\rangle$, occur only in the combination $\langle T^*B_\mu B_\nu\rangle$. We obtain the expression

$$M_s = i^2\{-2em_1 b^{*\mu}(k') + ieq^\mu\phi^*(k')\}i(g_{\mu\nu} + m_1^{-2}q_\mu q_\nu) \times (s - m_1^2)^{-1}\{-2em_1 b^\nu(k) - ieq^\nu\phi(k)\},$$

where $q = k + p$ and $s = -q^2$, and a similar expression for M_u. In the t channel a scalar boson is exchanged, and we find that

$$M_t = i^2\{-3fm_0\}i(t - m_0^2)^{-1}\{-2em_1 b_\mu^*(k')b^\mu(k) + em_1^{-1}(t - m_0^2)\phi^*(k')\phi(k)\},$$

where $t = -(k - k')^2$. Finally, the contribution of the quartic vertices is given by

$$M_{\text{direct}} = i\{-2e^2[b_\mu^*(k') - im_1^{-1}k_\mu'\phi^*(k')] \times [b^\mu(k) + im_1^{-1}k^\mu\phi(k)] - f^2\phi^*(k')\phi(k)\}.$$

Again the four contributions sum to the invariant expression

$$\begin{aligned} M_{\text{total}} &= -2im_1^2\{2e^2(s - m_1^2)^{-1}[b_\mu^*(k')b^\mu(k) + m_1^{-2}p_\mu' b^{*\mu}(k')p_\nu b^\nu(k)] \\ &\quad + 3f^2(t - m_0^2)^{-1}b_\mu^*(k')b^\mu(k) \\ &\quad + 2e^2(u - m_1^2)^{-1}[b_\mu^*(k')b^\mu(k) + m_1^{-2}p_\mu b^{*\mu}(k')p_\nu' b^\nu(k)]\} \\ &\quad - 2ie^2 b_\mu^*(k')b^\mu(k). \end{aligned} \quad (19)$$

A similar matrix element may be written down for the process, vector pair $\leftrightarrow$ scalar pair, by making appropriate interchanges of incoming and outgoing momenta and wave functions.

iv. Scalar Boson-Scalar Boson Scattering

This is the only simple process in which no invariance problems arise in lowest order: The vertices which are involved contain no vector boson factors. We find that

$$M_{\text{total}}=M_s+M_t+M_u+M_{\text{direct}}$$
$$=-9if^2m_0^2\{(s-m_0^2)^{-1}+(t-m_0^2)^{-1}$$
$$+(u-m_0^2)^{-1}+(3m_0^2)^{-1}\}. \quad (20)$$

IV. EQUIVALENT LAGRANGIAN

In the previous section we have illustrated the Lorentz and gauge invariance of the model by the somewhat unsophisticated device of performing a few lowest order calculations. From a more sophisticated point of view we remark that Lorentz invariance may be proved by constructing the generators of the Lorentz group and verifying their commutation relations. Provided that the Lagrangian (1) is first properly symmetrized, the proof goes through as in quantum electrodynamics[37]; spontaneous breakdown of the internal symmetry is irrelevant to the argument, which depends only on the equal-time commutators of products of field operators.

Concerning gauge invariance we remark that our result, that (in lowest order at least) decay and scattering amplitudes depend only on the gauge-invariant vector wave functions $b_\mu(k,\sigma)$, suggests that it must be possible to rewrite the theory in a form in which only gauge-invariant variables appear. Indeed, if one were shown only the expressions (17)-(20), he would guess that they had been derived from an interaction Lagrangian given by

$$\mathcal{L}_{\text{int}}'=-em_1B_\mu B^\mu\chi-\tfrac{1}{2}fm_0\chi^3-\tfrac{1}{2}e^2B_\mu B^\mu\chi^2-\tfrac{1}{8}f^2\chi^4. \quad (21)$$

We shall now show that the expressions (7) and (21) are equivalent by finding a transformation of the total Lagrangian (1) which takes the one into the other.

We note that the gauge dependence of the classical dynamic variables may be expressed in the form

$$\Phi_1(x)=R(x)\cos\Theta(x),$$
$$\Phi_2(x)=R(x)\sin\Theta(x), \quad (22)$$
$$A_\mu(x)=B_\mu(x)-e^{-1}\partial_\mu\Theta(x),$$

where $R(x)$ and $B_\mu(x)$ are gauge invariant variables and the transformations (4) take the simple form, $\Theta(x)\to\Theta(x)-\Lambda(x)$. The classical Lagrangian (1), expressed in terms of the new variables, takes the form

$$\mathcal{L}'=-\tfrac{1}{4}F_{\mu\nu}F^{\mu\nu}-\tfrac{1}{2}g^{\mu\nu}\partial_\mu R\partial_\nu R$$
$$-\tfrac{1}{2}e^2B_\mu B^\mu R^2+\tfrac{1}{4}m_0^2R^2-\tfrac{1}{8}f^2R^4. \quad (23)$$

Gauge invariance here has ensured the disappearance of the variable Θ from the scene. [What we have done is to exploit the freedom which *local* $U(1)$ invariance gives us to "rotate" the entire two-component field $\Phi_a(x)$ onto one of the "axes." In a theory with only *global* $U(1)$ invariance this rotation cannot be performed on the entire field but only on the static solution (3).] The existence of the solution $B_\mu=0$, $R^2=m_0^2/f^2$ suggests the substitution $R(x)=\eta+\chi(x)$. In this way, we find immediately that, apart from an additive constant, $\mathcal{L}'=\mathcal{L}_0'+\mathcal{L}_{\text{int}}'$, where

$$\mathcal{L}_0'=-\tfrac{1}{4}F_{\mu\nu}F^{\mu\nu}-\tfrac{1}{2}m_1^2B_\mu B^\mu-\tfrac{1}{2}g^{\mu\nu}\partial_\mu\chi\partial_\nu\chi-\tfrac{1}{2}m_0^2\chi^2. \quad (24)$$

The expression (24) is the same as (6), except that the exactly gauge-invariant variables B_μ and χ which we have just defined replace their interaction picture counterparts.

We conjecture that the equivalence demonstrated here between the classical Lagrangians (1) and (23) may be extended to the corresponding quantum mechanical operators, provided that careful attention is given to the ordering problems which may arise, for example, in the definition of the current j_μ.[38]

V. DISCUSSION

The foregoing considerations illustrate our contention in Refs. 15 and 27 that the extension of a spontaneously broken internal symmetry of a Lorentz-invariant Lagrangian from global to local transformations not only may but actually does change zerons into the longitudinal states of massive vector bosons. Since we believe the value of this simple model to lie in the insight which it may give into this phenomenon when one looks at it as simple-mindedly as we have here, we shall not go into more difficult questions, such as radiative corrections and renormalization, in the present paper.

We note that in this model the original symmetry is almost unrecognizable in the physical states. Even without the gauge field coupling, the invalidation of the usual argument leading to the conservation of $Q(t)$[39] by the asymptotic behavior of the term $\langle\Phi_2\rangle\partial_\mu\Phi_1$ in j_μ destroys the commutativity of Q with the Hamiltonian: Consequently, the one-particle states are not eigenstates of Q and the masses within the Φ_a multiplet are not degenerate, but at least the multiplet structure remains. The gauge field coupling obscures even the multiplet structure: The scalar doublet is now incomplete, having lost its formerly massless member to form the longitudinal polarization of the vector singlet.

[37] See B. Zumino, J. Math. Phys. 1, 1 (1960).

[38] J. Schwinger, Phys. Rev. Letters 3, 296 (1959); K. Johnson, Nucl. Phys. 25, 431 (1961).

[39] In passing, we remark that this feature of spontaneous breakdown theories seems to call into question the validity of the results obtained on the basis of chirality conservation by Nambu and his collaborators. See Y. Nambu and D. Lurié, Phys. Rev. 125, 1429 (1962); Y. Nambu and E. Shrauner, *ibid.* 128, 862 (1962); E. Shrauner, *ibid.* 131, 1847 (1963).

In view of the rather drastic nature of the symmetry breakdown which we have just summarized, it is of interest to inquire what happens when this system is coupled to a second in a Lagrangian which retains local $U(1)$ invariance and contains no additional mechanisms for spontaneous breakdown. To be specific, let us take the second system to be a pair of "baryons" of "charges" $\pm\frac{1}{2}$, together with their antiparticles, and let us assume that the Φ-baryon interaction is of the Yukawa type. The total Lagrangian is then given by

$$\mathcal{L}_{\text{total}}=\mathcal{L}(A,\Phi)-\bar{\psi}_a(\gamma^\mu\nabla_\mu+M)\psi_a+g[\Phi_1(\bar{\psi}_1\psi_2+\bar{\psi}_2\psi_1)+\Phi_2(\bar{\psi}_2\psi_2-\bar{\psi}_1\psi_1)], \quad (25)$$

in which $\mathcal{L}(A,\Phi)$ is the expression (1), $\nabla_\mu\psi_1=\partial_\mu\psi_1-\frac{1}{2}eA_\mu\psi_2$, $\nabla_\mu\psi_2=\partial_\mu\psi_2+\frac{1}{2}eA_\mu\psi_1$ and we have, without loss of generality, made a choice of a phase angle in writing down the invariant Yukawa term. But for the presence of this last term, the Lagrangian would be invariant under global $U(1)$ transformations on Φ and ψ *independently*; that is, the symmetry would be $U(1)\times U(1)$ and the currents $j_\mu(\Phi)$ and $j_\mu(\psi)$ would be *separately* conserved. Thus, despite the nonconservation of $Q(\Phi)$ brought about by the structure of the first term in (25), there would still be conservation of $Q(\psi)$.

The Yukawa term reduces the symmetry to $U(1)$, the divergences of the individual currents now being given by

$$\partial_\mu j^\mu(\psi)=g[\Phi_1(\bar{\psi}_1\psi_1-\bar{\psi}_2\psi_2)+\Phi_2(\bar{\psi}_1\psi_2+\bar{\psi}_2\psi_1)] = -\partial_\mu j^\mu(\Phi). \quad (26)$$

We observe that spontaneous breakdown of the symmetry in the Φ system breaks the symmetry of the ψ system to an extent which depends on the coupling constant g. In the spirit of the perturbative approach which we have been using, we may evaluate the major part of the effects on the ψ system by replacing Φ by its vacuum expectation value. Then in Eq. (25) the term $g\eta(\bar{\psi}_2\psi_2-\bar{\psi}_1\psi_1)$ removes the baryon mass degeneracy, and Eq. (26) becomes

$$\partial_\mu j^\mu(\psi)=g\eta(\bar{\psi}_1\psi_2+\bar{\psi}_2\psi_1)+\text{higher order terms}. \quad (27)$$

If we were to modify the Lagrangian (25) by adding to it such $U(1)$ invariant baryon-antibaryon interactions as would produce a doublet of low-mass scalar bound states with wave functions transforming as $\bar{\psi}_1\psi_2+\bar{\psi}_2\psi_1$ and $\bar{\psi}_2\psi_2-\bar{\psi}_1\psi_1$, then Eq. (27) would be approximately a partial conservation law of the type which has proved so successful for the axial currents of the weak interactions.[40] Moreover, the current $j_\mu(\psi)$ interacts with itself via the massive intermediate vector boson which the $\Phi-A_\mu$ coupling produces.

There appears to be some hope that the basic ingredients of our model, namely the combination of spontaneous symmetry breakdown with the gauge principle, may provide the basis for an understanding of the broken symmetries of high-energy physics. In a subsequent paper we shall discuss models in which the breakdown of higher symmetries such as $SU(3)$ is treated in the same fashion.

ACKNOWLEDGMENTS

I have learned much about the Goldstone theorem from discussions with Dr. G. S. Guralnik, Dr. C. R. Hagen, Dr. T. W. B. Kibble, and Dr. R. F. Streater. I wish to thank Professor Bryce and Professor Cécile DeWitt for the hospitality of the Institute of Field Physics at the University of North Carolina, where this work was completed.

[40] It would be the type of partial conservation law proposed by Y. Nambu, Phys. Rev. Letters 4, 380 (1960) and by M. Gell-Mann, Phys. Rev. 125, 1067 (1962), rather than that proposed by M. Gell-Mann and M. Lévy, Nuovo Cimento 16, 705 (1960), in which the low-lying scalar or pseudoscalar states are treated as elementary rather than composite in the context of a field theory.

General Theory of Broken Local Symmetries*

Steven Weinberg
Laboratory for Nuclear Science and Department of Physics,
Massachusetts Institute of Technology, Cambridge, Massachusetts 02139
(Received 4 October 1972)

A general formalism for theories with spontaneously broken local symmetries is developed in the unitarity gauge. The canonical quantization procedure is carried out, leading to a set of Lorentz-covariant Feynman rules. Various special topics are discussed.

I. INTRODUCTION AND SUMMARY

Models with spontaneously broken local symmetries have been suggested[1] as a solution to two of the major problems of elementary-particle theory:

(a) the unification[2] of the weak and electromagnetic interaction;

(b) the elimination of ultraviolet divergences appearing in higher-order effects of the weak interactions.

Recently there have been indications[3] that such models may also elucidate one other outstanding problem:

(c) the explanation of the weak breaking of intrinsic symmetries such as isospin.

The purpose of the present paper is to provide a formal foundation for general theories with spontaneously broken local symmetries.

The formalism described here is based on choice of a particular gauge, the "unitarity gauge," in which the absence of Goldstone bosons and the order-by-order unitarity of these theories is manifest. This is the gauge which was originally used to show that, instead of Goldstone bosons appearing when local symmetries are spontaneously broken, the Yang-Mills fields in the theory acquire a mass.[4] Also, it is this gauge that was originally used to suggest the occurrence of this "Higgs-Kibble phenomenon" in a combined theory of weak and electromagnetic interactions.[1] However, the renormalizability of the unitarity gauge is extremely obscure, arising as it does from miraculous cancellations among different diagrams.[5] (The term "cryptorenormalizable" has been coined to describe such theories.) Indeed, the proof that these theories are renormalizable had to wait for four years after the original suggestion of renormalizability, when a new gauge, the "renormalizable" gauge, was invented by 't Hooft[6] and Lee.[7]

Despite these considerations, there are a number of reasons for wishing to see a systematic development of the theory of spontaneously broken local symmetries in the unitarity gauge. The unitarity gauge has an obvious heuristic value, in immediately revealing the particle content of the theory. It is more convenient than the renormalizable gauge for calculation of physical processes in the tree approximation, just because it is not necessary to cancel out unphysical poles. It can be used in some simple one-loop calculations.[5] Finally, the canonical quantization of theories with spontaneously broken local symmetries has so far only been possible in the unitarity gauge. This last is important, because once we know that a set of Feynman rules can be derived by canonical quantization from an Hermitian Hamiltonian, we know that the S matrix is unitary, at least in the perturbative sense.

This paper begins in Sec. II with the description of a general notation for renormalizable gauge-invariant theories. The unitarity gauge is then introduced in Sec. III by imposing the condition

$$(\theta_\alpha\lambda, \phi(x))=0 \quad \text{for all } \alpha, x, \tag{1.1}$$

where $\phi(x)$ is a vector whose components comprise all the real scalar fields of the theory, θ_α is the matrix representing the action of the αth generator of the gauge group on this scalar-field vector, and λ is the value of ϕ for which the scalar-meson part of the Lagrangian is stationary in ϕ. This is a "unitarity gauge," because the scalar-field components corresponding to unphysical Goldstone bosons of zero mass are just those in the $\theta_\alpha\lambda$ directions[8] which are here made to vanish. We usually think of a gauge as being chosen through conditions on the vector fields $A_{\alpha\mu}$, but of course the scalar as well as the vector fields respond to gauge transformations, so a gauge can be specified by conditions on ϕ as well as by conditions on $A_{\alpha\mu}$. A simple proof is given, showing that (1.1) can always be satisfied by a suitable change of gauge.[9]

The canonical quantization procedure is carried out in Secs. IV–VI, leading to the main result of this paper, the general Feynman rules for theories in which all local symmetries are spontaneously broken. It turns out that the correct Feynman rules are given by the naive prescription of using covariant propagators for $A_{\alpha\mu}$ and $\partial_\mu\phi$ and using the negative of the interaction Lagrangian as the interaction Hamiltonian, provided we add to the effective interaction Hamiltonian a term

$$i\delta^4(0)\ln \mathrm{Det}[\mu^{-2}\Phi(x)], \tag{1.2}$$

where μ^2 and Φ are the matrices

$$\mu^2{}_{\beta\alpha} \equiv -(\theta_\beta\lambda, \theta_\alpha\lambda),$$

$$\Phi_{\beta\alpha} \equiv -(\theta_\beta\lambda, \theta_\alpha\phi).$$

This result is the same as would be found by a path-integral quantization in the unitarity gauge,[10] and since the path-integral formalism manifestly leads to a gauge-invariant S matrix, the unitarity of the S-matrix derived by path-integral methods in any gauge, renormalizable as well as unitary, is now assured.

The last two sections deal with special topics. In Sec. VII the class of "simple" theories, in which (1.1) forces ϕ to point in the direction of λ, is introduced and briefly discussed. Section VIII deals with the properties of theories with unbroken local subgroups, and shows how some previous results for photon–neutral-vector-meson mixing can be rederived in a more general way. The modifications in the Feynman rules required in the presence of massless vector mesons are discussed, but not fully worked out.

The formalism developed here has already been used in Ref. 3; and will be employed in future papers on symmetry breaking and other topics.

II. GENERAL GAUGE-INVARIANT RENORMALIZABLE LAGRANGIANS

We shall consider the general class of renormalizable field theories possessing gauge invariance with respect to some compact semisimple Lie group $\mathcal{G}$. For the sake of renormalizability, the contents of the Lagrangian will be limited to a set of spin-zero fields $\phi_b(x)$ transforming according to a representation D_B of $\mathcal{G}$, a set of spin-one-half fields $\psi_n(x)$ transforming according to a representation D_F of $\mathcal{G}$, and a set of spin-one gauge fields $A_{\alpha\mu}(x)$ transforming according to the adjoint representation of $\mathcal{G}$. (Either D_F or D_B or both may be reducible.) The most general gauge-invariant Lagrangian which satisfies the usual power-counting conditions for renormalizability is of the Yang-Mills form[11]

$$\mathcal{L} = -\tfrac{1}{4}F_{\alpha\mu\nu}F^{\mu\nu}_{\alpha} - \tfrac{1}{2}(D_\mu\phi, D^\mu\phi)$$
$$-\bar\psi\gamma^\mu D_\mu\psi - \bar\psi m_0\psi - \mathcal{P}(\phi) - \bar\psi(\Gamma,\phi)\psi. \quad (2.1)$$

Our notation and conventions are as follows:

(A) The gauge-covariant curl is

$$F_{\alpha\mu\nu} \equiv \partial_\mu A_{\alpha\nu} - \partial_\nu A_{\alpha\mu} - C_{\alpha\beta\gamma}A_{\beta\mu}A_{\gamma\nu}, \quad (2.2)$$

where $C_{\alpha\beta\gamma}$ is the structure constant, defined by the commutation relations for the generators T_α of the abstract group $\mathcal{G}$:

$$[T_\alpha, T_\beta] = iC_{\alpha\beta\gamma}T_\gamma. \quad (2.3)$$

(Repeated indices are summed over unless otherwise noted.) Without loss of generality, we are choosing the generators T_α and the corresponding gauge fields $A_{\alpha\mu}$ to be Hermitian,

$$T_\alpha^\dagger = T_\alpha, \quad (2.4)$$

$$A_{\alpha\mu}^\dagger = A_{\alpha\mu} \quad (2.5)$$

so that the structure constants are real,

$$C^*_{\alpha\beta\gamma} = C_{\alpha\beta\gamma}. \quad (2.6)$$

Also, we have normalized the generators and gauge fields so that $T_\alpha T_\alpha$ and $F_{\alpha\mu\nu}F^{\mu\nu}_\alpha$ are $\mathcal{G}$-invariant, and therefore the structure constants are totally antisymmetric,

$$C_{\alpha\beta\gamma} = -C_{\alpha\gamma\beta} = -C_{\beta\alpha\gamma} = -C_{\gamma\beta\alpha}. \quad (2.7)$$

This does not entirely specify the normalization of the generators and structure constants – there is still a free normalization constant for each simple subgroup into which $\mathcal{G}$ may be decomposed. In our Lagrangian, this freedom has been employed to absorb into the generators T_α all gauge coupling constants, so that $C_{\alpha\beta\gamma}$ is of first order in the gauge coupling constants.

(B) The gauge-covariant derivative of the scalar field $\phi_p(x)$ is

$$(D_\mu\phi)_p = \partial_\mu\phi_p + i(\theta_\alpha)_{pq}\phi_q A_{\alpha\mu}, \quad (2.8)$$

where θ_α is the matrix representing the abstract generator T_α in the representation D_B of $\mathcal{G}$ furnished by the boson fields:

$$[\theta_\alpha, \theta_\beta] = iC_{\alpha\beta\gamma}\theta_\gamma. \quad (2.9)$$

Note that θ_α, like T_α and $C_{\alpha\beta\gamma}$, is taken to be proportional to the various gauge coupling constants. If the ϕ_p were not Hermitian, we could form a real representation of $\mathcal{G}$ by separating the Hermitian and anti-Hermitian parts of ϕ_p; hence we shall simply assume from the beginning here that ϕ_p is Hermitian,

$$\phi_p^\dagger = \phi_p, \quad (2.10)$$

so that $i\theta_\alpha$ is real,

$$(\theta_\alpha)^*_{pq} = -(\theta_\alpha)_{pq}. \quad (2.11)$$

In (2.1), we are using a scalar-product notation

$$(a,b) \equiv a_p b_p. \quad (2.12)$$

The ϕ_p fields are normalized so that scalar products like $(D_\mu\phi, D^\mu\phi)$ are $\mathcal{G}$-invariant, so therefore θ_α must be antisymmetric

$$(\theta_\alpha)_{pq} = -(\theta_\alpha)_{qp} \quad (2.13)$$

and hence Hermitian.

(C) The gauge-covariant derivative of the spin-one-half field $\psi_n(x)$ is

$$(D_\mu\psi)_n \equiv \partial_\mu\psi_n + i(t_\alpha)_{nm}\psi_m A_{\alpha\mu}, \quad (2.14)$$

where t_α is the matrix representing the generator T_α in the representation D_F of $\mathcal{G}$ furnished by the fermion fields:

$$[t_\alpha, t_\beta] = iC_{\alpha\beta\gamma}t_\gamma. \quad (2.15)$$

Once again, t_α, like T_α, $C_{\alpha\beta\gamma}$, and θ_α, is taken to be proportional to the various gauge coupling constants. The t_α matrices will in general contain terms proportional to the Dirac matrix γ_5 as well as the unit Dirac matrix. They may be decomposed into left- and right-handed parts

$$t_\alpha = t_\alpha^L + t_\alpha^R, \quad (2.16)$$

$$t_\alpha^L \equiv \tfrac{1}{2}(1+\gamma_5)t_\alpha, \qquad t_\alpha^R \equiv \tfrac{1}{2}(1-\gamma_5)t_\alpha, \quad (2.17)$$

each of which separately satisfy the commutation relations of our group:

$$[t_\alpha^L, t_\beta^L] = iC_{\alpha\beta\gamma}t_\gamma^L, \quad (2.18)$$

$$[t_\alpha^R, t_\beta^R] = iC_{\alpha\beta\gamma}t_\gamma^R. \quad (2.19)$$

Correspondingly, the field ψ may be decomposed into left- and right-handed parts

$$\psi = \psi^L + \psi^R, \quad (2.20)$$

$$\psi^L \equiv \tfrac{1}{2}(1+\gamma_5)\psi, \qquad \psi^R \equiv \tfrac{1}{2}(1-\gamma_5)\psi, \quad (2.21)$$

which furnish representations D_L and D_R of $\mathcal{G}$, in which the generators T_α are represented respectively by t_α^L and t_α^R. The fields $\psi_n(x)$ are normalized in such a way as to make $\bar\psi\gamma^\mu D_\mu\psi$ invariant under $\mathcal{G}$, where $\bar\psi$ is the usual Lorentz invariant adjoint

$$\bar\psi = \psi^\dagger\gamma_4. \quad (2.22)$$

Since t_α commutes with $\gamma_4\gamma^\mu$, this choice of normalization requires that t_α be Hermitian

$$t_\alpha^\dagger = t_\alpha \quad (2.23)$$

and therefore also

$$t_\alpha^{L\dagger} = t_\alpha^L, \quad t_\alpha^{R\dagger} = t_\alpha^R. \tag{2.24}$$

In the notation used here, the Dirac matrices satisfy the anticommutation relations

$$\{\gamma_\mu, \gamma_\nu\} = 2g_{\mu\nu},$$

where $g_{\mu\nu}$ is a diagonal matrix, with elements +1, +1, +1, −1 for $\mu=\nu=1,2,3,0$. Also, γ_4 and γ_5 are defined by

$$\gamma_4 = -i\gamma_0, \quad \gamma_5 = \gamma_1\gamma_2\gamma_3\gamma_4.$$

The matrices γ_1, γ_2, γ_3, γ_4, and γ_5 are Hermitian, and have unit squares.

(D) The bare-mass matrix m_0 must be $\mathcal{G}$-invariant, in the sense that

$$[t_\alpha, \gamma_4 m_0] = 0. \tag{2.25}$$

Decomposing (2.25) into terms proportional to $\gamma_4(1\pm\gamma_5)$ gives

$$t_\alpha^L m_0 - m_0 t_\alpha^R = t_\alpha^R m_0 - m_0 t_\alpha^L = 0. \tag{2.26}$$

Thus m_0 is absent unless the representations $D_L^* \otimes D_R$ and $D_R^* \otimes D_L$ of $\mathcal{G}$ contain among their irreducible components the identity representation. When it is present, the matrix $\gamma_4 m_0$ must be Hermitian,

$$\gamma_4 m_0 \gamma_4 = m_0^\dagger. \tag{2.27}$$

In general, m_0 might contain terms proportional to the Dirac matrix γ_5 as well as 1, though such terms could always be eliminated by a suitable choice of ψ fields.

(E) The function $\mathcal{P}(\phi)$ is a real 4th-order polynomial in ϕ, and is $\mathcal{G}$-invariant in the sense that

$$\frac{\partial \mathcal{P}(\phi)}{\partial \phi_p}(\theta_\alpha)_{pq}\phi_q = 0. \tag{2.28}$$

In contrast to the case of the fermions, here there is no way that $\mathcal{G}$ invariance can prevent the appearance of a boson mass term quadratic in ϕ and a boson-boson interaction quartic in ϕ, while terms in $\mathcal{P}(\phi)$ linear or cubic in ϕ may or may not be allowed.

(F) The $\mathcal{G}$ invariance of the Yukawa interaction in (1) requires that $\gamma_4\Gamma_p$ transform under $\mathcal{G}$ according to the representation D_B, in the sense that

$$[t_\alpha, \gamma_4\Gamma_p] = -(\theta_\alpha)_{pq}\gamma_4\Gamma_q. \tag{2.29}$$

The matrices Γ_p include any coupling constants that may appear in the Yukawa interaction, and may contain terms proportional to the Dirac matrix γ_5 as well as 1. Decomposing (2.29) into terms proportional to $\gamma_4(1\pm\gamma_5)$ gives

$$t_\alpha^L \Gamma_p^{LR} - \Gamma_p^{LR} t_\alpha^R = -(\theta_\alpha)_{pq}\Gamma_q^{LR},$$

$$t_\alpha^R \Gamma_p^{RL} - \Gamma_p^{RL} t_\alpha^L = -(\theta_\alpha)_{pq}\Gamma_q^{RL},$$

where

$$\Gamma_p^{LR} = \tfrac{1}{2}(1-\gamma_5)\Gamma_p,$$

$$\Gamma_p^{RL} = \tfrac{1}{2}(1+\gamma_5)\Gamma_p.$$

Thus Γ_p^{LR} or Γ_p^{RL} vanish unless the representations $D_L^* \otimes D_R$ or $D_R^* \otimes D_L$ contain one or more of the irreducible components of D_B. The matrix $\gamma_4\Gamma_p$ must be Hermitian, so

$$\Gamma_p^\dagger = \gamma_4\Gamma_p\gamma_4 \tag{2.30}$$

or in other words

$$\Gamma_p^{LR\dagger} = \Gamma_p^{RL}.$$

III. THE UNITARITY GAUGE

In consequence of the gauge invariance of our Lagrangian, the fields ψ_n, ϕ_p, and $A_{\alpha\mu}$ do not all represent independent physical degrees of freedom. This redundancy stands in the way of a straightforward canonical quantization of the theory, so before quantizing we shall choose a gauge.

Many choices are possible, including gauges defined by covariant conditions on the $A_{\alpha\mu}$ (such as $\partial_\mu A_\alpha^\mu = 0$), which would require quantization according to an indefinite metric. However, in this paper we shall instead choose a gauge designed to make manifest the nature of the Hilbert space of physical state vectors. That is, we shall eliminate the fields which do not correspond to physical particles, and we shall avoid an indefinite metric, so that the unitarity of the S matrix will be apparent in each order of perturbation theory.

Our gauge is defined as follows: First, define a vector λ_p in the representation space of the scalar fields by the condition

$$\frac{\partial \mathcal{P}(\lambda)}{\partial \lambda_p} = 0, \tag{3.1}$$

where $\mathcal{P}(\phi)$ is the quartic polynomial in the Lagrangian (2.1). The unitarity gauge is then defined by the condition that the scalar field components in the directions $\theta_\alpha\lambda$ should vanish

$$(\theta_\alpha\lambda, \phi(x)) = 0. \tag{3.2}$$

Before exploring the properties of this gauge, we must first convince ourselves that it exists, and then consider to what extent it is unique.

A. Existence

We must verify that there always exists an appropriate gauge in which the scalar fields $\phi_p(x)$ obey the condition (3.2). That is, starting with a field $\phi(x)$ in any gauge, not necessarily satisfying Eq. (3.2), we wish to find the gauge transformation $\phi(x) \to \overline{\phi}(x)$ to a new field $\overline{\phi}(x)$ which *does* satisfy Eq. (3.2):

$$(\theta_\alpha\lambda, \overline{\phi}(x)) = 0 \quad \text{for all } \alpha, x.$$

Under an arbitrary gauge transformation, the field $\phi(x)$ transforms according to the rule

$$\phi(x) \to O(x)\phi(x),$$

where $O(x)$ is an arbitrary x-dependent orthogonal matrix belonging to the representation D_B of $\mathcal{G}$. Hence we want to show that for any field $\phi(x)$ we can find an $O(x)$ such that

$$(\theta_\alpha\lambda, O(x)\phi(x)) = 0 \quad \text{for all } \alpha, x.$$

We are working on a classical level, prior to quantization, so $\phi(x)$ may be treated here as a c-number field. Also, since $O(x)$ can have an arbitrary x dependence; the x dependence of $O(x)$ and $\phi(x)$ is irrelevant.[12] Thus, our task is to prove that for an arbitrary vector ϕ, we can choose an orthogonal matrix O_ϕ belonging to the representation D_B, such that

$$(\theta_\alpha\lambda, O_\phi\phi) = 0 \quad \text{for all } \alpha.$$

To prove that this *is* possible, consider the quantity

$$(\lambda, O\phi),$$

where O sweeps over all orthogonal matrices belonging to the representation D^B. Since D^B is a real representation of a compact Lie group, this quantity is a real bounded differentiable function defined in the closed manifold of the group $\mathcal{G}$. Therefore it has at least one maximum and at least one minimum. Let us tentatively choose O_ϕ to be any one of the O matrices for which $(\lambda, O\phi)$ is an extremum. At any point on the group manifold, the variation of O is of the form

$$\delta O = i\epsilon_\alpha\theta_\alpha O,$$

where the ϵ_α are arbitrary infinitesimal parameters. Since $(\lambda, O\phi)$ is stationary at O_ϕ, we have

$$\begin{aligned} 0 &= (\lambda, \delta O\phi)|_{O=O_\phi} \\ &= i\epsilon_\alpha(\lambda, \theta_\alpha O_\phi\phi) \\ &= -i\epsilon_\alpha(\theta_\alpha\lambda, O_\phi\phi) \end{aligned}$$

and, since ϵ_α is arbitrary,

$$(\theta_\alpha\lambda, O_\phi\phi) = 0$$

as was to be proven. Thus our desired gauge may be constructed by starting with $\phi(x)$ in any arbitrary gauge and then performing a gauge transformation $\phi(x) \to O(x)\phi(x)$, with $O(x)$ chosen at each x to give the quantity $(\lambda, O(x)\phi(x))$ an extremal value.[13]

B. Uniqueness

There are two different ways in which the unitarity gauge can fail to be unique. First, there may be several vectors λ, with different directions, satisfying the defining Eq. (3.1). Also, given any vector λ, there may be several choices of gauge satisfying the gauge condition (3.2).

If λ is a solution of Eq. (3.1) then so is $O\lambda$, where O is any orthogonal matrix belonging to the representation D_B of $\mathcal{G}$. However, since we start with a completely $\mathcal{G}$-invariant theory, the solutions λ and $O\lambda$ are physically indistinguishable. On the other hand, we might find two different solutions λ_1 and λ_2 which are *not* related by a $\mathcal{G}$ transformation. With the method of quantization to be developed here, two such solutions will in general lead to two physically inequivalent Hilbert spaces.

Given a vector λ satisfying Eq. (3.1), and a gauge in which the scalar field $\phi(x)$ satisfies Eq. (3.2), we may be able to find gauge transformations $\phi(x) \to O(x)\phi(x)$ such that the new field $O(x)\phi(x)$ also satisfies Eq. (3.2). In particular, this will be the case if $O(x)$ leaves λ invariant:

$$O(x)\lambda = \lambda \tag{3.3}$$

for then

$$\begin{aligned} (\theta_\alpha\lambda, O(x)\phi(x)) &= (O^{-1}(x)\theta_\alpha\lambda, \phi(x)) \\ &= \mathcal{A}_{\alpha\beta}(x)(\theta_\beta\lambda, \phi(x)) = 0, \end{aligned}$$

where $\mathcal{A}_{\alpha\beta}$ is the adjoint representation of D^B:

$$O^{-1}(x)\theta_\alpha O(x) = \mathcal{A}_{\alpha\beta}(x)\theta_\beta.$$

The elements of $\mathcal{G}$ satisfying Eq. (3.3) form a subgroup $\mathcal{S}$, which as we shall see, consists precisely of all the unbroken symmetries of the theory. (In the real world, $\mathcal{S}$ is the one-dimensional group of electromagnetic gauge transformations, although for some purposes it is convenient to consider approximations in which $\mathcal{S}$ is larger.[14]) Thus, if our original group $\mathcal{G}$ is N-dimensional, and there is an unbroken M-dimensional subgroup $\mathcal{S}$, then the condition (3.2) really imposes only $N-M$ independent constraints on ϕ, leaving us with complete freedom to perform gauge transformations in $\mathcal{S}$.

Conversely, the conditions (3.2) will in general provide N independent functional conditions on ϕ for an N-dimensional Lie algebra θ_α, unless there is one or more linear relations among the vectors $\theta_\alpha\lambda$:

$$C_\alpha\theta_\alpha\lambda = 0.$$

If Eq. (3.2) really provides only $N-M$ constraints on ϕ, then there must be M independent linear relations of this form, and the M linear combinations $C_\alpha\theta_\alpha$ will then generate an M-dimensional subgroup of $\mathcal{G}$ which leaves λ invariant.

We see that the *continuous* degrees of freedom in the gauge, which are not removed by the gauge condition (3.2), are precisely those associated with the unbroken subgroup $\mathcal{S}$ which leaves λ invariant. It is not clear to me whether there could also be a discrete choice of gauges not associated with $\mathcal{S}$ transformations, or what would be the physical significance of such a choice.

C. Physical Significance

Now we must show that the gauge condition (3.2) really accomplishes the purpose for which it was designed, of eliminating all Goldstone bosons from the theory. Let n_a be an orthonormal set of vectors spanning the subspace (in the representation space of the scalar fields) orthogonal to all $\theta_\alpha\lambda$:

$$(n_a, \theta_\alpha\lambda) = 0\,, \qquad (3.4)$$

$$(n_a, n_b) = \delta_{ab}\,, \qquad (3.5)$$

$$n_a(n_a, u) = u\,, \quad \text{if } (u, \theta_\alpha\lambda) = 0\,. \qquad (3.6)$$

Any scalar field vector satisfying Eq. (3.2) may be written

$$\phi_p = \phi_a (n_a)_p\,. \qquad (3.7)$$

The ϕ_a are then the independent scalar fields of the theory.

At this point we must recognize that the defining condition for λ, Eq. (3.1), is to be imposed *before* we choose our gauge, so that $\mathcal{P}(\phi)$ must be stationary at $\phi = \lambda$ with respect to variations of ϕ not only in the directions n_a, but also in the directions $\theta_\alpha\lambda$ perpendicular to the n_a. Thus Eq. (3.1) really gives us two different conditions: For all a and α,

$$\frac{\partial \mathcal{P}(\phi)}{\partial \phi_p} n_{ap} = 0 \quad \text{at} \quad \phi = \lambda\,, \qquad (3.8)$$

$$\frac{\partial \mathcal{P}(\phi)}{\partial \phi_p} (\theta_\alpha\lambda)_p = 0 \quad \text{at} \quad \phi = \lambda\,. \qquad (3.9)$$

However, Eq. (3.9) is automatically satisfied by virtue of the $\mathcal{G}$-invariance condition (2.28). Thus the full content of Eq. (3.1) is contained in Eq. (3.8), or in other words, in the statement that $\mathcal{P}(\phi)$ has no terms of first order in the "shifted field" ϕ'_a, where

$$\phi_a \equiv (n_a, \lambda) + \phi'_a\,. \qquad (3.10)$$

We may recognize this as the condition that ϕ'_a should have zero vacuum expectation value in the tree approximation,[8] and therefore conclude that λ is the vacuum expectation value of ϕ in the tree approximation:

$$\langle \phi_a \rangle_{\text{vac, tree}} = (n_a, \lambda) \qquad (3.11)$$

Now let us see what happens when we try to prove Goldstone's theorem. By differentiating Eq. (2.28) with respect to ϕ_r, we find for an arbitrary ϕ (not necessarily in the unitarity gauge) that

$$\frac{\partial^2 \mathcal{P}(\phi)}{\partial \phi_r \partial \phi_p} (\theta_\alpha)_{pq} \phi_q + \frac{\partial \mathcal{P}(\phi)}{\partial \phi_p} (\theta_\alpha)_{pr} = 0\,.$$

Now setting $\phi = \lambda$ and using Eq. (3.1), we have

$$\frac{\partial^2 \mathcal{P}(\lambda)}{\partial \lambda_r \partial \lambda_p} (\theta_\alpha)_{pq} \lambda_q = 0\,. \qquad (3.12)$$

In theories with a global rather than a local invariance group, this would require the tree approximation to the mass matrix of the scalar fields to have an eigenvector $\theta_\alpha\lambda$ with eigenvalue zero. Indeed there are proofs[8] that in such theories the propagator of the scalar field has poles which remain at zero mass to all orders in perturbation theory. However, in our present theory the second derivative appearing in Eq. (3.12) has nothing to do with the mass matrix of the scalar fields, which in the tree approximation is just the coefficient of the term in $\mathcal{P}(\phi)$ quadratic in the shifted fields ϕ'_a:

$$M^2{}_{ab} = \frac{\partial^2 \mathcal{P}(\lambda)}{\partial \lambda_p \partial \lambda_q} n_{ap} n_{bq}\,. \qquad (3.13)$$

There are no linear relations between the n_a and $\theta_\alpha\lambda$ vectors, so Eq. (3.12) tells us nothing whatever about the eigenvalues of the mass matrix (3.13). One of these eigenvalues may vanish, but there is no reason to expect it.

Our conclusions remain unchanged when we consider higher-order effects. If all eigenvalues of the scalar-boson mass matrix are nonzero in zeroth-order perturbation theory, then they must all be nonzero for sufficiently small values of the coupling constants. Of course, one of the eigenvalues of the mass matrix might vanish for some sufficiently large values of the coupling constants, but again, there is no particular reason to expect it, and certainly there is no Goldstone theorem here which requires a vanishing mass.

These remarks illuminate a difficulty that might be anticipated in some theories. When the representation D_B is sufficiently complicated we must expect that the true direction of the vacuum expectation value of the scalar field vector ϕ will differ from the direction of the vacuum expectation value λ in the tree approximation.[15] Should we then correct the gauge condition (3.2) in each order of perturbation theory by using in place of λ the true value of $\langle\phi\rangle_0$ calculated up to that order? Note that this *is* possible, because the above proof of the existence of a gauge satisfying (3.2) did not depend on λ having any particular direction. Nevertheless, it appears that the gauge defined by (3.2) is perfectly adequate for use in all orders.

IV. QUANTIZATION

Having chosen our gauge, we shall now quantize the theory. For simplicity, it will be assumed from now on that all gauge symmetries are broken, so that there is no nontrivial subgroup $\mathcal{S}$ of $\mathcal{G}$ which leaves the direction λ [defined by Eq. (3.1)] invariant. This way, we avoid the well-known complications of quantizing theories, like quantum electrodynamics, with massless spin-one particles. After quantization, we can always turn off the symmetry breaking of some subgroup $\mathcal{S}$ of $\mathcal{G}$, as discussed in Sec. VIII.

The canonical variables here will be chosen as the components $\psi_n(x)$ of the spin-one-half fields, the spatial components $A_{\alpha i}(x)$ of the gauge fields, and the independent components $\phi_a(x)$ of the scalar-field vector constrained by the unitarity gauge condition (3.2). The canonical "momenta" conjugate to these variables are

$$\chi_n \equiv \frac{\partial \mathcal{L}}{\partial(\partial_0 \psi_n)} = (\bar{\psi}\gamma_0)_n \,, \tag{4.1}$$

$$P_{\alpha i} \equiv \frac{\partial \mathcal{L}}{\partial(\partial_0 A_{\alpha i})} = F^{io}_{\alpha} \,, \tag{4.2}$$

$$\pi_a \equiv \frac{\partial \mathcal{L}}{\partial(\partial_0 \phi_a)} = (n_a, D_0\phi) \,. \tag{4.3}$$

The fields A^0_α are not independent dynamical variables, but rather may be expressed in terms of the above fields and conjugates by using the field equations. The field equation for A^ν_α reads

$$\begin{aligned} 0 &= \partial_\mu \left(\frac{\partial \mathcal{L}}{\partial(\partial_\mu A_{\beta\nu})}\right) - \frac{\partial \mathcal{L}}{\partial A_{\beta\nu}} \\ &= -\partial_\mu F^{\mu\nu}_\beta + F^{\mu\nu}_\alpha C_{\alpha\beta\gamma} A_{\gamma\mu} + i(\theta_\beta \phi, D^\nu \phi) \,. \end{aligned} \tag{4.4}$$

Setting $\nu = 0$ gives

$$0 = -\vec{\nabla}\cdot\vec{P}_\beta + \vec{P}_\alpha\cdot\vec{A}_\gamma C_{\alpha\beta\gamma} + i(\theta_\beta\phi, D^0\phi) \,. \tag{4.5}$$

To evaluate the last term, we note that our assumption that all gauge symmetries are broken means that no linear combination $c_\alpha\theta_\alpha$ leaves the vector λ invariant, so that the vectors $\theta_\alpha\lambda$ are *independent*. The vectors n_a and $\theta_\alpha\lambda$ form a complete set, so that

$$1 = n_a n_a^T - (\theta_\alpha\lambda)(\mu^{-2})_{\alpha\beta}(\theta_\beta\lambda)^T \,, \tag{4.6}$$

where μ^{-2} is the inverse of the matrix

$$\mu^2{}_{\alpha\beta} \equiv -(\theta_\alpha\lambda, \theta_\beta\lambda) \,. \tag{4.7}$$

(It will be seen later that μ is the zeroth-order vector-meson mass matrix.) Inserting (4.6) in (4.5) gives

$$\begin{aligned} 0 = &-\vec{\nabla}\cdot\vec{P}_\beta + \vec{P}_\alpha\cdot\vec{A}_\gamma C_{\alpha\beta\gamma} \\ &+ i(\theta_\beta\phi, n_a)(n_a, D^0\phi) \\ &- i(\theta_\beta\phi, \theta_\alpha\lambda)(\mu^{-2})_{\alpha\gamma}(\theta_\gamma\lambda, D^0\phi) \,. \end{aligned} \tag{4.8}$$

At this point, we use the unitarity gauge condition, which tells us that for all x,

$$(\theta_\gamma\lambda, \phi(x)) = 0$$

and therefore also

$$(\theta_\gamma\lambda, \partial_\mu\phi) = 0 \,.$$

It follows then that

$$(\theta_\gamma\lambda, D^0\phi) = i(\theta_\gamma\lambda, \theta_\delta\phi)A^0_\delta \,. \tag{4.9}$$

The unitarity gauge condition also lets us replace ϕ with $n_a\phi_a$. Using (4.3) and (4.9) in (4.8) gives our solution for $A_{\beta 0}$:

$$\begin{aligned} A_{\beta 0} = \Omega^{-1}{}_{\beta\delta}(\phi)[&-\vec{\nabla}\cdot\vec{P}_\delta + \vec{P}_\alpha\cdot\vec{A}_\gamma C_{\alpha\delta\gamma} \\ &- i(\theta_\delta)_{ab}\pi_a\phi_b] \,, \end{aligned} \tag{4.10}$$

where

$$(\theta_\alpha)_{ab} = (n_a, \theta_\alpha n_b)$$

and Ω^{-1} is the inverse of the symmetric matrix

$$\Omega_{\beta\delta}(\phi) = (\theta_\beta\phi, \theta_\alpha\lambda)(\mu^{-2})_{\alpha\gamma}(\theta_\gamma\lambda, \theta_\delta\phi) \,. \tag{4.11}$$

Our next task is to evaluate the Hamiltonian density:

$$\mathcal{H} \equiv \chi_n\dot{\psi}_n + \vec{P}_\alpha\cdot\dot{\vec{A}}_\alpha + \pi_a\dot{\phi}_a - \mathcal{L} \,. \tag{4.12}$$

The time derivatives $\dot{\vec{A}}_\alpha$ and $\dot{\phi}_a$ may be expressed in terms of canonical variables and their conjugates by using (4.2) and (4.3)

$$\dot{\vec{A}}_\alpha = \vec{\nabla}A_{\alpha 0} + C_{\alpha\beta\gamma}A_{\beta 0}\vec{A}_\gamma + \vec{P}_\alpha \,, \tag{4.13}$$

$$\dot{\phi}_a = -i(\theta_\alpha)_{ab}\phi_b A_{\alpha 0} + \pi_a \,. \tag{4.14}$$

Using (4.13) and (4.14) in (4.12), and noting the cancellation of terms involving $\dot{\psi}_n$, we have then

$$\begin{aligned} \mathcal{H} = \vec{P}_\alpha\cdot[&\vec{\nabla}A_{\alpha 0} + C_{\alpha\beta\gamma}A_{\beta 0}\vec{A}_\gamma + \vec{P}_\alpha] \\ &+ \pi_a[-i(\theta_\alpha)_{ab}\phi_b A_{\alpha 0} + \pi_a] - \tfrac{1}{2}\vec{P}_\alpha\cdot\vec{P}_\alpha \\ &+ \tfrac{1}{4}F_{\alpha ij}F_{\alpha ij} - \tfrac{1}{2}(D_0\phi, D_0\phi) + \tfrac{1}{2}(\vec{D}\phi, \vec{D}\phi) \\ &+ \bar{\psi}\vec{\gamma}\cdot\vec{D}\psi + i\bar{\psi}\gamma^0 t_\alpha\psi A_{\alpha 0} \\ &+ \bar{\psi}m_0\psi + \mathcal{P}(\phi) + \bar{\psi}(\Gamma, \phi)\psi \,. \end{aligned} \tag{4.15}$$

Using (4.6), (4.9), (4.11), and (4.3), the sixth term can be evaluated as

$$(D_0\phi, D_0\phi) = \pi_a\pi_a + \Omega_{\alpha\beta}(\phi)A_{\alpha 0}A_{\beta 0} \tag{4.16}$$

while the seventh term is

$$\begin{aligned} (\vec{D}\phi, \vec{D}\phi) = &\vec{\nabla}\phi_a\cdot\vec{\nabla}\phi_a + 2i\vec{A}_\alpha\cdot\vec{\nabla}\phi_a(\theta_\alpha)_{ab}\phi_b \\ &+ \vec{A}_\alpha\cdot\vec{A}_\beta[\Omega_{\alpha\beta}(\phi) - (\theta_\alpha)_{ac}(\theta_\beta)_{bc}\phi_a\phi_b] \,. \end{aligned}$$

Also, we can add a divergence $-\vec{\nabla}\cdot(P_\alpha A_{\alpha 0})$ to $\mathcal{H}$, so that the first term becomes

$$-A_{\alpha 0}\vec{\nabla}\cdot\vec{P}_\alpha \,.$$

Using (4.10) to express $A_{\alpha 0}$ in terms of independent fields and their canonical conjugates, we have at last

$$
\begin{aligned}
\mathcal{H} = {} & \tfrac{1}{2}\Omega^{-1}{}_{\alpha\beta}(\phi)[\vec{\nabla}\cdot\vec{\mathrm{P}}_\alpha - C_{\alpha\gamma\delta}\vec{\mathrm{P}}_\delta\cdot\vec{\mathrm{A}}_\gamma + i(\theta_\alpha)_{ab}\pi_a\phi_b][\vec{\nabla}\cdot\vec{\mathrm{P}}_\beta - C_{\beta\epsilon\zeta}\vec{\mathrm{P}}_\zeta\cdot\vec{\mathrm{A}}_\epsilon + i(\theta_\beta)_{cd}\pi_c\phi_d] \\
& + \tfrac{1}{2}\vec{\mathrm{P}}_\alpha\cdot\vec{\mathrm{P}}_\alpha + \tfrac{1}{2}\pi_a\pi_a + \tfrac{1}{4}[\partial_i A_{\alpha j} - \partial_j A_{\alpha i} - C_{\alpha\beta\gamma}A_{\beta i}A_{\gamma j}][\partial_i A_{\alpha j} - \partial_j A_{\alpha i} - C_{\alpha\delta\epsilon}A_{\delta i}A_{\epsilon j}] \\
& + \tfrac{1}{2}\vec{\nabla}\phi_a\cdot\vec{\nabla}\phi_a + i\vec{\mathrm{A}}_\alpha\cdot\vec{\nabla}\phi_a(\theta_\alpha)_{ab}\phi_b + \tfrac{1}{2}\vec{\mathrm{A}}_\alpha\cdot\vec{\mathrm{A}}_\beta[\Omega_{\alpha\beta}(\phi) - (\theta_\alpha)_{ac}(\theta_\beta)_{bc}\phi_a\phi_b] + \bar{\psi}\vec{\gamma}\cdot\vec{\nabla}\psi + i\bar{\psi}\vec{\gamma}t_\alpha\psi\cdot\vec{\mathrm{A}}_\alpha \\
& - i\bar{\psi}\gamma^0 t_\alpha\psi\,\Omega^{-1}{}_{\alpha\beta}(\phi)[\vec{\nabla}\cdot\vec{\mathrm{P}}_\beta - \vec{\mathrm{P}}_\delta\cdot\vec{\mathrm{A}}_\gamma C_{\beta\gamma\delta} + i(\theta_\beta)_{ab}\pi_a\phi_b] + \bar{\psi}m_0\psi + \mathcal{P}(\phi) + \bar{\psi}(\Gamma,\phi)\psi\,. \qquad (4.17)
\end{aligned}
$$

The quantization procedure is completed by writing down the canonical commutation relations:

$$[A_{\alpha i}(\vec{x},t), P_{\beta j}(\vec{y},t)] = i\delta_{\alpha\beta}\delta_{ij}\delta^3(\vec{x}-\vec{y})\,, \qquad (4.18)$$

$$[\phi_a(\vec{x},t), \pi_b(\vec{y},t)] = i\delta_{ab}\delta^3(\vec{x}-\vec{y}), \qquad (4.19)$$

$$\{\psi(\vec{x},t), \bar{\psi}(\vec{y},t)\gamma^0\} = i\delta^3(\vec{x}-\vec{y})1\,. \qquad (4.20)$$

V. THE INTERACTION REPRESENTATION

In order to develop a perturbation series, we must extract from ϕ its zeroth-order vacuum expectation value λ, and write

$$\phi = \lambda + \phi'\,, \qquad (5.1)$$

so that the shifted field ϕ' has zero vacuum expectation value in the tree approximation. Note that in this approximation, the Ω matrix (4.11) is just

$$\Omega_{\alpha\beta}(\lambda) = \mu^2{}_{\alpha\beta}\,. \qquad (5.2)$$

Also note that

$$(\theta_\alpha)_{ab}\lambda_b = (n_a, \theta_\alpha\lambda) = 0\,.$$

The free-field Hamiltonian $\mathcal{H}_0$ is just that part of (4.17) quadratic in the canonical variables and their conjugates:

$$
\begin{aligned}
\mathcal{H}_0 = {} & \tfrac{1}{2}\mu^{-2}{}_{\alpha\beta}(\vec{\nabla}\cdot\vec{\mathrm{P}}_\alpha)(\vec{\nabla}\cdot\vec{\mathrm{P}}_\beta) + \tfrac{1}{2}\vec{\mathrm{P}}_\alpha\cdot\vec{\mathrm{P}}_\alpha \\
& + \tfrac{1}{2}\pi_a\pi_a + \tfrac{1}{4}(\partial_i A_{\alpha j} - \partial_j A_{\alpha i})(\partial_i A_{\alpha j} - \partial_j A_{\alpha i}) \\
& + \tfrac{1}{2}\vec{\nabla}\phi'_a\cdot\vec{\nabla}\phi'_a + \tfrac{1}{2}\mu^2{}_{\alpha\beta}\vec{\mathrm{A}}_\alpha\cdot\vec{\mathrm{A}}_\beta \\
& + \bar{\psi}\vec{\gamma}\cdot\vec{\nabla}\psi + \bar{\psi}m\psi + \tfrac{1}{2}M^2{}_{ab}\phi'_a\phi'_b\,, \qquad (5.3)
\end{aligned}
$$

where

$$m \equiv m_0 + (\Gamma, \lambda)\,. \qquad (5.4)$$

We recognize this as the Hamiltonian for free particles with spin zero, one-half, and one, and with mass matrices M, m, and μ, respectively.

The interaction Hamiltonian here is

$$
\begin{aligned}
\mathcal{H}' \equiv {} & \mathcal{H} - \mathcal{H}_0 \\
= {} & \tfrac{1}{2}[\Omega^{-1}{}_{\alpha\beta}(\lambda+\phi') - \mu^{-2}{}_{\alpha\beta}]\vec{\nabla}\cdot\vec{\mathrm{P}}_\alpha\vec{\nabla}\cdot\vec{\mathrm{P}}_\beta + \Omega^{-1}{}_{\alpha\beta}(\lambda+\phi')\vec{\nabla}\cdot\vec{\mathrm{P}}_\beta\cdot[-C_{\alpha\gamma\delta}\vec{\mathrm{P}}_\delta\cdot\vec{\mathrm{A}}_\gamma + i(\theta_\alpha)_{ab}\pi_a\phi'_b] \\
& + \tfrac{1}{2}\Omega^{-1}{}_{\alpha\beta}(\lambda+\phi')[-C_{\alpha\gamma\delta}\vec{\mathrm{P}}_\delta\cdot\vec{\mathrm{A}}_\gamma + i(\theta_\alpha)_{ab}\pi_a\phi'_b][-C_{\beta\epsilon\zeta}\vec{\mathrm{P}}_\zeta\cdot\vec{\mathrm{A}}_\epsilon + i(\theta_\beta)_{cd}\pi_c\phi'_d] \\
& - \tfrac{1}{2}(\partial_i A_{\alpha j} - \partial_j A_{\alpha i})C_{\alpha\beta\gamma}A_{\beta i}A_{\gamma j} + \tfrac{1}{4}C_{\alpha\beta\gamma}C_{\alpha\delta\epsilon}A_{\beta i}A_{\gamma j}A_{\delta i}A_{\epsilon j} \\
& + i\vec{\mathrm{A}}_\alpha\cdot\vec{\nabla}\phi'_a(\theta_\alpha)_{ab}\phi'_b + \tfrac{1}{2}\vec{\mathrm{A}}_\alpha\cdot\vec{\mathrm{A}}_\beta[\Omega_{\alpha\beta}(\lambda+\phi') - \mu^2{}_{\alpha\beta} - (\theta_\alpha)_{ac}(\theta_\beta)_{bc}\phi'_a\phi'_b] \\
& + i\bar{\psi}\vec{\gamma}t_\alpha\psi\cdot\vec{\mathrm{A}}_\alpha - i\bar{\psi}\gamma^0 t_\alpha\psi\,\Omega^{-1}{}_{\alpha\beta}(\lambda+\phi')[\vec{\nabla}\cdot\vec{\mathrm{P}}_\beta - \vec{\mathrm{P}}_\delta\cdot\vec{\mathrm{A}}_\gamma C_{\beta\gamma\delta} + i(\theta_\beta)_{ab}\pi_a\phi'_b] + \mathcal{P}(\lambda+\phi') - \tfrac{1}{2}M^2{}_{ab}\phi'_a\phi'_b + \bar{\psi}(\Gamma,\phi')\psi
\end{aligned}
\qquad (5.5)
$$

We now perform a "unitary" transformation to the interaction representation, in which the time derivatives of all canonical variables are determined by $\mathcal{H}_0$ alone. Distinguishing interaction-representation variables by superscript I, we have for the vector fields

$$
\begin{aligned}
\dot{A}^I_{\alpha i} &= \frac{\partial\mathcal{H}^I_0}{\partial P^I_{\alpha i}} - \partial_j\left(\frac{\partial\mathcal{H}^I_0}{\partial(\partial_j P^I_{\alpha i})}\right) \\
&= P^I_{\alpha i} - \mu^{-2}{}_{\alpha\beta}\partial_i(\vec{\nabla}\cdot\vec{\mathrm{P}}^I_\beta), \qquad (5.6)
\end{aligned}
$$

$$
\begin{aligned}
\dot{P}^I_{\alpha i} &= -\frac{\partial\mathcal{H}^I_0}{\partial A^I_{\alpha i}} + \partial_j\left(\frac{\partial\mathcal{H}^I_0}{\partial(\partial_j A^I_{\alpha i})}\right) \\
&= -\mu^2{}_{\alpha\beta}A^I_{\beta i} - \partial_i\vec{\nabla}\cdot\vec{\mathrm{A}}^I_\alpha + \vec{\nabla}^2 A^I_{\alpha i}\,. \qquad (5.7)
\end{aligned}
$$

It proves extremely convenient to *define* the time component of the vector potential in the interaction representation by

$$A^I_{\alpha 0} \equiv -\mu^{-2}{}_{\alpha\beta}\vec{\nabla}\cdot\vec{\mathrm{P}}^I_\beta\,. \qquad (5.8)$$

[This is *not* what we would get by applying the general "unitary" transformation, that takes us from the Heisenberg to the interaction representation, to the Heisenberg representation quantity $A_{\alpha 0}$ given by (4.10).] For notational convenience, we also introduce the curl

$$f^I_{\alpha\mu\nu} \equiv \partial_\mu A^I_{\alpha\nu} - \partial_\nu A^I_{\alpha\mu} . \tag{5.9}$$

Equation (5.6) now reads simply

$$P^I_{\alpha i} = f^I_{\alpha 0 i} \tag{5.10}$$

and Eqs. (5.7) and (5.8) become just the space and time components of the free-field equation:

$$\partial_\mu f^{\mu\nu I}_\alpha - \mu^2{}_{\alpha\beta} A^{\nu I}_\beta = 0. \tag{5.11}$$

Similarly (but more simply) the equations for the time derivatives of the spin-zero and spin-one-half fields and their canonical conjugates yield the free-field equations

$$\Box^2 \phi'^I_a - M^2{}_{ab}\phi'^I_b = 0 , \tag{5.12}$$

$$\gamma^\lambda \partial_\lambda \psi^I_n + m_{nm}\psi^I_m = 0 \tag{5.13}$$

and the formulas

$$\pi^I_a = \dot{\phi}^I_a , \tag{5.14}$$

$$\chi^I_n = (\bar{\psi}^I \gamma_0)_n . \tag{5.15}$$

In writing down the propagators of the interaction-representation fields, it is convenient first to *define* covariant propagators:

$$\langle T^*\{A^I_{\alpha\mu}(x), A^I_{\beta\nu}(y)\}\rangle \equiv g_{\mu\nu}\Delta^F_{\alpha\beta}(x-y;\mu) - \partial_\mu\partial_\nu[\mu^{-2}\Delta^F(x-y;\mu)]_{\alpha\beta} , \tag{5.16}$$

$$\langle T^*\{f^I_{\alpha\lambda\mu}(x), A^I_{\beta\nu}(y)\}\rangle \equiv \langle T^*\{A^I_{\beta\nu}(y), f^I_{\alpha\lambda\mu}(x)\}\rangle = (\partial_\lambda g_{\mu\nu} - \partial_\mu g_{\lambda\nu})\Delta^F_{\alpha\beta}(x-y;\mu) , \tag{5.17}$$

$$\langle T^*\{f^I_{\alpha\lambda\mu}(x), f^I_{\beta\kappa\nu}(y)\}\rangle \equiv [-g_{\mu\nu}\partial_\lambda\partial_\kappa + g_{\lambda\nu}\partial_\mu\partial_\kappa + g_{\mu\kappa}\partial_\lambda\partial_\nu - g_{\lambda\kappa}\partial_\mu\partial_\nu]\Delta^F_{\alpha\beta}(x-y;\mu), \tag{5.18}$$

$$\langle T^*\{\phi'^I_a(x), \phi'^I_b(y)\}\rangle \equiv \Delta^F_{ab}(x-y;M) , \tag{5.19}$$

$$\langle T^*\{\partial_\mu\phi'^I_a(x), \phi'^I_b(y)\}\rangle \equiv \langle T^*\{\phi'^I_b(y), \partial_\mu\phi'^I_a(x)\}\rangle = \partial_\mu\Delta^F_{ab}(x-y;M), \tag{5.20}$$

$$\langle T^*\{\partial_\mu\phi'^I_a(x), \partial_\nu\phi'^I_b(y)\}\rangle \equiv -\partial_\mu\partial_\nu\Delta^F_{ab}(x-y;M) , \tag{5.21}$$

$$\langle T^*\{\psi^I_n(x), \bar{\psi}^I_m(y)\}\rangle \equiv [(-\gamma^\lambda\partial_\lambda + m)\Delta^F(x-y;m)]_{nm} , \tag{5.22}$$

with Δ^F defined for a general mass matrix $\mathfrak{M}$ by

$$\Delta^F(x-y;\mathfrak{M}) \equiv (2\pi)^{-3}\int d^4p\, \theta(p^0)\delta(p^2+\mathfrak{M}^2)[\theta(x^0-y^0)e^{ip\cdot(x-y)} + \theta(y^0-x^0)e^{-ip\cdot(y-x)}]. \tag{5.23}$$

A straightforward calculation then shows that the true vacuum expectation values of time-ordered products are given by

$$\langle T\{A^I_{\alpha\mu}(x), A^I_{\beta\nu}(y)\}\rangle_0 = \langle T^*\{A^I_{\alpha\mu}(x), A^I_{\beta\nu}(y)\}\rangle - i\,\mu^{-2}{}_{\alpha\beta}\delta^0_\mu\delta^0_\nu\delta^4(x-y) , \tag{5.24}$$

$$\langle T\{f^I_{\alpha\lambda\mu}(x), A^I_{\beta\nu}(y)\}\rangle_0 = \langle T^*\{f^I_{\alpha\lambda\mu}(x), A^I_{\beta\nu}(y)\}\rangle , \tag{5.25}$$

$$\langle T\{f^I_{\alpha\lambda\mu}(x), f^I_{\beta\rho\nu}(y)\}\rangle_0 = \langle T^*\{f^I_{\alpha\lambda\mu}(x), f^I_{\beta\rho\nu}(y)\}\rangle - i\delta_{\alpha\beta}\delta^4(x-y)[\delta^0_\rho\delta^0_\lambda g_{\mu\nu} - \delta^0_\nu\delta^0_\lambda g_{\mu\rho} - \delta^0_\mu\delta^0_\rho g_{\mu\nu} + \delta^0_\mu\delta^0_\nu g_{\lambda\rho}] , \tag{5.26}$$

$$\langle T\{\phi'^I_a(x), \phi'^I_b(y)\}\rangle_0 = \langle T^*\{\phi'^I_a(x), \phi'^I_b(y)\}\rangle , \tag{5.27}$$

$$\langle T\{\partial_\mu\phi'^I_a(x), \phi'^I_b(y)\}\rangle_0 = \langle T^*\{\partial_\mu\phi'^I_a(x), \phi'^I_b(y)\}\rangle , \tag{5.28}$$

$$\langle T\{\partial_\mu\phi'^I_a(x), \partial_\nu\phi'^I_b(y)\}\rangle_0 = \langle T^*\{\partial_\mu\phi'^I_a(x), \partial_\nu\phi'^I_b(y)\}\rangle - i\delta_{ab}\delta^0_\mu\delta^0_\nu\delta^4(x-y) , \tag{5.29}$$

$$\langle T\{\psi^I_n(x), \bar{\psi}^I_m(y)\}\rangle_0 = \langle T^*\{\psi^I_n(x), \bar{\psi}^I_m(y)\}\rangle . \tag{5.30}$$

The fermion propagators here have the same form as usual.

VI. DERIVATION OF THE FEYNMAN RULES

As a first step toward a Lorentz-invariant formulation of this theory, let us use Eqs. (5.8), (5.10), and (5.14) to eliminate canonical "momenta" from the interaction Hamiltonian $\mathcal{H}'^I$ in the interaction representation. Dropping the superscript I, Eq. (5.5) becomes

$$\begin{aligned}\mathcal{H}' = {} & \tfrac{1}{2}[\mu^2\Omega^{-1}(\lambda+\phi')\mu^2-\mu^2]_{\alpha\beta}A_{\alpha 0}A_{\beta 0}-[\Omega^{-1}(\lambda+\phi')\mu^2]_{\alpha\beta}A_{\beta 0}[-C_{\alpha\gamma\delta}f_{\delta 0 i}A_{\gamma i}+i(\theta_\alpha)_{ab}\dot{\phi}_a\phi_b] \\ & +\tfrac{1}{2}\Omega^{-1}{}_{\alpha\beta}(\lambda+\phi')[-C_{\alpha\gamma\delta}f_{\delta 0 i}A_{\gamma i}+i(\theta_\alpha)_{ab}\dot{\phi}_a\phi_b][-C_{\beta\epsilon\zeta}f_{\zeta 0 i}A_{\epsilon i}+i(\theta_\beta)_{cd}\dot{\phi}_c\phi_d] \\ & -\tfrac{1}{2}C_{\alpha\beta\gamma}f_{\alpha ij}A_{\beta i}A_{\gamma j}+\tfrac{1}{4}C_{\alpha\beta\gamma}C_{\alpha\delta\epsilon}A_{\beta i}A_{\gamma j}A_{\delta i}A_{\epsilon j} \\ & +i(\theta_\alpha)_{ab}\phi_b'\vec{\nabla}\phi_a'\cdot\vec{\mathbf{A}}_\alpha-\tfrac{1}{2}\vec{\mathbf{A}}_\alpha\cdot\vec{\mathbf{A}}_\beta[(\theta_\alpha)_{ab}(\theta_\beta)_{ac}\phi_b'\phi_c'+\mu^2{}_{\alpha\beta}-\Omega_{\alpha\beta}(\lambda+\phi')] \\ & +i\bar{\psi}\vec{\gamma}t_\alpha\psi\cdot\vec{\mathbf{A}}_\alpha-i\bar{\psi}\gamma^0t_\alpha\psi\,\Omega^{-1}{}_{\alpha\beta}(\lambda+\phi')[-\mu^2{}_{\beta\gamma}A_{\gamma 0}-f_{\delta 0 i}A_{\gamma i}C_{\beta\gamma\delta}+i(\theta_\beta)_{ab}\dot{\phi}_a\phi_b] \\ & +\mathcal{P}(\lambda+\phi')-\tfrac{1}{2}M^2{}_{ab}\phi_a'\phi_b'+\bar{\psi}(\Gamma,\phi')\psi\,.\end{aligned} \tag{6.1}$$

Now, we note that the objects appearing in (6.1) may be divided into two classes:

(A) The quantities $A_{\alpha i}$, $f_{\alpha ij}$, ϕ_a', $\vec{\nabla}\phi_a'$, ψ_n, and $\bar{\psi}_n$, all of whose propagators are given according to Eqs. (5.24)–(5.30) by the *covariant* T^* products defined in Eqs. (5.16)–(5.23).

(B) The quantities $A_{\alpha 0}$, $f_{\alpha 0 i}$, and $\dot{\phi}_a'$, which enter into propagators containing noncovariant local terms [given by Eqs. (5.24), (5.26), and (5.29)] in addition to the covariant T^* products.

We also note that the quantities of type (B) enter in the interaction (6.1) only linearly and quadratically. It is therefore possible to sum up the effects of the noncovariant terms in the propagators precisely, by using techniques developed long ago by Lee and Yang.[16]

Let us denote the general field variables of type (B) as $\beta_N(x)$, and write their propagators as

$$\langle T\{\beta_N(x),\beta_M(y)\}\rangle_0=\langle T^*\{\beta_N(x),\beta_M(y)\}\rangle+i\mathcal{B}_{NM}\,\delta^4(x-y)\,, \tag{6.2}$$

with T^* understood as the covariant propagators defined in Eqs. (5.16)–(5.23). The interaction here may be written as

$$\mathcal{H}'=\mathcal{E}(\alpha)+\mathcal{F}_N(\alpha)\beta_N+\tfrac{1}{2}\mathcal{G}_{NM}(\alpha)\beta_N\beta_M\,, \tag{6.3}$$

where α stands for all the fields of type (A), which have purely covariant propagators. Following Lee and Yang,[16] we can represent the T^* and $\mathcal{B}$ terms in (6.2) by straight and wiggly lines, respectively, and evaluate the effects of the $\mathcal{B}$ terms by summing chains of wiggly lines. The effect of inserting a chain of wiggly lines between two straight lines is to replace $\mathcal{G}(\alpha)$ with

$$\mathcal{G}(\alpha)+\mathcal{G}(\alpha)\mathcal{B}\mathcal{G}(\alpha)+\mathcal{G}(\alpha)\mathcal{B}\mathcal{G}(\alpha)\mathcal{B}\mathcal{G}(\alpha)+\cdots=\mathcal{G}(\alpha)[1-\mathcal{B}\mathcal{G}(\alpha)]^{-1}\,.$$

The effect of inserting a chain of wiggly lines between a straight line and an $\mathcal{F}$ vertex is to replace $\mathcal{F}(\alpha)$ with

$$\mathcal{F}(\alpha)+\mathcal{F}(\alpha)\mathcal{B}\mathcal{G}(\alpha)+\mathcal{F}(\alpha)\mathcal{B}\mathcal{G}(\alpha)\mathcal{B}\mathcal{G}(\alpha)+\cdots=\mathcal{F}(\alpha)[1-\mathcal{B}\mathcal{G}(\alpha)]^{-1}$$

or

$$\mathcal{F}(\alpha)+\mathcal{G}(\alpha)\mathcal{B}\mathcal{F}(\alpha)+\mathcal{G}(\alpha)\mathcal{B}\mathcal{G}(\alpha)\mathcal{B}\mathcal{F}(\alpha)+\cdots=[1-\mathcal{G}(\alpha)\mathcal{B}]^{-1}\mathcal{F}(\alpha)\,.$$

Note that these are equal, because $\mathcal{B}$ and $\mathcal{G}(\alpha)$ are symmetric matrices. Finally, it is possible for wiggly lines to form a ring, producing a contribution

$$\begin{aligned}\tfrac{1}{2}\delta^4(0)[\mathrm{Tr}(\mathcal{G}(\alpha)\mathcal{B})+\tfrac{1}{2}\mathrm{Tr}(\mathcal{G}(\alpha)\mathcal{B}\mathcal{G}(\alpha)\mathcal{B})+\tfrac{1}{3}\mathrm{Tr}(\mathcal{G}(\alpha)\mathcal{B}\mathcal{G}(\alpha)\mathcal{B}\mathcal{G}(\alpha)\mathcal{B})+\cdots] &= -\tfrac{1}{2}\delta^4(0)\,\mathrm{Tr}\ln[1-\mathcal{G}(\alpha)\mathcal{B}] \\ &= -\tfrac{1}{2}\delta^4(0)\ln\mathrm{Det}[1-\mathcal{G}(\alpha)\mathcal{B}]\,.\end{aligned}$$

Thus, we can drop the noncovariant $\mathcal{B}$ term in the propagators (6.2) if we replace the interaction (6.3) with an effective interaction:

$$\mathcal{H}'_{\mathrm{eff}}=-\mathcal{L}'-(i/2)\delta^4(0)\ln\mathrm{Det}[1-\mathcal{B}\mathcal{G}(\alpha)]\,, \tag{6.4}$$

with

$$-\mathcal{L}'\equiv\mathcal{E}(\alpha)+(\mathcal{F}(\alpha)[1-\mathcal{B}\mathcal{G}(\alpha)]^{-1})_N\beta_N+\tfrac{1}{2}(\mathcal{G}(\alpha)[1-\mathcal{B}\mathcal{G}(\alpha)]^{-1})_{NM}\beta_N\beta_M\,. \tag{6.5}$$

(It will turn out that $\mathcal{L}'$ is the interaction Lagrangian.)

In our case, the index N labeling the variables of class $\mathcal{B}$ may be considered to run over the values α, αi, and a, with

$$\beta_\alpha\equiv\mu_{\alpha\beta}A_{\beta 0},\quad \beta_{\alpha i}\equiv f_{\alpha 0 i},\quad \beta_a\equiv\dot{\phi}_a'\,. \tag{6.6}$$

With this normalization of β variables, Eqs. (5.24)–(5.30) give the nonvanishing elements of the $\mathcal{B}$ matrix here as

$$\mathcal{B}_{\alpha\beta}=-\delta_{\alpha\beta},\quad \mathcal{B}_{\alpha i,\beta j}=-\delta_{\alpha\beta}\delta_{ij},\quad \mathcal{B}_{ab}=-\delta_{ab}$$

or more compactly

$$\mathcal{B}=-1\,.$$

Also, inspection of (6.1) shows that the $\mathcal{G}$ matrix, the $\mathcal{F}$ vector, and the function $\mathcal{E}$ here are

$$\mathcal{G}_{\alpha\beta}=[\,\mu\Omega^{-1}(\lambda+\phi')\mu-1]_{\alpha\beta}\,, \tag{6.7}$$

$$\mathcal{G}_{\alpha i,\beta}=\mathcal{G}_{\beta,\alpha i}=[\Omega^{-1}(\lambda+\phi')\mu]_{\delta\beta}C_{\delta\gamma\alpha}A_{\gamma i}\,, \tag{6.8}$$

$$\mathcal{G}_{\alpha i,\beta j}=\Omega^{-1}{}_{\delta\zeta}(\lambda+\phi')C_{\delta\gamma\alpha}C_{\zeta\epsilon\beta}A_{\gamma i}A_{\epsilon j}\,, \tag{6.9}$$

$$\mathcal{G}_{\alpha a}=\mathcal{G}_{a\alpha}=-i(\theta_\beta)_{ab}\phi_b'[\Omega^{-1}(\lambda+\phi')\mu]_{\beta\alpha}\,, \tag{6.10}$$

$$\mathcal{G}_{\alpha i,a}=\mathcal{G}_{a,\alpha i}=-i(\theta_\beta)_{ab}\phi_b'\Omega^{-1}{}_{\delta\beta}(\lambda+\phi')C_{\delta\gamma\alpha}A_{\gamma i}\,, \tag{6.11}$$

$$\mathcal{G}_{ab}=-(\theta_\alpha)_{ac}\phi_c'(\theta_\beta)_{bd}\phi_d'\Omega^{-1}{}_{\alpha\beta}(\lambda+\phi')\,, \tag{6.12}$$

$$\mathcal{F}_\alpha=i\bar\psi\gamma^0 t_\beta\psi[\Omega^{-1}(\lambda+\phi')\mu]_{\beta\alpha}\,, \tag{6.13}$$

$$\mathcal{F}_{\alpha i}=i\bar\psi\gamma^0 t_\beta\psi\Omega^{-1}{}_{\beta\delta}(\lambda+\phi')C_{\delta\gamma\alpha}A_{\gamma i}\,, \tag{6.14}$$

$$\mathcal{F}_a=\bar\psi\gamma^0 t_\beta\psi\Omega^{-1}{}_{\beta\gamma}(\lambda+\phi')(\theta_\gamma)_{ab}\phi_b'\,, \tag{6.15}$$

$$\begin{aligned}\mathcal{E}=&-\tfrac{1}{2}C_{\alpha\beta\gamma}f_{\alpha ij}A_{\beta i}A_{\gamma j}+\tfrac{1}{4}C_{\alpha\beta\gamma}C_{\alpha\delta\epsilon}A_{\beta i}A_{\gamma j}A_{\delta i}A_{\epsilon j}\\&+i(\theta_\alpha)_{ab}\phi_b'(\vec\nabla\phi_a')\cdot\vec{A}_\alpha-\tfrac{1}{2}\vec{A}_\alpha\cdot\vec{A}_\beta[(\theta_\alpha)_{ab}(\theta_\beta)_{ac}\phi_b'\phi_c'-\Omega_{\alpha\beta}(\lambda+\phi')+\mu^2{}_{\alpha\beta}]\\&+i\bar\psi\vec\gamma t_\alpha\psi\cdot\vec{A}_\alpha+\mathcal{P}(\lambda+\phi')-\tfrac{1}{2}M^2{}_{ab}\phi_a'\phi_b'+\bar\psi(\Gamma,\phi')\psi\,.\end{aligned} \tag{6.16}$$

In calculating $\mathcal{H}'_{\text{eff}}$, it will be very convenient to introduce a supermatrix notation, with all α indices lumped together in the left column and/or upper row, and all α, i, *and* a indices in the right column and/or lower row. In this notation, the above formulas for $\mathcal{G}$ and $\mathcal{F}$ become simply

$$\mathcal{G}=\begin{bmatrix}\mu\Omega^{-1}\mu-1 & \mu\Omega^{-1}\mathcal{C}^T\\ \mathcal{C}\Omega^{-1}\mu & \mathcal{C}\Omega^{-1}\mathcal{C}^T\end{bmatrix},\quad \mathcal{F}=\begin{bmatrix}i\mu\Omega^{-1}\bar\psi\gamma^0 t\psi\\ i\mathcal{C}\Omega^{-1}\bar\psi\gamma^0 t\psi\end{bmatrix},$$

where

$$\mathcal{C}_{\alpha i,\beta}\equiv C_{\alpha\beta\gamma}A_{\gamma i},\quad \mathcal{C}_{a,\beta}\equiv -i(\theta_\beta)_{ab}\phi_b'\,.$$

It is easy to check that

$$\begin{aligned}[1-\mathcal{B}\mathcal{G}]^{-1}&=[1+\mathcal{G}]^{-1}\\&=\begin{bmatrix}\mu^{-1}(\Omega+\mathcal{C}^T\mathcal{C})\mu^{-1} & -\mu^{-1}\mathcal{C}^T\\ -\mathcal{C}\mu^{-1} & 1\end{bmatrix}.\end{aligned}$$

The last two terms in (6.5) are then

$$\begin{aligned}\tfrac{1}{2}[\mathcal{G}(1-\mathcal{B}\mathcal{G})^{-1}]_{NM}\beta_N\beta_M&=-\tfrac{1}{2}[(1+\mathcal{G})^{-1}-1]_{NM}\beta_N\beta_M\\&=-\tfrac{1}{2}[\Omega_{\beta\alpha}(\lambda+\phi')-\mu^2{}_{\beta\alpha}-(\theta_\beta)_{ab}\phi_b'(\theta_\alpha)_{ac}\phi_c'+C_{\delta\beta\gamma}C_{\delta\alpha\epsilon}\vec{A}_\gamma\cdot\vec{A}_\epsilon]A_{\beta 0}A_{\alpha 0}\\&\quad+C_{\alpha\beta\gamma}f_{\alpha 0i}A_{\beta 0}A_{\gamma i}-i(\theta_\beta)_{ab}\phi_b'\dot\phi_a'A_{\beta 0}\end{aligned}$$

and

$$\begin{aligned}[(1-\mathcal{B}\mathcal{G})^{-1}\mathcal{F}]_N\beta_N&=[(1+\mathcal{G})^{-1}\mathcal{F}]_N\beta_N\\&=i\bar\psi\gamma^0 t_\alpha\psi A_{\alpha 0}\,.\end{aligned}$$

Putting this together with (6.16), we find for the first term in (6.4)

$$\begin{aligned}-\mathcal{L}'=&-\tfrac{1}{2}A_{\alpha\mu}A^\mu_\beta[(\theta_\alpha)_{ab}(\theta_\beta)_{ac}\phi_b'\phi_c'-\Omega_{\alpha\beta}(\lambda+\phi')+\mu^2{}_{\alpha\beta}]\\&+\tfrac{1}{4}C_{\alpha\beta\gamma}C_{\alpha\delta\epsilon}A_{\beta\mu}A_{\gamma\nu}A^\mu_\delta A^\nu_\epsilon-\tfrac{1}{2}C_{\alpha\beta\gamma}f_{\alpha\mu\nu}A^\mu_\beta A^\nu_\gamma+i(\theta_\alpha)_{ab}\phi_b'\partial_\mu\phi_a'A^\mu_\alpha\\&+i\bar\psi\gamma^\mu t_\alpha\psi A_{\alpha\mu}+\mathcal{P}(\lambda+\phi')-\tfrac{1}{2}M^2{}_{ab}\phi_a'\phi_b'+\bar\psi(\Gamma,\phi')\psi\,,\end{aligned} \tag{6.17}$$

with Ω given by Eq. (4.11):

$$\Omega_{\alpha\beta}(\phi) \equiv (\theta_\alpha \phi, \theta_\gamma \lambda)\mu^{-2}{}_{\gamma\delta}(\theta_\delta \lambda, \theta_\beta \phi) \,.$$

This is covariant, and in fact is just what would be obtained by the naive procedure of using the negative of the interaction part of the Lagrangian as our interaction Hamiltonian. However, we also must deal with the second term in (6.4). To calculate the determinant, we note that

$$[1-\mathcal{B}\mathcal{G}]^{-1} = \begin{bmatrix} \mu^{-1}\Omega^{1/2} & 0 \\ 0 & 1 \end{bmatrix} \begin{bmatrix} 1+\Omega^{-1/2}\mathcal{C}^T\mathcal{C}\,\Omega^{-1/2} & -\Omega^{-1/2}\mathcal{C}^T \\ -\mathcal{C}\,\Omega^{-1/2} & 1 \end{bmatrix} \begin{bmatrix} \Omega^{1/2}\mu^{-1} & 0 \\ 0 & 1 \end{bmatrix} .$$

The matrix in the middle is of the form

$$\begin{bmatrix} 1+A^TA & A^T \\ A & 1 \end{bmatrix} , \quad A = -\mathcal{C}\Omega^{-1/2} ,$$

and therefore, according to Appendix A, has unit determinant. Thus, by the product rule,

$$\begin{aligned} \mathrm{Det}(1-\mathcal{B}\mathcal{G})^{-1} &= \mathrm{Det}(\mu^{-1}\Omega^{1/2})\,\mathrm{Det}(\Omega^{1/2}\mu^{-1}) \\ &= \mathrm{Det}(\mu^{-2}\Omega) \\ &= \mathrm{Det}(\mu^{-2}\Phi\,\mu^{-2}\Phi) \\ &= [\mathrm{Det}(\mu^{-2}\Phi)]^2 \,, \end{aligned}$$

where Φ is the matrix

$$\Phi_{\beta\alpha} \equiv -(\theta_\beta\lambda, \theta_\alpha\phi) \,. \tag{6.18}$$

The effective interaction Hamiltonian is then

$$\mathcal{H}'_{\text{eff}} = -\mathcal{L}' + i\,\delta^4(0)\ln \mathrm{Det}(\mu^{-2}\Phi) \,. \tag{6.19}$$

This is just the result that would be obtained by a path-integral quantization[10] of this theory in a gauge in which the propagators are given by the T^* products (5.16)–(5.22).

VII. SIMPLE THEORIES

As an example, suppose we are dealing with a scalar field representation D_B so simple that every vector in the representation can be rotated into a fixed direction by applying a gauge transformation in the group $\mathcal{G}$. For the gauge group SU(2) or U(2), this is the case if D_B consists of just a single complex doublet, as in the old model of leptons,[1] but it is *not* the case if D_B contains, say, a doublet and a triplet, or two doublets with different charge assignments.

If $\phi(x)$ at each point can be rotated into a fixed direction, then this must also be the direction of the vacuum expectation value of $\phi(x)$ in the tree approximation. Thus there is a "gauge" in which

$$\phi_p(x) = [1+\chi(x)]\lambda_p \,, \tag{7.1}$$

where $\chi(x)$ is a scalar field, the only one present in this gauge. This obviously is a "unitarity gauge," because the antisymmetry of θ_α gives

$$(\theta_\alpha\lambda, \phi(x)) = [1+\chi(x)](\theta_\alpha\lambda, \lambda) = 0 \,.$$

In these simple theories, the existence proof for the unitarity gauge presented in Sec. III is unnecessary.

Equation (7.1) allows an instant evaluation of the matrix (6.18)

$$\begin{aligned} \Phi_{\beta\alpha} &= -(1+\chi)(\theta_\beta\lambda, \theta_\alpha\lambda) \\ &= (1+\chi)\mu^2{}_{\beta\alpha} \,. \end{aligned} \tag{7.2}$$

The effective Hamiltonian (6.19) is then simply

$$\mathcal{H}'_{\text{eff}} = -\mathcal{L}'(x) + iN\delta^4(0)\ln[1+\chi(x)] \,, \tag{7.3}$$

where N is the number of θ_α generators.

VIII. THEORIES WITH UNBROKEN SUBGROUPS

The quantization program carried out in Secs. IV–VII is strictly applicable only to theories in which all gauge symmetries are broken. Let us now briefly return to the general case, and consider a theory with an unbroken subgroup $\mathcal{S}$.

The generators of $\mathcal{S}$ are simply those independent linear combinations $C_\alpha T_\alpha$ of the generators of $\mathcal{G}$ for which $1+i\epsilon C_\alpha\theta_\alpha$ leaves λ invariant:

$$C_\alpha\theta_\alpha\lambda = 0 \,. \tag{8.1}$$

Each such set of coefficients C_α forms an eigenvector of the vector-meson mass matrix with eigenvalue zero:

$$\mu^2{}_{\alpha\beta}C_\beta = -(\theta_\alpha\lambda, \theta_\beta\lambda)C_\beta = 0 \,.$$

Since μ^2 is a real symmetric matrix, we can find a complete set of orthonormal eigenvectors

$$\mu^2{}_{\alpha\beta} C_{\beta N} = \mu_N{}^2 C_{\alpha N} , \tag{8.2}$$

$$C_{\alpha N} C_{\alpha M} = \delta_{NM} , \tag{8.3}$$

$$\sum_N C_{\alpha N} C_{\beta N} = \delta_{\alpha\beta} , \tag{8.4}$$

among which must be all the vectors C_α satisfying (8.1). Thus, if we define new generators

$$\overline{T}_N \equiv C_{\alpha N} T_\alpha , \tag{8.5}$$

then the generators of $\mathcal{S}$ are just those $\overline{T}_N$ for which $\mu_N = 0$ and $C_{\alpha N}$ satisfies (8.1):

$$C_{\alpha N}\,\theta_\alpha \lambda = 0 \quad \text{for} \quad \overline{T}_N \in \mathcal{S} .$$

It is very convenient to use the $C_{\alpha N}$ also to define new canonically normalized vector fields

$$\overline{A}_{N\mu} \equiv C_{\alpha N} A_{\alpha\mu} . \tag{8.6}$$

Using (8.4), this gives

$$A_{\alpha\mu} = \sum_N C_{\alpha N} \overline{A}_{N\mu} . \tag{8.7}$$

Thus the vector-meson mass term in $\mathcal{H}_0$ may be written

$$\tfrac{1}{2}\mu^2{}_{\alpha\beta} A_{\alpha\mu} A^\mu_\beta = \tfrac{1}{2}\sum_N \mu_N{}^2 \overline{A}_{N\mu} \overline{A}^\mu_N . \tag{8.8}$$

The fields $\overline{A}_{N\mu}$ thus describe particles of definite mass. In particular, if $\overline{T}_N$ is one of the generators of the unbroken subgroup $\mathcal{S}$, then $\mu_N = 0$, so $\overline{A}_{N\mu}$ describes a particle of zero mass. Also, the interaction of gauge fields with spin-zero or spin-one-half fields is described by the matrices

$$t_\alpha A_{\alpha\mu} = \sum_N \overline{t}_N \overline{A}_{N\mu} , \tag{8.9}$$

$$\theta_\alpha A_{\alpha\mu} = \sum_N \overline{\theta}_N \overline{A}_{N\mu} , \tag{8.10}$$

where

$$\overline{t}_N \equiv C_{\alpha N} t_\alpha , \tag{8.11}$$

$$\overline{\theta}_N \equiv C_{\alpha N} \theta_\alpha . \tag{8.12}$$

This formalism is particularly useful when we do not have a complete model of the symmetry-breaking mechanism.[14] For instance, consider the gauge group SU(2)⊗U(1) used in the old model of leptons,[1] with gauge coupling constants g and g' associated with the SU(2) generators I_i and the U(1) generator Y, respectively. The charge is defined here by

$$Q = I_3 - Y . \tag{8.13}$$

Remember that we have agreed to absorb the gauge coupling constants into the generators, so this should be written

$$Q = \frac{1}{g} T_3 - \frac{1}{g'} T_Y , \tag{8.14}$$

where

$$\vec{T} \equiv g\vec{I}, \quad T_Y \equiv g'Y . \tag{8.15}$$

But Q is the generator of the unbroken subgroup $\mathcal{S}$, so we can read off from (8.14) the $C_{\alpha N}$ coefficients appearing in Eq. (8.5):

$$C_{3Q} \propto \frac{1}{g}, \quad C_{YQ} \propto -\frac{1}{g'} ,$$

or, recalling that C must be orthogonal,

$$C_{3Q} = \frac{g'}{(g^2+g'^2)^{1/2}} , \quad C_{YQ} = \frac{-g}{(g^2+g'^2)^{1/2}} . \tag{8.16}$$

The canonically normalized massless vector field coupled to Q is then

$$A_\mu = \frac{1}{(g^2+g'^2)^{1/2}} (g' A_{3\mu} - g A_{Y\mu}) . \tag{8.17}$$

Also, in this model the other neutral generator is defined by a vector orthogonal to (8.16):

$$C_{3Z} = \frac{g}{(g^2+g'^2)^{1/2}} , \quad C_{YZ} = \frac{g'}{(g^2+g'^2)^{1/2}} , \tag{8.18}$$

so the canonically normalized neutral field of nonzero mass is

$$Z_\mu = \frac{1}{(g^2+g'^2)^{1/2}} (g A_{3\mu} + g' A_{Y\mu}) . \tag{8.19}$$

The operators to which A_μ and Z_μ couple are respectively

$$A_\mu:\ C_{3Q} T_3 + C_{YQ} T_Y = \frac{gg'}{(g^2+g'^2)^{1/2}} (I_3 - Y) , \tag{8.20}$$

$$Z_\mu:\ C_{3Z} T_3 + C_{YZ} T_Y = \frac{(g^2 I_3 + g'^2 Y)}{(g^2+g'^2)^{1/2}} . \tag{8.21}$$

We see in particular that the electric charge of a particle with $I_3 - Y = +1$ is

$$e = \frac{gg'}{(g^2+g'^2)^{1/2}} . \tag{8.22}$$

These results are the same as those derived previously[1] under specific assumptions as to the mechanism of symmetry breaking.

The direct quantization of theories with an unbroken gauge symmetry subgroup $\mathcal{S}$ presents well-known difficulties. We can however break $\mathcal{S}$ weakly by adding an additional multiplet $\Delta\phi$ of scalar fields whose zeroth-order vacuum expectation value $\Delta\lambda$ is small, but is not annihilated by the $\overline{\theta}_N$ generators belonging to $\mathcal{S}$. Since the θ matrices act separately on ϕ and $\Delta\phi$, the matrix Φ now is

$$\Phi_{\beta\alpha}=(\theta_\beta\lambda,\theta_\alpha\phi)+(\theta_\beta\Delta\lambda,\theta_\alpha\Delta\phi)\ . \qquad (8.23)$$

This matrix can be put in a supermatrix form

$$\Phi=\begin{pmatrix}\Phi_{BB} & \Phi^T_{UB}\\ \Phi_{UB} & \Phi_{UU}\end{pmatrix},$$

where U refers to the generators $\overline{T}_N$ of $\mathcal{G}$ which span $\mathcal{S}$, and B refers to the other, strongly broken generators. (Note that Φ is symmetric). Both Φ_{UB} and Φ_{UU} receive contributions only from the second term in (8.23), so as long as $\Delta\lambda \ll \lambda$, we have

$$\Phi_{UB}\ll\Phi_{BB},\quad \Phi_{UU}\ll\Phi_{BB}\ ,$$

so to lowest nonvanishing order in terms of order $\Delta\lambda$,

$$\mathrm{Det}\Phi \to (\mathrm{Det}\Phi_{BB})(\mathrm{Det}\Phi_{UU})$$

and Eq. (6.19) becomes

$$\mathcal{H}'_{\text{eff}}\simeq -\mathcal{L}'+i\delta^4(0)\ln\mathrm{Det}\Phi_{BB}+i\delta^4(0)\ln\mathrm{Det}\Phi_{UU}\ . \qquad (8.24)$$

The last term involves $\Delta\phi$, not ϕ, and is presumably what is needed to provide a smooth limit as various gauge boson masses tend to zero. Thus, if we put aside such complications, we may conjecture that the determinant of Φ is to be evaluated taking account only of the *broken* generators for which $\overline{\theta}_N\lambda\neq 0$. In particular, in simple theories the number N in Eq. (7.3) is the number of *broken* generators[17] of $\mathcal{G}$.

The last remarks are intended as suggestive rather than conclusive. A more thorough study of gauge theories with unbroken gauge subgroups would be useful.

ACKNOWLEDGMENTS

I am grateful for valuable assistance received in conversations with Roman Jackiw and Ben Lee.

APPENDIX A: EVALUATION OF A DETERMINANT

We wish to calculate the determinant of a matrix S, expressed as a supermatrix

$$S=\begin{bmatrix}1+A^TA & A^T\\ A & 1\end{bmatrix},$$

where A is an arbitrary real matrix, not necessarily square. There is probably some easy way to do this calculation, but I have been able to manage it only by considering the eigenvalues λ of S. Writing the eigenvector Ψ as

$$\Psi=\begin{bmatrix}u\\ v\end{bmatrix},$$

the eigenvalue equation reads

$$S\Psi=\lambda\Psi$$

or in more detail

$$(1+A^TA)u+A^Tv=\lambda u\ ,$$

$$Au+v=\lambda v.$$

For every possible value of λ, other than $\lambda=1$, we may solve for v:

$$v=\frac{1}{\lambda-1}Au$$

and write the eigenvalue equation in terms of u alone

$$\lambda A^TAu=(\lambda-1)^2u\ .$$

Thus u must be an eigenvector of A^TA, with a necessarily positive eigenvalue $\alpha\geq 0$. For each such eigenvalue α, there will be two eigenvalues λ of S, given by the two roots of the quadratic equation

$$(\lambda-1)^2=\lambda\alpha\ .$$

These two roots have product unity. Thus, the eigenvalues S are either unity or come in reciprocal pairs λ_n, $1/\lambda_n$. Since S is real and symmetric, it can be put in the form

$$S=O^T\begin{bmatrix}1 &&&&&&&\\ & 1 &&&&&&\\ && \ddots &&&&&\\ &&& 1 &&&&\\ &&&& \lambda_1 &&&\\ &&&&& \lambda_1^{-1} &&\\ &&&&&& \lambda_2 &\\ &&&&&&& \lambda_2^{-1}\\ &&&&&&&& \ddots\end{bmatrix}O\ ,$$

where O is an orthogonal matrix. Inspection now yields our result

$$\mathrm{Det}\,S=1\ .$$

Note added in proof. M. Grisaru has pointed out to me that the result derived in Appendix A can be obtained much more easily by factoring the matrix S into a product of two matrices having ones on the main diagonal and zeros either above or below it.

*This work is supported in part through funds provided by the Atomic Energy Commission under Contract AT (11-1)-3069.

[1]S. Weinberg, Phys. Rev. Letters 19, 1264 (1967). For subsequent references see the rapporteur's talk by B. W. Lee, in Proceedings of the XVI International Conference on High Energy Nuclear Physics, Batavia, Ill., 1972 (to be published).

[2]Among early papers on the unification of weak and electromagnetic interactions, see J. Schwinger, Ann. Phys. (N.Y.) 2, 407 (1957); A. Salam and J. Ward, Nuovo Cimento 11, 568 (1959); S. L. Glashow, Nucl. Phys. 22, 579 (1961); A. Salam and J. Ward, Phys. Letters 13, 168 (1964); also see A. Salam, in *Elementary Particle Theory,* edited by N. Svartholm (Wiley, New York, 1969).

[3]S. Weinberg, Phys. Rev. Letters 29, 388 (1972); H. Georgi and S. L. Glashow, Phys. Rev. D 6, 2977 (1972).

[4]P. W. Higgs, Phys. Rev. Letters 12, 132 (1964); 13, 508 (1964); Phys. Rev. 145, 1156 (1966); F. Englert and R. Brout, Phys. Rev. Letters 13, 321 (1964); G. S. Guralnik, C. R. Hagen, and T. W. B. Kibble, *ibid.* 13, 585 (1965); T. W. Kibble, Phys. Rev. 155, 1554 (1967).

[5]S. Weinberg, Phys. Rev. Letters 27, 1688 (1971); For a survey of calculations carried out so far, see the report of J. Primack in the Proceedings of the XVI International Conference on High Energy Nuclear Physics, 1972 (to be published).

[6]G. 't Hooft, Nucl. Phys. B35, 167 (1971).

[7]B. W. Lee, Phys. Rev. D 5, 823 (1972).

[8]J. Goldstone, A. Salam, and S. Weinberg, Phys. Rev. 127, 965 (1962).

[9]A more complicated proof is given by Kibble, Ref. 4.

[10]For specific examples, see 't Hooft, Ref. 6, and Lee, Ref. 7. The general case has been worked out by R. Jackiw (unpublished). I will present the details in a forthcoming article on perturbative symmetry-breaking calculations.

[11]C. N. Yang and R. L. Mills, Phys. Rev. 96, 191 (1954); R. Utiyama, *ibid.* 101, 1597 (1956); M. Gell-Mann and S. Glashow, Ann. Phys. (N.Y.) 15, 437 (1961).

[12]It is at this point that we use our assumption that $\mathcal{G}$ is a *local* group. Kibble in Ref. 4 shows in detail how the gauge transformation transfers the Goldstone boson degrees of freedom to the gauge fields, but this is quite unnecessary in proving the existence of the unitarity gauge.

[13]It should be stressed that we did not have to assume in this proof that ϕ furnishes an irreducible representation of $\mathcal{G}$. Thus, we could add as many boson multiplets as we liked to theories like the model of leptons of Ref. 1, and still be confident of the existence of a gauge in which the linear combinations of these multiplets corresponding to Goldstone bosons were absent.

[14]S. Weinberg, Phys. Rev. D 5, 1962 (1972).

[15]This problem does not arise in the class of "simple" theories discussed here in Sec. VII, which includes most of the models considered in Refs. 1, 6, and 7.

[16]T. D. Lee and C. N. Yang, Phys. Rev. 128, 885 (1962); also see J. Honerkamp and K. Meetz, Phys. Rev. D 3, 1996 (1971); J. Charap, *ibid.* 3, 1998 (1971); I. S. Gerstein, R. Jackiw, B. W. Lee, and S. Weinberg, *ibid.* 3, 2486 (1971).

[17]The coefficient of the logarithm in the $SU(2)\times U(1)$ model of leptons was given as $6i\delta^4(0)$ in Ref. 5. This was a mistake, arising from neglect of the factor $\frac{1}{2}$ in Eq. (6.4). The number of broken generators in this model is three, and the correct coefficient of the logarithm is therefore $3i\delta^4(0)$.

HIGH ENERGY BEHAVIOUR OF TREE DIAGRAMS IN GAUGE THEORIES

J.S. BELL
CERN – Geneva

Received 14 May 1973

Abstract: Tree diagram scattering amplitudes may behave extra badly at high energy when vector mesons are involved. The cancellation, among bad terms, implied by gauge invariance is discussed. It is shown that in the renormalizable spontaneously broken gauge theories such cancellation is complete.

It is well known that in quantum field theories with vector mesons tree diagrams for scattering amplitudes may behave badly at high energy, and that this is related somehow to non-renormalizability [1]. It is well known that gauge invariance leads to cancellation among the bad terms [2] and it is natural to conjecture that these cancellations are complete in spontaneously broken gauge (SBG) theories* which are supposed to be renormalizable [3]. This will be demonstrated here. It seems that no one had bothered to do this already; but now Llewellyn Smith [4] has studied the (more interesting) inverse question: does the requirement of good high energy behaviour (for physical scattering amplitudes in the tree approximation) dictate SBG structure? The following may be a useful preface to his paper.

First, for ease of reference, it will be recalled how gauge invariance permits the complete elimination of the bad terms in the Abelian case – i.e., for "massive photons". Then the point at which such arguments fail in the non-Abelian case of massive Yang-Mills fields will be indicated. Finally it will be shown that related arguments again go through in the SBG case. Since we are concerned only with tree diagrams it is not necessary to invoke the full machinery of quantized gauge theories; classical field theory is sufficient [5].

Consider a Langragian

$$L = -\tfrac{1}{2}(\partial_\mu W_\nu - \partial_\nu W_\mu)^2 - \tfrac{1}{2}M^2 W_\mu^2 + J_\mu W_\mu - R\,, \tag{1}$$

where R is a polynominal residual Lagrangian not containing W in linear or bilinear terms. The equation of motion for W is

$$\partial_\mu \partial_\mu W_\nu - \partial_\nu \partial_\mu W_\mu - M^2 W_\nu = \frac{\delta R}{\delta W_\nu} - J_\nu\,. \tag{2}$$

* For a review see Lee, ref. [3].

With Feynman boundary conditions this can be cast into integral form

$$W_\mu = \int \mathrm{d}y\, G_{\mu\nu}(x,y)\left\{J_\nu(y) - \frac{\delta R}{\delta W_\nu}(y)\right\}, \tag{3}$$

where $G_{\mu\nu}$ is the Fourier transform of

$$(S_{\mu\nu} + k_\mu k_\nu/M^2)/(M^2 + k^2 - i\epsilon). \tag{4}$$

Treating all fields and J as small, an iterative solution of (3) yields a series of terms which can be represented in an obvious way *à la* Feynman by tree diagrams, such as

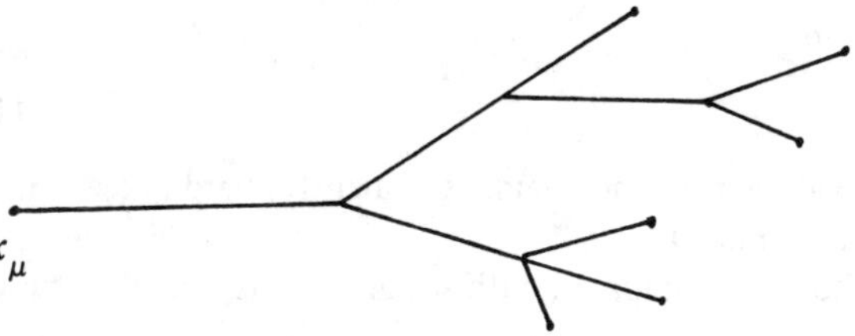

x_μ (5)

The end x_μ is associated with the point x_μ at which the field is evaluated, and the other ends are associated with various arguments $y, z, \ldots$ of the source J. In this way

$$W_\mu(x) = \int \mathrm{d}y\, G_{\mu\nu}(x,y)\, J_\nu(y) + \tfrac{1}{2!}\int \mathrm{d}y\,\mathrm{d}z\, G_{\mu\nu\lambda}(x,y,z)\, J_\nu(y)\, J_\lambda(z) + \ldots, \tag{6}$$

where

$$G_{\mu\nu\lambda\ldots}(x, y, z, \ldots) \tag{7}$$

is the sum of contributions from all tree diagrams with the appropriate end points. It is important to note that despite the unsymmetric construction described here, which seems to distinguish the arguments (μ, x), the functions (7) are actually fully symmetric in their end points. We identify the quantities (7) with the Feynman n-point functions (in the tree aproximation) of the quantum theory. Physical scattering matrix elements are then obtained by the following steps:

(i) Fourier transform:

$$\widetilde{G}_{\mu\nu\ldots}(k, l\ldots) = \int \mathrm{d}x\,\mathrm{d}y \ldots\ \mathrm{e}^{ikx+ily+\ldots}\, G_{\mu\nu\ldots}(x, y\ldots). \tag{8}$$

(ii) Cancel the propagator denominators $(M^2 + k^2)$ associated with external lines:

$$\Gamma_{\mu\nu\ldots}(k, l\ldots) = \widetilde{G}_{\mu\nu\ldots}(k, l\ldots)(M^2 + k^2)(M^2 + l^2)\ldots \tag{9}$$

(iii) Put the external momenta on their respective mass shells:

$$k^2 \to -M^2,\ l^2 \to -M^2, \ldots \tag{10}$$

(iv) Contract the free tensor indices with physical polarization vectors

$$\epsilon_\mu \epsilon'_\nu \ldots \Gamma_{\mu\nu\ldots}(k, l, \ldots) . \tag{11}$$

This is the scattering matrix element. Note that

$$k_\mu \epsilon_\mu = 0, \quad l_\mu \epsilon'_\mu = 0, \ldots, \tag{12}$$

and note that for zero helicity

$$\epsilon_\mu = \left(\frac{k}{|k|}\frac{k_0}{M}, i\frac{|k|}{M}\right) = \frac{k_\mu}{M} + \rho_\mu , \tag{13}$$

where ρ_μ is of order (M/k_0) and

$$k_\mu \rho_\mu = M \tag{14}$$

The extension of all this to allow for other fields is straightforward (except that the fermion sources must anticommute).

Associated with vector mesons, as compared with scalar mesons, there are two potential sources of bad behaviour:

(i) The propagator (4), because of the $k_\mu k_\nu/M^2$ in the numerator, does not fall off for large k_μ. (15)

(ii) The zero helicity polarization vector (14) increases for large k_μ. (16)

We have to study to what extent gauge invariance allows these dangers to be transformed away.

Actually it is possible (although not always very useful, as will be seen) to dispose of point (16) immediately by the Veltman trick. After step (10), in the formation of scattering matrix elements, there is still a factor

$$\delta_{\mu\nu} + k_\mu k_\nu/M^2 \tag{17}$$

associated with each external line. Note that

$$k_\mu (\delta_{\mu\nu} + k_\mu k_\nu/M^2) = k_\mu (M^2 + k^2)/M^2$$

is zero on mass shell. So the bad polarization vector (13) may systematically be replaced by the well-behaved ρ_μ:

$$\epsilon_\mu \to \rho_\mu . \tag{18}$$

We will initially suppose this to be done and concentrate on difficulty (15).

Suppose that only the mass and source terms in (1) spoil invariance under infinitesimal variations

$$W_\mu \to W_\mu - \partial_\mu \omega , \tag{19}$$

with associated variations of other fields. Then if this variation is made away from a solution of the equations of motion we must have*

* Such reasoning is used for example in ref. [6].

$$\delta \int \mathrm{d}x\, (J_\mu W_\mu - \tfrac{1}{2} M^2 W_\mu^2) = 0 \,,$$

because the action as a whole is stationary and any variation could come only from the gauge breaking part. So

$$\int \mathrm{d}x\, \partial_\mu \omega\, (J_\mu - M^2 W_\mu) = 0 \,,$$

whence

$$\partial_\mu W_\mu = M^{-2} \partial_\mu J_\mu \,. \tag{20}$$

Introducing this into (2)

$$(\partial_\mu \partial_\mu - M^2)\, W_\nu = \frac{\delta R}{\delta W_\nu} - J_\nu + M^{-2} \partial_\nu \partial_\mu J_\mu \,. \tag{21}$$

Starting from (21) rather than (2) we come to a diagram expansion involving the good propagator

$$\delta_{\mu\nu}/(M^2 + k^2 - i\epsilon) \,, \tag{22}$$

rather than the bad (4). However, an extra source term has appeared:

$$M^{-2}\, \partial_\nu \partial_\mu J_\mu \,. \tag{23}$$

When the current is finally replaced by a polarization vector this gives

$$-M^{-2} k_\nu k_\mu \epsilon_\mu = 0 \,,$$

or for the zero-helicity residual vector ρ of (13)

$$-M^{-2} k_\nu k_\mu \rho_\mu = -k_\nu / M \,,$$

i.e., the extra source term just restores the bad part of the zero helicity state (13), recreating difficulty (16) for all but one of the external lines. Accepting this, renouncing the Veltman trick, and anticipating (12), the extra source can be dropped.

In the present case one can go further. For this purpose it is useful to note that the modified equation

$$(\partial_\mu \partial_\mu - M^2) W_\nu = \frac{\delta R}{\delta W_\nu} - J_\nu \,, \tag{24}$$

follows directly from the modified Lagrangian

$$L - \tfrac{1}{2}\, (\partial_\mu W_\mu)^2 \,.$$

Here the extra terms, as well as the original mass term, breaks gauge invariance. The argument giving (20) now gives instead

$$(M^2 - \partial_\mu \partial_\mu)\, \partial_\nu W_\nu = \partial_\nu J_\nu \,, \tag{25}$$

or

$$\partial_\nu W_\nu(x) = \int \mathrm{d}y\, G(x,y)\, \partial_\nu J_\nu \,,$$

where G is the Feynman propagator. It follows that $\partial_\nu W_\nu$ is again just linear in the source, and so accounted for entirely by the trivial diagram

x_μ

For interesting diagrams we have effectively

$$\partial_\mu W_\mu = 0 \,.$$

Then

$$k_\mu \Gamma_{\mu\nu\ldots}(k, l \ldots) = 0 \,,$$

with $\Gamma_{\mu\nu\ldots}$ defined as in (9) but in the modified formalism of (25). By symmetry

$$l_\nu \Gamma_{\mu\nu\ldots}(k, l, \ldots) = 0$$

etc. So once again the bad terms in the zero helicity states can be dropped, and there are no bad terms left. So much for the Abelian case.

Note by the way that in this modified formalism the limit $M \to 0$ presents no problem. In particular scattering matrix elements involving zero helicity states go to zero. A massless theory can be defined in this way.

So far the only external lines considered were associated with a single species of vector meson. The role of other particles in general will be sufficiently illustrated by that of other vector mesons, which in any case are of particular interest. Suppose then that in the Lagrangian the quadratic and bilinear terms involving vector mesons are those displayed explicitely in

$$L = -\tfrac{1}{4}(\partial_\mu W^n_\nu - \partial_\nu W^n_\mu)^2 - \tfrac{1}{2}(MW^n_\mu)^2 + J^n_\mu W^n_\mu - R \,, \tag{26}$$

where the summation convention operates, and

$$MW^n_\mu \equiv M_n W^n_\mu \,.$$

Suppose that only the W mass and source terms in (26) spoil invariance under

$$\delta W^l_\mu = -\partial_\mu \omega^l + \Lambda^l_{nm}\, \omega^n W^m_\mu \,, \tag{27}$$

(and related variations of other fields) where the Λ's are a set of numbers antisymmetric for exchange of any pair of indices.

Consider now the application of the previous argument to a particular W^n, say W^0. The relevant gauge transformation is with

$$\omega^l = \delta_{l0}\, \omega \,. \tag{28}$$

Because of the antisymmetry of the Λ's, (27) then reduces for $l = 0$ to (19). The subsequent argument, however, must be modified because the extra source terms do, and the other mass terms may, also spoil gauge invariance. As a result (20) and (21) are replaced by

$$\partial_\mu W^0_\mu = M_0^{-2} \{\partial_\mu J^0_\mu + \Lambda^0_{lm} J^l_\mu W^m_\mu - \Lambda^0_{lm} (M^2 W^l_\mu) W^m_\mu\} \tag{29}$$

$= D$, say ,

$$(\partial_\mu \partial_\mu - M_0^2)\, W^0_\nu = \frac{\delta R}{\delta W^0_\nu} - J^0_\nu + \partial_\nu D\,. \tag{30}$$

The extra source term $\partial_\mu J_\mu$ can again be dropped at the expense of renouncing the Veltman trick (18). There is now another extra source in (29), linear in field strength W. This gives rise to diagrams in which at least one source is connected to the rest of the diagram by two lines rather than one. Such contributions do not have the one particle poles required to survive steps (9) and (10) above, and so do not contribute to physical scattering matrix elements. So reasoning for each vector meson in turn the original equations of motion

$$\partial_\mu \partial_\mu W^n_\nu - \partial_\nu \partial_\mu W^n_\nu - M^2 W^n_\mu = \frac{\delta R}{\delta W^n_\nu} - J^n_\nu\,, \tag{31}$$

can be replaced by

$$\partial_\mu \partial_\mu W^n_\nu - M^2 W^n_\nu = \frac{\delta R}{\delta W^n_\nu} - J^n_\nu - M_n^{-2}\, \partial_\nu \{\Lambda^n_{lm} (M^2 W^l_\mu) W^m_\mu\}. \tag{32}$$

Then danger (15) has again been eliminated at the expense of having some new vertices in the diagrams (and of restoring difficulty (16) for all but one external line). The extra vertex terms in (32) cancel against one another for given n (because of the antisymmetry of Λ) when the masses of all other mesons W^l, with $l \neq n$, are equal. In the completely symmetric case, all masses m_n equal, the extra vertices go away completely. More generally they do not pose a new threat of bad high energy behaviour, because their coupling constants

$$(M_l^2/M_n^2)\, \Lambda^n_{lm}$$

are dimensionless. .

To deal again with (16), one might note again that in the fully symmetric case the modified equations

$$(\partial_\mu \partial_\mu - M_n^2) W^n_\nu = \frac{\delta R}{\delta W^n_\nu} - J^n_\nu \tag{33}$$

again follow from the modified Lagrangian

$$L - \tfrac{1}{2} (\partial_\mu W^n_\mu)^2\,. \tag{34}$$

Then we again have

$$\delta \int \mathrm{d}x\, (-\tfrac{1}{2} (\partial_\mu W^n_\mu)^2 - \tfrac{1}{2} (M W^n_\mu)^2 + J^n_\mu W^n_\mu) = 0\,,$$

under gauge variations (27). In the particular case (28) for example this gives

$$0 = \int \mathrm{d}x \; \{M^2 W^0_\mu \partial_\mu \omega - J^0_\mu \partial_\mu \omega + J^l_\mu \Lambda^0_{lm} \omega W^m_\mu$$

$$+ \partial_\mu W^0_\mu \partial_\nu \partial_\nu \omega - \partial_\mu W^l_\mu \partial_\nu (\Lambda^0_{lm} \omega W^m_\nu)\} ,$$

whence

$$(M^2 - \partial_\nu \partial_\nu) \partial_\mu W^0_\mu = \partial_\mu J^0_\mu + \Lambda^0_{lm} J^l_\mu W^m_\mu + \Lambda^0_{lm} W^m_\nu \partial_\nu \partial_\mu W^l_\mu \; . \tag{35}$$

Because of the last term here, which is not a source term, this is much less immediately useful than (25). It does permit the conclusion that $\partial_\mu W_\mu$ is zero when all sources are finally on-mass shell and divergenceless (i.e., anticipating (12)). This permits the Veltman trick in one external line, but this we had already.

So for massive Yang-Mills fields we can eliminate the propagator difficulty (15), but have not entirely disposed of the zero helicity difficulty (16). There could remain as many bad powers of energy as there are zero helicity external lines less two. Less two, because the Veltman trick for one line disposes of two powers of k_0

$$\rho_\mu = O\left(\frac{M}{k_0}\right)^2 \epsilon_\mu \; .$$

It is notable that symmetry breaking mass terms do not weaken this conclusion. Note, however, that symmetry breaking Fermion masses, m, of spin-$\frac{1}{2}$ fields could be more serious, introducing extra vertices in (32) with dangerously dimensional coupling constants

$$\alpha \, m/M^2 \; .$$

The fact that we have not succeeded in explicitely eliminating the bad terms does not of course prove that they do not cancel among one another in some way that could be exhibited by being smarter. That they do not cancel has, however, been shown for particular examples by explicit calculation [4–7].

The persistence of powers of (k/M) when all momenta are scaled up implies of course the persistence of such powers when the masses are scaled down. With dimensionless coupling constants these limits are equivalent. Thus the zero mass limit of the Yang Mills theory does not exist, even at the tree diagram level [7, 8].

Consider finally SBG theories. We start from a Lagrangian

$$L = -\tfrac{1}{4}(\partial_\mu W^n_\nu - \partial_\nu W^n_\mu)^2 - \tfrac{1}{2}(M W^n_\mu + \partial_\mu \theta^n)^2 - R \; . \tag{36}$$

It is supposed that a scalar field θ occurs in association with each massive W; that the only linear or bilinear terms involving W or θ are those displayed explicitly; that the residual terms R are polynomial and have no dangerously dimensional couplings (i.e., no inverse masses). There is supposed to be exact gauge symmetry (i.e., unbroken even by masses) under

$$\delta W^n_\mu = -\partial_\mu \omega^n + \dots ,$$

$$\delta \theta^n = M\omega^n \quad + \dots , \tag{37}$$

$$\delta \phi = \quad + \dots ,$$

where the residual terms ... are field dependent, and ϕ stands for any other of the fields in the theory.

By a change of variables

$$W^n_\mu \to W^n_\mu - M_n^{-1} \partial_\mu \phi^n + \dots ,$$

$$\theta^n \to \theta^n + \dots \quad , \tag{38}$$

$$\phi \to \phi + \dots \quad ,$$

(where the residual terms ... are non-linear in fields) the variables θ can be eliminated from the Lagrangian. Thus these fields are considered as "ghost" fields, not assiciated with the asymptotic propagation of free particles. It is at this stage, with the ghost free Lagrangian, that we add source terms (for fields other than θ) and define n-point functions and scattering matrix elements in the usual way.

Add now to the starting Lagrangian (36), following 't Hooft [3], a term

$$-\tfrac{1}{2} (\partial_\mu W^n_\mu + M\theta^n)^2 . \tag{39}$$

At the "ghost-free" stage this would add

$$-\tfrac{1}{2} F_n F_n , \tag{40}$$

with

$$F_n = \partial_\mu W^n_\mu + M\theta^n - M_n^{-1} \partial_\mu \partial_\mu \theta^n + \dots . \tag{41}$$

This would restore some θ dependence to the otherwise ghost free Lagrangian. Varying (40) with respect to the θ's would yield the equations of motion

$$\sum_m \{(M^{-1}_n \partial_\mu \partial_\mu - M_n) \delta_{nm} + \dots\} F_m = 0 , \tag{42}$$

(not summed over n). Iterative solution of these along with the other equations of motion yields

$$F_n = 0 \tag{43}$$

Thus the extra term (40) serves only to determine the otherwise undetermined and unimportant θ, without changing the other fields – because (43) implies that the variation of the extra term (40) with respect to any field is zero.

An important special case is that in which there is a zero mass photon, say W^0, and a corresponding Abelian gauge invariance persisting at the "ghost free" stage.

In this case there is no field θ^0 which can be eliminated; the θ^0 terms in the above equations are omitted. The addition of the $n = 0$ term in (40) is not then justified as simply determining a particular redundant field θ, but it is justified as a legitimate choice of electromagnetic gauge.

With the extra term (39) the starting Lagrangian (36) becomes

$$L = -\tfrac{1}{4}(\partial_\mu W_\nu^n - \partial_\nu W_\mu^n)^2 - \tfrac{1}{2}(MW_\mu^n + \partial_\mu \theta^n)^2 - \tfrac{1}{2}(\partial_\mu W_\mu^n + M\theta^n)^2 - R\,. \tag{44}$$

In the action integral the $W\theta$ cross terms just cancel, so this is equivalent to

$$L = -\tfrac{1}{2}(\partial_\mu W_\nu^n)^2 - \tfrac{1}{2}(MW_\mu^n)^2 - \tfrac{1}{2}(\partial_\mu \theta^n)^2 - \tfrac{1}{2}(M\theta^n)^2 - R\,. \tag{45}$$

If we were to add source terms

$$J_\mu^n W_\mu^n + J\phi$$

to (45) and define n-point functions in the usual way, they would differ of course from those of the ghost free formalism. But they would give the same physical scattering matrix elements. The extra terms due to the non-linearities . . . in (38) would not have the right pole structure to survive steps (9) and (10). The linear term $\partial_\mu \theta$ would also be ineffective if we anticipate (12). Actually it is convenient not to anticipate (12), but to assume rather the Veltman trick (18). Therefore we choose to define n-point functions in terms of the combination

$$U_\mu^n = W_\mu^n + M_n^{-1} \partial_\mu \theta^n\,, \tag{46}$$

which differs from the ghost free W only by non-linear terms. So we use the source terms

$$J_\mu^n U_\mu^n + J\phi\,. \tag{47}$$

From (45) with (47),

$$\begin{aligned} (\partial_\mu \partial_\mu - M_n^2)\, W_\nu^n &= \frac{\delta R}{\delta W_\nu^n} - J_\nu^n\,, \\ (\partial_\mu \partial_\mu - M_n^2)\, \theta^n &= \frac{\delta R}{\delta \theta^n} + \frac{\partial_\mu J_\mu^n}{M_n}\,. \end{aligned} \tag{48}$$

Iterative solution yields a diagram expansion in which the only dangerous looking terms come from those explicitly containing reciprocal masses and gradients in (46) and (48). However, these dangers disappear when the currents are finally replaced by, and the field U contracted with polarization vectors ϵ or ρ. The $M^{-1}\,\partial\theta$ and $M^{-1}\,\partial J$ then give factors

$$M^{-1} k_\mu \epsilon_\mu = 0$$

for non-zero-helicity, or

$$M^{-1} k_\mu \rho_\mu = 1$$

for zero helicity. Thus the dangers associated with vector mesons are in SBG theories only apparent.

In connection with this work, I thank Professors G. 't Hooft, C.H. Llewellyn Smith, R. Stora and M.J.G. Veltman for guidance.

References

[1] M. Gell-Mann, M.L. Goldberger, N.M. Kroll and F.E. Low, Phys. Rev. 179 (1969) 1518.
[2] S. Weinberg, Phys. Rev. Letters 27 (1971) 1688.
[3] S. Weinberg, Phys. Rev. Letters 19 (1967) 1264;
A. Salam, Elementary particle theory ed. N. Svartholm (Almquist and, Stockholm, 1968);
G. 't Hooft, Nucl. Phys. B35 (1971) 167;
B.W. Lee, Proc. of 16th Int. Conf. on high energy physics, NAL 1972, vol. 4.
[4] C.H. Llewellyn Smith, Phys. Letters, to appear;
J.M. Cornwall, D.N. Levin and G. Tiktopoulos, UCLA Preprint 73/TEP/75, April 1973.
[5] Y. Nambu, Phys. Letters 26B (1966) 626;
L.S. Brown and D. Boulware, Phys. Rev. 172 (1968) 1628.
[6] L. Quaranta, A. Rouet, R. Stora and E. Tirapegui, Report at C.N.R.S. meeting on the renormalization of Yang-Miils fields, Marseille, June 1972.
[7] A.I. Vainshtein and I.B. Khriplovich, Sov. J. Nucl. Phys. 13 (1972) 111.
[8] H. Van Dam and M.J.G. Veltman, Nucl. Phys. B22 (1970) 397.

HIGH ENERGY BEHAVIOUR AND GAUGE SYMMETRY

C.H. LLEWELLYN SMITH
CERN, Geneva, Switzerland

Received 13 May 1973

The imposition of unitarity bounds is shown to lead to a Yang-Mills structure in a wide class of theories involving vector mesons. Scalar fields are needed and, at least in simple cases, the unique unitary theory is of the Higgs type.

Intuition, based on dispersive calculations of Feynman diagrams, suggests that there is an intimate connection between the high energy behaviour of amplitudes and renormalizability. In the standard renormalizable field theories (QED, ϕ^3, etc.) it is obvious that the behaviour of the invariant amplitudes for N point tree diagrams is bounded by E^{4-N} when the energy $E \to \infty$ with the ratio of all invariants fixed; Bell has recently proved [1] that this is also true in spontaneously broken gauge theories. Whether or not this "good" high energy behaviour is necessary for renormalizability, it is perhaps reasonable to insist that the four-point functions do not grow as $E \to \infty$. Otherwise the corresponding partial wave amplitudes would exceed their (constant) unitarity limits above some energy, at which perturbation theory would necessarily fail†1.

In this note we study the necessary conditions that in Born approximation all four-point functions are "well behaved" (do not grow as $E \to \infty$) in theories in which massive vector mesons interact with fermions. We assume that there are no interactions characterized by coupling constants with inverse mass dimensions, as is the case in all known renormalizable theories†2. Weinberg has stressed [2] that cancellations improve the high energy behaviour in the model invented by himself [3] and Salam [4]. Subsequently, several authors have "derived" various properties of simple models by insisting that certain terms cancel and it is known that, in general, the removal of the terms which are worst behaved in fermion-antifermion annihilation to two vector mesons ($F\bar{F} \to WW$) requires the F-W interaction to be that of a Yang-Mills theory [5, 6]. The constraints required to cancel the next to leading (but still growing) terms in $F\bar{F} \to WW$ and to ensure good behaviour in $WW \to WW$ had not been studied previously. After completing the work described here†3, however, I received a preprint from Cornwall, Levin and Tiktopoulos [8] who have independently addressed the same problem. Their work differs from mine mainly in that they also studied some five-point functions and deduced information about the self-interactions of their scalar fields, but they did not study fermions nor did they go so far in revealing the underlying symmetry in the general case (there are some other differences and some disagreements which will be noted below).

Let us represent all fundamental fermions (whose number is not specified) by a vector†4

$$\psi = \begin{pmatrix} \nu_e \\ e \\ \nu_\mu \\ \mu \\ \vdots \\ ? \end{pmatrix}$$

†1 Even if the invariant amplitude is constant asymptotically the partial wave amplitude may grow logarithmically. This occurs in some renormalizable theories but the energy at which unitarity is violated [$s \sim M^2 \cdot \exp(1/g^2)$] is very large for a small coupling constant g.

†2 Without this assumption we can get nowhere; "bad" high energy behaviour can be cancelled in an ad hoc way be brutally adding new contact interactions to the Lagrangian.

†3 The results were described very briefly in very general terms in ref. [7] and in somewhat more detail in lectures delivered at the Advanced School of Physics, Frascati, March 1973.

†4 ν_μ, μ may be replaced by $\bar{\nu}_\mu$, $\bar{\mu}$ depending on what form of conservation laws we wish to embody in the theory. We do not consider the more general possibility of putting (e.g.) both μ^- and μ^+ in ψ, which would allow processes such as $W^{--} \to \mu^-\mu^-$.

We suppose that there are g vector mesons represented by real fields W^i_μ $(i = 1, \dots, g)$ and write the F$\overline{\text{F}}$-W interaction in the form:

$$\mathcal{L}_1 = \overline{\psi}_\alpha \gamma^\mu \left[L^i_{\alpha\beta} \left(\frac{1-\gamma_5}{2} \right) + R^i_{\alpha\beta} \left(\frac{1+\gamma_5}{2} \right) \right] \psi_\beta W^i_\mu \qquad (1)$$

where L and R are Hermitean matrices, a summation over repeated indices is implied (here and henceforth) and the notation is that employed by Bjorken and Drell in their book. Since one of the W's will eventually be identified with the photon, there must be a W selfinteraction (e.g., a γW^+W^- vertex must exist). The most general form which satisfies our assumptions is:

$$\begin{aligned} \mathcal{L}_2 = \; & \tfrac{1}{2} D_{ij,k} W^k_\mu (W^j_\alpha \partial_\mu W^i_\alpha - W^i_\alpha \partial_\mu W^j_\alpha) \\ & + \tfrac{1}{2} D^+_{ij,k} W^k_\mu (W^j_\alpha \partial_\mu W^i_\alpha + W^i_\alpha \partial_\mu W^j_\alpha) \\ & + \tfrac{1}{2} G_{ij,k} \epsilon^{\alpha\beta\gamma\delta} W^i_\alpha W^j_\beta \partial_\gamma W^k_\delta \qquad (2) \\ & + \tfrac{1}{4} C_{ij,kl} W^i_\mu W^i_\mu W^k_\nu W^l_\nu \\ & + \tfrac{1}{4} S_{ijkl} \epsilon^{\alpha\beta\gamma\delta} W^i_\alpha W^j_\beta W^k_\gamma W^l_\delta \end{aligned}$$

where the couplings are all real and are supposed to have been fully (anti) symmetrized in an appropriate way.

For the moment we suppose that all the W's are massive. Straightforward but tedious calcualtions show that the conditions

$$\begin{aligned} & D^+ = S = G = 0, \quad D_{ij,k} = -D_{ik,j} (\equiv D_{ijk}), \\ & D_{kac} D_{kbd} - D_{kab} D_{kcd} - D_{kad} D_{kbc} = 0, \qquad (3) \\ & 2C_{ij,kl} = D_{ilp} D_{kjp} + D_{ikp} D_{ljp} \end{aligned}$$

are both sufficient and necessary to remove all the badly behaved pieces of the Born terms for WW → WW, except for terms which grow like E^2 when all four W's are longitudinal and like E when three are longitudinal and one transverse (we return to these residual terms below). Eqs. (3) specify a Yang-Mills theory.

Consider now FF → WW. Given that the 3W vertex has the Yang-Mills form, the leading (E^2) pieces cancel if and only if the coupling constants represent a Lie algebra [5, 6]:

$$[L^i, L^j] = \mathrm{i} D_{ijk} L^k, \qquad [R^i, R^j] = \mathrm{i} D_{ijk} R^k. \qquad (4)$$

The relevant diagrams are shown in fig. 1, where the

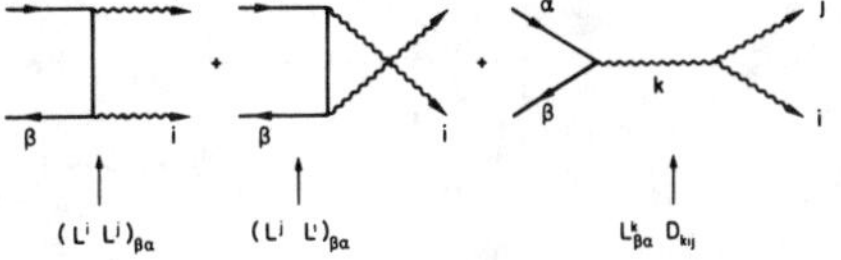

Fig. 1

origin of each term is indicated for the case of left-handed leptons. Non-leading (~E) terms necessarily remain unless either all fermions are massless or all fermions in a given irreducible multiplet are degenerate and parity is conserved [this can be inferred from eq. (6) below], which would not be interesting for physics. Additional particles must therefore be exchanged and, if we wish to avoid the vicious problems associated with particles with spin ⩾ $\frac{3}{2}$, they must have spin zero†5.

Using real scalar fields, $\phi^i (i = 1, \dots, N)$ the most general F-ϕ-W interaction is†6:

$$\begin{aligned} \mathcal{L}_3 = \; & \overline{\psi}_\alpha \left[C^b_{\alpha\beta} \left(\frac{1+\gamma_5}{2} \right) + C^{+b}_{\alpha\beta} \left(\frac{1-\gamma_5}{2} \right) \right] \psi_\beta \phi^b \qquad (5) \\ & + \tfrac{1}{2} W^i_\mu W^j_\mu \phi^b K^b_{ij} + \tfrac{1}{2} T^i_{ba} W^i_\mu (\phi^a \partial_\mu \phi^b - \phi^b \partial_\mu \phi^a) \\ & + \tfrac{1}{4} M^{ij}_{ab} W^i_\mu W^i_\mu \phi^a \phi^b + H^i_{ba} W^i_\mu (\phi^a \partial_\mu \phi^b + \phi^b \partial_\mu \phi^a). \end{aligned}$$

In simple models it can be shown that $H^i_{ba} = 0$ but I have not yet succeeded in showing this in general (it is sufficient, but not necessary, to ensure that $\phi\phi \to \phi\phi$

†5 We introduce these scalar particles in connection with the process F$\overline{\text{F}}$ → WW where good high energy behaviour seems to be necessary to ensure that the box diagram for F$\overline{\text{F}}$ → F$\overline{\text{F}}$ is renormalizable. We then adjust their contributions to WW → WW to give good high energy behaviour there (but note that this is not necessary for renormalizability at the one loop level, at which massive Yang-Mills theory is known to be renormalizable [9]). The rôle of scalar particles in cancelling divergences in one-loop diagrams was (to my knowledge) first discussed in detail in a model of the Higgs type in ref. [10]; the associated improvement of high energy behaviour was noted explicitly in ref. [11].

†6 According to our general assumptions there are only ϕ^3 and ϕ^4 interactions - but we learn nothing about them (nor about the masses of these particles) from studying four-point functions in Born approximation.

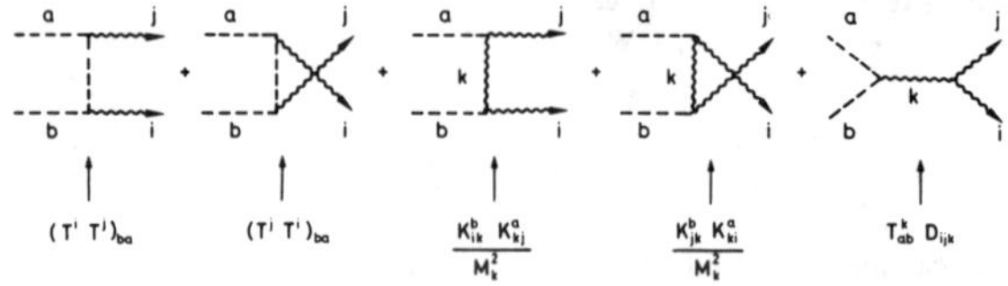

Fig. 2.

and $\phi\phi \to \phi$W are well behaved). For the ensuing argument we must assume that $H^i_{ba} = 0$ (noting that H^i_{ba} is effectively equivalent to a contact interaction with inverse mass dimensions, which we have already assumed to be absent). The coupling in eq. (5) are real and are supposed to have been fully (anti) symmetrized.

It is straightforward but tedious to derive the necessary and sufficient conditions which guarantee that $\mathrm{F\bar{F}} \to$ WW, WW $\to$ WW, $\phi\phi \to$ WW, ϕW $\to$ WW, and $\mathrm{F\bar{F}} \to \phi$W are well behaved in all cases. If we define quantities B, A and X by

$$2M_c B^a_{ci} = -K^i_{ac}\,, \qquad 2M_t M_c A^d_{tc} = D_{tcd}(M_d^2 - M_t^2 - M_c^2)\,,$$

$$M^a X^a_{ij} = m_t R^a_{ij} - L^a_{ij} m_j \qquad (a \leqslant g), \qquad (5)$$

$$X^a_{ij} = -\mathrm{i} C^{a-g}_{ij} \qquad (a = g+1, \dots, N)\,,$$

where M_i are the vector meson masses and m is the diagonal matrix of fermion masses, and construct real antisymmetric $(N{+}g) \times (N{+}g)$ matrices P^a which may be written in partitioned form

$$P^a = \left(\begin{array}{c|c} A^a & B^a \\ \hline -B^{a\mathrm{T}} & T \end{array}\right)\begin{array}{l}\updownarrow g \\ \updownarrow N\end{array} \qquad (7)$$
$$\qquad \overset{\longleftrightarrow}{g} \quad \overset{\longleftrightarrow}{N}$$

then these remaining "cancellation conditions" may be combined in the compact form†7

$$[P^a, P^b] = D_{abc} P^c \qquad (8)$$

$$L^i X^a - X^a L^i = \mathrm{i} P^i_{ca} X^c \qquad (9)$$

$$M^{ij}_{ab} = -\{P^i, P^j\}_{g+a,\, g+b} \qquad (10)$$

where $\{A, B\} = AB + BA$. To understand the origin of these conditions, it may be helpful to consider, for example, the constraints which follow from studying $\phi\phi \to \mathrm{W}_i \mathrm{W}_j$; they contribute a piece involving $[T^i, T^j]$ to eq. (8), whose origin is illustrated in fig. 2.

Remarks: 1. The only theories which have well behaved four-point Born amplitudes and satisfy our assumptions are those in which the W-W and W-F interactions have the Yang-Mills form.

2. There must be scalar particles coupled to fermion and W's in a way which is intimately connected with the masses. The coupling constant relations in eqs. (8) and (9) show that there is a "hidden" representation of the Lie algebra with dimension equal to the number of scalars (N) plus the number of vectors (g) (still assumed to be all massive).

3. Suppose there is one massless particle (the photon) and choose $W^1_\mu = A^\gamma_\mu$. The discussion above holds, except that

$$\begin{aligned} D_{1ab} &= 0 \quad \text{unless } M_a = M_b \\ T^1_{ab} &= 0 \quad \text{unless } M_a = M_b \\ R^1_{ab} &= L^1_{ab} = \delta_{ab} \lambda^a \\ K^b_{ij} &= 0 \end{aligned} \qquad (11)$$

(terms involving M_γ^{-2} can therefore never occur in the cancellation conditions). The cancellation conditions which are not satisfied identically are still summarized by eqs. (8)–(10), except that we now have:

$$P^a_{1b} = P^a_{b1} = X^1_{ij} = 0 \qquad (12)$$

i.e., the "hidden representation of the Lie algebra has dimensions $N + g - 1$ in this case.

†7 The piece of eq. (8) involving $[A^a, A^b]$ differs slightly from the corresponding equation in ref. [8] which is wrong. This, together with the retention of a $W^\mu(\varphi_a \delta_\mu \varphi_b + \varphi_b \delta_\mu \varphi_a)$ coupling, makes it impossible to recover the underlying symmetry from the equations in ref. [8] in the general case.

4. Eqs. (6)–(12) exhibit many of the features of the "spontaneously broken" gauge theories of Englert and Brout [12] and Higgs [13]. There is a class of solutions to these equations in which there exists a vector η_j ($j = 1, \ldots, N$) such that

$$T^a_{ij}\eta_j = 0, \quad \eta_i K^i_{ab} = 2\delta_{ab}M^2_a \tag{13}$$

which implies

$$\delta_{ab}m_a = M_{ab} - \eta_i C^{\,i}_{ab} \tag{14}$$

where

$$M_{ac}R^{\,i}_{cb} = L^i_{ac}M_{cb}.$$

Eqs. (6)–(14) define the general case of the non-Abelian Higgs model discussed by Kibble [14][†8] (apart from the ϕ mass and self interaction terms).

5. The question immediately arises whether there are solutions to the cancellation conditions other than those which specify a spontaneously broken gauge theory in the unitary gauge. If the Lie algebra is semi-simple the answer is no (it can be shown that the necessary vector η_i always exists). More generally, the answer is yes; an example is provided by massive QED ($M_\gamma \neq 0$) which is known to be a well behaved renormalizable theory. It might be that all solutions have the structure of generalized Higgs models except for certain gauge bosons corresponding to some Abelian subgroups, but I have been unable to find the complete answer to this question (I hope to discuss it further in a future publication containing full details of the work reported here). However, in models with only the known leptons and four vector mesons ($W^\pm$, γ, Z°) the unique (non-trivial) solution is the spontaneously broken gauge theory model of Weinberg [3] and Salam [4] (which is not based on a semi-simple Lie algebra) if we take the minimal number (one) of scalar particles. [Cornwall, Levin and Tiktopoulos [8] studied the W-ϕ system in models with $W^\pm$, γ, Z° and ϕ° and found two non-trivial solutions, one corresponding to the Weinberg-Salam model and the other, in which the Z° decouples, to the (semi-simple) Georgi-Glashow model [15]. Their study of five-point tree diagram uniquely fixes the ϕ^3 and ϕ^4 interactions in this case.]

Among many colleagues with whom I have had useful discussions, I am particularly grateful to J.S. Bell, J. Prentki, M. Veltman and J.D. Bjorken (who introduced me to the approach used here and in collaboration with whom this work was originally begun).

†8 Recall that, before spontaneous symmetry breaking occurs, there are $(N+g)$ ϕ's which transform as an $N+g$ dimensional representation of the Lie group; the ϕ interactions and the "bare" fermion mass term have the form:

$$\mathcal{L}^1 = \bar{\psi}_i\left[M_{ij} + X^a_{ij}\left(\frac{1+\gamma_5}{2}\right) + X^{+a}_{ij}\left(\frac{1+\gamma_5}{2}\right)\right]\psi_j\phi^a + \tfrac{1}{2}\,|\partial_\mu\phi_i + P^a_{ij}\,W^a_\mu\phi_j|^2$$

where M is Hermitean, P is real and antisymmetric and $[P^a, P^b] = D_{abc}P^c$, $L^c X^b - X^b R^c = iP^c_{ab}X^a$, $L^a M = MR^a$, where L and R are the $WF\bar{F}$ coupling matrices defined above. When the symmetry is completely "spontaneously broken" the Lagrangian in the "unitary gauge" is obtained by putting $\phi_i \to 0$, $i = 1, \ldots, g$; $\phi_i \to \eta_i + \phi_i$, $i = g+1, \ldots, g+N$; $P^a_{i+g,g+j}\eta_j = 0$, ($i = 1, \ldots, N$). Choosing (without loss of generality) a representation in which the vector meson masses are diagonal and the fermion masses are diagonal and independent of γ_5, we see that the conditions in eqs. (12)–(17) specify the theory in this "physical" gauge.

References

[1] J.S. Bell, CERN preprint TH. 1669 (1973).
[2] S. Weinberg, Phys. Rev. Lett. 27 (1971) 1688.
[3] S. Weinberg, Phys. Rev. Lett. 19 (1967) 1264.
[4] A. Salam, in Elementary particle physics, ed. N. Svartholm (Almqvist and Wiksells, Stockholm, 1968), p. 367.
[5] J.D. Bjorken, unpublished manuscript (May, 1972).
[6] J. Schwinger, Phys. Rev. D7 (1973) 908, footnote 4).
[7] C.H. Llewellyn Smith, An introduction to renormalizable models of weak interactions and their experimental consequences, to be published in Proc. of the meeting on Links between weak and electromagnetic interactions, Rutherford High Energy Laboratory, February 1973.
[8] J.M. Cornwall, D.N. Levin and G. Tiktopoulos, UCLA preprint UCLA/73/TEP/75, April 1973.
[9] M. Veltman, Nuclear Phys. B7 (1968) 637.
[10] T. Appelquist and H.R. Quinn, Phys. Lett. 39B (1972) 675.
[11] J.C. Taylor, Oxford University preprint: The physical rôle of scalar particles in convergent theories of charged vector particles (1972).
[12] F. Englert and R. Brout, Phys. Rev. Lett. 13 (1964) 321.
[13] P.W. Higgs, Phys. Lett. 12 (1964) 132.
[14] T.W.B. Kibble, Phys. Rev. 155 (1967) 1554;
See also: G.S. Guralnick, C.R. Hagen and T.W.B. Kibble, in Advances in particles physics, ed. Cool and Marshak (Interscience, 1968) p. 567.
[15] H. Georgi and S.L. Glashow, Phys. Rev. Lett. 28 (1972) 1494.

IV. THE STANDARD MODEL

PARTIAL-SYMMETRIES OF WEAK INTERACTIONS

SHELDON L. GLASHOW †

Institute for Theoretical Physics, University of Copenhagen, Copenhagen, Denmark

Received 9 September 1960

Abstract: Weak and electromagnetic interactions of the leptons are examined under the hypothesis that the weak interactions are mediated by vector bosons. With only an isotopic triplet of leptons coupled to a triplet of vector bosons (two charged decay-intermediaries and the photon) the theory possesses no partial-symmetries. Such symmetries may be established if additional vector bosons or additional leptons are introduced. Since the latter possibility yields a theory disagreeing with experiment, the simplest partially-symmetric model reproducing the observed electromagnetic and weak interactions of leptons requires the existence of at least four vector-boson fields (including the photon). Corresponding partially-conserved quantities suggest leptonic analogues to the conserved quantities associated with strong interactions: strangeness and isobaric spin.

1. Introduction

At first sight there may be little or no similarity between electromagnetic effects and the phenomena associated with weak interactions. Yet certain remarkable parallels emerge with the supposition that the weak interactions are mediated by unstable bosons. Both interactions are universal, for only a single coupling constant suffices to describe a wide class of phenomena: both interactions are generated by vectorial Yukawa couplings of spin-one fields ††. Schwinger first suggested the existence of an "isotopic" triplet of vector fields whose universal couplings would generate both the weak interactions and electromagnetism — the two oppositely charged fields mediate weak interactions and the neutral field is light [2]). A certain ambiguity beclouds the self-interactions among the three vector bosons; these can equivalently be interpreted as weak or electromagnetic couplings. The more recent accumulation of experimental evidence supporting the $\Delta I = \frac{1}{2}$ rule characterizing the non-leptonic decay modes of strange particles indicates a need for at least one additional neutral intermediary [3]).

The mass of the charged intermediaries must be greater than the K-meson mass, but the photon mass is zero — surely this is the principal stumbling block in any pursuit of the analogy between hypothetical vector mesons and photons. It is a stumbling block we must overlook. To say that the decay intermediaries

† National Science Foundation Post-Doctoral Fellow. Present Address: Physics Department, California Institute of Technology, Pasadena.

†† A scalar intermediary is also conceivable. See ref. [1])

together with the photon comprise a multiplet leads to no more than an excessively obscure notation unless a principle of symmetry is discovered which can relate the forms of weak and electromagnetic couplings. Because of the large mass splittings among the vector mesons, only a very limited symmetry among them may be anticipated. The purpose of this note is to seek such symmetries among the interactions of leptons in order to make less fanciful the unification of electromagnetism and weak interactions.

2. Partially-Symmetric Interactions

In the conventional Lagrangian formulation of quantum field theory the relation between symmetries of the Lagrange function and conservation laws is well known. We recently introduced the notion of "partial-symmetry" — invariance of only part of the Lagrange function under a group of infinitesimal transformations [4]). The part of the Lagrange function bilinear in the field variables which produces masses of the elementary particles need not be invariant under a partial-symmetry. Corresponding "partial-conservation laws" become conservation laws only with the neglect of appropriate masses or mass differences. This is the only sort of symmetry which could relate the massive decay intermediaries to the massless photon.

The most familiar example of a partial-symmetry is produced by an infinitesimal change of scale. If we change the coordinates, $x_\mu \rightarrow (1+\lambda)x_\mu$, and at the same time replace each field variable χ by $(1-\lambda)\chi$, that part of the integrated Lagrange function not involving dimensional parameters will be left unchanged. As long as all the interactions involve only dimensionless coupling constants the scale transformation will be a partial-symmetry. To require this partial-symmetry excludes such interactions as ps—pv meson theory and direct four-Fermion couplings.

Another kind of partial-symmetry has been recently examined by Gell-Mann and his collaborators [5]). They are led to a proportionality between the divergence of the axial-vector weak interaction current and the pion field. This they recognize as a partial-conservation law. (They must neglect both weak and electromagnetic interactions. We shall demand partial-symmetry of these interactions themselves.) It can result from the invariance of strong interactions under the infinitesimal unitary transformation,

$$\psi_N \rightarrow (1+ia\gamma_5\tau^{\pm})\psi_N,$$

where ψ_N are the nucleon fields, together with appropriate accompanying transformations of meson and hyperon fields. Requiring this partial-symmetry in order to generate a partially-conserved axial-vector current provides them a powerful restriction on the acceptable form of the strong interactions. If we may invert historical sequence for the sake of pedagogy, in the same fashion the assertion of Feynman and Gell-Mann [6]) that the vector weak-interaction current is conserved could have led to the discovery that the strong interactions are charge-independent.

A last prerequisite to our discussion is the possibility of constructing partially-symmetric interactions among a triplet of vector mesons, Z_μ^1, Z_μ^2 and Z_μ^3. We shall assume that under the CP transformation, $Z_0^1 \to -Z_0^1$, $Z_0^2 \to +Z_0^2$, and $Z_0^3 \to +Z_0^3$. The most general self-interaction tri-linear in the fields and consistent with CP invariance † is

$$g_1 Z^1_{\mu\nu} Z_\mu^2 Z_\nu^3 + g_2 Z^2_{\mu\nu} Z_\mu^3 Z_\nu^1 + g_3 Z^3_{\mu\nu} Z_\mu^1 Z_\nu^2$$
$$+ f_1 Z^1_{\mu\nu} Z^2_{\nu\lambda} Z^3_{\lambda\mu} + f_2 Z^2_{\mu\nu} Z^3_{\nu\lambda} Z^1_{\lambda\mu} + f_3 Z^3_{\mu\nu} Z^1_{\nu\lambda} Z^2_{\lambda\mu}.$$

To obtain a three-parameter group of partial-symmetries, we must choose $g_1 = g_2 = g_3$ and $f_1 = f_2 = f_3$. The resulting partial-conservation laws are

$$\partial_\mu (Z^1_{\mu\nu} Z_\nu^2 - Z^2_{\mu\nu} Z_\nu^1) = (M_1^2 - M_2^2) Z_\mu^1 Z_\mu^2,$$
$$\partial_\mu (Z^2_{\mu\nu} Z_\nu^3 - Z^3_{\mu\nu} Z_\nu^2) = (M_2^2 - M_3^2) Z_\mu^2 Z_\mu^3,$$
$$\partial_\mu (Z^3_{\mu\nu} Z_\nu^1 - Z^1_{\mu\nu} Z_\nu^3) = (M_3^2 - M_1^2) Z_\mu^3 Z_\mu^1,$$

corresponding to the infinitesimal transformations

$$\mathbf{Z} \to (1 + i\mathbf{a} \cdot \mathbf{t})\mathbf{Z},$$

where the $\mathbf{t}$ are the conventional anti-symmetric imaginary 3×3 matrices.

Without the quadrupole couplings (they involve implicit cubes of momenta), and with $M_1 = M_2$ and $M_3 = 0$, this partially-symmetric interaction describes the electrodynamics (with Z_μ^3 the vector potential) of a charged pair of vector bosons, $Z^\pm = (Z^1 + iZ^2)/\sqrt{2}$, with gyromagnetic ratio of two and with no electric quadrupole coupling. Some properties of this version of spin-one electrodynamics we have discussed elsewhere [4]).

3. Interactions of Leptons

We consider the interactions of a multiplet of vector bosons with a triplet of real Majorana fields, $\psi_1 = (\psi^+ + \psi^-)/\sqrt{2}$, $\psi_2 = i(\psi^+ - \psi^-)/\sqrt{2}$ and ψ_3. The mass-producing term has the form $m\psi\beta t_3^2 \psi$, so that we may regard $\psi^\pm$ as the positon and negaton and ψ_3 as that variety of neutrino produced in association with negatons and positons. To escape the $\mu \to e + \gamma$ difficulty [7]), we assume (with Schwinger [2])) that a quite distinct triplet of fields describes muons and those neutrinos produced with muons. No interaction shall be introduced that couples the one triplet to the other. The interactions between the Fermions and

† Our choice of the CP behaviour of the Z-meson triplet is such as to exclude the interactions $\varepsilon^{\mu\nu\lambda\sigma} Z^1_{\mu\nu} Z^2_\lambda Z^3_\sigma$ and $\varepsilon^{\mu\nu\lambda\sigma} Z^1_{\mu\nu} Z^2_{\lambda\beta} Z^3_{\beta\sigma}$.

the vector mesons should include both electromagnetism and such weak interactions as are necessary to produce observed decay phenomena.

The interaction Lagrangian,

$$g\mathbf{Z}_\mu \cdot \mathbf{J}_\mu = g\mathbf{Z}_\mu \cdot [(\mathbf{Z}_{\mu\nu} \times \mathbf{Z}_\nu) + i\psi\beta\gamma_\mu \mathbf{O}\psi], \tag{3.1}$$

includes both a symmetrical self-interaction of the Z-triplet and the Yukawa interaction of the bosons to the Fermions. The common coupling strength is in accord with the universality both of the electric charge and of the weak interaction coupling constant. For partial-symmetry, the three imaginary anti-symmetric matrices $\mathbf{O}$ must satisfy the commutation relations of an angular momentum,

$$\mathbf{O} \times \mathbf{O} = i\mathbf{O}. \tag{3.2}$$

In that case three partially-conserved currents

$$\mathbf{J}_\mu = \mathbf{Z}_{\mu\nu} \times \mathbf{Z}_\nu + \psi\beta\gamma_\mu \mathbf{O}\psi \tag{3.3}$$

result from the three infinitesimal transformations (i.e., partial-symmetries)

$$\psi \to (1 + i\mathbf{a} \cdot \mathbf{O})\psi, \quad \mathbf{Z} \to (1 + i\mathbf{a} \cdot \mathbf{t})\mathbf{Z}. \tag{3.4}$$

The $\mathbf{O}$ must commute with $\beta\gamma_\mu$ in order to leave invariant the kinematic terms of the Lagrange function. They must have the form $\mathbf{O} = \mathbf{A} + i\gamma_5 \mathbf{S}$, where the $\mathbf{A}$ are anti-symmetric, the $\mathbf{S}$ are symmetric, and both are Hermitean 3×3 matrices acting between the ψ_i. Further restrictions upon the $\mathbf{O}$ arise if we demand CP invariance and limit the currents to neutral and singly charged. The simplest choice is evidently $\mathbf{O} = \mathbf{t}$. Certainly J_μ^3 is the total electrical current, so that Z_μ^3 may be interpreted as the electromagnetic vector potential. But the charged currents do not display symmetric parity violation, hence they cannot reproduce the observed weak interactions.

Other matrices satisfying (3.2) but violating parity conservation are obtained from the $\mathbf{t}$ by a unitary transformation,

$$U\mathbf{t}U^{-1} = e^{-(\gamma_5 t_3{}^2\theta)}\,\mathbf{t}\,e^{+(\gamma_5 t_3{}^2\theta)}$$

Clearly U commutes with t_3 so that the leptons' electrical current $j_\mu{}^3$ remains unchanged. To obtain symmetric parity violation in the charged currents we take $\theta = \frac{1}{4}\pi$, whereupon

$$'O_1 = (t_1 + i\gamma_5\{t_2, t_3\})/\sqrt{2}, \qquad 'O_2 = (t_2 - i\gamma_5\{t_1, t_3\})/\sqrt{2}, \tag{3.5}$$

where curly brackets signify anti-commutators. Unfortunately the charged currents $j_\mu{}^{1,2} = \psi\beta\gamma_\mu {}'O_{1,2}\psi$ are not acceptable weak interaction currents. The parity-violating unitary transformation U is not invariant under the CP transformation so that, under CP,

$$'O_1 \to \tfrac{1}{2}(t_1 - i\gamma_5\{t_2, t_3\}).$$

In terms of charge eigenstates these currents involve the interaction of both negaton and positon to the same handed neutrino. The theory generated with '**O**, partially-symmetric and parity violating though it may be, is not a faithful model of the weak and electromagnetic interactions of leptons.

Recent experiments have determined the form of the charged leptonic currents. We shall choose O_1 and O_2 in accordance with these experiments. Negatons are produced in elementary-particle decays only in association with left-handed neutrinos whereas positons are accompanied by right-handed ones. The correct charged currents $j_\mu^{1,2} = \psi\beta\gamma_\mu O_{1,2}\psi$ are produced with [2])

$$O_1 = (t_1 + i\gamma_5\{t_1, t_3\})/\sqrt{8}, \qquad O_2 = (t_2 + i\gamma_5\{t_2, t_3\})/\sqrt{8}. \tag{3.6}$$

The decay interaction is partially-symmetric only when all three vector bosons participate, $g\mathbf{Z}_\mu \cdot \mathbf{j}_\mu$, where the neutral current is given by $j_\mu^3 = \psi\beta\gamma_\mu O_3\psi$ and

$$O_3 = -i[O_1, O_2] = \tfrac{1}{4}(t_3 + i\gamma_5(3t_3^2 - 2)). \tag{3.7}$$

It is readily seen that these three O_i satisfy (3.2). Since j_μ^3 is not the lepton electric current, Z_3 cannot be interpreted as the electromagnetic field. Thus the theory containing only the necessary weak interactions of two oppositely charged decay intermediaries together with the electromagnetic interactions of both the leptons and the bosons is not partially-symmetric.

4. Partially-Symmetric Synthesis

In order to achieve a partially-symmetric theory of weak and electromagnetic interactions, we must go beyond the hypothesis of only a triplet of vector bosons and introduce an additional neutral vector boson Z_S. It will have the same behaviour under CP as Z_3 and it is coupled to its own neutral lepton current $J_\mu^S = \psi\beta\gamma_\mu S\psi$. The three partial symmetries of section 3 are undisturbed by this new interaction provided that

$$[\mathbf{O}, S] = 0. \tag{4.1}$$

We use the **O** that yield correct weak interactions of charged currents, (3.6) and (3.7), and we define

$$S = \tfrac{3}{4}(t_3 - i\gamma_5(t_3^2 - \tfrac{2}{3})). \tag{4.2}$$

Note that (4.1) is satisfied, and moreover,

$$O_1^2 + O_2^2 + O_3^2 + S^2 = 1 \tag{4.3}$$

and

$$Q = t_3 = O_3 + S. \tag{4.4}$$

The last relation suggests the analogous expression relating strangeness and the third component of isobaric spin to the electrical charge of strongly interacting particles. Far more transparent expressions for $\mathbf{O}$ and S and for the relations among them emerge in a notation wherein the handedness of leptons is diagonal †.

The interaction Lagrange function including the couplings of four vector bosons is

$$e \sec\theta \; \mathbf{Z}_\mu \cdot [(\mathbf{Z}_{\mu\nu}\times\mathbf{Z}_\nu)+\psi\beta\gamma_\mu\mathbf{O}\psi]+e\csc\theta \; Z_\mu{}^S\psi\beta\gamma_\mu S\psi. \tag{4.5}$$

The parameter θ appears in order to permit an arbitrary choice of the strengths of the triplet and singlet interactions. The three partial-symmetries (3.4) have been preserved and an additional partial-symmetry yielding a partial-conservation law for $J_\mu{}^S$ is obtained,

$$\psi \to (1+ibS)\psi, \qquad \mathbf{Z}\to\mathbf{Z}, \qquad Z^S \to Z^S. \tag{4.6}$$

The reader may wonder what has been gained by the introduction of another neutral vector meson. Neither Z_3 nor Z_S interacts with the electrical current so that neither interaction may be identified with electromagnetism. (To have chosen $J_\mu{}^S$ to be the electrical current would have violated (4.1) and lost partial-symmetry.) However, both the neutral fields have the same CP property so that linear combinations of the two fields may correspond to "particles." The most general form for the boson mass producing part of the Lagrange function is a positive-definite bilinear expression in $Z_\mu{}^3$ and $Z_\mu{}^S$ whose diagonalization identifies those linear combinations of the fields which display unique masses (i.e., the "particles"). Most generally we may have

$$L_M = \tfrac{1}{2}M_A{}^2(Z_\mu{}^3\cos\theta'+Z_\mu{}^S\sin\theta')^2+\tfrac{1}{2}M_B{}^2(Z_\mu{}^S\cos\theta'-Z_\mu{}^3\sin\theta')^2,$$

in which the fields

$$A_\mu = Z_\mu{}^3\cos\theta'+Z_\mu{}^S\sin\theta', \qquad B_\mu = Z_\mu{}^S\cos\theta'-Z_\mu{}^3\sin\theta' \tag{4.7}$$

describe spin-one particles with masses M_A and M_B. In terms of these fields the interaction Lagrange function (4.5) becomes

$$\begin{aligned} &eA_\mu J_\mu{}^Q+eF_{\mu\nu}\mathbf{Z}_\mu t_3\mathbf{Z}_\nu+e\sec\theta(Z_\eta{}^1 j_\mu{}^1+Z_\mu{}^2 j_\mu{}^2)\\ &\quad -e\tan\theta(B_{\mu\nu}\mathbf{Z}_\mu t_3\mathbf{Z}_\nu+B_\mu\mathbf{Z}_{\mu\nu}t_3\mathbf{Z}_\nu)+eB_\mu\psi\beta\gamma_\mu(S\cot\theta-O_3\tan\theta)\psi, \end{aligned} \tag{4.8}$$

when θ' is put equal to θ. The total electrical current of leptons and bosons is denoted by $J_\mu{}^Q$,

$$J_\mu{}^Q = \mathbf{Z}_{\mu\nu}t_3\mathbf{Z}_\nu+\psi\beta\gamma_\mu(S+O_3)\psi.$$

The interaction of A_μ is precisly the electrodynamic interaction of the charged

† We are indebted to Professor M. Gell-Mann for this observation, and for presenting part of our work, in his notation, at the 1960 Conference on High Energy Physics in Rochester, New York.

Fermions and of the charged vector mesons (with gyromagnetic ratio of two). We may identify A_μ with the electromagnetic field if we put $M_A = 0$. This isolation of the electromagnetic interaction has been possible only because of (4.4) and our apparently arbitrary choice of S is now justified. The interactions of Z_1 and Z_2 generate the correct parity-violating decay interactions with coupling constant $g_W = e \sec \theta$. Remaining terms in (4.8) comprise the interaction of B with a neutral partially-conserved current. They are the price we must pay for partial-symmetry. The symmetries of (4.5) under the four-parameter group of transformations (3.4 and 4.6) are unaffected by the re-definition of fields required by the mass-producing part of the Lagrange function. Expressed in terms of A and B one of these symmetries yields the conservation of electrical charge; the other three shuffle the photon with decay-intermediaries and lead only to partial-conservation laws for the three weak interaction currents. For no choice of θ is the interaction of the neutral current small compared with weak interactions involving charged currents. The masses of the charged intermediaries M_Z and of the neutral M_B are as yet arbitrary.

5. Discussion

It seems remarkable that both the requirement of partial-symmetry and quite independent experimental considerations indicate the existence of neutral weakly interacting currents. It would be gratifying if the introduction of only a single neutral vector-meson field B could secure both partial-symmetry and the $\Delta I = \frac{1}{2}$ rule. Whether this is possible depends upon the extension of our work to the interactions of the vector-meson multiplet with strongly interacting particles. But the roles of B and of $Z^\pm$ are far from symmetrical in the leptonic decays of strange particles. Indeed, the modes $K \to \pi + \nu_R + \nu_L$ and $K \to \pi + e^+ + e^-$ have never been seen. Since the coupling strengths of Z to charged leptonic currents and of B to its neutral current are limited by the requirement of partial-symmetry, the absence of neutral leptonic decay modes must be attributed to a mass splitting between $Z^\pm$ and B. The unobserved modes are then suppressed by the factor $(M_Z/M_B)^4$. Does not this mass splitting prevent the symmetrical participation of the three decay intermediaries prerequisite to a $\Delta I = \frac{1}{2}$ rule? While in a leptonic decay mode, the momentum of the decay intermediary is just the sum of the momenta of the emerging leptons, in non-leptonic modes one must integrate over all momenta of the intermediary. Thus the matrix element for non-leptonic modes is expected to be less sensitive a function of the vector-meson mass than the matrix element for non-leptonic modes. Vector meson theory is less well-behaved than quantum-electrodynamics or pseudoscalar meson theory, so that it is not unreasonable to suppose that the non-leptonic modes are dominated by contributions at virtual momenta far beyond M_B. If this is so the mass difference between

$Z^{\pm}$ and B would be without significant effect and a $\Delta I = \frac{1}{2}$ rule might be secured.

To assure the experimentally observed selection rules of strangeness as well as of isotopic spin, two neutral decay intermediaries are needed [3]). One of these has CP behaviour opposite to that of A and B. It is not discouraging that this particle has not yet appeared in the interactions of the vector-meson multiplet with leptons, since no such interaction exists consistent with CP invariance.

We have argued that any underlying symmetries relating weak interactions and electromagnetism are obscured by the masses of elementary particles. Without a theory of the origins of these masses, any study based upon the analogy between decay-intermediaries and photons may make use only of partial-symmetries. The simplest partially-symmetric system exhibiting all known interactions of the leptons, the weak and the electromagnetic, has been determined. Although we cannot say *why* the weak interactions violate parity conservation while electromagnetism does not, we have shown *how* this property can be embedded in a unified model of both interactions. Unfortunately our considerations seem without decisive experimental consequence. For this approach to be more than academic, a partially-symmetric system correctly describing all decay modes of all elementary particles should be sought.

Appendix

A.1. ANOTHER ALTERNATIVE

In section 4 we said it is necessary to go beyond the framework of a triplet of vector mesons and a triplet of leptons in order to obtain a partially-symmetric system including the known interactions. This is true, but it is not entirely obvious that the introduction of additional lepton fields rather than an additional boson would not also do the trick. Two triplets of real Fermion fields were needed to describe all the leptons — one triplet including the electrons and one the muons, each with its own neutrino. Let us keep our original triplet of vector mesons while introducing two new (i.e., unobserved) Fermion fields. They will correspond to a massive neutral muon and to its distinct antiparticle. Altogether there are now eight kinds of Fermions. Half of them are defined to be leptons: μ^+, μ^0, e^-, and ν; and their antiparticles are anti-leptons. Conservation of lepton number will assure the absence of unwanted modes of muon decay. Four by four matrices which describe isobaric spin $\frac{1}{2}$ are the most convenient to use †. The doublet (μ^+, μ^0) is characterized by $\zeta_3 = +1$, while (e^-, ν) has $\zeta_3 = -1$. Similarly, the doublet (μ^+, ν) has $\tau_3 = +1$, while (e^-, μ^0) has $\tau_3 = -1$. The electrical charge has the form $Q = \frac{1}{2}(\tau_3+\zeta_3)$. We assume that the μ^0 is sufficiently massive so that the possibility of $\pi \to \mu^0+e$ is avoided. The problem is again to find a set of $\mathbf{O}$ satisfying (3.2) which generates charged

† The notation of J. Schwinger is employed. See ref. [2]).

currents embodying symmetric parity violation yet conserving CP. Since we have introduced only a triplet of vector mesons, the neutral must be the photon and $O_3 = \frac{1}{2}(\tau_3+\zeta_3)$. Such a set of **O** is

$$O_1 = [\tau_1(1+i\gamma_5\zeta_3)+\zeta_1(1-i\gamma_5\tau_3)]/\sqrt{8},$$
$$O_2 = [\tau_2(1+i\gamma_5\zeta_3)+\zeta_2(1-i\gamma_5\tau_3)]/\sqrt{8}.$$

With these **O** we may construct a partially-symmetric theory including both weak and electromagnetic interactions of the leptons. Unfortunately it is the wrong theory. If the electron is associated with a left-handed neutrino, the μ^- will be associated with a right-handed one. (A change of sign produces a converse assignment.) But it cannot be arranged for both electron and μ^- to be coupled with the same handed neutrino. Experiments indicate that the handedness of the neutrino is determined solely by the lepton's charge. With this theory muons would decay according to the schemes: $\mu^+ \to e^+ + 2\nu_R$ and $\mu^- \to e^- + 2\nu_L$. The electron spectrum from muon decay would necessarily be of the $\rho = 0$ shape and is now probably excluded by experiment. A more acceptable set of just three **O** has not been found.

A.2. OTHER INTERACTIONS

One might conclude from existing experimental determinations of the electrical charge and of the weak interaction strength g_W^2/M_Z^2 that $M_Z > \approx 137$ nucleon masses. But whatever mechanism is responsible for the large mass of the intermediaries should also produce a large wave-function renormalization of the massive vector fields not shared by the photon. This results in a reduction of the weak coupling strength compared with the electric charge and consequently relaxes this lower limit to the Z-meson mass.

Both its mass and its charge renormalization would arise if there were strong interactions quadratic in the Z-field. Should these couplings involve strongly interacting particles they would generate effective six-Fermion interactions of comparable strength to the existing four-Fermion couplings responsible for decay phenomena. A possible experimental consequence of these interactions is the stimulated decay of a muon:

$$\mu^- + N = e^- + \nu_R + \nu_L + N,$$

in the presence of nuclear matter Observable effects upon the decay rate of negative muons bound to heavy nuclei might result, but at this time we cannot exclude the possible existence of strong interactions of the decay intermediaries. Of course these interactions could also be detected in a search for the real production of decay intermediaries.

A.3. ANOTHER SYMMETRY

If only for completeness, we exhibit one remaining symmetry of (4 5)

Defining $W = 1 - t_3^2 + i\gamma_5 t_3$, we discover $W\mathbf{O} = \mathbf{O}W = \mathbf{O}$ and $WS = SW$. The infinitesimal transformation $\psi \rightarrow (1 + ia\gamma_5 W)\psi$ is not merely a partial symmetry, but it is a complete symmetry providing the neutrino mass is zero. The corresponding conserved current,

$$J_\mu^W = \psi\beta\gamma_\mu(t_3 - i\gamma_5(1 - t_3^2))\psi,$$

describes the "neutrinic" quantum number of Schwinger [2]). It is the leptons' electrical charge plus the number of left-handed neutrinos less the number of right-handed ones.

The author wishes to thank Professor Niels Bohr and Professor Aage Bohr for the hospitality extended to him at this Institute during the past two years. He is also very grateful to the CERN laboratories in Geneva for their hospitality while much of this work was done. Conservations with Professor M. Gell-Mann and with members of both Institutes are gratefully acknowledged.

References

1) C. Fronsdal and S. Glashow, Phys. Rev. Letters **3** (1959) 570
2) J. Schwinger, Ann. of Phys. **2** (1957) 407
3) M. Gell-Mann, private communication;
T. D. Lee and C. N. Yang, preprint
4) S. Glashow, Nuclear Physics **10** (1959) 107
5) M. Gell-Mann and M. Lévy, Nuovo Cim. **16** (1960) 705;
J. Bernstein, M. Gell-Mann and L. Michel, Nuovo Cim. **16** (1960) 560
6) R. Feynman and M. Gell-Mann, Phys. Rev. **109** (1958) 193
7) e. g., G. Feinberg, Phys. Rev. **110** (1958) 1482

ELECTROMAGNETIC AND WEAK INTERACTIONS

A. SALAM and J. C. WARD *
Imperial College, London

Received 24 September 1964

One of the recurrent dreams in elementary particles physics is that of a possible fundamental synthesis between electro-magnetism and weak interactions [1]. The idea has its origin in the following shared characteristics:

1) Both forces affect equally all forms of matter-leptons as well as hadrons.
2) Both are vector in character.
3) Both (individually) possess universal coupling strengths. Since universality and vector character are features of a gauge-theory these shared characteristics suggest that weak forces just like the electromagnetic forces arise from a gauge principle.

There of course also are profound differences:

1) Electromagnetic coupling strength is vastly different from the weak. Quantitatively one may state it thus: if weak forces are assumed to have been mediated by intermediate bosons (W), the boson mass would have to equal 137 M_p, in order that the (dimensionless) weak coupling constant $g_W{}^2/4\pi$ equals $e^2/4\pi$.

In the sequel we assume just this. For the outrageous mass value itself ($M_W \approx 137\ M_p$) we can offer no explanation. We seek however for a synthesis in terms of a group structure such that the remaining differences, viz:

2) Contrasting space-time behaviour (V for electromagnetic versus V and A for weak).
3) And contrasting ΔS and ΔI behaviours both appear as aspects of the same fundamental symmetry. Naturally for hadrons at least the group structure must be compatible with SU_3.

Lepton interactions define both the unit of the electric charge and (from μ-decay) the (bare) value of weak coupling constant. Leptons therefore must be treated first.

There is only one genuine lepton multiplet (in the limit $m_e = m_\mu = 0$) which really treats the neutrino field on the same footing ** as μ and e. This is the Konopinski-Mahmoud multiplet.

$$L = \begin{pmatrix} \nu \\ e^- \\ \mu^+ \end{pmatrix} \qquad (1)$$

In terms of SU_3 generators ***, the electric charge clearly equals:

$$Q_l = \begin{pmatrix} 0 & & \\ & -1 & \\ & & +1 \end{pmatrix} = -2U_3 = -\tfrac{2}{3}\sqrt{3}\,(I_0 - V_0) = 2(I_3 - V_3), \qquad (2)$$

while the weak interaction (with no neutral currents) has the unique form †

* Permanent address, John Hopkins University, Baltimore.

** There are other schemes where one postulates multiplets consisting of a two-component neutrino field together with a four-component electron or muon. These do not satisfy even the most elementary requirement of a genuine group-structure, i.e. that in some limit at least, the particles concerned should be transformable, one into the other.

***
$$[T^i, T^j] = i f^{ijk} T^k$$
$$\{T^i, T^j\} = \tfrac{1}{3}\delta^{ij}\,\mathbf{1} + d^{ijk}\,T^k$$

$$I_3 = T^3,\ I^{\pm} = \tfrac{1}{2}\sqrt{2}\,(T^1 \mp iT^2),\ I_0 = T^8,\ [I_0, \boldsymbol{I}] = 0$$

$$U_3 = \tfrac{1}{2}\sqrt{3}\,T^8 - \tfrac{1}{2}T^3,\ U^{\pm} = \tfrac{1}{2}\sqrt{2}\,(T^6 \mp iT^7),\quad U_0 = \tfrac{1}{2}\sqrt{3}\,T^3 + \tfrac{1}{2}T^8$$

$$V_3 = \tfrac{1}{2}\sqrt{3}\,T^8 + \tfrac{1}{2}T^3,\ V^{\pm} = \tfrac{1}{2}\sqrt{2}\,(T^4 \mp iT^5),\quad V_0 = \tfrac{1}{2}\sqrt{3}\,T^3 - \tfrac{1}{2}T^8$$

Note

$$Q_h = T^3 + \tfrac{1}{3}\sqrt{3}\,T^8 = \tfrac{2}{3}\sqrt{3}\ U_0 = \tfrac{2}{3}\sqrt{3}\,(I_0 + V_0) = \tfrac{2}{3}(I_3 + V_3)$$
$$Q_l = T^3 - \sqrt{3}\,T^8 = -2U_3 = -\tfrac{2}{3}\sqrt{3}\,(I_0 - V_0) = +2\,(I_3 - V_3)$$

Explicitly,

$$Q_h = \begin{pmatrix} \tfrac{2}{3} & & \\ & -\tfrac{1}{3} & \\ & & -\tfrac{1}{3} \end{pmatrix},\quad Q_l = \begin{pmatrix} 0 & & \\ & -1 & \\ & & +1 \end{pmatrix}$$

Q_h is the conventional hadron charge operator, Q_l gives lepton-charge.

† Define

$$\psi_L = \tfrac{1}{2}(1 + \gamma_5)\,\psi \qquad \psi_R = \tfrac{1}{2}(1 - \gamma_5)\,\psi$$
$$(A^+B)_L = \tfrac{1}{2}A^+\gamma_4\gamma_\mu(1 + \gamma_5)B$$

$$\mathcal{L}_{weak} = [((e^-)^\dagger \nu)_L + (\nu^\dagger \mu)_R]\, W^- + \text{h.c.}$$

$$= (I_{1L} + V_{1R})\, W_1 + (I_{2L} - V_{2R})\, W_2. \quad (3)$$

Here $W^\pm = \frac{1}{2}\sqrt{2}\,(W_1 + iW_2)$ and $I_{1L} = (\psi^+ I_1 \psi)_L$ etc. Now $I_{1L} + V_{1R}$, $I_{2L} - V_{2R}$, $I_{3L} - V_{3R}$ generate an SU_2 sub-group. Since two of the group generators $I_{1L} + V_{1R}$, $V_{2L} - V_{2R}$ give the weak currents, ideally one would have liked the neutral component [2] $(I_{3L} - V_{3R})$ to represent electromagnetism. This unfortunately is not the case. $I_{3L} - V_{3R}$ equals $\frac{1}{4}(Q_l + 3\gamma_5 Q_h)$ so that in addition to the "correct' electromagnetic current $\frac{1}{2}Q_l$ there appears also unwanted parity violating term $\gamma_5 Q_h$. We are therefore forced to extend the group structure as follows:

The free Lagrangian $(\psi^+ \partial \psi)$ is invariant for the following $SU_2 \otimes U_1$ transformation

$$L' = UL$$

where

$$U = \exp ie[(I_L + V'_R)\varepsilon - \sqrt{3}\,l\,(I_{0L} - V_{0R})\epsilon_0]. \quad (4)$$

Here l is an arbitrary constant determining the relative strength of the Abelian gauge $(I_{0L} - V_{0R})$ compared to the gauge $(I_L + V'_R)$ $(V' = V_1, -V_2, -V_3)$. Following the well-known procedure, construct $\mathcal{L}_{int}$ by replacing $(\partial\psi)$ in $\mathcal{L}_f$ with the co-variant derivative

$$\mathcal{D}L = \partial L + ie[(I_L + V'_R)\cdot \boldsymbol{W} - \sqrt{3}\,l\,(I_{0L} - V_{0R}) W_0]\, L \quad (5)$$

In terms of the two orthogonal combinations of fields,

$$A^0 = W_0 \cos\theta + W_3 \sin\theta$$

$$X^0 = -W_0 \sin\theta + W_3 \cos\theta$$

$$(l = \tan\theta)$$

one can write the neutral component of (4) in the form

$$(\cos\theta\, Q_x X^0 + \sin\theta\, Q_l A^0) \quad (6)$$

where $Q_x = \frac{1}{4}\sec^2\theta\,(3\gamma_5 Q_h + (2\cos 2\theta - 1)Q_l)$. The full interaction Lagrangian equals

$$\mathcal{L}_{int} = -ieL^+\gamma_4\gamma_\mu[(I_L^- + V'^+)\, W^+ + \text{h.c.} +$$

$$+ \sin\theta\, Q_l A^0 + \cos\theta\, Q_x X^0] L\,. \quad (7)$$

This Lagrangian has the following characteristics
1. The electromagnetic interaction $ie\,\sin\theta$ $(\bar{\mu}^+\mu^+ - \bar{e}\,e)$ is necessarily accompanied by a neutral current term $-ie\cos\theta\, Q_x X^0$. For the special case $\theta = 30^\circ$ $(3l^2 = 1)$ this neutral current conserves parity and is pure axial-vector, with the same sign of coupling for e^- and μ^+. For $\theta = 45^\circ$ $(l^2 = 1)$ the orthogonal fields A^0 and X^0 equal $\frac{1}{2}\sqrt{2}\,(W_3 \pm W_0)$ and the full neutral Lagrangian reads

$$\tfrac{1}{2}\sqrt{2}\, ie\,(\mu^+\mu - e^+e) A_0 \quad (8)$$

$$- \tfrac{1}{2}\sqrt{2}\, ie\,[\nu^+\gamma_5\nu - \tfrac{1}{2}e^+(1+\gamma_5)\,l + \tfrac{1}{2}\mu^+(1-\gamma_5)\mu] X_0\,.$$

Within Lepton physics the only assumption one need make is that X^0 is not mass-less (in order that the neutral X^0 current does not contribute to what is experimentally called the electric charge). When we come to consider hadrons, the absence of neutral leptonic currents interacting with heavy particles however requires that we assume X^0 particles are at least as massive as W^+ or W^-. The appearance of the X^0 current is about the minimum price one must pay to achieve the synthesis of weak and electromagnetic interactions we are seeking.

Since the interaction of W_3 with W^+, W^- (including the coupling strength) is completely determined by the gauge principle, it is crucial to check that the electric charge carried by W^+, W^- identically equals the charge on μ^+ end e^-. Group-theoretically this is equivalent to making sure that the W's belong to the *same* group representation $(SU_2 \times U_1)$ as L itself.

Concretely, in accordance with the transformation properties of the W's, one may define the field strengths

$$\boldsymbol{W}_{\mu v} = \partial_\mu \boldsymbol{W}_v - \partial_v \boldsymbol{W}_\mu - e\, \boldsymbol{W}_\mu \times \boldsymbol{W}_v \quad (9)$$

$$W_{0\mu v} = \partial_\mu W_{0v} - \partial_v W_{0\mu}\,. \quad (10)$$

The gauge-Lagrangian for the W fields equals

$$\mathcal{L}(W) = \tfrac{1}{4}(\boldsymbol{W}_{\mu v}\cdot \boldsymbol{W}_{\mu v} + W_{0\,\mu v} W_{0\mu v})\,. \quad (11)$$

In terms of A^0 and X^0, rewrite (9) and (10) in the form

$$W^\pm_{\mu v} = \partial_\mu W^\pm_v - \partial_\nu W^\pm_\mu \mp ieW^\pm_\mu\,(\cos\theta\, X^0 + \sin\theta\, A^0)_\nu$$

$$\pm ie\, W^\pm_v\,(\cos\theta\, X^0 + \sin\theta\, A^0)_\mu$$

$$W^0_{\mu v} = \partial_\mu X^0 - \partial_\nu X^0_\mu - ie\cos\theta\,(W^-_\mu W^+_\nu - W^+_\mu W^-_\nu)$$

$$A^0_{\mu v} = \partial_\mu A^0_v - \partial_v A^0_\mu - ie\sin\theta\,(W^-_\mu W^+_v - W^+_\mu W^-_v).$$

Clearly,

$$\mathcal{L}(W) = \tfrac{1}{4}(A^0_{\mu v}A^0_{\mu v} + X^0_{\mu v}X^0_{\mu v} + W^+_{\mu v}W^-_{\mu v} + W^-_{\mu v}W^+_{\mu v}) \quad (12)$$

Comparing the coupling constants in (7) and (9), it is obvious that the dynamical charge carried by W^- and W^+ equals respectively the charge on e^- and μ^+ both in sign and magnitude.

Note that the magnetic moment of $W^\pm$ equals 2 Bohr magnetons irrespective of the value of θ.

For leptons

$$Q_l = T^3 - \sqrt{3}\, T^8$$

while for hadrons the Gell-Mann - Nishijima formula gives

$$Q_h = T^3 + \tfrac{1}{3}\sqrt{3}\, T^8 .$$

To correspond to this fundamental difference, clearly the appropriate infinitesimal unitary transformation for hadrons equals

$$U = 1 + i(\varepsilon \cdot \boldsymbol{I} + \tfrac{1}{3}\sqrt{3}\, l\, I_0)$$

instead of the leptonic transformation

$$U = 1 + i(\varepsilon \cdot \boldsymbol{I} - \sqrt{3}\, l\, I_0).$$

Now weak interactions for hadrons experimentally appear to exhibit the pattern

$$F^V + D^A$$

It was pointed out [3,4] in an earlier paper that this precisely is the consequence of assuming that the 9-fold of baryons B transforms as a representation of the group [5] structure $(SU_3)_L \otimes (SU_3)_R$. Specialising for weak and electromagnetic interactions to the sub-group $(SU_2 \times U_1)$ in place of the full SU_3 group, consider the double gauge transformation

$$B' = [\exp i\,(I_L + V_R)]\; B\; [\exp -i\,(I_R + V_L)]$$

where I, stands for,

$$I = I_1\epsilon_1 + I_2\epsilon_2 + I_3\epsilon_3 + \tfrac{1}{3}\sqrt{3}\, l\, I_0\epsilon_0$$

and likewise for V.

With the standard gauge procedure, this gives rise to the currents

$$-\tfrac{1}{2}B^+\gamma_4\gamma_\mu\,[I + V, B]$$

$$-\tfrac{1}{2}B^+\gamma_4\gamma_\mu\gamma_5\,\{I - V, B\}$$

The neutral components of $\mathcal{L}_{int}$ are

$$\mathcal{L}_{int} = -ie\ \sin\theta\, A^0\ \mathrm{Tr}\ B^+[Q_h, B] A^0_\mu - ie\ \cos\theta\, X^0 J_x \qquad (13)$$

where

$$J_x = \tfrac{1}{4}\sec^2\theta\ \mathrm{Tr}\, B^+((1 + 2\cos 2\theta)[Q_h, B] - \gamma_5\{Q, B\}). \qquad (14)$$

Once again, note that for $\theta = 60^\circ$ ($l^2 = 3$) the new neutral current is purely axial vector *. It is gratifying that relative electric (as well as weak) charges on baryons are the same in magnitude as on leptons. This is not true in any obvious manner for the new neutral charge appearing in J_x.

Summarising the full Lagrangian equals

$$\mathcal{L}_{weak} = -ie\, W^- \Big(((e^-)^+\nu)_L + (\nu_\mu^+)_R + \tfrac{1}{2}\mathrm{Tr}\, B^+[I^+ + V^+, B] + \tfrac{1}{2}\mathrm{Tr}\ B^+\gamma_5\{I^+ - V^+, B\} + \text{meson terms}\Big) \qquad (15)$$

$$\mathcal{L}_{em} = -ie\ \sin\theta\, A^0\,(p^+p + \ldots + \mu\ \mu + \ldots) \qquad (16)$$

$$\mathcal{L}_{neutral} = -ie\ \cos\theta\ X^0\ (L^+ Q_x L + J_x) \qquad (17)$$

A number of problems remain

1) Even though the neutral X^0 current conserves strangeness and even though it is only just the lepton pairs e^+e^-, $\mu^+\mu^-$, $\bar{\nu}\nu$ which make their appearance.

The upper limit established in the CERN neutrino experiment on $\nu + p \rightarrow \nu + p$ (1% of σ_{weak}) implies that the effective coupling of X^0 particles must be much smaller than $g_{W^\pm}$. We speculate that there possibly may exist a geometrical relation among the effective coupling constants like

$$e^2 g^2_{x^0} = (g^2_{W^\pm})^2 .$$

We have no convincing reasons for this relation. This however appears to be the most plausible conjecture for g_x^2, once we do accept that $g^2_{W^\pm} \neq e^2$.

2) The weak hadron Lagrangian $[I^+ + V^+]_F + \gamma_5\{I^+ - V^+\}_D$ shows no sign of the Cabibbo suppression for the strangeness-changing currents $V^\pm$; it also predicts the combination $(V+A)$ in $\epsilon^- \rightarrow n$ + leptons compared to $(V-A)$ for β-decay.

There are two distinct views about Cabibbo's phenomenological Lagrangian $[I^\pm\cos\theta + V^\pm\sin\theta]_F + \gamma_5\,(a[I^\pm\cos\theta + V^\pm\sin\theta]_F + b\{I^\pm\cos\theta + V^\pm\sin\theta\}_D)$. One is that the Cabibbo suprression is a consequence of the symmetry-breaking mechanism operating in strong-interaction physics. The second view postulates that the suppression is intrinsic and results from a rotation of the weak relative to the strong i-spin and hypercharge axes. Now the most general charge and CP-conserving rotations one may consider are the following

* To incorporate the mesons, the natural assumption [3] appears to be to consider the P.S. particles M, together with a set of yet undiscovered scalar mesons M_2, to form a 9-fold $M = M_1 + iM_2$. The appropriate gauge transformation is

$$M' = \exp(iI)\, M \exp(-iV)$$

giving the gauge Lagrangian (see ref. 5, eq. (33) -

$$\mathcal{L} = -\tfrac{1}{4}\mathrm{Tr}\ [(\partial M_1 + \tfrac{1}{2}[I+V, M_1] - \tfrac{1}{2}\{I - V, M_2\})^2 + (\partial M_2 + \tfrac{1}{2}[I + V, M_2)] + \tfrac{1}{2}\{I - V, M_2\})^2$$

$$B'_L = X(\theta_1)\, B_L X^{-1}(\theta_2)$$
$$B'_R = X(\theta_3)\, B_R X^{-1}(\theta_4)$$

where

$$X = \exp\,(2i\theta\; T^7).$$

The net-effect of these is to replace I and V in (18) by the appropriate combinations of $I(\theta) = X^{-1}(\theta)\; I(\theta)$ and $V(\theta)$. For $I^{\pm}$, $V^{\pm}$, this may indeed give the Cabibbo suppression but for neutral currents (since $Q(\theta) = Q \cos 2\theta + \frac{1}{2} \sin 2\theta\; T^6$) any value of θ other than 0° or 90° inevitably gives rise to neutral strangeness-changing currents ($K^o \to \mu^+ + \mu^-$). For such currents there appears to be no experimental evidence. The "rotation" view is therefore incompatible with the present theory.

In a subsequent paper we consider the problem of starting with $[I + V]_F + \gamma_5\{I - V\}_D$, to show that the strong symmetry-breaking mechanisms do lead to a relative suppression of $[V^{\pm}]_F$ and $\gamma_5\{V^{\pm}\}_D$ currents, as well as the appearance of $\gamma_5[V^{\pm}]_F$ terms.

The authors would like to thank Prof. J. Prentki for a discussion.

References

1. A. Salam and J. C. Ward, Il Nuovo Cimento 11 (1959) 568;
S. Glashow, Nuclear Phys. 10 (1959) 103;
J. Schwinger, Ann. of Phys. 2 (1957) 407.
2. The group structure $I_L + V'_R$ for weak interactions was studied by R. Gatto (Il Nuovo Cimento 28 (1963) 567; and Y. Neeman, Il Nuovo Cimento 27 (1963) 567.
3. A. Salam and J. C. Ward, A gauge theory of elementary interactions, Phys, Rev., to be published
4. Y. Nambu and P. Freund, Phys. Rev. Letters 12 (1964) 714;
M. Gell-Mann, Physics Letters 12 (1964) 63;
M. Gell-Mann and Y. Neeman, to be published.
R. Makunda, R. E. Marshak and S. Okubu, to be published.
5. This group was introduced in elementary particle physics by A. Salam and J. C. Ward, Il Nuovo Cimento 27 (1961) 922;
M. Gell-Mann, Phys. Rev. 125 (1962) 1067;
and in slightly different form by R. E. Marshak and S. Okubo, 19 (1961) 1226.

* * * * *

A MODEL OF LEPTONS*

Steven Weinberg†
Laboratory for Nuclear Science and Physics Department,
Massachusetts Institute of Technology, Cambridge, Massachusetts
(Received 17 October 1967)

Leptons interact only with photons, and with the intermediate bosons that presumably mediate weak interactions. What could be more natural than to unite[1] these spin-one bosons into a multiplet of gauge fields? Standing in the way of this synthesis are the obvious differences in the masses of the photon and intermediate meson, and in their couplings. We might hope to understand these differences by imagining that the symmetries relating the weak and electromagnetic interactions are exact symmetries of the Lagrangian but are broken by the vacuum. However, this raises the specter of unwanted massless Goldstone bosons.[2] This note will describe a model in which the symmetry between the electromagnetic and weak interactions is spontaneously broken, but in which the Goldstone bosons are avoided by introducing the photon and the intermediate-boson fields as gauge fields.[3] The model may be renormalizable.

We will restrict our attention to symmetry groups that connect the observed electron-type leptons only with each other, i.e., not with muon-type leptons or other unobserved leptons or hadrons. The symmetries then act on a left-handed doublet

$$L \equiv [\tfrac{1}{2}(1+\gamma_5)]\binom{\nu_e}{e} \tag{1}$$

and on a right-handed singlet

$$R \equiv [\tfrac{1}{2}(1-\gamma_5)]e. \tag{2}$$

The largest group that leaves invariant the kinematic terms $-\overline{L}\gamma^{\mu}\partial_\mu L - \overline{R}\gamma^\mu\partial_\mu R$ of the Lagrangian consists of the electronic isospin $\vec{T}$ acting on L, plus the numbers N_L, N_R of left- and right-handed electron-type leptons. As far as we know, two of these symmetries are entirely unbroken: the charge $Q = T_3 - N_R - \frac{1}{2}N_L$, and the electron number $N = N_R + N_L$. But the gauge field corresponding to an unbroken symmetry will have zero mass,[4] and there is no massless particle coupled to N,[5] so we must form our gauge group out of the electronic isospin $\vec{T}$ and the electronic hyperchange $Y \equiv N_R + \frac{1}{2}N_L$.

Therefore, we shall construct our Lagrangian out of L and R, plus gauge fields $\vec{A}_\mu$ and B_μ coupled to $\vec{T}$ and Y, plus a spin-zero doublet

$$\varphi = \binom{\varphi^0}{\varphi^-} \tag{3}$$

whose vacuum expectation value will break $\vec{T}$ and Y and give the electron its mass. The only renormalizable Lagrangian which is invariant under $\vec{T}$ and Y gauge transformations is

$$\mathcal{L} = -\tfrac{1}{4}(\partial_\mu \vec{A}_\nu - \partial_\nu \vec{A}_\mu + g\vec{A}_\mu \times \vec{A}_\nu)^2 - \tfrac{1}{4}(\partial_\mu B_\nu - \partial_\nu B_\mu)^2 - \overline{R}\gamma^\mu(\partial_\mu - ig'B_\mu)R - L\gamma^\mu(\partial_\mu ig\vec{t}\cdot\vec{A}_\mu - i\tfrac{1}{2}g'B_\mu)L$$

$$-\tfrac{1}{2}|\partial_\mu\varphi - ig\vec{A}_\mu\cdot\vec{t}\varphi + i\tfrac{1}{2}g'B_\mu\varphi|^2 - G_e(\overline{L}\varphi R + \overline{R}\varphi^\dagger L) - M_1^{\,2}\varphi^\dagger\varphi + h(\varphi^\dagger\varphi)^2. \tag{4}$$

We have chosen the phase of the R field to make G_e real, and can also adjust the phase of the L and Q fields to make the vacuum expectation value $\lambda \equiv \langle\varphi^0\rangle$ real. The "physical" φ fields are then φ^-

and

$$\varphi_1 \equiv (\varphi^0 + \varphi^{0\dagger} - 2\lambda)/\sqrt{2} \quad \varphi_2 \equiv (\varphi^0 - \varphi^{0\dagger})/i\sqrt{2}. \tag{5}$$

The condition that φ_1 have zero vacuum expectation value to all orders of perturbation theory tells us that $\lambda^2 \cong M_1^2/2h$, and therefore the field φ_1 has mass M_1 while φ_2 and φ^- have mass zero. But we can easily see that the Goldstone bosons represented by φ_2 and φ^- have no physical coupling. The Lagrangian is gauge invariant, so we can perform a combined isospin and hypercharge gauge transformation which eliminates φ^- and φ_2 everywhere[6] without changing anything else. We will see that G_e is very small, and in any case M_1 might be very large,[7] so the φ_1 couplings will also be disregarded in the following.

The effect of all this is just to replace φ everywhere by its vacuum expectation value

$$\langle\varphi\rangle = \lambda\binom{1}{0}. \tag{6}$$

The first four terms in $\mathcal{L}$ remain intact, while the rest of the Lagrangian becomes

$$-\tfrac{1}{8}\lambda^2 g^2[(A_\mu{}^1)^2 + (A_\mu{}^2)^2] - \tfrac{1}{8}\lambda^2(gA_\mu{}^3 + g'B_\mu)^2 - \lambda G_e \bar{e}e. \tag{7}$$

We see immediately that the electron mass is λG_e. The charged spin-1 field is

$$W_\mu \equiv 2^{-1/2}(A_\mu{}^1 + iA_\mu{}^2) \tag{8}$$

and has mass

$$M_W = \tfrac{1}{2}\lambda g. \tag{9}$$

The neutral spin-1 fields of definite mass are

$$Z_\mu = (g^2 + g'^2)^{-1/2}(gA_\mu{}^3 + g'B_\mu), \tag{10}$$

$$A_\mu = (g^2 + g'^2)^{-1/2}(-g'A_\mu{}^3 + gB_\mu). \tag{11}$$

Their masses are

$$M_Z = \tfrac{1}{2}\lambda(g^2 + g'^2)^{1/2}, \tag{12}$$

$$M_A = 0, \tag{13}$$

so A_μ is to be identified as the photon field. The interaction between leptons and spin-1 mesons is

$$\frac{ig}{2\sqrt{2}}\bar{e}\gamma^\mu(1+\gamma_5)\nu W_\mu + \text{H.c.} + \frac{igg'}{(g^2+g'^2)^{1/2}}\bar{e}\gamma^\mu e A_\mu + \frac{i(g^2+g'^2)^{1/2}}{4}\left[\left(\frac{3g'^2-g^2}{g'^2+g^2}\right)\bar{e}\gamma^\mu e - \bar{e}\gamma^\mu\gamma_5 e + \bar{\nu}\gamma^\mu(1+\gamma_5)\nu\right]Z_\mu. \tag{14}$$

We see that the rationalized electric charge is

$$e = gg'/(g^2 + g'^2)^{1/2} \tag{15}$$

and, assuming that W_μ couples as usual to hadrons and muons, the usual coupling constant of weak interactions is given by

$$G_W/\sqrt{2} = g^2/8M_W{}^2 = 1/2\lambda^2. \tag{16}$$

Note that then the e-φ coupling constant is

$$G_e = M_e/\lambda = 2^{1/4}M_e G_W{}^{1/2} = 2.07\times10^{-6}.$$

The coupling of φ_1 to muons is stronger by a factor M_μ/M_e, but still very weak. Note also that (14) gives g and g' larger than e, so (16) tells us that $M_W > 40$ BeV, while (12) gives $M_Z > M_W$ and $M_Z > 80$ BeV.

The only unequivocal new predictions made by this model have to do with the couplings of the neutral intermediate meson Z_μ. If Z_μ does not couple to hadrons then the best place to look for effects of Z_μ is in electron-neutron scattering. Applying a Fierz transformation to the W-exchange terms, the total effective e-ν interaction is

$$\frac{G_W}{\sqrt{2}}\bar{\nu}\gamma_\mu(1+\gamma_5)\nu\left\{\frac{(3g^2-g'^2)}{2(g^2+g'^2)}\bar{e}\gamma^\mu e + \tfrac{3}{2}\bar{e}\gamma^\mu\gamma_5 e\right\}.$$

If $g \gg e$ then $g \gg g'$, and this is just the usual e-ν scattering matrix element times an extra factor $\frac{3}{2}$. If $g \simeq e$ then $g \ll g'$, and the vector interaction is multiplied by a factor $-\frac{1}{2}$ rather than $\frac{3}{2}$. Of course our model has too many arbitrary features for these predictions to be taken very seriously, but it is worth keeping in mind that the standard calculation[8] of the electron-neutrino cross section may well be wrong.

Is this model renormalizable? We usually do not expect non-Abelian gauge theories to be renormalizable if the vector-meson mass is not zero, but our Z_μ and W_μ mesons get their mass from the spontaneous breaking of the symmetry, not from a mass term put in at the beginning. Indeed, the model Lagrangian we start from is probably renormalizable, so the question is whether this renormalizability is lost in the reordering of the perturbation theory implied by our redefinition of the fields. And if this model is renormalizable, then what happens when we extend it to include the couplings of $\vec{A}_\mu$ and B_μ to the hadrons?

I am grateful to the Physics Department of MIT for their hospitality, and to K. A. Johnson for a valuable discussion.

---

*This work is supported in part through funds provided by the U. S. Atomic Energy Commission under Contract No. AT(30-1)2098).

†On leave from the University of California, Berkeley, California.

[1]The history of attempts to unify weak and electromagnetic interactions is very long, and will not be reviewed here. Possibly the earliest reference is E. Fermi, Z. Physik 88, 161 (1934). A model similar to ours was discussed by S. Glashow, Nucl. Phys. 22, 579 (1961); the chief difference is that Glashow introduces symmetry-breaking terms into the Lagrangian, and therefore gets less definite predictions.

[2]J. Goldstone, Nuovo Cimento 19, 154 (1961); J. Goldstone, A. Salam, and S. Weinberg, Phys. Rev. 127, 965 (1962).

[3]P. W. Higgs, Phys. Letters 12, 132 (1964), Phys. Rev. Letters 13, 508 (1964), and Phys. Rev. 145, 1156 (1966); F. Englert and R. Brout, Phys. Rev. Letters 13, 321 (1964); G. S. Guralnik, C. R. Hagen, and T. W. B. Kibble, Phys. Rev. Letters 13, 585 (1964).

[4]See particularly T. W. B. Kibble, Phys. Rev. 155, 1554 (1967). A similar phenomenon occurs in the strong interactions; the ρ-meson mass in zeroth-order perturbation theory is just the bare mass, while the A_1 meson picks up an extra contribution from the spontaneous breaking of chiral symmetry. See S. Weinberg, Phys. Rev. Letters 18, 507 (1967), especially footnote 7; J. Schwinger, Phys. Letters 24B, 473 (1967); S. Glashow, H. Schnitzer, and S. Weinberg, Phys. Rev. Letters 19, 139 (1967), Eq. (13) et seq.

[5]T. D. Lee and C. N. Yang, Phys. Rev. 98, 101 (1955).

[6]This is the same sort of transformation as that which eliminates the nonderivative $\vec{\pi}$ couplings in the σ model; see S. Weinberg, Phys. Rev. Letters 18, 188 (1967). The $\vec{\pi}$ reappears with derivative coupling because the strong-interaction Lagrangian is not invariant under chiral gauge transformation.

[7]For a similar argument applied to the σ meson, see Weinberg, Ref. 6.

[8]R. P. Feynman and M. Gell-Mann, Phys. Rev. 109, 193 (1957).

Weak and electromagnetic interactions

By Abdus Salam

Imperial College of Science and Technology, London, England, and International Centre for Theoretical Physics, Trieste, Italy

One of the recurrent dreams in elementary particle physics is that of a possible fundamental synthesis between electromagnetism and weak interaction. The idea has its origin in the following shared characteristics:

1. Both forces affect equally all forms of matter-leptons as well as hadrons.
2. Both are vector in character.
3. Both (individually) possess universal coupling strengths.

Since universality and vector character are features of a gauge theory those shared characteristics suggest that weak forces just like the electromagnetic forces arise from a gauge principle. There is of course also a profound difference: electromagnetic coupling strength is vastly different from the weak. Quantitatively one may state it thus: if weak forces are assumed to have been mediated by intermediate bosons (W), the boson mass would have to equal from 50 to as large as 137 M_p, in order that the (dimensionless) weak coupling constant $g_w^2/4\pi$ equals $e^2/4\pi$.

I shall approach this synthesis from the point of view of renormalization theory. I had hoped that I would be able to report on weak and electromagnetic interactions throughout physics, but the only piece of work that is complete is that referring to leptonic interactions, so I will present only that today. Ward and I worked with these ideas intermittently (1, 2), particularly the last section on renormalization of Yang-Mills theories. The material I shall present today, incorporating some ideas of Higgs & Kibble, was given in lectures (unpublished) at Imperial College. Subsequently I discovered that an almost identical development had been made by Weinberg (3) who apparently was also unaware of Ward's and my work.

The renormalization point of view has become increasingly more and more important as time has gone on. In fact, if I have heard the reports rightly, the major activity in the U.S.A. last year, at least according to the rumours we heard about it, was the attempt to renormalize the radiative corrections to beta-decay, particularly at Princeton, by Gell-Mann, Goldberger, Kroll, Low and others. I will not say that the problem of renormalization of weak interactions, together with their radiative corrections, has acquired the same importance now as in 1947 it had for electrodynamics in connection with the Lamb shift, but I think it is a pretty serious problem, which must be treated at a fundamental level.

Today it is not the renormalization to second order that one is speaking about. I have in mind a renormalization program that should apply for the complete theory just as the program applied for electrodynamics itself. From this point of view we shall see that this unification of electromagnetism and weak interactions becomes more and more of a necessity. Further, it places a large number of restraints on the type of theory one can have.

For leptonic weak interactions the basic problem is of course that one is dealing with vector and axial vector interactions of spin half particles. The four-fermion interaction has not the slightest vestige of renormalizability. We thus consider intermediate boson theory and what we wish to do is to formulate the theory such that the photon appears as the neutral intermediate boson. Before I do this I shall briefly review what we know about the renormalization problem of vector mesons. As is well known, the lack of renormalizability of the theory arises from massive vector mesons from the $k_\mu k_\nu/m^2(k^2-m^2)$ term in the propagator.

(*a*) Consider *neutral vector* mesons, where there is hope that currents can be conserved. In this case Feynman's old proof, in which he showed that this $k_\mu k_\nu$ term latches on to the currents J_μ and J_ν at either of the vertices and disappears because of the conservation law, applies. The theory is renormalizable even if the vector meson is massive, as was first shown by Matthews. Thus current conservation must be preserved at all costs as a minimum requirement.

(*b*) Clearly, for an *axial* current the mass of the source Fermi particle (electron or meson), which makes the current non-conserved, will destroy any hope of renormalizability. Therefore the second restriction is that we must not have a fundamental Fermi mass in the theory.

(*c*) If one is dealing with charged vector or axial mesons the currents must include these particles themselves. To construct a conserved current this time we must have a recipe which includes contributions of the vector and axial vector mesons themselves within the current. The only known method for constructing such currents is the Yang-Mills recipe which arises when one is dealing with gauge symmetries associated with non-abelian symmetry groups.

(*d*) Now what about the renormalizability of Yang-Mills theories? Around 1962 the following was proved. Take $U(3)$ symmetry; then out of the, say, nine currents of $U(3)$, three of these, J_0, J_3 and J_8, will allow the corresponding $k_\mu k_\nu/m^2$ terms in the propagator to be transformed away. But, for the remaining currents the corresponding meson propagator must contain these terms and they cannot cancel *even though the currents are conserved.* Thus Yang-Mills theory with *non-zero meson mass is not renormalizable even if the currents are fully conserved.*

Therefore the only hope for vector mesons is (1) a Yang-Mills theory but also (2) that mesons should be massless. *To state it more generally, it is not current conservation that is the criterion of renormalizability, but gauge invariance.* The mass of the vector meson is not gauge invariant even though it does not interfere with current conservation. Nothing that will destroy the gauge invariance is allowed. Thus the only hope for achieving a renormalizable theory involving vector currents is that the mass of the meson should come through a spontaneous symmetry breaking. Further, since the second source of lack of (axial) gauge invariance is going to be the Fermi mass term, what would be ideal is if the Fermi mass term could also come from spontaneous symmetry breaking and preferably the *same* symmetry breaking. The Fermi mass coming from symmetry breaking was an idea started by Nambu, and has been worked on particularly by Johnson, Baker and Willey in electrodynamics. One adds the mass term in the free part of the Hamiltonian, sets up an interaction representation, computes the self-mass and sets it equal to the physical mass. This gives a self-consistent equation. For the meson case the propagator has no $k_\mu k_\nu/m^2$ terms but just $\delta_{\mu\nu}/(k^2-m^2)$ and the theory from the outset is renormalizable. (See for more details ref. 2.)

A second proposal and a related one is not to do this rather brutal addition and subtraction of mass terms but to work more gently. This is the method of letting the vector mesons interact with a set of scalar particles and to let them acquire physical masses by assuming self-consistently that these scalar particles possess non-zero expectation values. Consider, for example, the charged mesons ϕ and ϕ^* interacting with a neutral vector meson. Compute self-consistently from the theory an expectation value $\langle\phi_0\rangle$ for the ϕ field. The term $e^2\phi^*\phi A_\mu^2$ in the Lagrangian now appears to have a piece $e^2(\phi_0)^2A_\mu^2$ which acts like the vector meson mass.

Now all this is beautiful, but there was the suspected difficulty with this theory which held people back for a long time. This was the fear of Goldstone mesons. What one feared was that whenever you have such a theory Goldstone boson sits like a snake in the grass ready to strike. According to Goldstone's theorem a number of massless particles must arise in any such theory. To see these massless objects it is simplest to take the same example as before of charged scalar meson; set

$$\phi=\varrho e^{i\theta},\quad \phi^*=\varrho e^{-i\theta} \tag{1}$$

The Lagrangian equals

$$\mathcal{L}=\partial_\mu\phi^*\partial_\mu\phi+m^2\phi^*\phi=(\partial_\mu\varrho)^2+\varrho^2(\partial_\mu\theta)^2+m^2\varrho^2 \tag{2}$$

It clearly has a massless field θ. It is this field θ which for spontaneous symmetry-breaking theories appears as the Goldstone boson.

A big advance in resolving this difficulty has recently been made but I do not think that many people are aware of it. This is due to Higgs and, following him, Kibble who have shown that, true, there may be massless scalar mesons when you have done the spontaneous symmetry breaking. *But if you have a gauge theory* such objects can be transformed away as gauges and they have no coupling to the physical particles. To see this let us go back to the same example. Consider the Lagrangian

$$\mathcal{L} = (\partial_\mu - ieA_\mu)\phi^*(\partial_\mu + ieA_\mu)\phi + F_{\mu\nu}F_{\mu\nu} + m^2\phi^*\phi \tag{3}$$

Now write

$$(\partial_\mu + ieA_\mu)\phi = e^{i\theta}(\partial_\mu \varrho + ieA'_\mu \varrho) \tag{4}$$

where $A'_\mu = A_\mu + \partial_\mu\theta$. Clearly the term $\partial_\mu \theta$ appears in such a way where it can be transformed away through the gauge invariance, and no trace of the θ field is left. Also, if we assume that the expectation value of ϱ could be non-zero, i.e. $\varrho = \langle\varrho_0\rangle + \varrho'$ where ϱ' is the quantized part of ϱ, then the term $e^2\varrho^2 A_\mu^2$ now has a piece $e^2(\varrho_0)^2 A_\mu^2$, i.e. the vector meson acquires a mass term in this particular gauge specified by $\langle\varrho_0\rangle$.

What has happened? We started with the two fields ϱ and θ. One of the fields, ϱ, I can use for spontaneous symmetry breaking. The other field was massless. The symmetry breaking gave mass to the vector meson and the massless objects got transformed away. So you have the best of all possible worlds in this theory. Since the mass term has come in gently through symmetry breaking, our claim is that the theory is renormalizable. It clearly is in the original formulation with ϕ, ϕ^* fields.

Let us now make use of this in the case of weak and electromagnetic interactions. I will try to get a Yang–Mills theory of weak and electromagnetic interactions combined because that is the only way of keeping currents conserved. Since there is a mass of the fermion, the Yang–Mills theory will want to be broken spontaneously. If the same symmetry breaking can be arranged to give masses to the charged intermediate bosons which mediate weak current but not to the neutral meson representing the electromagnetic field we would have removed the major objection to the unification of weak and electromagnetic fields of having in one multiplet massive and massless objects.

The application of the Yang-Mills ideas to weak and electromagnetic interaction has been carried through by Gatto, Ne'eman, Ward, Salam and recently Weinberg. I will write it in an *SU*(3) notation and I hope that this will prove to be more than a notational nicety. What you do is that you have a four-component neutrino, a negative electron and positive muon, and define the lepton triplet

$$L=\begin{pmatrix}\nu\\ e^-\\ \mu^+\end{pmatrix} \tag{5}$$

Consider I-spin and V-spin subgroups of $SU(3)$ where I define the scalars in I-space and V-space as

$$I_0=T^8,\; V_0=\frac{\sqrt{3}}{2}T^3-\tfrac{1}{2}T^8 \tag{6}$$

The lepton charge equals

$$Q_l=\begin{pmatrix}0 & & \\ & -1 & \\ & & 1\end{pmatrix} \tag{7}$$

Notice that the hadronic charge in the same notation is

$$Q_h=T^3+\frac{i}{\sqrt{3}}T^8=\begin{pmatrix}\frac{2}{3} & & \\ & -\frac{1}{3} & \\ & & -\frac{1}{3}\end{pmatrix} \tag{8}$$

so that Q_h and Q_l are "orthogonal" combinations in T^3, T^8 space. Define right and left particles, conventionally,

$$\psi_{L,R}=\tfrac{1}{2}(1\pm\gamma_5)\psi \tag{9}$$

and note that

$$\begin{aligned} I_{3L}-V_{3R}&=\tfrac{1}{4}(Q_l+3\gamma_5 Q_h)\\ I_{0L}-V_{0R}&=\tfrac{1}{2}(Q_l-\gamma_5 Q_h)\end{aligned} \tag{10}$$

Now the weak interaction can be written as

$$\mathcal{L}_{\text{weak}}=[(\bar{e}^-\nu)_L+(\bar{\nu}\mu^+)_R]\,W^-+\text{h.c.}=(I_{L1}+V_{R1})\,W_1+(I_{L2}-V_{R2})\,W_2 \tag{11}$$

where the intermediate bosons are W^+ and W^-. Note that the anti-neutrino right come in with muon and neutrino left with the electron so that all four components of the neutrino field are used up. This is a minor merit of the formulation. If one has a theory of symmetries it is undesirable to have to work with two-component neutrinos and four-component electrons and muons, because in some conceptual limit you would like to transform them into each other.

To complete the Yang-Mills gauge group we must introduce the third component of the vector **W**, adding on the term $(I_{3L} - V_{3R}) W_3$ to the Lagrangian. Ideally one would have liked this to be nothing but the electromagnetic field. Unfortunately the $I_{3L} - V_{3R}$ term contains the $\gamma_5 Q_h$ piece which has nothing to do with electromagnetism. One must therefore introduce in this theory an additional neutral vector meson W_0 and add in a term

$$\mathcal{L}_{\text{neutral}} = (I_{3L} - V_{3R}) W_3 - \sqrt{3} \tan\theta (I_{0L} - V_{0R}) W_0 \tag{12}$$

to the Lagrangian (11). The factor $\sqrt{3}\tan\theta$ gives an arbitrary scale factor between the two neutral currents. In terms of the quantities

$$\begin{aligned} A^0 &= W_0 \cos\theta + W_3 \sin\theta \\ X^0 &= -W_0 \sin\theta + W_3 \cos\theta \end{aligned} \tag{13}$$

one can write the neutral currents as

$$(\cos\theta Q_X X^0 + \sin\theta Q_l A^0) \tag{14}$$

where

$$Q_X = \tfrac{1}{4} \sec^2\theta (3 Q_h \gamma_5 + (2\cos\theta - 1) Q_l) \tag{15}$$

Identify A^0 with the Maxwell field. In addition one has to buy—as about the minimum price for this unification—another field X^0 which is neutral and has the current Q_X. Note that the current Q_X is diagonal in all the particles and does not transform electrons to muons.

Let us not worry about muon mass for the moment and concentrate on the electron. We shall now introduce one common source of breaking the symmetry introduced above which will give (*a*) mass to the electron (*b*) mass for W^+, W^- and for X^0 (but not to A^0). To bring out the structure of this term let us rewrite the symmetry introduced above in a more familiar language, define a lepton hypercharge and write

$$\tfrac{1}{2} Y = -\sqrt{3}\, T^8 \tag{16}$$

All that I described above in the $SU(3)$ notation can be re-expressed in the following way: we make a doublet out of the electron-left and neutrino-left with hypercharge -1 and let the electron-right correspond to $I=0$ with $Y=-2$. This can be indicated in this way

	I_3	Y	$Q = I_3 + \frac{1}{2}Y$
V_L^0	$\frac{1}{2}$	-1	0
e_L^-	$-\frac{1}{2}$	-1	-1
e_R^-	0	-2	-1

To bring out the analogy with hadron physics, it would be as if we were dealing with Ξ and Ω^-.

Write

$$\mathcal{L}_m = m\bar{e}_R e_L + \text{h.c.} \overset{\text{analogy}}{=} f\bar{\Omega}^-(\langle\mathcal{K}\rangle\tau_2\Xi) + \text{h.c.}; \quad \mathcal{K} = \begin{pmatrix}\bar{K}^0\\ K\end{pmatrix} \tag{17}$$

Since I would now want to get this mass term as an expectation value of some scalar entity, I have written it in a suggestive form in analogy with a sort of $\Delta I = \frac{1}{2}$ rule for I-spin breaking. To do this I have exhibited an analogous interaction of Ω^- and Ξ particles. The idea is that we introduce into the theory an object which transforms like the $\bar{K}$-meson in *leptonic I-space* and that this object—or rather the K_{01} component of it—possesses a non-zero expectation value, $f\langle K_{01}\rangle m$. The $\bar{K}$-meson thus introduced must itself have a gauge interaction plus a mass term (plus a four-field interaction to allow it to acquire a non-zero expectation value). Thus its Lagrangian is

$$\mathcal{L} = (\partial_\mu - ieW_\mu)\mathcal{K}^+(\partial_\mu + ieW_\mu)\mathcal{K} + m^2\mathcal{K}^+\mathcal{K} + \lambda(\mathcal{K}^+\mathcal{K})^2 \tag{18}$$

If one does a very simple-minded calculation the expectation value of K^0 comes out of the magnitude

$$\langle K_{01}\rangle \approx m^2/2\lambda \tag{19}$$

where m is the mass of the K-particle. We now use the Higgs ansatz and go into a special gauge where all objects except K_{01} are transformed away. Higgs would tell us that of the four real fields contained in $\mathcal{K}$, K_{01} to which we have hopefully assigned the spontaneously introduced non-zero expectation value, will survive in the Lagrangian while K_{02}, K^+ and K^- can be transformed away. What is further remarkable is that this will give mass terms proportional to $e^2\langle K_{01}\rangle^2$ to the charged intermediate bosons W^+, W^- and to X^0 but not A^0. The simplest way to see this is to remark that the vector meson field corresponding to the symmetry which survives (electric charge conservation) started with zero mass and will end up in the special gauge with zero mass. There is only one symmetry which we are preserving in the whole business and that is the symmetry corresponding to the electromagnetic field. We had no mass for this field to start with and no mass when we finish. The other three are the symmetries we break; the corresponding vector fields acquire masses which are $e^2\langle K_{01}\rangle^2$. The whole thing has therefore worked out in a unified and beautiful manner; a renormalizable theory of massive vector mesons, in which electromagnetism is built in as part of the theory and the Lagrangian is perfectly symmetrical. Only when one goes to a special gauge, where

$$\mathcal{K} = e^{i(\mathbf{I}\cdot\boldsymbol{\theta}\frac{1}{2}Y\theta_3)}\begin{pmatrix}\varrho\\ 0\end{pmatrix} \tag{20}$$

and where $\langle\varrho\rangle = K_{01}$, only in this gauge does the Lagrangian look ugly.

Like all suggestions in physics this one too has some associated difficulties. The first difficulty of course is "what shall we do with this expectation value?" We would like the mass of the vector meson

$$m_V^2 = e^2 \langle K_{01} \rangle^2 \tag{21}$$

to be larger than 50 BeV or so to get the weak interaction constant to equal the electromagnetism. If we want the coupling constant e to be the electromagnetic coupling constant, then $\langle K_{01} \rangle$ must be of the order of 500–1000 BeV. We can, of course, arrange for such a large value of $\langle K_{01} \rangle$ by making the mass m in front of the term $\mathcal{K}^+\mathcal{K}$ large. But this is really shifting the problem of a large mass for the intermediate boson to a (still) larger mass for the scalar K-meson.

This is not so pleasant. There may be another way out. Suppose we were using similar ideas for the baryon case, with something like a fundamental quark field which we assume to be very massive. If such a quark mass were also to arise from spontaneous symmetry breaking, the scalar object similar to what we called the K-meson above, coupling to such quarks, would need to possess an enormous value for its vacuum expectation value. Such a scalar object would couple with the intermediate bosons just like the K-particle above and give them mass. It may be more pleasant to find the source of large vector boson mass thus through its association with baryons rather than leptons. The point is that since weak bosons couple both to leptons *and* baryons, their large mass may come from the baryon side rather than leptons since all contributions of the type $\langle K_{01} \rangle^2$ which determine both the baryon masses and boson masses come additively in the latter.

All I am trying to say is that you must have a complete theory. And one must not be allowed to break symmetries in the fundamental Lagrangian ad hoc if vector meson renormalizability is to be preserved. I feel that this is a line worth pursuing just because it is so restrictive.

Now some remarks on the construction of a similar theory for hadrons. This Ward and myself tried very hard to do in 1964 but we failed. The main source of the failure was always the Cabibbo angle. One does not want the neutral current introduced into the theory to be strangeness changing. In every theory with a non-zero Cabibbo angle this will inevitably happen.

One line we tried was to use the same combination of I_L and V_R for the hadrons with two types of intermediate bosons

$$[I_L \cos\theta + V_R \sin\theta]\, W_1 + [-I_L \sin\theta + V_R \cos\theta]\, W_2 \tag{22}$$

as far as hadrons are concerned. Let only W_1-particles interact with the leptons and require that the charge values on leptons and hadrons given by the appropriate neutral current be Q_l and Q_h,

$$Q_l = T^3 - \sqrt{3}\,T^8, \quad Q_h = T^3 + \frac{1}{\sqrt{3}}T^8 \tag{23}$$

This gives a unique value for $\theta = 15°$ which, it so happens, is the right numerical value for Cabibbo suppression of leptonic decays of hadrons. It is easy to check that there are no neutral strangeness-changing currents. I am not presenting this as the solution for the hadron problem, firstly because in such a theory we would get a $V+A$ charged strangeness-changing quark weak current; secondly, to get any non-leptonic strangeness-changing decays at all we shall need to use a further spontaneous symmetry breaking which must employ something like a non-zero expectation value for the physical K_0-particles ($\langle K_{01}\rangle \neq 0$) in addition to the intermediate boson theory above. The hadron problem is still unsolved.

References

1. J. C. Ward & A. Salam, Nuovo Cimento *11*, 568 (1959); Phys. Lett. *13*, 168 (1964); Phys. Rev. *136*, 763 (1964).
2. A. Salam, Phys. Rev. *127*, 331 (1962).
3. S. Weinberg, Phys. Rev. Lett. *19*, 1264 (1967).

Discussion

Pais

Suppose you have by some mechanism or other a range for the effective mass of the charged W. Suppose I calculate some leptonic process, in which more than a single W-meson goes across between leptons. Is it not seriously divergent?

Salam

My claim is that in the theory I discussed there is no $k_\mu k_\nu$ term in the propagator.

Pais

In that case I want to ask a second question namely the following one. You have auxiliary scalar fields. I thought what you said is the following. The purpose of this field is by some self-consistency condition to generate the mass. At the same time you have no Goldstone particles. Suppose I have now retransformed the theory into this form. In what way does that theory now present itself to me in a different fashion from a theory in which I had an ordinary mass from the very beginning?

Salam

Formally none. Except that this mass does not have the same origin. All your troubles are now computing this mass.

Sudarshan
I would like to pursue this point a little further. Suppose, after the theory has reached its final stage, one attempts to write down rules for computing diagrams. Would I then simply substitute the $g_{\mu\nu}/(k^2-m^2)$ propagator?

Salam
Let us have it clear. There are two distinct formulations of the theory; the fields in the two formulations are being transformed one into the other by a non-linear (gauge) transformation. One is the perfect theory which contains all of the scalar particles; the gauge is preserved, etc. In that version of the formulation there is no $k_\mu k_\nu$ term and there are no divergences. If you ask me for a dictionary to the conventional theory obtained by transforming to the new set of fields where not all the scalar particles are present I haven't got this dictionary worked out yet. I conjecture that this will be a very simple dictionary but I have not evolved it yet.

Pais
So you have not proved that the $k_\mu k_\nu$ term can be dropped?

Salam
It can clearly be dropped in the symmetrical version of the theory. I believe this will also happen in the normal theory because I know there are no divergences; and also I know the two theories are equivalent so far as the final results are concerned.

Sudarshan
But in that case if I look at the propagator then the effective propagator for this particular particle does it not have a scalar negative norm particle because of the $g_{\mu\nu}$?

Salam
No. It cannot have negative norms because the symmetrical version of the theory does not have them. A local transformation of fields cannot introduce negative norms if there are none before.

Stech
I have some difficulty in understanding such a high mass for the intermediate boson. If one wants to calculate, in a very rough way, an absolute rate for non-leptonic decays as Glashow, Weinberg and Schnitzer did some time ago or when one wants to compare the electromagnetic mass differences with non-leptonic matrix elements of the current–current type one sees that a high mass for the intermediate particle is not likely. There are several arguments against a high value of a mass (Λ) representing the structure of weak interaction. My

own one is based on a comparison of electromagnetic mass differences with one-particle matrix elements of the weak interaction describing non-leptonic decay processes.

One has for an example:

$$\int \frac{d^4q}{-q^2} \int d^4x\, e^{iqx} \langle n | T(V^{\mu}_{1-i2}(x) V^{8}_{\mu}(0)) | p \rangle = \frac{\sqrt{3}\, m_p(m_n - m_p)}{e^2/4\pi}$$

and

$$\int \frac{d^4q}{-q^2+\Lambda^2} \frac{\Lambda^2}{m_p^2} \int d^4x \left\langle n \,|\, T(V^{\mu}_{1-i2}(x)\, V^{4-i5}_{\mu}(0) + A^{\mu}_{1+i2}(x)\, A^{4-i5}_{\mu}(0) \,|\, \Lambda^0 + \frac{1}{\sqrt{3}} \Sigma^0 \right\rangle$$

$$= \frac{(2\pi)^4 \left\langle n \,|\, \mathcal{H}_{\text{weak}} \,|\, \Lambda^0 + \frac{1}{\sqrt{3}} \Sigma^0 \right\rangle}{\sin\theta \cos\theta G m_p^2/\sqrt{2}}$$

A comparison of the two integrals can be made using the assumption that the high momentum transfer region (Björken limit) decisively determines the (dominant) octet parts of the two interactions. Defining a factor x by $x = \langle n | VV + AA | \Lambda^0 + (1/\sqrt{3})\Sigma^0 \rangle / 2\sqrt{2} \langle n | VV | p \rangle$ one gets from the known numerical values of the right hand sides of the above equations (using octet parts only) $|x| \approx 0.16\, m_p^2/\Lambda^2$. By this relation high values of Λ are excluded even if a sizeable suppression of the weak interaction matrix element (in breaking $SU(3) \times SU(3)$) occurs. This conclusion is, of course, not valid if the high momentum transfer region is unimportant for the value of the integrals. In this case $|x| \approx 0.16 \overline{(1/q^2)}\, m_p^2$ i.e. rather low values of q^2 determine the integrals. The Björken limit would be irrelevant and could not be held responsible for the octet dominance observed in both interactions.

Salam (remark added after discussion)

The two integrals can be compared only if one neglects Λ^2 in the denominator $1/(-q^2 + \Lambda^2)$. Clearly for $\Lambda^2 \approx (50\, m_p)^2$ this is questionable. So the argument based on neglecting Λ^2 in the denominator to prove that Λ^2 is small is somewhat circular.

Physical Processes in a Convergent Theory of the Weak and Electromagnetic Interactions*

Steven Weinberg
Laboratory for Nuclear Science and Department of Physics, Massachusetts Institute of Technology, Cambridge, Massachusetts 02139
(Received 20 October 1971)

A previously proposed theory of leptonic weak and electromagnetic interactions is found to be free of the divergence difficulties present in conventional models. The experimental implications of this theory and its extension to hadrons are briefly discussed.

Several years ago I proposed a unified theory[1] of the weak and electromagnetic interactions of leptons, and suggested that this theory might be renormalizable. This theory is one of a general class of models which may be constructed by a three-step process[2]: (A) First write down a Lagrangian obeying some exact gauge symmetry, in which massless Yang-Mills fields interact with a multiplet of scalar fields[3] and other particle fields. (B) Choose a gauge in which all the scalar field components vanish, except for a few (in our case one) real scalar fields. (C) Allow the gauge group to be spontaneously broken by giving the remaining scalar field a nonvanishing vacuum expectation value. Redefine this field by subtracting a constant λ, so that the "shifted" field φ has zero vacuum expectation value. In the resulting perturbation theory, all vector mesons acquire a mass, except for those (in our case, the photon) associated with unbroken symmetries.

In the proposed theory, this procedure was applied to the gauge group $SU(2)_L \otimes Y$, and resulted in a model involving electrons, electron-type neutrinos, charged intermediate bosons (W_μ), neutral intermediate bosons (Z_μ), photons (A_μ), and massive neutral scalar mesons (φ), with an interaction of the form[1]

$$\begin{aligned}\mathcal{L}' = {} & \frac{ig}{(g+g'^2)^{1/2}}[gZ^\nu - g'A^\nu][W^\mu(\partial_\mu W_\nu{}^\dagger - \partial_\nu W_\mu{}^\dagger) - W^{\mu\dagger}(\partial_\mu W_\nu - \partial_\nu W_\mu) + \partial^\mu(W_\mu W_\nu{}^\dagger - W_\nu W_\mu{}^\dagger)] \\ & - \frac{g^2}{(g^2+g'^2)} W_\mu W_\nu{}^\dagger (gZ_\rho - g'A_\rho)(gZ_\sigma - g'A_\sigma)(\eta^{\mu\nu}\eta^{\rho\sigma} - \eta^{\mu\rho}\eta^{\nu\sigma}) + \frac{g^2}{2}[|W_\mu W^\mu|^2 - (W_\mu W^{\mu\dagger})^2] \\ & + F(\varphi) - \frac{m_e}{\lambda}\varphi\bar{e}e - \tfrac{1}{8}(\varphi^2 + 2\lambda\varphi)[(g^2+g'^2)Z_\mu Z^\mu + 2g^2 W_\mu W^{\mu\dagger}] \\ & + i(g^2+g'^2)^{-1/2}\bar{e}\gamma^\mu\left[\left(\frac{1-\gamma_5}{2}\right)g'^2 + \frac{1}{2}\left(\frac{1+\gamma_5}{2}\right)(g'^2 - g^2)\right]eZ_\mu + \frac{i}{2}(g^2+g'^2)^{1/2}\bar{\nu}\gamma^\mu\left(\frac{1+\gamma_5}{2}\right)\nu Z_\mu \\ & + \frac{igg'}{(g^2+g'^2)^{1/2}}\bar{e}\gamma^\mu e A_\mu + \frac{ig}{\sqrt{2}}\bar{\nu}\gamma^\mu\left(\frac{1+\gamma_5}{2}\right)eW_\mu{}^\dagger + \frac{ig}{\sqrt{2}}\bar{e}\gamma^\mu\left(\frac{1+\gamma_5}{2}\right)\nu W_\mu. \end{aligned} \quad (1)$$

Here $F(\varphi)$ is a fourth-order polynomial in φ (chosen so that $\langle\varphi\rangle_0 = 0$), and g and g' are independent coupling constants. The electronic charge e, weak coupling constant G, and vector meson masses are given by the formulas

$$e = gg'/(g^2+g'^2)^{1/2}, \quad G/\sqrt{2} = \tfrac{1}{2}\gamma^2, \quad (2)$$

$$m_W = \lambda g/2, \quad m_Z = \lambda(g^2+g'^2)^{1/2}/2. \quad (3)$$

At the time that this theory was proposed, its renormalizability was still a matter of conjecture. It is well known[4] that the Yang-Mills theory with which we start in step A above is indeed renormalizable if quantized in the usual way. However, the shift of the scalar field performed in step C amounts to a rearrangement of the perturbation series, so that the S matrix calculated in perturbation theory corresponds to a representation of the algebra of field operators inequivalent to that with which we started in step A. There is no obvious way to tell that renormalizability is preserved in this shift.

Recently several studies have indicated that various models of this general class actually are renormalizable. By choosing a different gauge in step B, 't Hooft[5] derived effective Lagrangians which appear manifestly renormalizable, but which involve fictitious massless scalar mesons of both positive and negative norm. Subsequently, Lee[6] showed in one case that the renormalization program does actually work in this gauge, and that the spurious singularities associated with the fictitious particles all cancel. I would suggest, as an explanation of these results, that, although the shift of fields in step C really does generate an inequivalent representation of the field operators, the choice of gauge in step B does not, so that the S matrix calculated in the manifestly renormalizable gauges of 't Hooft should agree with the S matrix calculated in the "manifestly unitary" gauge used to derive Eq. (1).

If this is correct, then it ought to be possible to carry out calculations of higher-order weak interactions using the interaction (1) directly. This has obvious advantages over the use of Lagrangians of the 't Hooft–Lee form, because (1) involves only physical particles. This paper will explore the results of this model in certain physical processes, both in order to test its renormalizability, and also to gain some insight into its general properties.

First, it is necessary to derive the Feynman rules for this theory. The interaction Hamiltonian here is given by $-\mathcal{L}'$ plus noncovariant terms which are canceled by the noncovariant parts of the propagators of the vector meson fields and their derivatives. However, after this cancelation, there is left over a covariant effective interaction[7]

$$\delta\mathcal{L} = -6i\delta^4(0)\ln(1+\varphi/\lambda).$$

The correct Feynman rules are thus generated by using $-\mathcal{L}' - \delta\mathcal{L}$ as an effective interaction Hamiltonian, keeping only the "naive" covariant parts of the various propagators. No "ghost loops"[4] appear here.

Now let us consider some specific physical processes:

$\nu+\bar{\nu}\to W^+ + W^-$.—This is the reaction used by Gell-Mann, Goldberger, Kroll, and Low[8] to exhibit the difficulties associated with conventional intermediate boson theories. In such theories, the amplitude[9] for production of zero-helicity $W^\pm$ is given in lowest order by

$$f_{\rm GGKL} = \frac{-iGp^{3/2}\sin\theta e^{-i\varphi}}{2\pi(2E)^{1/2}}\left\{\frac{2E^2[1-(E/p)\cos\theta]-m_W^2}{2E^2[1-(p/E)\cos\theta]-m_W^2+m_e^2}\right\}, \qquad (4)$$

where E, p, θ, and φ are the energy, momentum, and scattering angles, respectively, of the W^+ in the center of mass system. For $E\to\infty$ this is dominated by a pure $J=1$ term which grows like E, so that in order to save unitarity it is necessary to introduce a cutoff[8] at energies of order $1/\sqrt{G}$. In the theory proposed here, there is an additional term produced by Z exchange in the s channel:

$$f_Z = \frac{iGp^{3/2}\sin\theta e^{-i\varphi}}{2\pi(2E)^{1/2}}\left[\frac{4E^2+2m_W^2}{4E^2-m_Z^2}\right]. \qquad (5)$$

Inspection of (4) and (5) shows that the total scattering amplitude now grows only near threshold and falls off like $1/E$ for $E\gg m_W$. This natural cutoff at $E\sim m_W$ obviates the need for any special unitarity cutoff. To the extent that this is a general phenomenon, we can expect that the perturbation series in G is really an expansion in powers of $Gm_W^2\sim g^2$, which may be as small as e^2.

$\nu+\nu\to\nu+\nu$.—This is a good process to use as a test of the performance of our theory in loop diagrams, because, as pointed out by Low,[10] the exchange of a pair of W bosons generates a quadratic divergence, related to the failure of unitarity bounds in Eq. (4). In our present theory, there are two additional fourth-order diagrams in which a pair of Z's is exchanged, plus a large number of fourth-order diagrams in which a single Z is exchanged with second-order self-energy or vertex insertions. The former diagrams contain quadratic divergences which cancel among themselves. The latter diagrams contain quartic divergences, which cancel among themselves, plus a large number of quadratically divergent terms. Some of these quadratic divergences can be grouped together as renormalizations of m_Z and the Z-ν coupling constant (and probably cancel) but there remain quadratic divergences in the Z self-energy proportional to $(t-m_Z^2)^2$, and in the Z-ν vertices proportional to $(t-m_Z^2)$. These terms generate an effective quadratically divergent neutrino Fermi interaction, *which turns out to cancel the quadratic divergence found by Low.* (I have not yet checked what happens to the logarithmically divergent terms.)

W-photon interactions.—The first term in Eq. (1) gives the W an "anomalous" magnetic moment, with gyromagnetic ratio $g_W=2$. This is just the value required[11] if the amplitude for Compton scattering on a W behaves well enough at high energies to satisfy a Drell-Hearn sum rule.

$e^+ + e^- \to W^+ + W^-$.—Because the W has an "anomalous" magnetic moment, the electromagnetic pair-production amplitude[9] grows like E as $E \to \infty$. Further, the neutrino t-channel and Z s-channel exchange diagrams do *not* cancel here, so that the weak pair-production amplitude also grows like E. However, *the weak and electromagnetic amplitudes cancel each other as* $E \to \infty$, leaving a scattering amplitude which vanishes like $1/E$ as $E \to \infty$, as required by unitarity bounds. This cooperation between the weak and electromagnetic interactions in solving each other's problems is one of the most satisfying features of this theory.

The weak and electromagnetic interactions of the leptons appear to be in good shape, so let us consider how to incorporate the hadrons. In order to preserve renormalizability, it is necessary to couple Z, W, and A to the currents of an *exact* $SU(2)_L \otimes U(1)$ symmetry of the strong-interaction Lagrangian. This poses a problem, because, apart from any spontaneous symmetry-breaking mechanisms responsible for the baryon masses, it is usually presumed that the nonzero masses of the π and K arise from an *intrinsic* breaking of $SU(2)\otimes SU(2)$ or $SU(3)\otimes SU(3)$. The only way that I can see to save renormalizability is to suppose instead that the π and K masses arise from the same purely spontaneous symmetry-breaking mechanism responsible for the W and Z masses. The problem then is whether it is natural for the strong interactions to conserve parity and isospin.

Leaving aside strange particles, the simplest way to couple the scalar doublet[3] $(\varphi^+, \varphi^0+\lambda)$ of our model to the hadrons is to find some $(\frac{1}{2}, \frac{1}{2})$ $SU(2)\otimes SU(2)$ multiplet $(\sigma, \vec{\pi})$ of hadronic field operators, and write an $SU(2)_L$-invariant interaction,

$$-if\varphi^{+\dagger}(\pi_1 - i\pi_2) + f(\varphi^{0\dagger}+\lambda)(\sigma + i\pi_3) + \text{H.c.}$$

The rest of the strong-interaction Lagrangian is assumed to conserve $SU(2)_R$ as well as $SU(2)_L$, so by an $SU(2)_R$ rotation we can define σ and $\vec{\pi}$ so that f is real. After eliminating φ^+ and $\text{Im}\varphi^0$ in step B, the only remaining symmetry-breaking term is $2f(\lambda+\varphi)\sigma$, which does conserve parity and isospin. Thus the spontaneous symmetry breaking of the weak interactions can act as a seemingly intrinsic symmetry-breaking mechanism for the strong interactions, which in turn is amplified if σ develops a large vacuum expectation value. A particularly attractive aspect of this approach is that the requirement of renormalizability provides the rationale for the conservation or partial conservation of the hadronic weak currents.

The most direct verification of this theory would be the discovery of W's and Z's with the predicted properties. However, the lower limits on m_W and m_Z are, respectively, $\lambda e/2 = 37.3$ GeV and $\lambda e = 74.6$ GeV, so this discovery will take a while.[12] The most accessible effect of the Z's is to change the cross sections for scattering of neutrinos and antineutrinos on electrons. We know nothing about the mass of the scalar meson φ, but its field might contribute to the level shifts in muonic atoms. Higher-order weak interactions produce various "radiative" corrections, including a change of order Gm_μ^2 in the gyromagnetic ratio of the muon.[13] The extension of this theory to strange particles appears to require both strangeness-changing and strangeness-conserving neutral hadronic currents, but the former can be eliminated in an $SU(4)\otimes SU(4)$–invariant model.[14] These matters will be dealt with at greater length in future papers.

I am deeply grateful to Francis Low, both for his indispensable advice and encouragement during the course of this work, and also for discussions over the last several years on the divergence difficulties of the weak interactions.

---

*Work supported in part through funds provided by the the U. S. Atomic Energy Commission under Contract No. AT(30-1)-2098.

[1]S. Weinberg, Phys. Rev. Lett. 19, 1264 (1967).

[2]P. W. Higgs, Phys. Rev. Lett. 12, 132 (1964), and 13, 508 (1964), and Phys. Rev. 145, 1156 (1966); F. Englert and R. Brout, Phys. Rev. Lett. 13, 321 (1964); G. S. Guralnik, C. R. Hagen, and T. W. B. Kibble, Phys. Rev. Lett. 13, 585 (1965); T. W. B. Kibble, Phys. Rev. 155, 1554 (1967). Also see A. Salam, in *Elementary Particle Physics*, edited by N. Svartholm (Almqvist and Wiksells, Stockholm, 1968), p. 367.

[3]There is a mistake in Eq. (3) of Ref. 1. The upper and lower members of the scalar doublet should have charges +1 and 0, not 0 and −1.

[4]R. P. Feynman, Acta Phys. Pol. 24, 697 (1963); B. S. De Witt, Phys. Rev. 162, 1195 (1967); L. D. Fadeev and V. N. Popov, Phys. Lett. 25B, 29 (1967); S. Mandelstam, Phys. Rev. 175, 1580 (1968); E. S. Fradkin and I. V. Tyutin, Phys. Rev. D 2, 2841 (1970); R. Mills, Phys. Rev. D 3, 2969 (1971).

[5]G. 't Hooft, to be published.

[6]B. W. Lee, to be published.

[7]The methods used to derive this result are based on the work of T. D. Lee and C. N. Yang, Phys. Rev. 128, 885 (1962); also see J. Honerkamp and K. Meetz, Phys. Rev. D 3, 1996 (1971); J. Charap, Phys. Rev. D 3, 1998 (1971); I. S. Gerstein, R. Jackiw, B. W. Lee, and

S. Weinberg, Phys. Rev. D 3, 2486 (1971).

[8]M. Gell-Mann, M. L. Goldberger, N. M. Kroll, and F. E. Low, Phys. Rev. 179, 1518 (1969).

[9]The scattering amplitude f is normalized so that the differential cross section is $|f|$.

[10]F. E. Low, *Lectures on Weak Interactions* (Tata Institute of Fundamental Research, Bombay, 1970), p. 11.

[11]S. Weinberg, in *Lectures on Elementary Particles and Quantum Field Theory*, edited by S. Deser, M. Grisaru, and H. Pendleton (Massachusetts Institute of Technology Press, Cambridge, Mass., 1970), p. 234.

[12]A value of precisely 37.3 GeV has been suggested for the intermediate boson mass by J. Schechter and Y. Ueda, Phys. Rev. D 2, 736 (1970) and more recently by T. D. Lee, Phys. Rev. Lett. D 3, 801 (1971). Lee also gets a gyromagnetic ratio $g_W = 2$. In the present theory, a W mass of 37.3 GeV is only possible if $g = e$, $g' \gg e$, and $m_Z \gg m_W$.

[13]R. Jackiw and S. Weinberg, to be published.

[14]S. L. Glashow, J. Iliopoulos, and L. Maiani, Phys. Rev. D 2, 1285 (1970).

Weak Interactions with Lepton-Hadron Symmetry*

S. L. Glashow, J. Iliopoulos, and L. Maiani†
Lyman Laboratory of Physics, Harvard University, Cambridge, Massachusetts 02139
(Received 5 March 1970)

We propose a model of weak interactions in which the currents are constructed out of four basic quark fields and interact with a charged massive vector boson. We show, to all orders in perturbation theory, that the leading divergences do not violate any strong-interaction symmetry and the next to the leading divergences respect all observed weak-interaction selection rules. The model features a remarkable symmetry between leptons and quarks. The extension of our model to a complete Yang-Mills theory is discussed.

INTRODUCTION

Weak-interaction phenomena are well described by a simple phenomenological model involving a single charged vector boson coupled to an appropriate current. Serious difficulties occur only when this model is considered as a quantum field theory, and is examined in other than lowest-order perturbation theory.[1] These troubles are of two kinds. First, the theory is too singular to be conventionally renormalized. Although our attention is not directed at this problem, the model of weak interactions we propose may readily be extended to a massive Yang-Mills model, which may be amenable to renormalization with modern techniques. The second problem concerns the selection rules and the relationships among coupling constants which are carefully and deliberately incorporated into the original phenomenological Lagrangian. Our principal concern is the fact that these properties are not necessarily maintained by higher-order weak interactions.

Weak-interaction processes, and their higher-order weak corrections, may be classified[2] according to their dependence upon a suitably introduced cutoff momentum Λ. Contributions to the S matrix of the form

$$\sum_{n=1}^{\infty} A_n(G\Lambda^2)^n$$

(where G is the usual Fermi coupling constant and A_n are dimensionless parameters) are called zeroth-order

* Work supported in part by the Office of Naval Research, under Contract No. N00014-67-A-0028, and the U. S. Air Force under Contract No. AF49(638)-1380.

† On leave of absence from the Laboratori di Fisica, Istituto Superiore di Santa, Roma, Italy.

[1] B. L. Ioffe and E. P. Shabalin, Yadern. Fiz. 6, 828 (1967) [Soviet J. Nucl. Phys. 6, 603 (1968)]; Z. Eksperim. i Teor. Fiz. Pis'ma v Redaktsiyu 6, 978 (1967) [Soviet Phys. JETP Letters 6, 390 (1967)]; R. N. Mohapatra, J. Subba Rao, and R. E. Marshak, Phys. Rev. Letters 20, 1081 (1968); Phys. Rev. 171, 1502 (1968); F. E. Low, Comments Nucl. Particle Phys. 2, 33 (1968); R. N. Mohapatra and P. Olesen, Phys. Rev. 179, 1917 (1969).

[2] T. D. Lee, Nuovo Cimento 59A, 579 (1969).

weak effects, terms of the form

$$G \sum_{n=0}^{\infty} B_n (G\Lambda^2)^n$$

are called first-order weak effects, and generally, terms of the form

$$G^l \sum_{n=0}^{\infty} C_{ln} (G\Lambda^2)^n$$

are called lth order. (We are disregarding possible logarithmic dependences on the cutoff.) The zeroth-order terms present us with the dangerous possibility of serious violations of parity and hypercharge in strong interactions. First-order terms include the usual lowest-order contributions (order G) to leptonic and semileptonic processes. However, other first-order terms may yield violations of observed selection rules: There can be $\Delta S=2$ amplitudes, yielding a K_1-K_2 mass splitting, beginning at order $G(G\Lambda^2)$, as well as contributions to such unobserved decay modes as $K_2 \to \mu^+ + \mu^-$, $K^+ \to \pi^+ + l + \bar{l}$, etc., involving neutral lepton pairs, or departures from the leptonic $\Delta S = \Delta Q$ law. We shall say of a model that its divergences are properly ordered if it is true that the zeroth-order terms *do not* yield violations of parity or hypercharge, and if the first-order terms *do* satisfy the observed selection rules of weak-interaction phenomena.

In most conventional formulations of a weak-interaction field theory (say, a vector boson coupled to a quark triplet), the divergences are not properly ordered. Defenders of such theories must argue that there is an effective weak-interaction cutoff which guarantees that the induced higher-order effects are as small as experiment indicates. A remarkably small cutoff,[1] not greater than 3 or 4 GeV, seems necessary. Should such a cutoff be justified, the problem of higher-order departures from known selection rules is solved; all such departures are small.

Others feel that such a small cutoff is implausible and unrealistic, and that one must confront the possibility that $G\Lambda^2$ is large—perhaps obtaining sensible results in the limit $G\Lambda^2 \to \infty$. In this case, one may regard all the first-order terms as having the same general magnitude, that of observed weak phenomena, and nth-order terms as having the magnitude naively expected of nth-order weak interactions.

An elegant solution to the problem of the zeroth-order terms was recently discovered, removing the specter of strong violations of parity and hypercharge.[3,4] One assumes a particular form for the breakdown of chiral $SU(3)$: The symmetry-breaking term must transform like the $(3,\bar{3})+(\bar{3},3)$ representation[5]; in a quark model, like the quark mass term. In this case, the zeroth-order weak interactions may be identified as an object belonging to the same representation as the symmetry-breaking term. After an appropriate $SU(3) \times SU(3)$ transformation, their only effect is to cause a renormalization of the symmetry-breaking terms, giving renormalized quark masses.[4] There is no violation of hypercharge or parity. Indeed, from a speculative stability requirement of the symmetry-breaking term under weak and electromagnetic corrections, the correct value of the Cabibbo angle may be deduced.[4]

Although the zeroth-order terms are controlled with an appropriate model of strong interactions, the first-order terms remain troublesome. Indeed, with a quark model, we immediately encounter strangeness-violating couplings of neutral lepton currents and contributions to the neutral kaon mass splitting to order $G(G\Lambda^2)$.[6] (In such a model, departures from $\Delta S = \Delta Q$ first appear at second order.) For this reason, it appears necessary to depart from the original phenomenological model of weak interactions. One suggestion[7] involves the introduction of a large number of intermediaries of spins one and zero, so coupled that the leading divergences are associated with only the diagonal symmetry-preserving interactions; in this fashion a proper ordering of divergences is readily obtained. But this model is an awkward one involving many intermediaries with different spins but degenerate coupling strengths. Few would concede so much sacrifice of elegance to expediency.[8]

We wish to propose a simple model in which the divergences are properly ordered. Our model is founded in a quark model, but one involving four, not three, fundamental fermions; the weak interactions are mediated by just one charged vector boson. The weak hadronic current is constructed in precise analogy with the weak lepton current, thereby revealing suggestive lepton-quark symmetry. The extra quark is the simplest modification of the usual model leading to the proper ordering of divergences. Just as importantly, we argue that universality is preserved, in the sense that the

[3] C. Bouchiat, J. Iliopoulos, and J. Prentki, Nuovo Cimento **56A**, 1150 (1968); J. Iliopoulos, *ibid.* **62A**, 209 (1969); R. Gatto, G. Sartori, and M. Tonin, Phys. Letters **28B**, 128 (1968); Nuovo Cimento Letters **1**, 1 (1969).

[4] N. Cabibbo and L. Maiani, Phys. Letters **28B**, 131 (1968); Phys. Rev. D **1**, 707 (1970).

[5] S. L. Glashow and S. Weinberg, Phys. Rev. Letters **20**, 224 (1968); M. Gell-Mann, R. J. Oakes, and B. Renner, Phys. Rev. **175**, 2195 (1968).

[6] Of course, one cannot exclude *a priori* the possibility of a cancellation in the sum of the relevant perturbation expansion in the limit $\Lambda \to \infty$.

[7] M. Gell-Mann, M. L. Goldberger, N. M. Kroll, and F. E. Low, Phys. Rev. **179**, 1518 (1969).

[8] For other departures from the conventional theory, see, for example, C. Fronsdal, Phys. Rev. **136B**, 1190 (1964); W. Kummer and G. Segrè, Nucl. Phys. **64**, 585 (1965); G. Segrè, Phys. Rev. **181**, 1996 (1969); L. F. Li and G. Segrè, *ibid.* **186**, 1477 (1969); N. Christ, *ibid* **176**, 2086 (1968). It should be understood that the ingenious conjecture of T. D. Lee and G. C. Wick [Nucl. Phys. **B9**, 209 (1969)] for removing divergences is logically independent of our analysis. If their hypothesis is correct, the role of the cutoff momentum is played by M_W. Only if M_W is small (~3–4 GeV) would the problems associated with ordering of divergences be solved; otherwise, a modification of the coupling scheme, such as ours, is still necessary.

leading divergent corrections (i.e., the first-order terms) yield a *common* renormalization to each of the various observed coupling constants.

The new model is discussed in Sec. I. Since Cabibbo's algebraic notion of universality[9] is maintained, that is to say, the entire weak charges generate the algebra of $SU(2)$, we observe in Sec. II that an extension to a three-component Yang-Mills model may be feasible. In contradistinction to the conventional (three-quark) model, the couplings of the neutral intermediary—now hypercharge conserving—cause no embarrassment. The possibility of a synthesis of weak and electromagnetic interactions is also discussed.

In Sec. III we briefly note some of the implications of the existence of a fourth quark, and finally, in Sec. IV we discuss some of the experimental tests of our model of weak interactions.

I. NEW MODEL

We begin by introducing four quark fields.[10] The three quarks $\mathcal{P}$, $\mathfrak{N}$, and λ form an $SU(3)$ triplet, and the fourth, $\mathcal{P}'$, has the same electric charge as $\mathcal{P}$ but differs from the triplet by one unit of a new quantum number $\mathcal{C}$ for charm. The strong-interaction Lagrangian is supposed to be invariant under chiral $SU(4)$, except for a symmetry-breaking term transforming, like the quark masses, according to the $(4,\bar{4})+(\bar{4},4)$ representation. This term may always be put in real diagonal form by a transformation of $SU(4)\times SU(4)$, so that B, Q, Y, $\mathcal{C}$, and parity are necessarily conserved by these strong interactions.

The extra quark completes the symmetry between quarks and the four leptons ν, ν', e^-, and μ^-. Both quadruplets possess unexplained unsymmetric mass spectra, and consist of two pairs separated by one in electric charge.

The weak lepton current may be expressed as

$$J_\mu{}^L=\bar{l}C_L\gamma_\mu(1+\gamma_5)l, \tag{1}$$

where l is a column vector consisting of the four lepton fields (ν, ν', e^-, μ^-) and the matrix C_L is given by

$$C_L=\begin{bmatrix}0&0&1&0\\0&0&0&1\\0&0&0&0\\0&0&0&0\end{bmatrix}. \tag{2}$$

This is a convenient way to rewrite the conventional current. In analogy with this expression, we define the weak hadron current to be

$$J_\mu{}^H=\bar{q}C_H\gamma_\mu(1+\gamma_5)q, \tag{3}$$

where q is the quark column vector $(\mathcal{P}',\mathcal{P},\mathfrak{N},\lambda)$ and the matrix C_H must be of the form

$$C_H=\left[\begin{array}{cc|c}0&0&\\0&0&U\\\hline 0&0&0\quad 0\\0&0&0\quad 0\end{array}\right] \tag{4}$$

in order for $J_\mu{}^H$ to carry unit charge. Pursuing the analogy further, we demand that the 2×2 submatrix U be unitary, so that the matrix C_H is equivalent to C_L under an $SU(4)$ rotation. Of course, it is not convenient to carry out the transformation making C_H and C_L coincide, for this would destroy the diagonalization of the $SU(4)$-breaking term, the quark masses. Nevertheless, suitable redefinitions of the relative phases of the quarks may be performed in order to make U real and orthogonal, so without loss of generality we write

$$U=\begin{bmatrix}-\sin\theta & \cos\theta\\ \cos\theta & \sin\theta\end{bmatrix}. \tag{5}$$

This is just the form of the weak current suggested in an earlier discussion of $SU(4)$ and quark-lepton symmetry.[10] What is new is the observation that this model is consistent with the phenomenological selection rules and with universality even when all divergent first-order terms [i.e., $G(G\Lambda^2)^n$] are considered.

To see this, we proceed diagrammatically in the quark model ignoring the strong $SU(4)$-invariant interactions.[11] Zeroth-order terms occur only in diagrams with only one external quark line, and give contributions to the quark mass operator of the form

$$\delta M(\gamma k)=\sum A_n(G\Lambda^2)^n\bar{q}M_n\gamma\cdot k(1+\gamma_5)q. \tag{6}$$

The A_n are dimensionless parameters, and the matrix M_n is a symmetric homogeneous polynomial of order n in C_H and of order n in $C_H{}^\dagger$. From the definition of C_H, it is seen that M_n must be a multiple of the unit matrix—again in contradistinction to the $SU(3)$ situation. Now, the zeroth-order terms are $SU(4)$ invariant.

There remains an apparent zeroth-order violation of parity, which may be transformed away because of the simple fashion of chiral $SU(4)$ breaking we have assumed. We simply define new quark fields

$$q_i'=(\alpha+\beta\gamma_5)q_i \tag{7}$$

with the real cutoff-dependent parameters α and β chosen so that the entire (bare plus zeroth-order) mass operator, in terms of q_i', is diagonal and parity conserving. The $SU(4)\times SU(4)$-invariant strong interactions are left unchanged. The procedure is analogous

[9] N. Cabibbo, Phys. Rev. Letters **10**, 531 (1963).

[10] B. J. Bjorken and S. L. Glashow, Phys. Letters **11**, 255 (1964).

[11] All our results about the zero- and first-order selection rules are trivially extended to the case of an $SU(4)$-invariant strong interaction which consists of a neutral vector boson coupled to quark number, the so-called "gluon" model. The only results of this paper which might be affected by such an interaction are the universality conditions in Eq. (9).

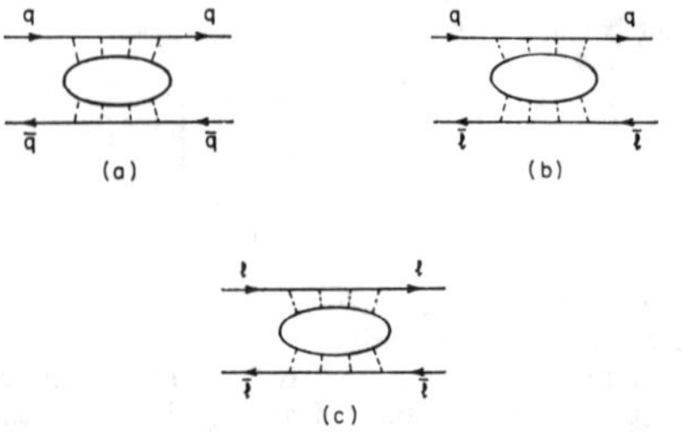

FIG. 1. (a) Connected part of the $q\bar{q}\rightarrow q\bar{q}$ amplitude. The crossed (annihilation) channel is also understood. (b) Connected part of the $ql\rightarrow ql$ amplitude. (c) Connected part of the $l\bar{l}\rightarrow l\bar{l}$ amplitude.

to that of Ref. 4, with the difference that the transformation (7) is $SU(4)$ invariant and does not change the definition of strangeness (or charm), or of the Cabibbo angle. An important consequence of the fact that M_n does not depend on the Cabibbo angle is that, unlike the situation in Ref. 4, it is impossible in our case to evaluate the Cabibbo angle by imposing a condition on the leading divergences. We conclude that zeroth-order weak effects are not significant.

We now consider the first-order $G(G\Lambda^2)^n$ terms which are of four types: (i) next-to-the-leading contributions to the quark and lepton mass operators, (ii) leading contributions to quark-quark or quark-antiquark scattering, (iii) leading contributions to quark-lepton scattering, and (iv) leading contributions to lepton-lepton scattering. Graphs with more than two external fermion lines yield no larger than second-order effects. Terms of type (i) are harmless: They contribute to observable nonleptonic $\Delta I=\frac{1}{2}$ processes, but since they cannot give $\Delta Y=2$, they do not produce a K_1K_2 mass splitting. On the other hand, type-(ii) diagrams could lead to $\mathfrak{N}\bar{\lambda}\rightarrow\bar{\mathfrak{N}}\lambda$, possibly giving rise to first-order contributions to the K_1K_2 mass difference, contrary to experiment. Let us show that they do not.

Graphs contributing to type (ii) effects are of the general form shown in Fig. 1(a), where the bubble includes any possible connections among the boson lines, and any number of closed fermion loops. The leading divergent contributions to q-$\bar{q}$ scattering from these graphs have the form

$$T_{HH}=G\sum_{n=2}^{\infty}B_n(G\Lambda^2)^{n-1}[\bar{q}\gamma_\mu(1+\gamma_5)\times B_H{}^{(n)}q\bar{q}\gamma^\mu(1+\gamma_5)B_H{}^{(n)\dagger}q], \quad (8)$$

where the B_n are finite dimensionless parameters independent of masses or momenta. It is clear that these first-order terms are independent of all external momenta. The matrix $B_H{}^{(n)}$ is a polynomial in C_H and $C_H{}^\dagger$ of order k and l, respectively, with $k+l\leq n$. Furthermore, the charge structure of the quark multiplets allows a change of charge no greater than unity, so that $|k-l|$ must be zero or one, and the matrices $B_H{}^{(n)}$ are easily computed (see the Appendix) to be

$$B_H{}^{(n)}=C_H \text{ or } C_H{}^\dagger \quad (k=l\pm1) \quad (8')$$
$$=[C_H,C_H{}^\dagger] \quad (k=l)\,. \quad (8'')$$

Thus, T_{HH} gives rise to contributions with $|\Delta Y|\leq1$ and, in particular, it does not yield a first-order K_1K_2 mass splitting. Of course, the next-to-the-leading divergences of these graphs will give $\Delta Y=2$, and do contribute to a second-order K_1K_2 mass difference, agreeing with experiment.

The leading divergences of types (iii) and (iv) give first-order contributions T_{HL} and T_{LL}, to semileptonic and leptonic processes. There will be a 1-to-1 correspondence among the graphs contributing to T_{LL}, T_{HL} [Figs. 1(b) and 1(c)], and T_{HH}. Because the algebraic properties of C_H and C_L are identical, we construct T_{HL} and T_{LL} from T_{HH} by the appropriate substitutions of $q\rightarrow L$ and $C_H\rightarrow C_L$.

In processes where the lepton charge changes, no violations of observed selection rules occur, but the first-order terms cause a renormalization of observed coupling constants. It is important to note that these renormalizations are common to leptonic and semileptonic processes, so that the relations

$$G_V(\Delta S=0)=G_\mu\cos\theta\,,\qquad G_V(\Delta S=1)=G_\mu\sin\theta \quad (9)$$

remain true when all first-order terms are included. This renormalization is given by the factor $1+\sum B_n(G\Lambda^2)^{n-1}$. A sufficient condition for these renormalizations to be common is the algebraic version of universality—a condition which is satisfied by our model, as well as by the usual three-quark model.

Next, we turn to the induced first-order couplings of hadrons to neutral lepton currents and self-couplings of neutral lepton currents. The neutral lepton currents are generated by the matrix $C_L{}^0$ and the neutral hadron currents by the matrix $C_H{}^0$, where

$$C_L{}^0=[C_L,C_L{}^\dagger]=\begin{bmatrix}1 & 0\\ 0 & -1\end{bmatrix}=[C_H,C_H{}^\dagger]=C_H{}^0. \quad (10)$$

Evidently, there are no induced couplings of neutral lepton currents to strangeness-changing currents. The induced couplings involve the strangeness-conserving current

$$\begin{aligned}J_\mu{}^0&=\bar{q}\gamma_\mu C_H{}^0(1+\gamma_5)q+\bar{l}\gamma_\mu C_L{}^0(1+\gamma_5)l\\&=\bar{\mathcal{P}}'\gamma_\mu(1+\gamma_5)\mathcal{P}'+\bar{\mathcal{P}}\gamma_\mu(1+\gamma_5)\mathcal{P}-\bar{\mathfrak{N}}\gamma_\mu(1+\gamma_5)\mathfrak{N}\\&\quad-\bar{\lambda}\gamma_\mu(1+\gamma_5)\lambda+\bar{\nu}'\gamma_\mu(1+\gamma_5)\nu'+\bar{\nu}\gamma_\mu(1+\gamma_5)\nu\\&\quad-\bar{e}\gamma_\mu(1+\gamma_5)e-\bar{\mu}\gamma_\mu(1+\gamma_5)\mu\,. \quad (11)\end{aligned}$$

The coupling constant for this new neutral current-current interaction is a first-order expression of the form

$$G\sum_{n=2}^{\infty} C_n(G\Lambda^2)^{n-1}.$$

We anticipate that its strength should be comparable to the strength of the charged leptonic interactions. The new coupling plays no role in observed decay modes, but is should be detectable in accelerator experiments.

In Sec. II we discuss the possible extension of our model to a Yang-Mills model, where the coupling strength of the neutral W to its current is uniquely determined. These neutral lepton couplings constitute the most characteristic and interesting feature of our model. Relevant experimental evidence is discussed in Sec. IV.

II. YANG-MILLS MODEL OF WEAK INTERACTIONS

Divergences appear in our model of weak interactions, but they are properly ordered; observed selection rules are broken only in order $G^2(G\Lambda^2)^n$. But, the model is certainly not renormalizable. There is at least a possibility that a Yang-Mills model of weak interactions may be less singular.[12] In this section, we show how our model can be extended to include a symmetrically coupled triplet of W's. It is possible that W self-couplings can be introduced to give a complete Yang-Mills theory.

The Lagrangian with which we work may be written, in the four-quark model, without electromagnetism,

$$\mathfrak{L}=\mathfrak{L}_{\text{kin}}+\mathfrak{L}_s+\mathfrak{L}_M+\mathfrak{L}_w\,, \tag{12}$$

where $\mathfrak{L}_{\text{kin}}$ is the purely kinematic term

$$\mathfrak{L}_{\text{kin}}=\bar{q}\gamma\cdot pq+\bar{l}\gamma\cdot pl+G_{\mu\nu}G^{\mu\nu}+W_{\mu\nu}{}^\dagger W^{\mu\nu} \tag{13}$$

describing four free massless quarks, four leptons, and their strong and weak intermediaries ($X_{\mu\nu}$ denotes the antisymmetric curl of X_μ). $\mathfrak{L}_s$ denotes the $SU(4)$-invariant strong interaction, most simply

$$\mathfrak{L}_s=fG_\mu\bar{q}\gamma^\mu q\,, \tag{14}$$

and $\mathfrak{L}_w$ is the weak interaction

$$\mathfrak{L}_w=gW_\mu{}^\dagger[\bar{q}C_H\gamma^\mu(1+\gamma_5)q+\bar{l}C_L\gamma^\mu(1+\gamma_5)l+\text{H.a.}]\,. \tag{15}$$

The bare-mass term $\mathfrak{L}_M$ produces the observed masses of the leptons, the masses of W and G, and gives rise to the observed hierarchy of hadron symmetry,

$$\mathfrak{L}_M=\bar{q}M_Hq+\bar{l}M_Ll+m^2G_\mu G^\mu+M^2W_\mu W^\mu\,, \tag{16}$$

where M_H and M_L are 4×4 matrices. This model gives a complete description of weak-interaction phenomena. The most important new feature is the appearance of neutral currents generated by the most divergent parts of diagrams containing an exchange of W^+, W^- pairs between two fermion lines. The effective coupling strength of these currents is expected to be of order G but, at this stage, we cannot predict its precise numerical value since we are unable to sum the perturbation series. In order to extend this model to a more symmetric one, we introduce an additional weak intermediary W_0 with appropriate couplings.

The couplings of W_0 to hadrons and leptons must be taken to be

$$2^{-1/2}gW_0{}^\mu\{\bar{q}[C_H{}^\dagger,C_H]\gamma_\mu(1+\gamma_5)q+\bar{l}[C_L{}^\dagger,C_L]\gamma_\mu(1+\gamma_5)l\}\,. \tag{17}$$

We emphasize that the introduction of W_0 is by no means necessary in our model; however, we think that it gives a much more symmetric and aesthetically appealing theory.

In the conventional model of weak interactions, the extension to a three-component vector-meson theory cannot be made without contradicting experiment: The neutral boson leads to strangeness-changing decays involving neutral-lepton currents and to $\Delta S=2$ at order G. This is because the commutator of the conventional weak charge with its adjoint yields a strangeness-violating neutral charge. In our case, the corresponding operator is diagonal, and these difficulties are absent.

It is straightforward to show that the introduction of the neutral current does not spoil the proper ordering of divergences: The observed selection rules are preserved by all terms of order $G(G\Lambda^2)^n$. This is shown in the Appendix.

We note that W_0 is coupled to precisely the same neutral current appearing in the last section as an induced coupling. In the symmetric three-W model, its strength is uniquely predicted. Universality now applies to both charged and neutral couplings. That is to say, the leading divergent corrections to each are the same. The bare relationship

$$G_0=\tfrac{1}{2}G \tag{18}$$

is preserved by the renormalizations, to first order [i.e., including all terms of order $G(G\Lambda^2)^n$]. This assertion is proved in the Appendix.

The introduction of a neutral W opens the possibility of formulating the weak interactions into a Yang-Mills theory. Self-couplings must be introduced among the W triplet in order to ensure the gauge symmetry. This is accomplished if we choose the Lagrangian in a manifestly gauge-invariant fashion:

$$\mathfrak{L}=\bar{q}\gamma\Pi_Hq+\bar{l}\gamma\Pi_Ll+\mathbf{W}_{\mu\nu}\mathbf{W}^{\mu\nu}+G_{\mu\nu}G^{\mu\nu}+\mathfrak{L}_M+\mathfrak{L}_s\,, \tag{19}$$

where

$$\Pi_H{}^\mu=\partial^\mu+ig(\mathbf{C}_H\cdot\mathbf{W}^\mu)(1+\gamma_5)\,, \tag{19$'$}$$

$$\Pi_L{}^\mu=\partial^\mu+ig(\mathbf{C}_L\cdot\mathbf{W}^\mu)(1+\gamma_5)\,, \tag{19$''$}$$

[12] See, for example, S. Mandelstam, Phys. Rev. **175**, 1580 (1969); M. Veltman, Nucl. Phys. **B7**, 637 (1968); H. Reiff and M. Veltman, *ibid.* **B13**, 545 (1969); D. Boulware, Ann. Phys. (N. Y.) **56**, 140 (1970); A. A. Slavnor, University of Kiev Report No. ITP 69/20 (unpublished); E. S. Fradkin and I. V. Tyutin, Phys. Letters **30B**, 562 (1969). Notice, however, that none of these references consider the far more difficult case of vector mesons coupled to nonconserved currents.

TABLE I. Quark quantum numbers.

	Fractional assignment			Integral assignment		
	Q	Y	$\mathfrak{C}$	Q	Y	$\mathfrak{C}$
$\mathcal{P}'$	$\frac{2}{3}$	$-\frac{2}{3}$	1	0	0	0
$\mathcal{P}$	$\frac{2}{3}$	$\frac{1}{3}$	0	0	0	-1
$\mathfrak{N}$	$-\frac{1}{3}$	$\frac{1}{3}$	0	-1	0	-1
λ	$-\frac{1}{3}$	$-\frac{2}{3}$	0	-1	-1	-1

and

$$W^{\mu\nu}=\Pi_W{}^{\mu}W^{\nu}-\Pi_W{}^{\nu}W^{\mu}, \qquad (19''')$$

where

$$(\Pi_W{}^{\mu})_{ij}=\delta_{ij}\partial^{\mu}+ig2^{-1/2}(\mathbf{t}\cdot\mathbf{W}^{\mu})_{ij}. \qquad (19'''')$$

The matrix-valued vectors $\mathbf{C}_H$ and $\mathbf{C}_L$ have components $(C,C^{\dagger},2^{-1/2}[C^{\dagger},C])$ in a basis where charge is diagonal, and $\mathbf{t}$ are the usual 3×3 generators of $O(3)$, with t_3 diagonal. The gauge group thus introduced is an exact symmetry of the entire Lagrangian excepting both $\mathcal{L}_M$ and electromagnetism.

The Yang-Mills model is undoubtedly the most attractive way to couple a triplet of vector mesons and the only one for which people have expressed some hope of constructing a renormalizable theory. The massless case has been proved to be renormalizable[12]; however, very little is known about the physically more interesting massive theory. In fact, the naive power counting shows that the highest divergence in a Yang-Mills theory is $g^{2n}\Lambda^N$ with $N=6n$. Notice that in the absence of the self-couplings the corresponding divergences are given, as we have already seen, by $N=2n$. So, at first sight, the Yang-Mills theory seems to be much more divergent than the ordinary coupling of the vector mesons with the currents. However, one can show that the naive limit $N=6n$ can be considerably lowered. We have already been able to show that $N\leq3n$ and we believe that one can still lower this limit to at least $N=2n$. In other words, we believe that the introduction of the self-couplings does not make the theory more divergent.

Let us briefly consider a more daring speculation. It has long been suspected[13] that there may be a fundamental unity of weak and electromagnetic interactions, reflected phenomenologically by the common vectorial character of their couplings. For this reason, it may have been wrong for us to introduce a gauge symmetry for the weak interactions not shared by electromagnetism. As a more speculative alternative, consider the possibility of a four-parameter gauge group involving $\mathbf{W}$, and an additional Abelian singlet W_S, broken only by the mass term $\mathcal{L}_M$. Suppose, however, that a one-parameter gauge symmetry, corresponding to a linear combination A of W_0 and W_S remains unbroken. Then A must be massless, and may be identified as the photon. The orthogonal neutral combination B is massive, and acts as an intermediary of weak interactions along with $W^{\pm}$. This model could be correct only if the weak bosons are very massive (100 GeV) so that the weak and electromagnetic coupling constants could be comparable. With this model, the relation (18) would not persist, and the weak neutral current would involve $(1-\gamma_5)$ as well as $(1+\gamma_5)$ currents. The precise form of the model would depend on what linear combination of W_0 and W_S is the photon.

[13] J. Schwinger, Ann. Phys. (N. Y.) 2, 407 (1957); S. L. Glashow, Nucl. Phys. 10, 107 (1959); 22, 579 (1961)

III. ANOTHER QUARK MAKES $SU(4)$

Having introduced four quarks, we must consider strong interactions which admit the algebra of chiral $SU(4)$. Does this mean we should expect $SU(4)$ to be an approximate symmetry of nature? Nothing in our argument depends on how much $SU(4)$ is broken; the divergences are necessarily properly ordered. However, for the higher-order nonleading divergences to be as small as they must be, the breaking of $SU(4)$ cannot be too great: The limit on the cutoff Λ is replaced by a limit on Δ, a parameter measuring $SU(4)$ breaking; and from the observed K_1K_2 mass difference we now conclude that Δ must be not larger than 3–4 GeV. Thus, some residue of $SU(4)$ symmetry should persist.

We expect the appearance of charmed hadron states.[10] Meson multiplets, made up of a quark-antiquark pair, must belong to the 15-dimensional adjoint representation of $SU(4)$, consisting of an uncharmed $SU(3)$ singlet and octet, as well as two $SU(3)$ triplets of charm ±1. The structure of baryons depends on the quantum numbers assigned to the quarks. The two simplest possibilities are shown in Table I. For the more conventional fractional charge assignment, the baryons are made up of three quarks, and must belong to one of the representations contained in $4\times4\times4$. The only possibility is a 20-dimensional representation, which contains, besides the baryon octet, a triplet and sextet of charmed states and a doubly charmed triplet. The $j=\frac{3}{2}^+$ baryon decuplet belongs to another 20-dimensional representation with a charmed sectet, a doubly charmed triplet, and a triply charmed singlet.

With the integral-charge assignment, the baryon octet must be made of two quarks and an antiquark, the decuplet of three quarks and two antiquarks. The lepton and quark charged spectra now coincide, and the synthesis of weak and electromagnetic interactions appears more plausible. Moreover, there is no difficulty in obtaining the correct value for the π^0 lifetime.

Why have none of these charmed particles been seen? Suppose they are all relatively heavy, say ~2 GeV. Although some of the states must be stable under strong (charm-conserving) interactions, these will decay rapidly ($\sim10^{+13}$ sec^{-1}) by weak interactions into a very wide variety of uncharmed final states (there are about a hundred distinct decay channels). Since the charmed particles are copiously produced only in associated production, such events will necessarily be of very complex topology, involving the plentiful decay products of both charmed states. Charmed particles could easily have escaped notice.

Finally, we briefly comment on the leptonic decay rates of ρ, ω, and ϕ (Γ_ρ, Γ_ω, and Γ_ϕ). Our electric current contains $SU(3)$ singlet as well as octet terms, so that the inequality

$$m_\omega\Gamma_\omega+m_\phi\Gamma_\phi\geq\tfrac{1}{3}m_\rho\Gamma_\rho \tag{20}$$

may be deduced from the Weinberg spectral function sum rules and ω, ϕ, ρ dominance.[14] A stronger result is obtained if we extend Weinberg's Schwinger-term hypothesis to the vector currents of $SU(4)$:

$$m_\omega\Gamma_\omega+m_\phi\Gamma_\phi\geq m_\rho\Gamma_\rho. \tag{21}$$

This result is in poor agreement with experiment, which favors the equality in (20). A resolution of this difficulty that does not abandon the Schwinger-term symmetry requires the introduction of a third $Y=T=0$ vector meson, another partner of ω and ϕ, corresponding to the $SU(4)$ singlet vector current.

IV. EXPERIMENTAL SUGGESTIONS

In this section, we discuss some of the observable effects characteristic of our picture of strong and weak interactions. Firstly, consider the experimental implications of the existence of a new quantum number—charm—broken only by weak interactions. The charmed particles, because they are heavy, are too short lived to give visible tracks. However, they should be copiously produced in hardonic collisions at accelerator energies:

$$(\text{hadron or }\gamma)+(\text{hadron})\rightarrow X^{(+)}+X^{(-)}+\cdots,$$

where $X^{(\pm)}$ are oppositely charmed particles, each rapidly decaying into uncharmed hadrons with or without a charged lepton pair. The purely hadronic decay modes could provide illusory violations of hypercharge conservation in strong interactions. The leptonic decay modes provide a mechanism for the seemingly direct production of one or two charged leptons in hadron-hadron collisions.[15] Conceivably, muons thus produced may be responsible for the anomalous observed angular distribution of cosmic-ray muons in the 10^{12}-eV range,[16] where these directly produced muons may dominate the sea-level muon flux.

Should this last speculation about cosmic rays be correct, we need to revise radically estimates of the flux of ν and $\bar\nu$ in this energy range. We expect the charmed particle decays to yield equal numbers of each lepton variety; this gives a flux of electron neutrinos and antineutrinos equal to the muon flux, and 10–100 times greater than other estimates. This fact is of crucial importance to the possible detection of the resonance scattering[17]

$$\bar\nu+e^-\rightarrow\bar\nu'+\mu^-.$$

Charmed particles may be produced singly by neutrinos in such reactions as

$$\nu'+N\rightarrow\mu^-+X,\quad \bar\nu'+N\rightarrow\mu^++X,$$

where the charmed particle X would have a variety of decay modes, including leptonic ones. With the fractional charge assignment, the neutrino processes are suppressed by $\sin^2\theta$ and the antineutrino processes are forbidden. On the other hand, with the integral-charge assignment, the neutrino processes are again proportional to $\sin^2\theta$ while the antineutrino processes are proportional to $\cos^2\theta$.

The second new feature of our model is the appearance of neutral leptonic and semileptonic couplings involving a specified ($Y=0$) current and with a coupling constant comparable with the Fermi constant. Without the introduction of a W_0, we may say only $G_0\sim G$. To be more definite, we shall phrase our arguments in terms of the value $G_0=\frac{1}{2}G$ of Eq. (18).

Let us summarize the presently available data about these interactions.[18] Consider the following three reactions induced by muon neutrinos:

$$\text{(i)}\quad \nu'+e^-\rightarrow\nu'+e^-,$$

$$\text{(ii)}\quad \nu'+p\rightarrow\nu'+p,$$

$$\text{(iii)}\quad \nu'+p\rightarrow\nu'+\pi^++n.$$

None of these neutral couplings have been observed; experimentally, we can only quote limits. From the absence of observed forward energetic electrons in the CERN bubble-chamber experiments, we may conclude

$$G_0\lesssim G,$$

a limit which is close to but consistent with our prediction.

For reaction (ii), it is found that

$$R=\sigma(\nu'p\rightarrow\nu'p)/\sigma(\bar\nu'p\rightarrow\mu^+n)\lesssim0.5.$$

Because our neutral current contains both $I=0$ and $I=1$ parts, we cannot unambiguously predict this ratio. In a naive quark model, where the proton consists of only $\mathfrak{N}$ and $\mathcal{P}$ quarks, we find $R=\frac{1}{4}$, again close but consistent.

Finally, we quote the experimental limit on reaction (iii):

$$R'=\sigma(\nu'+p\rightarrow\pi^++n+\nu')/\sigma(\nu'+p\rightarrow\pi^++p+\mu^-)\lesssim0.08.$$

---

[14] S. Weinberg, Phys. Rev. Letters **18**, 507 (1967); T. Das, V. Mathur, and S. Okubo, *ibid.* **19**, 470 (1967).

[15] In a recent experiment, P. J. Wanderer *et al.* [Phys. Rev. Letters **23**, 729 (1969)] have performed a search for W's by measuring the intensity and polarization of prompt energetic muons from the interaction of 28-GeV protons with nuclei. Their results are compatible with the assumption that all 25-GeV prompt muons have electromagnetic origin. There is no indication of the existence of W's. However, the published evidence does not seem to be relevant to the existence of charmed particles, which are produced in pairs, decay into many final states, and are not expected to produce many very energetic muons.

[16] H. E. Bergeson *et al.*, Phys. Rev. Letters **21**, 1089 (1968).

[17] M. G. K. Menon *et al.*, Proc. Roy. Soc. (London) **A301**, 137 (1967).

[18] See D. H. Perkins, in Proceedings of the Topical Conference in Weak Interactions, CERN, 1969 [CERN Report No. 69-7], pp. 1–42 (unpublished).

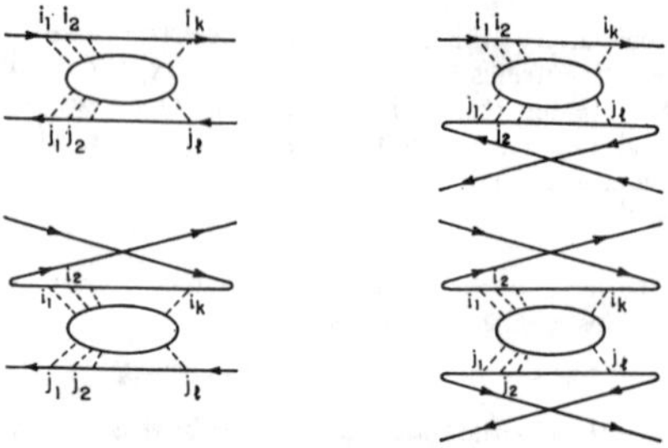

FIG. 2. Decomposition of the $q\bar{q} \to q\bar{q}$ connected amplitude by crossing the external fermion lines.

Because this transition is $\Delta I=1$, we unambiguously predict $R'=\frac{1}{9}$ under the hypothesis of $\Delta(1238)$ dominance. In each of these three reactions, experiment is very close to a decisive test of our model.

In our model, the parity-violating nonleptonic interaction is also changed. In particular, the parity-violating one-pion-exchange nuclear force is no longer suppressed by $\sin^2\theta$.

Next we consider some experiments which could discover the existence of W_0. A simple and attractive possibility is the search for muon tridents in the semiweak reaction[19]

$$\mu^- + Z \to \mu^- + W_0 + Z\,,$$

with the subsequent muonic decay of W_0. Another possibility is the reaction[20]

$$e^+e^- \to \mu^+\mu^-\,.$$

The interference between the W^0 and photon contributions causes an asymmetry of the μ^+ angular distribution relative to the momentum of the incident e^+ given by

$$\delta = \frac{N_F - N_B}{N_F + N_B} = \frac{3M_W{}^2}{16\sqrt{2}} \frac{G}{\alpha\pi} \frac{s}{s - M_W{}^2}\,,$$

where

$$G = 10^{-5} M_p{}^{-2}\,, \quad \alpha = 1/137\,, \quad \text{and} \quad s = 4E_e{}^2\,.$$

Away from the W^0 pole, the effect is rather small (less than 1% for $E_e = 3.5$ GeV) and it is masked by a similar effect due to the two-photon contribution. However, the factor $s/(s-M_W{}^2)$ makes the asymmetry increase sharply and change sign near M_W. Therefore, this reaction is an excellent tool to sweep a substantial mass range looking for W's. Another effect, much harder to detect, would be the direct observation of parity violation in $e^+e^- \to \mu^+\mu^-$. This requires the measurement of μ polarization.

Finally, we recall from Sec. III that the $SU(4)$ description of leptonic decays of vector mesons suggested the existence of another strongly coupled neutral $I=0$ vector meson with considerable coupling to lepton pairs. Evidence for its existence could come from colliding beam experiments.

[19] M. Tannenbaum (private communication).

[20] N. Cabibbo and R. Gatto, Phys. Rev. 129, 1577 (1961).

APPENDIX

In this appendix we determine the form of the leading weak corrections to the q-$\bar{q}$, q-$\bar{l}$, and l-$\bar{l}$ amplitudes.

We have already shown that the wave-function renormalization of spinors is the same for both quarks and leptons and contributes a common factor to T_{HH}, T_{HL}, and T_{LL}. Therefore we need consider only the q-$\bar{q}$ amplitude T_{HH}. The other amplitudes T_{HL} and T_{LL} can be obtained from T_{HH} by appropriate substitutions. In the following, we shall omit the common wave-function renormalization factors.

For the sake of clarity, let us first consider our model of weak interactions, where we have three vector bosons symmetrically coupled.

The graphs of Fig. 1(a) can be decomposed into four classes of terms, obtained by keeping the boson lines fixed and reversing the fermion lines, as shown in Fig. 2. We then obtain for the contribution to T_{HH} corresponding to these four classes of diagrams

$$\begin{aligned} T_{HH}{}^{(n,k,l)} &= \bar{q}\gamma_\mu(1+\gamma_5)[C_{i_1}C_{i_2}\cdots C_{i_k} - (-1)^k C_{i_k}C_{i_{k-1}}\cdots C_{i_1}]q \\ &\quad \times P_{j_1\cdots j_l;\, i_1\cdots i_k}\bar{q}\gamma^\mu(1+\gamma_5) \\ &\quad\quad \times [C_{j_1}C_{j_2}\cdots C_{j_l} - (-1)^l C_{j_l}C_{j_{l-1}}\cdots C_{j_1}]q\,, \\ &\quad\quad\quad k+l \le n\,, \quad k,l \ge 1\,. \end{aligned} \tag{A1}$$

All the i's and j's go from 1 to 3 and the sum over all indices is understood. $P_{j_1\cdots j_l;\, i_1\cdots i_k}$ is a tensor made out of the invariant tensors δ_{ij} and ϵ_{ijk}.

It is easy to show that for any k

$$\mathrm{Tr}[C_{i_1}C_{i_2}\cdots C_{i_k} - (-)^k C_{i_k}C_{i_{k-1}}\cdots C_{i_1}] = 0\,. \tag{A2}$$

Therefore, since the interaction is $O(3)$ invariant, the connected part of T_{HH} has the form

$$T_{HH} = G\sum_{n=0}^{\infty} b_n(G\Lambda^2)^n \times (\bar{q}\gamma_\mu(1+\gamma_5)\mathbf{C}_H q)\cdot(\bar{q}\gamma^\mu(1+\gamma_5)\mathbf{C}_H q)\,. \tag{A3}$$

In the case where we have only charged bosons, the argument is even simpler. Each of the indices $i_1\cdots i_k$, $j_1\cdots j_l$ appearing in Eq. (A1) takes only two possible values. With the relations

$$\begin{aligned} (C_H)^2 &= (C_H{}^\dagger)^2 = 0\,, \\ (C_H C_H{}^\dagger)^2 &= C_H C_H{}^\dagger\,, \\ (C_H{}^\dagger C_H)^2 &= C_H{}^\dagger C_H\,, \end{aligned} \tag{A4}$$

Eq. (A1) explicitly reads

$$T_{HH}{}^{(n;k,l)} = (\bar{q}\gamma_\mu(1+\gamma_5)[C_H,C_H{}^\dagger]q) \times (\bar{q}\gamma^\mu(1+\gamma_5)[C_H,C_H{}^\dagger]q)\,, \quad k=l$$

$$T_{HH}{}^{(n,k,l)} = (\bar{q}\gamma_\mu(1+\gamma_5)C_H q)(\bar{q}\gamma^\mu(1+\gamma_5)C_H{}^\dagger q)\,, \quad k=l+1\,. \quad \text{Q.E.D.}$$

V. QUANTIZATION OF GAUGE FIELDS, FEYNMAN RULES

FEYNMAN DIAGRAMS FOR THE YANG-MILLS FIELD

L. D. FADDEEV and V. N. POPOV
Mathematical Institute. Leningrad. USSR

Received 1 June 1967

Feynman and De Witt showed, that the rules must be changed for the calculation of contributions from diagrams with closed loops in the theory of gauge invariant fields. They suggested also a specific recipe for the case of one loop. In this letter we propose a simple method for calculation of the contribution from arbitrary diagrams. The method of Feynman functional integration is used.

It is known, that one can associate the field of the Yang-Mills type with an arbitrary simple group G [1-3]. It is appropriate to describe this field by means of the matricies $B_\mu(x)$ with values in the Lie algebra of this group.

The gauge group consits of the transformations

$$B_\mu \to B_\mu^\Omega = \Omega B_\mu \Omega^{-1} + \epsilon^{-1}\partial_\mu \Omega\Omega^{-1}$$

where $\Omega(x)$ is an arbitrary function with values in the group G.

The Lagrange function

$$\mathcal{L}(x) = -\tfrac{1}{4}\mathrm{Sp}G_{\mu\nu}G^{\mu\nu}; \quad G_{\mu\nu} = \partial_\nu B_\mu - \partial_\mu B_\nu + \epsilon[B_\mu, B_\nu]$$

is invariant with respect to these transformations. It is clear, that

$$\mathcal{L} = \mathcal{L}_0 + \mathcal{L}_1$$

where $\mathcal{L}_0$ is a quadratic form, and $\mathcal{L}_1$ is the sum of trilinear and quartic forms in B. In the quantization of the Feynman type $\mathcal{L}_1$ generates vertices with three and four external lines and $\mathcal{L}_0$ is to define the propagator function. However the form $\mathcal{L}_0$ is singular and the longitudinal part of the propagator can not be found unambiguously. This ambiguity does not influence the physical results in quantum electrodynamics. It seems that Feynman [4] was the first to show that the matter is not so simple in the cases of Yang-Mills and gravitational fields. Namely the contribution of the closed loop diagrams depends essentially on the longitudinal part of the propagator and spoils the transversality and unitarity properties of scattering amplitudes. Feynman himself described the necessary change of rules for calculation the contribution from diagrams with one closed loop. A more detailed derivation of the new rules was given by De Witt [5]. However it seems that nobody gave the generalization of these rules for arbitrary diagrams.

The formal considerations below are to give a simple explanation of the described difficulties and a quite workable recipe to circumvent them.

We know from Feynman [6] that every element of the S-matrix can be written down as the functional integral

$$\langle \mathrm{in}|\mathrm{out}\rangle \sim \int \exp\{\mathrm{i}S[B]\} \prod_x \mathrm{d}B(x)$$

up to an (infinite) normalizing factor. Here $S[B] = \int \mathcal{L}(x)\mathrm{d}x$ is the action functional and one is to integrate over all fields $B(x)$ with the asymptote at $t = x_0 \to \pm\infty$ prescribed by in-and out-states. The diagrams appear naturally in the perturbative calculation of this integral.

In the case of gauge invariant theory it is necessary to transform this integral a little. In fact, we can say, using the natural geometrical language, that the integrand is constant on the "orbits" $B_\mu \to B_\mu^\Omega$ of the gauge group in the manifold of all fields $B_\mu(x)$. It follows that the integral itself is proportional to the volume of this orbit which can be expressed as the integral $\int \prod_x d\Omega(x)$ over all matrices $\Omega(x)$. This integral should be factorized before using the perturbation theory.

There exist several methods for this purpose. The idea of one of them is to integrate over the orbits and some transversal surface. It is appropriate to choose for the latter the "plane" $\partial_\mu B^\mu = 0$. Then the integral reduces to the following

$$\int \exp\{iS[B]\}\ \Delta[B]\prod_x \delta\left(\partial_\mu B^\mu(x)\right)dB(x) \int\prod_x d\Omega(x)$$

where the factor $\prod_x \delta\left(\partial_\mu B^\mu\right)$ symbolizes that we integrate over transverse fields and $\Delta[B]$ is to be chosen such that the condition

$$\Delta[B] \int \prod_x \delta\left(\partial_\mu (B^\mu)^\Omega\right)\ d\Omega = \text{const}$$

holds. It is the nontriviality of $\Delta[B]$ which distinguishes the theories of Yang-Mills and gravitational fields from quantum electrodynamics.

We must know $\Delta[B]$ only for transverse fields and in this case all contribution to the last integral is given by the neighbourhood of the unit element of the group. After appropriate linearization we come to the condition

$$\Delta[B]\ \int\prod_x \delta\left(\Box u - \epsilon[B^\mu, \partial_\mu u]\right)\ du(x) = \text{const}$$

where $\Box$ is the D'Alembert operator and $u(x)$ are functions with values in the Lie algebra of the group G.

Formally $\Delta[B]$ is equal to the determinant of the operator

$$Au = \Box u - \epsilon[B^\mu, \partial_\mu u] \equiv A_0 u - \epsilon V(B)u$$

After extracting the trivial infinite factor $\det A_0$ we obtain the following expression for $\ln\Delta[B]$

$$\ln\Delta[B] = \ln(\det A/\det A_0) = \mathrm{Sp}\ln(1-\epsilon A_0^{-1}V(B))$$

Developing the right hand side in a power series in ϵ we have the following expressions for the coefficients

$$\int dx_1 \ldots \int dx_n\ \mathrm{Sp}\left(B^{\mu_1}(x_1)\ldots B^{\mu_n}(x_n)\right) \times$$
$$\times\ \partial_{\mu_1}G(x_1-x_2)\ldots\partial_{\mu_n}G(x_n-x_1)$$

where $G(x)$ is a Green function of the D'Alembert operator. This expression corresponds to the closed loop with the scalar particle propagating along it and interacting with the transverse vector particles according to the law $\sim \epsilon\ \mathrm{Sp}(\varphi[B^\mu\partial_\mu\varphi])$.

There results the diagram technique with the following features:

1. The pure transversal Green function is to be used as a propagator for the vector particles (Landau gauge).

2. It is necessary to take into account the new vertex with two scalar and one vector external line in addition to the ordinary vertices with three and four lines.

Concrete calculations with these changes in the rules give transverse and unitary expressions for the scattering amplitudes.

It must be stressed that the Landau gauge is essential for the new rules. It is connected with the chosen method of extracting the fact or $\int\prod_x d\Omega(x)$.

An other method leads to the expression

$$\int \exp\{iS[B] - \tfrac{1}{2}i\int \mathrm{Sp}(\partial_\mu B^\mu)^2\, dx\}\ \varphi[B]\prod_x dB$$

where the factor $\varphi[B]$ must be found from the condition

$$\varphi[B] \int \exp\{-\tfrac{1}{2}i\int \mathrm{Sp}\left(\partial_\mu (B^\mu)^\Omega\right)^2\, dx\}\prod_x d\Omega = \text{const}$$

This integral gives the perturbation series with Feynman propagator, but the calculation of $\varphi[B]$ is more cumbersome than that of $\Delta[B]$.

We conclude with the comment that one can proceed in an analogous way with gravitation theory. The analog for $\Delta[B]$ is the determinant of the Beltrami - Laplace operator in a harmonic coordinate system.

References

1. C.N.Yang and R.L.Mills, Phys. Rev. 96 (1954) 191.
2. R.Utiyama, Phys. Rev. 101 (1956) 1597.
3. S.L.Glashow and M.Gell-Mann, Ann. of Phys. 15 (1961) 437.
4. R.P.Feynman, Acta Physica Polonica, 24 (1963) 697.
5. B.S.De Witt, Relativity, groups and topology. (Blackie and Son Ltd 1964) pp 587-820.
6. R.P.Feynman, Phys. Rev. 80 (1950) 440.

PERTURBATION THEORY FOR GAUGE-INVARIANT FIELDS

V. N. POPOV* and L. D. FADDEEV*
Academy of Sciences of Ukrainian SSR
Institute for Theoretical Physics

Edited by David Gordon and Benjamin W. Lee, May 1972

PREFACE

In 1967 an excellent paper by V. N. Popov and L. D. Faddeev appeared as a Kiev Report No. ITP 67-36. We regret that this remarkable paper has not been available to a wider audience. We have undertaken the task of making their work more accessible to the English reading audience.

Due to recent developments in both dual resonance models and the intense theoretical activity in constructing a renormalizable theory of weak interactions we believe that this work is not only "timely" but merits thoughtful consideration for those theorists involved in constructing theories which possess gauge symmetries.

The present text is based on an informal translation of the original Russian version that Professor M. T. Veltman arranged in Paris in 1968. We have also consulted Dr. Yu. K. Pilipenko for clarification of some passages. None of them are responsible for any errors. *Caveat emptor!*

The reader is also referred to Faddeev's article "The Feynman Integral for Singular Lagrangians" in Teoreticheskaya i Matematicheskaya Fizika, Vol. 1, No. 1, p. 3 (1969) [English translation: Theoretical and Mathematical Physics, Vol. 1, p. 1 (1970), Copyright by Consultants Bureau], which was written after the present paper.

Finally we thank Professor L. D. Faddeev for permitting this venture.

May 1972 DAVID GORDON and BENJAMIN W. LEE

ABSTRACT: A method is developed for the manifestly convariant quantization of gauge-invariant fields by means of a functional integration. It is shown that for the fields with non-Abelian gauge groups (the Yang-Mills and gravitational fields) fictitious particles appear naturally in the diagram technique, which are not present in the initial Lagrangian. An appearance of these particles restores the transversality of scattering amplitudes and the unitarity of the S-matrix.

Reported at the Seminar ın Institute for Theoretical Physics, Kiev, Ukrainian SSR.

*Permanent Address: Mathematical Institute, Leningrad, USSR.

I. INTRODUCTION

It is well known that certain difficulties are involved in the quantization of gauge-invariant fields, and that they are due to the *singular nature of the Lagrangian concerned.* There are many artificial tricks invented in order to circumvent this difficulty in the case of quantum electrodynamics; one of the best known is Fermi's method which makes use of an indefinite metric.[1]

But this technique does not work in the case of a non-abelian gauge group. It seems that Feynman was the first to observe this in trying to develop a diagrammatic perturbation theory for the gravitational and the Yang-Mills fields.[2] He noted, for instance, that diagrams with closed loops depend non-trivially on the longitudinal parts of Green's functions (in internal lines), and the scattering amplitudes obtained were neither unitary nor transverse.

In order to get rid of these difficulties Feynman proposed some modified rules of computing one-loop diagrams; similar and more detailed descriptions of these Feynman rules were given by de Witt.[3] As far as we are aware, however, these rules have not been generalized for arbitrary diagrams.

In this work we are going to analyze a quantization method which permits the possibility of such a generalization (cf. de Witt, Ref. 3). The method is based on a redefinition of the Feynman path integral. When developed in perturbation series, this integral gives a relativistic invariant diagram technique, which in turn leads to scattering amplitudes fulfilling the unitarity condition. The striking feature of this technique is the *presence of some new diagrams* which *restore unitarity* and which have no analogue in the case where no gauge fields are involved.

In the case of Yang-Mills fields we shall also show the connection of our formalism with the Hamiltonian one; such a connection was also developed by Schwinger.[4] It is this possibility of passing to the Hamiltonian formalism which assures the unitarity of the S-matrix. At the end we will try to employ the formalism to the theory of gravitation.

Throughout this work we use $\delta_{\mu\nu}$ to denote the Minkowski tensor, in which the only non-vanishing elements are given by: $\delta_{00}=-\delta_{11}=-\delta_{22}=-\delta_{33}=1$..Vector indices everywhere, with the exception of Section V, do not recognize any distinction between contravariant and covariant indices. Repeated Greek indices imply summation over the pseudo-Euclidean metric; repeated Latin indices imply summation over 1, 2, and 3. For example: $k^2 = k_\mu k_\mu = k_0^2 - k_1^2 - k_2^2 - k_3^2$

$k_i k_i = k_1^2 + k_2^2 + k_3^2$, $\partial_\mu A_\mu = \partial_0 A_0 - \partial_1 A_1 - \partial_2 A_2 - \partial_3 A_3$

II. QUANTIZATION BY PATH INTEGRAL

Following Feynman, we write[5] an element of the S-matrix, up to an infinite normalization factor, as a functional integral:

$$\langle in | out \rangle = \int exp\{iS[B]\} \prod_x dB(x) \tag{1.1}$$

where $S[B] = \int \mathcal{L}(x)\, dx$ is the action functional, and the integration is performed over all fields which converge to the in (out) states for $x_0 \to \mp\infty$. Elements of the S-matrix can be expressed by Green's functions which can be written also in the form of the functional integrals over all the fields B(x) vanishing at infinity, the integrand being a product $B(x_1) \ldots B(x_n)$ with a weighting factor exp (iS[B]).

The Lagrangian as a rule is always the sum $\mathcal{L} = \mathcal{L}_0 + \mathcal{L}_I$, where $\mathcal{L}_0$ is a quadratic form, $\mathcal{L}_I$ is a sum of higher powers of the field B, and ϵ is some small parameter. After developing such an integral in a series in powers of ϵ, we obtain the perturbation series, for which there is a Feynman diagram corresponding to each term. The vertices of a diagram are produced by perturbing term $\mathcal{L}_I$, whereas $\mathcal{L}_0$ gives the form of the propagator corresponding to a line in a diagram. The propagator is the inverse of the matrix of the quadratic form $\mathcal{L}_0$. For *gaugeless fields the form $\mathcal{L}_0$ is non-degenerate, and the propagator is uniquely defined.*

When there are gauge fields, the form $\mathcal{L}_0$ is degenerate and the propagator is not uniquely defined. In quantum electrodynamics, however, this ambiguity does not affect the physical results of the theory. Feynman[2] was the first to remark that this is not true in the case of non-abelian gauge fields (Yang-Mills fields, gravitation). The trouble is that *there are diagrams with closed loops, which do affect unitarity, and which depend non-trivially on the choice of the propagator.*

In what follows we proceed with some formal considerations which give an explanation of these difficulties, and give a prescription for circumventing them.

In gauge-invariant theories the action functional should not be changed after replacing B(x), by $B^\Omega(x)$, where $B^\Omega(x)$ is a result of applying the element Ω of the gauge group to the field B(x). In other words the action is constant on the orbits of the gauge group, which are formed by all $B^\Omega(x)$ for fixed B and Ω ranging over the group G. Hence the integral is proportional to the "volume of an orbit" $\prod_x d\Omega(x)$ where the integration is performed over all the elements of the group G. It seems natural to extract this infinite factor before proceeding to develop a perturbation series.

There are different possibilities for giving a recipe for such an extraction. The idea of one of them is to pass from integrating over all possible fields to the integral over a "hypersurface" in the manifold of all fields, which intersect any orbit only once. This means that if the equation of our hypersurface is f[B] = 0, then the equation $f[B^\Omega] = 0$ has only one solution in $\Omega(x)$ for any B(x).

We define a functional $\Delta_f[B]$ from the condition,

$$\Delta_f[B]\int \prod_x \delta(f[B^\Omega])\, d\Omega(x) = \text{const.}, \tag{1.2}$$

by integrating the "infinite-dimensional" δ -function $\prod_x \delta(f[B^\Omega])$ over the group G. Below we shall give a few examples of computing such an integral. Note that the functional $\Delta_f[B]$ is gauge-invariant, i.e. $\Delta_f[B^\Omega] = \Delta_f[B]$

Now put under the integral (1.1) the left-hand side of (1.2) which does not depend on B. In the resulting integral over the variables B and Ω perform the substitution $B^\Omega \to B$ The new expression obtained for Eq. (1.1) is a product of the orbit volume $\prod_x d\Omega(x)$ and the integral

$$\int \exp\{iS[B]\}\, \Delta_f[B] \prod_x \delta(f[B(x)])\, dB(x) \tag{1.3}$$

It is this redefined Feynman functional integral (1.3) that we propose to utilize in developing a perturbation theory for gauge-invariant fields.

We define the generating functional of Green's functions:

$$Z[\eta] = \int \exp\left\{ iS[B] + i\int \eta(x)\, B(x)\, dx \right\} \Delta_f[B] \prod_x \delta(f[B])\, dB(x) \tag{1.4}$$

The Green's functions themselves are logarithmic variational derivatives of this functional:

$$\Gamma_f(x_1, x_2, \cdots, x_n) = \left. \frac{\delta^n \ln Z_f[\eta]}{\delta\eta(x_1) \cdots \delta\eta(x_n)} \right|_{\eta=0}$$

The notations Z_f, Γ_f show that these expressions depend on the choice of a hypersurface f[B] = 0.

There is also another possibility of extracting the "orbit volume". Let $\Delta S[B]$ be some term non-invariant with respect to G, and such that the quadratic form $\mathcal{L}_0$ corresponding to the action $S + \Delta S$ shall be non-degenerate. Defining a gauge-invariant functional $\Phi[B]$ by the definition

$$\Phi[B]\int \exp\{i\Delta S[B^\Omega]\} \prod_x d\Omega(x) = \text{const.}, \tag{1.5}$$

then, in the same way as before, we shall express the integral (1.1) as a product of an orbit volume $\prod_x d\Omega(x)$ and the factor

$$\int \exp\{i(S+\Delta S)\}\, \Phi[B] \prod_x dB(x) \tag{1.6}$$

Both procedures described leave great freedom in choosing the hypersurface f[B] = 0 or an addition to the action, which amounts to the choice of a particular gauge.

In quantum electrodynamics one of the variants of the first procedure leads to the perturbation theory with purely transverse photon Green's functions (Landau gauge), whereas proceeding as in second case mentioned, we arrive at the perturbation theory with non-transverse Green's functions (in particular, the Feynman gauge), but in both cases the factors $\Delta_f[B]$ and $\phi[B]$ do not depend on B(x).

On the contrary, in the case of a non-abelian gauge group (Yang-Mills fields, gravitation) these factors depend non-trivially on B(x) and give rise to some new diagrams for Green's functions. We shall also remark that the computation of $\phi[B]$ in higher orders of perturbation theory is much more difficult than that of $\Delta[B]$.

So our main idea is to extract from the integral (1.1) the factor which is proportional to the orbit volume. This can be done in different ways and leads to the integrals of the type (1.3) or (1.6) where the factors $\Delta[B]$ or $\phi[B]$ appear, which depend on B in the case of a non-abelian gauge fields. The following will illustrate this procedure on concrete examples.

III. QUANTUM ELECTRODYNAMICS

This section has a methodological character. Here we show how all known results of quantum electrodynamics can be obtained by means of the general scheme of part I.

It is well known that the Lagrangian of electrodynamics:

$$\mathcal{L}(x) = \bar{\psi}(i\hat{\partial} - m + e\hat{A})\psi - \frac{1}{4}(\partial_\nu A_\mu - \partial_\mu A_\nu)^2 = \mathcal{L}_0 + \mathcal{L}_I \tag{2.1}$$

where $\mathcal{L}_I = \bar{\psi}\hat{A}\psi$ is invariant with respect to the abelian gauge group:

$$A_\mu(x) \to A_\mu(x) + \partial_\mu C(x) \; ;$$
$$\psi(x) \to e^{ieC(x)}\psi(x) \; ; \qquad \bar{\psi}(x) \to e^{-ieC(x)}\bar{\psi}(x) \tag{2.2}$$

The first recipe of Sec. I is to choose a hypersurface $f = 0$. There are two well known special cases: the Lorentz gauge and the Coulomb (radiation) gauge:

$$f_L[A] \equiv \partial_\mu A_\mu = 0 \, , \qquad f_R[A] \equiv \operatorname{div}\vec{A} = 0$$

which gives us the equations of the "hyperplanes" in the manifold of all fields A(x). The corresponding integrals of the type (1.2)

$$\int \prod_x \delta(\partial_\mu A_\mu(x) - \Box C(x))\, dC(x) \qquad \text{(Lorentz gauge)}$$

and

$$\int \prod_x \delta(\operatorname{div}\vec{A}(x) - \Delta C(x))\, dC(x) \qquad \text{(Coulomb gauge)}$$

where Δ and $\Box$ mean the Laplace and d'Alembert operators respectively, do not depend on A(x), hence we can put without loss of generality $\Delta_L = \Delta_R = 1$.

In order to proceed to relativistic perturbation theory it is useful to find a generating functional of the form (1.4) for the non-perturbed Green's functions:

$$Z^{\circ}[\eta] = \int \exp\{iS + i\int(\bar{\eta}\psi + \bar{\psi}\eta + \eta_\mu A_\mu)\,dx\} \prod_x \delta(\partial_\mu A_\mu)\,dA_\mu\, d\psi\, d\bar{\psi}$$

Here $\bar{\eta}, \eta, \eta_\mu$ are the sources of the fields $\psi, \bar{\psi}, A_\mu$ This functional can be computed by means of a translation

$$\psi \to \psi + \psi_0 \; ; \quad \bar{\psi} \to \bar{\psi} + \bar{\psi}_0 \quad ; \quad A_\mu \to A_\mu + A_\mu^{(0)}$$

that does not affect $\delta(\partial_\mu A_\mu)$, i.e. $\partial_\mu A_\mu^{(0)} = 0$, and is equal to

$$\exp\left\{-i\int\left[\bar{\eta}(x)\,G(x-y)\,\eta(y) + \frac{1}{2}\eta_\mu(x)\,\mathcal{D}^T_{\mu\nu}(x-y)\,\eta_\nu(y)\right]dx\,dy\right\} \tag{2.3}$$

where

$$G(x) = (2\pi)^{-4}\int \exp\{ipx\}\,(\not{p} - m)^{-1}\,dp$$

and

$$\mathcal{D}^T_{\mu\nu}(x) = (2\pi)^{-4}\int \exp\{ikx\}\,(k^2 + i\epsilon)^{-2}(-k^2\delta_{\mu\nu} + k_\mu k_\nu)\,dk \tag{2.4}$$

From (2.3) and (2.4) the Wick theorem follows, and hence, all usual diagrammatic techniques with well known expressions for the electron lines and vertices and purely transverse photon Green's functions (Landau gauge).

In the second procedure we have to choose the additional term ΔS. Choosing it to be

$$\Delta S = -\frac{1}{2\beta}\int [\partial_\mu A_\mu(x)]^2\,dx$$

we see the corresponding factor $\phi[A]$ (cf. 1.5) does not depend on A and therefore we can put $\phi = 1$ Hence we come once more to the diagrammatic techniques with the transverse photon Green's functions of the form

$$(k^2 + i\epsilon)^{-2}(-k^2\delta_{\mu\nu} + (\beta - 1)k_\mu k_\nu) \qquad (\text{in } k\text{-space})$$

which for the special case $\beta = 1$ is equal to the Feynman function:

$$\delta_{\mu\nu}\,(k^2 + i\epsilon)^{-1}$$

Both frameworks of perturbation theory obtained are well known and lead to the same results for physical quantities.

Exact relations among Green's functions of electrodynamics can be obtained by a change of variables in the functional integrals. As an illustration of this we shall derive the Ward identity[6] and the relations between Green's functions in different gauges:

$$G_L(x-y) = \langle \psi(x)\bar{\psi}(y)\rangle_L = N_L^{-1}\int \psi(x)\bar{\psi}(y)\, e^{iS} \prod_x \delta(\partial_\mu A_\mu)\, dA\, d\psi\, d\bar{\psi} \qquad (2.5)$$

where

$$N_L = \int \exp\{iS\} \prod_x \delta(\partial_\mu A_\mu)\, dA\, d\psi\, d\bar{\psi}$$

A rotation of the spinor field into:

$$\psi(x) \to e^{ic(x)}\psi(x) \quad ; \quad \bar{\psi}(x) \to e^{-ic(x)}\bar{\psi}(x)$$

in the numerator of (2.5) gives rise to the factor:

$$\exp\{ ie[c(x) - c(y)] - \int c(z)\, \partial_\mu j_\mu\, dz \,]\}$$

where $j_\mu = \bar{\psi}\gamma_\mu\psi$ Differentiating with respect to c(z) and putting afterwards $c = 0$ we obtain

$$G(x-y)[\delta(x-z) - \delta(y-z)] = \langle \psi(x)\bar{\psi}(y)\, \partial_\mu j_\mu(z)\rangle_L$$

from which the Ward identity follows:

$$G^{-1}(p) - G^{-1}(q) = (p-q)_\mu \Gamma_\mu(p,q;p-q)$$

connecting the Green's function G(p) and the irreducible vertex part $\Gamma_\mu(p,q;p-q)$. It is clear that this identity is valid in any gauge of the photon Green's function, because in the derivation the change of variables was made only on the spinor fields.

Let us look at the transition from thė Coulomb gauge to the Lorentz gauge for the electron Green's function:

$$G_R(x-y) = \langle \psi(x)\bar{\psi}(y)\rangle_R = N_R^{-1}\int \psi(x)\bar{\psi}(y)\, e^{iS} \prod_x \delta(\mathrm{div}\vec{A})\, dA\, d\psi\, d\bar{\psi} \qquad (2.6)$$

where

$$N_R = \int \exp\{iS\} \prod_x \delta(\mathrm{div}\vec{A})\, dA\, d\psi\, d\bar{\psi}$$

In order to accomplish the transition we put under the integral signs in the numerator and in the denominator of (2.6) the expression $\int \prod_x \delta(\Box c - \partial_\mu A_\mu)\, dc(x)$ which does not depend on A. Then we perform a transformation of the type (2.2) which does not affect the action S. Under this transformation the arguments of the delta functions change as $\delta(\Box c - \partial_\mu A_\mu) \to \delta(\partial_\mu A_\mu), \delta(\mathrm{div}\,\vec{A}) \to \delta(\mathrm{div}\,\vec{A} + \Delta c)$ In the factor $\exp\{ie[c(x) - c(y)]\}$ which emerges in the numerator we can put for c(x) a solution of the equation $\Delta c(x) + \mathrm{div}\,\vec{A} = 0$, i.e.,

$$c(x) = \frac{1}{4\pi}\int |\vec{x} - \vec{z}|^{-1}\, \mathrm{div}\,\vec{A}(z)\, d^3z = \int \phi_i(x-z)\, A_i(z)\, dz$$

where

$$\phi_i(x) = -\delta(x_0)\, \partial_i \left(4\pi\, |\vec{x} - \vec{z}|\right)^{-1}$$

When this substitution is made the integrals over c(x) in the numerator and denominator cancel each other. The formula that results is:

$$G_R(x-y) = \langle \psi(x)\bar{\psi}(y) \int \exp\{ie[\phi_i(x-z) - \phi_i(y-z)] \quad . \; A_i(z)\, dz \rangle$$

It expresses the Green's function in Coulomb gauge as a power series of Green's functions $\langle \psi(x)\bar{\psi}(y)\prod_k A_k(z_k)\rangle_{L}$ in the Lorentz gauge.[7]

IV. THE YANG-MILLS FIELDS

1. Yang-Mills field theories[8] are the simplest example of fields associated with non-abelian gauge groups. It is very convenient to describe Yang-Mills vector fields related to any simple compact Lie group, G,[9,10] in terms of matrices $B_\mu(x)$ which form a definite representation of the Lie algebra characterizing the group. It is clear that such a representation may be defined by

$$B_\mu(x) = \sum_{a=1}^{n} b_\mu^a(x)\, \tau_a$$

where τ_a are linearly independent matrices representing the Lie algebra, are normalized according to the relation $Tr(\tau_a \tau_b) = 2\,\delta_{ab}$. n is the number of group parameters, and $b_\mu^a(x)$ is a function with vector index μ and isotopic index a. As is known one may represent the latter index to denote a matrix element of $B_\mu(x)$ by means of the relation, $(B_\mu)_{ab} = \sum_c (\tau_c)_{ab} b_\mu^c = \sum_c t_{abc} b_\mu^c$, where t_{abc} are the totally anti-symmetric structure constants of the group G.

The gauge transformations are:

$$B_\mu \to B^\Omega \equiv \Omega B_\mu \Omega^{-1} + \frac{1}{2}\partial_\mu \Omega\, \Omega^{-1}$$

where $\Omega(x)$ is an arbitrary matrix-function with the values in G. The Lagrangian

$$\mathcal{L}(x) = \frac{1}{8}\, Tr\, F_{\mu\nu} F_{\mu\nu} \;,$$

where
$$F_{\mu\nu} \equiv \partial_\nu B_\mu - \partial_\mu B_\nu + \epsilon [B_\mu, B_\nu] \tag{3.1}$$

is invariant with respect to such transformations. It is also clear that $\mathcal{L} = \mathcal{L}_0 + \mathcal{L}_I$ where $\mathcal{L}_0$ is a quadratic form and $\mathcal{L}_I$ is a sum of trilinear and quadratic forms in the fields B(x).

2. Now we shall proceed with the formalism developed in the part 1, making use of the first recipe. As in electrodynamics the most convenient gauges seem to be the Lorentz or Coulomb gauges:

$$f_L[B] \equiv \partial_\mu B_\mu = 0 \quad , \quad f_R[B] \equiv \operatorname{div} \vec{B} = 0$$

Both equations are matrix equations and represent really n conditions (n = dim G) instead of only one condition as in electrodynamics.

In order to compute the factor $\Delta_L[B]$ we shall remark that we are interested only in its value for the transverse fields ($\partial_\mu B_\mu = 0$). For such fields the only solution of the equation $\partial_\mu B_\mu^\Omega = 0$ in Ω will be the identity element of the group G, and so the entire contribution to the integral (1.2) is given by the neighborhood of the identity element. Therefore we can put

$$\Omega \simeq 1 + u(x)$$

where u(x) is an element of the Lie algebra of the group, and retain in $\partial_\mu B_\mu^\Omega$ only the terms linear in u(x):

$$\begin{aligned}\partial_\mu B_\mu^\Omega &\simeq \partial_\mu (B_\mu + \epsilon [u, B_\mu] + \partial_\mu u) \\ &= \Box u - \epsilon [B_\mu, \partial_\mu u]\end{aligned}$$

where $\Box$ is the d'Alembert operator.

The condition (1.2) reads as

$$\Delta_L[B] \int \prod_x \delta (Au)\, du(x) = \text{Const.},$$

where

$$Au = \Box u - \epsilon [B_\mu, \partial_\mu u]$$

Formally $\Delta_L[B]$ is the determinant of the operator A. It is useful to have also another realization of this operator. We introduce instead of the matrices u(x) the column vector $u_a(x)$ in such a way that $u = u_a \tau_a$, τ_a being the matrices of the adjoined representation of G. In this new representation the operator A is given by the matrix

$$(A)_{ab} = \Box \delta_{ab} - \epsilon (B_\mu)_{ab} \partial_\mu$$

Extracting the trivial infinite factor det $\Box$ we are lead to a following expression for $\ln \Delta_L[B]$:

$$\ln \Delta_L[B] = \ln \det(\Box^{-1} A) = \operatorname{tr} \ln (1 - \epsilon \Box^{-1} B_\mu \partial_\mu) \tag{3.2}$$
$$= \sum_{n=2}^{\infty} \frac{\epsilon^n}{n} \int dx_1 \cdots dx_n \operatorname{tr}(B_{\mu_1}(x_1) \cdots B_{\mu_n}(x_n))$$
$$\cdot \partial_{\mu_1} \mathcal{D}(x_1 - x_2) \cdots \partial_{\mu_n} \mathcal{D}(x_n - x_1)$$

where $\mathcal{D}(x)$ is the Feynman Green's function for the d'Alembert operator

$$\mathcal{D}(x) = (2\pi)^{-4} \int \exp\{ikx\} (k^2 + i\epsilon)^{-1} dk$$

On the left-hand side of (3.2) the trace is understood in the operator sense, whereas on the right-hand side the trace is taken only in the matrix sense.

Analogous computations in the Coulomb gauge give the factor $\Delta_R[B]$. It is given by the expression,

$$\ln \Delta_R[B] = \operatorname{tr} \ln (1 - \epsilon \Delta^{-1} B_i \partial_i) \tag{3.3}$$
$$= \sum_{n=2}^{\infty} \frac{\epsilon^n}{n} \int d^3x_1 \cdots d^3x_n \operatorname{tr}(B_{i_1}(x_1) \cdots B_{i_n}(x_n))$$
$$\partial_{i_1} \tilde{\mathcal{D}}(\vec{x}_1 - \vec{x}_2) \cdots \partial_{i_n} \tilde{\mathcal{D}}(\vec{x}_n - \vec{x}_1),$$

where $\tilde{\mathcal{D}}(\vec{x})$ is a Green's function for the Laplace operator:

$$\tilde{\mathcal{D}}(\vec{x}) = (2\pi)^{-3} \int \exp\{\vec{k} \cdot \vec{x}\} (k_i k_i)^{-1} d^3k$$
$$= -(4\pi |\vec{x}|)^{-1}$$

and the indices $i_1 \ldots i_n$ run over the values 1, 2, 3.

3. Relativistic invariant perturbation theory for computing the Green's functions in the Lorentz gauge, viz,

$$\langle B_{\mu_1}(x_1) \cdots B_{\mu_n}(x_n) \rangle_L = N_L^{-1} \int B_{\mu_1}(x_1) \cdots B_{\mu_n}(x_n) \exp\{iS[B]\}$$
$$\cdot \Delta_L[B] \prod_x \delta(\partial_\mu B_\mu) \, dB(x)$$

where

$$N_L = \int \exp\{iS[B]\} \Delta_L[B] \prod_x \delta(\partial_\mu B_\mu) \, dB(x)$$

arises naturally in the Lorentz gauge when developing the functional

$$\exp\{iS[B]\}\,\Delta_L[B] = \exp\{iS[B] + \ln \Delta_L[B]\}$$

in a power series in the parameter ϵ . The expression $\ln \Delta_L[B]$ can be interpreted as an addition to the action, but this addition does not have the form $\int \mathcal{L}(x)\,dx$ associated with some Lagrangian. The term of degree ϵ^n of this series in a striking way looks like a contribution of a closed loop along which a scalar particle is propagating and interacting with a vector field in a tri-linear way. This circumstance permits a description of the perturbation theory in terms of additional graphs containing the propagation of a fictitious scalar particle and a vertex with one vector and two scalar ends. The elements of this diagrammatic technique are shown in Fig. 1, the corresponding analytic expressions are given by:

$$\begin{aligned}
G^{ab}_{\mu\nu}(p) &= \delta_{ab}\,(p^2+i0)^{-2}\,(-p^2\delta_{\mu\nu} + p_\mu p_\nu) \\
G^{ab}(p) &= -\,\delta_{ab}\,(p^2+i0)^{-1} \\
V^{abc}_{\mu,\nu\rho} &= i\epsilon\, t_{abc}\,(p_{1\nu}\,\delta_{\mu\rho} - p_{1\rho}\,\delta_{\mu\nu}) \\
V^{abc}_{\mu} &= \frac{i\epsilon}{2}\, t_{abc}\,(p_3 - p_2)_\mu
\end{aligned} \tag{3.4}$$

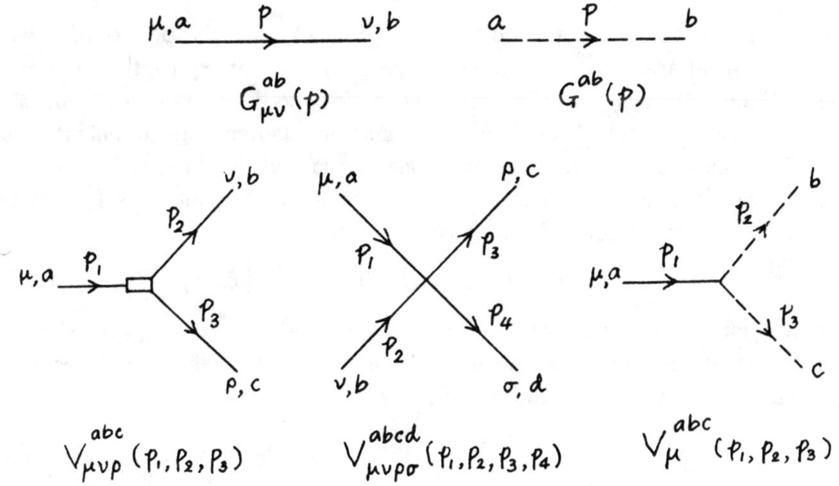

Figure 1.

In order to calculate the contribution of a given diagram we ought to integrate the product of the expressions for its elements over the independent momenta and sum over discrete indices, and then multiply the result by $R^{-1}[(2\pi)^{-4} i]^{l_i - v + 1}$ $(-)^S$, where v is the number of vertices, l_i the number of internal lines, S – the number of closed scalar loops, and R is the rank of a symmetry group of the diagram concerned.

In this way all the effects of the presence of a factor $\Delta_L[B]$ can be interpreted as the introduction of a fictitious scalar particle. The propagation lines of this particle should be taken into consideration only inside the diagrams so that they form closed loops. Moreover these scalar particles behave in a way as fermions. The factor $(-1)^S$, indeed, where S is the number of loops is characteristic of fermions. The last fact becomes less queer if we remark that for $\Delta_L[B]$ one can rewrite the integral representation,

$$\Delta_L[B] = \text{const} \int \exp\{i \operatorname{tr} \bar{u}(\Box u - \epsilon[B_\mu, \partial_\mu u])\} \, du \, d\bar{u}$$

in terms of the anticommuting fields u and $\bar{u}$. Putting this expression in the integral (1.3) we are lead to the usual Feynman integral for a two field system: The transverse vector field B_μ, $\partial_\mu B_\mu = 0$ and scalar fermion field $(\bar{u}, u)$

4. In the second-type approach to Yang-Mills fields it is natural to choose an additional term to the action as in electrodynamics, i.e.,

$$\Delta S = \frac{1}{4\beta} \int \operatorname{tr} (\partial_\mu B_\mu)^2 \, dx$$

Then after expanding the functional $\Phi[B]\exp\{i(S+\Delta S)\}$ in the powers of ϵ we are led to a non-transverse Green's function, in particular, for $\beta = 1$, the Feynman function. Nevertheless in this case it is no longer possible to account for all additional diagrams generated by introducing into our diagrammatic technique just one new line and one new vertex as was done in the previous case.

5. Sometimes the so-called first order formalism is more useful. This formalism can be obtained if the Lagrangian (3.1) is re-expressed as follows:

$$\mathcal{L}(x) = -\frac{1}{8} \operatorname{tr} F_{\mu\nu} F_{\mu\nu} + \frac{1}{4} \operatorname{tr} [(\partial_\nu B_\mu - \partial_\mu B_\nu) + \epsilon[B_\mu, B_\nu]] F_{\mu\nu} \qquad (3.5)$$

and to integrate in the functional integral over the fields B_μ and $F_{\mu\nu}$ as if they were independent, extracting, as usual the "orbit volume". Then for instance, for the Lorentz gauge we arrive at the integral,

$$\int \exp\{iS[B,F]\} \Delta_L[B] \prod_x \delta(\partial_\mu B_\mu) \, dB \, dF \qquad (3.6)$$

Here the integral over F can be computed exactly, thus leading us to the formerly investigated formalism. However, if we do not do that, but yet in the integral (3.6) expand the expression $\exp\{i S[B,F]\}\Delta_L[B]$ in powers of ϵ, we will obtain a new variant of the diagrammatic perturbation theory with three lines, correspon-ding to functions $\langle B, B\rangle$, $\langle B, F\rangle$, $\langle F, F\rangle$ and one vertex describing tri-linear interaction

FBB. The elements of this diagrammatic technique are given in Fig. 2 and formulae (3.7); the field B is represented by a single line, whereas F is represented by a double line:

$$
\begin{aligned}
G^{ab}_{\mu\nu}(p) &= \delta_{ab}(p^2+i0)(-\delta_{\mu\nu}p^2 + p_\mu p_\nu) \\
G^{ab}_{\mu\nu,\rho}(p) &= i\,\delta_{ab}(p^2+i0)^{-1}(p_\nu \delta_{\mu\rho} - p_\mu \delta_{\nu\rho}) \\
G^{ab}_{\mu\nu,\rho\sigma}(p) &= \delta_{ab}\big[\delta_{\mu\rho}\delta_{\nu\sigma} - \delta_{\mu\sigma}\delta_{\nu\rho} \\
&\quad + (p^2+i0)^{-1}(\delta_{\mu\sigma}p_\nu p_\rho + \delta_{\nu\rho}p_\mu p_\sigma \\
&\qquad - \delta_{\mu\rho}p_\nu p_\sigma - \delta_{\nu\sigma}p_\mu p_\rho)\big] \\
V^{abc}_{\rho\sigma\mu\nu} &= i\epsilon\, t_{abc}\,\delta_{\rho\mu}\delta_{\sigma\nu}
\end{aligned}
\tag{3.7}
$$

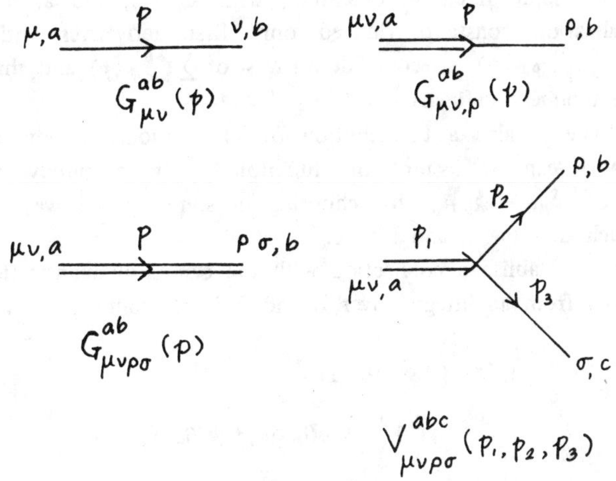

Figure 2.

The lines and vertices describing the propagation of the fictitious scalar particles and their interaction with the vector ones are the same as in the case of the second-order formalism, because the factor depends only on B, but not on F.

6. In the first-order formalism discussed above we have three full one-particle Green's functions which can be expressed in terms of the corresponding non-perturbed functions (3.7) and three self-energy parts, which in second order perturbation theory take the form:

$$\Sigma^{ab}_{\mu\nu}(k) = \frac{\delta_{ab}\,\epsilon^2}{12\pi^2}\Big[(\delta_{\mu\nu}k^2 - k_\mu k_\nu)\ln\Big(\frac{-k^2}{k_0^2}\Big) - (ak^2+b)\,\delta_{\mu\nu} + c\,k_\mu k_\nu\Big]$$

$$\Sigma^{ab}_{\mu\nu,\sigma}(k) = \frac{\delta_{ab}\,3i\epsilon^2}{16\pi^2}\big[k_\mu\delta_{\nu\sigma} - k_\nu\delta_{\mu\sigma}\big]\Big[\ln\Big(\frac{-k^2}{k_0^2}\Big) + d\Big]$$

$$\Sigma^{ab}_{\mu\nu,\rho\sigma}(k) = \frac{\delta_{ab}\,\epsilon^2}{16\pi^2}\Big\{(\delta_{\mu\rho}\delta_{\nu\sigma} - \delta_{\mu\sigma}\delta_{\nu\rho})\Big[\ln\Big(\frac{-k^2}{k_0^2}\Big) + e\Big] + \frac{1}{2}(k^2+i0)^{-1}\big(k_\mu k_\rho\delta_{\nu\sigma} + k_\nu k_\sigma\delta_{\mu\rho} - k_\mu k_\sigma\delta_{\nu\rho} - k_\nu k_\rho\delta_{\mu\sigma}\big)\Big\}$$

Here k_0 denotes some fixed 4-momentum, with $k_0^2 > 0$; and a, b, c, d, e are some renormalization constants (indeed only first derivatives with respect to momentum of $\Sigma^{ab}_{\mu\nu,\rho\sigma}(p)$, second derivatives of $\Sigma^{ab}_{\mu\nu,\sigma}(p)$ and third derivatives of $\Sigma^{ab}_{\mu\nu}(p)$ are uniquely defined).

In $\Sigma^{ab}_{\mu\nu,\sigma}$ there is also a contribution of an additional diagram and only due to its appearance can we assume the function $\Sigma^{ab}_{\mu\nu}$ to be purely transverse, i.e., proportional to $k^2\delta_{\mu\nu} - k_\mu k_\nu$, by choosing in some special way the constants a, b,, and c such that $b = 0$, and $a = -c$.

7. In order to establish a connection with the canonical quantization method it is useful to start from an integral over B and F in Coulomb gauge with sources:

$$Z[\eta] = \int \exp\Big\{iS[B,F] + i\int tr\,(\eta_\mu B_\mu + \tfrac{1}{2}\eta_{\mu\nu}F_{\mu\nu})\,dx\Big\} \cdot \Delta_R[B]\prod_x \delta(\operatorname{div}\vec{B})\,dB\,dF \tag{3.8}$$

Following Schwinger[4] we shall choose for the dynamical variable the transverse (in three-dimensional sense) components of the field B_i, and F_{io}, $i = 1, 2, 3$. We also assume the sources present only correspond to our dynamical variables; this means that

$$\eta_0 = \eta_{ik} = \partial_i\eta_i = \partial_i\eta_{oi} = 0\,.$$

In three-dimensional notations our Lagrangian (3.5) takes the form

$$\mathcal{L}(x) = \mathrm{tr}\Big\{ -\tfrac{1}{2} F_{ik} F_{ik} + \tfrac{1}{4} F_{io} F_{io}$$

$$+ \tfrac{1}{4} F_{ik} (\partial_k B_i - \partial_i B_k + \epsilon [B_i, B_o])$$

$$- \tfrac{1}{2} F_{io} \partial_o B_i - \tfrac{1}{2} B_o (\partial_i F_{io} + \epsilon [B_i, F_{io}]) \Big\}$$

The absence of the sources corresponding to F_{ik} and B_o permits the integration in (3.8) over these fields, which is equivalent to changing in the resulting integral over the remaining variables B_i, F_{io} the matrix F_{ik} into

$$H_{ik} = \partial_k B_i - \partial_i B_k + \epsilon [B_i, B_k]$$

and also the δ -function into

$$\prod_x \delta(\partial_i F_{io} + \epsilon [B_i, F_{io}])$$

Now, put into the integral (3.8) the factor $\int \prod_x \delta(\Delta c - \partial_i F_{io}) dc(x)$ that really does not depend on F_{io} and perform a translation $F_{io} \to F_{io} + \partial_i c$, thus inducing the transformations:

$$\prod_x \delta(\Delta c - \partial_i F_{io}) \to \prod_x \delta(\partial_i F_{io}) \prod_x \delta(\partial_i F_{io} + \epsilon [B_i, F_{io}])$$

$$\to \prod_x \delta(\Delta c + \epsilon [B_i, \partial_i c] + \epsilon [B_i, F_{io}])$$

Let c_o be a solution of the equation:

$$\Delta c + \epsilon [B_i, \partial_i c] = \epsilon [F_{io}, B_i] ,$$

which can be expressed by a Green's function depending on B (cf. Ref. 4)

$$c_o(x) = \epsilon \int \mathcal{D}(\vec{x}, \vec{y}; B) [F_{io}(y), B_i(y)] \, d^3y \quad ;$$

$$\Delta \mathcal{D}(\vec{x}, \vec{y}; B) + \epsilon [B_i(x), \partial_i \mathcal{D}(\vec{x}, \vec{y}; B)] = \delta^3(\vec{x} - \vec{y}) .$$

After performing a translation $c \to c + c_o$ the δ -function appears as $\prod_x \delta(\Delta c + \epsilon [B_i, \partial_i c])$ and we can put $c = 0$ everywhere except when equal to the argument of this function. The resulting integral

$$\int \prod_x \delta(\Delta c + \epsilon [B_i, \partial_i c]) \, dc(x)$$

cancels with $\Delta_R [B]$.

Hence the integral (3.8) can be transformed, up to an infinite factor, to the form

$$Z_R[\eta] = \int \exp\{ i S_R + i \int tr[\eta_i B_i + \eta_{io} F_{io}]dx \} \cdot \prod_x \delta(\partial_i B_i)\,\delta(\partial_i F_{io})\, dB_i\, dF_{oi} \tag{3.9}$$

where

$$S_R = \int dx_o \{ \int f^a_{io}\, \partial_o b^a_i\, d^3x \; - H(f, h)\}$$

and

$$H = \frac{1}{2}\int d^3x\, [\frac{1}{2} h^a_{ik} h^a_{ik} + f^a_{io} f^a_{io} + \partial_i c^a_o\, \partial_i c^a_o]$$

The relations obtained look like the usual formulae of quantization of a classical Hamiltonian system using Feynman method (cf. Ref. 11). The transverse variables $f^a_{oi}(\vec{x})$ and $b^a_i(\vec{x})$ are playing the role of canonically conjugate variables, whereas the functional H[f, b] that of the Hamiltonian function. Following the general quantization scheme, we ought to introduce the operators $\hat{b}^a_i(\vec{x}), \hat{f}^a_{oi}(\vec{x})$ fulfilling the transversality conditions:

$$\partial_i \hat{f}_{oi} = \partial_i \hat{b}_i = 0$$

as well as the commutation relations

$$[\hat{f}^a_{oi}(\vec{x}), \hat{b}^c_j(\vec{y})] = i\,\delta_{ac}\,\delta^T_{ij}(\vec{x}-\vec{y})$$

where

$$\delta^T_{ij}(\vec{x}) = (2\pi)^{-3}\int \exp\{i\vec{k}\cdot\vec{x}\}\,(\delta_{ij} - \frac{k_i k_j}{k_s k_s})\, d^3k$$

The energy operator $\hat{H}$ can be optained by putting the operators $\hat{b}$ and $\hat{f}$ instead of the functions b and f into the functional H. The canonical quantization of Yang-Mills fields given above is equivalent to the one proposed by Schwinger in Ref. 4.

It is also known that the Feynman and the canonical quantization procedure are equivalent. In particular the functional integral (3.9) is the generating functional for the Green-Schwinger function. Strictly speaking, the following relations occur:

$$\langle \psi_o | T(\hat{B}(x) \cdots \hat{F}(y) \cdots) | \psi_o \rangle$$

$$= \langle B(x) \cdots F(y) \cdots \rangle_R$$

$$\equiv N_R^{-1} \int B(x) \cdots F(y) \cdots e^{iS_R} \prod \delta(\partial_i F_{oi})\,\delta(\partial_i B_i)\, dB\, dF$$

where

$$N_R = \int e^{iS_R} \prod \delta(\partial_i F_{oi})\, \delta(\partial_i B_i)\, dB\, dF$$

The left-hand side is an average over the physical vacuum of the chronological product of the Heisenberg operators B(x).... F(y)....

As a closing remark we indicate that the presence of the factor $\Delta_R[B]$ in the primary integral (3.8) is necessary for casting it into canonical form.

8. After having established the equivalence of our quantization with more conventional one in terms of operators, we can define the unitary S-matrix using well known reduction formulae (cf. Ref. 12). For the sake of brevity we return to the second-order formalism, so that the Coulomb-gauge Green's functions are given by the following expressions:

$$\langle b_{i_1}^{a_1}(x_1) \cdots b_{i_n}^{a_n}(x_n) \rangle_R \tag{3.10}$$

$$= N_R^{-1} \int b_{i_1}^{a_1}(x_1) \cdots b_{i_n}^{a_n}(x_n) \exp\{iS[B]\}$$

$$\cdot\, \Delta_R[B] \prod \delta(\operatorname{div} \vec{B})\, dB$$

where

$$N_R = \int \exp\{iS[B]\}\, \Delta_R[B] \prod \delta(\operatorname{div} \vec{B})\, dB$$

Take now their Fourier transforms;

$$G_{R\, i_1 \cdots i_n}^{a_1 \cdots a_n}(p_1, \cdots, p_n) = -(2\pi)^{-4} \int \langle b_{i_1}^{a_1}(x_1) \cdots b_{i_n}^{a_n}(x_n) \rangle\, e^{i\sum k\cdot x}\, dx_1 \cdots dx_n$$

which are proportional to $\delta(\sum p)$ due to translational invariance. In order to obtain the transition amplitudes between an initial state containing r particles and a final state containing s particles, r + s = n, we have to go to the mass shell (i.e., to put $p^2 = 0$ for all the momenta p) in the product of $G_{R\, i_1 \cdots i_n}^{a_1 \cdots a_n}(p_1, \cdots, p_n)$ and n factors of the form $e_i(p)\, u(p)\, Z_R^{-1}$ for every particle. Here $e_i(p)$ is a unit transverse polarization vector, $p_i e_i = 0$ and $u(p) = (2\pi)^{-3/2} p^2 |2p_0|^{-\frac{1}{2}} \theta(\pm p_0)$ where the last sign takes account of the difference between the in- and outgoing particles; the renormalization constant Z_R is defined by the residuum of the one-particle Green's function $G_{ij}^{ab}(p)$ for $p^2 = 0$. More exactly, for $p^2 = 0$ this function has the asymptotic form:

$$G_{ij}^{ab}(p) \simeq \frac{Z_R\, \delta_{ab}}{p^2 + i0} \left(\delta_{ij} - \frac{p_i p_j}{p_s p_s} \right)$$

up to infrared singularities.

From the practical point of view it is useful to express the matrix elements of the S-matrix in terms of the Green's functions in the Lorentz gauge, for which we have developed above a covariant perturbation theory. As a matter of fact this can be done and the corresponding formulae have exactly the same form as the ones developed above in the terms of Coulomb-gauge Green's functions. We need only change the Coulomb gauge Green's functions to the Lorentz gauge Green's functions, and put instead of the factor Z_R the residuum of $G_L^{ab}(p^2)$ for $p^2 = 0$.

For a more detailed explanation we consider the connection between these two types of Green's functions corresponding to different gauges. We put into the denominator and numerator of (3.10) the factor not depending on B:

$$\Delta_L[B] \int \prod_x \delta(\partial_\mu B_\mu^{\Omega^{-1}})\, d\Omega$$

and then perform a translation $B \to B^\Omega$ In the numerator the following integral over Ω appears:

$$\int B_{i_1}^{\Omega}(x_1) \cdots B_{i_n}^{\Omega}(x_n) \prod_x \delta(\mathrm{div}\, \vec{B}^{\Omega})\, d\Omega$$

from which can be extracted a factor $B_{i_1}^{\Omega_0} \cdots B_{i_n}^{\Omega_0}$, where Ω_0 is a gauge transformation changing the field B satisfying the Lorentz condition $\partial_\mu B_\mu = 0$ into a field transverse in the three-dimensional sense, i.e., div $\vec{B} = 0$ This transformation depends on B, so that B^{Ω_0} is a complicated functional of B. Here are the first terms of this functional developed in powers of ϵ :

$$B^{\Omega_0} = B^T - \frac{\epsilon}{2}\left[\Delta^{-1}\, \mathrm{div}\, \vec{B},\; B + B^T\right]^T + \cdots \tag{3.11}$$

Here T denotes the three-dimensional transverse part of the corresponding vector. The rest of the integral cancels with $\Delta_R[B]$ and we are lead to the expression of the form:

$$N_L^{-1} \int B_{i_1}^{\Omega_0}(x_1) \cdots B_{i_n}^{\Omega_0}(x_n)\, e^{iS}\, \Delta_L[B] \prod_x \delta(\partial_\mu B_\mu)\, dB$$

that can be used to calculate the Coulomb Green's functions in terms of the Lorentz Green's functions. When expanding in terms of perturbation series we can ascribe the extra vertices to the terms of (3.11), so that any Coulomb Green's function is really an infinite series containing the integrals of Lorentz-gauge Green's functions with any number of external lines.

In this way the connection between the Green's functions in different gauges appears to be very complicated. Our scheme becomes much simpler, however, if we go to the mass-shell: We ought to compare only the terms having the asymptotic singular behaviour when $p^2 \to 0$. It is clear that the pole for a given p will remain only if the whole effect of extra vertices reduced to terms of the type of self energy insertions into a corresponding external line. In the limit $p^2 \to 0$, the resulting Lorentz and Coulomb Green's functions will differ by the factor σ^n where σ is certain constant, and n denotes the number of external lines. By comparing the one-particle functions we obtain $\sigma^2 = Z_R Z_L^{-1}$ so that $\sigma = (Z_R/Z_L)^{1/2}$ This ends our explanatory remarks concerning the elements of the S-matrix.

All these considerations have an explicitly invariant character and therefore apply to any Green's functions. In other words, the expressions for the elements of S-matrix have the same form in any gauge. In particular we can make use of the following formula for computing the S-matrix of the Green's functions:

$$\langle b_{i_1}^{a_1}(x_1) \cdots b_{i_n}^{a_n}(x_n) \rangle_{\Delta S} = N_{\Delta S}^{-1} \int b_{i_1}^{a_1}(x_1) \cdots b_{i_n}^{a_n}(x_n)\, e^{i\{S+\Delta S\}} \phi_{\Delta S}[B] \prod dB$$

where

$$N_{\Delta S} = \int e^{i\{S+\Delta S\}} \phi_{\Delta S}[B] \prod dB$$

for which the relativistic invariant perturbation theory could be developed as well. It can be shown that the recipe given by Feynman in (2) for computing one-loop diagrams is equivalent to taking into account some of the terms appearing in the expansion of the factor $\phi_{\Delta S}[B]$ in powers of ϵ . In the general case this expansion gives rise to the infinite number of extra vertices, whereas in the formalism of the first type it is sufficient to introduce only one extra vertex.

As a closing remark: The computation of S-matrix elements in perturbation theory without taking into account factors of the type $\Delta_L[B]$ or $\phi[B]$ would lead to the formally invariant amplitudes, which would not, however, satisfy the unitarity condition. This fact was indeed revealed by Feynman in Ref. 2.

V. THE THEORY OF GRAVITATION

Now we will examine one proposed scheme for quantizing the theory of gravitation. Here we limit ourselves only to an outline of the perturbation theory and compute the analogue of $\Delta_L[B]$

The Lagrangian of the gravitation field,

$$\mathcal{L}(x) = (2\kappa^2)^{-1} R(x) = (2\kappa^2)^{-1} \sqrt{-g}\, g^{\mu\nu} \cdot (\partial_\nu \Gamma^\sigma_{\mu\sigma} - \partial_\sigma \Gamma^\sigma_{\mu\nu} + \Gamma^\rho_{\mu\sigma} \Gamma^\sigma_{\nu\rho} - \Gamma^\sigma_{\mu\nu} \Gamma^\rho_{\sigma\rho})$$

is invariant with respect to general coordinate transformations which we write here in an infinitesimal form:

$$\delta g^{\mu\nu} = -\delta x^\lambda \partial_\lambda g^{\mu\nu} + g^{\mu\lambda} \partial_\lambda \delta x^\nu + g^{\lambda\nu} \partial_\lambda \delta x^\mu$$

$$\delta \Gamma^{\sigma}_{\mu\nu} = - \delta x^{\lambda} \partial_{\lambda} \Gamma^{\sigma}_{\mu\nu} - \Gamma^{\sigma}_{\mu\lambda} \partial_{\nu} \delta x^{\lambda} - \Gamma^{\sigma}_{\nu\lambda} \partial_{\mu} \delta x^{\lambda}$$
$$+ \Gamma^{\lambda}_{\mu\nu} \partial_{\lambda} \delta x^{\sigma} - \partial_{\mu} \partial_{\nu} \delta x^{\sigma}$$

Here δx^{λ} are arbitrary infinitesimal functions.

The coordinate transformation depend on four arbitrary functions. Therefore in our scheme a hypersurface in the manifold of all possible fields should be given by four conditions, and we choose the harmonicity conditions:[13]

$$\partial_{\nu} (\sqrt{-g}\, g^{\mu\nu}) = 0$$

If for the variables of functional integration we choose $\sqrt{-g}\, g^{\mu\nu}$ and $\Gamma^{\sigma}_{\mu\nu}$ we will arrive at an integral of the form

$$\int \exp \{ i S [g, \Gamma] \} \Delta_{h}[g] \prod \delta (\partial_{\nu} \sqrt{-g}\, g^{\mu\nu})\, dg\, d\Gamma \tag{4.1}$$

where

$$dg\, d\Gamma = \prod_{\mu<\nu} d(\sqrt{-g}\, g^{\mu\nu}) \prod_{\sigma, \mu<\nu} d\Gamma^{\sigma}_{\mu\nu}$$

The notation $\Delta_{h}(g)$ recalls the origin of this factor from the harmonicity conditions.

In spite of the fact that the integral over $\Gamma^{\sigma}_{\mu\nu}$ in (4.1) can be taken explicitly we will prefer not to do it and work in the first-order formalism.

The diagrammatic perturbation technique arises in a natural way if we put

$$\sqrt{-g}\, g^{\mu\nu} = \delta^{\mu\nu} + \kappa h^{\mu\nu}, \quad \Gamma^{\sigma}_{\mu\nu} = \kappa \Pi^{\sigma}_{\mu\nu}$$

and then expand the integral (4.1) in the powers of κ

At that moment we get two kinds of vertices: One of them corresponding to interaction $\kappa h^{\mu\nu}(\Pi^{\rho}_{\mu\sigma}\Pi^{\sigma}_{\nu\rho} - \Pi^{\rho}_{\mu\nu}\Pi^{\sigma}_{\rho\sigma})$ and the other being generated by the factor $\Delta_{h}[g]$ which we shall now evaluate.

We must know $\Delta_{h}[g]$ only for the harmonic fields. For such fields

$$\partial_{\nu} \delta (\sqrt{-g}\, g^{\lambda\nu}) = \sqrt{-g}\, g^{\mu\nu} \partial_{\mu} \partial_{\nu} \delta x^{\lambda} = - \Box \delta x^{\lambda} + \kappa h^{\mu\nu} \partial_{\mu} \partial_{\nu} \delta x^{\lambda}$$

and the general formula (1.2) for the factor we are seeking takes the form:

$$\Delta_{h}[g] \int \prod_{x} \delta (\sqrt{-g}\, g^{\mu\nu} \partial_{\mu} \partial_{\nu} (\delta x^{\lambda}))\, d(\delta x^{\lambda}) = \text{Const.}$$

Proceeding thereafter as in the case for Yang-Mills fields we obtain the result:

$$\ln \Delta_h[g] = 4 \, \mathrm{tr} \ln (1 - \Box^{-1} h_{\mu\nu} \partial_\mu \partial_\nu)$$

$$= -4 \sum_{n=2}^{\infty} \frac{\kappa^n}{n} \int dx_1 \cdots dx_n \tag{4.2}$$

$$h^{\mu_1\nu_1} \partial_{\mu_1} \partial_{\nu_1} \mathcal{D}(x_1 - x_2) \cdots h^{\mu_n\nu_n} \partial_{\mu_n} \partial_{\nu_n} \mathcal{D}(x_n - x_1)$$

A general term in the expansion (4.2) describes a fictitious vector particle propagating along a loop with n external lines and interacting with the "harmonic" field $h^{\mu\nu}$ following the coupling rule $\kappa h^{\mu\nu} \partial_\mu x^\rho \partial_\nu x^\rho$.

This paper was received
by the Publishing Department
July 5, 1967.

REFERENCES

[1] Bogoliubov and Shirkov, *An Introduction to the Theory of Quantized Fields,* Moscow 1957 (in Russian).
[2] Feynman, R. P. *Acta Phys, Polonica,* **24,** 6 (12), (1963).
[3] De Witt, B. S., *Relativity, Groups and Topology, London,* (1964).
[4] Schwinger, J., *Phys. Rev.* **125,** 1043, (1962); *Phys. Rev.* **127,** 321, (1962).
[5] Feynman, R. P., *Phys. Rev.* **80,** 440, (1950).
[6] Ward, J. C., *Phys. Rev.* **77,** 293L, (1950); *Phys. Rev.* **78,** 182L, (1950).
[7] Schwinger, J., *Phys. Rev.,* **115,** 721, (1959).
[8] Yang, C. N. and Mills, R. L., *Phys. Rev.* **96,** 191, (1959).
[9] Glashow, S. L., Gell-Mann, M., *Ann. Phys.* **15,** 437, (1961).
[10] Utiyama, R., *Phys. Rev.* **101,** 1957, (1956).
[11] Garrod, C., *Rev. Mod. Phys.* **38,** 483, (1966).
[12] Matthews, P. T., *Relativistic Quantum Theory of the Interactions of the Elementary Particles,* Russian translation (1959).
[13] Fock, V., *The Theory of Space, Time and Gravitation,* issued by the State Publishing Department, Moscow, 1955 (in Russian).

Generalized Renormalizable Gauge Formulation of Spontaneously Broken Gauge Theories

Kazuo Fujikawa*
The Enrico Fermi Institute, The University of Chicago, Chicago, Illinois 60637

and

Benjamin W. Lee† and A. I. Sanda
National Accelerator Laboratory,‡ Batavia, Illinois 60510
(Received 29 June 1972)

The spontaneously broken gauge theory is formulated in the generalized renormalizable gauge (R_ξ gauge). A parameter ξ can be adjusted to include existing gauges, U gauge, R gauge, and 't Hooft-Feynman gauge as special cases. Three applications of the R_ξ-gauge formulation are given. First we compute the weak correction to the muon magnetic moment unambiguously in the existing models for leptons. Secondly, we discuss the large-momentum-transfer limit of the Pauli magnetic form factor of the muon. Finally, we discuss the static charge of the neutrino, and show that an appropriate regularization makes it vanish.

I. INTRODUCTION

The possibility of constructing a unified theory of weak and electromagnetic interactions in terms of a spontaneously broken gauge symmetry has attracted a great deal of attention lately, following the works of Weinberg[1] and 't Hooft.[2] In this paper we shall present a formulation of spontaneously broken gauge theories (SBGT) which is particularly suited for practical calculations. In this formulation the gauge condition one adopts is a generalization of the one used by 't Hooft and depends on a parameter ξ which can vary continuously from 0 to ∞. In this gauge, which we shall call generically the R_ξ gauge, the massive-vector-boson propagator is precisely the one invented by Lee and Yang in their discussion[3] of the ξ-limiting process:

$$\begin{aligned}\Delta_{\mu\nu}(p,\xi) &= -i\left[g_{\mu\nu} - \left(1-\frac{1}{\xi}\right)\frac{p_\mu p_\nu}{p^2 - M^2/\xi}\right]\frac{1}{p^2-M^2} \\ &= -i\left(g_{\mu\nu} - \frac{1}{M^2}p_\mu p_\nu\right)\frac{1}{p^2-M^2} \\ &\quad - i\frac{1}{M^2}p_\mu p_\nu \frac{1}{p^2-M^2/\xi}. \end{aligned} \tag{1.1}$$

The difference between the R_ξ-gauge formulation of SBGT and the ξ-limiting process applied to the electrodynamics of massive vector bosons is this:

In the former, the negative-metric scalar-boson pole of the vector-boson propagator at $p^2 = M^2/\xi$ is canceled by the pole of the unphysical-scalar-boson propagator

$$\frac{i}{p^2 - M^2/\xi} \tag{1.2}$$

in the S matrix, and the S matrix of the former is independent of the parameter ξ and is unitary, whereas in the latter, one recovers the unitarity of the S matrix only in the limit $\xi \to 0$. The ξ independence of the S matrix[4] in the former is a direct consequence of the non-Abelian gauge invariance of the relevant Lagrangian.

It is worthwhile to note the connection between the R_ξ gauge and other gauges discussed in the literature.

(1) The R Gauge: In the proof of renormalizability of SBGT by Lee and Zinn-Justin,[5] and also in the discussion of Salam and Strathdee,[6] a generalization of the Landau gauge in quantum electrodynamics, the so-called R gauge, was used. The R gauge is obtained from the R_ξ gauge for $\xi = \infty$.

(2) The 't Hooft–Feynman Gauge: This gauge, which was discussed by 't Hooft, is obtained when we set $\xi = 1$. In this gauge the vector-boson propagator is proportional to $g_{\mu\nu}$, and the unphysical-scalar-boson propagator of Eq. (1.2) has a pole at $p^2 = M^2$.

(3) The U Gauge: In this formulation, the unphysical scalar bosons are absent and the vector-boson propagator is the canonical one:

$$\Delta_{\mu\nu}(p) = -i\left(g_{\mu\nu} - \frac{p_\mu p_\nu}{M^2}\right)\frac{i}{p^2 - M^2}. \tag{1.3}$$

In this gauge, the unitarity of the S matrix is manifest since there are no spurious singularities at $p^2 = M^2/\xi$. However, Green's functions are unrenormalizable in this gauge: It is only the S matrix that can be defined in this gauge. The U-gauge is *formally* equivalent to the R_ξ gauge in the limit $\xi \to 0$. The equivalence here is "formal," in the sense that Feynman amplitudes in the two formulations are equal if the limit $\xi \to 0$ is taken before the Feynman integral is performed.

The U-gauge formulation of SBGT deserves some more discussion. Because the quantization of SBGT in this gauge is most straightforward, most of the existing calculations were performed in this formulation, despite the divergence difficulties unique to this gauge. The cancellation of divergences in the S matrix (but not in Green's functions) has been demonstrated by various authors in a number of cases.[7] However, isolation of the finite part of an S-matrix element in this gauge may prove ambiguous. In fact, Jackiw and Weinberg and Bars and Yoshimura[8] have commented on an ambiguity that exists in the calculation of the weak-interaction contribution to the anomalous magnetic moment of the muon. We claim that, based on our own experiences, computation of Feynman amplitudes is enormously simplified in the R_ξ gauge. It is also easier to check the ξ independence of the S matrix (thereby verifying the unitarity of the S matrix) in the R_ξ gauge, than to establish the cancellation of higher-order divergences. When there are ambiguities in fixing the finite part of an S-matrix element in the U gauge, the R_ξ-gauge formulation provides a gauge-invariant (with respect to the non-Abelian gauge group) way of circumventing such difficulties. In fact, we shall resolve the ambiguity in the computation of the anomalous magnetic moment of the muon by evaluating it in the R_ξ gauge. Our study explains also why the ξ-limiting process used by Jackiw and Weinberg and by Bars and Yoshimura yields the correct result.[9]

This paper is organized as follows: In Sec. II we formulate the generalized renormalizable gauge (R_ξ gauge). In Sec. III, we apply the R_ξ gauge to the calculation of weak correction to the magnetic moment of the muon. We will present unambiguous answers for three existing models of Weinberg,[1] of Georgi and Glashow,[10] and of Lee,[11] and Prentki and Zumino.[11] In Sec. IV, we show that the naive calculation of the neutrino static charge gives a nonvanishing result and we discuss how to remedy this situation. In Appendix A, we give details of Sec. III. In Appendix B we point out the reasons for ambiguities present in the U-gauge calculations of the weak correction to the muon magnetic moment. Finally, in Appendix C, we give the Lagrangians and necessary Feynman rules for our calculations.

After the completion of this paper, we received a paper by Yao[11a] in which a formulation similar to ours is discussed in the context of an *Abelian* gauge theory.

II. FORMULATION OF THE R_ξ GAUGE

In this section we shall discuss the formulation of SBGT in a general class of covariant linear gauge conditions. We shall consider, for definiteness, the Georgi-Glashow model based on the O(3) gauge symmetry without fermions. In Appendix C, we will extend our considerations of this section to all three models mentioned in the Introduction, with fermions.

In the absence of fermions, the Georgi-Glashow model consists of a triplet of gauge bosons and a triplet of scalar mesons. The Lagrangian is of the form

$$\mathcal{L} = -\tfrac{1}{4}(\partial_\mu \vec{B}_\nu - \partial_\nu \vec{B}_\mu + g\vec{B}_\mu \times \vec{B}_\nu)^2 + \tfrac{1}{2}[(\partial_\mu + g\vec{B}_\mu \times)\vec{\phi}]^2 - V(\vec{\phi}), \tag{2.1}$$

where $V(q)$ is an isospin-invariant quartic polynomial of the scalar fields $\vec{\phi}$. The potential is assumed to have an absolute minimum at $\vec{\phi} = \vec{v} \neq 0$. We can always choose the isospin z axis to coincide with the direction of $\vec{v}$.

It is convenient to define a unit vector $\hat{\eta}$ along the z axis:

$$\vec{v} = v\hat{\eta}.$$

We also define $\vec{\phi}_t$ and ϕ by

$$\vec{\phi} = \vec{\phi}_t + \hat{\eta}(v + \phi), \quad \vec{\phi}_t \cdot \hat{\eta} = 0. \tag{2.2}$$

The gauge condition we shall adopt is (see also Appendix B)

$$\partial^\mu \vec{B}_\mu - \frac{1}{\xi} g v \hat{\eta} \times \vec{\phi} = 0, \tag{2.3}$$

where ξ is a non-negative real parameter.

It was shown by 't Hooft[12] that the gauge condition (2.3) may be taken care of by defining the effective action[13]:

$$S[\vec{B}_\mu, \vec{\phi}] = \int d^4x \left[\mathcal{L}(x) - \frac{\xi}{2}\left(\partial^\mu \vec{B}_\mu - \frac{1}{\xi} g \vec{v} \times \vec{\phi}\right)^2 - \frac{1}{2}\left(\frac{1}{\alpha} - \xi\right)(\partial^\mu B_\mu^3)^2 \right] - i \operatorname{Tr} \ln[1 + g\mathcal{G}\gamma], \tag{2.4}$$

where $\mathcal{G}$ is defined by

$$\left[(-\partial^2 + i\epsilon)\delta_{ab} - \frac{1}{\xi}(gv)^2(\delta_{ab} - \eta_a\eta_b)\right]\langle b, x | \mathcal{G} | c, y\rangle = \delta_{ac}\delta^4(x-y) \tag{2.5}$$

and γ is defined by

$$\langle a, x | \gamma | b, y\rangle = \left(\epsilon_{abc} \frac{\partial}{\partial x_\mu} B_\mu^c(x) + \frac{1}{\xi} g v [\phi_{ta}(x)\eta_b - \phi(\delta_{ab} - \eta_a\eta_b)]\right)\delta^4(x-y) \tag{2.6}$$

and Tr denotes the trace operation over the space-time variables x, y, as well as over the isospin indices a, b. In Eq. (2.4), ξ and α are in general arbitrary non-negative real numbers.

The effective action S of Eq. (2.4) is to be used in defining the generating functional of connected Green's functions. Thus, if we define

$$\exp(iZ[\vec{J}_\mu, \vec{k}]) = \int [d\vec{B}_\mu] \int [d\vec{\phi}] \exp\left\{ iS[\vec{B}_\mu, \vec{\phi}] + i \int d^4x [\vec{k}(x)\cdot\vec{\phi}(x) - \vec{J}_\mu(x)\cdot\vec{B}^\mu(x)]\right\}, \tag{2.7}$$

functional derivatives of Z at $\vec{J}_\mu = \vec{k} = 0$ give connected Green's functions of the theory. The generating functional Z depends on two parameters α and ξ. Note that the choice $\alpha^{-1} = \xi = \infty$ leads to the R gauge discussed in Refs. 5 and 6.

The effective Lagrangian can be written as

$$\begin{aligned}
\mathcal{L} - \frac{\xi}{2}\left(\partial^\mu \vec{B}_\mu - \frac{1}{\xi} g\vec{v}\times\vec{\phi}\right)^2 - \frac{1}{2}\left(\frac{1}{\alpha} - \xi\right)(\partial^\mu B_\mu^3)^2 = {} & \frac{1}{2}[\partial_\mu \vec{\phi} + g\vec{B}_\mu \times (\vec{\phi}_t \times \phi\hat{\eta})]^2 - \frac{(gv)^2}{2\xi}\vec{\phi}_t^{\,2} \\
& + g^2 v\hat{\eta}\times\vec{B}_\mu \cdot \vec{B}^\mu \times (\vec{\phi}_t + \phi\hat{\eta}) - V(\vec{\phi}) \\
& - \tfrac{1}{4}(\partial_\mu\vec{B}_\nu - \partial_\nu\vec{B}_\mu + g\vec{B}_\mu\times\vec{B}_\nu)^2 + \frac{(gv)^2}{2}(\vec{B}_\mu^t)^2 \\
& - \frac{\xi}{2}(\partial^\mu \vec{B}_\mu^t)^2 - \frac{1}{2\alpha}(\partial^\mu B_\mu^3)^2 .
\end{aligned} \tag{2.8}$$

The terms proportional to $\vec{B}_\mu \cdot \partial^\mu \vec{\phi}$ have disappeared from Eq. (2.8). The propagators for various fields are obtained by inverting the matrix of the quadratic form $\mathcal{L}_0$ of the above expression:

$$\mathcal{L}_0 = \frac{1}{2}\left[(\partial_\mu\vec{\phi})^2 - \frac{M^2}{\xi}\vec{\phi}_t^{\,2} - \mu_\phi^{\,2}\phi^2\right] - \tfrac{1}{4}(\partial_\mu\vec{B}_\nu - \partial_\nu\vec{B}_\mu)^2 - \frac{\xi}{2}(\partial^\mu\vec{B}_\mu^t)^2 + \frac{M^2}{2}(\vec{B}_\mu^t)^2 - \frac{1}{2\alpha}(\partial^\mu B_\mu^3)^2, \tag{2.9}$$

where

$$M^2 = gv, \quad \mu_\phi^{\,2} = 2\partial V(v\hat{\eta})/\partial v^2,$$

and they are[14]

$$s^\pm \equiv \frac{1}{\sqrt{2}}(\phi^1 \pm i\phi^2): \; i\frac{1}{k^2 - M^2/\xi + i\epsilon},$$

$$\psi \equiv \phi^3 - v:\ i\frac{1}{k^2-\mu_\psi{}^2+i\epsilon},$$

$$W_\mu \equiv \frac{1}{\sqrt{2}}(B^1_\mu \mp iB^2_\mu):\ -i\left[g_{\mu\nu} - k_\mu k_\nu \frac{1}{k^2-M^2/\xi}\left(1-\frac{1}{\xi}\right)\right]\frac{1}{k^2-M^2+i\epsilon}, \tag{2.10}$$

and

$$A_\mu \equiv B^3_\mu:\ -i\left(g_{\mu\nu} - \frac{k_\mu k_\nu}{k^2}(1-\alpha)\right)\frac{1}{k^2+i\epsilon}.$$

We see that our gauge interpolates between the R gauge ($\xi \to \infty$) and the U gauge ($\xi \to 0$). For $\xi = 1$, we recover the 't Hooft-Feynman gauge, in which the vector-boson propagators are proportional to $g_{\mu\nu}$.

In the Weinberg model (and also in the model of Lee and of Prentki and Zumino) we have another gauge boson Z_μ. We fix the gauge for this boson by adding the following term to the Lagrangian:

$$\mathcal{L}^{c\prime} = -\frac{\eta}{2}\left(\partial_\mu Z^\mu + \frac{Gv}{\eta}\chi\right)^2, \qquad G = (g^2+g'^2)^{1/2} \tag{2.11}$$

where η is a parameter that can vary over the range

$$0 \leqslant \eta \tag{2.12}$$

and G and χ are the coupling constant of Z_μ and the corresponding unphysical neutral scalar boson, respectively. We must also modify the last term $-i\,\mathrm{Tr}\ln[1+g\mathcal{G}\gamma]$ in Eq. (2.4) accordingly. The propagators for Z_μ and χ are given by

$$Z_\mu:\ -i\frac{[g_{\mu\nu} - k_\mu k_\nu(1-\eta)/(M_Z{}^2-\eta k^2)]}{k^2-M_Z{}^2+i\epsilon}, \tag{2.13}$$

$$\chi:\ i\frac{1}{k^2-(1/\eta)M_Z{}^2+i\epsilon}, \tag{2.14}$$

where $M_Z \equiv Gv$.

In an S-matrix element, the pole at $k^2 = M^2/\xi$ of the vector-boson propagator is canceled by the similar pole at M^2/ξ of the $s^\pm$ propagator. Neither of the scalar particles implied by these poles are physical. (In the R-gauge formulation the $s^\pm$ are the would-be Goldstone fields.) In fact, the couplings of $s^\pm$ to other particles can be determined based on the above considerations. As an example, let us determine the coupling of s^- to the $e\nu$ pair. We write the coupling of $W^-_\mu[\equiv(1/\sqrt{2})(B^1_\mu + iB^2_\mu)]$ to the $e\nu$ pair as

$$\mathcal{L}_{We\nu} = g\bar{e}(1-\gamma_5)\nu W^-_\mu.$$

Now consider the T-matrix element for the process

$$e(p)+\nu(q) \to \nu(q')+e(p').$$

To lowest order, the W exchange gives

$$(ig)^2(-i)[\bar{e}(p')\gamma^\mu(1-\gamma_5)\nu(q)][\bar{\nu}(q')\gamma^\lambda(1-\gamma_5)e(p)]\left[\left(g^{\mu\lambda} - \frac{k^\mu k^\lambda}{M^2}\right)\frac{1}{k^2-M^2} + \frac{k^\mu k^\lambda}{M^2}\frac{1}{k^2-M^2/\xi}\right],$$

where $k = p'-q = p-q'$, and we have used the vector-boson propagator of Eq. (2.10). The pole term at $k^2 = M^2/\xi$,

$$\left(\frac{igm_e}{M}\right)^2(-i)[\bar{e}(p')(1-\gamma_5)\nu(q)][\bar{\nu}(q')(1+\gamma_5)e(p)]\frac{1}{k^2-M^2/\xi}, \qquad M^2 = (gv)^2$$

must be canceled by the s^- exchange contribution. This requires the $s^- e\nu$ coupling to be

$$\mathcal{L}_{s^-e\nu} = \pm\frac{m_e}{v}\bar{e}(1-\gamma_5)\nu s^-.$$

(The sign ambiguity is superficial, since the sign of v is indeterminate. Once a definite sign convention is made here, all other couplings are uniquely determined.)

We note that the above cancellation is one of the consequences of the following two fundamental relations:

(i) The S matrix is gauge-independent, namely

$$\frac{\partial S}{\partial \xi} = 0.$$

(ii) The propagator $D_{\mu\nu}$ for W_μ and the propagator D for $s^\pm$ satisfy the identity

$$\frac{\partial}{\partial\xi}D_{\mu\nu}(k) \equiv -\frac{k_\mu k_\nu}{M^2}\frac{\partial}{\partial\xi}D(k).$$

III. MAGNETIC FORM FACTOR OF THE MUON

A. Weak Correction to the Magnetic Moment of the Muon

In a unified theory of weak and electromagnetic interactions, one-loop contributions to the anomalous magnetic moment a_μ of the muon are formally of order α, whether they derive from photon exchange or weak-vector-boson exchanges. The anomalous magnetic moment to this order can be written as

$$a_\mu = \frac{\alpha}{2\pi}\left[1+\left(\frac{\mu}{M}\right)^2 f\left(\left(\frac{\mu}{M}\right),\left(\frac{m_{Y^0}}{\mu}\right)\right)\right], \tag{3.1}$$

where μ is the muon mass, M is the W-boson mass, f is a function of the mass ratios (μ/M) and (m_{Y^0}/μ), and m_{Y^0} is the mass of a neutral heavy lepton Y^0, that might exist in such a theory. The second term in Eq. (3.1) is *in magnitude* of order $(\alpha/M^2)\mu^2 \sim G_F\mu^2$ and we shall call it the weak correction to a_μ and denote it by

$$a_\mu^W = \frac{\alpha}{2\pi}\left(\frac{\mu}{M}\right)^2 f\left(\left(\frac{\mu}{M}\right),\left(\frac{m_{Y^0}}{\mu}\right)\right). \tag{3.2}$$

In addition, there are contributions of massive Higgs scalar bosons in such a theory to a_μ. However, they are of order $(\mu/m_\phi)^2$ compared to Eq. (3.2), where m_ϕ is a typical Higgs scalar mass, and since the masses of these scalars are presumably very large, we shall ignore them in the following discussion.

The weak correction to a_μ^W has previously been computed by several authors[8] in the U gauge. In this gauge, the electromagnetic vertex of the muon is quadratically divergent, so that its separation into the electric and anomalous (Pauli) magnetic form factors is ambiguous. As a consequence, one finds that a_μ^W computed in this gauge depends on the way the internal momentum is routed in a diagram, even though it is finite.

In the R_ξ gauge, the electromagnetic vertex of the muon is only logarithmically divergent, and there is no such ambiguity in evaluating a_μ^W. In order to verify the gauge invariance we have evaluated it in three different gauges: $\xi=\infty$, 1, and 0. In this section we will present the results of these calculations. In Appendix A, we will present a general proof that the value of a_μ^W is independent of ξ. In the following we shall refer to the result obtained in the limit $\xi\to 0$ after the Feynman integration as the U-gauge result. For those diagrams involving unphysical scalars in this particular case, the limit $\xi\to 0$ and the integration commute. This explains why the procedure used by Jackiw and Weinberg[8] and by Bars and Yoshimura,[8] of replacing the vector propagator (1.3) by the regularized one, (1.1), yielded the correct result, even though this replacement *per se* is not a gauge-invariant procedure.

Our results are given for three different models. These are the model of Weinberg[9] based on SU(2)×U(1), that of Georgi and Glashow[10] based on O(3), and that of Lee and Prentki and Zumino (LPZ) based on O(3) ×O(2). The diagrams shown in Figs. 1 and 2 contribute to the models given by Weinberg and LPZ, and the diagrams shown in Figs. 1 and 3 contribute to the model of Georgi and Glashow. In these figures s^- and χ are unphysical scalars and Y^0 is a neutral lepton. For the purpose of illustration let us evaluate the diagram shown in Fig. 2b. It gives the contribution

$$-e\frac{\mu^2}{M_Z^2}\frac{(g^2+g'^2)}{4}\int\frac{d^4k}{(2\pi)^4}\frac{\bar\mu(p')\gamma_5(\not p'-\not k+\mu)\gamma_\mu(\not p-\not k+\mu)\gamma_5\mu(p)}{[(p-k)^2-\mu^2][(p'-k)^2-\mu^2][k^2-M_Z^2/\eta]}, \tag{3.3}$$

where p and p' are the incident and final muon momenta, respectively, and k is the internal momentum of χ. We separate this expression into the charge and anomalous magnetic form factors:

$$-ie\bar\mu(p')\left[F_1(q^2)\gamma_\mu+F_2(q^2)\frac{i}{2\mu}\sigma_{\mu\nu}q^\nu+\text{parity-violating terms}\right]\mu(p),\quad q=p'-p. \tag{3.4}$$

After the k integration $F_2(q^2)$ of Eq. (3.4) is found to be

$$F_2(q^2) = -\left(\frac{g\mu^2}{4M^2}\right)\frac{\mu^2}{8\pi^2}\int_0^1 dx\int_0^{1-x}dy\frac{(x+y)^2}{\mu^2(x+y)^2-q^2xy+(M_Z^2/\eta)(1-x-y)}. \tag{3.5}$$

Integrations over x and y yield the desired result in the gauge characterized by η. For example, to obtain the R-gauge result, we take the limit $\eta \to \infty$. Then

$$a_\mu^w[\text{Fig. 2(b)}]_R = -\frac{G_F\mu^2}{8\pi^2\sqrt{2}}. \tag{3.6}$$

To get from Eq. (3.3) to Eq. (3.6) we have used the relations

$$g^2 + g'^2 = g^2\frac{M_Z^2}{M^2} \text{ and } \frac{g^2}{8M^2} = G_F/\sqrt{2}.$$

To obtain U-gauge result, we let $\eta \to 0$ and we find that the diagram does not contribute:

$$a_\mu^w[\text{Fig. 2(b)}]_U = 0. \tag{3.7}$$

Finally in a 't Hooft–Feynman gauge, we let $\eta = 1$ and we see that

$$a_\mu^w[\text{Fig. 2(b)}]_{\text{'t H-F}} = O(\mu^2/M_Z^2). \tag{3.8}$$

Details of all other diagrams can be found in Appendix A. Tables I, II, and III give contributions from Figs. 1, 2, and 3, respectively. For example, the contribution of Fig. 2(b) can be found in Table II, in the column labeled Fig. 2(b). We neglected terms of order μ/M. In Table III, terms of order m_{Y^0}/M, μ/m_{Y^0} were also neglected. It is amusing to see that individual diagrams are quite gauge-dependent, but, as they must, the diagrams always add up to give a gauge-independent result not only to the leading order in $(\mu/M)^2$, $(\mu/m_Y)^2$, but to all orders. (See Appendix A.)

To obtain the result of Weinberg's model, we add the results of Tables I and II:

$$a_\mu^w = \frac{G_F\mu^2}{8\pi^2\sqrt{2}}\{\tfrac{10}{3} + \tfrac{1}{3}[(3 - 4\cos^2\theta)^2 - 5]\}. \tag{3.9}$$

Note that this result is the same as that obtained previously in the U gauge by the ξ-limiting regularization.[8]

The result for the model proposed by LPZ can be obtained from that of Weinberg's model by merely changing the definition for the coupling constants. We obtain

$$a_\mu^w = \frac{G_F\mu^2}{8\pi^2\sqrt{2}}\left(\frac{10}{3} - \frac{4}{3}\frac{u^2+v^2}{u^2}(1 + \sin^2\theta\cos^2\theta)\right), \tag{3.10}$$

where θ is related to the physical quantities by

$$\frac{G_F}{\sqrt{2}} = \frac{e^2}{4M^2\sin^2\theta} \text{ and } \frac{u^2+v^2}{u^2} = 2\frac{M^2}{M_Z^2}(1 + \tan^2\theta). \tag{3.11}$$

u and v are the vacuum expectation values of Higgs scalars in this model. In Eq. (3.10), a_μ^w can be about the same order of magnitude as that of Eq. (3.9) if $(M^2/M_Z^2)\tan^2\theta \sim O(1)$. In that case both of these models predict the weak correction to the muon magnetic moment to be of order 10^{-8}, below the experimental detectability. In any case, the strong-interaction correction a_μ^s has been estimated from the colliding-beam experiment to be of order $(6.5 \pm 0.5) \times 10^{-8}$,[15] so that the latter seems to be bigger than the former.

The Georgi-Glashow model receives contributions from Figs. 1 and 2. The results of the calculation in the special case $m_{Y^0} \ll M$ are given in Table III and Table I. Note that Fig. 3 gives contributions which are much larger than the previous two models by the factor m_{Y^+}/μ. We will thus concentrate our attention on the contributions of order $G_F\mu m_{Y^0}$ in Fig. 3. Evaluating for arbitrary value of m_{Y^0}/M we obtain

$$a_\mu^w = -\frac{G_F(m_{Y^+} + \mu)\mu}{8\sqrt{2}\pi^2\sin^2\theta}\left[1 + \frac{6}{(1-y)^3}(\tfrac{1}{2} - 2y + \tfrac{3}{2}y^2 - y^2\ln y)\right], \tag{3.12}$$

where $y = m_{Y^0}^2/M^2$ and $\sin^2\theta = 4G_F M^2/\sqrt{2}e^2$. Note also the relation

$$2m_{Y^0}\cos\theta = m_{Y^+} + \mu.$$

In Fig. 4 we have plotted the prediction of Eq. (3.12). Two sets of curves correspond to contours of constant a_μ^w and m_{Y^0} at various values of m_{Y^+} and M. The present status on experimental and theoretical uncertainties in a_μ places the limit[15]

$$2 \times 10^{-8} \le a_\mu^w + a_\mu^s \le 6 \times 10^{-7}, \tag{3.13}$$

where a_μ^s corresponds to the hadronic correction to the muon magnetic moment. Using the value for a_μ^s quoted above, it seems quite safe to guess that

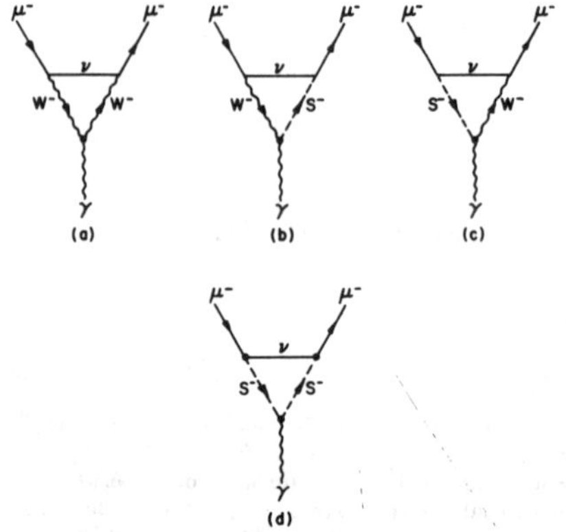

FIG. 1. W-boson and scalar-boson contributions to weak correction to the muon magnetic moment. These give contributions in all three models.

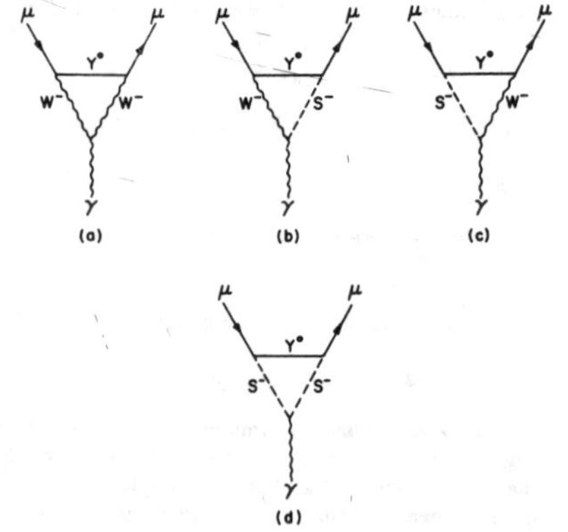

FIG. 3. Y^0-lepton contribution to the weak correction to the muon magnetic moment. These diagrams are for the Georgi-Glashow model.

$$-2\times 10^{-7} \leqslant a_\mu^w \leqslant 0\,.$$

If we further demand, for example, that $m_{Y^+} \geqslant 0.5$ GeV, then Fig. 4 readily gives the allowed range of m_{Y^0} and M. The generous lower limit $-2\times 10^{-7} = a_\mu^w$ gives $m_{Y^+} \leqslant 5$ GeV. If we use $|a_\mu^w| \leqslant +1.1 \times 10^{-7}$, we get $m_{Y^+} \leqslant 2$ GeV. In these estimates we assume $m_{Y^+} m_\mu / M_\Psi{}^2 \ll 1$, where Ψ is the physical scalar boson. A charged heavy lepton of this mass range can be detected in the near future. A pair of Y^+ and Y^- can be produced in reactions such as

$$\gamma + (Z) \to Y^+ + Y^- + (Z)$$

or

$$e^+ + e^- \to Y^+ + Y^-\,.$$

The detection of coincident $e^-\mu^+$ from the decays

$$Y^- \to \bar\nu_\mu + \nu_e + e^-\,,$$

$$Y^+ \to \bar\nu_\mu + \nu_\mu + \mu^+$$

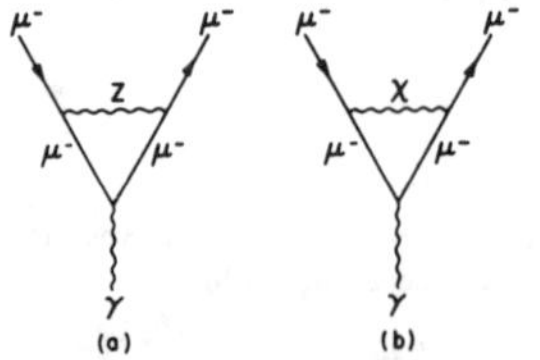

FIG. 2. Z- and χ-boson contribution to the weak correction to the muon magnetic moment. These diagrams are for Weinberg's model. Similar sets of diagrams exist for the LPZ model.

is a signature of the Y^+Y^- pair production.

It is important to recognize that the a_μ^w in Eq. (3.12) does not vanish even in the limit $m_{Y^0} \to \infty$. If one performs a naive U-gauge calculation the first term in Eq. (3.12) is absent and $a_\mu^w \to 0$ for $m_{Y^0} \to \infty$ (i.e., we can make a_μ^w arbitrarily small by letting m_{Y^0} be large).

B. Asymptotic Behavior of $F_2(q^2)$

In this subsection we discuss the behavior of the Pauli magnetic form factor $F_2(q^2)$ as $|q^2| \to \infty$. We caution the reader that $F_2(q^2)$ for $q^2 \neq 0$ is not an on-shell S-matrix element (i.e., not measurable) and is not invariant under non-Abelian gauge group (i.e., depends on the gauge). Figure 5(a) gives a process in which $F_2(q^2)$ is relevant. But Fig. 5(b) is of the same order. We obtain the gauge-invariant answer only when Figs. 5(a) and 5(b) and all other diagrams of the same order are added. $F_2(q^2=0) = a_\mu^w$ is available for experimental measurements only because of the pole due to the photon propagator. Still, the knowledge of the asymptotic behavior of $F_2(q^2)$, though gauge-dependent, is important in the question of renormalizability when diagrams of the type Fig. 5(a) are inserted in more complicated diagrams.

Our conclusion is that $F_2(q^2) \to 0$ as $|q^2| \to \infty$ in all gauges except the U gauge, i.e., for all combinations $\eta \neq 0$, $\xi \neq 0$. This can be easily seen, at least for off-mass-shell muons, as follows: For $\eta \neq 0$, $\xi \neq 0$, the triangle diagram that we consider has the degree of divergence at most zero. Thus due to the

TABLE I. Contributions of diagrams shown in Fig. 1. To obtain the answer these numbers should be multiplied by $G_F\mu^2/8\pi^2\sqrt{2}$.

Gauge \ Diagram	Fig. 1(a)	Fig. 1(b) + 1(c)	Fig. 1(d)	Total
U gauge	$\frac{10}{3}$	0	0	$\frac{10}{3}$
't Hooft-Feynman gauge	$\frac{7}{3}$	1	0	$\frac{10}{3}$
R gauge	$\frac{4}{3}$	1	1	$\frac{10}{3}$

kinematical factor, $\sigma_{\mu\nu}q_\nu$, the integral for $F_2(q^2)$ has the degree of divergence -1. Therefore, by Weinberg's theorem,[16] $F_2(q^2) \leq O(1/\sqrt{q^2})$.

We have also done the calculation for the on-mass-shell muon amplitude and verified that $F(q^2) \to 0$ as $|q^2| \to \infty$ in all gauges except the U gauge in the Weinberg model. In order to obtain the result for the U gauge, we let $\xi \to 0$ and then let $q^2 \to -\infty$. The result is (for the Weinberg model)

$$F_2(q^2) \to \frac{G_F\mu^2}{\sqrt{2}8\pi^2}\left(4 - \frac{\mu^2}{M^2}\right)\ln(-q^2)$$

$$+\text{constant for } q^2 \to -\infty\,.$$

These results indicate the gauge dependence of the off-shell amplitude. In particular, for the renormalizable gauge (i.e., for $\xi \neq 0$), $F_2(q^2)$ shows a manifestly renormalizable behavior. On the other hand, $F(q^2)$ in the U gauge exhibits a divergent behavior at $q^2 = -\infty$. As remarked in Ref. 17, the logarithmic growth of $F_2(q^2)$ for large q^2 does not necessarily imply any trouble with S-matrix elements for physical processes. When all diagrams of the same order for a physical process are added the bad behavior of $F_2(q^2)$ can be canceled by those of other diagrams.

The dispersion relation for F_2 in the U gauge requires a subtraction (which cannot be determined *a priori*) while its absorptive part may be computed by the standard Landau-Cutkosky rule. On the other hand, F_2 in the R_ξ gauge has an absorptive contribution from unphysical states, while it requires no unknown subtraction.

Note added. After the completion of this paper, we received a preprint of Bardeen *et al.*,[18] in which they evaluate a_μ in the Weinberg model using the n regularization method of 't Hooft and Veltman.[19] Their answer agrees with ours. Quinn and Primack[20] have computed a_μ for the Georgi-Glashow model. We appreciate Professor Quinn's explaining their result to us.

IV. STATIC CHARGE OF THE NEUTRINO

As in the case of Pauli magnetic form factor discussed in Sec. III, the notion of the electric form factor of a (muon) neutrino is an unphysical one in the present theory: The electric form factor, for nonzero momentum transfer, is not an element of the S matrix and, in the present theory, is neither gauge-invariant nor unitary for arbitrary ξ. However, the electric charge, i.e., the value of the electric form factor F_1 at zero momentum transfer, is an element of the S matrix and measurable. It must be zero if due care is exercised in evaluating Feynman integrals.

In the Georgi-Glashow model, there are altogether 10 diagrams contributing to the electric charge

TABLE II. Contributions of diagrams shown in Fig. 2. To obtain the answer these numbers should be multiplied by $G_F\mu^2/8\pi^2\sqrt{2}$.

Gauge \ Diagram	Fig. 2(a)	Fig. 2(b)	Total
U gauge	$\frac{1}{3}[(3-4\cos^2\theta)^2-5]$	0	$\frac{1}{3}[(3-4\cos^2\theta)^2-5]$
't Hooft-Feynman gauge	$\frac{1}{3}[(3-4\cos^2\theta)^2-5]$	0	$\frac{1}{3}[(3-4\cos^2\theta)^2-5]$
R gauge	$\frac{1}{3}[(3-4\cos^2\theta)^2-5]$ $+1$	-1	$\frac{1}{3}[(3-4\cos^2\theta)^2-5]$

TABLE III. Contributions of diagrams shown in Fig. 3. To obtain the answer these numbers should be multiplied by $G_F m_{Y^+} \mu/2\pi^2\sqrt{2}\sin^2\theta$.

Gauge \ Diagram	Fig. 3(a)	Fig. 3(b) +3(c)	Fig. 3(d)	Total
U gauge	-1	0	0	-1
't Hooft-Feynman gauge	$-\frac{3}{4}$	$-\frac{1}{4}$	0	-1
R gauge	$-\frac{1}{2}$	$-\frac{1}{4}$	$-\frac{1}{4}$	-1

of the neutrino. In Fig. 6, we show five of them which involve internal muon lines. The other five are similar and involve internal Y^+ lines. We shall evaluate the Feynman integrals in the R gauge ($\xi=\infty$) for convenience. (We have also checked the ξ independence of our results.) The contribution of each of the five diagrams in Fig. 6 is as follows.

$$\text{(a):}\quad F_1^{(a)}(0) = -i\frac{3}{2}\left(\frac{g}{2}\right)^2 \int \frac{d^4k}{(2\pi)^4}\frac{1}{(k^2-M^2)^2} - i\frac{3}{2}\left(\frac{g}{2}\right)^2 \int \frac{d^4k}{(2\pi)^4}\frac{\mu^2}{k^2-\mu^2}\frac{1}{(k^2-M^2)^2},$$

$$\text{(b):}\quad F_1^{(b)}(0) = i\frac{3}{2}\left(\frac{g}{2}\right)^2 \int \frac{d^4k}{(2\pi)^4}\frac{1}{k^2}\frac{\mu^2}{k^2-\mu^2}\frac{1}{k^2-M^2},$$

$$\text{(c):}\quad F_1^{(c)}(0) = -i\frac{1}{2}\left(\frac{f_\mu}{2}\right)^2 \int \frac{d^4k}{(2\pi)^4}\frac{1}{k^2}\frac{1}{k^2-\mu^2}, \qquad (4.1)$$

$$\text{(d):}\quad F_1^{(d)}(0) = -i\frac{3}{2}\left(\frac{g}{2}\right)^2 \int \frac{d^4k}{(2\pi)^4}\frac{\mu^2}{(k^2-M^2)(k^2-\mu^2)^2},$$

$$\text{(e):}\quad F_1^{(e)}(0) = -i\frac{1}{2}\left(\frac{f_\mu}{2}\right)^2 \int \frac{d^4k}{(2\pi)^4}\frac{k^2-2\mu^2}{k^2(k^2-\mu^2)^2},$$

where μ and M are the masses of the muon and the W boson, and f_μ is the coupling constant of the scalar meson to the muon and neutrino:

$$f_\mu = g\mu/M.$$

In Eq. (4.1) we have written $F_1^{(a)}(0)$ as a sum of logarithmically divergent and convergent integrals. A simple computation shows that the sum of the second term of $F_1^{(a)}$, $F_1^{(b)}$, and $F_1^{(d)}$ is zero after the integration, so that *if* the sum $F_1^{(c)}+F_1^{(e)}$ vanishes, then the muon contribution to the electric charge of the neutrino is

$$-\frac{3}{2}\left(\frac{g}{2}\right)^2 \int \frac{d^4k}{(2\pi)^4}\frac{1}{(k^2-M^2)^2}$$

and is independent of the muon mass. The Y^+ contribution to the electric charge of the neutrino is exactly opposite to the above-mentioned μ contribution, so the net charge of the neutrino is zero as it should be.

Thus, the matter hinges entirely on whether the sum $F_1^{(c)}+F_1^{(e)}$ is zero. A naive evaluation of these

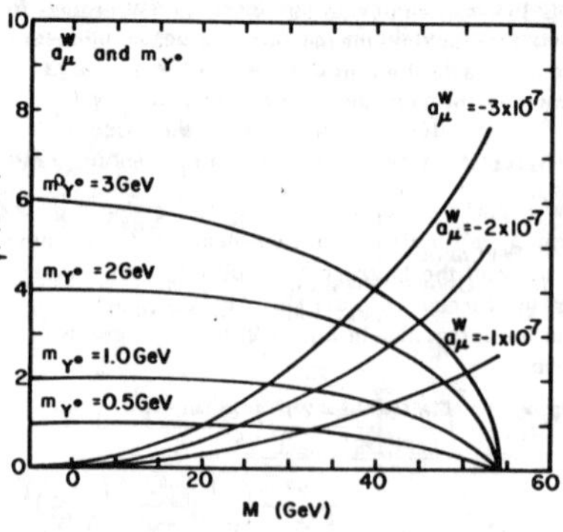

FIG. 4. Predictions of Eq. (3.12). Sets of contours correspond to constant a_μ^W and constant m_{Y^0} on the (m_{Y^+}, M) plane. If we take $-2\times10^{-7} \le a_\mu^W \le 0$, for example, the experimentally allowed region lies below the line labeled -2×10^{-7}. If we further take $m_{Y^+} \ge 0.5$ GeV, the allowed region is bounded from below. The upper bound for m_{Y^+}, in any case, is approximately 5 GeV.

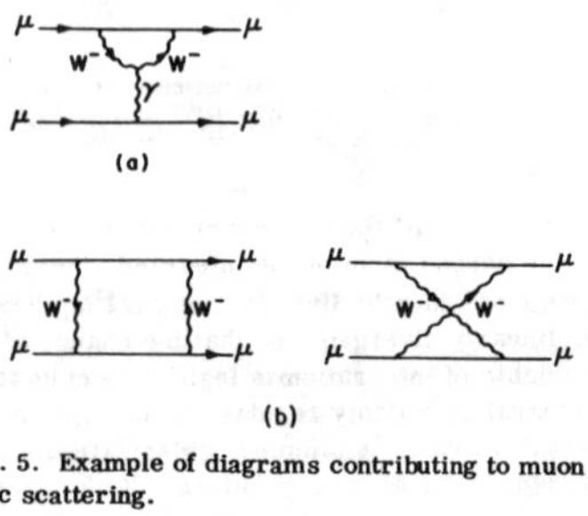

FIG. 5. Example of diagrams contributing to muon elastic scattering.

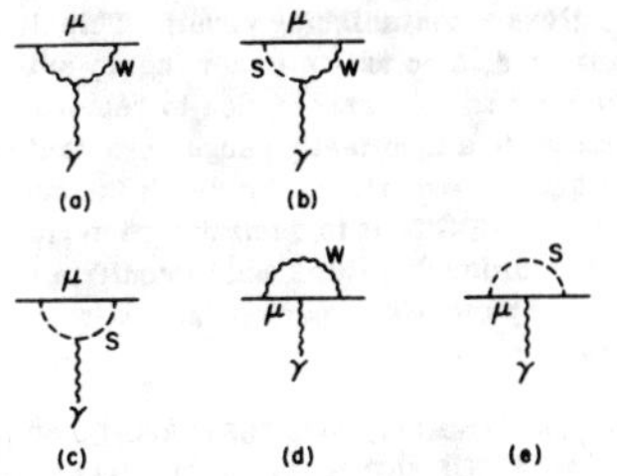

FIG. 6. Diagrams which contribute to the static charge of the neutrino in the Georgi-Glashow model.

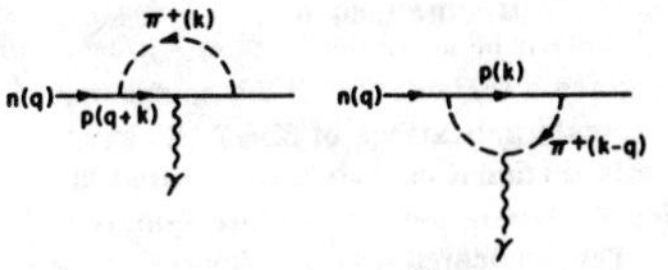

FIG. 7. Diagrams contributing to the neutron static charge.

two terms gives

$$F_1^{(c)}+F_1^{(e)}=\frac{f_\mu^{\,2}}{8}I,$$

$$I=i\int\frac{d^4k}{(2\pi)^4}\left[\frac{k^2-2\mu^2}{k^2(k^2-\mu^2)^2}-\frac{1}{k^2(k^2-\mu^2)}\right] \qquad (4.2)$$

$$=-\frac{1}{8\pi^2},$$

which is not zero; it is significant that the value of I is independent of the muon mass μ. Note that $F_1^{(c)}+F_1^{(e)}$ cannot be canceled by the similar contribution of Y^+, since the latter is proportional to $(f_{Y^+})^2$, f_{Y^+} being the coupling constant of the scalar meson to the muon-neutrino pair. That is, a naive evaluation of Feynman integrals leads to a nonzero electric charge of the neutrino.

The above paradox has nothing to do with the non-Abelian gauge invariance of the theory or the massless nature of the neutrino. The offending diagrams, Figs. 5(c) and 5(e), are characteristic of a theory in which fermions are coupled to a scalar meson. The sum of the two diagrams shown in Fig. 7 may be written as

$$F_\mu\sim-\frac{\partial}{\partial q^\mu}\int\frac{d^4k}{(2\pi)^4}\gamma_5\left[\frac{1}{(\not q+\not k-m_p)(k^2-m_\pi^{\,2})}-\frac{1}{(\not k-m_p)[(k-q)^2-m_\pi^{\,2}]}\right]\gamma_5, \qquad (4.3)$$

and if we can shift the contour of integration $k\to k+q$ in the second term of the integrand, the integral vanishes identically. The integral is, however, linearly divergent, so that the change of the variable of integration is legitimate only after the integral is suitably regularized in a gauge-invariant manner. A simple regularization scheme is to replace the pion propagators in (4.3) by

$$\frac{1}{k^2-m_\pi^{\,2}}\to\frac{1}{k^2-m_\pi^{\,2}}-\frac{1}{k^2-\Lambda_0^{\,2}},$$
$$\frac{1}{(k-q)^2-m_\pi^{\,2}}\to\frac{1}{(k-q)^2-m_\pi^{\,2}}-\frac{1}{(k-q)^2-\Lambda_0^{\,2}}. \qquad (4.4)$$

In this case, the naive evaluation of the charge of the neutron gives a result independent of the mass-$m_\pi^{\,2}$ internal-pion lines so that the regularization implied by Eq. (4.4) yields to a zero neutron charge. The result here is counter to the folklore which says that convergent Feynman integrals need not be regulated: If we perform the differentiation with respect to q prior to integration in Eq. (4.3), as one would to recover the original Feynman integral, then the integral becomes convergent and the conventional wisdom would say that it is not necessary to regulate the integral. What we have learned is that to *keep the charge of a neutral fermion zero, it is necessary to regularize Feynman integrals in a gauge-invariant way, even if the integrals are convergent.*

Let us return now to our problem. We can regularize the scalar-meson line in a gauge-invariant manner as in the σ model: We insert in the Lagrangian the regulator term

$$-\tfrac{1}{2}[(D_\mu\vec\phi')^2-\Lambda_0^{\,2}\vec\phi'^2]=-\tfrac{1}{2}\{[(\partial_\mu+g\vec B_\mu\times)\vec\phi']^2-\Lambda_0^{\,2}\vec\phi'^2\}$$

and replace the untranslated scalar fields ϕ by the sum $\phi+\phi'$ in all interaction terms.[21,22] This modification of the Lagrangian is clearly gauge-invariant (with respect to the non-Abelian gauge group), and the integral I in Eq. (4.2) is now regulated to read

$$I_{reg}=i\int\frac{d^4k}{(2\pi)^4}\left[\frac{k^2-2\mu^2}{(k^2-\mu^2)^2}\left(\frac{1}{k^2}-\frac{1}{k^2-\Lambda_0^{\,2}}\right)-\frac{k^2}{k^2-\mu^2}\left(\frac{1}{(k^2)^2}-\frac{1}{(k^2-\Lambda_0^{\,2})^2}\right)\right],$$

which is zero for all values of $\Lambda_0^{\,2}$. Thus, a gauge-invariant regularization of the Feynman integral does give the physically correct result $F_1^{(c)}+F_1^{(e)}=0$. We remind the reader that $F_1^{(c)}+F_1^{(e)}$ is non-Abelian gauge-invariant by itself. Thus the regularization procedure stated above is sufficient to remove the neutrino static charge for arbitrary gauge.

V. CONCLUSION

We have given a formulation of the convenient gauge for actual applications of SBGT (R_ξ gauge). Based on this formulation, we have verified the gauge independence of several simple S-matrix elements. This indicates that the ghost-eliminating mechanism is indeed working in examples we have considered. It is important to show that the gauge-independence properties of the S matrix are preserved at every stage of the renormalization program.

In our formulation, the finite part of the S matrix is uniquely determined. Results of our calculation of weak correction to the muon magnetic moment agree with U-gauge calculations with the ξ-limiting regularization procedure. An experimental implication of our results is that the charged heavy lepton in the Georgi-Glashow model is required to be small (of the order of 0.5–5 GeV). It is therefore worthwhile searching for this lepton in the existing accelerators.

A naive calculation of the neutrino static charge in SBGT gives a nonvanishing result. This difficulty, however, also exists in any theory with neutral spinor fields. A prescription to remove the static charge in a manifestly gauge-invariant (non-Abelian) manner was given. To check the self-consistency of SBGT, it is desirable to evaluate other lower-order diagrams and to confirm the absence of any other unexpected "anomalous" behaviors.

APPENDIX A: FORMULAS FOR THE MUON MAGNETIC FORM FACTOR AND PROOF OF THE GAUGE INDEPENDENCE OF THE g_μ FACTOR

In this appendix we give general formulas for the weak corrections to the anomalous magnetic moment of the muon based on our R_ξ gauge in Sec. II. We calculate the neutral-vector-meson contribution in the Weinberg model and the massive-neutral-lepton contribution in the Georgi-Glashow model. The Feynman rules are given in Appendix C. From these two results one can easily derive the magnetic moment for other schemes of lepton interactions.

1. Neutral Vector Meson in the Weinberg Model

We have two diagrams shown in Figs. 2(a) and 2(b) for the neutral vector- and scalar-meson contribution in the Weinberg model. Figure 2(b) has been discussed in Sec. III [see Eq. (3.5)].

For Fig. 2(a) we have

$$-e\bar{\mu}(p')\Big\}\gamma_\beta\left[a\left(\frac{1+\gamma_5}{2}\right)+b\left(\frac{1-\gamma_5}{2}\right)\right](\not{p}'+\mu)\gamma_\mu(\not{p}+\mu)\gamma_\alpha\left[a\left(\frac{1+\gamma_5}{2}\right)+b\left(\frac{1-\gamma_5}{2}\right)\right]\Big\}\mu(p)\,\frac{P^{\alpha\beta}(k)}{k^2-M_Z^{\,2}}, \tag{A1}$$

where

$$a=\frac{g'^2}{(g^2+g'^2)^{1/2}}, \qquad b=\frac{1}{2}\,\frac{g'^2-g^2}{(g^2+g'^2)^{1/2}}, \tag{A2}$$

$$P^{\alpha\beta}(k)=g^{\alpha\beta}-\frac{k^\alpha k^\beta(1-\eta)}{M_Z^{\,2}-\eta k^2}. \tag{A3}$$

The result is given by

$$\begin{aligned}F(q^2)=&-\frac{\mu^2}{8\pi^2}(a^2+b^2)\int dx\,dy\,\frac{(x+y-2)(x+y-1)}{Q^2+M_Z^{\,2}(1-x-y)}\\&-\frac{4\mu^2}{8\pi^2}ab\int dx\,dy\,\frac{(x+y-1)}{Q^2+M_Z^{\,2}(1-x-y)}\\&+\frac{\mu^2}{8\pi^2}\,\frac{(a-b)^2}{2M_Z^{\,2}}\Big\{\int dx\,dy[3(x+y)-2]\ln\left[\frac{Q^2+\Lambda(1-x-y)}{Q^2+M_Z^{\,2}(1-x-y)}\right]\\&-\int dx\,dy[Q^2(x+y)+2q^2xy]\left[\frac{1}{Q^2+M_Z^{\,2}(1-x-y)}-\frac{1}{Q^2+\Lambda(1-x-y)}\right]\Big\},\end{aligned} \tag{A4}$$

where

$$Q^2=\mu^2(x+y)^2-q^2xy, \qquad \Lambda=M_Z^{\,2}/\eta.$$

Note the relation

$$\frac{(a-b)^2}{M_Z^{\,2}}=\frac{1}{4}\,\frac{g^2}{M^2}, \tag{A5}$$

with M the mass of the W boson.

Based on Eqs. (3.5), (A4), and (A5) we prove the gauge-independence condition for the *physical* S-matrix element, $(\partial/\partial\eta)F_2(0)\equiv 0$ or equivalently

$$\frac{\partial}{\partial\Lambda}F_2(0)\equiv 0\,. \tag{A6}$$

Equation (A6) demands the following relation:

$$\int_0^1 dt\,t\int_{-1}^1 dz\left\{(3t-2)\frac{(1-t)}{\mu^2t^2+\Lambda(1-t)}-\frac{\mu^2t^3(1-t)}{[\mu^2t^2+\Lambda(1-t)]^2}+\frac{2\mu^2t^2(1-t)}{[\mu^2t^2+\Lambda(1-t)]^2}\right\}\equiv 0\,, \tag{A7}$$

where we changed the Feynman variables to

$$t=(x+y),\qquad z=(x-y)/t\,. \tag{A8}$$

Equation (A7) is indeed satisfied if one notices the relation

$$\frac{\partial}{\partial t}\left[\frac{1-t}{\mu^2t^2+\Lambda(1-t)}\right]=-\mu^2\frac{2t-t^2}{[\mu^2t^2+\Lambda(1-t)]^2}\,. \tag{A9}$$

Therefore the sum of the neutral vector and scalar meson contributions is gauge-independent, and consequently it is free of ghost contributions. Equation (A6) allows us to use any gauge we want to calculate the matrix element. In particular if one takes the limit $\eta\to 0$ in Eq. (A4) one recovers the results based on the ξ-limiting procedure (scalar-meson contribution vanishes in this limit).

In passing we note that the Higgs neutral scalar meson (which is independent of the gauge) in the Weinberg model (see Fig. 7) gives the magnetic form factor

$$-\frac{\mu^2}{8\pi^2}\left(\frac{g^2\mu^2}{4M^2}\right)\int dx\,dy\,\frac{(x+y)(x+y-2)}{\mu^2(x+y)^2-q^2(xy)+M_\phi{}^2(1-x-y)}\,. \tag{A10}$$

This gives a small contribution to $F_2(0)$ for $(\mu^2/M_\phi{}^2)\ll 1$.

2. Neutral Massive Lepton in the Georgi-Glashow Model

The neutral massive lepton in the Georgi-Glashow scheme contains four diagrams; see Figs. 3(a)–3(d). Figure 3(d) gives

$$\frac{e^3}{2M^2}[(\mu-m\cos\theta)^2+(\mu\cos\theta-m)^2]\bar{u}(p')\not{k}u(p)\frac{(l+l')_\mu}{(k^2-m^2)(l^2-\Lambda)(l'^2-\Lambda)}$$
$$+\frac{e^3}{M^2}(\mu-m\cos\theta)(\mu\cos\theta-m)\bar{u}(p')\not{k}u(p)\frac{(l+l')_\mu}{(k^2-m^2)(l^2-\Lambda)(l'^2-\Lambda)}\,, \tag{A11}$$

and the result is

$$F(q^2)=-\frac{e^2m\mu}{8\pi^2M^2}[(\mu-m\cos\theta)(\mu\cos\theta-m)]\int dx\,dy\frac{(1-x-y)}{f(\Lambda,\Lambda)}$$
$$-\frac{e^2\mu^2}{16\pi^2M^2}[(\mu-m\cos\theta)^2+(\mu\cos\theta-m)^2]\int dx\,dy\,\frac{(x+y)(1-x-y)}{f(\Lambda,\Lambda)}\,, \tag{A12}$$

where we defined

$$\begin{aligned}&\Lambda=M^2/\xi\,,\\&m=\text{mass of the neutral massive lepton }Y^0,\\&f(a,b)=Q^2+x(a-\mu^2)+y(b-\mu^2)+(1-x-y)m^2\,,\\&Q^2=\mu^2(x+y)^2-q^2xy\,.\end{aligned} \tag{A13}$$

Note that $F(q^2)$ in Eq. (A12) vanishes in the U-gauge limit, $\xi\to 0$ (or $\Lambda\to\infty$).

Figures 3(b) and 3(c) give rise to

$$-\frac{e^3}{2}[\mu(1+\cos^2\theta)-2m\cos\theta]\left\{\bar{u}(p')\not{k}\gamma_\alpha u(p)\frac{P_\mu^\alpha(l)}{(k^2-m^2)(l'^2-\Lambda)(l^2-\Lambda)}+\bar{u}(p')\gamma_\beta\not{k}u(p)\frac{P_\mu^\beta(l')}{(k^2-m^2)(l^2-\Lambda)(l'^2-\Lambda)}\right\}$$
$$+\frac{e^3}{2}m[m(1+\cos^2\theta)-2\mu\cos\theta]\left\{\bar{u}(p')\gamma_\alpha u(p)\frac{P_\mu^\alpha(l)}{(k^2-m^2)(l'^2-\Lambda)(l^2-\Lambda)}+\bar{u}(p')\gamma_\beta u(p)\frac{P_\mu^\beta(l')}{(k^2-m^2)(l^2-\Lambda)(l'^2-\Lambda)}\right\}, \tag{A14}$$

where

$$P^{\alpha}_{\mu}(l)=g^{\alpha}_{\mu}+\frac{k_\mu k^{\alpha}(1/\xi-1)}{l^2-\Lambda}\,.$$

The result is given by

$$F(q^2)=\frac{e^2\mu}{16\pi^2M^2}[\mu(1+\cos^2\theta)-2m\cos\theta]\Big\{2M^2\int dx\,dy\,\frac{y}{f(\Lambda,M^2)}+\int dx\,dy[3(x+y)-2]\ln\left(\frac{f(\Lambda,\Lambda)}{f(\Lambda,M^2)}\right)$$
$$-\int dx\,dy(x+y-1)[Q^2-\mu^2(x+y)]\left[\frac{1}{f(\Lambda,M^2)}-\frac{1}{f(\Lambda,\Lambda)}\right]\Big\} \tag{A15}$$

for the first group in Eq. (A14). The second group in Eq. (A14) gives

$$F(q^2)=\frac{1}{16\pi^2}\frac{e^2m\mu^2}{M^2}[m(1+\cos^2\theta)-2\mu\cos\theta]\int dx\,dy(x+y-1)^2\left[\frac{1}{f(\Lambda,M^2)}-\frac{1}{f(\Lambda,\Lambda)}\right]. \tag{A16}$$

Note that $F(q^2)$ in Eqs. (A15) and (A16) vanishes in the U-gauge limit $\Lambda\to\infty$.

Finally Fig. 3(a) gives the expression

$$-e^3m\cos\theta\bar{\mu}(p')\gamma_\beta\gamma_\alpha\mu(p)\frac{P^{\beta\tau}(l')V_{\tau\mu\sigma}P^{\sigma\alpha}(l)}{(k^2-m^2)(l'^2-M^2)(l^2-M^2)}-\frac{e^3}{2}(1+\cos^2\theta)\bar{\mu}(p')\gamma_\beta\not{k}\gamma_\alpha\mu(p)\frac{P^{\beta\tau}(l')V_{\tau\mu\sigma}P^{\sigma\alpha}(l)}{(k^2-m^2)(l'^2-M^2)(l^2-M^2)}, \tag{A17}$$

where

$$V_{\tau\mu\sigma}\equiv g_{\sigma\tau}(l+l')_\mu-l_\tau g_{\mu\sigma}-l'_\sigma g_{u\tau}+g_{\sigma\mu}q_\tau-g_{\tau\mu}q_\sigma \tag{A18}$$

and

$$q=l'-l\,.$$

The result is rather lengthy. The first group in Eq. (A17) gives rise to

$$F(q^2)=-\frac{1}{8\pi^2}\frac{e^2\mu m\cos\theta}{M^2}\Big\{3M^2\int dx\,dy\,\frac{x+y}{f(M^2,M^2)}$$
$$+3\int dx\,dy(x-y)\ln\frac{f(\Lambda,M^2)}{f(M^2,M^2)}$$
$$+\int dx\,dy\{(1-x+y)[Q^2-2\mu^2(x+y)+\mu^2]+2q^2y(x-1)\}\left[\frac{1}{f(M^2,M^2)}-\frac{1}{f(\Lambda,M^2)}\right]$$
$$-\frac{1}{2}\frac{q^2}{M^2}\int dx\,dy[3(x+y)-2]\ln\left(\frac{f(\Lambda,M^2)f(M^2,\Lambda)}{f(M^2,M^2)f(\Lambda,\Lambda)}\right)$$
$$+\frac{1}{2}\frac{q^2}{M^2}\int dx\,dy(x+y-1)[Q^2-2\mu^2(x+y)+\mu^2]$$
$$\times\left[\frac{1}{f(M^2,M^2)}-\frac{1}{f(M^2,\Lambda)}-\frac{1}{f(\Lambda,M^2)}+\frac{1}{f(\Lambda,\Lambda)}\right]\Big\} \tag{A19}$$

and the second group in Eq. (A17) gives

$$F(q^2)=\frac{1}{8\pi^2}\frac{e^2\mu^2}{4M^2}(1+\cos^2\theta)\Big\{2M^2\int dx\,dy\,\frac{(x+y)[2(x+y)+1]}{f(M^2,M^2)}$$
$$-2\int dx\,dy[4(x+y)^2-9(x+y)+2+6y]\ln\left(\frac{f(\Lambda,M^2)}{f(M^2,M^2)}\right)$$
$$+2\int dx\,dy\{\mu^2(x+y-1)^2[2y+(x+y)(x+y-1)]-q^2xy[2y+(1-x-y)(2-x-y)]\}$$
$$\times\left[\frac{1}{f(M^2,M^2)}-\frac{1}{f(\Lambda,M^2)}\right]$$
$$+4\int dx\,dy\;q^2(xy-y)\left[\frac{1}{f(M^2,M^2)}-\frac{1}{f(\Lambda,M^2)}\right]$$

$$+\left(\frac{q^2}{M^2}\right)\int dxdy[4(x+y)^2-9(x+y)+4]\ln\left(\frac{f(\Lambda,M^2)f(M^2,\Lambda)}{f(M^2,M^2)f(\Lambda,\Lambda)}\right)$$

$$-\left(\frac{q^2}{M^2}\right)\int dxdy(x+y-1)[(x+y-2)Q^2+\mu^2(x+y)]$$

$$\times\left[\frac{1}{f(M^2,M^2)}-\frac{1}{f(M^2,\Lambda)}-\frac{1}{f(\Lambda,M^2)}+\frac{1}{f(\Lambda,\Lambda)}\right]\Bigg\}. \tag{A20}$$

We would like to discuss the gauge-independence condition $(\partial/\partial\xi)F(0)\equiv 0$ or

$$\frac{\partial}{\partial\Lambda}F(0)\equiv 0. \tag{A21}$$

From Eqs. (A12), (A15), (A16), (A19), and (A20) we readily recognize that $F(q^2)$ consists of two groups, one of which is proportional to $m\cos\theta$ and the other proportional to $(1+\cos^2\theta)$. These two groups separately satisfy Eq. (A21). The proof of Eq. (A21) can be made as in Eqs. (A7)–(A9) by the repeated use of partial integration. We note that the contributions with a $f(\Lambda,M^2)$ factor and the contribution with a $f(\Lambda,\Lambda)$ factor in Eqs. (A12)–(A20) *separately* satisfy Eq. (A12). We do not write down this lengthy but straightforward proof.

Equation (A21) ensures the absence of the ghost contribution to the anomalous magnetic moment. Equation (A21) also allows us to use the most convenient gauge when we calculate numerical values. We also note that in the limit $\xi\to 0$ we recover the result based on the ξ-limiting process.

The neutrino contribution is obtained from the above result by setting $m=0$. The coupling constant should be adjusted according to the specific model one uses.

3. Large-q^2 Behavior of $F(q^2)$

From the above general results for $F(q^2)$ it is easy to see that all the contributions to $F(q^2)$ from scalar mesons vanish in the U-gauge limit, $\xi=0$ and $\eta=0$. They also vanish at $q^2=-\infty$ (i.e., large spacelike momentum transfer). It is also not difficult to see that the Z contribution in Eq. (A4) also vanishes at $q^2=-\infty$ independently of the value of η.

In the following we discuss the large-q^2 behavior of the W contribution in Eqs. (A19) and (A20). Those equations show that all the contributions to $F(q^2)$ vanish at $q^2=-\infty$ for $\xi=1$ (i.e., 't Hooft – Feynman gauge). However the gauge independence of the off-shell amplitude cannot be proved. We expect that $F(q^2)$ at $q^2=-\infty$ may depend on the gauge one chooses. We show this explicitly in the case of the neutrino contribution in the Weinberg model. We thus put $m=0$ in Eqs. (A19) and (A20).

We first note the following relations:

$$-q^2\int dxdy\frac{y}{Q^2+x(\Lambda-\mu^2)+y(M^2-\mu^2)}\underset{q^2\to-\infty}{\sim}\ln(-q^2)+\int_0^1 dt\ln\frac{t}{\mu^2t+M^2-\mu^2} \tag{A22}$$

and

$$-q^2\int dxdy\,\rho(x,y)\ln f(M^2,\Lambda)\underset{q^2\to-\infty}{\sim}-q^2\int_0^1 dt\,\rho(t)t\ln t$$

$$+(\Lambda+M^2-2\mu^2)\int_0^1 dt\,\rho(t)\ln(-q^2t)+\tfrac{1}{2}(\Lambda-M^2)\int_0^1 dt\,\rho(t)\ln\left[\frac{\mu^2t+M^2-\mu^2}{\mu^2t+\Lambda-\mu^2}\right]$$

$$-\tfrac{1}{2}\int_0^1 dt\,\rho(t)\left[\mu^2t-\mu^2+\frac{\Lambda+M^2}{2}\right]\ln[(\mu^2t+M^2-\mu^2)(\mu^2t+\Lambda-\mu^2)]$$

$$+(\Lambda+M^2-2\mu^2)\int_0^1 dt\,\rho(t), \tag{A23}$$

where

$$\rho(x,y)\equiv 4(x+y)^2-9(x+y)+4$$

and

$$\rho(t)\equiv 4t^2-9t+4.$$

Using these two relations in Eq. (A20) we can readily show that

$$F(q^2) \to 0 \quad \text{for } q^2 \to -\infty \text{ and } \xi \neq 0, \tag{A24}$$

$$F(q^2) \sim \frac{1}{8\pi^2}\frac{g\mu^2}{8M^2}\left(4-\frac{\mu^2}{M^2}\right)\ln(-q^2) + \text{cost} \quad \text{for } q^2 \to -\infty \text{ after } \xi = 0. \tag{A25}$$

Equations (A24) and (A25) indicate the gauge dependence of the off-shell amplitude.

We also point out an interesting large-q^2 behavior obtained if one uses (incorrect) regularization schemes other than the ξ-limiting procedure in the U gauge. The last two terms in Eq. (A20) show that the linear divergence in q^2 at large q^2 could exist. For the gauge-invariant calculation this linear divergence cancels. But if one uses other regularization schemes such as the "proper-time" (see Appendix B) or the "Pauli-Villars" regularization with a massive neutrino in the U gauge, this linear divergence indeed survives.

APPENDIX B: AMBIGUITIES IN THE U GAUGE

In this appendix we briefly review the ambiguity that Jackiw and Weinberg and also Bars and Yoshimura[8] encountered in their calculation of the muon g factor.

The logarithmic term in the parametric integral for $F(0)$ in the U-gauge limit (i.e., $\xi \to 0$) causes an ambiguity: Equation (A20) in Appendix A contains the logarithmic term

$$F(0) = -\frac{1}{8\pi^2}\frac{g^2\mu^2}{4M^2}\int dx\,dy[4(x+y)^2 - 9(x+y) + 2 + 6y]\ln\left[\frac{x}{\mu^2(x+y)^2 + (x+y)(M^2-\mu^2)}\right] + \text{other terms}. \tag{B1}$$

This is the correct answer. On the other hand if one regulates the neutrino propagator we get

$$F(0) = -\frac{1}{8\pi^2}\frac{g^2\mu^2}{4M^2}\int dx\,dy[4(x+y)^2 - 9(x+y) + 2 + 6y]\ln\left[\frac{1-x-y}{\mu^2(x+y)^2 + (x+y)(M^2-\mu^2)}\right] + \text{other terms}. \tag{B2}$$

If one first exponentiates the Feynman amplitude (a sort of "proper time") and performs a loop integral, the following result is obtained:

$$F(0) = -\frac{1}{8\pi^2}\frac{g^2\mu^2}{4M^2}\int dx\,dy[4(x+y)^2 - 9(x+y) + 2 + 6y]\ln\left[\frac{1}{\mu^2(x+y)^2 + (x+y)(M^2-\mu^2)}\right] + \text{other terms}. \tag{B3}$$

All these expressions give rise to different answers. This kind of ambiguity is absent in the general R-gauge calculation. Naive U-gauge calculations are plagued by this kind of ambiguity. Some existing proofs of cancellations of divergences in higher-order diagrams are based on the exponential parametrization of propagators that leads to (B3). While such a method is acceptable in establishing the absence of divergences, it will not provide a reliable finite part.

APPENDIX C: LAGRANGIANS AND FEYNMAN RULES

In this appendix we present Lagrangians for the existing models of leptons mentioned in the Introduction. We also write down necessary Feynman rules for our calculations in Secs. III and IV. In the R_ξ gauge discussed in Sec. II, we have unphysical scalars in the Lagrangian. The Feynman rules are therefore more complicated than those in the U-gauge limit. The Feynman rules for those unphysical scalars and also the relative signs for various amplitudes can be conveniently checked based on the gauge independence of the T-matrix element as we discussed in Sec. II.

1. Georgi-Glashow Model

This model is based on the group O(3). We have a triplet of leptons and also a singlet of neutral massive lepton. The mass is generated by a triplet of real scalars. A part of this Lagrangian has been given in Sec. II. (We follow the notation of Bjorken and Drell[16].) The total Lagrangian has the following form:

$$\mathcal{L} = \mathcal{L}_{\text{I}} + \mathcal{L}_{\text{II}} + \mathcal{L}_{\text{III}} - V,$$

$$\begin{aligned}\mathcal{L}_{\text{I}} \equiv{}& \bar{Y}^+(i\not\partial - m_{Y^+})Y^+ + \bar{Y}^0(i\not\partial - m_{Y^0})Y^0 \\ &+ \bar{\mu}(i\not\partial - m_\mu)\mu + \bar{\nu}_L i\not\partial\nu_L + e\bar{L}\gamma^\mu(\vec{T}\cdot\vec{B})_\mu L + e\bar{R}\gamma^\mu(\vec{T}\cdot\vec{B})_\mu R - \frac{e(m_{Y^0}\cos\theta - m_\mu)}{M}[\bar{L}(\vec{T}\cdot\vec{s})R + \text{H.c.}] \\ &- \frac{em_{Y^0}\sin\theta}{M}\{[\sin\theta\,\bar{Y}^0_L - \cos\theta\,\bar{\nu}_L][s^-Y^+_R + s^+\mu_R + \psi Y^0_R] + \text{H.c.}\},\end{aligned} \tag{C1}$$

where

$$L \equiv \begin{bmatrix} Y_L^+ \\ \cos\theta\, Y_L^0 + \sin\theta\, \nu_L \\ \mu_L \end{bmatrix}, \qquad R \equiv \begin{bmatrix} Y_R^+ \\ Y_R^0 \\ \mu_R \end{bmatrix}, \tag{C2}$$

with

$$\mu_R \equiv \left(\frac{1+\gamma_5}{2}\right)\mu\,, \qquad \mu_L \equiv \left(\frac{1-\gamma_5}{2}\right)\mu\,, \quad \text{etc.,} \tag{C3}$$

$$\vec{T}\cdot\vec{B}_\mu = \begin{bmatrix} A_\mu & -W_\mu^+ & 0 \\ -W_\mu^- & 0 & W_\mu^+ \\ 0 & W_\mu^- & -A_\mu \end{bmatrix}, \qquad \vec{T}\cdot\vec{s} = \begin{bmatrix} \psi & -s^+ & 0 \\ -s^- & 0 & s^+ \\ 0 & s^- & -\psi \end{bmatrix}, \tag{C4}$$

where

$$\begin{aligned} &\psi = \text{Higgs scalar}\,, \\ &s^\pm = \text{unphysical scalars}\,, \\ &G_F/\sqrt{2} = (e^2 \sin^2\theta/4M^2)\,. \end{aligned} \tag{C5}$$

There is a constraint:

$$2m_{Y^0}\cos\theta = m_{Y^+} + m_\mu\,. \tag{C6}$$

Note that the covariant derivative is given by

$$\nabla_\mu \psi \equiv [\partial_\mu + ig(\vec{T}\cdot\vec{B})_\mu]\psi\,,$$

with $g \equiv -e$.

$\mathcal{L}_{\rm II}$ is given by

$$\mathcal{L}_{\rm II} \equiv \tfrac{1}{2}|\partial_\mu\psi + ie[W_\mu^- s^+ - W_\mu^+ s^-]|^2 + |\partial_\mu s^+ + iMW_\mu^+ - ieA_\mu s^+ + ie\psi W_\mu^+|^2\,. \tag{C7}$$

The quadratic term of $\mathcal{L}_{\rm II}$ is given by

$$\mathcal{L}_{\rm II}^{\rm quad} = |\partial_\mu s^+|^2 + \tfrac{1}{2}(\partial_\mu\psi)^2 + M^2|W_\mu^+|^2 - iM[\partial_\mu s^+ W^{-\mu} - \partial_\mu s^- W^{+\mu}]\,. \tag{C8}$$

This $\mathcal{L}_{\rm II}^{\rm quad}$ suggests the following gauge term:

$$\mathcal{L}^c = -\frac{1}{2\alpha}(\partial_\mu A^\mu)^2 - \xi\left|\partial_\mu W^{+\mu} + \frac{iM}{\xi}s^+\right|^2; \tag{C9}$$

see also Sec. II. $\mathcal{L}_{\rm III}$ is given by

$$\mathcal{L}_{\rm III} = -\tfrac{1}{2}|\partial_\mu W_\nu^+ - \partial_\nu W_\mu^+ + ie[W_\mu^+ A_\nu - W_\nu^+ A_\mu]|^2 - \tfrac{1}{4}|\partial_\mu A_\nu - \partial_\nu A_\mu - ie[W_\mu^+ W_\nu^- - W_\mu^- W_\nu^+]|^2\,. \tag{C10}$$

The mass of $W_\mu^\pm$ is given in Eq. (C8).

The potential is given by

$$\begin{aligned} V(\phi) &= \tfrac{1}{2}\mu_0^2|\phi|^2 + \lambda|\phi|^4 \\ &= \tfrac{1}{2}m_\psi^2\psi^2 + \lambda[4v\psi(2s^+s^- + \psi^2) + (2s^+s^- + \psi^2)^2] + [\mu_0^2 + 4v^2\lambda][s^+s^- + v\psi]\,, \end{aligned} \tag{C11}$$

where

$$m_\psi^2 = \mu_0^2 + 12\lambda v^2\,, \tag{C12}$$

$\mu_0^2 < 0$, and we have the condition $\mu_0^2 + 4v^2\lambda = 0$. The field ϕ is the unshifted field, and it is expressed as

$$\phi = \begin{pmatrix} \phi^+ \\ \phi^0 \\ \phi^- \end{pmatrix} = \begin{pmatrix} s^+ \\ v+\psi \\ s^- \end{pmatrix}, \qquad \langle\phi\rangle_0 = \begin{pmatrix} 0 \\ v \\ 0 \end{pmatrix}. \tag{C13}$$

The gauge-compensating term for Eq. (C9) is given by (see also Sec. II)

$$-i\,\mathrm{Tr}\ln(1 + \mathcal{G}\cdot\gamma)\,, \tag{C14}$$

where

$$\begin{bmatrix} -(\partial^2+M^2/\xi-i\epsilon) & 0 & 0 \\ 0 & -\partial^2+i\epsilon & 0 \\ 0 & 0 & -(\partial^2+M^2/\xi-i\epsilon) \end{bmatrix} \mathcal{G}=\delta^4(x-y) \tag{C15}$$

and

$$\gamma\equiv\begin{bmatrix} ie\partial_\mu A^\mu-(eM/\xi)\psi & -ie\partial_\mu W^{+\mu}+(eM/\xi)s^+ & 0 \\ -ie\partial_\mu W^{-\mu} & 0 & ie\partial_\mu W^{+\mu} \\ 0 & ie\partial_\mu W^{-\mu}+(eM/\xi)s^- & -ie\partial_\mu A^\mu-(eM/\xi)\psi \end{bmatrix}\delta^4(x-y). \tag{C16}$$

The divergence in this matrix stands for an operator, e.g., $\partial_\mu A^\mu\equiv(\partial_\mu A^\mu)+A^\mu\partial_\mu$. The lowest-order contribution from Eq. (C14) is a ψ-vacuum tadpole diagram. For $\xi=\infty$ (Landau gauge), for example, Eq. (C14) becomes

$$-i\,\mathrm{Tr}\ln\left[1+\frac{1}{-\partial^2+i\epsilon}\partial_\mu(ie\vec{T}\cdot\vec{B}^\mu)\right]. \tag{C17}$$

2. Weinberg Model

The detailed form of this Lagrangian is found in Refs. 1 and 6. We briefly summarize it in the following:

$$\mathcal{L}\equiv\mathcal{L}_{\mathrm{I}}+\mathcal{L}_{\mathrm{II}}+\mathcal{L}_{\mathrm{III}}-V,$$

$$\mathcal{L}_{\mathrm{I}}=\bar{\nu}_L i\partial\!\!\!/\nu_L+\bar{\mu}(i\partial\!\!\!/-m_\mu)\mu-e\bar{\mu}A\!\!\!/\mu-\frac{g}{\sqrt{2}}[\bar{\nu}_L W\!\!\!\!/^+\mu+\mathrm{H.c.}]-\frac{G}{2}\bar{\nu}_L Z\!\!\!/\nu_L$$
$$+G\bar{\mu}Z\!\!\!/\left[\frac{\cos2\theta}{2}\left(\frac{1-\gamma_5}{2}\right)-\sin^2\theta\left(\frac{1+\gamma_5}{2}\right)\right]\mu-\frac{gm_\mu}{2M}[(\sqrt{2}\bar{\mu}\nu_L s^-+\mathrm{H.c.})+\bar{\mu}\mu\psi+\bar{\mu}i\gamma_5\mu\chi], \tag{C18}$$

where

$s^\pm$ and χ = unphysical scalar, ψ = Higgs scalar,

$$G=(g^2+g'^2)^{1/2},$$

$$\cos\theta=\frac{g}{G}=\frac{M}{M_Z},\qquad e=-\frac{gg'}{G},\qquad \frac{G_F}{\sqrt{2}}=\frac{g^2}{8M^2}. \tag{C19}$$

For notational convenience we defined the charge e with an extra minus sign.

$\mathcal{L}_{\mathrm{II}}$ is given by

$$\mathcal{L}_{\mathrm{II}}=\left|\partial_\mu s^++iMW_\mu^++\frac{i}{\sqrt{2}}gW_\mu^+s^0+i\left(-eA_\mu+\frac{G\cos2\theta}{2}Z_\mu\right)\right|^2+\left|\partial_\mu s^0-i\frac{M_Z}{\sqrt{2}}Z_\mu-i\frac{GZ_\mu}{2}s^0+\frac{i}{\sqrt{2}}gW_\mu^-s^+\right|^2, \tag{C20}$$

where

$$s^0\equiv\frac{1}{\sqrt{2}}(\psi+i\chi).$$

The quadratic part of $\mathcal{L}_{\mathrm{II}}$ is given by

$$\mathcal{L}_{\mathrm{II}}^{\mathrm{quad}}=|\partial_\mu s^+|^2+\tfrac{1}{2}[(\partial_\mu\psi)^2+(\partial_\mu\chi)^2]+M^2|W_\mu^+|^2$$
$$+\tfrac{1}{2}M_Z^2(Z_\mu)^2-iM[\partial_\mu s^+W^{-\mu}-\partial_\mu s^-W^{+\mu}]-M_Z\partial_\mu\chi Z^\mu. \tag{C21}$$

The gauge term is given as

$$\mathcal{L}^c=-\frac{1}{2\alpha}(\partial_\mu A^\mu)^2-\xi\left|\partial_\mu W^{+\mu}+i\frac{M}{\xi}s^+\right|^2-\frac{\eta}{2}\left(\partial_\mu Z^\mu+\frac{M_Z}{\eta}\chi\right)^2. \tag{C22}$$

The gauge-compensating term will be discussed later.

For $\mathcal{L}_{\mathrm{III}}$,

$$\mathcal{L}_{\mathrm{III}}=-\tfrac{1}{2}|\partial_\mu W_\nu^+-\partial_\nu W_\mu^++ie[W_\mu^+A_\nu-W_\nu^+A_\mu]-iG\cos^2\theta[W_\mu^+Z_\nu-W_\nu^+Z_\mu]|^2$$
$$-\tfrac{1}{4}|\partial_\mu A_\nu-\partial_\nu A_\mu-ie[W_\mu^+W_\nu^--W_\nu^+W_\mu^-]|^2-\tfrac{1}{4}|\partial_\mu Z_\nu-\partial_\nu Z_\mu+iG\cos^2\theta[W_\mu^+W_\nu^--W_\nu^+W_\mu^-]|^2. \tag{C23}$$

The masses for W_μ and Z_μ are given in Eq. (C21);

$$V=\lambda\left[s^+s^-+\left(s^0+\frac{v}{\sqrt{2}}\right)\left(\bar{s}^0+\frac{v}{\sqrt{2}}\right)-\frac{v^2}{2}\right]^2$$
$$=\tfrac{1}{2}m_\psi{}^2\psi^2+\lambda\{(2/\lambda)^{1/2}m_\psi\psi(s^+s^-+s^0\bar{s}^0)+[s^+s^-+s^0\bar{s}^0]^2\}. \tag{C24}$$

The gauge-compensating term is given by

$$-i\,\mathrm{Tr}\ln[1+\mathcal{G}\cdot\gamma], \tag{C25}$$

where

$$\begin{bmatrix} -(\partial^2+M^2/\xi-i\epsilon) & 0 & 0 & 0 \\ 0 & -(\partial^2+M^2/\xi-i\epsilon) & 0 & 0 \\ 0 & 0 & -\partial^2+i\epsilon & 0 \\ 0 & 0 & 0 & -(\partial^2+M^2/\eta-i\epsilon)\end{bmatrix}\mathcal{G}=\delta^4(x-y) \tag{C26}$$

and the matrix γ is defined by

$$\begin{bmatrix}\delta f^+\\ \delta f^-\\ \delta f^A\\ \delta f^Z\end{bmatrix}\equiv[\gamma]\begin{bmatrix}\Omega^+\\ \Omega^-\\ \Omega^A\\ \Omega^Z\end{bmatrix}\equiv[\gamma]\begin{bmatrix}\frac{1}{\sqrt{2}}(\Omega^1-i\Omega^2)\\ \frac{1}{\sqrt{2}}(\Omega^1+i\Omega^2)\\ \frac{g'}{g}\Omega^3+\frac{2g}{g'}\Omega^0\\ \Omega^3-2\Omega^0\end{bmatrix}, \tag{C27}$$

with

$$\delta f^\pm\equiv\mp ig\partial^\mu[(Z_\mu\cos\theta+A_\mu\sin\theta)\Omega^\pm]-\frac{gM}{2\xi}(\psi\pm i\chi)\Omega^\pm$$
$$\mp ie\cos\theta\,\partial^\mu[W_\mu^\pm(\Omega^A+\cot\theta\,\Omega^Z)]+\frac{eM\cos\theta}{2\xi}s^\pm\left(2\Omega^A-\frac{G}{e}\cos2\theta\,\Omega^Z\right),$$
$$\delta f^A\equiv iG\sin\theta\,\partial^\mu(W_\mu^-\Omega^+-W_\mu^+\Omega^-), \tag{C28}$$
$$\delta f^Z\equiv ig\partial^\mu(W_\mu^-\Omega^+-W_\mu^+\Omega^-)+\frac{GM_Z}{2\eta}(-\psi\Omega^Z+s^+\Omega^-+s^-\Omega^+).$$

Another way to take care of this compensating term is to introduce four auxiliary complex scalar fields[12] ϕ_a, $a=(+,-,A,Z)$. We add the following extra piece to the effective Lagrangian:

$$\tilde{\mathcal{L}}\equiv\phi_a^\dagger(\mathcal{G}^{-1}+\gamma)^{a,b}\phi_b. \tag{C29}$$

The ordinary perturbative treatment of $\tilde{\mathcal{L}}$ with an extra (−) sign for each closed loop of the ficticious scalar particles ϕ_a gives rise to Eq. (C25).

3. The Lepton Model of Lee, Prentki, and Zumino

The group structure of this model is close to that of the Weinberg model. We need two sets of scalar triplets to accomodate "quarks" in this model. However only one triplet of complex scalars is sufficient for the lepton model. We present here this simplified lepton model:

$$\mathcal{L}=\mathcal{L}_\mathrm{I}+\mathcal{L}_\mathrm{II}+\mathcal{L}_\mathrm{III}-V,$$
$$\mathcal{L}_\mathrm{I}=\bar{M}^+(i\not\partial-m_{M^+})M^++\bar{\nu}_Li\not\partial\nu_L+\bar{\mu}(i\not\partial-m_\mu)\mu$$
$$-\frac{gm_{M^+}}{M}[(\bar{L}\cdot\mathfrak{s})M_R^++\mathrm{H.c.}]-\frac{gm_\mu}{M}[(\bar{L}\cdot s)\mu_R+\mathrm{H.c.}]-e[\bar{\mu}\not A\mu-\bar{M}^+\not AM^+]-g[(\bar{\nu}\not W^+\mu-\bar{M}^+\not W^+\nu)+\mathrm{H.c.}]$$
$$+G\bar{\mu}\not Z\left[\cos^2\theta\left(\frac{1-\gamma_5}{2}\right)-\sin^2\theta\left(\frac{1+\gamma_5}{2}\right)\right]\mu-G\bar{M}^+\not Z\left[\cos^2\theta\left(\frac{1-\gamma_5}{2}\right)-\sin^2\theta\left(\frac{1+\gamma_5}{2}\right)\right]M^+, \tag{C30}$$

where

$$L \equiv \begin{bmatrix} M_L^+ \\ \nu_L \\ \mu_L \end{bmatrix}, \quad s \equiv \begin{bmatrix} s^{++} \\ s^+ \\ s^0 \end{bmatrix}, \quad \bar{s} \equiv \begin{bmatrix} \bar{s}^0 \\ s^- \\ s^{--} \end{bmatrix},$$

$$G = (g^2 + g'^2)^{1/2}, \quad e = (-)\frac{gg'}{G}, \quad \frac{G_F}{\sqrt{2}} = \frac{g^2}{4M^2}, \quad \cos\theta = \frac{g}{G}, \quad M = gu, \quad M_Z = \sqrt{2}Gu. \tag{C31}$$

For $\mathcal{L}_{II}$,

$$\mathcal{L}_{II} = |\partial_\mu s^{++} - igW_\mu^+ s^+ + i(-2eA_\mu + G\cos 2\theta Z_\mu)s^{++}|^2$$
$$+ |\partial_\mu s^+ + iMW_\mu^+ + igW_\mu^+ s^0 - igW_\mu^- s^{++} - i[eA_\mu + G\sin^2\theta Z_\mu]s^+|^2 + \left|\partial_\mu s^0 - i\frac{M_Z}{\sqrt{2}}Z_\mu + igW_\mu^- s^+ - iGZ_\mu s^0\right|^2. \tag{C32}$$

The quadratic part of $\mathcal{L}_{II}$ is given by

$$\mathcal{L}_{II}^{\text{quad}} = |\partial_\mu s^{++}|^2 + |\partial_\mu s^+|^2 + \tfrac{1}{2}[(\partial_\mu\psi)^2 + (\partial_\mu\chi)^2] + M^2|W_\mu^+|^2 + \tfrac{1}{2}M_Z^2(Z_\mu)^2 - iM[\partial^\mu s^+ W_\mu^- - \partial^\mu s^- W_\mu^+] - M_Z\partial_\mu\chi\cdot Z^\mu, \tag{C33}$$

where we defined

$$s^0 \equiv \frac{1}{\sqrt{2}}(\psi + i\chi),$$
$s^\pm$ and χ = unphysical scalars, (C34)

$s^{\pm\pm}$ and ψ = Higgs scalars.

The gauge term is the same as Eq. (C22) in the Weinberg model. The Yang-Mills Lagrangian is also given by Eq. (C23).

Finally we discuss the potential:

$$V = \mu_0(\vec{\xi}^* \cdot \vec{\xi}) + \frac{\nu}{4}(\vec{\xi}^* \times \vec{\xi})^2 + \lambda(\vec{\xi}^* \cdot \vec{\xi})^2$$
$$= \tfrac{1}{2}m_\psi^2\psi^2 + \tfrac{1}{2}m_{s^{++}}^2|s^{++}|^2 + 2\sqrt{2}\lambda u\psi(s^0\bar{s}^0 + s^+s^- + s^{++}s^{--}) + \lambda(s^0\bar{s}^0 + s^+s^- + s^{++}s^{--})^2$$
$$-\tfrac{1}{4}\nu[2\sqrt{2}u\psi(s^0\bar{s}^0 + s^{++}s^{--}) + 2us^+(s^0s^- - s^+s^{--}) + 2us^-(\bar{s}^0s^+ - s^-s^{++})$$
$$+ (s^0\bar{s}^0 - s^{++}s^{--})^2 + 2(\bar{s}^0s^+ - s^-s^{++})(s^0s^- - s^+s^{--})], \tag{C35}$$

where ξ is the unshifted complex scalar field

$$\xi = \begin{bmatrix} s^{++} \\ s^+ \\ u + s^0 \end{bmatrix}, \quad \langle\xi\rangle_0 = \begin{bmatrix} 0 \\ 0 \\ u \end{bmatrix}, \tag{C36}$$

$$m_\psi^2 = -2\mu_0 > 0, \quad m_{s^{++}}^2 = \nu u^2 > 0.$$

There is the following constraint:

$$\lambda = \frac{\nu}{4} + \left(-\frac{\mu_0}{2u}\right) = \frac{1}{4u^2}[m_\psi^2 + m_{s^{++}}^2]. \tag{C37}$$

We can make m_ψ and $m_{s^{++}}$ arbitrarily large. The gauge-compensating term is also similar to that of the Weinberg model. We do not discuss it here.

4. Feynman Rules

We summarize several Feynman rules we use in Secs. III and IV. Propagators for vector bosons and also for scalars are found in Sec. II. The fermion propagator has the standard form

$$\frac{i}{\not{p} - m}. \tag{C38}$$

In the following we give Feynman rules for the lepton models due to Georgi and Glashow (GG), Weinberg (W), and Lee, Prentki, and Zumino (LPZ). We write the Feynman rules for the R_ξ gauge of Sec. II.

All of these models give the identical Feynman rules for Figs. 8(a)–8(d). They are

$$8(\text{a}):\ (-ie)\bar{\mu}\gamma_\alpha\mu\,; \tag{C39}$$

$$8(\text{b}):\ (-ie)(l'+l)_\mu\,; \tag{C40}$$

$$8(\text{c}):\ (ie)[(g_{\alpha\beta}(l'+l)_\mu - l_\beta g_{\mu\alpha} - l'_\alpha g_{\mu\beta}) + (g_{\alpha\mu}q_\beta - g_{\beta\mu}q_\alpha)],\ \text{where } q=l'-l\,; \tag{C41}$$

$$8(\text{d}):\ (-ie)Mg_{\alpha\mu}\,. \tag{C42}$$

Note that we normalized the sign of the charge by Fig. 8(a). The magnetic form factor appears as a coefficient of $(-ie)\bar{\mu}(i\sigma_{\mu\nu}q^\nu/2\mu)\mu$. For other diagrams Feynman rules depend on the model. We just list them below:

$$8(\text{e}):\ \text{GG}:\ (ie\sin\theta)\bar{\nu}\gamma_\alpha\left(\frac{1-\gamma_5}{2}\right)\mu\,;$$
$$\text{W}:\ \left(-\frac{ig}{\sqrt{2}}\right)\bar{\nu}\gamma_\alpha\left(\frac{1-\gamma_5}{2}\right)\mu\,; \tag{C43}$$
$$\text{LPZ}:\ (-ig)\bar{\nu}\gamma_\alpha\left(\frac{1-\gamma_5}{2}\right)\mu\,;$$

$$8(\text{f}):\ \text{GG}:\ \left(\frac{ie\sin\theta\, m_\mu}{M}\right)\bar{\nu}\left(\frac{1+\gamma_5}{2}\right)\mu\,;$$
$$\text{W}:\ \left(\frac{-igm_\mu}{\sqrt{2}M}\right)\bar{\nu}\left(\frac{1+\gamma_5}{2}\right)\mu\,; \tag{C44}$$
$$\text{LPZ}:\ \left(\frac{-igm_\mu}{M}\right)\bar{\nu}\left(\frac{1+\gamma_5}{2}\right)\mu\,;$$

$$8(\text{g}):\ \text{GG}:\ (ie)\bar{Y}^0\left[\cos\theta\left(\frac{1+\gamma_5}{2}\right)+\left(\frac{1-\gamma_5}{2}\right)\right]\gamma_\alpha\mu\,; \tag{C45}$$

$$8(\text{h}):\ \text{GG}:\ \left(\frac{ie}{M}\right)\bar{Y}^0\left[(m_\mu\cos\theta - m_{Y^0})\left(\frac{1+\gamma_5}{2}\right) + (m_\mu - m_{Y^0}\cos\theta)\left(\frac{1-\gamma_5}{2}\right)\right]\mu\,; \tag{C46}$$

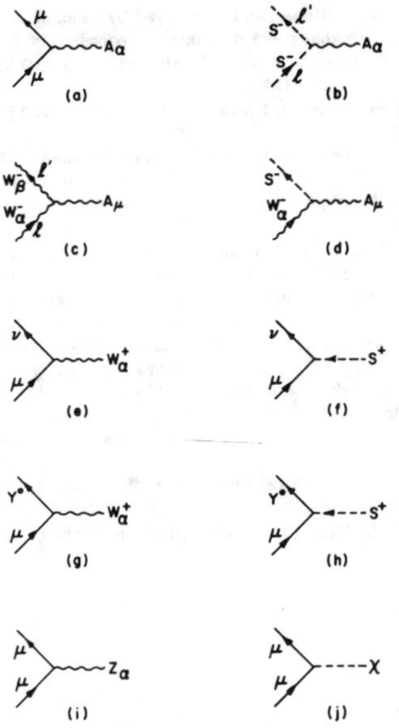

FIG. 8. Several vertex diagrams for lower-order calculations.

$$8(\text{i}):\ \text{W}:\ iG\bar{\mu}\left[\frac{\cos 2\theta}{2}\left(\frac{1+\gamma_5}{2}\right)-\sin^2\theta\left(\frac{1-\gamma_5}{2}\right)\right]\gamma_\alpha\mu\,; \tag{C47}$$
$$\text{LPZ}:\ iG\bar{\mu}\left[\cos^2\theta\left(\frac{1+\gamma_5}{2}\right)-\sin^2\theta\left(\frac{1-\gamma_5}{2}\right)\right]\gamma_\alpha\mu\,;$$

$$8(\text{j}):\ \text{W}:\ \left(-i\frac{gm_\mu}{2M}\right)\bar{\mu}i\gamma_5\mu\,; \tag{C48}$$
$$\text{LPZ}:\ \left(-i\frac{gm_\mu}{\sqrt{2}M}\right)\bar{\mu}i\gamma_5\mu\,.$$

*Work supported in part by the NSF under Contract No. NSF GP 32904X.

†On leave of absence from Institute for Theoretical Physics, SUNY, Stony Brook, New York 11790.

‡Operated by Universities Research Association, Inc., under contract with the United States Atomic Energy Commission.

[1]S. Weinberg, Phys. Rev. Letters 19, 1264 (1967); 27, 1688 (1972).

[2]G. 't Hooft, Nucl. Phys. B35, 167 (1971).

[3]T. D. Lee and C. N. Yang, Phys. Rev. 128, 885 (1962).

[4]The R_ξ gauge was formulated independently by Zinn-Justin (unpublished). The proof of renormalizability of SBGT in this gauge is presently being pursued by Zinn-Justin and one of the present authors (B.W.L). What is to be proved is that divergences in higher orders of

Green's functions can be removed by renormalization in such a way that the S matrix is independent of ξ.

[5]B. W. Lee and J. Zinn-Justin, Phys. Rev. D 5, 3121 (1972); 5, 3137 (1972); 5, 3155 (1972).

[6]A. Salam and J. Strathdee, Nuovo Cimento 11A, 397 (1972).

[7]S. Weinberg, Phys. Rev. Letters 27, 1688 (1972); S. Y. Lee, Phys. Rev. D 6, 1701 (1972); 6, 1803 (1972); T. W. Appelquist and H. R. Quinn, Phys. Letters 39B, 229 (1972).

[8]R. Jackiw and S. Weinberg, Phys. Rev. D 5, 2396 (1972); I. Bars and M. Yoshimura, *ibid.* 6, 374 (1972).

[9]Some of the numerical values for the weak correction to the muon magnetic moment have been previously evaluated without using the lepton models based on SBGT. T. Burnett and J. J. Levine, Phys. Letters 24, 467 (1967); S. J. Brodsky and J. Sullivan, Phys. Rev. 156, 1644 (1967); T. Kinoshita *et al.*, Phys. Rev. D 2, 910 (1970).

[10]H. Georgi and S. L Glashow, Phys. Rev. Letters 28, 1494 (1972).

[11]B. W. Lee, Phys. Rev. D 6, 1188 (1972); J. Prentki and B. Zumino, Nucl. Phys. B47, 99 (1972).

[11a]Y.-P. Yao, Phys. Ref. D (to be published).

[12]G. 't Hooft, Nucl. Phys. B33, 173 (1971).

[13]For details see, for example, Ref. 5.

[14]The physical scalar bosons are called Higgs scalars, and denoted by Ψ, as opposed to unphysical scalars which are denoted by s^*.

[15]For summary and review of the recent advances in this field, see S. Brodsky, in *Proceedings of the International Symposium on Electron and Photon Interactions at High Energies, 1971*, edited by N. B. Mistry (Cornell Univ. Press, Ithaca, N. Y., 1971).

[16]J. D. Bjorken and S. D. Drell, *Relativistic Quantum Fields* (McGraw-Hill, New York, 1965).

[17]G. Altarelli, N. Cabibbo, and L. Maiani, Phys. Letters 40B, 415 (1972).

[18]W. A. Bardeen *et al.*, Nucl. Phys. B46, 319 (1972).

[19]G. 't Hooft and M. Veltman, Nucl. Phys. (to be published).

[20]J. Primack and H. R. Quinn, Phys. Rev. D (to be published).

[21]J. L. Gervais and B. W. Lee, Nucl. Phys. B12, 627 (1969).

[22]S. L. Adler and W. A. Bardeen, Phys. Rev. 182, 1517 (1969).

WARD IDENTITIES AND CHARGE RENORMALIZATION OF THE YANG-MILLS FIELD

J.C.TAYLOR
Department of Theoretical Physics,
Oxford University, Oxford, England

Received 25 June 1971

Abstract: We derive, for the massless Yang-Mills field, a generalized Ward-Takahashi identity which implies that Z_1/Z_2 has a common value for the vector field and for the "fictitious" scalar field, thus ensuring that the renormalized charges are equal. The identity may also be used to avoid ambiguities in working with quadratically divergent integrals.

1. INTRODUCTION

The Feynman rules for the massless Yang-Mills [1] field are known [2–4]. They involve "fictitious scalar particles" which occur only in closed loops*.

The resulting theory** appears to be renormalizable, and this has been discussed in detail by 't Hooft [6]. He finds some Ward-Takahashi identities which must be respected in extracting the convergent parts. If this is done, he argues, the Ward identities are sufficient to guarantee unitarity.

However, 't Hooft tacitly assumes the renormalized charges*** of the vector and of the "fictitious" scalar particles to be equal, the bare charges being equal by construction. This is reminiscent of the equality of the renormalized electric charges of, say, the electron and the proton, which is a consequence of the relation $Z_1/Z_2 = 1$. In the Yang-Mills theory, therefore, we may seek an analogue of this relation, as a formal justification for equating the renormalized charges.

The relations found may be expressed in terms of the following definitions. Let

* In the Feynman gauge, Faddeev and Popov [3] give different rules from the authors of refs. [2, 4]. Presumably both sets of rules give the same S-matrix elements. In this paper, we refer to the simpler set of rules given by the majority of the authors.

** We leave aside all questions of infra-red divergences. For mention of these problems in the Yang-Mills case, see Weinberg [5].

*** The use of the work "charge" for the "fictitious" field is perhaps unjustified. We mean simply the constant appearing in the effective coupling. In fact, in the Landau gauge, the "fictitious" field is couplied as if it had charge $\frac{1}{2}g$ (with the conventions of ref. [6]).

the wave-function renormalization constants for the Yang-Mills field be Z_3^Y, for the "fictitious" spin-zero field be Z_3^F, and for a spin-$\frac{1}{2}$, isospin-$\frac{1}{2}$ matter field be Z_2^M; and let the vertex part renormalization constants be Z_1^Y, Z_1^F and Z_1^M respectively. Then we prove, first of all, that

$$Z_1^Y/Z_3^Y = Z_1^F/Z_3^F . \tag{1}$$

This implies the identity of the renormalized charges.

We also find that

$$Z_1^M/Z_2^M = Y/Z_3^F \tag{2}$$

where Y is a constant which is defined in terms of a Feynman integral not normally appearing in the theory. In the Landau gauge, it can be shown further that

$$Z_1^F = Y = 1 \quad \text{(Landau gauge)} . \tag{3}$$

It then follows from eqs. (1) and (2) that

$$Z_1^Y/Z_3^Y = Z_1^M/Z_2^M \tag{4}$$

and hence that the renormalized charges of the Yang-Mills and matter fields are equal. That this must be so follows from a general argument of Weinberg [7]. Therefore the eq. (4) must be true in any gauge, and so also

$$Z_1^F = Y . \tag{5}$$

However, we have not found a direct proof of relation (5) except in the Landau gauge.

Note that eq. (1) does *not* follow from Weinberg's paper [7], since the "fictitious particles" do not occur on external lines.

These results are derived in sect. 2 from a generalized Ward-Takahashi identity, which is based on 't Hooft's [5] methods. A proof of the Ward identities using path-integral methods is given in the appendix. We believe that this identity can be used to resolve ambiguities in dealing with quadratically divergent integrals, even *without* the use of a gauge-invariant regulator. This is argued in a little more detail in sect. 3.

2. THE GENERALIZED WARD-TAKAHASHI IDENTITIES

We are concerned here with a modification of an identity proved by 't Hooft [6], expressed in his eq. (6.12). We shall follow his combinatorial technique and use his graphical conventions. The results are common to any gauge so we use the general propagator

$$[g_{\mu\nu} + (\lambda - 1) k_\mu k_\nu / k^2] \, (k^2 + i\epsilon)^{-1} \,. \tag{6}$$

The Landau gauge is given by $\lambda = 1$*. The results may also be obtained by path-integral methods, as shown in the appendix.

Graphically, 't Hooft's Ward identity is expressed by fig. 1. Here continuous lines stand for Yang-Mills particles and dotted lines for "fictitious" ones. The symbol . . . . . = stands for the replacement of a polarization vector $e_\mu(k)$ by k_μ, and k^2 need not be zero. The symbol ——— denotes an external particle which is on-shell ($q^2 = 0$) but not transverse ($q \cdot e \neq 0$). The circle represents the set of all relevant Feynman graphs, and there may be other external lines provided that they are all on-shell and transverse.

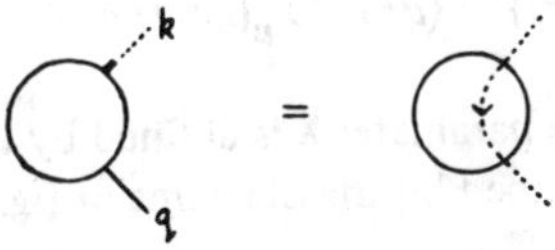

Fig. 1. 't Hooft's Ward identity.

't Hooft establishes fig. 1 in the renormalized theory. In order to deduce eq. (1) we have to consider the corresponding formal steps also in the unrenormalized theory. Fig. 1 must then be modified because of self-energy insertions on the external lines. We therefore separate the self-energy graphs explicitly and obtain, for $q^2 \neq 0$, fig. 2. Here the circles on external lines represent self-energy parts, which are excluded from the main circle. Where a "fictitious particle" line runs to the wall of a

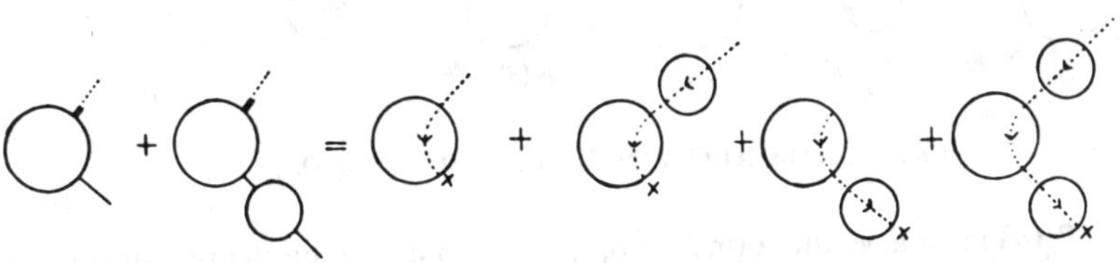

Fig. 2. Ward identity allowing for self-energy graphs. The symbol X denotes the action of the tensor $(q_\mu q_\nu - q^2 g_{\mu\nu})$.

* 't Hooft's proof of unitarity employs the Feynman gauge in the renormalized theory. It is perhaps worth mentioning however, that, since the renormalized value of λ is given by $\lambda_R = Z_3 \lambda$, $\lambda_R = 1$ does not imply $\lambda = 1$

circle but does not emerge, it means that the factor q_μ which normally accompanies an emerging "fictitious" line is absent. Thus there is a free Lorentz index μ at such a point. The crosses X represent the presence of the tensor $(q_\mu q_\nu - q_{\mu\nu} q^2)$, contracted into the free μ-index.

The proof of fig. 2 follows by the same methods as those with which 't Hooft proved fig. 1. In particular, the factors $(q_\mu q_\nu - q_{\mu\nu} q^2)$ come from eq. (4.3c) of ref. [6], when the q-line is an external, off-shell, non-transverse line. There are no self-energy insertions on the top line of the left-hand side of fig. 2 because the Yang-Mills self-energy part is proportional to $(k^2 g_{\lambda\rho} - k_\lambda k_\rho)$ (see eq. (3.8) of ref. [6]).

For the case in which there is just one other on-shell, transverse external line (momentum p), fig. 2 may be expressed

$$e^\sigma(p)\, k^\lambda\, \Gamma_{\lambda\sigma\mu}(k,p,q)\, [\{1-\Pi(q^2)\}^{-1}(g^{\mu\nu} - q^\mu q^\nu/q^2) + \lambda\, q^\mu q^\nu/q^2]$$

$$= \{1-P(k^2)\}^{-1} e^\sigma(p)\, [G_{\sigma\mu}(k,p,q)$$

$$+\, G_\sigma(k,p,q)\{1-P(q^2)\}^{-1}(q^2)^{-1} P_\mu(q)]\, (q^\mu q^\nu - q^2 g^{\mu\nu})\,, \tag{7}$$

where $p^2 = 0, p_\sigma e^\sigma(p) = 0$. The parameter λ is defined by eq. (6). The functions appearing in this equation are defined by the diagrams of fig. 3 as follows:

$$\text{(a) } \Gamma_{\lambda\sigma\mu}\,, \quad \text{(b) } \Pi(q^2)(q^2 g_{\mu\nu} - q_\mu q_\nu)\,, \quad \text{(c) } G_{\sigma\mu}\,,$$
$$\text{(d) } G_\sigma = q^\mu G_{\sigma\mu}\,, \quad \text{(e) } P_\mu\,, \quad \text{(f) } q^2 P(q^2) = q^\mu P_\mu(q^2)\,. \tag{8}$$

The obvious isotopic spin tensors are omitted from these definitions. All the graphs shown are proper graphs.

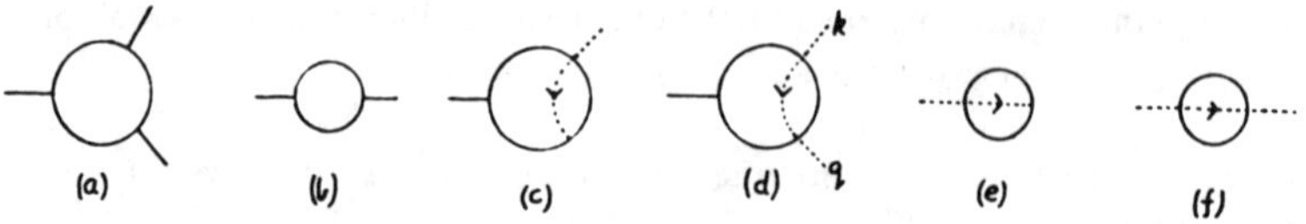

Fig. 3. Graphs defining the functions listed in eq. (8).

Since $P_\mu(q)$ must have the form $P(q^2) q_\mu$, the second term on the right-hand side of eq. (7) vanishes. This enables one to put $q^2 = 0$ there, obtaining

$$\lim_{q^2\to 0}\; e^\sigma(p)\, k^\lambda\, \Gamma_{\lambda\sigma\mu}(k,p,q)\, [\{1-\Pi(q^2)\}^{-1}(g^{\mu\nu} - q^\mu q^\nu/q^2) + \lambda\, q^\mu q^\nu/q^2]$$

$$= \{1-P(k^2)\}^{-1} e^\sigma(p)\, G_\sigma(k,p,q; q^2=0)\, q^\nu\,. \tag{9}$$

Note that in this equation it is $\Pi(q^2)$ and $P(k^2)$ that survive; so 't Hooft's relation, fig. 1, is applicable, provided it is understood that there are self-energy insertions on the bottom line on the left and on the top line on the right only.

To obtain the result (1) from eq. (9) all that is now required is to set $k^2 = 0$ and use the following definitions

$$\Gamma_{\lambda\sigma\mu}(k^2 = p^2 = q^2 = 0) = -igZ_1^Y\{g_{\sigma\mu}(p-q)_\lambda + g_{\mu\lambda}(q-k)_\sigma + g_{\lambda\sigma}(k-p)_\mu\}\,, \quad (10)$$

$$e^\sigma(p)\,G_\sigma(k^2 = p^2 = q^2 = 0) = -igZ_1^F e^\sigma(p) q_\sigma\,, \quad (11)$$

$$\Pi(0) = 1 - Z_3^Y\,, \qquad P(0) = 1 - Z_3^F\,. \quad (12)$$

We next prove that $Z_1^F = 1$ in the Landau gauge. The graphs contributing to G_σ may be split up as shown in fig. 4, where, in the last term, the "first" and "last" vertices on the "fictitious" line have been exposed. The symbol r denotes a variable of integration. The "last" vertex contributes a factor q, and the "first" a factor r, but in the Landau gauge r may be replaced by k. Therefore G_σ must have the form

$$-igq_\sigma + q^\nu k^\lambda F_{\sigma\nu\lambda}(k, p, q)\,. \quad (13)$$

Evaluating at $k^2 = p^2 = q^2 = 0$, the second term vanishes and so, with eq. (11), $Z_1^F = 1$.

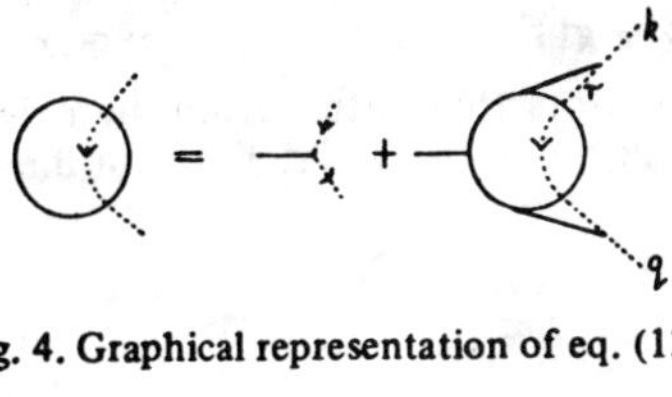

Fig. 4. Graphical representation of eq. (13).

We now go on to consider the interaction with a spin-$\frac{1}{2}$, isospin-$\frac{1}{2}$ matter field (ψ). The interaction is

$$g\bar{\psi}\gamma_\mu\tau_a\psi W_a^\mu\,.$$

In Feynman graphs, we shall write dashed lines for the spin-$\frac{1}{2}$ particles. The analogue of fig. 2 is now shown in fig. 5, where the symbol X represents the action of the matrix $(m - \gamma \cdot q)$. The graphs on the right-hand side of fig. 5 are not ones which normally occur in the theory, since the "fictitious" field does not couple directly to the matter field. Nevertheless, the contributions to fig. 5 are to be calculated as if there were a coupling

$$g\bar{\psi}\tau_a\psi\phi_a\,,$$

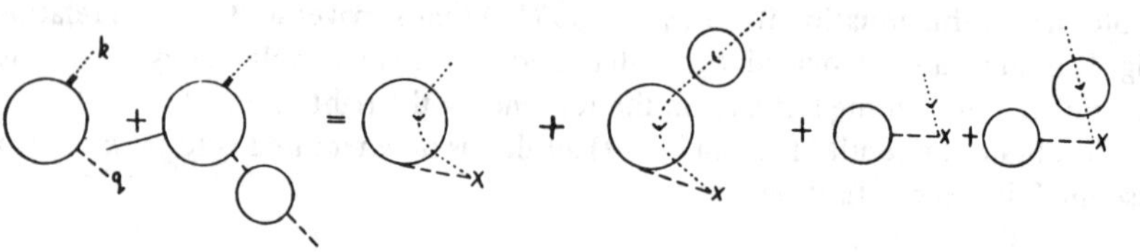

Fig. 5. Ward identity for spin-$\frac{1}{2}$ particles (dashed lines). The symbol X here denotes the action of the matrix $(m - \gamma \cdot q)$.

(ϕ_a being the "fictitious field"), acting at the right-most vertex of each term on the right-hand side. The combinatorial proof of fig. 5 follows the methods laid down by 't Hooft [6].

To apply fig. 5 in the case where there is just one other external line (an on-shell spin-$\frac{1}{2}$ line) we define $\Gamma_\lambda(p, k, q)$ to be the spin-$\frac{1}{2}$ vertex part and $\Sigma(q)(\gamma \cdot q - m)$ to be the self-energy function (after mass renormalization). Then fig. 5 gives

$$\bar{u}(p) k^\lambda \Gamma_\lambda(p, k, q) \{1 - \Sigma(q)\}^{-1}$$

$$= \bar{u}(p) [(1 - P(k^2)\}^{-1} \{g + H(p, k, q)\}] (m - \gamma \cdot q) , \qquad (14)$$

where $\bar{u}(p)$ is a Dirac spinor and H is defined by fig. 6. Eq. (2) is got from (14) by differentiating with respect to k_λ (keeping p fixed) and then setting $k = 0$, with Y defined by

$$\bar{u}(p) H(p, 0, p) = g(Y - 1) \bar{u}(p) , \qquad (p^2 = m^2) . \qquad (15)$$

Finally, for the Landau gauge, one finds that $Y = 1$, in the same way that fig. 4 was used to show that $Z_1^F = 1$.

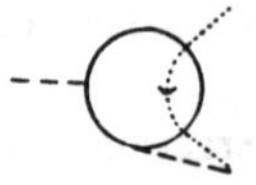

Fig. 6. Graph defining the quantity H in eq. (14).

We make one final comment on charge renormalization. The Yang-Mills Lagrangian contains a four-field coupling with strength g^2. One of 't Hooft Ward identities (eq. (3.10) of ref. [6]) assures that this constant g is renormalized in the same way as the three-field coupling constant g.

3. WARD IDENTITIES AND PRACTICAL CALCULATIONS

In quantum electrodynamics [8], the Ward identity connecting the convergent part of the electron self-energy Σ and the vertex function Γ_λ turns out to be satisfied, even if a non-gauge-invariant cut-off is used. The photon self-energy function, $\Pi_{\mu\nu}$, on the other hand, is quadratically divergent, and much more care is necessary to produce a finite part obeying the Ward identity $k^\mu \Pi_{\mu\nu}(k) = 0$. On the basis of this experience, then, it seems likely that logarithmically divergent integrals are relatively safe, whereas quadratically divergent ones require gauge-invariant regulators.

If this is so, eq. (7) may be useful, as it expresses the Yang-Mills self-energy function $\Pi(q^2)$ (which is defined from quadratically divergent graphs) in terms of other integrals which are only logarithmically divergent. Of these, the quantity $G_{\sigma\mu}$ does not normally occur in the Feynman rules. But it is well defined by fig. 3(c), and may be used as an intermediate step in the calculation of $\Pi(q^2)$. The result, of course, satisfies the Ward identity by construction.

From this point of view, one may question whether the existence of a gauge invariant regularization procedure is of practical importance, assuming that the formally derived Ward identities are in fact correct.

I am indebted to Mr. D.A.Ross for valuable help.

APPENDIX

We give here a path-integral derivation of the generalized ward identities. The method is an extension of appendix B of ref. [6].

If W_μ^a is the Yang-Mills field, we define

$$W_\mu^{ab} = \epsilon^{abc} W_\mu^c , \tag{A.1}$$

$$D_\mu^{ab} = \delta^{ab} \partial_\mu - g\, W_\mu^{ab} , \tag{A.2}$$

$$\Delta(W) = \det(\partial^\mu D_\mu^{ab}) . \tag{A.3}$$

In terms of source functions η, j_μ^a, J^a we write the path-integral

$$Z(\eta, j_\mu^a, J^a) = \int \mathrm{d}\{W\}\, \Delta(W) \exp(iS) , \tag{A.4}$$

where

$$S = S_0 - (2\lambda)^{-1} (\partial \cdot W^a)^2 - \lambda^{-1} J^a \partial \cdot W^a + \bar{\psi}\eta + \bar{\eta}\psi + j_\mu^a W^{\mu a} \tag{A.5}$$

Here S_0 is the action for the interacting Yang-Mills and matter fields. The parameter

λ determines the gauge, giving the propagator of eq. (6). The functional $\Delta(W)$ is responsible for the "fictitious particle" loops. The source J^a is 't Hooft's device for generating Ward identities. The sources η, $\bar{\eta}$ and j^a_μ enable one to *define* off-shell amplitudes. However, they destroy the gauge invariance of eq. (A.4).

Following ref. [6], we take J^a to be infinitesimal, and make an infinitesimal gauge transformation

$$W^a_\mu \to W^a_\mu - g^{-1} D^{ab}_\mu \Lambda^b , \tag{A.6}$$

with Λ^b given by

$$\partial \cdot D^{ab} \Lambda^b = g J^a , \tag{A.7}$$

or

$$\Lambda = g(\partial \cdot D)^{-1} J . \tag{A.8}$$

As proved explicitly in sect. IIC of ref. [4], the measure $\mathrm{d}\{W\}\,\Delta(W)$ in eq. (A.4) is invariant under this W-dependent gauge transformation. But S is replaced by

$$\begin{aligned} S' = S_0 &- (2\lambda)^{-1} (\partial \cdot W^a)^2 + \bar{\eta}(1 + i\Lambda_a \tau_a) \psi \\ &+ \bar{\psi}(1 - i\Lambda_a \tau_a)\eta + j^a_\mu W^{\mu a} - g^{-1} j^a_\mu D^{\mu ab} \Lambda^b . \end{aligned} \tag{A.9}$$

The equivalence of the two theories defined by S and S' in eqs. (A.5) and (A.9) provided the Ward identities. S gives the left-hand sides of the identities, S' gives the right-hand sides. The relation (7) (or fig. 2) is obtained from

$$\left[\frac{\delta^3 Z}{\delta J^a \delta j^b_\sigma \delta j^c_\mu}\right]_{J=j=\eta=0} , \tag{A.10}$$

and relation (14) (fig. 5) from

$$\left[\frac{\delta^3 Z}{\delta J^a \delta\eta \delta\bar{\eta}}\right]_{J=j=\eta=0} \tag{A.11}$$

The external lines corresponding to j^b_μ in (A.10) and to η in (A.11) are put on the mass shell (and transverse in the former case). The Λ-dependent terms in eq. (A.9) provide, because of (A.8), the open "fictitious particle" lines in figs. 2 and 5.

In order to understand the factors $(q_\mu q_\nu - q^2 \delta_{\mu\nu})$ on the right-hand side of eq. (7), we use in (A.9) the identity

$$g^{-1} j^{\mu a} D^{ab}_\mu \Lambda^b = j^{\mu a} \partial_\mu \Box^{-1} J^a - j^{\mu a}(g_{\mu\nu} - \partial_\mu \partial_\nu \Box^{-1})\, W^{\nu ab} \Lambda^b , \tag{A.12}$$

which is a consequence of eqs. (A.2) and (A.7). Then the second term on the right-hand side of eq. (A.12) provides the final vertices on the right-hand side of fig. 2. The first term on the right-hand side of eq. (A.12) merely cancels that contribution from S (eq. (A.5)) which results from contraction of the W-fields in $j^{\mu a} W^a_\mu$ and in $-\lambda^{-1} J^a \partial \cdot W^a$ (using the propagator of eq. (6)).

REFERENCES

[1] C.N.Yang and R.L.Mills, Phys. Rev. 96 (1954) 191;
R.Shaw (Cambridge thesis 1955, unpublished).

[2] R.P.Feynman, Acta Phys. Pol. 24 (1963) 697;
B.S.de Witt, Phys. Rev. 162 (1967) 1195, 1239;
S.Mandelstam, Phys. Rev. 175 (1968) 1580, 1604.

[3] L.D.Faddeev and V.N.Popov, Phys. Letters 25B (1967) 29.

[4] E.S.Fradkin and I.V.Tyutin, Phys. Rev. D2 (1970) 2841.

[5] S.Weinberg, Phys. Rev. 140 (1965) B516.

[6] G.'t Hooft, Nucl. Phys. B33 (1971) 173.

[7] S.Weinberg, Phys. Rev. 135 (1964) B1049.

[8] J.M.Jauch and S.Rohrlich, The theory of photons and electrons (Addison-Wesley) p. 178.

WARD IDENTITIES IN GAUGE THEORIES

A. A. Slavnov

V. A. Steklov Mathematics Institute, Academy of Sciences of the USSR, Moscow.

Translated from Teoreticheskaya i Matematicheskaya Fizika, Vol. 10, No. 2, pp. 153–161, February, 1972. Original article submitted June 23, 1971.

Generalized Ward-Takahashi identities are obtained for gauge theories of the type of the massless Yang-Mills field. It is shown that all the divergences of the Yang-Mills theory can be removed by means of a renormalization of the charge and the wave function.

A relativistically invariant formalism for gauge theories of the type of the massless Yang-Mills field was first constructed in [1,2]. In this formalism the S matrix is constructed from an effective Lagrangian, which is a nonlocal function of the fields. Despite this, the gauge invariance ensures unitarity of the S matrix. The perturbation series contains a finite number of types of divergent diagram and the theory can therefore be renormalized by means of the Bogoliubov – Parasuik R operation [3]. However, to ensure unitarity of the renormalized S matrix it is necessary to take into account the Ward identities, which guarantee gauge invariance of the theory, when carrying out the renormalization.

In the present paper we obtain Ward identities that relate diagrams with a different number of external lines and accordingly renormalization constants of different vertices. We shall show that all divergences can be eliminated by a renormalization of the charge and the wave function of a vector particle (for simplicity we consider a selfinteracting Yang-Mills field. Allowance for an interaction with other fields does not lead to additional difficulties). Ward identities have been considered in [4]. However, in [4] only relations of the type of transversality conditions for the matrix elements are obtained and there are no explicit relations between the renormalization constants.

The method of obtaining the Ward identities developed in the present paper is applicable to any gauge theory, including the gravitational field. To illustrate the method we derive the well-known Ward identities in electrodynamics in the first section. In the second section we consider the Yang-Mills field.

In this investigation we have been concerned solely with the problem of ultraviolet divergences and have ignored the difficulties associated with infrared divergences. To give the expressions we have obtained a well-defined mathematical meaning it may be assumed, for example, that the subtractions are not made on the mass shell. Of course, this does not solve the problem of infrared divergences and this question remains open at the present time.

1. ELECTRODYNAMICS

The generating functional for the Green's functions in quantum electrodynamics is determined by the functional Feynman integral

$$Z(\eta_\mu, \xi, \bar{\xi}) = \int \exp i \Big\{ \int \Big[\mathcal{L}(x) + \frac{1}{2\alpha}(\partial_\mu A_\mu)^2 + \eta_\mu A_\mu + \xi \bar{\psi} + \bar{\xi}\psi \Big] dx \Big\} dA\, d\bar{\psi}\, d\psi \tag{1}$$

Here $\mathcal{L}(x)$ is the gauge-invariant Lagrangian of the electromagnetic and electron-position fields. The term $(1/2\alpha)(\partial_\mu A_\mu)^2$ fixes the gauge. In particular, $\alpha = 1$ corresponds to the transverse gauge; $\alpha = 0$, to the Feynman gauge. The Green's functions can be expressed in terms of the generating functional (1):

$$(i)^n \langle T A_\mu(x_1) \cdots \bar{\psi}(x_i) \cdots \psi(x_n) \rangle = \left. \frac{\delta^n Z}{\delta \eta_\mu(x_1) \cdots \delta \xi(x_i) \cdots \delta \bar{\xi}(x_n)} \right|_{\eta = \xi = \bar{\xi} = 0} \tag{2}$$

The gauge invariance of the Lagrangian $\mathcal{L}(x)$ leads immediately to generalized Takahashi-Ward identities. To obtain them we make a gauge transformation of the variables in the integral (1):

$$A_\mu \to A_\mu + \frac{\partial \varphi}{\partial x_\mu} \quad \psi \to e^{ig\varphi} \psi \, ; \quad \bar{\psi} \to e^{-ig\varphi} \bar{\psi} \tag{3}$$

Since $\mathcal{L}(x)$ is invariant, all that changes under this transformation are the coefficients of the external sources and the term that fixes the gauge. The change in the latter is

$$\frac{1}{\alpha} \partial_\mu A_\mu \, \Box \varphi(x) \equiv \frac{1}{\alpha} \partial_\mu A_\mu \, \chi(x) \tag{4}$$

Since the change of variables does not alter the value of the integral, we can equate the coefficient of the arbitrary function $\chi(x)$ to zero:

$$\int \exp i \left\{ \int \left[\mathcal{L}(x) + \frac{1}{2\alpha} (\partial_\mu A_\mu)^2 + \eta_\mu A_\mu + \xi \bar{\psi} + \bar{\xi} \psi \right] dx \right\} \tag{5}$$
$$\left\{ \frac{1}{\alpha} (\partial_\mu A_\mu) - \Box^{-1} \partial_\mu \eta_\mu + ig \Box^{-1} \xi(x) \bar{\psi}(x) - ig \Box^{-1} \bar{\xi}(x) \psi(x) \right\} dA \, d\bar{\psi} \, d\psi = 0$$

Equation (5) can be rewritten in terms of the variational derivatives of the functional $Z(\eta_\mu, \xi, \bar{\xi})$

$$\frac{1}{\alpha} \frac{\partial}{\partial x_\mu} \left(\frac{\delta Z}{\delta \eta_\mu(x)} \right) - \int D_0^c(x-y) \, \partial_\mu \eta_\mu(y) \, dy \, Z \tag{6}$$
$$+ ig \int D_0^c(x-y) \left\{ \xi(y) \frac{\delta Z}{\delta \xi(y)} - \bar{\xi}(y) \frac{\delta Z}{\delta \bar{\xi}(y)} \right\} dy = 0$$

where $D_0^c(x)$ is the Feynman Green's function of the d'Alembert operator (the rules for avoiding the pole are, as always, chosen in such a way that a transition to Euclidean variables is possible).

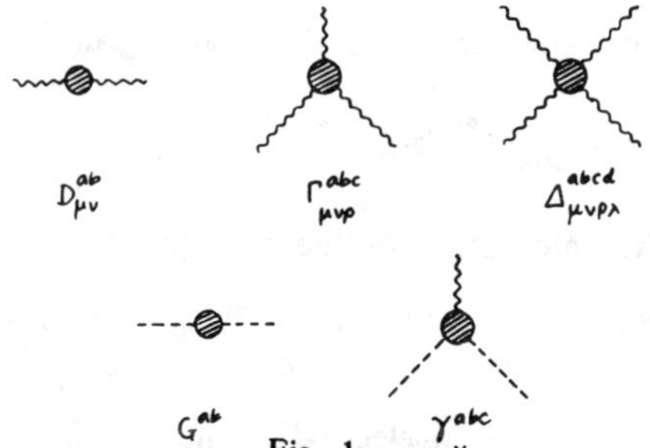

Fig. 1

The relation (6) is a compact expression of the Ward-Takahashi identities. Differentiating (6) with respect to η_ν and setting $\eta_\alpha = \xi = \bar{\xi} = 0$ we obtain

$$-\frac{1}{\alpha}\frac{\partial}{\partial x_\mu}\left.\frac{\delta^2 Z}{\delta\eta_\mu(x)\,\delta\eta_\nu(y)}\right|_{\eta=0} = \frac{1}{\alpha}\frac{\partial}{\partial x_\mu} D_{\mu\nu}(x-y) = \partial_\nu D_0^c(x-y) \qquad (7)$$

which expresses the fact that there is no renormalization of the longitudinal part of the Green's function. Differentiating Eq. (6) with respect to ξ and $\bar{\xi}$, we obtain

$$\frac{1}{\alpha}\frac{\partial}{\partial x_\mu}\langle T A_\mu(x)\bar{\psi}(y)\psi(z)\rangle = g\left\{ D_0^c(x-y) - D_0^c(x-z)\right\}\langle\bar{\psi}(y)\psi(z)\rangle \qquad (8)$$

Using the definition of the proper vertex part,

$$\langle T A_\mu(x)\bar{\psi}(y)\psi(z)\rangle = g\int D_{\mu\nu}(x-x_1)\,G(y-y_1)\,G(z-z_1)\,\Gamma_\nu(x_1,y_1,z_1)\,dx_1 dy_1 dz_1 \qquad (9)$$

and taking into account (7), we obtain the well-known Ward identity

$$\frac{\partial}{\partial x_\mu}\Gamma_\mu(x,y,z) = G^{-1}(y-z)\left\{\delta(x-y) - \delta(x-z)\right\} \qquad (10)$$

Note, as one would expect, that the dependence on the gauge-fixing parameter α has disappeared and the relation (10) has an invariant form.

Of course, the identity (10) can also be readily obtained by the usual methods: in particular, by making a direct analysis of the diagrams that describe the vertex and selfenergy parts. However, in more complicated gauge theories, for example, in the Yang-Mills theory, such an analysis is very laborious. The method described here can be transferred immediately to any gauge theory.

2. YANG-MILLS FIELD

The generating functional for the Green's functions in the Yang-Mills theory has the form [1, 2, 3]

$$Z(\eta_\mu) = \int \exp i\left\{\int\left[\mathcal{L}(x) + \eta_\mu^a B_\mu^a + \frac{1}{2\alpha}(\partial_\mu B_\mu^a)^2\right]dx\right\}\Delta(B)\,dB \qquad (11)$$

where $\mathcal{L}(x)$ is the gauge-invariant Yang-Mills Lagrangian:

$$\mathcal{L} = -\frac{1}{4}f_{\mu\nu}^a f_{\mu\nu}^a - \frac{1}{2}g\,\epsilon^{abc} f_{\mu\nu}^a B_\mu^b B_\nu^c - \frac{1}{4}g^2\epsilon^{abc}\epsilon^{ade} B_\mu^b B_\nu^c B_\mu^d B_\nu^e \qquad (12)$$

the factor $\Delta(B)$ is the determinant of the operator $\Box^{-1}M$

$$M^{ab}\varphi^b = \Box\varphi^a(x) + g\,\epsilon^{abc}\partial_\mu\left\{B_\mu^b(x)\varphi^c(x)\right\} \qquad (13)$$

the term $(1/2\alpha)(\partial_\mu B_\mu^a)^2$, as in electrodynamics, fixes the guage. From the point of view of Feynman diagrams, the determinant $\Delta(B)$ is a sum of closed loops with respect to which a scalar fermion of vanishing mass propagates,

$$\mathrm{Tr}\,\ln\Delta(B) = -\sum_{n=2}^{\infty}\frac{g^n}{n}\int\partial_{\mu_1}D_0^c(x_1-x_2)\,B_{\mu_1}^{a_1}(x_1)$$
$$\dots\partial_{\mu_n}D_0^c(x_n-x_1)\,B_{\mu_n}^{a_n}(x_n)\;\epsilon^{aa_1b}\,\epsilon^{ba_2c}\dots\epsilon^{ea_na}\,dx_1\cdots dx_n \tag{14}$$

The functional $Z(\eta_\mu)$ can also be expressed as an integral of a local action by introducing an auxiliary Fermi field C [1],

$$Z(\eta_\mu) = \int \exp i\Big\{\int\Big[\mathcal{L}(x) + \eta_\mu^a B_\mu^a + \frac{1}{2\alpha}(\partial_\mu B_\mu^a)^2 - \bar{C}^a\Box C^a + g\epsilon^{abc}B_\mu^a C^b\partial_\mu\bar{C}^c\Big]dx\Big\}dB\,dC\,d\bar{C} \tag{15}$$

As is readily seen, the theory contains five types of divergent diagram; these are shown in Fig. 1, in which the way line denotes the propagation function of the vector particle; the dashed line, the propagation function of the C particles. The Ward identities which we shall obtain establish relationships between the corresponding counter-terms; ultimately, only two arbitrary constants remain in the theory – the renormalizations of the wave function and charge.

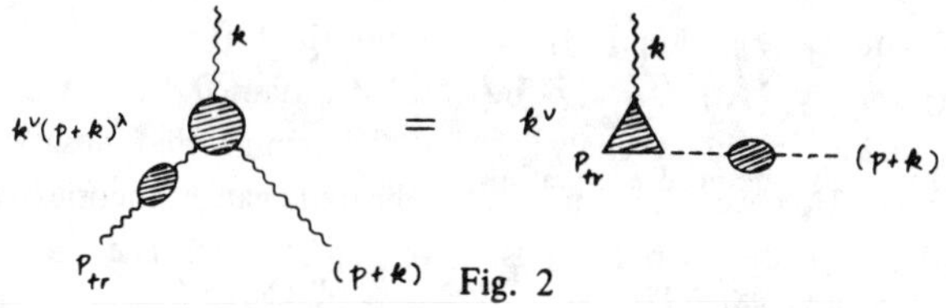

Fig. 2

To obtain the Ward identities, we shall proceed as in electrodynamics. In the integral (11) we make an infinitesimal gauge transformation of the variables:

$$B_\mu^a(x) \to B_\mu^a(x) + g\epsilon^{abc}B_\mu^b(x)\varphi^c(x) + \partial_\mu\varphi^a(x) \tag{16}$$

where the function $\varphi^a(x)$ satisfies the equation

$$M^{ab}\varphi^b(x) = \chi^a(x) \tag{17}$$

and $\chi^a(x)$ is an arbitrary function. The operator M was determined by Eq. (13) (in electrodynamics the analog of M was simply the d'Alembert operator). The Lagrangian $\mathcal{L}(x)$ is obviously invariant under such a transformation. One can show that the integration measure $\Delta(B)dB$ is also invariant under this transformation. To do this it is sufficient to calculate directly the Jacobian of the transformation (16) and the variation $\Delta(B)$

$$\delta\Delta(B) = \delta\,\mathrm{Tr}\ln M = \delta M^{ab}M^{-1ba} \tag{18}$$

where the inverse operator M_{xy}^{-1ba} satisfies the equation

$$\Box M^{-1ab}_{xy} + g\epsilon^{acd}\partial_\mu\{B^c_\mu(x)M^{-1db}_{xy}\} = \delta^{ab}\delta(x-y) \tag{19}$$

The operator M^{-1ab}_{xy} is obviously the Green's function of the C particle in the external vector field B_μ

Equating the coefficient of $\chi(x)$ to zero, we obtain

$$\int \exp i\{\int[\mathcal{L}(x) + \eta^a_\mu(x)B^a_\mu(x) + \frac{1}{2\alpha}(\partial_\mu B^a_\mu)^2]\,dx\}$$
$$\{\frac{1}{\alpha}\frac{\partial}{\partial y_\mu}B^d_\mu(y) + \int[\eta^a_\mu(z)\frac{\partial}{\partial z_\mu}M^{-1ad}_{zy} \tag{20}$$
$$+ g\epsilon^{abc}\eta^a_\mu(z)B^a_\mu(z)M^{-1cd}_{zy}]\}\,\Delta(B)\,dB = 0$$

Equation (20) is the system of generalized Ward identities. Differentiating this equation with respect to $\eta^b_\nu(x)$ and setting $\eta^i_\alpha = 0$ we obtain

$$-\frac{1}{\alpha}\langle T\frac{\partial B^a_\mu(x)}{\partial x_\mu}B^b_\nu(x)\rangle = \langle T\frac{\partial}{\partial y_\nu}M^{-1ba}_{yx}\rangle + g\epsilon^{bcd}\langle TB^c_\nu(y)M^{-1da}_{yx}\rangle \tag{21}$$

Differentiating (21) with respect to y and using Eq. (19), we obtain

$$\frac{1}{\alpha}\langle T\frac{\partial B^a_\mu}{\partial x_\mu}\frac{\partial B^b_\nu}{\partial y^\nu}\rangle = -\delta^{ab}\delta(x-y) \tag{22}$$

From this we obviously obtain the equation

$$\frac{1}{\alpha}\frac{\partial}{\partial x_\mu}D^{ab}_{\mu\nu}(x,y) = \delta^{ab}\partial_\nu D^0_c(x-y) \tag{23}$$

which is the exact analog of the condition (7) in electrodynamics and shows there is no renormalization of the longitudinal part of the Green's function. It follows from the condition (23) that the counter-term of the mass renormalization vanishes.

To obtain the analog of the Ward identity (10), we differentiate Eq. (20) twice with respect to η^k_ν and set $\eta^i_\alpha = 0$. We obtain

$$-\frac{i}{\alpha}\langle TB^a_\mu(x)B^b_\nu(y)B^c_\rho(z)\rangle = \langle T\partial_\mu M^{-1ab}_{xz}B^b_\nu(y)\rangle + \langle TB^a_\mu(x)\partial_\nu M^{-1bc}_{yz}\rangle$$
$$+ g\epsilon^{ade}\langle TB^d_\mu(x)M^{-1ec}_{xz}B^b_\nu(y)\rangle + g\epsilon^{bd\cdot}\langle TB^a_\mu(x)B^d_\nu(y)M^{-1ec}_{yz}\rangle \tag{24}$$

In contrast to electrodynamics, the divergence of the vertex function can no longer be expressed solely in terms of the two-point Green's function. To obtain from (24) a relationship between the renormalization constants, we differentiate this equation with respect to $D^c_0(x)$ nd separate out in the results a structure that is transverse with respect to the momentum p conjugate to the coordinate x. Going over to Fourier transforms we have

$$D^{tr}_{\mu\lambda}(p)\frac{k^\nu}{k^2}\frac{(p+k)^\rho}{(p+k)^2}\Gamma^{abc}_{\lambda\nu\rho}(p,k) = \frac{k^\nu}{k^2}G(p+k)\gamma^{abc\,tr}_{\mu\nu}(p,k) \tag{25}$$

Here $\Gamma^{abc}_{\lambda\nu\rho}$ is the proper vertex function of three vector fields and we used Eqs. (19) and (23); G(q) is the Green's function of the C particle,

$$\delta^{ab}(2\pi)^{-4}\int e^{ik(x-y)}G(k)\,d^4k = \langle TM^{-1\,ab}_{xy}\rangle \tag{26}$$

and the function $\gamma^{abc}_{\mu\nu}$ is related to the proper vertex function γ^{abc}_{ν}, corresponding to the transition of two C particles into a single vector particles by the relation

$$\gamma^{abc}_{\nu} = i p_\mu \gamma^{abc}_{\mu\nu}(p,k) \tag{27}$$

The symbol tr means that in this equation the structure transverse with respect to the variable p has been separated out. Equation (25) is expressed graphically in Fig. 2. The triangle stands for the function γ_ν, which is related to the vertex function $\gamma^{abc}_{\mu\nu}$ by Eq. (27). Note that the function $\gamma^{abc}_{\mu\nu}$ cannot have a single-particle pole in the variable p since the corresponding term is proportional to p_μ for reasons of relativistic invariance and does not therefore make a contribution to the transverse structure.

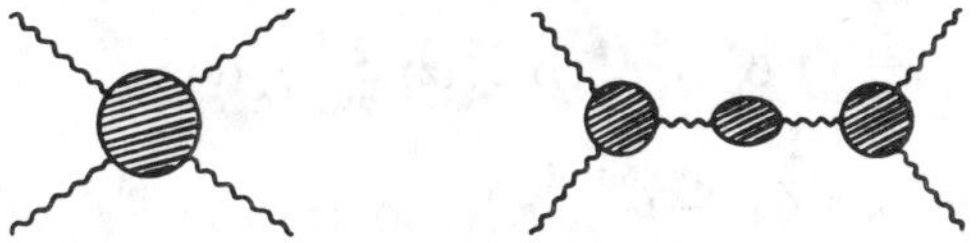

Fig. 3

The integral that determines $\Gamma^{abc}_{\lambda\nu\rho}$, diverges linearly. Therefore, with allowance for the symmetry properties, $\Gamma^{abc}_{\lambda\nu\rho}$ can be represented in the form

$$\Gamma^{abc}_{\lambda\nu\rho}(p,k,q) = i\epsilon^{abc}\{Z_1^{-1}[g_{\nu\lambda}(p-k)_\rho + g_{\nu\rho}(k-q)_\lambda + g_{\lambda\rho}(q-p)_\nu] + O_{\lambda\nu\rho} \tag{28}$$

where $O_{\lambda\nu\rho}$ is the finite part which vanishes as a third power of the momentum in the limit $p, k, q \to 0$ The other functions in (25) can be represented similarly:

$$D^{tr}_{\mu\nu}(p) = \left\{g_{\mu\nu} - \frac{p_\mu p_\nu}{p^2}\right\}\frac{1}{p^2}\{Z_2 + O(p^2)\}, \tag{29}$$

$$G(q) = \frac{\{\tilde{Z}_2 + O(q^2)\}}{q^2} \tag{30}$$

$$\gamma^{abc}_{\mu\nu}(p,q) = \epsilon^{abc}\{g_{\mu\nu}\tilde{Z}_1^{-1} + O_{\mu\nu}(p,q)\} \tag{31}$$

By virtue of (27) the constant $\tilde{Z}_1$ in Eq. (31) is the renormalization constant of the vertex function γ^{abc}_{ν}. Substituting all these expressions into (25) and making the passage to the limit $p, k, q \to 0$ we obtain

$$Z_2^{-1} Z_1 = \tilde{Z}_2^{-1} \tilde{Z}_1 \tag{32}$$

This equation replaces the relation $Z_1 = Z_2$ in electrodynamics. By virtue of this equation the renormalization of the charge that appears next to the triple vector vertex and the triple vertex that describes the interaction with the auxiliary scalar field is realized with one and the same constant. To see this we introduce the renormalized fields

$$B^a_{\mu R} = Z^{-1/2} B^a_\mu \; ; \quad C^a_R = \tilde{Z}_2^{-1/2} C^a \; ; \quad \bar{C}^a_R = \tilde{Z}_2^{-1/2} \bar{C}^a \tag{33}$$

and note that the corresponding renormalization constants are $Z_2^{-3/2} Z$ and $\tilde{Z}_2^{-1} \tilde{Z}_1 Z_2^{-1/2}$. By virtue of (32) these quantities are equal.

We now turn to the fourfold vertex. From (20) we obtain an identity that expresses the divergence of the four-line diagram with respect to one of the variables in terms of the remaining Green's functions:

$$\begin{aligned} &\frac{1}{\alpha} \langle T B^a_\mu(x) B^b_\nu(y) B^c_\rho(z) \frac{\partial}{\partial u_\lambda} B^d_\lambda(u) \rangle \\ &= \langle T \frac{\partial}{\partial x_\mu} M^{-1\,ad}_{xy} B^b_\nu(y) B^c_\rho(z) \rangle \\ &+ g \epsilon^{aef} \langle T B^e_\mu(x) M^{-1\,fd}_{xu} B^b_\nu(y) B^c_\rho(z) \rangle \\ &+ (x, a, \mu \leftrightarrow y, b, \nu) \\ &+ (x, a, \mu \leftrightarrow z, c, \rho) \end{aligned} \tag{34}$$

To obtain the corresponding relations between the renormalization constants, it is sufficient to consider the expression for the vacuum expectation value of the T product of the divergences of four fields. Differentiating (34) with respect to x_μ, y_ν, z_ρ and using Eq. (19), we obtain

$$\begin{aligned} &\langle T \frac{\partial B^a_\mu}{\partial x_\mu} \frac{\partial B^b_\nu}{\partial y_\nu} \frac{\partial B^c_\rho}{\partial z_\rho} \frac{\partial B^d_\lambda}{\partial u_\lambda} \rangle \\ &= \delta(x-u)\, \delta(y-z)\, \delta^{ab} \delta^{cd} + (x, a \leftrightarrow y, b) + (x, a \leftrightarrow z, c) \end{aligned} \tag{35}$$

It follows from this that the connected part of the Green's function under consideration vanishes. The connected part is described by the two diagrams shown in Fig. 3. The first of these diagrams is the proper fourfold vertex; the second, two triple vertices joined by a single-particle Green's function. The fourfold vertex has the structure

$$\Delta^{abcd}_{\mu\nu\rho\lambda}(p,k,q) = P g^2 \{ Z_3^{-1} \epsilon^{eab} \epsilon^{ecd} g_{\mu\rho} g_{\nu\lambda} + Z_4^{-1} \delta^{ab} \delta^{cd} g_{\mu\nu} g_{\lambda\rho} \} + O^{abcd}_{\mu\nu\rho\lambda}(p,k,q) \tag{36}$$

where P is the symmetrization operator with respect to the index pairs $a\mu$, $b\nu$, $c\rho$, $d\lambda$ and $O^{abcd}_{\mu\nu\rho\lambda}$, as before, stands for the finite part that vanishes as the second power of the momentum in the limit of small p, k, and q. Substituting this expression and the expressions (28) for the triple vertices into Eq. (34) and going to the limit $p, k, q \to 0$ we obtain

$$Z_4^{-1} = 0 \; ; \qquad Z_3 = Z_1^2 Z_2^{-1} \tag{37}$$

Thus, with allowance for the counter-terms, the effective Lagrangian has the form

$$\begin{aligned} \mathcal{L} = & -\frac{1}{4} Z_2 f^a_{\mu\nu R} f^a_{\mu\nu R} - \frac{g_R}{2} Z_1 \epsilon^{abc} f^a_{\mu\nu R} B^b_{\mu R} B^c_{\nu R} \\ & - \frac{g_R^2}{4} Z_3 \epsilon^{abc} \epsilon^{aed} B^b_{\mu R} B^c_{\nu R} B^e_{\mu R} B^d_{\nu R} \\ & - \tilde{Z}_2 \bar{C}^a_R \Box C^a_R + g_R \tilde{Z}_1 \epsilon^{abc} B^a_{\mu R} \bar{C}^b_R \partial_\mu C^c_R \end{aligned} \tag{38}$$

where g_R is the physical renormalized charge. Going over to the renormalized fields and taking into account (33) and (37), we have

$$\begin{aligned} \mathcal{L} = & -\frac{1}{4} f^a_{\mu\nu} f^a_{\mu\nu} - \frac{(Z_1 Z_2^{-3/2} g_R)}{2} \epsilon^{abc} f^a_{\mu\nu} B^b_\mu B^c_\nu \\ & + (Z_1 Z_2^{-3/2} g_R) \epsilon^{abc} B^a_\mu \bar{C}^b \partial_\mu C^c \\ & - \frac{1}{4} (Z_1 Z_2^{-3/2} g_R)^2 \epsilon^{abc} \epsilon^{ade} B^b_\mu B^c_\nu B^d_\mu B^e_\nu \\ & - \bar{C}^a \Box C^a \end{aligned} \tag{39}$$

It can be seen that the renormalization does not alter the structure of the effective Lagrangian and does not violate the gauge invariance of the theory. The Lagrangian (38) actually contains two independent renormalization constants – the renormalization of the charge and the wave function of the vector particle, $Z_1 Z_2^{-3/2}$ and Z_2, respectively. The renormalization constant of the C-particle wave function does not occur in the final expression for the matrix elements. Since the C particles are absent in the asymptotic states, this constant can be eliminated by a change of variables in the functional integral that defines the S matrix.

In deriving the Ward identities we operated with divergent integrals. To give our arguments a well-defined meaning, we must first carry out an invariant regularization. An invariant regularization for the Yang-Mills field is constructed in our paper. The results obtained enable one to use any regularization in practical calculations, it only being necessary to establish that the Ward identities are fulfilled.

Finally, we note that one of the renormalization constants can be made finite by an appropriate choice of the gauge; for suppose the invariant regularization has been performed. It then follows from Eq. (11) that

$$\frac{d}{d\alpha} G^{ab}_{\mu\nu\Lambda}(x,y) = -\frac{1}{2\alpha^2}\langle T B^a_\mu(x) B^b_\nu(y) \int (\partial_\lambda B^c_\lambda(z))^2 dz\rangle_\Lambda. \tag{40}$$

The subscript Λ means that all the quantities are regularized. Since the free Green's function is a linear function of α, we obtain from (40) a linear differential equation for the constant Z_2:

$$\frac{dZ_{2\Lambda}}{d\alpha} = \sum_{n=0}^{\infty} c_n^\Lambda \alpha^n \tag{41}$$

A direct calculation readily shows that the constants c_n^Λ do not vanish. In any finite order g the sum on the right-hand side terminates at some n and therefore

$$Z^{(k)}_{2\Lambda} = Z^{(k)}_{20} + \sum_{n=1}^{k} \frac{c_n^\Lambda}{n+1} \alpha^{n+1} \tag{42}$$

(until the regularization is lifted, all the c_n^Λ are finite). We set $Z^{(k)}_{2\Lambda}$ equal to a finite, Λ -independent quantity. Then Eq. (42) can be regarded as an algebraic equation for α. This equation is always solvable. Consequently, $\alpha(\Lambda)$ can be chosen such that Z_2 remain finite when the regularization is lifted. At the same time, α tends, in general, to infinity. However, this is not disastrous, since the perturbation theory series for the S matrix does not depend at all on α and, in particular, admits the passage to the limit $\alpha \to \infty$. Thus, as in electrodynamics, there remains only one essentially divergent constant in the Yang-Mills theory.

The Ward identities obtained here enable one to eliminate in an invariant manner all the ultraviolet divergences and, thus, to construct a unitary renormalized S matrix. A similar procedure can be applied to other gauge theories, in particular, gravitation. However, the situation as regards the renormalization constants is then much more complicated since this theory is, at least according to a formal calculation of the degrees of the divergence, unrenormalizable. The question of the extent to which the Ward identities restrict the renormalization ambiguity in this theory is as yet open.

I am grateful to L. D. Faddeev for stimulating discussions.

LITERATURE CITED

1. L. D. Faddeev and V. N. Popov, *Phys. Lett.*, **25B**, 29 (1967); Preprint *ITF*-67-36 [in Russian], Institute for Theoretical Physics, Kiev (1967).
2. B. S. De Witt, *Phys. Rev.*, **162**, 1195, 1239 (1967).
3. N. N. Bogoliubov and O. S. Parasiuk, *Acta Math.*, **97**, 227 (1957).
4. G. 't Hooft, Preprint of *Utrecht University* (1971).
5. E. S. Fradkin and I. V. Tyutin, *Phys. Lett.*, **30B**, 562 (1969).

VI. RENORMALIZATION OF GAUGE THEORIES

RENORMALIZATION OF MASSLESS YANG-MILLS FIELDS

G.'t HOOFT
Institute for Theoretical Physics, University of Utrecht, Utrecht, The Netherlands

Received 12 February 1971

Abstract: The problem of renormalization of gauge fields is studied. It is observed that the use of non-gauge invariant regulator fields is not excluded provided that in the limit of high regulator mass gauge invariance can be restored by means of a finite number of counter-terms in the Lagrangian. Massless Yang-Mills fields can be treated in this manner, and appear to be renormalizable in the usual sense.

Consistency of the method is proved for diagrams with non-overlapping divergencies by means of gauge invariant regulators, which however, cannot be interpreted in terms of regulator fields. Assuming consistency the S-matrix is shown to be unitary in any order of the coupling constant. A restriction must be made: no local, parity-changing transformations must be contained in the underlying gauge group. The interactions must conserve parity.

1. INTRODUCTION

In recent years the Feynman rules for massless Yang-Mills fields have been established [1–5]. Naive power counting suggests a renormalizable theory; however, in order to carry through a renormalization procedure one must first define a cut-off procedure. And if the cut-off procedure breaks the gauge-invariance of the theory then it is no more clear what the Feynman rules are. The reason is that gauge-invariance, through Ward identities, is essential for the S-matrix to be unitary.

Thus the problem poses itself as follows: how to find a gauge invariant cut-off procedure. This problem is of course quite the same in quantum electrodynamics. There the problem was solved by Pauli, Villars [6] and Gupta [7] who succeeded in finding a set of regulator fields that could be coupled in a gauge invariant way. Now in the case of massless Yang-Mills fields a gauge invariant regularizing procedure also seems to exist. Unfortunately, however, this procedure cannot be interpreted in terms of fields with indefinite metric and/or wrong statistics, like in the case of electrodynamics. Hence, unitarity and causality are no longer evident.

However, it must be realized that the whole renormalization procedure involves also the addition of counterterms in the Lagrangian. And in fact the important point is that the total effect of regulator fields and counterterms is to be gauge invariant, at least in the limit of infinite regulator masses. Thus let us suppose now that we have found a set of regulator fields, that makes the various amplitudes finite but destroys the gauge invariancc. If we are to restore gauge invariance by means of a finite number of counterterms in the Lagrangian then the gauge-invariance breaking terms in the above mentioned amplitudes must be polynomials of a definite degree in the external momenta, order by order in perturbation theory. But this is precisely the same problem as with the ultra-violet infinities in perturbation theory: the cut-off dependent terms must be polynomials of a definite degree in order for the theory to be renormalizable. Thus the usual proofs of the renormalizability of quantum electrodynamics also guarantee that the unwanted effects of a non-gauge invariant regulator procedure may be off-set by suitably chosen counterterms. Our aim with this procedure is twofold: first, causality is evident, and unitarity can be proven using Cutkosky relations. Secondly, actual calculations are much easier this way, because the counterterms can be fixed easily by applying Ward identities, whereas gauge-invariant regulators become rather complicated particularly at higher orders.

The above point may be illustrated in quantum electrodynamics; in sect. 2 our cut-off procedure is applied to the lowest order photon self energy diagram. Here the unwanted effects of a non-gauge invariant regulator procedure are seen to be such that they can be cured by means of counterterms, one of which has the form of a photon mass term.

One may argue that the method is equivalent with a dispersion relation technique, where the subtraction constants are determined by generalized Ward identities; and that is then sufficient to have a completely gauge invariant theory.

In sect. 3 the situation for massless Yang-Mills fields is outlined. First we use non-gauge invariant regulators, and require that counterterms that remove divergencies are such that Ward identities hold*.

Consequently, three important questions must be answered:

(i) Do the Ward identities determine the hitherto arbitrary coefficients uniquely? Indeed, we will show that only one arbitrary physical constant remains, being the renormalized coupling constant. Two other arbitrary numbers are unobservable and can be chosen by some convention.

(ii) Are there no internal inconsistencies, like in the PCAC case [8, 9], where no renormalizable counterterm could be found in such a way that PCAC and gauge invariance hold at the same time? In sect. 4 we show a combinatorial proof of the Ward identities, and it appears that many shifts of integration variables are necessary for this proof. Nevertheless, there are no inconsistencies, and for the case of one closed loop we prove this by deriving the gauge invariant set of regulators already referred to (sect. 5). Extension of a similar regulator technique to higher-orders

* The method of removing infinities by the use of Ward identities and counterterms for Quantum Electrodynamics is described in Jauch and Rohrlich, Theory of photons and electrons, p. 189.

seems possible, but complicated and tricky; and we shall not bother about it in this article.

(iii) Is the resulting S-matrix unitary? In sect. 5 we generalize the Ward identities, in order to show that the ghost particle intermediate states cancel the intermediate states with non-physically polarized W-particles. Thus in the unitarity equation only physically (i.e. transversely) polarized W-particles occur in the intermediate states.

In appendix A a simple formal path integral derivation of the Feynman rules for Yang-Mills fields and the generalized Ward identities is given for both Landau and Feynman gauge. The rules are listed in appendix B.

We use the notation $k_\mu = (\boldsymbol{k}, ik_0)$; $k^2 = \boldsymbol{k}^2 - k_0^2$. Throughout the paper we confine ourselves to the perturbation expansion. The underlying group here is SU(2), though this is not essential. For simplicity also, no other particles with isospin are taken into account, but introducing them does not give rise to any serious difficulty, as long as the matrix γ^5 and the tensor $\epsilon_{\kappa\lambda\mu\nu}$ do not occur in the Lagrangian.

2. QUANTUM ELECTRODYNAMICS

In this section we review the situation in quantum electrodynamics. We calculate the contribution of the diagram in fig. 1 to the photon self-energy: a spin $\frac{1}{2}$ particle forms a closed loop. We do this calculation in order to show the procedure, which can readily be extended to non-Abelian gauge fields.

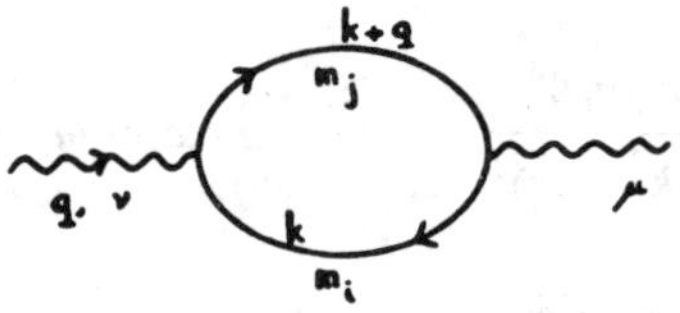

Fig. 1.

The integral diverges quadvutically. Now suppose we regularize by replacing the propagator $(m + i\gamma k)^{-1}$ by

$$\sum_i c_i(m_i + i\gamma k)^{-1} , \tag{2.1}$$

with

$$\sum_i c_i = 0 , \qquad \sum_i c_i m_i = 0 , \qquad \sum_i c_i m_i^2 = 0 , \qquad c_0 = 1 , \qquad m_0 = m , \tag{2.2}$$

and let ultimately m_i go to infinity for $i \neq 0$ (c_i remain finite)

For finite m_i the integral now converges and we may shift the integration variable and integrate symmetrically. Then we have

$$\Pi_{\mu\nu} = -\frac{ie^2}{(2\pi)^4}\int \mathrm{d}^4k \sum_{ij} c_i c_j \frac{\mathrm{Tr}(m_i - i\gamma k)\gamma_\mu(m_j - i\gamma(k+q))\gamma_\nu}{(k^2+m_i^2)((k+q)^2+m_j^2)}$$

$$= -\frac{4ie^2}{(2\pi)^4}\int_0^1 \mathrm{d}x \int \mathrm{d}^4k \sum_{ij} c_i c_j \frac{(m_i m_j + \frac{1}{2}k^2 - x(1-x)q^2)\delta_{\mu\nu} + 2x(1-x)q_\mu q_\nu}{[k^2 + m_i^2 x + m_j^2(1-x) + q^2 x(1-x)]^2}\,. \tag{2.3}$$

Let us define

$$\mu_{ij}^2 \equiv m_i^2 x + m_j^2(1-x) + q^2 x(1-x)\,, \tag{2.4}$$

then we also have

$$\sum_{ij} c_i c_j \mu_{ij}^2 = 0\,, \tag{2.5}$$

and we can evaluate the convergent integral using

$$\int \sum_{ij} c_i c_j \frac{\mathrm{d}^4k}{(k^2+\mu_{ij}^2)^2} = -i\pi^2 \sum_{ij} c_i c_j \log \mu_{ij}^2\,,$$

$$\int \sum_{ij} c_i c_j \frac{m_i m_j\, \mathrm{d}^4k}{(k^2+\mu_{ij}^2)^2} = -i\pi^2 \sum_{ij} c_i c_j m_i m_j \log \mu_{ij}^2\,,$$

$$\int \sum_{ij} c_i c_j \frac{k^2\, \mathrm{d}^4k}{(k^2+\mu_{ij}^2)^2} = 2i\pi^2 \sum_{ij} c_i c_j \mu_{ij}^2 \log \mu_{ij}^2\,, \tag{2.6}$$

so that (2.3) becomes

$$\left(\frac{e}{2\pi}\right)^2 \int_0^1 \mathrm{d}x \sum_{ij} c_i c_j \{\delta_{\mu\nu}(2x(1-x)q^2 + m_i^2 x + m_j^2(1-x) - m_i m_j)$$

$$- 2x(1-x)q_\mu q_\nu\} \log\,[m_i^2 x + m_j^2(1-x) + q^2 x(1-x)]\,. \tag{2.7}$$

To see what happens if for $i \neq 0$ m_i goes to infinity while the c_i remain finite, we split off the term $i = j = 0$ and ignore contributions of order q^2/m_i^2 for $i \neq 0$:

$$\Pi_{\mu\nu} = \left(\frac{e}{2\pi}\right)^2 \int_0^1 \mathrm{d}x \left\{ 2x(1-x)(q^2\delta_{\mu\nu} - q_\mu q_\nu) \left[\log(m^2 + q^2 x(1-x)) \right. \right.$$
$$\left. + \sum_{ij}{}' c_i c_j \log(m_i^2 x + m_j^2(1-x)) \right]$$
$$+ \sum_{ij}{}' c_i c_j \delta_{\mu\nu}(m_i^2 x + m_j^2(1-x) - m_i m_j) \left[\log(m_i^2 x + m_j^2(1-x)) \right.$$
$$\left. \left. + \frac{q^2 x(1-x)}{m_i^2 x + m_j^2(1-x)} \right] + \text{terms } \mathrm{O}\left(\frac{q^2}{m_{i\neq 0}^2}\right) \right\} , \tag{2.8}$$

where Σ'_{ij} denotes the sum over all i and j except the term with both $i = j = 0$. This result does not satisfy the usual gauge condition

$$q_\mu \Pi_{\mu\nu}(q) = 0 , \tag{2.9}$$

and the renormalized mass of the photon is not evidently zero.

Of course, the reason is that our regulators are not gauge invariant; a vertex where a photon line is attached to particle lines with different masses is not allowed. If we had used Pauli-Vilars-Gupta regulator fields instead of the propagators (2.1), that is, if in formulae (2.3)–(2.8) $\Sigma_{ij}\, c_i c_j$ is replaced by

$$\sum_{ij} c_i \delta_{ij} ,$$

then the second term in (2.8) would vanish identically and eq. (2.9) would be fulfilled [6, 7].

However, it is important to note that the gauge non-invariant term in (2.8) is only a polynomial of rank one as a function of q^2. Let us abbreviate it by

$$\left(\frac{e}{2\pi}\right)^2 (M + Lq^2)\delta_{\mu\nu} . \tag{2.10}$$

It can be removed from expression (2.8) if we add a simple counterterm into the Lagrangian*

$$\Delta\mathcal{L} = -\tfrac{1}{2}\left(\frac{e}{2\pi}\right)^2 (MA_\mu^2 + L(\partial_\nu A_\mu)^2) . \tag{2.11}$$

* This implies that terms in the Lagrangian are renormalized, not the fields, as is often done. The difference is merely a scale transformation of the bare quantities.

These terms are local and have dimension less than or equal to four, so that causality and renormalizability are not destroyed.

This can be seen to be a very general feature: instead of the gauge invariant Pauli-Villars-Gupta regularization technique we could just as well regularize with the revised propagator (2.1) (which is a non-gauge invariant procedure) and add to the Lagrangian as many local counterterms with dimension less than or equal to four, as desirable. All arbitrary coefficients can then be fixed by requiring the validity of identities like (2.9).

Equations like (2.9) will be called generalized Ward identities* from now on. They are derived from the usual Ward-Takahashi identity

$$(p'-p)_\mu \Gamma_\mu(p',p) = S_F'^{-1}(p') - S_F'^{-1}(p)\,, \tag{2.12}$$

which can be symbolized as

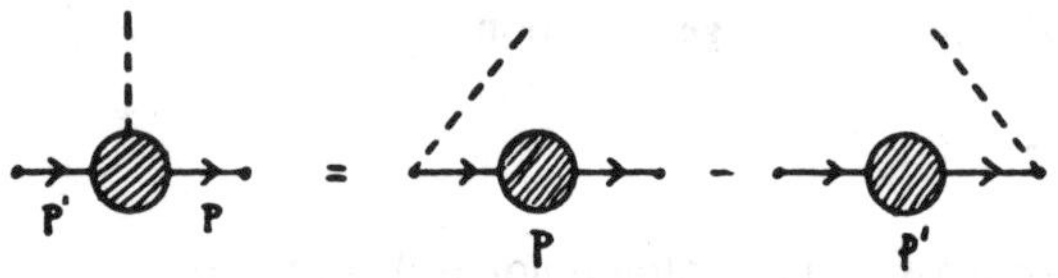

Here the dashed line denotes a "scalar photon" (a photon line with polarization vector proportional to its own momentum). This identity can be used to derive other equalities for diagrams. For instance

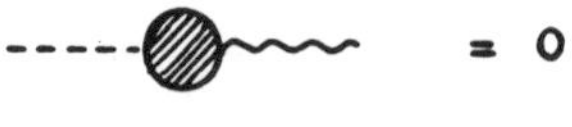

which is precisely eq. (2.9).

In our example we see that the coefficient in front of $(q^2\delta_{\mu\nu} - q_\mu q_\nu)$ is still unspecified. This is because we can add freely counter terms proportional to $F_{\mu\nu}F_{\mu\nu}$ to the Lagrangian because they are gauge invariant themselves. It corresponds to a scale transformation in our definition of the field A_μ. So the freedom we have is only a freedom in definition. The most convenient choice is to keep the matrix element of $A_\mu(x)$ between the vacuum and the one-photon state fixed:

$$\langle 0|\, A_\mu(x)\, |k,\epsilon\rangle = \epsilon_\mu e^{ikx}\,. \tag{2.13}$$

The renormalized propagator must then have a pole with residue unity at $k^2 = 0$, just as the bare propagator.

* See e.g. J.D.Bjorken and S.D.Drell, Relativistic quantum fields.

So (2.8) must vanish on the mass shell:

$$\Pi_{\mu\nu}(q^2 = 0) = 0 \,, \tag{2.14}$$

and we derive finally

$$\Pi_{\mu\nu}(q^2) = \left(\frac{e}{2\pi}\right)^2 \int_0^1 \mathrm{d}x\, 2x(1-x)(q^2\delta_{\mu\nu} - q_\mu q_\nu)$$

$$\times \left[\log(m^2 + q^2 x(1-x)) - \log m^2\right] \tag{2.15}$$

Once we know that the above mentioned procedure works well, we can go even further and leave the particular set of regulator fields or propagators altogether unspecified. Instead of the identities (2.6) we may use the symbolic expressions:

$$\int \frac{\mathrm{d}^4 k}{(k^2+\mu^2)^2} = -i\pi^2 \log \mu^2 + D_1 \,,$$

$$\int \frac{k^2 \mathrm{d}^4 k}{(k^2+\mu^2)^2} = 2i\pi^2\mu^2 \log\mu^2 + D_2 + D_3\mu^2 \,, \tag{2.16}$$

indicating only the terms $i = j = 0$ in eq. (2.6) explicitly.

The constants $D_{1,2,3}$ depend on the diagram for which the integral is evaluated, but do not depend on μ. Of course, expressions like (2.15) must be handled with great care, but in general they give a very clear idea of where arbitrary numbers enter in the theory. The arbitrariness can only be removed if some additional symmetry property of the system is known, like gauge invariance.

3. MASSLESS YANG-MILLS FIELDS

We now consider the Lagrangian of the massless Yang-Mills theory [10]:

$$\mathcal{L}_{\mathrm{YM}} = -\tfrac{1}{4} G_{\mu\nu} G_{\mu\nu} \,, \tag{3.1}$$

$$G^a_{\mu\nu} = \partial_\mu W^a_\nu - \partial_\nu W^a_\mu + g\,\epsilon_{abc} W^b_\mu W^c_\nu \,, \tag{3.2}$$

which is invariant under the local gauge transformation

$$W'^a_\mu(x) = f_{ab}(x)\, W^b_\mu(x) - \frac{1}{2g}\epsilon_{abc}(\partial_\mu f(x) f^{-1}(x))_{cb} \,. \tag{3.3}$$

If one wants to apply conventional field theory to this model one encounters difficulties [1]. Mandelstam [2] derived Feynman rules for the system using path dependent Green's functions. DeWitt, Faddeev and Popov [3, 4] derived the same rules using a path integral method. We sketch a simple path integral derivation for different gauges in appendix A, and the resulting rules are listed in appendix B.

An auxiliary "ghost particle" appears. In fact it will be seen to cancel the third polarization direction of the W-particles. There is an arbitrariness in gauge, expressed in the parameter λ in the propagator

$$\frac{\delta_{\mu\nu} - \lambda \frac{k_\mu k_\nu}{k^2}}{k^2}.$$

Other gauges, like the transversal, can be described in the same way [5].

A path integral derivation of generelized Ward identities is also given in appendix A. A "scalar" W-line

$$-\!-\!-\!\overset{k}{-}\!-\!-\!\blacksquare$$

is defined as a W-line with polarization vector $-ik_\mu$:

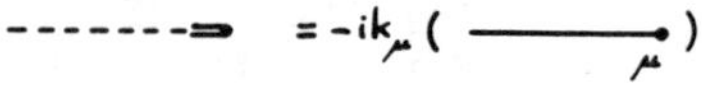

$$-\!-\!-\!-\!-\!-\!\blacksquare \;=\; -ik_\mu \left(\underset{\mu}{\xrightarrow{\hspace{1.2cm}}} \right) \tag{3.4}$$

A "transversal line" has a polarization vector ϵ_μ satisfying

$$k_\mu \epsilon_\mu = 0 ,$$

$$\epsilon_4 = 0 . \tag{3.5}$$

A generalized Ward identity is then:

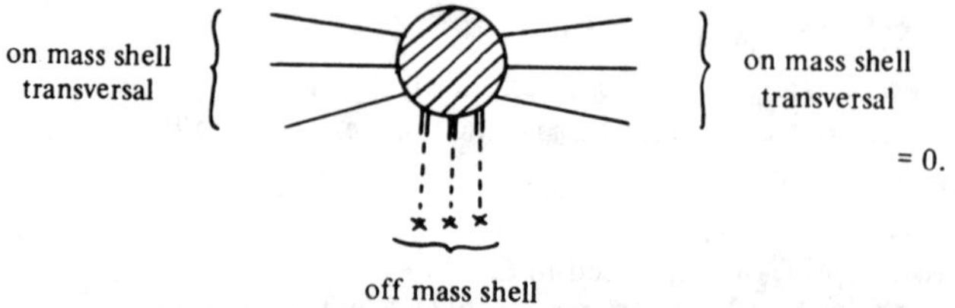

$$\left. \begin{array}{c} \text{on mass shell} \\ \text{transversal} \end{array} \right\} \qquad \left\{ \begin{array}{c} \text{on mass shell} \\ \text{transversal} \end{array} \right.$$

$$= 0. \tag{3.6}$$

off mass shell

Amplitudes with "longitudinal W-lines" ($\epsilon_\mu = (-1)^{\delta_{\mu 4}} k_\mu$) satisfy more complicated Ward identities (cf. sect. 6).

These identities are seen to express the gauge invariance of the theory. For example, the equivalence of the Feynman ($\lambda = 0$) and the Landau gauge ($\lambda = 1$) can be proven using (3.6).

Without much effort one now can verify that the Ward identities are sufficient to prescribe all subtraction constants uniquely, except for the coupling constant. The only needed (and allowed) counterterms are of the following type

 $\delta_{ab}[\delta_{\mu\nu}(C_o + C_1 k^2) + C_2 k_\mu k_\nu]$, (3.7a)

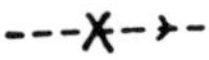 $\delta_{ab} C_3 k^2$, (3.7b)

 $-igC_4\,\epsilon_{abc}[\delta_{\beta\gamma}(q-p)_\alpha + \delta_{\gamma\alpha}(k-q)_\beta + \delta_{\alpha\beta}(p-k)_\gamma]$, (3.7c)

$$-g^2 C_5\,[\epsilon_{gac}\epsilon_{gbd}\delta_{\alpha\beta}\delta_{\gamma\delta} + \text{permutations}]$$
$$+ g^2 C_6\,[\delta_{ab}\delta_{cd}(\delta_{\alpha\beta}\delta_{\gamma\delta} + \delta_{\alpha\delta}\delta_{\gamma\beta}) + \text{permutations}] \;, \quad (3.7d)$$

$-igC_7 q_\alpha$, (3.7e)

(vertices with more φ-lines do not occur because any amplitude must contain as a factor the momenta of the *outgoing* φ-particles (or ingoing φ-antiparticles) as can be seen from the rules (B.1)–(B.6).

The numbers C_1, C_3 and C_4 may be chosen freely, using some convention for the physical amplitude of the W- and φ-fields, and the definition of the physical coupling constant $g^{\text{renormalized}}$. In the Landau gauge moreover, C_2 is immaterial.

According to the Ward identity for the self-energy correction one must have:

 +  = 0 (3.8)

where the counterterm is indicated explicitly, while

 $= \delta_{ab}(C_o k^2 + C_1 k^4 + C_2 k^4)$.

So C_o is fixed and C_2 is expressed in C_1.

Indeed, an actual calculation of the second-order self energy diagram in the Feynman gauge using the symbolic expressions (2.16) shows:

$$\Pi^{ab}_{\mu\nu} = -\frac{g^2}{(4\pi)^2}\delta_{ab}[\tfrac{10}{3}k^2\delta_{\mu\nu} - \tfrac{10}{3}k_\mu k_\nu]\log k^2 + \delta_{ab}[\delta_{\mu\nu}(C_o + C_1 k^2) + C_2 k_\mu k_\nu] \;, \quad (3.9)$$

so indeed the Ward identity (3.8) can be satisfied:

$$C_0 = 0\,, \qquad C_1 + C_2 = 0\,.$$

The renormalized mass, depending on C_0, turns out to be zero. Note that the coefficients in front of the terms $k^2\delta_{\mu\nu}\log k^2$ and $k_\mu k_\nu \log k^2$ would not be the same if the φ-particle loop had been left out.

For the four-point function we have,

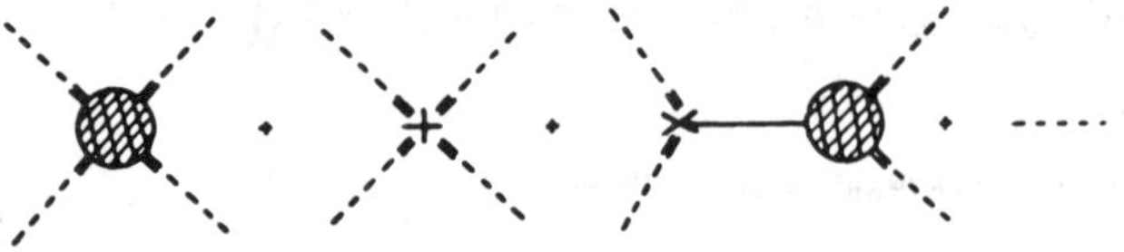 $= 0$ (3.10)

while

 $\neq 0$ if C_5 or $C_6 \neq 0$,

so C_5 and C_6 are expressed in terms of the other subtraction constants.

Finally, C_7 can be determined by applying the Ward identity (3.8) for the higher order self-energy diagram of the W-particle, using for instance the BPH procedure of renormalization [11], and the above mentioned observation that

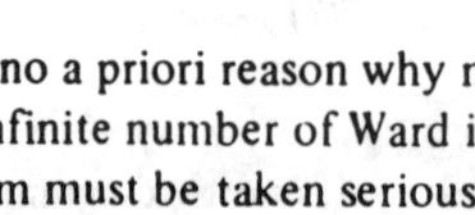 $\neq 0$

4. COMBINATORIAL PROOF OF THE WARD IDENTITIES

There is no a priori reason why no conflict situation could emerge if we try to satisfy an infinite number of Ward identities using a finite number of counter terms. This problem must be taken seriously, because the algebraic proof of the Ward identities, which will be given below, involves many shifts of integration variables. A proof of the absence of such a conflict will be given only for one closed loop.

Let us introduce some conventions:

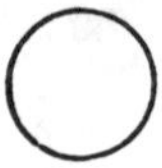

stands for the set of diagrams of a given order in g, and a given number of external transversal W-lines (cf. (3.5)) on mass-shell. There are no longitudinal or scalar external lines. They are denoted explicitly:

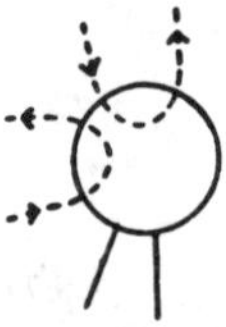

stands for the set of diagrams of a given order in g, and a given number of external transversal W-lines, as above, and in addition a number of external ghost lines and W-lines with arbitrary polarization and momentum, as drawn. The ghost lines are followed inside the graph, which is possible because (B.5) is the only kind of vertex for the ghost particle. The graphs may be disconnected.

The combinatorial proof of the validity of the Ward identities is as follows. From now on we use the Feynman gauge.

Let us perform an infinitesimal gauge transformation in the Lagrangian (3.1):

$$\mathcal{L}_{\mathrm{YM}} = -\tfrac{1}{4} G_{\mu\nu} G_{\mu\nu} = \mathcal{L}'_{\mathrm{YM}} = -\tfrac{1}{4} G'_{\mu\nu} G'_{\mu\nu}\,, \tag{4.1}$$

$$W'^{a}_{\mu}(x) = W^{a}_{\mu}(x) + g\,\epsilon_{abc}\Lambda^{b}(x)\,W^{c}_{\mu}(x) - \partial_\mu \Lambda^{a}(x)\,, \tag{4.2}$$

Λ is some external source which, according to (4.1), remains uncoupled.

Then we must add to all vertices (B.3) and (B.4) all vertices we get from (B.3) and (B.4) if one of the W-lines  has been substituted by

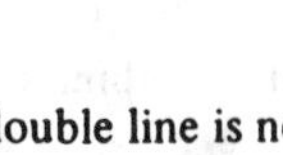

$$-g\,\epsilon_{abc}\,, \tag{4.3a}$$

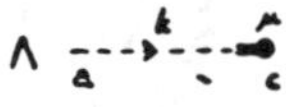

$$-\delta_{ac} i k_\mu\,. \tag{4.3b}$$

(Note: the double line is not meant to be a propagator; (4.3a) is a part of one vertex). Also from the free part of $\mathcal{L}_{\mathrm{YM}}$ we derive an extra vertex term in $\mathcal{L}'_{\mathrm{YM}}$, which appears to be

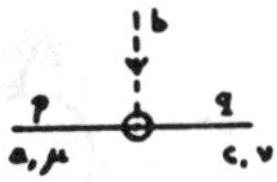

$$-g\,\epsilon_{abc}(\delta_{\mu\nu}p^2 - p_\mu p_\nu - \delta_{\mu\nu}q^2 + q_\mu q_\nu)\,. \tag{4.3c}$$

The ghost particle resulting from the use of a certain gauge condition, is not included in our gauge transformation (4.2). Hence, its vertices and propagators are unchanged.

Now it is easy to verify that up to first order in Λ all extra vertices cancel, which they should do. In diagrams:

(4.4a)

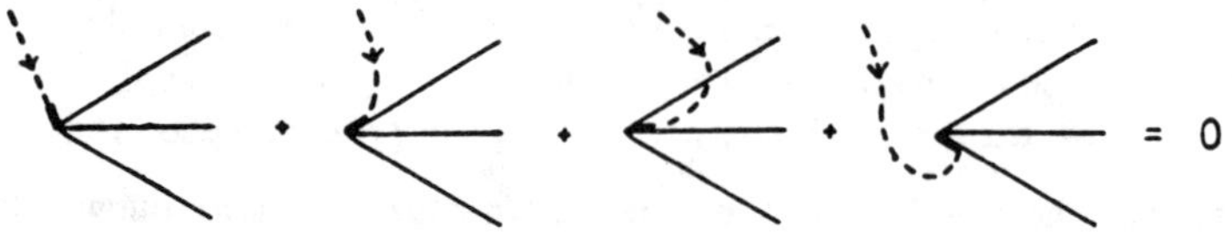

(4.4b)

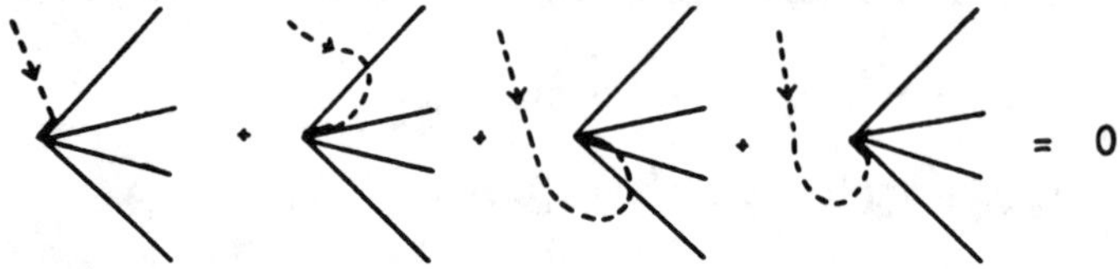

(4.4c)

(4.3b) is of the type which occurs in our Ward identities. We now see that it can be replaced by (4.3a) and (4.3c) using eqs. (4.4), except for the connections with the ghost particle. So as

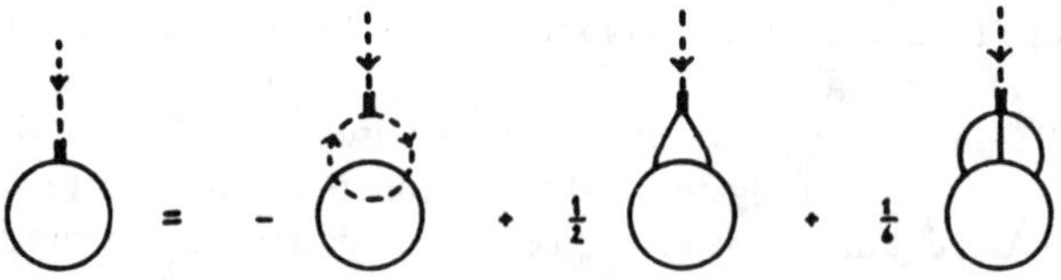

(4.5)

(Note the explicitly written minus sign for the φ-loop and the combinatory factors, because the blobs are already symmetrized) we have

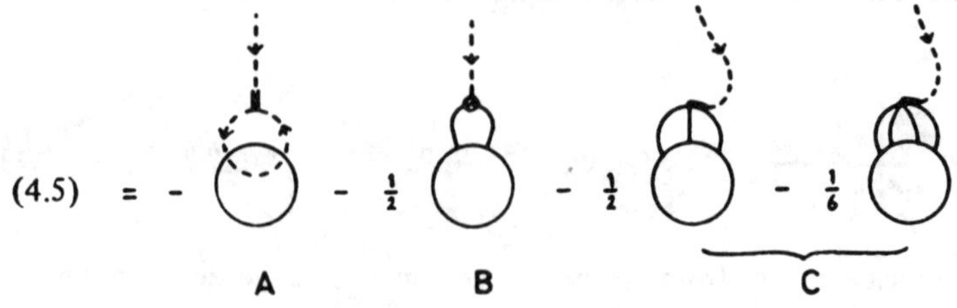

(4.6)

eq. (4.6) may be written as

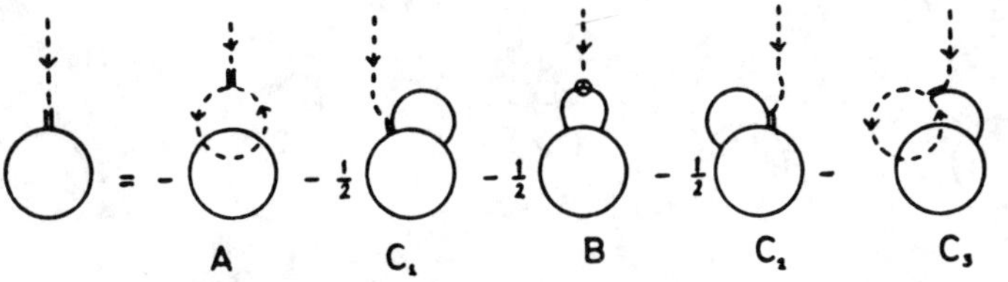

(4.7)

(Of course, C_1 equals C_2.)

Note that C_3 cancels those diagrams contained in C_1 and C_2 where the double line is attached to a ghost vertex.

The next step is a propagator identity which is related to invariance of the gauge condition under special gauge transformations Λ with $\partial_\mu(\partial_\mu \Lambda^a + g\epsilon_{abc} W_\mu^b \Lambda^c) = 0$:

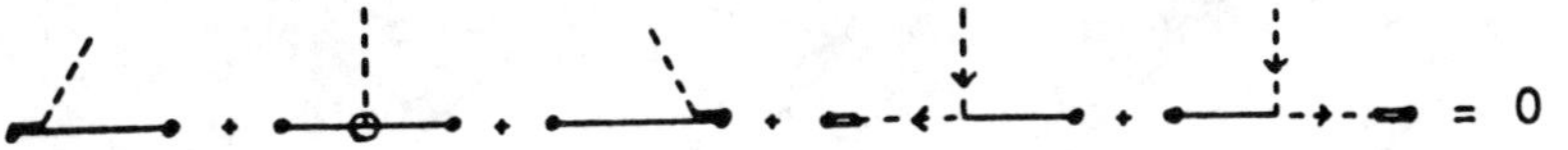

(4.8a)

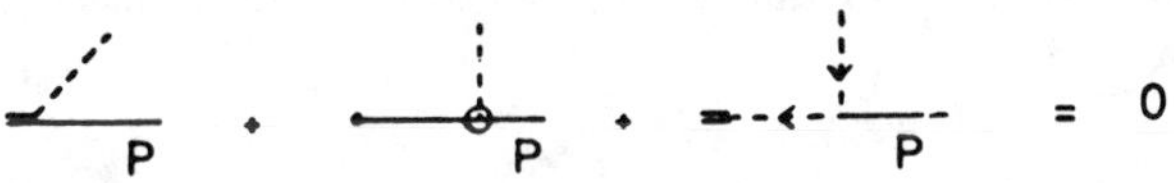

(4.8b)

The P denotes a transversal W-line on mass shell (cf. (3.5)). Note again that the double line is no propagator.

Eq. (4.8a) is the Yang-Mills counterpart of the usual Ward-Takahashi identity (2.12) for bare electron propagators and vertex functions. In the last two terms the dashed line ("Λ-line") has the same vertices and propagators as the ghost particle ("φ-line", compare (B.2) and (B.5)). If some of the lines in (4.8) are parts of a closed loop these identities are true provided one may shift integration variables. This is the reason why subtraction constants must be chosen carefully.

Applying eqs. (4.8) to eq. (4.7) we find

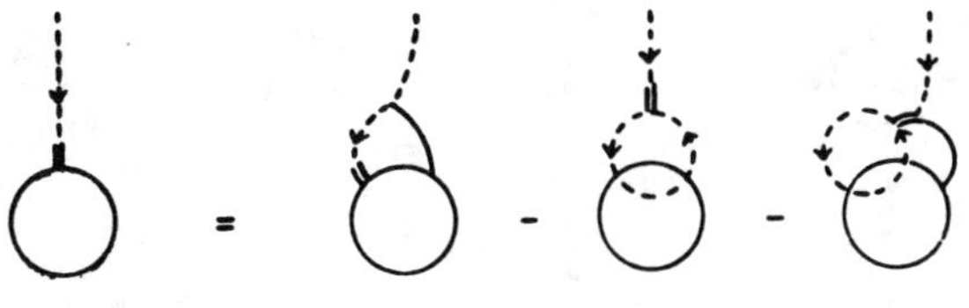

(4.9)

Eq. (4.9) can now be iterated, but then we must include the possibility that the

Λ-line forms a closed loop and is attached to itself. The result is:

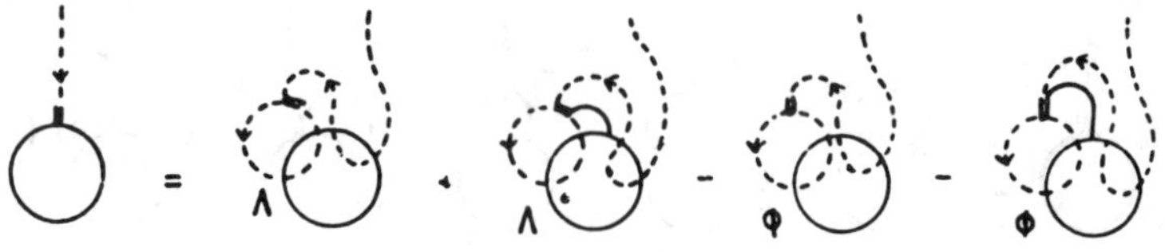

(4.10)

Using one more identity

(4.11)

we have

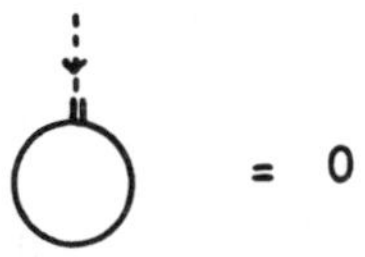

(4.12)

Substituting (4.3b) into (4.3a) one obtains another vertex, for which the following equation holds:

(4.13)

Consequently the derivation remains valid even if there are more off-mass shell scalar W-lines:

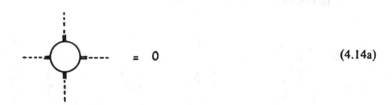

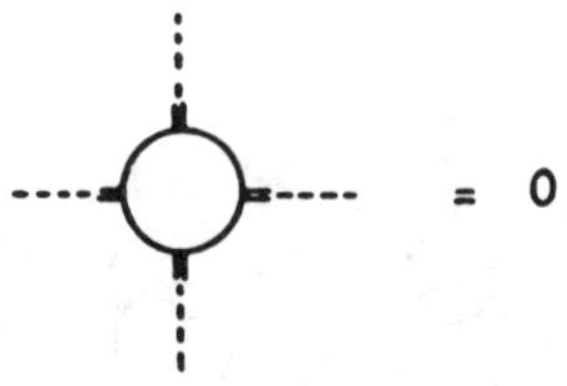

(4.14a)

which is the graphical notation for the formula

$$\frac{\partial}{\partial x^1_{\mu_1}} \cdots \frac{\partial}{\partial x^N_{\mu_N}} \langle \mathrm{out} | T^*(W^{a_1}_{\mu_1}(x^1) \ldots W^{a_N}_{\mu_N}(x^N)) | \mathrm{in} \rangle = 0 \,, \tag{4.14b}$$

in conventional field theory.

From this algebraic derivation of the Ward identities we draw the following conclusion: if we succeed in regularizing graphs containing one of the auxiliary vertices (4.3a)–(4.3c) in such a way that eqs. (4.8), (4.11) and (4.13) remain valid *also inside closed loops*, then we acquire gauge invariant amplitudes (amplitudes satisfying (4.14)).

5. GAUGE INVARIANT REGULATORS

In this section we construct a set of regulators satisfying all requirements formulated in the previous section, but we confine ourselves to the one closed-loop case. The mere existence of these regulators implies that no conflict situation arises if one uses Ward identities for calculating subtraction constants in the first quantum-mechanical correction, instead of gauge invariant regulators.

The procedure is as follows. Note that the identities (4.8), (4.11) and (4.13) are not only valid in a four-dimensional Minkowsky space, but we may add another dimension. Then the momenta k_μ have five components, and the fields W^a_μ have 15 components. Let for all diagrams with one closed loop the external momenta be in the Minkowsky space, that is, only their first four components differ from zero. Let the momenta inside the closed loop have one more component of fixed length M in a fixed fifth direction. Because of conservation of momentum, M is the same for all propagators of the closed loop. With this interpretation in mind, we may now reformulate the Feynman rules, which now contain an extra parameter M. Furthermore, they depend on which of the propagators belong to the closed loop; those propagators will be denoted by a *.

The W- and φ-propagators inside the closed loop are replaced by:

$$\frac{\delta_{ab}\delta_{\mu\nu}}{k^2 + M^2} \,, \tag{5.1a}$$

$$\frac{\delta_{ab}}{k^2 + M^2} \,. \tag{5.1b}$$

The vertices (B.3)–(B.5) remain the same, as well as the propagators (B.1) and (B.2) in the tree parts of a graph. In (5.1a) we let the indices μ, ν run from 1 to 4 as usual. The fifth polarization direction of the W-field is treated as a new particle, which only occurs inside the closed loop:

$$\frac{\delta_{ab}}{k^2 + M^2} \, . \tag{5.1c}$$

It has the vertices:

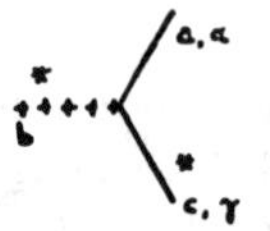

$$Mg\, \epsilon_{abc} \delta_{\alpha\gamma} \, , \tag{5.1d}$$

$$-Mg\, \epsilon_{abc} \, , \tag{5.1e}$$

(note that the factors $\pm\, i$ at each end of a crossed line have cancelled), and

$$-ig\, \epsilon_{abc} (q-p)_\alpha \, , \tag{5.1f}$$

$$-g^2 (\epsilon_{gac} \epsilon_{gbd} + \epsilon_{gad} \epsilon_{gbc}) \delta_{\alpha\beta} \, . \tag{5.1g}$$

Now with vertices (5.1f) and (5.1g) one may have closed loops of crossed lines, but these contributions are gauge invariant themselves, since the vertices (5.1f) and (5.1g) are precisely those of an ordinary isospin one scalar particle. So we may exclude diagrams with closed loops of crossed lines without invalidating the Ward identities. The above vertices with the rule of no closed loop of crossed lines define a set of diagrams which, up to one loop, satisfy the Ward identities. For $M = 0$ we have the diagrams of the massless theory. For M non-zero we have diagrams that may be used as regulator diagrams.

Consider now the sum of diagrams of the massless theory and regulator diagrams. Choosing the appropriate integration variables (remember that each individual contribution may be infinite, and relative shifts of integration variables may give different results) and furthermore regulators with masses M_i and signs e_i, in such a way that

$$\sum e_i = 0\,, \qquad e_0 = 1\,,$$
$$\sum e_i M_i^2 = 0\,, \qquad M_0 = 0\,, \tag{5.2}$$

we obtain a finite result.

One may choose convenient, finite values for

$$\sum_{i \neq 0} e_i \log M_i^2 = -A\,, \qquad \sum e_i M_i^2 \log M_i^2 = B\,. \tag{5.3}$$

In the limit $M_{i\neq 0} \to \infty$ we find the desired gauge invariant amplitudes.

Let us demonstrate this regulator technique for the second order self-energy contributions to the W-propagator:

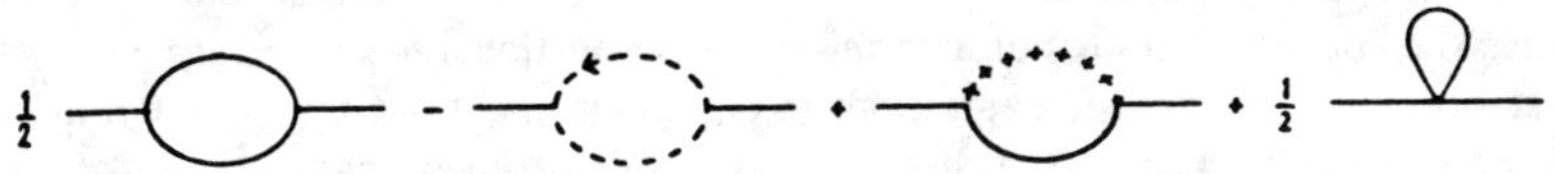
(5.4)

Using expressions (2.6) we find

$$\Pi^{ab}_{\mu\nu}(k) = \frac{-g^2}{(4\pi)^2}\delta_{ab}\int_0^1 \mathrm{d}x \sum_i e_i [\{k^2(5-10x(1-x))\delta_{\mu\nu}$$
$$-k_\mu k_\nu(2+8x(1-x))\}\log(M_i^2+x(1-x)k^2)$$
$$-6M_i^2\delta_{\mu\nu}\log(M_i^2+x(1-x)k^2)+6M_i^2\delta_{\mu\nu}\log M_i^2]\,. \tag{5.5}$$

Indeed, one may convince oneself that this satisfies the Ward identity

$$k_\mu k_\nu \Pi^{ab}_{\mu\nu}(k) = 0\,. \tag{5.6}$$

In the limit $M_{i\neq 0} \to \infty$ we have

$$\Pi^{ab}_{\mu\nu} = -\frac{g^2}{(4\pi)^2}\delta_{ab}(k^2\delta_{\mu\nu}-k_\mu k_\nu)\,[\tfrac{10}{3}\log k^2 - \tfrac{10}{3}A - \tfrac{56}{9}]\,. \tag{5.7}$$

The number A is the logarithm of a suitably chosen reference mass. It must have the same value for all graphs with one closed loop.

It must be emphasized that even if our regulator method appears very similar to the Pauli-Villars method it is in fact very different. The regulators do not correspond to fields in Lagrangians etc., and the procedure works only for one closed loop. In fact the above is just a convenient way of implementing the scheme proposed in the beginning. Tentative investigation shows that probably a modification of this regulator technique can produce finite gauge invariant amplitudes at higher orders. As yet we shall consider this as a conjecture. It is important to note that this technique of introducing more dimensions only works if the matrix γ^5 and the tensor $\epsilon_{\kappa\lambda\mu\nu}$ do not occur in the Lagrangian.

6. UNITARITY

In proving unitarity of the S-matrix one has to deal with on mass-shell amplitudes. We are then confronted with infrared difficulties. Now if we add a very small mass term κ^2 in the propagators, then the on mass-shell amplitudes (in finite order of g) are proportional to some power of $\log \kappa^2$. The Ward identities however, are violated with terms proportional to κ^2, $\kappa^2 \log \kappa^2$, etc. So we can still use these Ward identities keeping $\log \kappa^2$ finite, but ignoring terms proportional to κ^2, $\kappa^2 \log \kappa^2$ etc. For instance, in the regularized expressions in sect. 5 we might put $M_0 = \kappa \neq 0$, but ignore the crossed line with mass κ, because it is coupled with strength κ^2.

We shall not go into the problems of the physical interpretation of these infrared divergencies.

To compute imaginary parts we shall make use of the well-known Cutkosky rules [12]:

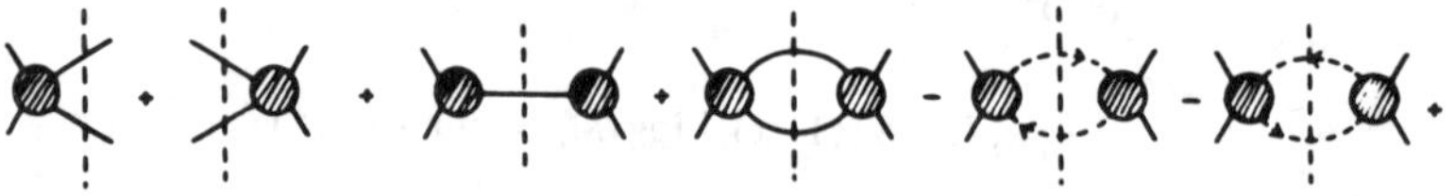

+ (graphs with more than two lines cut through) = 0 , (6.1)

where at the right-hand side of the dashed line the $i\epsilon$ in the propagators is replaced by $-i\epsilon$, and an extra minus sign is introduced for each propagator and each vertex. The blobs are at least of order one in g. Now, if in the blobs of (6.1) all graphs are added, including disconnected ones, such that the total order in g is kept fixed, then equation (6.1) is an identity, whatever the choice of our subtraction coefficients may be, provided that we use the following rules:

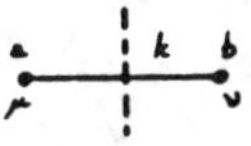

$$2\pi\delta(k^2)\theta(k_0)\delta_{\mu\nu}\delta_{ab}\,, \tag{6.2}$$

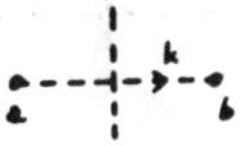

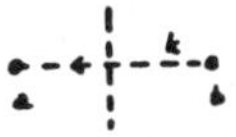

$$\left.\right\}\; 2\pi\delta(k^2)\theta(k_0^2)\delta_{ab}\,, \tag{6.3}$$

(a dashed line going through an external particle-line has no special meaning, except that it separates the ingoing lines from the outgoing lines).

Now if we can prove a slightly different equation,

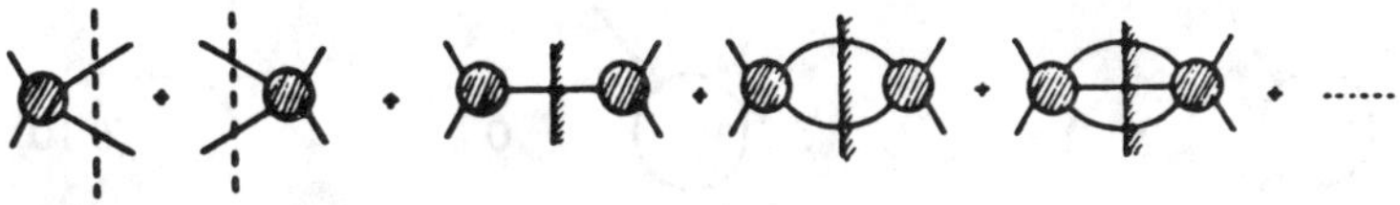

$= 0$ (6.4)

with

standing for $2\pi\delta(k^2)\theta(k_0)\delta_{ab}\left(\delta_{\mu\nu}-\dfrac{k_\mu k_\nu}{|\boldsymbol{k}|^2}\right)(1-\delta_{\mu4})(1-\delta_{\mu4})\,,$ (6.5)

then unitarity has bee.. proven, for the case that bosons with a given isospin have only two helicity stat..., like the photons. We shall prove eq. (6.4) from eq. (6.1) provided that we only look at the transverse components of the other outgoing lines. Let us first consider the case of only two intermediate particles. Define

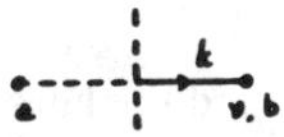

$$2\pi\delta(k^2)\theta(k_0)\delta_{ab}\frac{-i\bar{k}_\nu}{2|\boldsymbol{k}|^2}\,, \qquad \bar{k}_\nu \equiv (-1)^{\delta_{\nu 4}}\,k_\nu$$

(6.6)

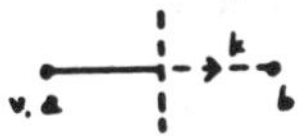

$$2\pi\delta(k^2)\theta(k_0)\delta_{ab}\frac{i\bar{k}_\nu}{2|\boldsymbol{k}|^2}\,.$$

A useful equation is:

$$\delta_{\mu\nu} = \tfrac{1}{2}\,\frac{k_\mu\bar{k}_\nu + \bar{k}_\mu k_\nu}{|\boldsymbol{k}|^2} + \left(\delta_{\mu\nu}-\frac{k_\mu k_\nu}{|\boldsymbol{k}|^2}\right)(1-\delta_{\mu4})(1-\delta_{\nu4}) \qquad \text{if } k^2 = 0\,. \tag{6.7}$$

Symbolically:

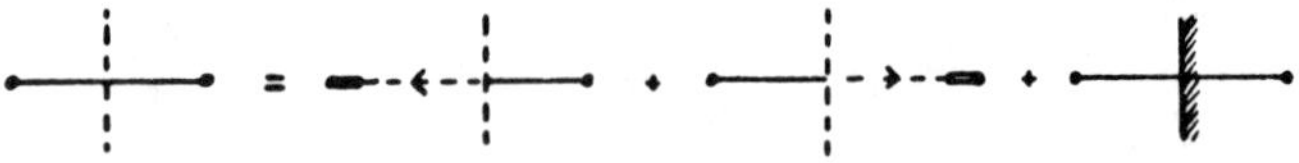 (6.8)

Also we have

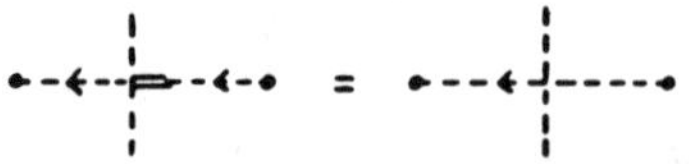 (6.9)

We shall apply the Ward identities

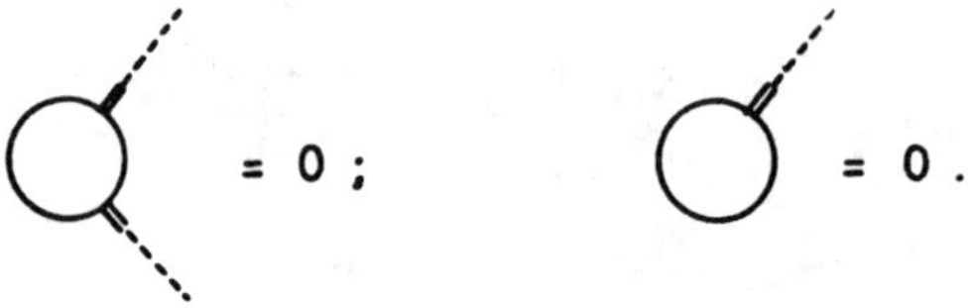 (6.10)

Moreover, we need a generalization of the Ward identities (4.14) for amplitudes with on mass-shell ghost particles and non-physically polarized W-particles, in particular W-particles with polarization vector e_μ not satisfying $k_\mu e_\mu = 0$. Formula (4.8b) is extended to

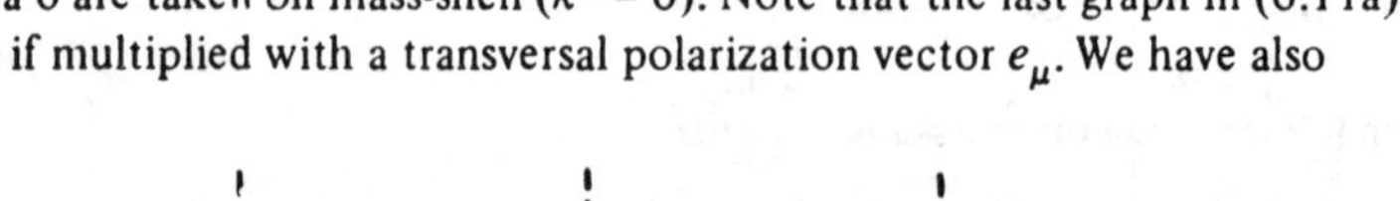 (6.11a)

where the arrow in $\dashrightarrow \mu$ stands for multiplication with $-ik_\mu$, and the lines with a o are taken on mass-shell ($k^2 = 0$). Note that the last graph in (6.11a) vanishes if multiplied with a transversal polarization vector e_μ. We have also

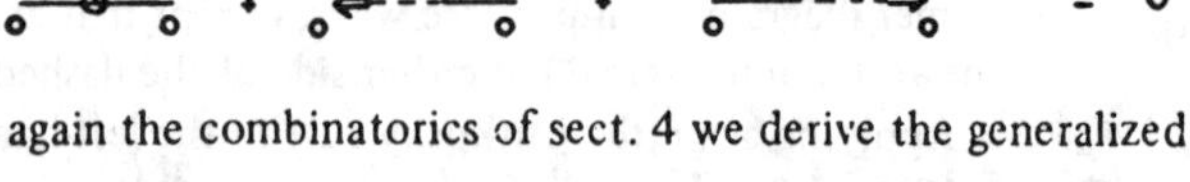 (6.11b)

Applying again the combinatorics of sect. 4 we derive the generalized Ward identity

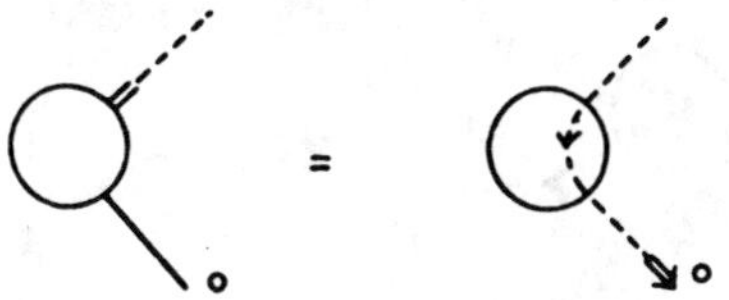

(6.12)

(This identity is not altered if other gauge invariant interactions are introduced. The other isospin particles must then be on mass-shell).

Equipped with eqs. (6.8), (6.9), (6.10) and (6.12) we derive

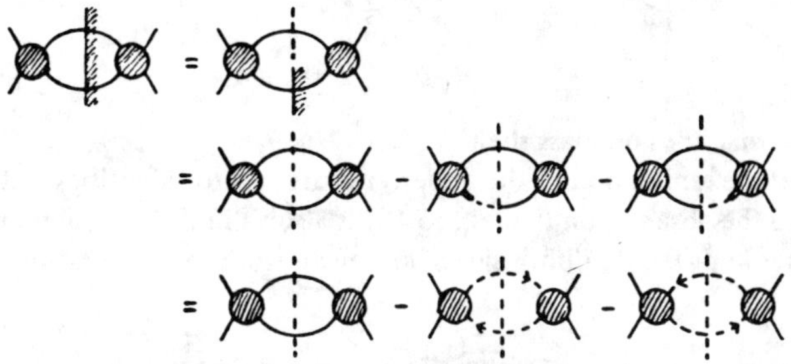

(6.13)

from which eq. (6.4) follows, as long as we confine ourselves to the contributions with at most two particles in the intermediate states.

In the same way it can be shown for intermediate states with more than two particles that the ghost particles cancel the non-physical polarization directions of the W-bosons. In principle this can be verified by writing down further generalizations of the Ward identity (6.12), but a more straightforward proof of this cancellation goes as follows. We apply induction with respect to the number of particles in the intermediate states.

Suppose we have a diagram

(the external lines being on mass-shell). Let then

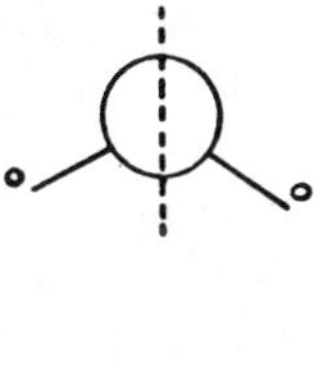

stand for the sum of all graphs acquired by cutting the former diagram in all possible ways, except that at least one vertex must remain at either side of the dashed line.

Applying again the Cutkosky rule to the left-hand side of (6.12):

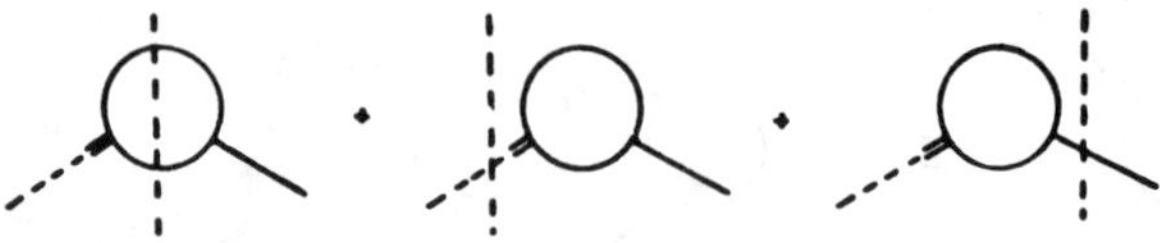

$= 0$ (6.14)

one derives easily:

(6.15)

with the external lines on mass shell.

Now careful examination of the underlying propagator identities and combinatorics leads to the observation that eq. (6.15) is also valid if the total number of cut propagators is kept fixed at both sides. So if we introduce the notation

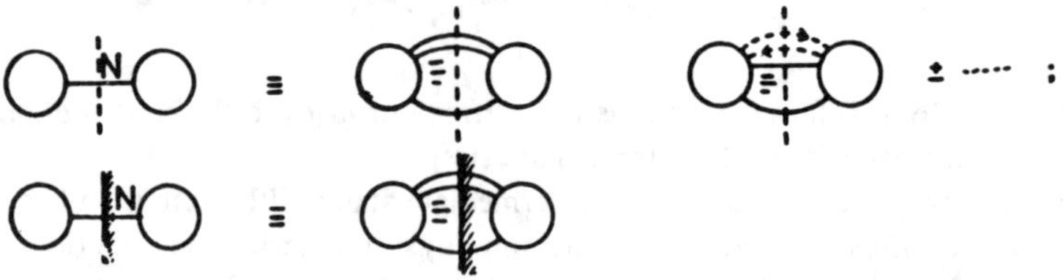

(6.16)

N denoting the total number of cut propagators, then (6.15) reads:

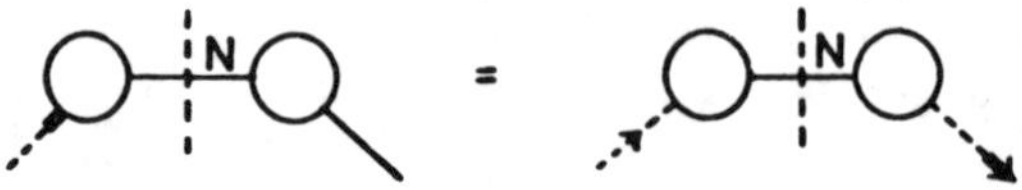

(6.17)

for all *N*. Moreover, one can impose the restriction that the cutting line must pass through both of the explicitly denoted external lines in (6.17), and then we get:

(6.18)

Now suppose that for a certain value of *N*

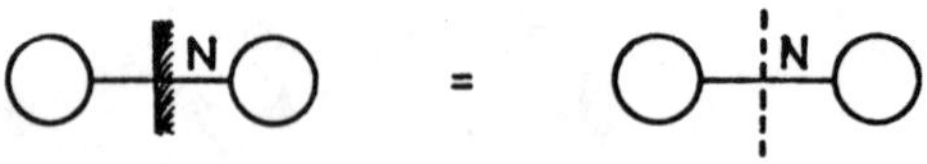

(6.19)

then we have

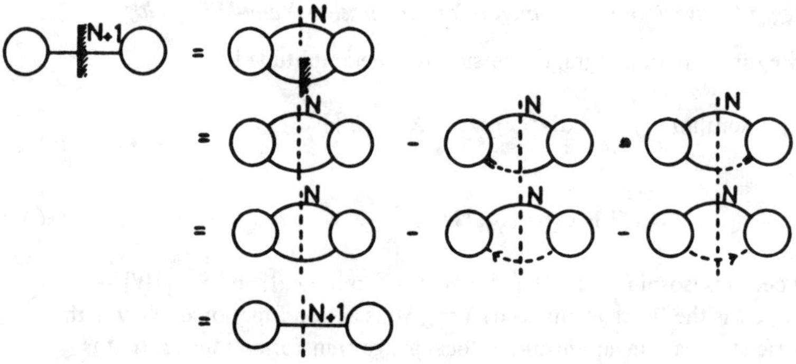

(6.20)

which completes the proof by induction.

So the S-matrix is unitary in a Hilbert space with only plane wave W-particle states, in which each particle has helicity ± 1. A necessary condition is that subtraction constants are chosen in such a way that all generalized Ward identities are satisfied.

7. CONCLUSION

Massless YM fields can be renormalized. A formal regulator procedure exists, at least for diagrams with one closed loop, but the simplest way to deal with the divergencies is to use the subtracted expressions (2.16) for divergent integrals, calculating subtraction constants by means of the Ward identities. In this article we have not gone into the details of a regulator technique for diagrams with more loops, so as yet a consistency proof of the Ward identity method for removing overlapping divergencies, is lacking.

With this restriction, we have proven that the resulting S-matrix is unitary, if infrared divergencies are dealt with in a proper way. There is only one physical parameter in the theory, which is the coupling constant g. The renormalized mass of the bosons is zero (at least, in perturbation theory).

The author is greatly indebted to Prof. M.Veltman for many helpful discussions and critical remarks.

APPENDIX A

Path integral derivation of Feynman rules for massless Yang-Mills fields

The Feynman path integral expression for the amplitude is

$$\langle \text{out}|\text{in}\rangle = \int \prod_{x,\mu,a} \mathrm{d}W_\mu^a(x) \exp\{iS_{\text{YM}}[W]\}$$

$$\equiv \int \mathcal{D}W \exp iS_{\text{YM}}[W] \,, \tag{A.1}$$

where a denotes isospin, μ the Lorentz vector component, and $S_{\text{YM}}[W] = \int \mathcal{L}_{\text{YM}}(x)\,\mathrm{d}x$ is the (unrenormalized) Yang-Mills action functional. Now if the asymptotic states are invariant under local gauge transformations Ω, that is

$$\Omega|\text{in}\rangle = |\text{in}\rangle \,, \qquad \Omega|\text{out}\rangle = |\text{out}\rangle \,,$$

then the integrand, as well as the measure $\mathcal{D}W$, are invariant under local gauge transformations.

In order to extract the infinite constant arising from this invariance we alter expression (A.1) by multiplying with a delta-function $\delta(\log \Omega)$ (defined in terms of the same measure $\mathcal{D}W$) where Ω is defined such, that the field

$$W' = \Omega^{-1}(W)$$

satisfies a special gauge condition. We choose the gauge

$$\partial_\mu W_\mu'^a(x) = C^a(x) \,, \tag{A.2}$$

with $C^a(x)$ a fixed function. Then expression (A.1) becomes

$$\int \mathcal{D}W \delta(\log \Omega) \exp iS_{\text{YM}}[W] = \int \mathcal{D}W \delta(\partial_\mu W_\mu^a - C^a) \times \det\left(\frac{\partial}{\partial \Omega(x')} \partial_\mu W_\mu^a(x)\right) \exp iS_{\text{YM}}[W] \tag{A.3}$$

In order to calculate the determinant we only need to know the change of $\partial_\mu W_\mu^a(x)$ under an infinitesimal gauge transformation $\Lambda^b(x)$:

$$\partial_\mu W_\mu^{a'} = \partial_\mu W_\mu^a + \epsilon_{abc}\, \partial_\mu(\Lambda^b W_\mu^c) - g^{-1}\partial^2 \Lambda^a$$

$$= \partial_\mu W_\mu^a - g^{-1}\partial_\mu (D_\mu \Lambda)^a \tag{A.4}$$

(D_μ is the covariant derivative and g is the coupling constant).

So we must calculate the determinant of the operator $g^{-1}\partial_\mu D_\mu$. This we do with the following trick. Note that even for a non-hermitean matrix A_{ij} the identity

$$\frac{1}{\det A} = C \int \prod_i \mathrm{d}\,\mathrm{Re}\, z_i\, \mathrm{d}\,\mathrm{Im}\, z_i \exp i(z^*, Az) \tag{A.5}$$

holds, where C is a trivial constant. So we write in a symbolic notation eq. (A.3) as

$$\int \mathcal{D}W\, \delta(\partial_\mu W_\mu^a - C^a) \int \mathcal{D}'\varphi \exp\{iS_{\mathrm{YM}}[W] + i\int \varphi^*(x)\,\partial_\mu D_\mu \varphi(x)\,\mathrm{d}x\}\,. \tag{A.6}$$

$\varphi^a(x)$ is a complex scalar particle field. The notation is symbolic because the determinant in eq. (A.3) stands in the numerator and not in the denominator like in eq. (A.5). But this only means that we have to add a factor -1 for each closed loop of φ's, as can easily be established. It is denoted by the prime in $\mathcal{D}'\varphi$.

If C^a is put equal to zero, we get the rules derived by Faddeev and Popov [4]. The transversal propagators

$$\delta_{ab} \frac{\delta_{\mu\nu} - \dfrac{k_\mu k_\nu}{k^2}}{k^2}\,, \tag{A.7}$$

emerge (Landau gauge)†. We can get rid of the annoying $k_\mu k_\nu$ term by noting that expression (A.6) is completely independent of the choice of $C^a(x)$. So we may integrate over all values of C, together with an arbitrary weight function $\exp iS'[C]$. We then get

$$\int \mathcal{D}W \int \mathcal{D}'\varphi \exp\{iS_{\mathrm{YM}}[W] - i\int (\partial_\mu\varphi)^*\, D_\mu\varphi\,\mathrm{d}x + iS'[\partial_\mu W_\mu]\}\,. \tag{A.8}$$

$S'[\partial_\mu W_\mu]$ may be chosen such that it cancels the corresponding term in $S_{\mathrm{YM}}[W]$ and we then find the Feynman gauge, with propagators

$$\frac{\delta_{ab}\delta_{\mu\nu}}{k^2}\,. \tag{A.9}$$

The resulting Feynman rules are listed in appendix B.

Ward identities

We first derive Ward identities in the Landau gauge. Let us treat C^a in expression (A.6) as a source function and make an expansion with respect to it. Even with out-

† The $i\epsilon$ in a propagator is not found by the path integral method. Its sign is dictated by unitarity and is essential for derivation of the Cutkosky rules (sect. 6).

or ingoing particles at plus or minus infinity expression (A.6) is independent of C^a. So all expansion terms with respect to C^a must be zero except the first.

In order to derive the Feynman rules for the expansion terms we must treat the transversal and longitudinal parts of the W-field separately. Integration over the transversal part leads to the Feynman rules (B.1)–(B.6), with $\lambda = 1$, but the fact that $\partial_\mu W_\mu$ now is C and not zero gives us the additional C-lines:

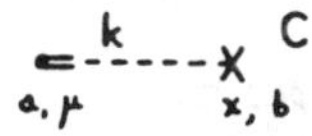

$$\delta_{ab} \frac{-ik_\mu}{k^2} \mathrm{e}^{-ikx} \qquad \text{(A.10)}$$

where the cross denotes the action of the "source" $C^b(x)$, and the double line simply acts as a normal Yang-Mills boson. (The derivation is done by making $\partial_\mu W_\mu$ variable and adding $-\alpha(\partial_\mu W^a_\mu - C^a)^2$ to the Lagrangian, which gives rise to a delta function for $\alpha \to \infty$.)

We can now formulate our Ward identity in the Landau gauge: *The total contribution of all diagrams with a given (non-zero) number of C-lines off mass-shell, and a given number of in- or outgoing lines on mass-shell, is zero.*

This rule is visualized in the diagram notation (3.6), and corresponds to formula (4.14b).

Eq. (3.6) greatly resembles the corresponding Ward identities in quantum electrodynamics, the only difference being that we have to contract *all* off mass-shell lines with their own momentum (that is, choose a polarization vector proportional to their own momentum). The outgoing lines must be physical, that is, their polarization vector must be orthogonal to their own momentum.

In the Feynman gauge we can do something similar. In expression (A.8) we made the choice

$$S'[C] = \int \mathrm{d}x \{-\tfrac{1}{2} C^2(x)\} .$$

Now we add a source function $J(x)$:

$$S'[C] = \int \mathrm{d}x \{-\tfrac{1}{2}(C(x) - J(x))^2\} . \qquad \text{(A.11)}$$

Again, the result must be independent of $J(x)$.

The Feynman rules are those of appendix B, with $\lambda = 0$, together with a J-source contribution which is the same as (A.10) except for the (immaterial) factor $1/k^2$. So the Ward identities in this case are again those of eqs. (3.6) and (4.14).

APPENDIX B

Feynman rules for massless Yang-Mills fields

W: 

$$\frac{\delta_{ab}}{k^2 - i\epsilon}\left(\delta_{\mu\nu} - \lambda \frac{k_\mu k_\nu}{k^2 - i\epsilon}\right) \qquad \begin{array}{l}\lambda = 1 \text{ Landau gauge ,} \\ \lambda = 0 \text{ Feynman gauge ,}\end{array} \tag{B.1}$$

ϕ: 

$$\frac{\delta_{ab}}{k^2 - i\epsilon}, \tag{B.2}$$

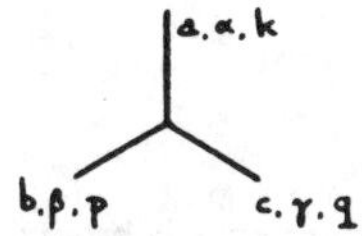

$$-ig\,\epsilon_{abc}[\delta_{\beta\gamma}(q-p)_\alpha + \delta_{\gamma\alpha}(k-q)_\beta + \delta_{\alpha\beta}(p-k)_\gamma] \,, \tag{B.3}$$

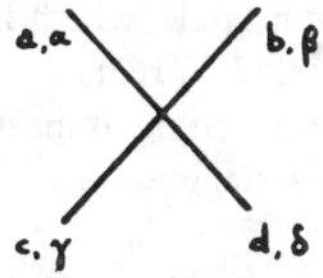

$$\begin{aligned} &-g^2 \epsilon_{gac}\epsilon_{gbd}(\delta_{\alpha\beta}\delta_{\gamma\delta} - \delta_{\alpha\delta}\delta_{\gamma\beta}) \\ &-g^2 \epsilon_{gad}\epsilon_{gbc}(\delta_{\alpha\beta}\delta_{\delta\gamma} - \delta_{\alpha\gamma}\delta_{\delta\beta}) \\ &-g^2 \epsilon_{gab}\epsilon_{gcd}(\delta_{\alpha\gamma}\delta_{\beta\delta} - \delta_{\alpha\delta}\delta_{\beta\gamma}) \,, \end{aligned} \tag{B.4}$$

$$-ig\,\epsilon_{abc}q_\alpha \,, \tag{B.5}$$

(at the vertices all momenta are defined to be inwards).

For each closed loop of φ particles: -1 . (B.6)

As usual: a factor $1/(2\pi)^4 i$ for each propagator and $(2\pi)^4 i$ for each vertex.

REFERENCES

[1] R.P.Feynman, Acta Phys. Polon. 24 (1963) 697.
[2] S.Mandelstam, Phys. Rev. 175 (1968) 1580; 1604.
[3] B.S.DeWitt, Phys. Rev. 162 (1967) 1195; 1239.
[4] L.D.Faddeev and V.N.Popov, Phys. Letters 25B (1967) 29.
[5] E.S.Fradkin and I.V.Tyutin, Phys. Rev. D2 (1970) 2841.
[6] W.Pauli and F.Villars, Rev. Mod. Phys. 21 (1949) 434.
[7] S.N.Gupta, Proc. Phys. Soc. 66 (1953) 129.
[8] J.S.Bell and R.Jackiw, Nuovo Cimento 60A (1969) 47.
[9] St.L.Adler, Phys. Rev. 177 (1969) 2426.
[10] C.N.Yang and R.L.Mills, Phys. Rev. 96 (1954) 191.
[11] K.Hepp, Comm. Math. Phys. 2 (1966) 301.
[12] R.E.Cutkosky, J. Math. Phys. 1 (1960) 429;
M.Veltman, Physica 29 (1963) 186.

RENORMALIZABLE LAGRANGIANS FOR MASSIVE YANG-MILLS FIELDS

G.'t HOOFT
Institute for Theoretical Physics, University of Utrecht

Received 13 July 1971

Abstract: Renormalizable models are constructed in which local gauge invariance is broken spontaneously. Feynman rules and Ward identities can be found by means of a path integral method, and they can be checked by algebra. In one of these models, which is studied in more detail, local SU(2) is broken in such a way that local U(1) remains as a symmetry. A renormalizable and unitary theory results, with photons, charged massive vector particles, and additional neutral scalar particles. It has three independent parameters.

Another model has local SU(2)$\otimes$U(1) as a symmetry and may serve as a renormalizable theory for ρ-mesons and photons.

In such models electromagnetic mass-differences are finite and can be calculated in perturbation theory.

1. INTRODUCTION

In a preceding article [1], henceforth referred to as I, it has been shown that, owing to their large symmetry, mass-less Yang-Mills fields may be renormalized, provided that a certain set of Ward identities is not violated by renormalization effects. With this we mean that anomalies like those of the axial current Ward identities in nucleon-nucleon interactions [2–4], which are due to an unallowed shift of integration variables in the "formal" proof, must not occur. In I it is proved that such anomalies are absent in diagrams with one closed loop, if there are no parity-changing transformations in the local gauge group. We do know an extension of this proof for diagrams with an arbitrary number of closed loops, but it is rather involved and we shall not present it here.

Thus, our prescription for the renormalization procedure is consistent, so the ultraviolet problem for mass-less Yang-Mills fields has been solved. A much more complicated problem is formed by the infrared divergencies of the system. Weinberg [5] has pointed out that, contrary to the quantum electrodynamical case, this problem cannot merely be solved by some closer contemplation of the measuring process. The disaster is such that the perturbation expansion breaks down in the infrared region, so we have no rigorous field theory to describe what happens.

However, although the Lagrangian is invariant under local gauge transformations, the physical solutions we are interested in may provide us with a certain preference gauge, in which these solutions take a simple form. If this is the case, then the local gauge invariance is hidden, and it is very well possible that all Yang-Mills bosons become massive vector particles [6]. We do not know whether such a thing can happen with mass-less Yang-Mills fields alone, but it surely can happen in other models, of which we present some.

In all these models additional scalar fields are introduced, which are representations of the local gauge group. If, in some gauge, these fields have a non-zero vacuum expectation value, then they may fix the gauge, either completely, or partly. In the latter case, invariance under transformations of a local subgroup of the original invariance group remains evident, and some of the Yang-Mills bosons remain mass-less.

The transition from a "symmetric" to a "non-symmetric" representation is done in a way analogous to the treatment of the σ-model by Lee and Gervais [7, 8]. The difference is of course that we have a local invariance, and we have no symmetry breaking term in the Lagrangian.

Our result is a large set of different models with massive, charged or neutral, spin one bosons, photons, and massive scalar particles. Due to the local symmetry our models are renormalizable, causal, and unitary. They all contain a small number of independent physical parameters.

A nice feature is that in certain models the electromagnetic mass-differences are finite and can be expressed in terms of the other parameters.

In sect. 2 we give a short review of the results in the preceding paper (I) on massless Yang-Mills fields. A general procedure appears to exist for deriving Feynman rules for models with a local gauge invariance. One statement must be made on our use of path integrals here: we only apply path integral techniques in order to get some idea of what the Feynman rules and Ward-identities might be. Consistency and unitarity of the renormalized expressions must always be checked later on. This has been done for the models described in this paper.

In sect. 3 we consider SU(2) gauge fields and an additional scalar isospin one boson. We show how the vacuum expectation value of this boson field can become non-zero due to dynamical effects, and how two of the Yang-Mills bosons become massive, oppositely charged, vector particles, while the third becomes an ordinary photon. Of the original scalar fields one component survives in the form of a neutral spinless particle. Interaction and gauge are formulated in such a way that the theory remains renormalizable. In sect. 4 a renormalization scheme is presented, but for a more elaborate description of the renormalization procedures we refer to I. In sect. 5 we prove that the model of sect. 3 is unitary and it is easily seen that the proof applies also to the other models.

In sect. 6 we describe an example where local invariance seems broken, while global invariance remains evident. All Yang-Mills particles get equal mass, and the model resembles very much the massive Yang-Mills field studied by other authors

[9, 10] except for the presence of one extra neutral scalar boson with arbitrary mass. The model can be used to describe ρ-mesons as elementary particles.

In sect. 7 it is shown that our "ρ-meson model" can be enriched with electromagnetic interactions without destroying renormalizability or unitarity. $\rho^0-\gamma$ mixture leads to phenomena like vector-dominance.

In the appendix we formulate the Feynman rules for the various models.

2. RESUME MASSLESS YANG-MILLS FIELDS

The massless Yang-Mills field has been discussed in I. The Lagrangian is

$$\mathcal{L}_{\text{YM}} = -\tfrac{1}{4} G^a_{\mu\nu} G^a_{\mu\nu} + \mathcal{L}^c(\partial_\mu W^a_\mu) , \tag{2.1}$$

with

$$G^a_{\mu\nu} \equiv \partial_\mu W^a_\nu - \partial_\nu W^a_\mu + g f_{abc} W^b_\mu W^c_\nu , \tag{2.2}$$

where W^a_μ are the Yang-Mills field components and f_{abc} are the structure constants of the underlying gauge group. g is a coupling constant.

$\mathcal{L}^c$ is an extra term, only depending on the divergence of the field W^a_μ, and it may be chosen in an arbitrary way, thus fixing the gauge.

From (2.1) the Feynman rules may be constructed by ordinary Feynman path integral methods: the procedure is clarified in the appendix. But, because of the gauge non-invariance of $\mathcal{L}^c$, an extra complex ghost particle field φ^a must be introduced, described by the Lagrangian

$$\mathcal{L}_\varphi = -\partial_\mu \varphi^{*a} (D_\mu \varphi)^a , \tag{2.3}$$

where D_μ is the covariant derivative, defined as

$$(D_\mu X)^a \equiv \partial_\mu X^a + g f_{abc} W^b_\mu X^c . \tag{2.4}$$

Furthermore, an extra factor -1 must be inserted in the amplitude for each closed loop of φ's.

The φ particles (and antiparticles) do not occur in the intermediate states in the unitarity condition, because they cancel the helicity-0 states in the W-field.

The Lagrangian (2.3) is related to the behaviour under local gauge transformations of the gauge non-invariant part $\mathcal{L}^c$ of the Lagrangian (2.1):

$$\mathcal{L}^c = \mathcal{L}^c(\partial_\mu W^a_\mu) .$$

Under an infinitesimal gauge transformation generated by $\Lambda^a(x)$, the quantity $\partial_\mu W^a_\mu$

transforms as:

$$\partial_\mu W^{a'}_\mu = \partial_\mu W^a_\mu - g^{-1}\partial_\mu (D_\mu \Lambda)^a . \tag{2.5}$$

Because the fields W^a_μ occur explicitly in the covariant derivative D_μ in eq. (2.5), a non-trivial Jacobian factor is needed in the gauge dependent expressions for the amplitude, which is precisely the φ-particle contribution. In appendix A of I it is shown how to derive the Lagrangian (2.3) from eq. (2.5).

The choice

$$\mathcal{L}^C = -\alpha(\partial_\mu W^a_\mu - C^a)^2 ; \qquad \alpha \to \infty \tag{2.6}$$

leads to the Landau gauge for $C = 0$, and from the fact that the amplitudes must be independent of $C^a(x)$, one can derive Ward identities. The most appropriate choice however is

$$\mathcal{L}^C = -\tfrac{1}{2}(\partial_\mu W^a_\mu(x) - J^a(x))^2 . \tag{2.7}$$

For $J = 0$ we have the Feynman gauge, in which the propagators are:

$$\frac{\delta_{\mu\nu}}{k^2 - i\epsilon} \tag{2.8}$$

and again one can derive Ward identities.

These Ward identities supply a unique prescription for the subtraction constants in a renormalization procedure, and from them unitarity of the system follows.

If we introduce other fields which are representations of the gauge group, then all derivatives in their Lagrangian parts must be replaced by covariant derivatives, thus ensuring local gauge invariance and unitarity.

3. SELECTION OF A PREFERENCE GAUGE BY INTRODUCING AN ISOSPIN 1 SCALAR FIELD

Consider the case that the local gauge group is SU(2), and suppose that an isospin one scalar field or current $X^a(x)$ exists which has (in a certain gauge) a non-zero vacuum expectation value [6]. Then this isovector is apt to select a preference gauge, which may be taken to be the gauge in which

$$X^1(x) = X^2(x) = 0 \tag{3.1}$$

for all x. However, a gauge transformation of the original system to this "X-field gauge" would in general involve a non-polynomial Jacobian factor (cf. I), thus destroying renormalizability of our model. So, in general we shall abandon the gauge (3.1), but its formal possibility indicates clearly that the components X^1 and X^2 are unobservable, and X^3 acts as a "schizon": our symmetry *seems* to be broken.

In our renormalizable model, X is simply a boson field, and we fix the gauge by adding some functional $\mathcal{L}^c$ to the Lagrangian as in sect. 2. In the symmetric representation the Lagrangian is

$$\mathcal{L} = \mathcal{L}_{\mathrm{YM}} - \tfrac{1}{2}(D_\mu X)^2 - \tfrac{1}{2}\mu^2 X^2 - \tfrac{1}{8}\lambda(X^2)^2 \,. \tag{3.2}$$

This Lagrangian corresponds to a renormalizable theory. The last term is necessary for fixing the counterterm in divergent graphs with four X-lines. Thus we have three independent parameters g, μ and λ.

In order to get some insight in what might happen let us consider the tree-approximation, that is, we disregard all graphs with closed loops. In this approximation all fields may be considered as being classical, and the vacuum corresponds to the equilibrium state, where all fields are constant and the total energy has a minimum. (If we specify the gauge, then this energy can be written as an integral over space pf an Hamiltonian density $\mathcal{H}(x)$.) In order for this vacuum to exist, the parameter λ must be positive, but μ^2 may have a negative value. In the latter case we expect that the X-field is non-zero in the equilibrium state: for slowly varying X^a, and $W_\mu^a \simeq 0$, we have

$$\mathcal{H} \simeq \int_V \mathrm{d}x(\tfrac{1}{2}\mu^2 X^2 + \tfrac{1}{8}\lambda(X^2)^2) \,.$$

This has a minimum for

$$X^a(x) = e^a \sqrt{\frac{-2\mu^2}{\lambda}} \,, \tag{3.3}$$

with e^a an arbitrary vector with modulus unity. By a global gauge transformation this vector e^a can always be pointed in the z-direction.

If we do quantum mechanics things do not change drastically. Eq. (3.3) must be replaced by

$$\langle 0|X^a(x)|0\rangle = F\begin{pmatrix}0\\0\\1\end{pmatrix} \,, \qquad F = \sqrt{\frac{-2\mu^2}{\lambda}} + \mathrm{O}(g) \,, \tag{3.4}$$

(the parameter λ is of order g^2).

We now proceed as Lee did for the σ-model [7]. We write

$$X^a(x) \equiv F \begin{pmatrix} 0 \\ 0 \\ 1 \end{pmatrix} + A_a(x) , \tag{3.5}$$

with

$$\langle 0|A_a(x)|0\rangle \equiv 0 . \tag{3.6}$$

The Lagrangian (3.2) then becomes

$$\begin{aligned} \mathcal{L} = \mathcal{L}_{\mathrm{YM}} &- \tfrac{1}{2}(D_\mu A)^2 - \tfrac{1}{2}\lambda F^2 A_3^2 - \tfrac{1}{2}g^2F^2(W_\mu^{1\,2} + W_\mu^{2\,2}) - \tfrac{1}{2}\lambda F A^2 A_3 - \tfrac{1}{8}\lambda (A^2)^2 \\ &- g^2 F A_3 (W_\mu^{1\,2} + W_\mu^{2\,2}) + g^2 F W_\mu^3 (A_1 W_\mu^1 + A_2 W_\mu^2) - \beta(\tfrac{1}{2}A^2 + FA_3) \\ &+ gF(A_1 \partial_\mu W_\mu^2 - A_2 \partial_\mu W_\mu^1) , \end{aligned} \tag{3.7}$$

where

$$\beta \equiv \mu^2 + \tfrac{1}{2}\lambda F^2 .$$

Note, that the "tadpole condition" (3.6) implies that $\beta = 0$ in first order of g and λ, in accordance with eq. (3.4).

We deliberately have not yet specified the local gauge. We have seen that the gauge (3.1) is no good, because it renders the theory unrenormalizable. One could very well proceed like in sect. 2 and choose the local gauge by adding

$$\mathcal{L}^c = \mathcal{L}^c(\partial_\mu W_\mu) = -\tfrac{1}{2}(\partial_\mu W_\mu^a - J_a(x))^2 ,$$

but the resulting Feynman rules are rather complicated and a massless ghost remains. It is more convenient to choose:

$$\begin{aligned} \mathcal{L}^c = &-\tfrac{1}{2}(\partial_\mu W_\mu^3 - J_3(x))^2 - \tfrac{1}{2}(\partial_\mu W_\mu^1 - gFA_2 - J_1(x))^2 \\ &-\tfrac{1}{2}(\partial_\mu W_\mu^2 + gFA_1 - J_2(x))^2 . \end{aligned} \tag{3.8}$$

In here the functions $J_a(x)$ will be put equal to zero, but the fact that all physical amplitudes are independent of them enables us to formulate Ward-identities.

The fields in eq. (3.8) transform as follows under infinitesimal local gauge transformations:

$$\partial_\mu W^{3'}_\mu = \partial_\mu W^3_\mu - g^{-1}\partial_\mu (D_\mu \Lambda)^3 ;$$

$$(\partial_\mu W^1_\mu - gFA_2)' = \partial_\mu W^1_\mu - gFA_2 - g^{-1}\partial_\mu (D_\mu \Lambda)^1 + gF^2\Lambda^1 - gF\epsilon_{2bc}\Lambda^b A_c ; \qquad (3.9)$$

$$(\partial_\mu W^2_\mu + gFA_1)' = \partial_\mu W^2_\mu + gFA_1 - g^{-1}\partial_\mu (D_\mu \Lambda)^2 + gF^2\Lambda^2 + gF\epsilon_{1bc}\Lambda^b A_c .$$

With the same procedure as in sect. 2 we derive the φ-ghost Lagrangian‡:

$$\mathcal{L}_\varphi = -\partial_\mu \varphi^{*a}(D_\mu \varphi)^a - g^2F^2(\varphi^{*1}\varphi^1 + \varphi^{*2}\varphi^2)$$
$$+ g^2F(\varphi^{*1}\epsilon_{2bc}\varphi^b A_c - \varphi^{*2}\epsilon_{1bc}\varphi^b A_c) . \qquad (3.10)$$

Because of our choice (3.8) for $\mathcal{L}^c$ the last term in (3.7) is cancelled; likewise the term

$$+\tfrac{1}{2}(\partial_\mu W^a_\mu)^2 \quad \text{in} \quad -\tfrac{1}{4}G^a_{\mu\nu}G^a_{\mu\nu} .$$

Let us finally replace the three parameters by

$$M \equiv gF > 0 ,$$
$$\alpha \equiv \lambda/g^2 > 0 , \qquad (3.11)$$

and g.

The resulting theory has two massive, charged vector particles $W^1_\mu \pm i W^2_\mu$, with the propagator

$$\frac{\delta_{\mu\nu}}{k^2 + M^2 - i\epsilon} ; \qquad (3.12)$$

a massless photon W^3_μ, with the propagator

$$\frac{\delta_{\mu\nu}}{k^2 - i\epsilon} \qquad (3.13)$$

and a neutral, scalar particle A_3, with mass $M\sqrt{\alpha}$.

There are two different types of ghosts:

‡ Note that this expression is not Hermitian; the φ-ghost restores unitarity. Because of these features Feynman rules must be derived by path integral methods. The heuristic Feynman rules are here the correct ones, as is shown in the appendix.

First: the complex φ-ghost with the oriented propagators: $(k^2+M^2-i\epsilon)^{-1}$ for $\varphi^{1,2}$ and $(k^2-i\epsilon)^{-1}$ for φ^3.

A minus sign must be inserted for each closed loop of φ-ghosts.
Second: the real $A^{1,2}$ ghosts, with no minus sign, and mass M^2 (resulting from a contribution of $\mathcal{L}^c$, eq. (3.8)).

These, and all other Feynman rules, except for the above mentioned minus sign, may be derived with ordinary Feynman path-integral techniques, from the Lagrangian

$$\mathcal{L} = -\tfrac{1}{4}G^a_{\mu\nu}G^a_{\mu\nu} - \tfrac{1}{2}(\partial_\mu W^a_\mu)^2 + \mathcal{L}_\varphi - \tfrac{1}{2}(D_\mu A)^2 - \tfrac{1}{2}M^2(W^{1 2}_\mu + W^{2 2}_\mu)$$
$$-\frac{\alpha M^2}{2}A_3^2 - \frac{M^2}{2}(A_1^2+A_2^2) - \frac{\alpha}{2}gMA_a^2A_3 - \frac{\alpha g^2}{8}(A_a^2)^2 - gMA_3(W^{1 2}_\mu + W^{2 2}_\mu)$$
$$+ gMW^3_\mu(A_1W^1_\mu + A_2W^2_\mu) - \beta(\tfrac{1}{2}A^2 + \frac{M}{g}A_3) + J_3(x)\,\partial_\mu W^3_\mu + J_1(x)$$
$$\times\,[\partial_\mu W^1_\mu - MA_2] + J_2(x)\,[\partial_\mu W^2_\mu + MA_1] - \tfrac{1}{2}J_a^2(x) \tag{3.14}$$

with

$$\mathcal{L}_\varphi = -\partial_\mu\varphi^{*a}(D_\mu\varphi)^a - M^2(|\varphi^1|^2 + |\varphi^2|^2) + gM(\varphi^{*1}\epsilon_{2bc} - \varphi^{*2}\epsilon_{1bc})\varphi^bA_c\,. \tag{3.15}$$

The functions $J_a(x)$ are arbitrary, which enables us to formulate Ward identities. They may be chosen to be zero. The constant β must be adjusted in such a way, that the tadpole condition (3.6) holds.

The complex φ-particles with their unphysical "Fermi statistics", and the $A^\pm$ particles, with positive definite metric, must all be considered as ghosts. The most compelling reason for this is the unitarity condition; in sect. 5 we derive that these particles cancel the unphysical polarisation directions of the W-fields in the intermediate states, resulting from the anomalous propagators (3.12) and (3.13).

4. RENORMALIZATION

The expression (3.8) for $\mathcal{L}^c$, has especially been chosen in order to acquire the simple, quadratically convergent propagators (3.12) and (3.13), and to arrive at Feynman rules which are those of a renormalizable theory.

However, renormalization requires the introduction of a cut-off procedure, and in general this modifies the theory such that the symmetry, and therefore also unitarity, get lost. Thus, the cut-off procedure must be chosen in accordance with our symmetry requirement. To this purpose we can use the observation that in all orders of g the physical amplitudes must be independent of the source function $J_a(x)$. The Feynman rules for the contribution of their Fourier transform $J_a(k)$ are given in

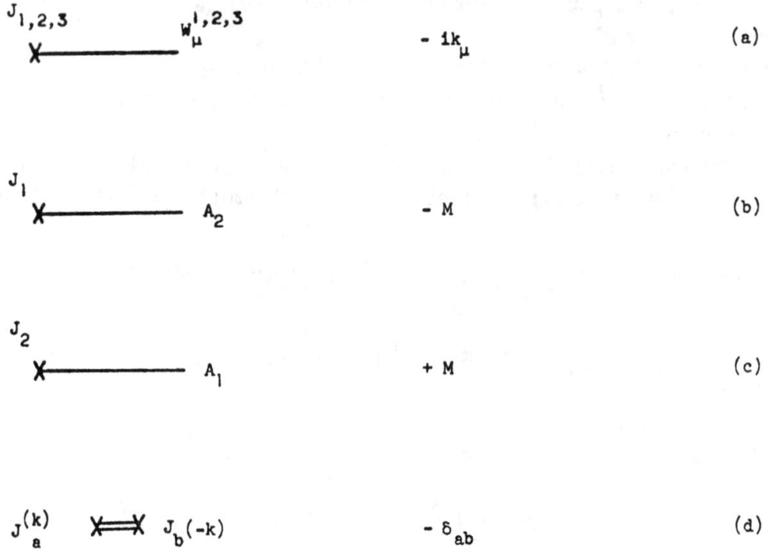

Fig. 1. Feynman rules for the contribution of the source function J_a to the amplitude.

fig. 1 (compare eq. (3.14)). (Note that (d) is cancelled by contributions of bare W and A particles.)

A graphical notation for the Ward identities is shown in fig. 2. The number of "J-lines" must be non-zero. A combinatorial proof of these Ward identities can be given in the same way as in the case of the massless Yang-Mills fields.

Further, also in this model, a variation upon these identities can be found for the case that one of the external W-lines on mass shell has a non-physical polarization direction (fig. 3). The identity in fig. 3 can be proven either by combinatorics, or by using a formulation in terms of path integrals: one must consider an infinitesi-

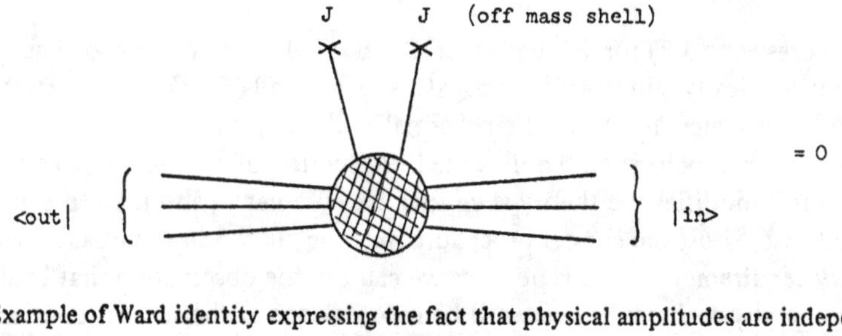

Fig. 2. Example of Ward identity expressing the fact that physical amplitudes are independent of the J-source.

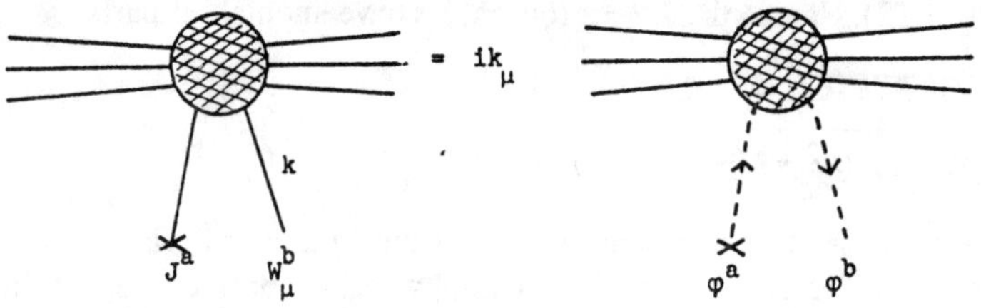

Fig. 3. Ward identity for the case that one of the external W-lines on mass shell has a non-physical polarization direction.

mal gauge transformation generated by $\Lambda^a(x)$ with the property

$$\partial_\mu D_\mu \Lambda(x) = J(x)\,, \tag{4.1}$$

($J(x)$ being also infinitesimal).

One of the simplest ways to use these Ward identities to calculate the renormalized higher order corrections to the amplitudes, is to apply subtracted dispersion relations. The behaviour of the amplitudes for the momenta going to infinity is prescribed by the condition that renormalizability must not be destroyed at higher orders: hence, the amplitude for a diagram with N outgoing boson lines must behave at infinity like $(k)^{4-N} L(k)$, where (k) stands for the momenta of the outgoing lines, and L is a logarithmic factor. Thus the number of subtractions has been determined, whereas the Ward identities give a large restriction on the possible values of the subtraction constants. The procedure sketched here can be proven to be consistent as soon as some gauge covariant set of regulators is found. Such a set can indeed be formulated for diagrams with one closed loop by the introduction of a fifth dimension in Minkowsky space (cf. I). By introducing more dimensions one can give a consistency proof for all orders, but we shall not present it here.

5. UNITARITY

The equations shown in figs. 2 and 3 may be used to prove unitarity of the model. The procedure is analogous to the proof given in (I). The proof that the contributions of the φ^3 and $\bar{\varphi}^3$ intermediate states cancel those of the unphysical W^3_μ-states is the same as in I and will not be repeated here. Actually, it is enough to show that the residues of the poles at $k^2 = 0$ of these propagators cancel.

As to the charged particle states, a unitary field theory of massive vector particles would have propagators

$$\frac{\delta_{\mu\nu} + \dfrac{k_\mu k_\nu}{M^2}}{k^2 + M^2 - i\epsilon} \tag{5.1}$$

instead of (3.12). Hence, the $W^{1,2}$ propagators have anomalous parts

$$-\frac{k_\mu k_\nu}{M^2(k^2+M^2-i\epsilon)}. \tag{5.2}$$

Now, indeed, we see that as a consequence of the equations in figs. 2 and 3, the residues of the poles at $k^2 = -M^2$ of the unphysical propagators cancel. In fig. 4 it is shown which combination of these propagators has to be considered in order to prove this cancellation. (Note the explicitly written minus sign for the φ-loops; the integration over k has not yet been carried out.) For more details of this proof we refer to the treatment of the analogous case in massless Yang-Mills fields, given in (I). It is because of this phenomenon that we can consider the anomalous part of the W-field, and the A_1, A_2 and φ-particles, as unphysical. They can be left out of the intermediate states without invalidating the unitarity equation.

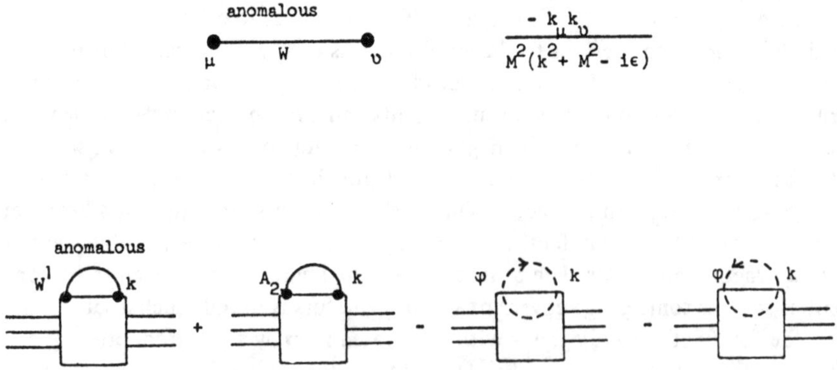

Fig. 4. Combination of unphysical propagators for fixed value of k.

6. ISOSPIN-$\frac{1}{2}$ SCALAR FIELD

Our most important conclusion from the foregoing is that a basic principle like gauge invariance can lead to renormalizable, unitary theories with massive, charged, vector particles. Many different models may be constructed this way. We like to mention in more detail a very interesting case: local SU(2)-gauge invariance with an isospin-$\frac{1}{2}$ "symmetry breaking" field

$$K(x) \equiv \begin{pmatrix} K_1' + iK_1'' \\ K_2' + iK_2'' \end{pmatrix}$$

Let the Lagrangian in the symmetric representation be

$$\mathcal{L} = \mathcal{L}_{\mathrm{YM}} - (D_\mu K)^* D_\mu K - \mu^2 K^* K - \tfrac{1}{2}\lambda (K^* K)^2 . \tag{6.1}$$

The covariant derivative of an isospin-$\frac{1}{2}$ field is:

$$D_\mu K \equiv \partial_\mu K - \tfrac{1}{2} i g \tau^a W_\mu^a K . \tag{6.2}$$

For negative values of μ^2, the field K is expected to have a non-zero vacuum expectation value, which by a suitable global rotation in isospin space can be taken to be:

$$\langle 0 | K(x) | 0 \rangle \equiv F \begin{pmatrix} 1 \\ 0 \end{pmatrix} . \tag{6.3}$$

Now, it appears to be convenient to express the complex spinor K in terms of a real isospin singlet and a real triplet:

$$\begin{aligned} K &\equiv F \begin{pmatrix} 1 \\ 0 \end{pmatrix} + \frac{1}{\sqrt{2}} (Z + i\psi_a \tau^a) \begin{pmatrix} 1 \\ 0 \end{pmatrix} , \\ &= F \begin{pmatrix} 1 \\ 0 \end{pmatrix} + \frac{1}{\sqrt{2}} \begin{pmatrix} Z + i\psi_3 \\ -\psi_2 + i\psi_1 \end{pmatrix} , \end{aligned} \tag{6.4}$$

and to introduce the independent parameters

$$\begin{aligned} M^2 &\equiv \tfrac{1}{2} g^2 F^2 , \\ \alpha &= \lambda / g^2 , \end{aligned} \tag{6.5}$$

and g.

In this representation, the Lagrangian is:

$$\begin{aligned} \mathcal{L} = & -\tfrac{1}{4} G_{\mu\nu}^a G_{\mu\nu}^a + \mathcal{L}^c - \tfrac{1}{2} M^2 W_\mu^2 - \tfrac{1}{2} (\partial_\mu Z)^2 - \tfrac{1}{2} (\partial_\mu \psi) D_\mu \psi - \tfrac{1}{2} 4\alpha M^2 Z^2 \\ & + \tfrac{1}{2} g W_\mu^a (Z \partial_\mu \psi_a - \psi_a \partial_\mu Z) - \tfrac{1}{8} g^2 W_\mu^2 (\psi^2 + Z^2) - \tfrac{1}{2} g M W^2 Z - \alpha M g Z (\psi^2 + Z^2) \\ & - \tfrac{1}{8} \alpha g^2 (\psi^2 + Z^2)^2 - \beta \left[\tfrac{1}{2} (Z^2 + \psi^2) + \frac{2M}{g} Z \right] - M \psi_a \partial_\mu W_\mu^a , \end{aligned} \tag{6.6}$$

where the parameter $\beta \equiv \mu^2 + \lambda F^2$ must be chosen in such a way that all tadpoles cancel. It is of order g^2.

$\mathcal{L}^c$ is chosen to be

$$\mathcal{L}^c = -\tfrac{1}{2}(\partial_\mu W^a_\mu - M\psi_a - J_a)^2 \,, \tag{6.7}$$

so that again the Feynman propagator (3.12) for the W-field emerges, and the last term of eq. (6.6) cancels.

By studying the behaviour of $\mathcal{L}^c$ under gauge transformations we derive the Lagrangian for the ghost field φ (compare sects. 2 and 3)

$$\mathcal{L}_\varphi = -\partial_\mu \varphi^* D_\mu \varphi - M^2 \varphi^* \varphi - \tfrac{1}{2} M g \varphi^* \varphi Z + \tfrac{1}{2} M g \epsilon_{abc} \varphi^{*a} \varphi^b \psi^c \,. \tag{6.8}$$

The Feynman rules for the source function contributions are now those of fig. 5.

Now we observe that the Lagrangian (6.6) as well as the Ward identities remain invariant under global isospin transformations, if the ψ fields are considered as a triplet and the Z as a singlet. Only local gauge invariance has been broken. Here the ψ fields act as additional ghosts, and all three isospin components of the W^a_μ-fields have become massive. The Z is an additional physical particle.

Many authors [9, 10] have considered the massive Yang-Mills theory described by the Lagrangian

$$\mathcal{L}_{\mathrm{YM}} - \tfrac{1}{2} M^2 W_\mu^2 \,. \tag{6.9}$$

That model appears to be non renormalizable, although many of the divergencies can be seen to cancel by the introduction of two kinds of ghost fields [11, 12]. In our model also two ghosts appear: the complex φ-field, with Fermi statistics, and the ψ-field. But now we see that the introduction of one physical isospin-zero particle Z can render the model renormalizable. Its mass is a new independent parameter. For large values of this mass we get something very similar to the old model (6.9).

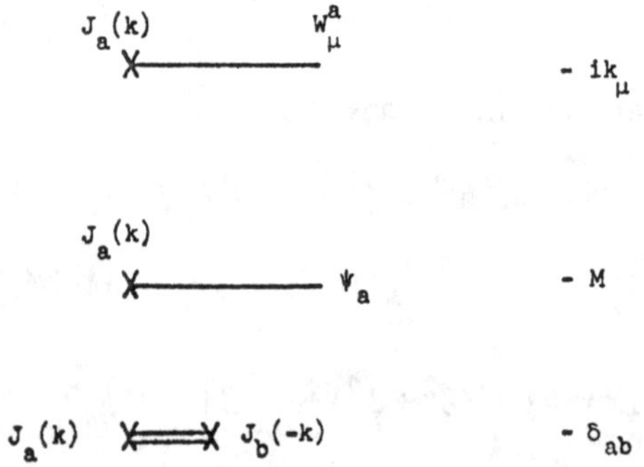

Fig. 5. Feynman rules for the J-source contribution to the amplitude.

7. ISOSPIN AND ELECTROMAGNETISM; VECTOR DOMINANCE

In the previous model, electromagnetism can be introduced in an elegant way† Consider first the symmetric representation (6.1). Let us assume the presence of a "hyperelectromagnetic" field, $\tilde{A}_\mu$, which does not break isospin. Let in (6.1) only the K particle have a "hypercharge" q. The Lagrangian is then:

$$\mathcal{L} = \mathcal{L}_{\mathrm{YM}} - \tfrac{1}{4}\tilde{F}_{\mu\nu}\tilde{F}_{\mu\nu} - (\tilde{D}_\mu K)^* \tilde{D}_\mu K - \mu^2 K^* K - \tfrac{1}{2}\lambda(K^*K)^2 ,$$

with

$$\begin{aligned} \tilde{F}_{\mu\nu} &\equiv \partial_\mu \tilde{A}_\nu - \partial_\nu \tilde{A}_\mu , \\ \tilde{D}_\mu K &\equiv D_\mu K + iq\tilde{A}_\mu K . \end{aligned} \tag{7.1}$$

The gauge group in this model is SU(2)$\otimes$U(1). Let us consider an infinitesimal gauge transformation:

$$\begin{aligned} K' &= (1 - \tfrac{1}{2} i \Lambda^a \tau^a - i\tilde{\Lambda})K , \\ W'^a_\mu &= W^a_\mu - g^{-1}(D_\mu \Lambda)^a , \\ \tilde{A}'_\mu &= \tilde{A}_\mu + q^{-1}\partial_\mu \tilde{\Lambda} , \end{aligned} \tag{7.2}$$

where $\Lambda^a(x)$, $\tilde{\Lambda}(x)$ are generators of an infinitesimal gauge transformation.

Now if the K field has a non-zero vacuum expectation value:

$$\langle 0|K(x)|0\rangle = F\begin{pmatrix}1\\0\end{pmatrix} , \tag{7.3}$$

then the physical fields will only appear to be invariant under those transformations (7.2) that leave the spinor

$$\begin{pmatrix}1\\0\end{pmatrix}$$

invariant; that are the transformations with

$$\Lambda^1 = \Lambda^2 = 0 ; \qquad \Lambda^3 = -2\tilde{\Lambda} \equiv \Lambda^{\mathrm{EM}} . \tag{7.4}$$

† The model of this section is due to Weinberg [13], who showed that it can describe weak interactions between leptons. His lepton model can be shown to be renormalizable.

Let us call such transformations electromagnetic gauge transformations. If we define

$$W_\mu^{1,2} \equiv \rho_\mu^{1,2} \qquad \text{(a)}$$

$$W_\mu^3 \equiv \rho_\mu^3 + \frac{2q}{g}\tilde{A}_\mu \qquad \text{(b)} \qquad (7.5)$$

$$K \equiv F\begin{pmatrix}1\\0\end{pmatrix} + \frac{1}{\sqrt{2}}\begin{pmatrix}Z + i\psi_3\\ -\psi_2 + i\psi_1\end{pmatrix} \qquad \text{(c)}$$

then these quantities transform under electromagnetic gauge transformations like:

$$\begin{aligned} Z' &= Z\,, \\ \psi_3' &= \psi_3\,, \\ \psi_1' \pm i\psi_2' &= \mathrm{e}^{\pm i\Lambda^{\mathrm{EM}}}(\psi_1 \pm i\psi_2)\,, \\ \rho_\mu^{3'} &= \rho_\mu^3\,, \\ \rho_\mu^{1'} \pm i\rho_\mu^{2'} &= \mathrm{e}^{\pm i\Lambda^{\mathrm{EM}}}(\rho_\mu^1 \pm i\rho_\mu^2)\,, \\ \tilde{A}_\mu' &= \tilde{A}_\mu - \frac{1}{2q}\partial_\mu \Lambda^{\mathrm{EM}}\,. \end{aligned} \qquad (7.6)$$

Finally, we make the substitutions:

$$\tilde{A}_\mu \equiv A_\mu\left(1 + \frac{4q^2}{g^2}\right)^{-\frac{1}{2}}; \qquad e \equiv 2q\left(1 + \frac{4q^2}{g^2}\right)^{-\frac{1}{2}}. \qquad (7.7)$$

In terms of these variables the Lagrangian (7.1) becomes:

$$\begin{aligned} \mathcal{L} = &-\tfrac{1}{4}\rho^a_{\mu\nu}\rho^a_{\mu\nu} - \tfrac{1}{4}F_{\mu\nu}F_{\mu\nu} - \frac{e}{2g}F_{\mu\nu}\rho^3_{\mu\nu} + \mathcal{L}^c - \tfrac{1}{2}M^2\rho_\mu^2 - \tfrac{1}{2}(\partial_\mu Z)^2 - \tfrac{1}{2}4\alpha M^2 Z^2 \\ &- \tfrac{1}{2}(\partial_\mu^{\mathrm{EM}}\psi)D_\mu^{\mathrm{EM}}\psi + \tfrac{1}{2}g\rho_\mu^a(Z\partial_\mu^{\mathrm{EM}}\psi_a - \psi_a\partial_\mu Z) - \tfrac{1}{2}gM\rho^2 Z - \tfrac{1}{8}g^2\rho^2(\psi^2 + Z^2) \\ &- \alpha MgZ(\psi^2 + Z^2) - \tfrac{1}{8}\alpha g^2(\psi^2 + Z^2)^2 - \beta\left(\tfrac{1}{2}(Z^2 + \psi^2) + \frac{2M}{g}Z\right) - M\psi_a\partial_\mu^{\mathrm{EM}}\rho_\mu^a\,, \end{aligned} \qquad (7.8)$$

where

$$\partial_\mu^{\mathrm{EM}} \psi_3 \equiv \partial_\mu \psi_3 ,$$
$$\partial_\mu^{\mathrm{EM}} (\psi_1 \pm i\psi_2) \equiv (\partial_\mu \pm ieA_\mu)(\psi_1 \pm i\psi_2) ,$$
$$(D_\mu^{\mathrm{EM}} \psi)_a \equiv \partial_\mu^{\mathrm{EM}} \psi_a + g\epsilon_{abc}\rho_\mu^b \psi_c ,$$
$$\rho_{\mu\nu}^a \equiv \partial_\mu^{\mathrm{EM}} \rho_\nu^a - \partial_\nu^{\mathrm{EM}} \rho_\mu^a + g\epsilon_{abc}\rho_\mu^b \rho_\nu^c . \tag{7.9}$$

Let us choose

$$\mathcal{L}^c = -\tfrac{1}{2}(\partial_\mu^{\mathrm{EM}} \rho_\mu^a - M\psi_a - J_a)^2 - \tfrac{1}{2}(\partial_\mu A_\mu - J^{\mathrm{EM}})^2 , \tag{7.10}$$

so that the last term in (7.8) is cancelled, while the photon, and the ρ-particle have the Feynman propagators (3.12) and (3.13) resp.

The contributions of the φ-ghosts is described by

$$\mathcal{L}_\varphi = -\partial_\mu^{\mathrm{EM}} \varphi^* D_\mu^{\mathrm{EM}} \varphi - M^2 \varphi^* \varphi - \tfrac{1}{2}Mg\varphi^* \varphi Z + \tfrac{1}{2}Mg\epsilon_{abc}\varphi^{*a}\varphi^b \psi^c , \tag{7.11}$$

again with the additional minus sign for each closed loop of φ's.

So, we have arrived at a renormalizable model containing photons, ρ-mesons and neutral Z-particles. There are four independent parameters: g, M, α and e. The parameter β is dictated by the tadpole condition:

$$\langle 0|Z(x)|0\rangle = 0 . \tag{7.12}$$

Note the third term in (7.8), which leads to phenomena like vector dominance. It is a consequence of the translation (7.5b).

8. CALCULATION OF ELECTROMAGNETIC MASS DIFFERENCES

One of the main virtues of the models presented here is, that there are no ambiguities due to infinities, and the number of independent parameters is small. It is interesting to calculate some "electromagnetic" mass differences. For instance, in the model of sect. 3 one may introduce an isospin one fermion†

$$\mathcal{L}_N = -\bar{N}(m + \gamma_\mu D_\mu)N . \tag{8.1}$$

† It appears that counterterms of the form $\bar{N}^a \epsilon_{abc} X^b N^c$, or, in the asymmetric representation, $\bar{N}^a \epsilon_{abc} A^b N^c$, are not needed for renormalization.

A direct computation of the difference in mass of the $N^{\pm}$ and N^{0} seems to lead to ambiguous results because the integrals for the self-energy corrections diverge. But, one of the Ward identities states that the graph of fig. 6a equals zero, if the charged particle N_2, and the neutral N_3 are both on mass shell. Let us consider all terms of this graph up to third order in g (fig. 6b).

The first term only contributes if $m^{\pm} \neq m^{0}$. Now the mass difference $m^{\pm} - m^{0}$ will be of order g^2, so q can be taken of order g^2. Hence the last four diagrams of fig. 6b will not contribute in the third order of g.

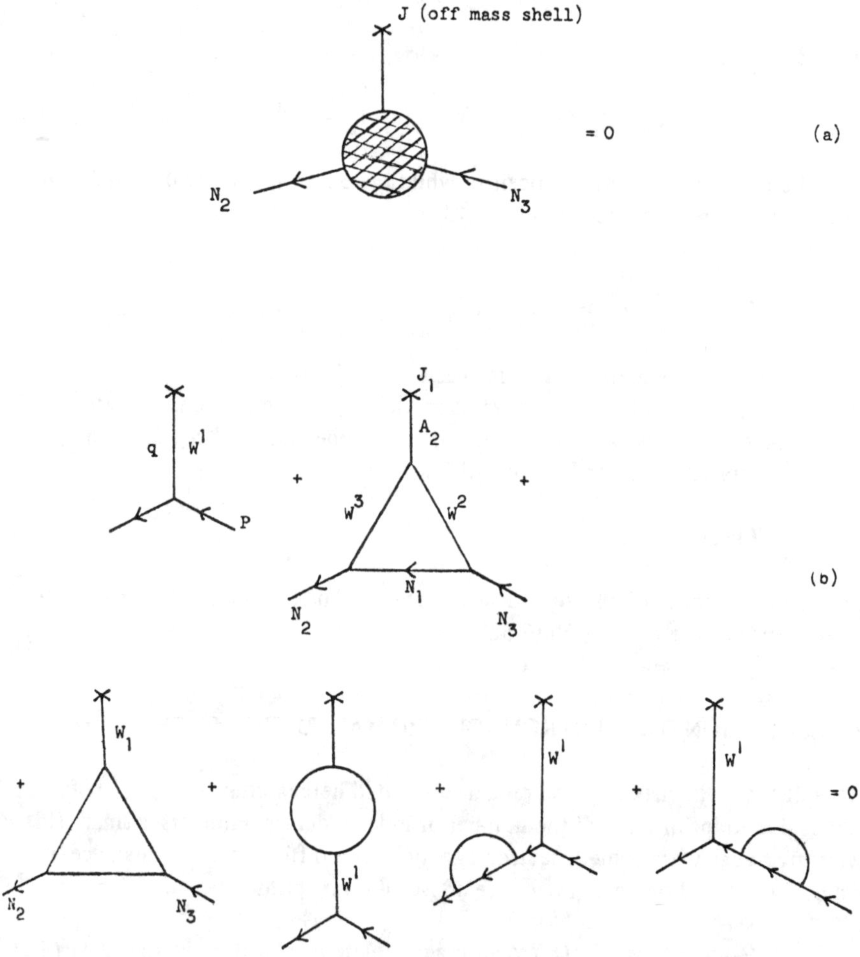

Fig. 6. a. Ward identity for the model of sect. 3, augmented with isospin-one fermions. b. All contributions up to order g^3.

Thus, the Ward identity reads

$$g(m_{\pm} - m_{o}) - M.B = 0 \tag{8.2}$$

where B is the second graph of fig. 6b.

The momentum transfer q may now be taken to be zero, and both external N-lines have momentum p with $p^2 = -m^2$. The integral in B converges, and the result is:

$$\frac{m^{\pm} - m^{o}}{m} = \frac{2g^2}{(4\pi)^2} \int_0^1 dx(1+x) \log\left[1 + \frac{M^2(1-x)}{m^2x^2}\right] \tag{8.3}$$

(in second order of g).

This mass-difference is always positive, and for small resp. large values of M^2/m^2 eq. (8.3) may be simplified to

$$\begin{aligned} m^{\pm} - m^{o} &= \frac{g^2M}{8\pi} \qquad (M^2 \ll m^2)\,; \\ m^{\pm} - m^{o} &= \frac{3g^2m}{(4\pi)^2} \log\frac{M^2}{m^2} \qquad (M^2 \gg m^2)\,. \end{aligned} \tag{8.4}$$

A negative mass-difference is found for the ρ-meson itself in the model of sect. 7. This mass-difference is of zeroth order, and originates from the "vector dominance" term in eq. (7.8). Diagonalisation of the bilinear terms in (7.8) leads to the mass formula for the ρ-mesons (in zeroth order):

$$M_o^2 = \frac{M^{\pm 2}}{1 - e^2/g^2}\,. \tag{8.5}$$

The author wishes to thank Prof. M.Veltman for his invaluable criticism and encouragement.

APPENDIX

Feynman rules for the various models

In the preceding article I a system with a complete local gauge invariance was quantized using a path integral technique. It was shown how to make the integrand gauge non-invariant without changing the total amplitude. The same procedure has been applied here. The term which breakes the gauge invariance is always called $\mathcal{L}^c$. As a consequence of this procedure, the Feynman rules must always be derived from the final Lagrangians by performing the path integral, and not by canonical quantization. This means that the propagators are always the inverse of the matrices in the bilinear terms of the Lagrangian, and the vertices are the coefficients in front of the remaining terms in the Lagrangian, regardless whether time derivatives occur or not. We mention here some of these Feynman rules.

Model of section 3. (compare (3.14) and (3.15)).
physical particles:

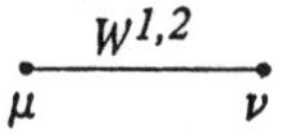

$$\frac{\delta_{\mu\nu}}{k^2 + M^2 - i\epsilon} \qquad \text{(A.1)}$$

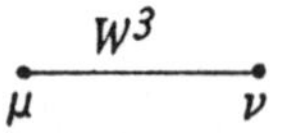

$$\frac{\delta_{\mu\nu}}{k^2 - i\epsilon} \qquad \text{(A.2)}$$

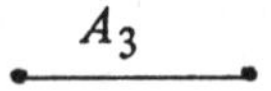

$$\frac{1}{k^2 + \alpha M^2 - i\epsilon} \qquad \text{(A.3)}$$

ghosts:

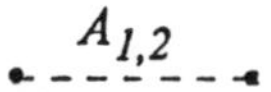

$$\frac{1}{k^2 + M^2 - i\epsilon} \qquad \text{(A.4)}$$

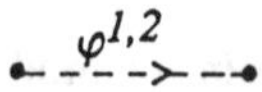

$$\frac{1}{k^2 + M^2 - i\epsilon} \qquad \text{(A.5)}$$

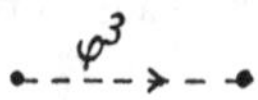

$$\frac{1}{k^2 - i\epsilon} \qquad \text{(A.6)}$$

for each closed loop of φ lines: $\quad -1$ (A.7)

some of the vertices:

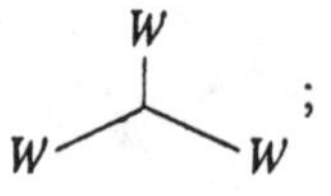

;

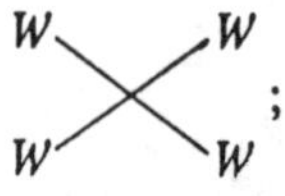

;

as in massless Yang-Mills fields.

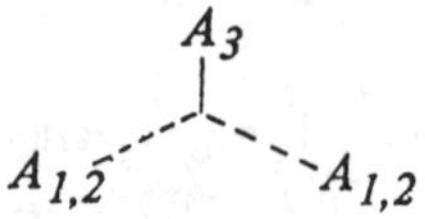

$$-\alpha g M \qquad \text{(A.8)}$$

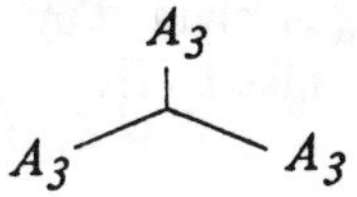

$$-3\alpha g M \tag{A.9}$$

etc.

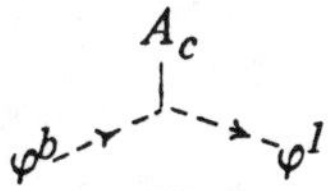

$$g M \epsilon_{2bc} \tag{A.10}$$

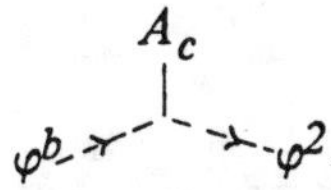

$$-g M \epsilon_{1bc} \tag{A.11}$$

$(k = 0)$

$$-\frac{\beta M}{g} \tag{A.12}$$

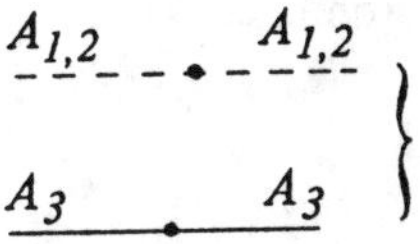

$$-\beta \tag{A.13}$$

The vertices (A.12) and (A.13) must be added to higher order tadpole- and self-energy diagrams, with β chosen such, that all tadpole contributions cancel.

Model of section 6. (compare (6.6), (6.7) and (6.8)).
physical particles:

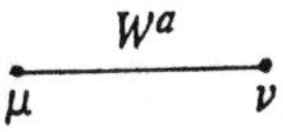

$$\frac{\delta_{\mu\nu}}{k^2 + M^2 - i\epsilon} \tag{A.14}$$

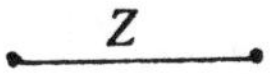

$$\frac{1}{k^2 + 4\alpha M^2 - i\epsilon} \tag{A.15}$$

ghosts:

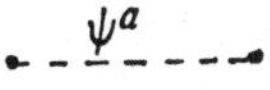

$$\frac{1}{k^2 + M^2 - i\epsilon} \tag{A.16}$$

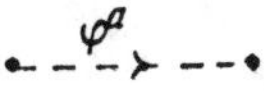

$$\frac{1}{k^2 + M^2 - i\epsilon} \tag{A.17}$$

some of the vertices:

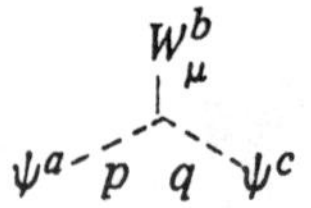

$$-\tfrac{1}{2} i g \epsilon_{abc} (p-q)_\mu \qquad \text{(A.18)}$$

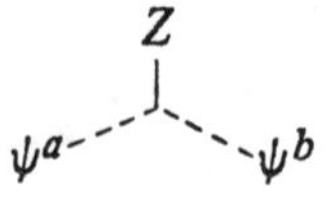

$$-2\alpha M g \delta_{ab} \qquad \text{(A.19)}$$

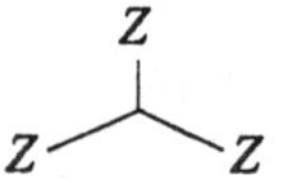

$$-6\alpha M g \qquad \text{(A.20)}$$

etc.

Model of section 7. (compare (7.8), (7.9), (7.10) and (7.11)).
Rules as in preceding model, but with additional photon lines:

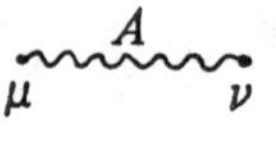

$$\frac{\delta_{\mu\nu}}{k^2 - i\epsilon} \qquad \text{(A.21)}$$

and vertices:

$$-\frac{e}{g}(k^2 \delta_{\mu\nu} - k_\mu k_\nu) \qquad \text{(A.22)}$$

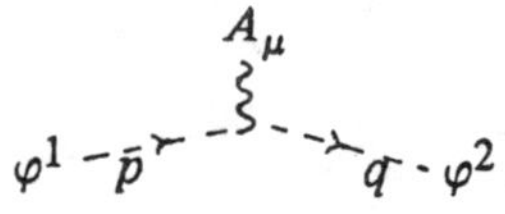

$$e(p-q)_\mu \qquad \text{(A.23)}$$

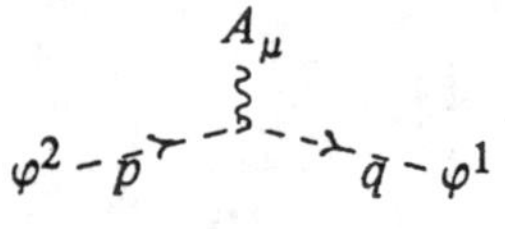

$$-e(p-q)_\mu \qquad \text{(A.24)}$$

etc.

The J-source can now emit a photon and a ρ-meson simultaneously:

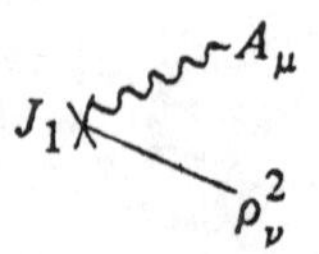

$$-e\,\delta_{\mu\nu} \qquad \text{(A.25)}$$

etc.

The vertex (A.22) is a consequence of the fact that we did not diagonalize the bilinear terms in (7.8) completely. The diagonalized propagators have a rather complicated form. For small e it is easier to leave (A.22) as it stands.

If necessary, vertices like (A.25) for the J-source contribution may be avoided by choosing another expression for $\mathcal{L}^c$ (eq. (7.10)).

REFERENCES

[1] G.'t Hooft, Nucl. Phys. B33 (1971) 173.
[2] J.S.Bell and R.Jackiw, Nuovo Cimento 60A (1969) 47.
[3] S.L.Adler, Phys. Rev. 177 (1969) 2426.
[4] I.S.Gerstein and R.Jackiw, Phys. Rev. 181 (1969) 1955.
[5] S.Weinberg, Phys. Rev. 140 (1965) B516.
[6] T.W.B.Kibble, Phys. Rev. 155 (1967) 1554.
[7] B.W.Lee, Nucl. Phys. B9 (1969) 649.
[8] J.L.Gervais and B.W.Lee, Nucl. Phys. B12 (1969) 627.
[9] M.Veltman, Nucl. Phys. B7 (1968) 637; Nucl. Phys. B21 (1970) 288.
[10] D.G.Boulware, Ann. of Phys. 56 (1970) 140.
[11] A.A.Slavnow, L.D.Faddeyev, Massless and massive Yang-Mills field, Moscow preprint 1970.
[12] E.S.Fradkin, U.Esposito and S.Termini, Rivista del Nuovo Cimento 2 (1970) 498.
[13] S.Weinberg, Phys. Rev. Letters 19 (1967) 1264.

REGULARIZATION AND RENORMALIZATION OF GAUGE FIELDS

G. 't HOOFT and M. VELTMAN
Institute for Theoretical Physics *, *University of Utrecht*

Received 21 February 1972

Abstract: A new regularization and renormalization procedure is presented. It is particularly well suited for the treatment of gauge theories. The method works for theories that were known to be renormalizable as well as for Yang-Mills type theories. Overlapping divergencies are disentangled. The procedure respects unitarity, causality and allows shifts of integration variables. In non-anomalous cases also Ward identities are satisfied at all stages. It is transparent when anomalies, such as the Bell-Jackiw-Adler anomaly, may occur.

1. INTRODUCTION

Recently it has been shown [1] that it is possible to formulate renormalizable theories of charged massive vector bosons. The derived Feynman rules involve ghost particles, and in order to establish unitarity and causality of the S-matrix Ward identities are needed. The necessary combinatorial techniques were given in ref. [2], in the treatment of massless Yang-Mills fields. It was emphasized that these same techniques work also in the case of massive vector boson theories obtained from the massless theory by means of the Higgs-Kibble [3] mechanism. Stated somewhat differently: the manifestly renormalizable set ** of Feynman rules involving ghosts may be transformed into a set of manifestly unitary and causal Feynman rules by means of Ward identities. Actually these manifestly unitary and causal Feynman rules are quite meaningless in view of the occurring divergencies, and a direct proof of unitary and causality starting from the manifestly renormalizable rules is to be preferred. This is precisely the program carried through in refs. [1, 2].

However, even with a set of manifestly renormalizable rules one cannot be sure that a consistent theory results unless a suitable cut-off and subtraction procedure has been defined. In particular, since unitarity depends crucially on the validity of the Ward indentities one must have a procedure that respects those Ward identities. In ref. [2] the existence of such a procedure was proven for diagrams containing at

* Postal address: Maliesingel 23, Utrecht, the Netherlands.
** i.e. renormalizable with respect to power counting.

most one closed loop; it is the aim of the present paper to extend the argument to arbitrary order of perturbation theory.

In connection with the question of renormalizability there is the problem of overlapping divergencies. This problem is of course not peculiar to gauge theories, and since it has been solved in other cases one could perhaps consider this as a not very urgent problem. However, our treatment deviates in several respects from the conventional procedures, and for this reason we have also considered this problem. It turns out that the techniques given below are particularly well suited to this purpose, and it will be shown that no difficulties arise.

The procedure suggested in ref. [2] was based on the observation that the Ward identities do hold irrespective of the dimension of the space involved. By introducing a fictitious fifth dimension and a very large fifth component of momentum inside the loop suitable gauge invariant regulator diagrams could be formulated. This procedure breaks down for diagrams containing two or more closed loops because then the "fifth" component of loop momentum may be distributed over the various internal lines. It was guessed that more dimensions would have to be introduced, and thus the idea of continuation in the number of dimensions suggested itself. This is the basic idea employed in this paper *.

In sect. 2 we define an analytic continuation of the S-matrix elements in the complex n-plane, where n is a variable that for positive integer values equals the dimension of the space involved with respect to loop quantities. The physical situation is obtained for $n = 4$. This definition is such that for finite diagrams the limit for $n = 4$ equals the conventional result. It turns out that the generalized S-matrix elements so defined are analytic in n and the infinities of perturbation theory manifest here as poles for $n = 4$.

Renormalization amounts to subtraction of these poles, and one must show that this subtraction procedure does not violate unitarity etc. In fact, in sect. 3 it will be shown that the generalized S-matrix elements satisfy Ward identities, unitarity and causality for all n. In sect. 4 we consider the question of renormalization and overlapping divergencies.

Since the whole subject of this paper is rather involved and technical, we have stripped the argument as much as possible of non-essential details. The arguments of sects. 2 to 4 are valid for theories containing scalar, vector etc. particles; in sect. 5 the extension to include fermions is indicated.

Sect. 6 is devoted to a discussion of the limitations of the method. It is indicated where there arise conflicts between the method and Ward identities; there seem to be no limitations with respect to the other properties. It appears that such conflicts happen only where there really are troubles, i.e. in cases where anomalies [4–6] occur. Even then the method is very suitable for practical evaluation of the anomalies, which is demonstrated in this section.

* Independently C.G. Bollini and J.J. Giambiagi [12] have also advanced and pursued the idea of continuation in the number of dimensions.

Infrared difficulties associated with zero mass particles are not considered in this paper.

For completeness we note that our method bears some resemblance to the method of analytical continuation [7]. The analytic continuation in the exponents of the propagators, as suggested by Bollini et al., amounts in the actual calculations to almost the same as continuation in the number of dimensions (see for instance eq. (A5) in appendix A). In several crucial respects, however, continuation in n gives less deformation of the structure of perturbation theory.

2. DEFINITIONS

As an example we take a photon interacting with charged pions. Vertices:

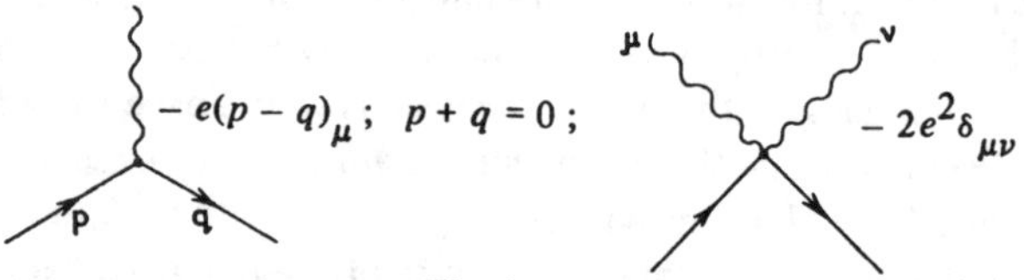

Fig. 1.

The arrows denote the direction of charge flow. Momenta are taken in the direction of the vertices.

The lowest order photon self-energy diagrams are:

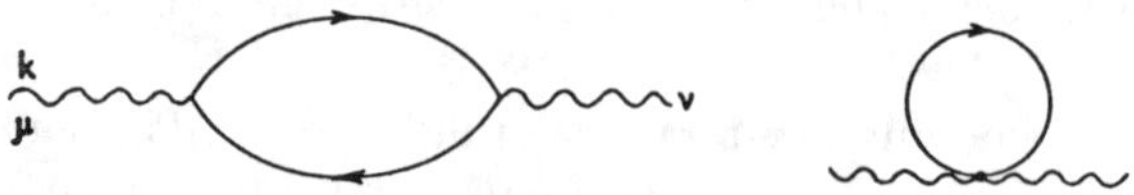

Fig. 2.

Assuming n component loop momentum p we have

$$e^2 \int \mathrm{d}_n p \left[\frac{(2p+k)_\mu (2p+k)_\nu}{(p^2+m^2)((p+k)^2+m^2)} - \frac{2\delta_{\mu\nu}}{p^2+m^2} \right] \tag{1}$$

Evaluating the integral without worrying about divergencies:

$$= e^2 \int_0^1 \mathrm{d}x \int \mathrm{d}_n p \, \frac{4p_\mu p_\nu + 2p_\mu k_\nu + 2k_\mu p_\nu + k_\mu k_\nu - 2((p+k)^2+m^2)\delta_{\mu\nu}}{(p^2+2pkx+k^2x+m^2)^2} \,.$$

Using the formulae of appendix A (note that in the end terms odd in $(1-2x)$ may be dropped):

$$(1) = e^2 i \pi^{\frac{1}{2}n}\, \Gamma(2-\tfrac{1}{2}n) \int_0^1 \mathrm{d}x \, \frac{(1-2x)^2 (k_\mu k_\nu - k^2 \delta_{\mu\nu})}{(m^2+k^2x(1-x))^{2-\frac{1}{2}n}} \,. \tag{2}$$

This expression is manifestly gauge invariant. In the complex n plane there are simple poles for $n = 4, 6, 8$ etc. Note that the n-dependence is such that gauge invariance holds for any n. This is the property referred to in the introduction: Ward identities do not involve the dimensionality of space.

In order to carry through renormalization we subtract from (2) the pole and its residue for $n = 4$

$$e^2 i\pi^2 \frac{2}{4-n}(k_\mu k_\nu - k^2\delta_{\mu\nu})\int_0^1 \mathrm{d}x(1-2x)^2 \,. \tag{3}$$

This subtraction term is a polynomial in the external momentum, and of course gauge invariant. Subtracting (3) from (2) and taking the limit $n = 4$ gives the customary result:

$$\begin{aligned} &-ie^2\pi^2(k_\mu k_\nu - k^2\delta_{\mu\nu})\int_0^1 \mathrm{d}x\,(1-2x)^2 \ln(m^2 + k^2 x(1-x)) \\ &\quad + C(k_\mu k_\nu - k^2\delta_{\mu\nu}) \,, \end{aligned} \tag{4}$$

where C is a constant related to the n dependence other then in the exponent of the denominator. Actually C is undetermined, which may be seen as follows. Suppose that in (2) we replace e^2 by e^2M^{4-n}, where M is an arbitrary mass. This gives (2) an n independent dimension of $(\text{mass})^2$. However C in (4) is changed by a term proportional to $\ln M$. It may be noted that the same arbitrariness results if one evaluates (1) with the help of Pauli-Villars regulators.

The above heuristic derivation shows many of the features of the method advocated in this paper. In practical calculations for one loop diagrams this provides a very simple scheme for computing gauge invariant results. It could for instance be used to show cancellation of divergencies in the manifestly unitary set of Feynman rules mentioned in the introduction and investigated by several authors [8].

There are several serious objections to the above manipulations. First of all, our starting point eq. (1) is meaningless for $n \geqslant 2$. In order to obtain a sensible result one must (i) change the Feynman rules such that for non-integer n all diagrams give rise to well-defined expressions, and (ii) define a suitable limiting procedure for $n = 4$, which restores originally convergent diagrams to their original values while originally divergent diagrams are given a meaning consistent with unitarity etc. Thus first of all a redefinition of the S-matrix is in order.

Consider again eq. (1). First we split the n-dimensional space in a 4 dimensional (physical) and an n–4 dimensional subspace:

$$\int \mathrm{d}_n p \to \int \mathrm{d}_4 \underline{p} \int \mathrm{d}_{n-4} P \,. \tag{5}$$

Multiplying (1) with two arbitrary physical four vectors $e_{1\mu}$ and $e_{2\nu}$ we see that (1) depends on the direction of $\underline{p}$ but not on the direction of P. Introducing polar coordinates in P space and integrating over angles one finds:

$$(1) = \int \mathrm{d}_4\underline{p} \int_{\cdot} \mathrm{d}\omega\, \omega^{n-5} \frac{2(\pi)^{\frac{1}{2}(n-4)}}{\Gamma(\frac{1}{2}(n-4))} f(\underline{p}, \omega^2)\,, \tag{6}$$

where ω is the length of P in the $n-4$ dimensional subspace. The dependence on the external vectors e_1, e_2 and k is not shown explicitly. Note that $(pk) = (\underline{p}k)$, $(e_1p) = (e_1\underline{p})$, $(e_2p) = (e_2\underline{p})$ and $p^2 = \underline{p}^2 + \omega^2$. (6) is still quite meaningless, and we continue our formal manipulations until we arrive at an expression that can be given a meaning. Note that the second integral in (6) contains an infrared divergence for $n \leqslant 4$. This divergence is superficial and may be removed by partial integration (throwing away surface terms):

$$\int_0^\infty \mathrm{d}\omega\, \omega^{n-5} f(\underline{p}, \omega^2) = -\frac{2}{n-4} \int_0^\infty \mathrm{d}\omega\, \omega^{n-3} \frac{\partial}{\partial\omega^2} f(\underline{p}, \omega^2)\,. \tag{7}$$

Doing this λ times on (6) gives

$$\frac{\pi^{\frac{1}{2}(n-4)}\, 2}{\Gamma(\frac{1}{2}(n-4)+\lambda)} \int \mathrm{d}_4\underline{p} \int_0^\infty \mathrm{d}\omega\, \omega^{n-5+2\lambda} \left(-\frac{\partial}{\partial\omega^2}\right)^\lambda f(\underline{p}, \omega^2) \tag{8}$$

For the (in 4 dimensional space) quadratically divergent diagrams of fig. 2 this is a well defined formula for $4 - 2\lambda < n < 2$. Note that $\omega \geqslant 0$. *Eq.* (8) *with sufficiently large λ defines the contribution of one loop diagrams to the generalized S-matrix elements in a finite region of the complex n-plane. This region is the domain of convergence of the integrals in* (8).

By taking a sufficiently large λ the domain of convergence extends to arbitrarily small n. Furthermore, the degree of convergence is $2-n$ as far as ultraviolet behaviour is concerned and $n-4+2\lambda$ for the infrared behaviour. Clearly, by choosing a suitable λ and n one has a representation of the generalized diagrams in some region of the n-plane in terms of arbitrarily convergent integrals.

If a diagram is convergent in 4-dimensional space then the redefinition (8) exists for $n < n_0$ with $n_0 > 4$. Moreover, for $n = 4$ (8) equals the result evaluated in the conventional way, as may be seen by taking $\lambda = 1$ and setting $n = 4$. Thus for finite diagrams our prescription gives the conventional result in the limit $n = 4$.

For divergent diagrams (8) will be meaningless for $n = 4$. However, as will be shown, (8) may be continued in the complex n-plane to larger n values. The result will in general have a pole at $n = 4$. In order to make sense in the limit $n = 4$ one must introduce counterterms in the perturbation expansion, and those counterterms must be chosen such that the poles at $n = 4$ disappear. Whether this can be done in a consistent manner is a separate problem, to be tackled in sect. 4.

For values of n outside the domain of convergence of the integrals in (8) *the contribution to the generalized S-matrix elements is defined as the analytic continuation of* (8).

It turns out to be possible to construct explicity this analytic continuation towards larger n values. The method is as follows. By means of partial integration, valid inside the domain of convergence of (8) we derive a new formula, identical to (8) inside the domain of convergence of (8), but analytic in n in an enlarged domain. By the principles of analytic continuation this new formula is then equal to the analytic continuation of (8) in this enlarged domain.

In view of the importance of this construction we will try to formulate it as clearly as possible. Consider the integral:

$$I = \int \mathrm{d}_\kappa p \frac{p_a^{\lambda_1} p_b^{\lambda_2} \ldots p_c^{\lambda_j}}{((p+k_1)^2+m_1^2)^{\alpha_1}((p+k_2)^2+m_2^2)^{\alpha_2}\ldots((p+k_l)^2+m_l^2)^{\alpha_l}} \tag{9}$$

p_a, p_b etc. are components a, b etc. of p. The exponents $\lambda_i \ldots \lambda_j$ are not necessarily integer. (8) is of the form (9) with $\kappa = 5$, where the integration over p_5 in (9) is nothing but the ω-integration in (8). Thus p_5 occurs with an n-dependent exponent in the numerator. Also $\underline{p}_1$, $\underline{p}_2$ etc. may occur in the numerator, they are contained in (8) in the function f. The differentiations with respect to ω^2 in (8) have as net effect an increase of the exponents of the factors in the denominator; the α_i will be integer, but they can be larger than 1.

The integral in (9) will be convergent if

$$\begin{gathered} \lambda_1 > -1, \lambda_2 > -1, \ldots, \lambda_j > -1 \\ \kappa + \lambda_1 + \lambda_2 + \ldots \lambda_j - 2(\alpha_1 + \alpha_2 + \ldots \alpha_l) < 0 . \end{gathered} \tag{10}$$

Next we insert in (9) the expression, identical to unity:

$$\frac{1}{\kappa}\sum_{i=1}^{\kappa}\frac{\partial p_i}{\partial p_i}. \tag{11}$$

Within the region (10) we may perform partial integration. After some trivial algebra we obtain:

$$I = -\frac{\lambda_1+\lambda_2+\ldots\lambda_j}{\kappa} I + \frac{2(\alpha_1+\alpha_2+\ldots\alpha_l)}{\kappa} I - \frac{1}{\kappa} I'$$

with

$$\begin{aligned} I' = \int \mathrm{d}_\kappa p p_a{}^{\lambda_1} \ldots p_c^{\lambda_j} \Bigg[& \frac{2\alpha_1(m_1^2+k_1^2+(pk_1))}{((p+k_1)^2+m_1^2)^{\alpha_1+1}(\ )^{\alpha_2}\ldots(\ )^{\alpha_l}} \\ & + \frac{2\alpha_2(m_2^2+k_2^2+(pk_2))}{(\ )^{\alpha_1}(\ )^{\alpha_2+1}\ldots(\ )^{\alpha_l}} + \ldots + \frac{2\alpha_l(m_l^2+k_l^2+(pk_l))}{(\ )^{\alpha_1}(\ )^{\alpha_2}\ldots(\ )^{\alpha_l+1}}\Bigg] \end{aligned} \tag{12}$$

or

$$I = -\frac{1}{(\kappa+\lambda_1+\lambda_2+\ldots\lambda_j-2\alpha_1-2\alpha_2-\ldots-2\alpha_l)} I' \tag{13}$$

The integral I' converges if

$$\lambda_1 > -1\,, \qquad \lambda_2 > -1\,, \ldots, \lambda_j > -1$$

$$\kappa + \lambda_1 + \lambda_2 + \ldots \lambda_j - 2(\alpha_1 + \alpha_2 + \ldots \alpha_l) < 1\,. \tag{14}$$

This is a larger domain than (10), and the right hand side of (13) is the explicit representation of the analytic continuation of I into this domain

For one loop diagrams the variable n appears linearly in some exponent λ. In that case one obtains an analytic continuation over a region of magnitude one in the direction of positive n.

The above operation will be called *partial p*. By repeated application of partial p and for sufficiently large λ one obtains an explicit representation valid in an arbitrarily large domain in the complex n-plane. Or, in a given region of the n-plane a representation in terms of arbitrarily convergent integrals.

With this prescription one may now evaluate the integrals in the example (1). The result is of course precisely (2).

For diagrams with two closed loops one may proceed in a similar way. There will be two n-fold integrals, and one writes:

$$\int \mathrm{d}_n p \int \mathrm{d}_n p' \to \int \mathrm{d}_4 \underline{p} \int \mathrm{d}_4 \underline{p}' \int \mathrm{d}_{n-4} P \int \mathrm{d}_{n-4} P'\,. \tag{15}$$

In the P' integral the fifth axis is taken in the direction of the $(n-4)$ vector P:

$$(15) \to \int \mathrm{d}_4 \underline{p} \int \mathrm{d}_4 \underline{p}' \int \mathrm{d}_{n-4} P \int \mathrm{d}p'_5 \int \mathrm{d}_{n-5} P'\,.$$

The integrands will be independent of the P and P' directions. The integration over angles may be performed:

$$(15) \to \frac{4\pi^{\frac{1}{2}(2n-9)}}{\Gamma(\frac{1}{2}(n-4))\,\Gamma(\frac{1}{2}(n-5))} \int \mathrm{d}_4 \underline{p} \int \mathrm{d}_4 \underline{p}' \int_0^\infty \mathrm{d}\omega\, \omega^{n-5} \int_{-\infty}^{\infty} \mathrm{d}p'_5 \int_0^\infty \mathrm{d}\omega'\, \omega'^{n-6} \tag{16}$$

The argument of such integrals will be a function of the components $\underline{p}$ and $\underline{p}'$, of ω^2, $p_5'^2 + \omega'^2$ and $(p'_5 + \omega)^2 + \omega'^2$ (arising from p^2, p'^2 and $(p+p')^2$).

(16) may be written in an elegant form by introducing a twodimensional space, and the vectors

$$q = \begin{pmatrix} \omega \\ 0 \end{pmatrix}, \qquad q' = \begin{pmatrix} p'_5 \\ \omega' \end{pmatrix} \tag{17}$$

With ϵ_{ij} the completely antisymmetric tensor in two dimensions ($\epsilon_{12} = 1$) we have:

$$\int \mathrm{d}_n p \int \mathrm{d}_n p' \, f(p^2, p'^2, (p+p')^2)$$

$$= \frac{2\pi^{\frac{1}{2}(2n-9)-1}}{\Gamma(\frac{1}{2}(n-4))\,\Gamma(\frac{1}{2}(n-5))} \int \mathrm{d}_4 \underline{p} \int \mathrm{d}_4 \underline{p}' \int \mathrm{d}_2 q \int \mathrm{d}_2 q' \, (\epsilon_{ij} q_i q_j')^{n-6} \theta(\epsilon_{ij} q_i q_j')$$

$$f(q^2, q'^2, (q+q')^2) \,. \tag{18}$$

The step-function θ is needed because of the lower limit 0 in the ω' integration in (16). We have dropped explicit indication of the dependence on the components of $\underline{p}$ and $\underline{p}'$.

The equivalent of (8) is also obtained by partial integration. To this purpose one observes that

$$(\epsilon_{ij} q_i q_j')^\alpha = \frac{1}{(\alpha+2)(\alpha+1)} \epsilon_{ab} \frac{\partial}{\partial q_a} \frac{\partial}{\partial q_b'} (\epsilon_{ij} q_i q_j')^{\alpha+1} \tag{19}$$

λ times application of (19) and subsequent partial integration leads to an expression similar to (8):

$$\frac{2\pi^{\frac{1}{2}(2n-11)}}{\Gamma(\frac{1}{2}(n-4)+\lambda)\,\Gamma(\frac{1}{2}(n-5)+\lambda)} \int \mathrm{d}_4 \underline{p} \int \mathrm{d}_4 \underline{p}' \int \mathrm{d}_2 q \int \mathrm{d}_2 q' \, (\epsilon_{ij} q_i q_j')^{n-6+2\lambda}$$

$$\times\, \theta(\epsilon_{ij} q_i q_j') \left(\frac{\partial^2}{\partial q^2\, \partial q'^2} + \frac{\partial^2}{\partial q^2\, \partial (q+q')^2} + \frac{\partial^2}{\partial q'^2\, \partial (q+q')^2} \right)^\lambda$$

$$\times\, f(q^2, q'^2, (q-q')^2) \,. \tag{20}$$

Again, (20) and its analytic continuation to larger n define the contribution of the two-loop diagrams to the generalized S-matrix elements. Explicit representations for large n may be obtained by operations similar to partial p in the one loop case. We need four such operations in the two loop case, and we will discuss them in sect. 4.

Definitions similar to (20) may be given for the three or more closed loop cases.

The above prescription applies if all loop particles are scalars. To complete our prescription to cover vector fields we note that indices that are part of the propagators contained in the loops now take the values 1 to n for integer n. This is because polarization vectors corresponding to internal lines become n-component vectors. The only practical consequence of this fact is that in doing the vector algebra of all occurring loop indices one must use the rule

$$\delta_{\mu\mu} = n \,. \tag{21}$$

After that one has expressions that can be used to define the diagrams for non-integer n. In establishing Ward-identities one notes an interplay between these factors

n and the factors n occurring in association with averaging over all directions in p space of factors like $p_\mu p_\nu$. See eqs. (A7), (A8). By virtue of (21) equations like

$$p_\mu(\delta_{\mu\nu}p^2 - p_\mu p_\nu) = 0$$

remain true even for continuous n in the sense defined above.

3. WARD IDENTITIES, UNITARITY, CAUSALITY

We must now establish that our generalized amplitudes satisfy Ward identities. These Ward identities involve (i) vector algebra and (ii) shifting of integration variables. In the sense defined in sect. 2 it is easy to see that the vector algebra goes through unchanged for any n. For example, consider the photon-pion vertex of fig. 1. One requires that this vertex when multiplied by the photon four momentum equals the difference of two inverse pion propagators. Thus, in 4 dimensions with the vertex of fig. 1 where $q = -p - k$:

$$k_\mu \; \{-(2p+k)_\mu\} = (p+k-p)_\mu \; \{-(p+k+p)_\mu\}$$

$$= -(p+k)^2 - m^2 + (p^2+m^2)\,.$$

In the n dimensional formulation, with the notation of eq. (5):

$$k_\mu \; \{-(2\underline{p}+k)_\mu\} = -(k, 2\underline{p}+k) = -(k, 2p+k)$$

$$= -(k+p-p, k+p+p) = -(p+k)^2 - m^2 + (p^2+m^2)\,,$$

with $(p+k)^2 = (\underline{p}+k)^2 + P^2$ and $p^2 = \underline{p}^2 + P^2$

It is this rather trivial type of vector algebra that is involved in proving the gauge invariance of eq. (1). In the case of vector particles things are slightly more complicated, and the rule (21) comes in. In that case one must demonstrate for instance that eq. (A7) from appendix A can be obtained from (A8) on multiplication with $\delta_{\mu\nu}$, which indeed happens to be the case. In general one must show that the vector algebra that must hold for the left hand side of eqs. (A5) – (A10) (and their generalizations) actually holds for the right hand sides for any n. One easily convinces oneself that this property holds keeping in mind that all necessary equations can be obtained from (A5) by differentiation with respect to k.

As for point (ii) the shifting of integration variables, we first note that any shift over an external (= physical) four vector is certainly allowed since we have kept the integrations over the first four components unchanged. Nothing else is required in the one loop case. In the two loop case we must have invariance for shifts like $p \to p + p'$, where p and p' are both loop momenta. From formula (20) this is evidently correct, due to the fact that the ϵ-product is invariant for such shifts.

Establishing unitarity and causality, or more precisely cutting rules [9, 10] is also very easy. With the usual $\pm i\epsilon$ prescriptions one needs only to establish the validity of the "largest time" equation of ref. [11] (or (2.8) of ref. [10]) which involves only the time components of the n-vectors, or after fourier transformation, the energy components. Since, as far as these components are concerned, we have not changed the structure of the propagators and integrations (as is evident from (8) and (20)), and since we can always take sufficiently convergent representations in some region of the n-plane, it is obvious that cutting rules hold in some region of the n-plane. Because any term in the cutting equations can be continued analytically to smaller and larger n values by means of the methods of sect. 3 we have the result that the cutting rules hold for any n.

It must be noted that in these rules the integration over intermediate states involves n-space. If all poles for $n = 4$ have been removed then in the limit $n = 4$ this integration over intermediate states reduces to the required integral over physical phase space. It is essential in this context that the phase space integrations are themselves finite, and do not introduce new poles.

In considering cut diagrams some care in handling the δ-functions is necessary. The following remarks may be of help in this respect;

(i) in (8) (and analogously in (20)) the various factors ω^2 occur in denominators simply in addition to the masses of the loop particles. E.g. $p^2 + m^2 = \underline{p}^2 + \omega^2 + m^2$. Thus one can see (8) as a superposition of diagrams where internal masses m_i^2 have been replaced by $\omega^2 + m_i^2$, with weight function

$$\omega^{n-5-2\lambda} \left(-\frac{\partial}{\partial\omega^2}\right)^{\lambda} .$$

One may go further and exchange the ω and $\underline{p}$ integrations in (8). For cut diagrams, where the $\underline{p}$ integration is convergent also one may further exchange differentiation with respect to ω^2 and the $\underline{p}$ integration. Then the calculation of cut diagrams becomes identical to the conventional calculation followed by differentiation and integration with respect to ω.

(ii) If one chooses a very small n the above mentioned weight function may induce strong threshold singularities.

4. RENORMALIZATION

In order to obtain a consistent theory it must be shown that the poles for $n = 4$ can be removed, order by order in perturbation theory. In a given order any new subtraction terms to be introduced must satisfy a stringent criterion: they may not have an imaginary part. This follows very simple from the fact that in a given order the imaginary part, through unitarity, is determined unambigeously by the lower order results. In practice this means that new subtraction terms must be finite polynomials in the external momenta. The demonstration that this is possible includes treatment of the overlapping divergencies. It is perhaps worthwhile to mention the fact that Ward identities and the problem of overlapping divergencies have nothing to do with each other, even though in quantum electrodynamics Ward identities have been of technical assistance in unraveling the divergence structure of perturbation theory.

In this section we will treat one and two closed loop diagrams. The generalization to more closed loops is obvious. In order to keep the discussion transparant we will omit most numerator factors as for instance occurring when there are vector particles. Yet the treatment will be such that these factors can be written without interference with the main argument.

The definitions (8) and (20) will form the basis of our discussion. Consider first one loop diagrams, eq. (8). We must show that the residues of eventual poles for $n = 4$ are finite polynomials in the external momenta. If the diagram is convergent in 4-space the expression (8) will be non-singular for $n = 4$. If the diagram is divergent than (8) will be well-defined only in some region to the left of $n = 4$. Subsequent analytic continuation by means of partial p shows a single pole for $n = 4$ multiplied by a finite and well-defined expression. On the other hand, (8) may be evaluated explicitly by means of Feynman parameters and the equations of appendix A. One obtains an expression of the form:

$$\Gamma(j - \tfrac{1}{2}n) \int \mathrm{d}y_1 \ldots \int \mathrm{d}y_i \frac{1}{(M^2)^{j-\frac{1}{2}n}} P(y, m, k) \tag{22}$$

independent of λ. In here j is some integer and M^2 and P are polynomials in the Feynman parameters y_i, the masses and the external momenta. The eventual pole for $n = 4$ is explicit in the Γ function; one has poles for $n \geqslant 2j$ in agreement with the results obtained by means of partial p. All this is to ensure that there is no trouble hidden in the Feynman parameter integrations *.

The residue of a pole for n an even integer $\geqslant 2j$ is:

$$C \int \mathrm{d}y_1 \ldots \int \mathrm{d}y_i (M^2)^l \, P(y, m, k) \tag{23}$$

with $l = \frac{1}{2}n - j \geqslant 0$. Obviously (23) is a finite polynomial, there are no terms of the form $\ln k^2$.

Thus, up to one closed loop, eventual poles for $n = 4$ (or for any other n) have as residue polynomials in the external momenta, and may be subtracted.

Next we turn to two closed loops. We assume that counter-terms of the form

$$\frac{1}{n-4} P(k, m)$$

* This is by no means trivial. For instance infrared divergencies are usually hidden in the Feynman parameter integrations. For two or more closed loops ultra-violet divergencies may also be transferred from the momentum integrations to the Feynman parameter integrations. For this reason one must be very careful in taking together propagators belonging to different loops by means of Feynman parameters.

have been introduced such as to make all one loop diagrams finite. Consider the general two loop diagram:

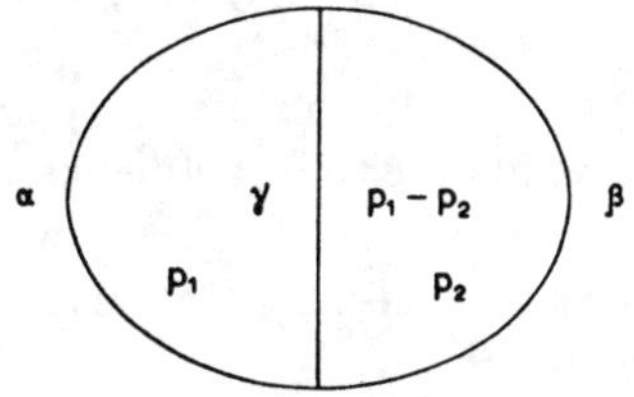

Fig. 3. $\alpha\beta\gamma$ diagram.

We have omitted all external lines. The corresponding expression is:

$$\int \mathrm{d}_n p_1 \, \mathrm{d}_n p_2 \frac{1}{(p_1^2 + m_1^2)^\alpha \, (p_2^2 + m_2^2)^\beta \, ((p_1 - p_2 + k)^2 + M^2)^\gamma} \tag{24}$$

and we will speak of the $\alpha\,\beta\,\gamma$ diagram.

In here we have assumed that all propagators that depend on p_1 and external momenta have been taken together by means of the Feynman parameter method. Similarly for p_2 and $p_1 - p_2$ dependent propagators. Furthermore we have suppressed all numerators, except insofar as power counting is concerned; thus

$$\frac{p_{1_\mu} \, p_{1_\nu}}{(p_1^2 + m_1^2)^5}$$

is in the above represented as

$$\frac{1}{(p_1^2 + m_1^2)^4}$$

Here, and in what follows, we write the integrals as if we are operating in an n-dimensional space with positive integer n, but this must be understood as symbolic for integrals like in (20) with sufficiently large λ. We work in a region of small n sufficiently far to the left of $n = 4$.

There are three one-loop diagrams contained in the above two-loop diagram. We will call them the $\alpha\gamma$, $\beta\gamma$ and $\alpha\beta$ sub-diagrams respectively:

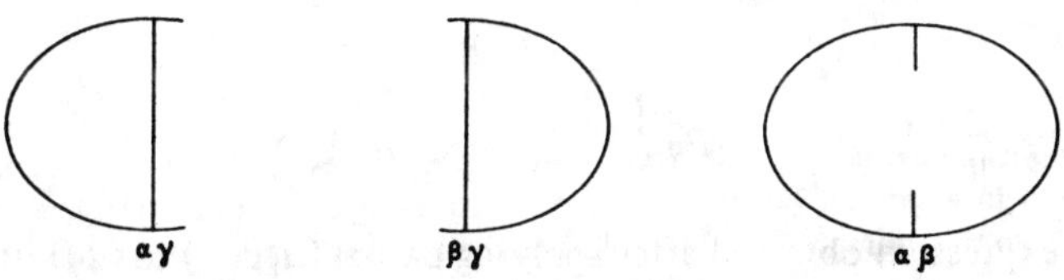

Fig. 4.

If any of these subdiagrams diverge we have counterterms associated with them. In addition to the $\alpha\beta\gamma$ diagram we must therefore also consider the subtraction diagrams:

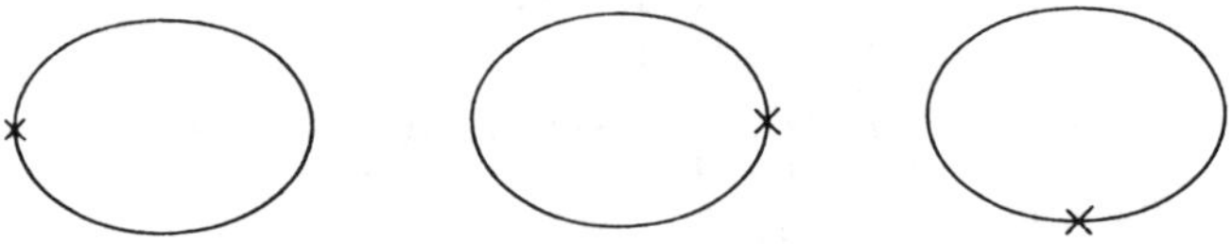

Fig. 5.

The crossed vertex in the first diagram refers to the pole with polynomial coefficient to be subtracted from the $\alpha\gamma$ diagram. Similarly for the other diagrams. It is clear that the subtraction diagrams contain double poles (1 pole from the crossed vertex and 1 pole from the loop integration) as well as single poles. In particular we have the single pole from the vertex multiplying the finite part of the loop integration. This will lead to terms of the form

$$\frac{1}{n-4}\ln(k^2)\,.$$

Such terms cannot be renormalized away, and the theory will be renormalizable only if these terms cancel against similar terms coming from the two loop ($\alpha\beta\gamma$) diagram. This means that we must unravel the pole structure of the $\alpha\beta\gamma$ diagram, eq. (24). In this way presents itself here the problem of the overlapping divergencies.

The expression (24) is well defined for sufficiently low n. The continuation to large n is slightly more complicated then in the one loop case. There are four operations that may be performed. First one may insert in (24) the expression, equal to unity:

$$\frac{1}{n}\sum_{i=1}^{n}\frac{\partial p_{1_i}}{\partial p_{1_i}}$$

and perform partial integration. The result is an equation similar to (13) showing a pole for $n = 2(\alpha+\gamma)$. Note that the denominators with the exponents α and γ involve p_1. We will call this operation partial (α, γ). Similarly one may define an operation partial (β, γ) by partial integration with respect to p_2. Further there is the operation partial (α, β) obtained by partial integration with respect to p_1 after the substitution $p_2' = p_1 - p_2$. Finally there is the operation partial (α, β, γ) showing a pole at $n = \alpha+\beta+\gamma$, and obtained by insertion of

$$\frac{1}{2n}\sum_{i=1}^{n}\left(\frac{\partial p_{1_i}}{\partial p_{1_i}} + \frac{\partial p_{2i}}{\partial p_{2i}}\right)$$

The explicit expression obtained after applying partial (α, β, γ) to (24) is:

$$(24) = -\frac{1}{(2n - 2\alpha - 2\beta - 2\gamma)} I' \tag{25}$$

with

$$I' = \int \mathrm{d}_n p_1\, \mathrm{d}_n p_2 \left[\frac{2\alpha\, m_1^2}{(\;)^{\alpha+1}(\;)^{\beta}(\;)^{\gamma}} + \frac{2\beta\, m_2^2}{(\;)^{\alpha}(\;)^{\beta+1}(\;)^{\gamma}} + \frac{2\gamma(M^2 + k^2 + (p_1 - p_2, k))}{(\;)^{\alpha}(\;)^{\beta}(\;)^{\gamma+1}} \right] \tag{26}$$

where the explicit form of the denominators is as in (24).

The procedure for continuation of (24) to large n is now clear. (24) has poles for $n = 2(\alpha+\beta)$, $n = 2(\alpha+\gamma)$, $n = 2(\beta+\gamma)$ and $n = \alpha + \beta + \gamma$. The integral representation will be valid for n less than the minimum of these four. Applying partial (α, β, γ) eventually followed by partial (α, β), (α, γ) and/or (β, γ) gives the desired continuation.

It would appear that (24) could contain four coincident poles, but actually there are at most second order poles.

We are now interested in poles with non-polynomial residues. *Definition.* A harmless pole is a pole with as residue a polynomial of finite order in the external momenta. *Definition.* The subintegral α, β is said to be divergent or convergent according to whether $\alpha + \beta \leqslant 2$ or $\alpha + \beta > 2$. *Definition.* The $\alpha\beta\gamma$ diagram (24) is said to be overall convergent or overall divergent according to whether $\alpha + \beta + \gamma > 4$ or $\alpha + \beta + \gamma \leqslant 4$.

More specifically we may call a subintegral α, β logarithmically, linearly etc. divergent if $\alpha + \beta = 2, \frac{3}{2}$, etc. Similar for the overall divergent cases $\alpha + \beta + \gamma = 4, 3, ...$ · etc.

First we will consider a simple situation, namely the case that the α, γ subintegral is logarithmically divergent while all other subintegrals are convergent. Then there is only one subtraction diagram, and we must show that (24) together with this subtraction diagram contains only harmless poles. More specifically consider the case $\alpha + \gamma = 2, \beta = 2$. Taking together the α and γ denominators of (24) by means of Feynman parameters and performing the integration over p_1 we obtain an expression of the form

$$\int \mathrm{d}_n p \frac{\Gamma(\alpha + \gamma - \frac{1}{2}n)}{(p^2 + m_3^2)^{\alpha+\gamma-\frac{1}{2}n}} \frac{1}{(p^2 + 2pq + m_4^2)^{\beta}}, \tag{27}$$

where we have written p instead of p_2 and suppressed various irrelevant factors. From this we must subtract the expression corresponding to the subtraction diagram

$$\int \mathrm{d}_n p \frac{2}{2\alpha + 2\gamma - n} \frac{1}{(p^2 + 2pq + m^2)^{\beta}}. \tag{28}$$

Main Theorem. The difference of (27) and (28) for $\alpha + \gamma = 2, \beta \geqslant 1$ contains only harmless poles. Proof: see appendix B.

This theorem is really the key theorem to our problem, because the general case may be reduced by means of partial operations to the case considered in the main theorem.

The general case may now be treated systematically. The case that any of the exponents α, β or γ is zero or negative is trivial and corresponds to a diagram containing no overlapping divergencies. Example:

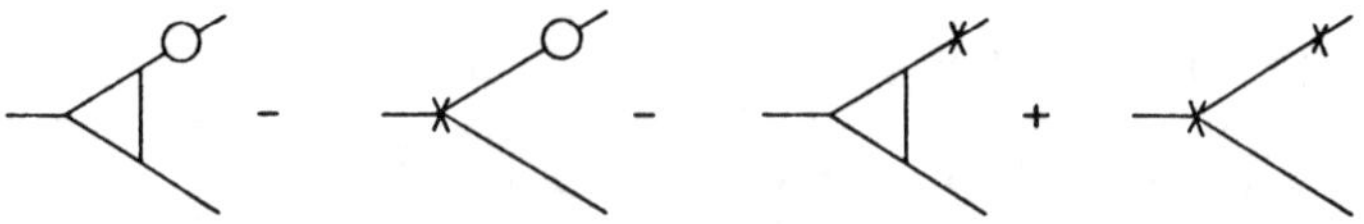

Fig. 6.

Such cases do not require new subtractions. Thus we assume now $\alpha > 0, \beta > 0, \gamma > 0$.
Theorem 1. If the integral (24) is logarithmically divergent and contains no divergent subintegrals than it contains a harmless single pole at $n = 4$.
Proof: by means of partial (α, β, γ) it is seen that the integral behaves at $n = 4$ as a single pole times a finite function. Actual calculation of (24) with the help of Feynman parameters exhibits this pole:

$$(24) \rightarrow \Gamma(\alpha + \beta + \gamma - n) \int \mathrm{d}y_1 \int \mathrm{d}y_2\, f(y_1, y_2)$$

and the integrals over the Feynman parameters $y_1\, y_2$ are well defined and finite. By actually setting $n = 4$ in those integrals one obtains the desired result.
Theorem 2. If the integral (24) is overall divergent and contains no divergent subintegrals then it contains a harmless single pole at $n = 4$.
Proof: by means of partial (α, β, γ) this case may be reduced to the case of theorem 1.
Theorem 3. If the integral (24) is overall convergent or logarithmically divergent then it contains at most one divergent sub-integral. The denominator not involved in the sub-integral has an exponent $\geqslant 2$.
Proof: if $\alpha + \beta + \gamma \geqslant 4$ and for instance $\alpha + \beta \leqslant 2$ then $\gamma \geqslant 2$.
Theorem 4. If the integral (24) is overall convergent or logarithmically divergent and contains one divergent sub-integral then the difference with the subtraction diagram containing the pole subtraction term corresponding to the divergent sub-integral has only harmless poles.
Proof: let α, β be the divergent subintegral. By means of partial (α, β) the divergent subintegral may be reduced to a logarithmically divergent subintegral. Since the remaining denominator has an exponent $\geqslant 2$ we are then in the case considered in the main theorem.

Finally we must consider the case that the integral (24) is linearly, quadratically etc. divergent. By means of partial (α, β, γ) this case can be reduced to the case of an overall convergent or logarithmically divergent integral, which has been considered above. This terminates the proof of renormalizability up to and including two loop diagrams.

As an example consider the case $\alpha = \beta = \gamma = 1$.
Thus:

$$I = \int \mathrm{d}_n p_1 \, \mathrm{d}_n p_2 \frac{1}{(p_1^2 + m_1^2)(p_2^2 + m_2^2)((p_1 - p_2 - k)^2 + m_3^2)} \tag{29}$$

The integral is overall quadratically divergent, and $\alpha\gamma$, $\beta\gamma$ and $\alpha\beta$ subintegrals are all logarithmically divergent. Thus we have three subtraction diagrams, and for instance the subtraction diagram corresponding to the $\alpha\gamma$ subintegral is:

$$I^{\mathrm{s}}_{\alpha\gamma} = -\int \mathrm{d}_n p_2 \, PP \left\{ \int \mathrm{d}_n p_1 \frac{1}{(p_1^2 + m_1^2)((p_1 - p_2 - k)^2 + m_3^2)} \; \frac{1}{(p_2^2 + m_2^2)} \right\} \tag{30}$$

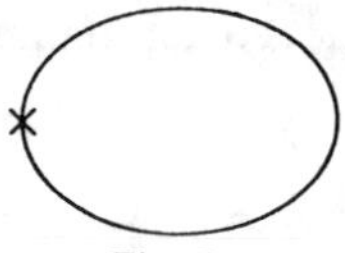

Fig. 7.

where $PP\{\ \}$ means pole part for $n = 4$. Because we have a logarithmic divergence the polynomial multiplying the pole $1/(n-4)$ will be simply a constant. Similarly for the other two subintegrals.

Applying partial (α, β, γ) to (29) we obtain:

$$I = -\frac{1}{2n-6} I'$$

$$I' = \int \mathrm{d}_n p_1 \, \mathrm{d}_n p_2 \left\{ \frac{2m_1^2}{(p_1^2 + m_1^2)^2 (p_2^2 + m_2^2)((p_1 - p_2 - k)^2 + m_3^2)} \right.$$

$$+ \frac{2m_2^2}{(p_1^2 + m_1^2)(p_2^2 + m_2^2)^2 ((p_1 - p_2 - k)^2 + m_3^2)}$$

$$\left. + \frac{2m_3^2 + 2k^2 - 2(p_1 - p_2, k)}{(p_1^2 + m_1^2)(p_2^2 + m_2^2)((p_1 - p_2 - k)^2 + m_3^2)^2} \right\} \tag{31}$$

Applying partial integration with respect to p_2 in (30) we obtain:

$$I^{s}_{\alpha\gamma} = + \frac{1}{n-2} \int \mathrm{d}_n p_2 \, PP\{\;\} \, \frac{2m_2^2}{(p_2^2 + m_2^2)^2} \tag{32}$$

In the limit $n = 4$ the second term of (31) combines together with the subtraction diagram such as to have precisely the case discussed in the main theorem. Similarly for the other terms, to be combined with the other subtraction diagrams.

This example demonstrates how the operation partial (α, β, γ) neatly separates out the various overlapping divergencies. After that the main theorem guarantees the absence of unwanted poles.

5. EXTENSION TO FERMIONS

The extension to fermions is straightforward and based on the following observation. Everything may be formulated such that only traces of strings of γ-matrices occur. If there are external fermion lines this may be done through the use of suitable projection operators. These traces must than be evaluated according to the rules (see appendix C)

$$\{\gamma^\mu, \gamma^\nu\} = 2\delta_{\mu\nu} \, , \tag{33}$$

$$\mathrm{Tr}\,(S) = 0 \text{ if } S \text{ is an odd string of } \gamma\text{'s} \, . \tag{34}$$

$$\mathrm{Tr}\,(I) = 4 \, . \tag{35}$$

Remember that $\delta_{\mu\mu} = n$. Any Ward identity relying as far as γ-matrices is concerned only on (33) (as in quantum-electrodynamics) will be satisfied for every n.

Note that there is no place for the pseudo-scalar γ^5 (in conventional notation) in (33), as there is no place for the pseudo tensor $\epsilon_{\mu\nu\alpha\beta}$. See sect. 6.

The rule (35) can be satisfied by finite matrices only for $n = 4$, but this is of no importance because we are only interested in a consistent algebra for $n \neq 4$. Or, in n-dimensional space one will have

$$\mathrm{Tr}\,(I) = f(n)$$

where f is a function of n only. We need only $f(n) = 4$, and the deviations of $f(n)$ from $f(4)$ are never important for Ward identities because one always compares diagrams with an equal number of traces.

As an example we consider the lowest order photon selfenergy diagram in quantum electrodynamics:

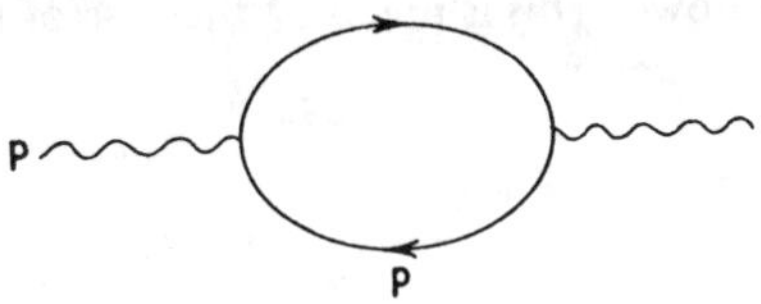

Fig. 8.

$$\int \mathrm{d}_n p \, \frac{\mathrm{Tr}\,[\gamma^\mu (i\gamma (p+k) + m) \gamma^\nu (i\gamma p + m)]}{((p+k)^2 + m^2)(p^2 + m^2)}$$

The trace may be evaluated using (33), (34) and (35). Taking denominators together one obtains:

$$4 \int_0^1 \mathrm{d}x \int \mathrm{d}_n p \, \frac{\delta_{\mu\nu}(m^2 + p^2 + pk) - 2p_\mu p_\nu - p_\mu k_\nu - k_\mu p_\nu}{(p^2 + 2pkx + k^2 x + m^2)^2} \, .$$

Using the equations of appendix A:

$$= \frac{4i\pi^{\frac{1}{2}n}}{\Gamma(2)} \Gamma(2 - \tfrac{1}{2}n) \int_0^1 \mathrm{d}x \, \frac{2x(1-x)(k_\mu k_\nu - \delta_{\mu\nu} k^2)}{(m^2 + k^2 x(1-x))^{2-\frac{1}{2}n}}$$

which is manifestly gauge invariant.

6. LIMITATIONS OF THE METHOD

The method fails if in the Ward identities there appear quantities that have the desired properties only in four dimensional space. An example is the completely antisymmetric tensor $\epsilon_{\mu\nu\alpha\beta}$. If the particular properties of this tensor are vital for the Ward identities to hold our method will fail because we cannot generalize $\epsilon_{\mu\nu\alpha\beta}$ to a tensor satisfying the required properties for non-integer n. Similarly for γ^5. One can write:

$$\gamma^5 = \frac{1}{4!} \epsilon_{\mu\nu\alpha\beta} \gamma^\mu \gamma^\nu \gamma^\alpha \gamma^\beta \, ,$$

insert this whenever γ^5 occurs and take the ϵ-tensor outside of the expression to be generalized to non-integer n. However, if we are dealing with Ward identities that rely on

$$\{\gamma^5, \gamma^\alpha\} = 0 \qquad \text{for} \qquad \alpha = 1, \ldots, n$$

$$\mathrm{Tr}\,(\gamma^5 \gamma^\mu \gamma^\nu \gamma^\alpha \gamma^\beta) = 4\epsilon_{\mu\nu\alpha\beta}$$

than this method breaks down. This is precisely the case of the well-known anomalies; to see this consider the vertex

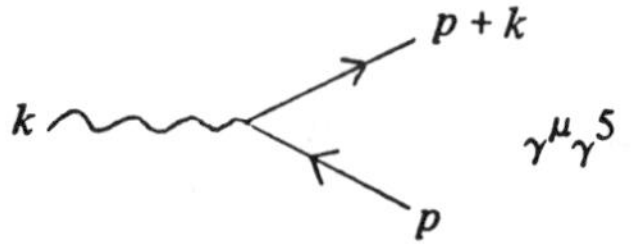

Fig. 9.

which is part of a linearly divergent loop with p as integration variable. One of the required Ward identities is that multiplication with k_μ should give something proportional to the fermion mass. In the combinatorial proof one writes:

$$\gamma k \gamma^5 = (\gamma(p+k) - \gamma p)\gamma^5 = \gamma(p+k)\gamma^5 + \gamma^5 \gamma p$$

where p is the loop momentum.

This is incorrect if $n \neq 4$. It is easily seen that the breakdown of the Ward identity is proportional to $\underline{p} - p$, and after integration to $n - 4$. When multiplied by a pole arising from the loop integration a non-zero part remains in the limit $n = 4$.

To exhibit all this explicitly we evaluate the anomaly for the well-known triangle graph. In the notation of Bell and Jackiw [4] the triangle graph is:

$$\Gamma^{\alpha\mu\nu} = \int \mathrm{d}_n r \, \frac{i\,\mathrm{Tr}[\gamma^5\gamma^\alpha \{i\gamma(p+r) - m\}\ \gamma^\mu (i\gamma r - m)\gamma^\nu \{i\gamma(r-q) - m\}]}{((p+r)^2 + m^2)(r^2+m^2)((r-q)^2+m^2)} \tag{36}$$

where p and q are the photon momenta. Multiplying by $k_\alpha = (p+q)_\alpha$ one writes:

$$\begin{aligned}\gamma^5 i\gamma k &= \gamma^5 \{-i\gamma(p+r) - m + i\gamma(r-q) - m + 2m\} \\ &= -\ \{i\gamma(r-q) + m\}\ \gamma^5 - \gamma^5 \{i\gamma(p+r) + m\}\ + 2m\gamma^5 \\ &\quad + 2i\gamma^5\gamma(r - \underline{r})\end{aligned}$$

where $\underline{r}$ coincides with r in the first four components, but is zero otherwise. The last term is the anomaly which we will evaluate. The four-vector $s = r - \underline{r}$ has the first four components zero. This greatly simplifies computation of the trace if one everywhere sets $r = \underline{r} + s$. The expression for the anomaly becomes

$$2i \int \mathrm{d}_n r \, \frac{\mathrm{Tr}[\gamma s \{i\gamma(p+\underline{r}) + i\gamma s - m\}\ \gamma^\mu\ \{i\gamma\underline{r} + i\gamma s - m\}\ \gamma^\nu \{i\gamma(\underline{r} - q) + i\gamma s - m\}\gamma^5]}{((p+r)^2 + m^2)(r^2+m^2)((r-q)^2+m^2)}$$

Remembering that $\gamma^5 = \gamma^1\gamma^2\gamma^3\gamma^4$, and that all vectors except s are physical (have zero components beyond the fourth) one reduces the trace to $4\, is^2 \epsilon_{\lambda\mu\nu\kappa} p_\lambda q_\kappa$ where one uses the fact that γ_5 anticommutes with all other γ-matrices. Finally we must evaluate the integral

$$\int \mathrm{d}_n r \frac{(r - \underline{r}, r - \underline{r})}{(r^2 + m^2)^3}$$

where we have omitted p and q in the denominator since we need only the pole for $n = 4$. With the help of (A8) multiplied with $\delta_{\mu\nu}$ where now the indices are taken to run from 5 to n only we obtain for this integral:

$$\frac{i\pi^{\frac{1}{2}n}}{(m^2)^{2-\frac{1}{2}n}} \frac{\Gamma(2 - \frac{1}{2}n)}{\Gamma(3)} \tfrac{1}{2}(n-4)\,.$$

In the limit $n = 4$, and taking into account that there are two graphs (the second obtained by the interchange $\mu \leftrightarrow \nu$, $p \leftrightarrow q$) we find for the anomaly:

$$8i\pi^2 \, \epsilon_{\lambda\mu\nu\kappa} p_\lambda q_\kappa$$

which agrees with the results of Bell and Jackiw [4], eq. (3.12c), and Adler [4], eq. (22).

Note that in our formulation the anomaly has nothing to do with the pecularities of shifts of integration variables.

The usual ambiguity of the choice of integration variables is replaced in our formalism by the ambiguity of the location of γ^5 in the trace in (36). If before generalization to $n \neq 4$ is done, γ^5 is (anti) commuted to the right, a different result emerges.

7. CONCLUSIONS

The method presented essentially completes the proof of the renormalizability of the theories presented in ref. [1]. The method fails if quantities particular to four-dimensional space, such as γ^5 or the tensor $\epsilon_{\mu\nu\alpha\beta}$ or scaling behaviour play an essential role in the Ward identities.

We have not considered infrared problems. Here we only wish to remark that the generalized S-matrix elements will have additional poles for integer n-values if infrared divergencies are present.

The authors are indebted to the participants of the discussion meeting at Orsay, Jan. 1972, for inspiring and constructive criticism.

APPENDIX A. SOME USEFUL FORMULAE

$$\int \mathrm{d}_n x f(x) = \int f(x) r^{n-1} \, \mathrm{d}r \sin^{n-2}\theta_{n-1} \, \mathrm{d}\theta_{n-1} \sin^{n-3}\theta_{n-2} \, \mathrm{d}\theta_{n-2} \ldots \mathrm{d}\theta_1 \quad \text{(A1)}$$

with $0 \leqslant \theta_i \leqslant \pi$, except $0 \leqslant \theta_1 \leqslant 2\pi$.

If $f(x)$ depends only on $r = \sqrt{x_1^2 + \ldots x_n^2}$ one may perform the integration over angles using

$$\int_0^\pi \sin^m \theta \, \mathrm{d}\theta = \sqrt{\pi}\, \frac{\Gamma(\frac{1}{2}(m+1))}{\Gamma(\frac{1}{2}(m+2))}, \tag{A2}$$

leading to

$$\int \mathrm{d}_n x f(r) = \frac{2\pi^{\frac{1}{2}n}}{\Gamma(\frac{1}{2}n)} \int f(r)\, r^{n-1}\, \mathrm{d}r \tag{A3}$$

$$\int_0^\infty \mathrm{d}x \frac{x^\beta}{(x^2+M^2)^\alpha} = \frac{1}{2} \frac{\Gamma(\frac{1}{2}(\beta+1))\,\Gamma(\alpha - \frac{1}{2}(\beta+1))}{\Gamma(\alpha)\,(M^2)^{\alpha-\frac{1}{2}(\beta+1)}}. \tag{A4}$$

Keeping the prescriptions and definitions of sect. 3 in mind, the following equations hold for arbitrary n:

$$\int \mathrm{d}_n p \frac{1}{(p^2+2kp+m^2)^\alpha} = \frac{i\pi^{\frac{1}{2}n}}{(m^2-k^2)^{\alpha-\frac{1}{2}n}} \frac{\Gamma(\alpha-\frac{1}{2}n)}{\Gamma(\alpha)}, \tag{A5}$$

$$\int \mathrm{d}_n p \frac{p_\mu}{(p^2+2kp+m^2)^\alpha} = \frac{i\pi^{\frac{1}{2}n}}{(m^2-k^2)^{\alpha-\frac{1}{2}n}} \frac{\Gamma(\alpha-\frac{1}{2}n)}{\Gamma(\alpha)} (-k_\mu), \tag{A6}$$

$$\int \mathrm{d}_n p \frac{p^2}{(p^2+2kp+m^2)^\alpha} = \frac{i\pi^{\frac{1}{2}n}}{(m^2-k^2)^{\alpha-\frac{1}{2}n}} \cdot \frac{1}{\Gamma(\alpha)} \{\Gamma(\alpha-\tfrac{1}{2}n)k^2 + \Gamma(\alpha-1-\tfrac{1}{2}n)\tfrac{1}{2}n(m^2-k^2)\}, \tag{A7}$$

$$\int \mathrm{d}_n p \frac{p_\mu p_\nu}{(p^2+2kp+m^2)^\alpha} = \frac{i\pi^{\frac{1}{2}n}}{(m^2-k^2)^{\alpha-\frac{1}{2}n}} \frac{1}{\Gamma(\alpha)} \{\Gamma(\alpha-\tfrac{1}{2}n)k_\mu k_\nu + \Gamma(\alpha-1-\tfrac{1}{2}n)\tfrac{1}{2}\delta_{\mu\nu}(m^2-k^2)\}, \tag{A8}$$

$$\int \mathrm{d}_n p \frac{p_\mu p_\nu p_\lambda}{(p^2+2kp+m^2)^\alpha} = \frac{i\pi^{\frac{1}{2}n}}{(m^2-k^2)^{\alpha-\frac{1}{2}n}} \frac{1}{\Gamma(\alpha)} \{-\Gamma(\alpha-\tfrac{1}{2}n)k_\mu k_\nu k_\lambda - \Gamma(\alpha-1-\tfrac{1}{2}n)\tfrac{1}{2}(\delta_{\mu\nu}k_\lambda + \delta_{\mu\lambda}k_\nu + \delta_{\nu\lambda}k_\mu)(m^2-k^2)\}, \tag{A9}$$

$$\int \mathrm{d}_n p \frac{p^2 p_\mu}{(p^2+2kp+m^2)^\alpha} = \frac{i\pi^{\frac{1}{2}n}}{(m^2-k^2)^{\alpha-\frac{1}{2}n}} \frac{1}{\Gamma(\alpha)} (-k_\mu)\{\Gamma(\alpha-\tfrac{1}{2}n)k^2$$

$$+\Gamma(\alpha-\tfrac{1}{2}n-1)\tfrac{1}{2}(n+2)(m^2-k^2)\} . \tag{A10}$$

The above equations contain indices μ, ν, λ. These indices are understood to be contracted with arbitrary n-vectors q_1, q_2 etc. In computing the integrals one first integrates over the part of n-space orthogonal to the vectors k, q_1, q_2 etc., using (A1) – (A4). After that the expressions are meaningful also for non-integer n. Note that formally (A6) – (A10) may be obtained from (A5) by differentiation with respect to k, or by using $p^2 = (p^2+2pk+m^2) - 2pk - m^2$. The Feynman parameter method for non-integer exponents:

$$\frac{1}{a^\alpha b^\beta} = \frac{\Gamma(\alpha+\beta)}{\Gamma(\alpha)\,\Gamma(\beta)} \int_0^1 \mathrm{d}x \frac{x^{\alpha-1}(1-x)^{\beta-1}}{(ax+b(1-x))^{\alpha+\beta}} \tag{A11}$$

valid for $\alpha > 0, \beta > 0$. If one needs this formula for α in the neighbourhood of 0 one may write

$$\frac{1}{a^\alpha b^\beta} = \frac{a}{a^{\alpha+1} b^\beta}$$

and then use (A11).
The generalization of eq. (A11) for many factors:

$$\frac{1}{a_1^{\alpha_1} a_2^{\alpha_2} \dots a_m^{\alpha_m}} = \frac{\Gamma(\alpha_1+\alpha_2+\dots\alpha_m)}{\Gamma(\alpha_1)\,\Gamma(\alpha_2)\dots\Gamma(\alpha_m)} \cdot \int_0^1 \mathrm{d}x_1 \int_0^{x_1} \mathrm{d}x_2 \dots \int_0^{x_{m-2}} \mathrm{d}x_{m-1}$$

$$\times \frac{x_{m-1}^{\alpha_1-1}(x_{m-2}-x_{m-1})^{\alpha_2-1}\dots(1-x_1)^{\alpha_m-1}}{[a_1 x_{m-1} + a_2(x_{m-2}-x_{m-1}) + \dots\dots + a_m(1-x_1)]^{\alpha_1+\alpha_2+\dots\alpha_m}} \tag{A12}$$

APPENDIX B. THE MAIN THEOREM

Consider the integral:

$$\int \mathrm{d}_n p \left\{ \frac{\Gamma(\frac{1}{2}(n_0-n))}{(p^2+m^2)^{\frac{1}{2}(n_0-n)}} - \frac{2}{n_0-n} \right\} \frac{A}{(p^2+2pk+M^2)^\alpha} . \tag{B1}$$

We must prove that (B1) contains only harmless poles for $n = n_0$ (where n_0 is the dimension of physical space, i.e. $n_0 = 4$) and $\alpha \geqslant 1$. For $\alpha > \frac{1}{2}n_0$ the theorem is trival because then the only poles for $n = n_0$ are those already explicit in (B1), and they obviously cancel if the rest is nonsingular for $n = n_0$. The nontrivial cases are $\alpha = \frac{1}{2}(n_0 - 1)$ and $\alpha = \frac{1}{2}n_0$.

Formula (B1) is symbolic insofar that the exponent α stands in fact for the difference of the powers of p in the numerator A and those in the denominator. We will prove (B1) for the case $\alpha = \frac{1}{2}n_0$ with $A = 1$; the case $A = (pq)$ and exponent $\alpha = \frac{1}{2}n_0$ in the denominator may be proven similarly, and the case that there are two or more p's in A may be reduced to the case of no or one p in the numerator.

Thus consider the special case:

$$\int \mathrm{d}_n p \left\{ \frac{\Gamma(\frac{1}{2}(n_0 - n))}{(p^2 + m^2)^{\frac{1}{2}(n_0 - n)}} - \frac{2}{n_0 - n} \right\} \frac{1}{(p^2 + 2pk + M^2)^{\frac{1}{2}n_0}} . \tag{B2}$$

The first term of (B2) may be worked out (see (A11)):

$$\Gamma(\tfrac{1}{2}(n_0 - n)) \int \mathrm{d}_n p \, \frac{p^2 + m^2}{(p^2 + m^2)^{\frac{1}{2}(n_0 - n + 2)} (p^2 + 2pk + M^2)^{\frac{1}{2}n_0}}$$

$$= \frac{\Gamma(\frac{1}{2}(n_0 - n))\, \Gamma(n_0 + 1 - \frac{1}{2}n)}{\Gamma(\frac{1}{2}(n_0 + 2 - n))\, \Gamma(\frac{1}{2}(n_0))} \int_0^1 \mathrm{d}x \, x^{\frac{1}{2}(n_0 - 2)} (1 - x)^{\frac{1}{2}(n_0 - n)} .$$

$$\int \mathrm{d}_n p \, \frac{p^2 + m^2}{(p^2 + 2pkx + M^2 x + m^2(1 - x))^{n_0 + 1 - \frac{1}{2}n}} . \tag{B3}$$

Use of eqs. (A5) and (A7) gives:

$$(\mathrm{B3}) = \mathrm{i}\pi^{\frac{1}{2}n} \frac{\Gamma(\frac{1}{2}(n_0 - n))\, \Gamma(n_0 + 1 - \frac{1}{2}n)}{\Gamma(\frac{1}{2}(n_0 + 2 - n))\, \Gamma(\frac{1}{2}(n_0))} \int_0^1 \mathrm{d}x \, x^{\frac{1}{2}(n_0 - 2)} (1 - x)^{\frac{1}{2}(n_0 - n)}$$

$$\times \left[\frac{\Gamma(n_0 + 1 - n)}{\Gamma(n_0 + 1 - \frac{1}{2}n)} \frac{m^2 + k^2 x^2}{(M^2 x + m^2(1 - x) - k^2 x^2)^{n_0 + 1 - n}} \right.$$

$$\left. + \frac{\Gamma(n_0 - n)}{2\Gamma(n_0 + 1 - \frac{1}{2}n)} \frac{n}{(M^2 x + m^2(1 - x) - k^2 x^2)^{n_0 - n}} \right] . \tag{B4}$$

The only singularities for $n = n_0$ are now located in the two Γ-functions. We are only interested in possible non-polynomial residues of the poles at $n = n_0$. To this purpose we may set everywhere $n = n_0$ except in the two Γ-functions and in the last exponent in (B4). Next, writing

$$x^{\frac{1}{2}(n_0 - 2)} = \frac{2}{n_0} \frac{\mathrm{d}}{\mathrm{d}x} x^{\frac{1}{2}n_0}$$

and performing partial integration with respect to x in the second term we obtain:

$$(\mathrm{B4}) \Rightarrow i\pi^{\frac{1}{2}n_0} \frac{\Gamma(\frac{1}{2}(n_0 - n))}{\Gamma(\frac{1}{2}n_0)} \int_0^1 \mathrm{d}x\, x^{\frac{1}{2}(n_0-2)} \cdot \frac{1}{(M^2 x + m^2(1-x) - k^2x^2)^{n_0-n}}$$

$$+ i\pi^{\frac{1}{2}n_0} \frac{\Gamma(\frac{1}{2}(n_0 - n))\,\Gamma(n_0 - n)}{\Gamma(\frac{1}{2}n_0)} \cdot \frac{1}{(M^2 - k^2)^{n_0-n}} . \qquad (\mathrm{B5})$$

The second term in (B5) is the surface term arising from the partial integ a ion.

The first term in (B5) displays a harmless single pole.

The second term contains the single pole

$$i\pi^{\frac{1}{2}n_0} \frac{\Gamma(\frac{1}{2}(n_0 - n))\,\Gamma(n_0 - n)}{\Gamma(\frac{1}{2}n_0)} (n_0 - n) \ln (M^2 - k^2) \qquad (\mathrm{B6})$$

in addition to harmless poles.

Consider next the contribution of the second term in (B2):

$$\frac{2}{n_0 - n} \int \mathrm{d}_n p \frac{1}{(p^2 + 2pk + M^2)^{\frac{1}{2}n_0}} =$$

$$= i\pi^{\frac{1}{2}n_0} \frac{2}{n_0 - n} \frac{\Gamma(\frac{1}{2}(n_0 - n))}{\Gamma(\frac{1}{2}n_0)} \frac{1}{(M^2 - k^2)^{\frac{1}{2}(n_0-n)}} . \qquad (\mathrm{B7})$$

(B7) contains harmless poles as well as the pole

$$i\pi^{\frac{1}{2}n_0} \frac{\Gamma(\frac{1}{2}(n_0 - n))}{\Gamma(\frac{1}{2}n_0)} \frac{2}{n_0 - n} \tfrac{1}{2}(n_0 - n) \ln (M^2 - k^2) \qquad (\mathrm{B8})$$

Since $\Gamma(x)$ behaves as $1/x$ for x in the neighbourhood of zero we see that the dangerous pole in (B6) is cancelled by (B8).

It may be noted that the difference of (B6) and (B8) contains a harmless double pole. Thus in general we may expect to need double pole counterterms at the two loop level.

APPENDIX C. TRACES MAY BE COMPUTED BY MEANS OF THE EQUATION

$$\mathrm{Tr}\,(\gamma^{\mu_1}\gamma^{\mu_2} \ldots \gamma^{\mu_m}) = -\,\mathrm{Tr}\,(\gamma^{\mu_m}\gamma^{\mu_1} \ldots \gamma^{\mu_{m-1}})$$

$$+ 2 \sum_{i=1}^{m-1} (-1)^{i+1}\, \mathrm{Tr}\,(\gamma^{\mu_1} \ldots \gamma^{\mu_{i-1}}\gamma^{\mu_{i+1}} \ldots \gamma^{\mu_{m-1}})\, \delta_{\mu_i \mu_m} \qquad (\mathrm{C1})$$

valid for even m. This equation is based solely on eq. (33). Since the first term on the right hand side equals minus the term on the left hand side we have a recursion formula relating traces of m γ-matrices to traces of $m-2$ matrices.

The requirement

$$\mathrm{Tr}\,(\gamma^{\mu_1} \ldots \gamma^{\mu_m}) = 0 \text{ for odd } m$$

excludes for instance for $n = 3$ the choice

$$\gamma^i = \sigma^i , \qquad i = 1, 2, 3$$

where the σ^i are the well-known Pauli spin matrices. Instead one may simply use the 4×4 matrices γ^1, γ^2 and γ^3. In fact, if one has a set of γ-matrices for some large n one may for lower n always use a subset of that set.

REFERENCES

[1] G. 't Hooft, Nucl. Phys. B35 (1971) 167;
See also B.W. Lee, Phys. Rev. D5 (1972) 823.
[2] G. 't Hooft, Nucl. Phys. B33 (1971) 173.
[3] P.W. Higgs, Phys. Letters 12 (1964) 132; Phys. Rev. Letters 13 (1964) 508; Phys. Rev. 145 (1966) 1156;
F. Englert and R. Brout, Phys. Rev. Letters 13 (1964) 321;
G.S. Guralnik, C.R. Hagen and T.W.B. Kibble, Phys. Rev. Letters 13 (1964) 585; T.W.B. Kibble, Phys. Rev. 155 (1967) 627.
[4] J.S. Bell and R. Jackiw, Nuovo Cimento 51 (1969) 47;
S.L. Adler, Phys. Rev. 177 (1969) 2426.
[5] D. Gross and R. Jackiw, M.I.T. preprint, Jan. 1972. It is demonstrated that anomalies can be cured only if the theory is modified.
[6] C. Bouchiat, J. Illiopoulos and Ph. Meyer, Orsay preprint January 1972. In this preprint and also in ref. [5] a modification of the theory such that no anomalies occur is presented.
[7] C.G. Bollini, J.J. Giambiagi and A. Gonzalez Dominguez, Nuovo Cimento 31 (1964) 550;
E.R. Speer, Generalized Feynman amplitudes, Princeton University Press, Princeton 1969.
[8] S. Weinberg, Phys. Rev. Letters 27 (1971) 1688;
Th. Appelquist and H. Quinn, Harvard University preprint, January 1972.
[9] R.E. Cutkosky, J. Math. Phys. 1 (1960) 429.
[10] M. Veltman, Physica 29 (1963) 186.
[11] M. Veltman, Marseille Conf. 1971.
[12] C.G. Bollini and J.J. Giambiagi, Preprint of the Universidad Nacional de la Plata, Argentina, February 1972. In this preprint attention is focused on applications to quantum electrodynamics.

Spontaneously Broken Gauge Symmetries. I. Preliminaries

Benjamin W. Lee
National Accelerator Laboratory, P. O. Box 500, Batavia, Illinois 60510
and Institute for Theoretical Physics, State University of New York at Stony Brook, Stony Brook, New York 11790

and

Jean Zinn-Justin*
Institute for Theoretical Physics, State University of New York at Stony Brook, Stony Brook, New York 11790
(Received 10 March 1972)

This is the first of a series of papers addressed to the renormalizability question of spontaneously broken gauge theories. We give a brief outline of the motivation for such an investigation and describe the manner in which the renormalizability of such theories will be proved in the sequel. Put briefly, we will show that in an appropriate gauge, ultraviolet divergences of a spontaneously broken gauge theory are removed completely by the gauge-invariant counterterms in the Lagrangian which would make the Green's functions of the corresponding unbroken gauge theory finite, that the *S* matrix computed in this gauge is unitary, and that the *S* matrix is independent of the gauge chosen. In this paper, the renormalizability question of the unbroken gauge theory is considered. We derive the Ward-Takahashi identities of the theory. We discuss several ways of regulating divergent Feynman integrals of the theory without destroying gauge invariance. Infrared divergences are avoided by the device of intermediate renormalization, wherein we choose as subtraction points some points where external momenta are Euclidean. This suffices to establish that the Bogoliubov-Parasiuk-Hepp renormalization will give renormalized Green's functions which satisfy the Ward-Takahashi identities. The existence of finite, renormalized Green's functions satisfying the Ward-Takahashi identities provides us with the means of proving the renormalizability of the spontaneously broken symmetry case. The Ward-Takahashi identities were previously derived for the gauge bosons by Slavnov. We present here a new derivation. The discussions on regularization methods and intermediate renormalization procedure and the renormalization conditions for matter fields, we believe, are new contributions of the present paper.

I. INTRODUCTION

This is the first of a series of papers which will deal with the renormalizability of spontaneously broken gauge symmetries. The intriguing possibilities of unifying electromagnetic and weak interactions in terms of Yang-Mills gauge bosons,[1-9] whose masses are generated by spontaneous breakdown of gauge invariance of the second kind,[4,5] and of constructing a finite theory of weak interactions[4,7-9] prompt a closer examination of the quantization and renormalization questions of theories of this genre.

In the sequel of this series, we wish to examine

the following questions: (1) We will discuss both the group-theoretic and field-theoretic problems associated with the Higgs phenomenon.[10-12] This entails a careful study of the stability of the physical system which possesses the freedom associated with the gauge invariance. (2) We will also study the perturbative treatment of such theories. Here our aim is to show that, in an appropriate gauge, ultraviolet divergences of a spontaneously broken gauge theory are removed completely by the counterterms in the Lagrangian which would make Green's functions of the corresponding unbroken gauge theory finite. Thus the renormalizability of the unbroken Yang-Mills theory (to be defined below) implies the same for the spontaneously broken gauge theory. The philosophy and methodology we shall follow are the same as those we employed in the study of the σ model.[13] (3) In the gauge in which the renormalizability can be proven, the unitarity of the S matrix is not manifest, since the quantization in that gauge implies the use of an indefinite-metric Hilbert space for the construction of Green's functions. We will show that the physical S matrix is nonetheless unitarity. (4) We shall also discuss the equivalence of the S matrix constructed in this gauge and in the gauge in which the unitarity of the S matrix is manifest (but not the renormalization).

In this paper, we shall give a discussion of the renormalization problem of the (unbroken) Yang-Mills field theory. It is not attempted in the present paper to establish that a renormalized Yang-Mills theory exists as a physically satisfactory theory of massless particles. Due to the infrared problem associated with massless quanta, such a theory may very well not exist at all. What we wish to demonstrate is that renormalized Green's functions of the theory exist (without implying the same for the S matrix), which satisfy the Ward-Takahashi identities which will be derived. The existence of renormalized Green's functions will prove to be a sufficient foundation for the discussion of the renormalizability of the spontaneously broken symmetry theory, which we shall discuss in the sequel.

We will proceed in the following manner. After a brief review of the quantization of the non-Abelian gauge theories, we shall derive the Ward-Takahashi identities. We will then discuss ways of regularizing divergent Feynman integrals in a gauge-invariant manner. The regularized Feynman amplitude then satisfies the identities automatically. The Bogoliubov-Parasiuk-Hepp-Zimmermann (BPHZ) renormalization procedure[14-17] requires specifying the values of primitively divergent vertices (A primitively divergent vertex is a proper vertex whose superficial degree of divergence is non-negative. Our definition here differs from the conventional usage of this term.) at some subtraction points. When these values are chosen in accordance with the Ward-Takahashi identities, and the cutoff parameters associated with the regularization are let go to infinity, the renormalized amplitudes are obtained which satisfy the Ward-Takahashi identities. Because of the infrared divergence, it is prudent to choose as subtraction points some points other than where all external momenta vanish. We shall describe in some detail this "intermediate" renormalization procedure.

The Ward-Takahashi identities were previously derived by Slavnov[18] for the gauge bosons. The derivation we shall present is somewhat different from his. The discussions on regularization methods and intermediate renormalization procedure, and renormalization conditions for matter fields, we believe, are new contributions of the present paper.

The organization of the paper is as follows: Sec. II: Quantization; Sec. III: Ward-Takahashi Identities I; Sec. IV: Ward-Takahashi Identities II; Sec. IV A: Two-Point Functions; Sec. IV B: Three-Point Vertices; Sec. IV C: Four-Point Vertices; Sec. V: Regularization; Sec. VI: Renormalization Conditions and Infrared Divergences; Sec. VII: Renormalization of Matter Fields; Appendix A: Derivation of Some Equations; Appendix B: Generating Functional of Proper Vertices; Appendix C: Ward-Takahashi Identity for the Generating Functional of Proper Vertices; Appendix D: Power Counting.

II. QUANTIZATION

Following the works of Feynman,[19] DeWitt,[20] Popov and Faddeev,[21] and 't Hooft[22] we define the generating functional $Z[\vec{J}_\mu]$ of connected Green's functions by

$$e^{Z[\vec{J}_\mu]} = \int [d\vec{A}]\,\Delta_L[\vec{A}]\exp i\left\{\mathcal{L}(x) - \frac{1}{2\alpha}(\partial_\mu \vec{A}^\mu)^2 - \vec{J}_\mu(x)\cdot\vec{A}^\mu(x)\right\}, \tag{2.1}$$

where $[d\vec{A}]$ is the canonical functional metric for the vector fields

$$[d\vec{A}] = \prod_{a,\mu,x} dA_\mu^a(x), \tag{2.2}$$

where a is the internal-symmetry label. We shall assume the internal symmetry to be SU(2), so that $a = 1, 2, 3$, but the generalizations to other groups are trivial and immediate. $\mathcal{L}(x)$ is the Lagrangian for the Yang-Mills fields

$$\mathcal{L}(x) = -\tfrac{1}{4}\vec{F}_{\mu\nu}\cdot\vec{F}^{\mu\nu}, \tag{2.3}$$

$$\vec{F}_{\mu\nu} = \partial_\mu\vec{A}_\nu - \partial_\nu\vec{A}_\mu - g\vec{A}_\mu\times\vec{A}_\nu, \tag{2.4}$$

which is invariant under local gauge transformations, the infinitesimal version of which is

$$A_\mu^a \to A_\mu^a + g(\vec{\omega}\times\vec{A}_\mu)^a + \partial_\mu \omega^a$$
$$= A_\mu^a + (D_\mu)^{ab}\omega^b\,, \qquad (2.5)$$

D_μ being the covariant derivative:

$$D_\mu[\vec{A}]^{ab} = \delta^{ab}\partial_\mu - g(\vec{t}\cdot\vec{A}_\mu)^{ab}\,, \qquad (2.6)$$

$$(l_c)^{ab} = \epsilon^{acb} \qquad (2.7)$$

The Jacobian $\Delta_L[\vec{A}_\mu]$ is essentially the determinant of the operator $\partial_\mu D^\mu$, and may be expressed as

$$\Delta_i[\vec{A}_\mu] = \exp\left[\operatorname{Tr}\ln\left(1 - g\frac{1}{\partial^2}\partial^\mu\vec{A}_\mu\cdot\vec{t}\right)\right]$$
$$= \exp\left\{-\sum_{n=2}^{\infty}\frac{(-g)^n}{n}\int dx_1\cdots dx_n\ \operatorname{tr}\tilde{D}_F(x_n - x_1)\partial_{x_1}^{\mu_1}\vec{A}_{\mu_1}(x_1)\cdot\vec{t}\cdots\tilde{D}_F(x_{n-1}-x_n)\partial_{x_n}^{\mu_n}\vec{A}_{\mu_n}(x_n)\cdot\vec{t}\right\}$$
$$= \exp\left[\operatorname{Tr}\ln\left(1 - g\vec{t}\cdot\vec{A}^\mu\partial_\mu\frac{1}{\partial^2}\right)\right]\,, \qquad (2.8)$$

where we have used the Feynman propagator $D_F(x-y)$ defined as

$$\tilde{D}_F(x-y) = \langle x|(-\partial^2 + i\epsilon)^{-1}|y\rangle\,. \qquad (2.9)$$

The symbol Tr denotes the trace operation over x and the isospin a; the trace operation over the isospin index is denoted by tr.

The Feynman rules for this theory are obtained in the usual manner if we regard

$$\int d^4x\left[\mathfrak{L}(x) - \frac{1}{2\alpha}[\partial_\mu\vec{A}^\mu(x)]^2\right]$$
$$-i\operatorname{Tr}\ln\left(1 - g\vec{t}\cdot\vec{A}^\mu\partial_\mu\frac{1}{\partial^2}\right) \qquad (2.10)$$

as the effective Lagrangian. The bare vector-boson propagator is

$$-i\Delta_{\mu\nu}(k^2;\alpha) = -i\left[g_{\mu\nu} - \frac{k_\mu k_\nu}{k^2+i\epsilon}(1-\alpha)\right]\frac{1}{k^2+i\epsilon}\,. \qquad (2.11)$$

In Eqs. (2.8) and (2.11) the $i\epsilon$ prescription is dictated by the unitarity considerations which we shall discuss in the sequel. The term $(1/2\alpha)(\partial_\mu\vec{A}^\mu)^2$ specifies the gauge one is employing and depends on a parameter which can vary from $-\infty$ to ∞. For $\alpha = 0$, we obtain the transverse or Landau gauge, and for $\alpha = 1$, the Feynman gauge. The last, nonlocal term in Eq. (2.10) is the new feature of non-Abelian gauge theories. It may be viewed[19] diagrammatically as the sum of closed-loop contributions from fictitious complex massless scalar fields (ghosts) obeying Fermi statistics which are coupled to the gauge fields through the interaction

$$g\bar{c}(x)\partial^\mu[\vec{t}\cdot\vec{A}_\mu(x)c(x)]\,. \qquad (2.12)$$

The connected Green's functions of $\vec{A}_\mu$'s are obtained as the functional derivatives of $Z[J_\mu]$:

$$\frac{i\delta^n Z[\vec{J}_\mu]}{\delta J_\mu^a(x)\delta J_\nu^b(y)\cdots} = (-i)^n\langle T^*(A_\mu^a(x)A_\nu^b(y)\cdots)\rangle_0^c\,. \qquad (2.13)$$

The Feynman rules are summarized in Fig. 1. In addition, the following rules should be kept in mind: The ghost-ghost-vector vertex is "dotted," the dot indicating which ghost line is differentiated; a ghost line cannot be dotted at both ends; a ghost loop carries an extra minus sign.

III. WARD-TAKAHASHI IDENTITIES-I

The invariance of the Lagrangian under the local gauge transformation (2.5) gives rise to a hierarchy of identities among the Green's functions (2.13). Alternatively, these relations may be expressed globally as an equation satisfied by the generating functional $Z[\vec{J}_\mu]$.

We will first rewrite Eq. (2.1) as

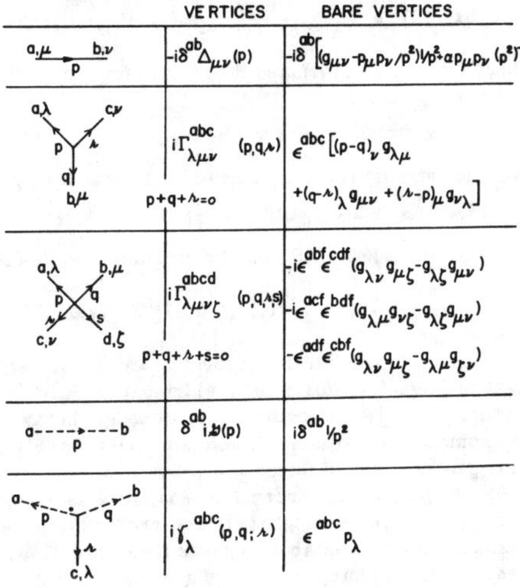

FIG. 1. Feynman rules of the Yang-Mills theory.

$$W[\vec{J}_\mu]=\exp iZ[\vec{J}_\mu]=\Delta_L[i\delta/\delta\vec{J}_\mu]W_0[\vec{J}_\mu]\,, \tag{3.1}$$

where

$$W_0[\vec{J}_\mu]=\int[d\vec{A}]\exp i\left\{\mathcal{L}(x)-\frac{1}{2\alpha}[\partial^\mu\vec{A}_\mu(x)]^2 -\vec{J}^\mu(x)\cdot\vec{A}_\mu(x)\right\}. \tag{3.2}$$

We perform the gauge transformation (2.5) on the variables of integration $\vec{A}_\mu(x)$. Due to the invariance of the Lagrangian and the metric $[d\vec{A}]$, this transformation will affect only the source term and the gauge defining term:

$$\delta\int d^4x\left\{\frac{1}{2\alpha}[\partial_\mu\vec{A}^\mu]^2+\vec{J}^\mu\cdot\vec{A}_\mu\right\} =\int d^4x\,\delta\vec{\omega}\cdot\left[\frac{1}{\alpha}D^\lambda\partial_\lambda\partial^\mu\vec{A}_\mu-D^\mu\vec{J}_\mu\right]. \tag{3.3}$$

Since a transformation of integration variables does not change the value of an integral, we may put the variation of W_0 with respect to $\delta\omega(x)$ equal to zero. We obtain thereby

$$\frac{1}{\alpha}D^\lambda[i\delta/\delta\vec{J}]^{ab}\partial_\lambda\partial_\mu\frac{i\delta W_0}{\delta J^b_\mu(x)}-D^\mu[i\delta/\delta\vec{J}]^{ab}J^b_\mu(x)W_0=0. \tag{3.4}$$

We note that (see Appendix A)

$$\Delta_L[i\delta/\delta\vec{J}]J^a_\lambda(x)\Delta_L^{-1}[i\delta/\delta\vec{J}] =J^a_\lambda(x)-ig\,\mathrm{tr}\,t^a[\partial_\lambda H(x,y;i\delta/\delta\vec{J})]|_{x=y}\,, \tag{3.5}$$

where H is the solution of the equation

$$D_\lambda[\vec{A}]^{ab}\partial^\lambda H^{bc}(x,y;\vec{A})=\delta^{ac}\delta^4(x-y), \tag{3.6}$$

satisfying the outgoing boundary condition; it has the representation

$$H^{ab}(x,y;\vec{A})=-\langle x,a|[-\partial^2+i\epsilon+g\vec{t}\cdot\vec{A}_\mu\partial^\mu]^{-1}|y,b\rangle\,. \tag{3.7}$$

Combining Eqs. (3.4) and (3.5), and recalling Eq. (3.1), we obtain

$$D^\lambda_x[i\delta/\delta\vec{J}]^{ab}\left\{\frac{i}{\alpha}\partial_\lambda\partial_\mu\frac{\delta}{\delta J^b_\mu(x)}-J^b_\lambda(x)+ig\,\mathrm{tr}\,t^b[\partial_\lambda H(x,y;i\delta/\delta\vec{J})]_{y=x}\right\}W=0. \tag{3.8}$$

The last term on the left-hand side is equal to (see Appendix A)

$$igD^\lambda[i\delta/\delta\vec{J}]^{ab}t^{cbd}[\partial_\lambda H^{dc}(x,y;i\delta/\delta\vec{J})]_{y=x}=-ig\int d^4y\,\mathrm{tr}\,t^a[\partial_\mu H(x,y;i\delta/\delta\vec{J})]\,D^\mu_y[i\delta/\delta\vec{J}]\,\delta^4(x-y). \tag{3.9}$$

In showing this, one makes use of Eq. (3.6) and the Jacobi identity of the matrices t^a. Equation (3.8) may now be written as

$$D^\lambda[i\delta/\delta\vec{J}]^{ab}\partial_\lambda\left\{\frac{i}{\alpha}\partial_\mu\frac{\delta}{\delta J^b_\mu(x)}-\int d^4y:H^{bc}(x,y;i\delta/\delta\vec{J})D^\mu_y[i\delta/\delta\vec{J}]^{cd}J^d_\lambda(y):\right\}W=0, \tag{3.10}$$

where the symbol : : denotes the normal product prescription that the $\delta/\delta J$ must stand to the right of the J.

We now define G by

$$\partial^\lambda D_\lambda[\vec{A}]^{ab}G^{bc}(x,y;\vec{A})=\delta^{ac}\delta^4(x-y), \tag{3.11}$$

with the outgoing boundary condition, so that it may be represented as

$$G^{ab}(x,y;\vec{A})=-\langle x,a|[-\partial^2+i\epsilon+g\vec{t}\cdot\vec{A}_\mu\partial^\mu]^{-1}|y,b\rangle=H^{ba}(y,x;\vec{A}). \tag{3.12}$$

In terms of G, Eq. (3.10) may be considerably simplified. We finally obtain the desired identity:

$$\frac{i}{\alpha}\partial_\mu\frac{\delta W}{\delta J^a_\mu(x)}-\int d^4y\,J^c_\lambda(y)D^\lambda_y[i\delta/\delta\vec{J}]^{cb}G^{ba}(y,x;i\delta/\delta\vec{J})W=0. \tag{3.13}$$

The above Ward-Takahashi identity was previously derived by Slavnov.[18] He considers a restricted class of gauge transformations which satisfy $\partial_\mu D^\mu\vec{\omega}=\vec{\chi}$ where $\vec{\chi}$ is an arbitrary function. He shows that the product $[d\vec{A}]\Delta_L[\vec{A}_\mu]$ remains invariant under the nonlinear gauge transformation generated by $\vec{\omega}=\vec{\omega}[\vec{A}_\mu,\vec{\chi}]$. The point of the above derivation is to show that Slavnov's form of the Ward-Takahashi identities is the most general form of the constraint on $W[\vec{J}_\mu]$ that follows from the gauge invariance of the Lagrangian.

For the purpose of renormalization, it is usually much more convenient to study the Ward-Takahashi identities connecting single particle irreducible (proper) vertices, as was done for the σ model[13,23] and the spontaneously broken Abelian gauge theories.[8] However, in the present instance, the Ward-Takahashi identities for the proper vertices are extremely complicated, being nonlinear relations among them. The Ward-Takahashi identity satisfied by the generating functional of the proper vertices is nevertheless derived and analyzed in Appendix C. The renormalization conditions will be analyzed on the basis of Eq. (3.13) in the following sections.

IV. WARD-TAKAHASHI IDENTITIES-II

We shall study the implications of Eq. (3.13) on the primitively divergent vertices.

A. Two-Point Functions

Differentiating Eq. (3.13) with respect to $J_\nu^b(y)$ and then letting $J_\mu = 0$, we obtain

$$\frac{i}{\alpha}\frac{\partial}{\partial x_\mu}\frac{\delta^2 W}{\delta J_\mu^a(x)\delta J_\nu^b(y)}\bigg|_{\mathfrak{J}_\mu=0} - D_y^\nu[i\delta/\delta\mathfrak{J}]^{bc}G^{ca}(y,x;i\delta/\delta\mathfrak{J})W\bigg|_{\mathfrak{J}_\mu=0} = 0. \tag{4.1}$$

Taking the divergence of the above equation with respect to y, and remembering the equation satisfied by G, we see that

$$\frac{i}{\alpha}\frac{\partial}{\partial x^\mu}\frac{\partial}{\partial y^\nu}\frac{\delta^2 W}{\delta J_\mu^a(x)\delta J_\nu^b(y)}\bigg|_{\mathfrak{J}_\mu=0} = \delta^{ab}\delta^4(x-y), \tag{4.2}$$

which shows that the longitudinal part of the propagator $\Delta_{\mu\nu}$,

$$\frac{\delta^2 Z[\mathfrak{J}_\mu]}{\delta J_\mu^a(x)\delta J_\nu^b(y)}\bigg|_{\mathfrak{J}_\mu=0} = \delta^{ab}\bar{\Delta}_{\mu\nu}(x-y), \tag{4.3}$$

is not renormalized: The vector propagator has the form

$$\bar{\Delta}_{\mu\nu}(x-y) = (g_{\mu\nu} - \partial_\mu\partial_\nu/\partial^2)\bar{D}(x-y) + \alpha(\partial_\mu\partial_\nu/\partial^2)\bar{D}_F(x-y). \tag{4.4}$$

In the momentum space, the inverse of the vector propagator, therefore, takes the form

$$[\Delta^{-1}(k)]_{\mu\nu} = (k^2 g_{\mu\nu} - k_\mu k_\nu)J(k^2) + \frac{1}{\alpha}k_\mu k_\nu\,. \tag{4.5}$$

Were it not for the n-particle thresholds at $k^2=0$, $J(k^2)$ would be regular at $k^2=0$ (at least in perturbation theory), and the transverse part of the vector-meson propagator would have a simple pole at $k^2=0$.

Equation (4.1), when combined with Eq. (4.4) gives

$$\left\{\delta^{ab}\partial_\mu\bar{D}_F(x-y) + \partial_\mu G^{ab}(x,y;i\delta/\delta\mathfrak{J}) - ig\left[\vec{t}\cdot\frac{\delta}{\delta\mathfrak{J}^\mu(x)}\right]^{ac}G^{cb}(x,y;i\delta/\delta\mathfrak{J})\right\}W\bigg|_{\mathfrak{J}_\mu=0} = 0 \tag{4.6}$$

or

$$\bar{\mathcal{G}}^{ab}(x-y) = \delta^{ab}\bar{D}_F(x-y) + ig\int d^4z\,\bar{D}_F(x-z)\frac{\partial}{\partial z_\mu}\left[\vec{t}\cdot\frac{\delta}{\delta\mathfrak{J}^\mu(z)}\right]^{ac}W^{-1}[\mathfrak{J}_\mu]G^{cb}(z,y;i\delta/\delta\mathfrak{J})W[\mathfrak{J}_\mu]|_{\mathfrak{J}_\mu=0} = 0, \tag{4.7}$$

where $\bar{\mathcal{G}}$ is the ghost propagator:

$$\bar{\mathcal{G}}^{ab}(x-y) = -\{W^{-1}[\mathfrak{J}_\mu]G^{ab}(x,y;i\delta/\delta\mathfrak{J})W[\mathfrak{J}_\mu]\}|_{\mathfrak{J}_\mu=0}.$$

We may define the self-energy part Σ_g of the ghost by

$$\delta^{ab}\mathcal{G}(k^2) = \int d^4x\,e^{ik\cdot x}\,\bar{\mathcal{G}}^{ab}(x),$$
$$\mathcal{G}(k^2) = (k^2+i\epsilon)^{-1}[1-\Sigma_g(k^2)\mathcal{G}(k^2)]\,. \tag{4.8}$$

The structure of Eq. (4.7) implies that $\Sigma_g(k^2)$ is of the form $-k_\mu\Sigma_g^\mu(k)$. Again, were it not for the fact that $k^2=0$ is the onsets of n-particle intermediate states, $\Sigma_g(k^2)$ would behave like k^2 near $k^2=0$, so that $\bar{\mathcal{G}}(k^2)$ would have just a simple pole there.

B. Three-Point Vertices

From Eq. (3.13) we obtain

$$\left(\frac{i}{\alpha}\right)^2\frac{\partial}{\partial x^\mu}\frac{\partial}{\partial y^\nu}\frac{\delta^3 W}{\delta J_\mu^a(x)\delta J_\nu^b(y)\delta J_\lambda^c(z)} + D_z^\lambda[i\delta/\delta\mathfrak{J}]^{cd}G^{da}(z,x;i\delta/\delta\mathfrak{J})\frac{i}{\alpha}\frac{\partial}{\partial y^\nu}\frac{\delta W}{\delta J_\nu^b(y)}\bigg|_{\mathfrak{J}_\mu=0} = 0. \tag{4.9}$$

We note that

$$gW^{-1}[\vec{t}\cdot\delta/\delta\vec{J}_\lambda(z)]^{cd}G^{da}(z,x;i\delta/\delta\vec{J})\delta W/\delta J^b_\nu(y)|_{\vec{J}_\mu=0}$$

$$=g\int d^4x'd^4y'\bar{\gamma}^{cab}_{\lambda\rho}(z,x';y')\mathcal{G}(x'-x)\bar{\Delta}^{\rho\nu}(y'-y)+g\int d^4z'd^4z''d^4x'd^4y'\bar{\Sigma}^\lambda_g(z-z')\mathcal{G}(z'-z'')\bar{\gamma}^{cab}_\rho(z'',x';y')$$
$$\times\,\mathcal{G}(x'-x)\bar{\Delta}^{\rho\nu}(y'-y), \tag{4.10}$$

where

$$\bar{\gamma}^{abc}_\rho(x,y;z)=-i\frac{\partial}{\partial x_\lambda}\bar{\gamma}^{abc}_{\lambda\rho}(x,y;z),$$

and $\gamma^{abc}_\rho(p,q;r)$, defined by

$$\gamma^{abc}_\rho(p,q;r)(2\pi)^4\delta^4(p+q+r)=\int d^4x\,d^4y\,d^4z\,e^{i(p\cdot x+q\cdot y+r\cdot z)}\bar{\gamma}^{abc}_\rho(x,y;z),$$

is the proper vertex for the coupling of a vector meson of momentum r, polarization ρ and isospin index c, with two ghosts, of momenta p and q and isospin indices a and b, respectively (we define the momenta outwardly from the vertex) (see Fig. 1). We have

$$\gamma^{abc}_\rho(p,q;r)=p^\lambda\gamma^{abc}_{\lambda\rho}(p,q;r). \tag{4.11}$$

The quantity $\bar{\Sigma}^\lambda_g$ is defined by the equation

$$\bar{\Sigma}_g(x-y)=-i\frac{\partial}{\partial x^\mu}\bar{\Sigma}^\mu_g(x-y), \tag{4.12}$$

where

$$\Sigma_g(k^2)=\int d^4x\,e^{-ik\cdot x}\bar{\Sigma}_g(x) \tag{4.13}$$

and satisfies the unrenormalized equation

$$i\delta^{ab}\bar{\Sigma}^\lambda_g(x-y)=g\epsilon^{adc}\int(-i)\bar{\Delta}^{\lambda\mu}(x-z)i\mathcal{G}(x-\omega)i\bar{\gamma}^{dbc}_\mu(w,y;z)d^4\omega d^4z. \tag{4.14}$$

We consider the part which is transverse with respect to the index λ of Eq. (4.9). Noting that

$$W^{-1}[\vec{J}_\mu]\left.\frac{\delta^3W[\vec{J}_\mu]}{\delta J^a_\mu(x)\delta J^b_\nu(y)\delta J^c_\lambda(z)}\right|_{\vec{J}_\mu=0}=-ig\int d^4x'd^4y'd^4z'\,\bar{\Delta}^{\mu\mu'}(x-x')\bar{\Delta}^{\nu\nu'}(y-y')\bar{\Delta}^{\lambda\lambda'}(z-z')\bar{\Gamma}^{abc}_{\mu'\nu'\lambda'}(x',y',z'),$$

where $\bar{\Gamma}^{abc}_{\mu\nu\lambda}(x,y,z)$ is the proper three-point vertex of vector mesons, and taking the Fourier transform we obtain

$$(p^\lambda/p^2)(q^\mu/q^2)[r^2J(r^2)]^{-1}(g^{\rho\nu}-r^\rho r^\nu/r^2)\Gamma^{abc}_{\lambda\mu\nu}(p,q,r)=(q^\mu/q^2)\mathcal{G}(p^2)(g^{\lambda\rho}-r^\lambda r^\rho/r^2)\gamma^{cab}_{\lambda\mu}(r,p;q),\quad p+q+r=0. \tag{4.15}$$

Equation (4.15) is a constraint among the propagators and three-point proper vertices. Equation (4.15) was first derived by Slavnov.[18]

C. Four-Point Vertices

It follows from Eq. (3.13) that

$$\frac{\partial}{\partial x^\lambda}\frac{\partial}{\partial y^\mu}\frac{\partial}{\partial z^\nu}\frac{\partial}{\partial w^\rho}\left.\frac{\delta^4Z[\vec{J}_\mu]}{\delta J^a_\lambda(x)\delta J^b_\mu(y)\delta J^c_\nu(z)\delta J^d_\rho(w)}\right|_{\vec{J}_\mu=0}=0, \tag{4.16}$$

from which one obtains a constraint on the four-point vertex $\Gamma^{abc}_{\lambda\mu\nu\rho}$:

$$p^\lambda q^\mu r^\nu s^\rho\{\Gamma^{abcd}_{\lambda\mu\nu\rho}(p,q,r,s)+[\Gamma^{abe}_{\lambda\mu\sigma}(p,q,-p-q)\Delta^{\sigma\zeta}(p+q)\Gamma^{cde}_{\nu\rho\zeta}(r,s,p+q)+\text{two more terms}]\}=0,$$
$$p+q+r+s=0, \tag{4.17}$$

where $\Gamma^{abcd}_{\lambda\mu\nu\rho}$ is defined as

$$i \frac{\delta^4 Z[\vec{J}_\mu]}{\delta J^a_\lambda(x)\delta J^b_\mu(y)\delta J^c_\nu(z)\delta J^d_\rho(w)} = (-i)^4 ig \int d^4x' d^4y' d^4z' d^4w' \bar{\Delta}^{\lambda\lambda'}(x-x')\bar{\Delta}^{\mu\mu'}(y-y')\bar{\Delta}^{\nu\nu'}(z-z')\bar{\Delta}^{\rho\rho'}(w-w')$$
$$\times \bar{\Gamma}^{abcd}_{\lambda'\mu'\nu'\rho'}(x',y',z',w') + \text{reducible parts} \tag{4.18}$$

and

$$\int d^4x d^4y d^4z d^4w \, e^{i(p\cdot x + q\cdot y + r\cdot z + s\cdot w)} \bar{\Gamma}^{abcd}_{\lambda\mu\nu\rho}(x,y,z,w) = (2\pi)^4 \delta^4(p+q+r+s)\Gamma^{abcd}_{\lambda\mu\nu\rho}(p,q,r,s). \tag{4.19}$$

The ghost-ghost-vector-vector vertex is superficially convergent and requires no discussion.

V. REGULARIZATION

The Feynman amplitudes constructed from the expression (2.1) would automatically satisfy the Ward-Takahashi identities discussed in the last two sections, if it were not for the ultraviolet divergences in their construction. A standard procedure of constructing the renormalized amplitudes satisfying the Ward-Takahashi identities is to regularize the Feynman integrals in a gauge-invariant manner; and then perform the R operation[14-17] of Bogoliubov, Parasiuk, and Hepp (BPH). The resulting amplitudes are cutoff independent, and if the values of primitively divergent vertices at subtraction points are chosen in accordance with the Ward-Takahashi identities, then the full amplitudes satisfy them too. Furthermore, under such circumstances, the R operation may be formally implemented by a gauge-invariant set of counter terms in the Lagrangian.

In this section we will discuss a few gauge-invariant regularization methods which can be implemented by adding gauge-invariant terms in the Lagrangian (e.g., Pauli-Villars regularization). 't Hooft[22] discussed a method which works for one-loop diagrams, but which does not appear to be implementable by modifying the Lagrangian.

We choose as regulator fields both scalar and spinor fields. They have all positive masses. They may belong to arbitrary representations (in general, reducible) of the symmetry group. They are coupled to the gauge fields by the minimal gauge-invariant coupling. They may, however, be quantized by the wrong spin-statistics connection (i.e., some scalar field multiplets may be quantized by the anticommutation relation). Let us show that the addition of these regulator fields to the Lagrangian renders the divergent Feynman integrals with one loop finite.

Let us first consider the self-energy of the gauge boson in the one-loop approximation. We will carry out the computations in the Feynman gauge. There are three diagrams (one is a ghost loop), and the sum of the contributions from these is

$$\Sigma_{\mu\nu}(p) = -4g_{\mu\nu}\frac{g^2}{16\pi^2}\int_0^\infty \frac{dz_1 dz_2}{(z_1+z_2)^3}\left(-1 + i\frac{z_1 z_2}{z_1+z_2}p^2\right)\exp\left(i\frac{z_1 z_2}{z_1+z_2}p^2\right)$$
$$+ \frac{2g^2}{16\pi^2} i(g_{\mu\nu}p^2 - p_\mu p_\nu)\int_0^\infty \frac{dz_1 dz_2}{(z_1+z_2)^2}\frac{z_1^2 + 6z_1 z_2 + z_2^2}{(z_1+z_2)^2}\exp\left(i\frac{z_1 z_2}{z_1+z_2}p^2\right). \tag{5.1}$$

The first term on the right-hand side is gauge-noninvariant and quadratically divergent; the second term, which is gauge-invariant, is logarithmically divergent. We shall regulate $\Sigma_{\mu\nu}$ by the replacement

$$\Sigma_{\mu\nu} \to \Sigma^{(\ )}_{\mu\nu} = \Sigma_{\mu\nu} + \Sigma^{(1)}_{\mu\nu} + \Sigma^{(2)}_{\mu\nu}, \tag{5.2}$$

where $\Sigma^{(1)}_{\mu\nu}$ is the sum of the scalar regulator contributions:

$$\Sigma^{(1)}_{\mu\nu} = \sum_i C_i\left\{-\frac{2g^2}{16\pi^2}g_{\mu\nu}\int_0^\infty \frac{dz_1 dz_2}{(z_1+z_2)^2}\exp\left[i\left(\frac{z_1 z_2}{z_1+z_2}p^2 - (z_1+z_2)m_i^2\right)\right]\left[-1 + i\left(\frac{z_1 z_2}{z_1+z_2}p^2 - (z_1+z_2)m_i^2\right)\right]\right.$$
$$\left. - \frac{g^2}{16\pi^2} i(p^2 g_{\mu\nu} - p_\mu p_\nu)\int_0^\infty \frac{dz_1 dz_2}{(z_1+z_2)^2}\exp\left(i\frac{z_1 z_2}{z_1+z_2}p^2 - i(z_1+z_2)m_i^2\right)\left(\frac{z_1 - z_2}{z_1+z_2}\right)^2\right\}$$

and $\Sigma^{(2)}_{\mu\nu}$ is the sum of the spinor regulator contributions:

$$\Sigma^{(2)}_{\mu\nu} = -\sum_i D_i \int \frac{dz_1 dz_2}{(z_1+z_2)^2} \exp\left(i\frac{z_1 z_2}{z_1+z_2}p^2 - i(z_1+z_2)\mu_i^2\right)$$
$$\times\left\{-\frac{2g^2}{16\pi^2} g_{\mu\nu}\left[\frac{-1}{z_1+z_2} + i\left(\frac{z_1 z_2}{(z_1+z_2)^2}p^2 - \mu_i^2\right)\right] - \frac{4g^4}{16\pi^2} i(g_{\mu\nu}p^2 - p_\mu p_\nu)\frac{z_1 z_2}{z_1+z_2}\right\}.$$

The coefficients C_i and D_i depend on the representations to which the regulators belong, and also on whether they obey the normal or abnormal statistics. In any case, if we choose

$$2 + \sum_i C_i - \sum_j D_j = 0 \tag{5.3}$$

and

$$\sum C_i m_i^2 - \sum D_j \mu_j^2 = 0 \tag{5.4}$$

then the gauge-noninvariant term vanishes identically. Furthermore if we choose

$$10 - \sum_i C_i + 2\sum_j D_j = 0 \tag{5.5}$$

the logarithmic divergence in the gauge-invariant part may be eliminated. The introduction of two kinds of regulators is necessitated by the requirement that both quadratic and logarithmic divergences be eliminated.

Next, we consider the three-point vertex $\Gamma_{\lambda\mu\nu}(p,q,r)$ of three gauge bosons in the one-loop approximation. The integral is linearly divergent and has the asymptotic structure, in the Feynman gauge,

$$\sim \int \frac{d^4 l}{(2\pi)^4} \frac{l_\lambda l_\mu l_\nu}{(l^2)^3}, \tag{5.6}$$

when all diagrams, including the one with a ghost loop, are added. Again, by taking a suitable combination of scalar and spinor regulators, it is possible to eliminate all divergences from the Feynman integral for the three-point vertex.

The four-point vertex is logarithmically divergent and offers no special difficulty. We have not verified that this method works for higher-order diagrams. In any case, when there are matter fields present in the Lagrangian, the method presented above is insufficient and it becomes necessary to dampen the high-energy behavior of the gauge boson propagator itself. The method described below will do just this, and when combined with the spinor-scalar regulators, will render all Feynman integrals finite.

We will add gauge-invariant higher-derivative terms to the Lagrangian. Consider, for example, the Lagrangian

$$\mathcal{L}_\Lambda = -\tfrac{1}{4}\vec{F}^{\mu\nu}\cdot\vec{F}_{\mu\nu} - \frac{\alpha}{4\Lambda^2}(D^\sigma \vec{F}^{\mu\nu})\cdot(D_\sigma \vec{F}_{\mu\nu})$$
$$-\frac{\beta}{4\Lambda^4}(D^\sigma D_\sigma \vec{F}_{\mu\nu})\cdot(D^\rho D_\rho \vec{F}^{\mu\nu}). \tag{5.7}$$

The vector boson propagator is now

$$\Delta_{\mu\nu}(k;\Lambda^2) = (g_{\mu\nu} - k_\mu k_\nu/k^2)(k^2+i\epsilon)^{-1} \times\left[1+\alpha\frac{k^2}{\Lambda^2} + \beta\left(\frac{k^2}{\Lambda^2}\right)^2\right]^{-1}$$
$$+\text{gauge-dependent term}$$

and behaves like $(p^2)^{-3}$ asymptotically. The maximum dimension of various new couplings (in powers of mass) is eight. A power-counting argument (see Appendix D) shows that in this case only the two-, three-, and four-point vertices with *one loop* are primitively divergent (quadratically, linearly, and logarithmically, respectively). Other proper vertices, including two-, three-, and four-point vertices with more than one loop are at least superficially convergent. As one adds still higher gauge-covariant derivatives, the propagator becomes more convergent at large momentum, but the maximum dimensions of the interaction terms increase also, in such a way that the two-, three-, and four-point vertices with one loop remain always divergent (see Appendix D). Note also that ghosts loops for these vertices remain divergent.

Therefore, by the addition of the last two terms to the Lagrangian (5.7), the divergences of the theory have now been isolated to those diagrams for which the spinor-scalar regularization was shown to work.

The BPH R operation is to be applied to the entire two-, three-, and four-point proper vertices. The resulting vertices are cutoff-independent, in the sense that the limits $\Lambda^2\to\infty$ of these amplitudes are finite and independent of α and β of Eq. (5.7). This can be seen as follows. A proper amplitude with two, three, or four external lines which is proportional to some powers of α and β has in general an ambiguous limit as $\Lambda^2\to\infty$. However, the finite part of such an integral vanishes like $\Lambda^{-2}(\ln\Lambda^2)^m$ as $\Lambda^2\to\infty$. A proper Feynman diagram with n external lines, $n>4$, with one or more vertices proportional to α or β which are not contained in any subdiagrams with two, three, or four external lines vanishes at least as fast as $\Lambda^{-(n-4)}\times(\ln\Lambda^2)^m$ as $\Lambda^2\to\infty$, after the R operations are applied to the subdiagrams. (The above is a summary of a rather lengthy analysis.)

The results of regulating the Feynman integral by the method described above, applying the R operation and then letting the cutoff Λ^2 go to infinity is identical to applying the R operation directly to the Feynman integral. This shows that the BPH R operation is in fact a gauge-invariant procedure.

A similar regularization procedure has been applied to nonlinear chiral Lagrangians by Slavnov.[23] We understand from Jackiw and Faddeev[23a] that Slavnov has considered the regularization method of Eq. (5.7) for the gauge fields also. (After the completion of this work we received a report by Slavnov.[24]) This possibility has also been known to Johnson.[25]

VI. RENORMALIZATION CONDITIONS AND INFRARED DIVERGENCES

Let us first discuss briefly how the values of primitively divergent vertices are determined from the considerations of Sec. IV, ignoring the problems associated with infrared divergences. Under such circumstances, we may choose as subtraction points the points at which all external momenta vanish. Later we will discuss the nature of infrared divergences in gauge theories and give a set of renormalization conditions which avoid the infrared difficulties.

We may by convention choose, in Eq. (4.5),

$$J(0)=1,$$

which amounts to

$$\lim_{k^2\to 0}\Delta_{\mu\nu}(k)\sim\left(g_{\mu\nu}-\frac{k_\mu k_\nu}{k^2}\right)\frac{1}{k^2}+\alpha\frac{k_\mu k_\nu}{(k^2)^2}\,. \tag{6.1}$$

The normalization of the ghost propagator is arbitrary. The ghost propagator has a simple pole at $k^2=0$ [see the discussion following Eq. (4.8)], and we write

$$\lim_{k^2\to 0}\mathcal{G}(k^2)\sim Z_g/k^2, \tag{6.2}$$

where Z_g is an arbitrary (finite) constant.

In the limit p, q, and $r=-p-q$ all go to zero, the three-point vertex $\Gamma^{abc}_{\lambda\mu\nu}(p,q,r)$ has the form

$$\lim_{p,q,r\to 0}\Gamma^{abc}_{\lambda\mu\nu}(p,q,r)\sim -iG\epsilon^{abc}[(p-q)_\nu g_{\lambda\mu}+(q-r)_\lambda g_{\mu\nu}+(r-p)_\mu g_{\nu\lambda}], \tag{6.3}$$

as follows from Lorentz covariance, isospin conservation, and Bose symmetry. Likewise, the low-energy form of the vertex $\gamma^{cab}_{\lambda\mu}(r,p;q)$ is given by

$$\lim_{p,q,r\to 0}\gamma^{cab}_{\lambda\mu}(r,p;q)=-iG'\epsilon^{abc}g_{\lambda\mu},$$

so that

$$\lim_{p,q,r\to 0}\gamma^{cab}_{\mu}(r,p;q)\sim -iG'\epsilon^{abc}r_\mu.$$

Equation (4.15) then tells us that

$$G=G'Z_g. \tag{6.4}$$

In the Green's functions, only the combination $G'Z_g$ enters, because the ghost never appears as an external line, so that it is convenient to set $Z_g=1$ and $G=G'$.

The low-energy form of the four-point vertex is given by

$$\lim_{p,q,r,s\to 0} i\Gamma^{abcd}_{\lambda\mu\nu\rho}(p,q,r,s)=-iF[\epsilon^{abe}\epsilon^{cde}(g_{\lambda\nu}g_{\mu\rho}-g_{\lambda\rho}g_{\mu\nu})+\epsilon^{ace}\epsilon^{bde}(g_{\lambda\mu}g_{\nu\rho}-g_{\lambda\rho}g_{\mu\nu})+\epsilon^{ade}\epsilon^{cbe}(g_{\lambda\nu}g_{\mu\rho}-g_{\lambda\mu}g_{\rho\nu})]$$
$$+F'[\delta^{ab}\delta^{cd}g_{\lambda\mu}g_{\nu\rho}+\delta^{ac}\delta^{bd}g_{\lambda\nu}g_{\mu\rho}+\delta^{ad}\delta^{bc}g_{\lambda\rho}g_{\mu\nu}]\,. \tag{6.5}$$

Equation (4.7) tells us that

$$F=G^2\quad\text{and}\quad F'=0. \tag{6.6}$$

The conditions (6.1), (6.4), and (6.6) allow us to express all primitively divergent vertices in terms of only one constant G.

As we have stated before, the foregoing discussion is of heuristic value only because of the infrared divergences that the Feynman integrals experience when all external momenta are set equal to zero. More precisely, a Feynman integral of the theory becomes divergent if two or more external lines are set on the mass shell. We assert that a Feynman integral suffers no infrared catastrophe if all of the external lines are kept off the mass shell. One can easily see this for diagrams with one loop. As long as all of the external lines are off the mass shell, infrared divergences in any subintegrations can occur only in the measure zero of the space of all integration variables where the rest of the integrand is nonsingular. Therefore, by choosing subtraction points to be somewhere other than where all external lines are on the mass shell, we can circumvent the infrared difficulties in the construction of Green's functions (but not the S matrix) altogether.

A convenient convention for the subtraction points is given by Symanzik.[26] We choose as such the points where the squares of external momenta are all equal to a negative number, say, $-a^2$. Defining all external momenta outwardly from the vertex, we have at the subtraction point $p_i^2=-a^2$, $p_i\cdot p_j=(n-1)^{-1}a^2$. As an example, we will work out the renormalization conditions for two- and three-point vertices.

We shall normalize the fields $\vec{A}_\mu$ and $\vec{c}$ so that

$$J(-a^2)=1,$$

and

$$\Sigma_g(-a^2)=0.$$

Then we have

$$\lim_{k^2\to -a^2} \Delta_{\mu\nu}(k) \sim (g_{\mu\nu}-k_\mu k_\nu/k^2)(-a)^{-2} + \text{gauge-dependent terms},$$

$$\mathcal{G}(a^2)=-a^{-2}.$$

At the symmetric point $p^2=q^2=r^2=-a^2$, the three-point vertices $\Gamma^{abc}_{\lambda\mu\nu}(p,q,r)$ and $\gamma^{cab}_{\lambda\mu}(r,p;q)$ have the structures

$$\lim_{p^2=q^2=r^2=-a^2} i\Gamma^{abc}_{\lambda\mu\nu}(p,q,r)=\{G[(p-q)_\nu g_{\lambda\mu}+(q-r)_\lambda g_{\mu\nu}+(r-p)_\mu g_{\nu\lambda}]+H[(r-q)_\lambda r_\mu q_\nu+(p-r)_\mu p_\nu r_\lambda+(q-p)_\nu q_\lambda p_\mu]$$
$$+J[r_\lambda p_\mu q_\nu - q_\lambda r_\mu p_\nu]\}\epsilon^{abc}, \tag{6.7}$$

$$\lim_{p^2=q^2=r^2=-a^2} i\gamma^{cab}_{\lambda\mu}(r,p;q)=\epsilon^{abc}\{G' g_{\lambda\mu}+K_1 p_\lambda r_\mu+K_2 p_\lambda p_\mu+K_3 p_\lambda q_\mu+L_1 q_\lambda r_\mu+L_2 q_\lambda p_\mu+L_3 q_\lambda q_\mu+r_\lambda(\cdots)\}, \tag{6.8}$$

where the omitted terms $(\cdots)$ are of no interest to the present problem. The quantities G and G' require subtractions whereas the other form factors $J, J, K_i, \ldots$ are superficially convergent. Substituting Eqs. (6.7) and (6.8) into Eq. (4.15) and taking the limit $p^2=q^2=r^2=-a^2$, we obtain

$$G-\tfrac{1}{2}a^2(H+J)=G'+\tfrac{1}{2}a^2(L_1+L_2-2L_3-K_1-K_2+2K_3), \tag{6.9}$$

which gives G' in terms of G.

VII. RENORMALIZATION OF MATTER FIELDS

So far our discussion was based on the Lagrangian (2.3) which contains only the Yang-Mills quanta. As an illustration of the renormalization procedure in the presence of matter fields, we consider the case in which a triplet of real scalar fields ϕ^a is added to the Lagrangian by the minimal gauge-invariant coupling.

Let $K^a(x)$ be the sources of the scalar fields ϕ^a. It is not difficult to derive the generalization of Eq. (3.13). It is

$$\frac{i}{\alpha}\partial_\mu\frac{\delta W}{\delta J^a_\mu(x)}-\int d^4y J^c_\lambda(y)D^\lambda_y[i\delta/\delta\vec{J}]^{cb}G^{ba}(y,x;i\delta/\delta\vec{J})W+ig\int d^4y K^c(y)t^{cdb}\frac{\delta}{\delta K^b(y)}G^{da}(y,x;i\delta/\delta\vec{J})W=0. \tag{7.1}$$

In this theory the $A_\mu\phi^2$, $A_\mu{}^2\phi^2$, ϕ^2, and ϕ^4 vertices are primitively divergent. The renormalization of the ϕ^4 vertex has nothing to do with the local gauge invariance and presents no problem. In order to regularize Feynman integrals for these vertices, it becomes necessary to regularize the gauge boson propagators in a gauge-invariant manner, for example, by the use of the Lagrangian (5.7). The renormalization conditions for the ϕ^2, $A_\mu\phi^2$, and $A_\mu{}^2\phi^2$ vertices are obtained from Eq. (7.1). First, though, let us ignore the infrared complications. Later, we will detail the renormalization conditions which take due account of the infrared difficulties.

From Eq. (7.1) we obtain

$$\left(\frac{i}{\alpha}\right)^2\frac{\partial}{\partial x^\mu}\frac{\partial}{\partial y^\nu}\frac{\delta^4 iZ[\vec{J}_\mu,\vec{K}]}{\delta J^a_\mu(x)\delta J^b_\nu(y)\delta K^c(z)\delta K^d(w)}\Bigg|_{\vec{J}_\mu=\vec{K}=0}$$
$$=-ig\epsilon^{cfe}W^{-1}\frac{\delta}{\delta K^e(z)}G^{fa}(z,x;i\delta/\delta\vec{J})\frac{\delta}{\delta K^d(w)}\frac{i}{\alpha}\frac{\partial}{\partial y^\nu}\frac{\delta W}{\delta J^b_\nu(y)}\Bigg|_{\vec{J}_\mu=\vec{K}=0}+(c\leftrightarrow d, z\leftrightarrow w). \tag{7.2}$$

If we amputate the scalar propagators and go to the mass shells of two scalar particles $q_1{}^2=q_2{}^2=\mu^2$ in Eq. (7.2), the right-hand side of the equation vanishes. Let us define the primitively divergent vertices:

$$\left.\frac{\delta^2 iZ}{\delta K^a(x)\delta K^b(y)}\right|_{\vec{J}_\mu=\vec{K}=0} = -\bar{\Delta}^{ab}(x-y) = -\delta^{ab}\bar{\Delta}(x-y),$$

$$\left.\frac{\delta^3 iZ}{\delta J^a_\mu(x)\delta K^b(y)\delta K^c(z)}\right|_{\vec{J}_\mu=\vec{K}=0} = -g\int d^4x'd^4y'd^4z'\bar{\Delta}_{\mu\nu}(x-x')\bar{\Delta}(y-y')\bar{\Delta}(z-z')\bar{C}^{abc}_\nu(x';y,z),$$

$$\left.\frac{\delta^4 iZ}{\delta J^a_\mu(x)\delta J^b_\nu(g)\delta K^c(z)\delta K^d(w)}\right|_{\vec{J}_\mu=\vec{K}=0} = ig^2\int d^4x'd^4y'd^4z'd^4w'\bar{\Delta}_{\mu\rho}(x-x')\bar{\Delta}_{\nu\sigma}(y-y')\bar{\Delta}(z-z')\bar{\Delta}(w-w')$$

$$\times\, \bar{C}^{abcd}_{\rho\sigma}(x',y';z',w'),$$

and

$$\int d^4x d^4y d^4z\, e^{i(px+q_1y+q_2z)}\bar{C}^{abc}_\nu(x;y,z) = C^{abc}_\nu(p;q_1,q_2)(2\pi)^4\delta^4(p+q_1+q_2),$$

$$C^{abc}_\nu(p;q_1,q_2) = \epsilon^{abc}C_\nu(p;q_1,q_2),$$

and similarly for $C^{abcd}_{\mu\nu}$. Then we have from Eq. (7.2) and the subsequent discussion that

$$\lim_{q_1{}^2=q_2{}^2=\mu^2} p_1^\mu p_2^\nu\{C^{abcd}_{\mu\nu}(p_1,p_2;q_1,q_2)+\Gamma^{abf}_{\mu\nu\lambda}(p_1,p_2,-p_1-p_2)\Delta^{\lambda\rho}(q_1+q_2)C^{fcd}_\rho(p_1+p_2;q_1,q_2)$$
$$-[C^{acf}_\mu(p_1;q_1,p_2+q_2)\Delta(p_2+q_2)C^{bdf}_\nu(p_2;q_2,p_1+q_1)+(c\leftrightarrow d,q_1\leftrightarrow q_2)]\}=0. \tag{7.3}$$

We shall now consider the limit $p_1, p_2\to 0$ while $Q=q_1-q_2$ is kept finite. The low-energy forms of the vertices above are

$$\lim_{q_1{}^2=q_2{}^2=\mu^2} C_\mu(p;q_1,q_2) = -iCQ_\mu, \tag{7.4}$$

$$\lim_{p_1,p_2\to 0}\Gamma^{abc}_{\mu\nu\lambda}(p_1,p_2,r)\sim -iG[(p_1-p_2)_\lambda g_{\mu\nu}+(p_2-r)_\mu g_{\nu\lambda}+(r-p_1)_\nu g_{\mu\lambda}], \tag{7.5}$$

where G is defined previously and $r+p_1+p_2=0$;

$$\lim_{\substack{q_1{}^2=q_2{}^2=\mu^2\\ p_1,p_2\to 0}} C^{abcd}_{\mu\nu}(p_1,p_2;q_1,q_2) = \delta^{ab}\delta^{cd}(Ag_{\mu\nu}+BQ_\mu Q_\nu)+(\delta^{ab}\delta^{cd}+\delta^{ad}\delta^{cb})(A'g_{\mu\nu}+B'Q_\mu Q_\nu). \tag{7.6}$$

The factors C, A, and A' require subtractions. We may renormalize the ϕ fields so that

$$\lim_{p^2\to\mu^2}\Delta(p^2)\sim(p^2-\mu^2)^{-1}. \tag{7.7}$$

Substituting Eqs. (7.4)–(7.7) in Eq. (7.3) and isolating the part antisymmetric in a and b, we get

$$GC=C^2$$

or

$$G=C. \tag{7.8}$$

Next looking at the part symmetric in a and b, we obtain

$$B=B'=0$$

and

$$A=-2G^2,\quad A'=G^2. \tag{7.9}$$

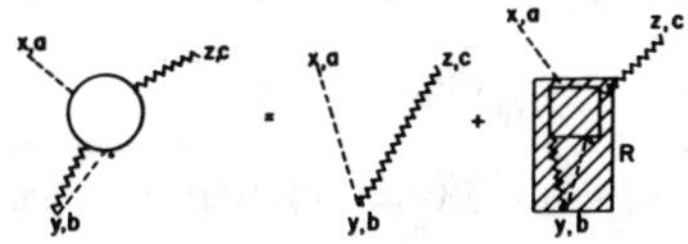

FIG. 2. Diagrammatic representation of Eq. (7.11).

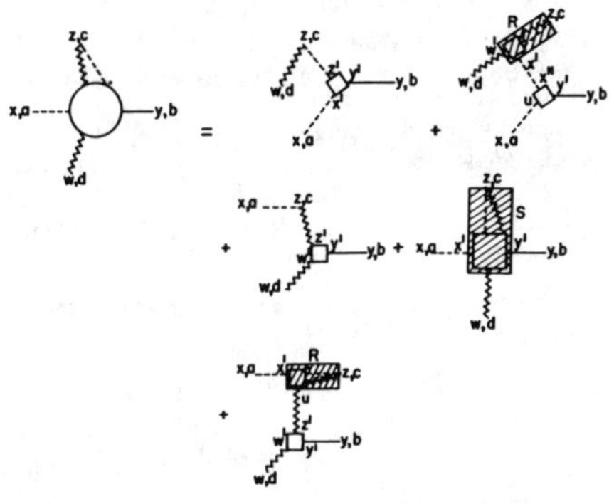

FIG. 3. Diagrammatic representation of Eq. (7.18).

The above treatment is careless, since the vertices $C_\mu, C_{\mu\nu}$ exhibit infrared divergence in the limit q_1^2 or $q_2^2=\mu^2$. We must, therefore, determine the renormalization conditions Δ, C_μ, and $C_{\mu\nu}$ without going to the mass shells of the scalar fields. For this purpose, we will return to Eq. (7.1). From Eq. (7.1) follows the relation

$$\frac{i}{\alpha}\frac{\partial}{\partial x^\mu}\frac{\delta^3 iZ}{\delta J_\mu^a(x)\delta K^b(y)\delta K^c(z)}\bigg|_{\vec{J}_\mu=\vec{K}=0}=-ig\left[W^{-1}\epsilon^{bfe}\frac{\delta}{\delta K^e(y)}\frac{\delta}{\delta K^c(z)}G^{fa}(y,x;i\delta/\delta\vec{J})W+(c\leftrightarrow d,\,y\leftrightarrow z)\right]_{\vec{J}_\mu=\vec{K}=0}. \tag{7.10}$$

In order to discuss this equation, it is necessary to define a new proper vertex R^{abc}. We write

$$\epsilon^{bfe}W^{-1}\frac{\delta^2}{\delta K^e(y)\delta K^c(z)}G^{fa}(y,x;i\delta/\delta\vec{J})W\bigg|_{\vec{J}_\mu=\vec{K}=0}=i\int d^4x'd^4y'\tilde{R}^{bac}(y,x',z')\bar{\mathcal{G}}(x'-x)\bar{\Delta}(z'-z)$$
$$+i\bar{\mathcal{G}}(y-z)\bar{\Delta}(y-z)\epsilon^{bac}, \tag{7.11}$$

$$R^{abc}(q,p,r)(2\pi)^4\delta(p+q+r)=\int d^4x\,d^4y\,d^4z\,\tilde{R}^{abc}(x,y,z)e^{i(px+qy+rz)},\quad R^{abc}(p,q,r)=\epsilon^{abc}R(p,q,r)$$

(see Fig. 2). The new vertex does not arise in the perturbative construction of the Green's functions. It is, however, relevant to our discussion of the renormalization of Eq. (7.2). The vertex R^{abc} has the superficial degree of divergence equal to zero, and therefore, requires one subtraction. The value of this vertex at the subtraction point is related to that of C_μ^{abc} through Eq. (7.10). Let us again choose as subtraction point the point where $p^2=q^2=r^2=-a^2$, $p\cdot q=q\cdot r=r\cdot p=\frac{1}{2}a^2$. The quantity $R^{abc}(q_1,p,q_2)$ may be written in the neighborhood of the subtraction point as

$$R^{bac}(q_1,p,q_2)=\epsilon^{bac}[R+(q_1^2+a^2)r_1+(q_2^2+a^2)r_2+(p^2+a^2)r_3+\cdots], \tag{7.12}$$

where R requires a subtraction and r_1, r_2, r_3 are finite.

We now transform Eq. (7.10) into the momentum space. It reads then

$$[p^2\mathcal{G}(p^2)]^{-1}(-ip^\mu)C_\mu(p;q_1,q_2)=\Delta^{-1}(q_1^2)-\Delta^{-1}(q_2^2)+[R(q_1,p,q_2)\Delta^{-1}(q_1^2)-R(q_2,p,q_1)\Delta^{-1}(q_2^2)]. \tag{7.13}$$

Adopting the field normalization conventions

$$[p^2\mathcal{G}(p^2)]\,|_{p^2+a^2=0}=i\,, \tag{7.14}$$

$$\lim_{q_1^2=-a^2}\Delta^{-1}(q_1^2)\sim q_1^2+a^2-M^2, \tag{7.15}$$

and defining C by

$$\lim_{p^2=q_1^2=q_2^2=-a^2}C_\mu(p;q_1,q_2)\sim -iQ_\mu C,\quad Q_\mu=(q_1-q_2)_\mu \tag{7.16}$$

we have from Eq. (7.13)

$$C=1+R-M^2(r_1-r_2), \tag{7.17}$$

which is the required relation.

Now we turn to Eq. (7.2). We need again to define a new proper vertex. We write (see Fig. 3)

$$\epsilon^{cfe}W^{-1}\frac{\delta}{\delta K_e(z)}G^{fa}(z,x;i\delta/\delta\vec{J})\frac{\delta}{\delta K^d(w)}\frac{\delta W}{\delta J_\nu^b(y)}\bigg|_{\vec{J}_\mu=\vec{K}=0}$$
$$=-ig\epsilon^{cfd}\bar{\Delta}(z-w)\int d^4z'd^4x'd^4y'\bar{\mathcal{G}}(z-z')\tilde{\gamma}_\lambda^{fab}(z',x';y')\bar{\Delta}_\nu^\lambda(y'-y)\bar{\mathcal{G}}(x'-x)$$
$$-ig\int d^4x'd^4x''d^4y'd^4w'd^4u\tilde{R}^{cfd}(z,x',w')\bar{\Delta}(w'-w)\bar{\mathcal{G}}(x'-x'')\tilde{\gamma}_\lambda^{fab}(x'',u;y')\bar{\Delta}_\nu^\lambda(y'-y)\bar{\mathcal{G}}(u-x)$$
$$+ig\epsilon^{cae}\bar{\mathcal{G}}(x-z)\int d^4y'd^4z'd^4w'\bar{\Delta}(z-z')\bar{\Delta}(w-w')\bar{\Delta}_\nu^\mu(y-y')\tilde{C}_\mu^{bed}(y';z',w')$$
$$+ig\int\tilde{R}^{cae}(z,x',u)\bar{\mathcal{G}}(x'-x)\bar{\Delta}(u-z')\bar{\Delta}(w-w')\bar{\Delta}_\nu^\mu(y-y')\tilde{C}_\mu^{bed}(y';z',w')d^4x'd^4ud^4y'd^4z'd^4w'$$
$$+g\int d^4w'd^4x'd^4y'\tilde{S}_\mu^{abcd}(z,w',x',y')\bar{\Delta}(w'-w)\bar{\mathcal{G}}(x'-x)\bar{\Delta}_\nu^\mu(y'-y), \tag{7.18}$$

which defines the vertex S_μ^{abcd}:

$$S_\mu^{abcd}(q_1,q_2,p_1,p_2)(2\pi)^4\delta^4(q_1+q_2+p_1+p_2)=\int d^4x d^4x' d^4y d^4y' \tilde{S}_\mu^{abcd}(y,y',x,x')e^{i(q_1y+q_2y'+p_1x+p_2x')}. \tag{7.19}$$

The fact that S_μ^{abcd} is superficially convergent is of importance in the ensuing discussion.

Now going to the momentum space, we rewrite Eq. (7.2) as

$$\begin{aligned}
i[p_1^2\mathcal{G}(p_1^2)]^{-1}p_1^\mu p_2^\nu\{C_{\mu\nu}^{abcd}(p_1,p_2;q_1,q_2)+\cdots\}
&= p_2^\nu\Delta^{-1}(q_1)\{\epsilon^{abf}\epsilon^{cfd}\mathcal{G}((p_1+p_2)^2)[1+R(q_1,p_1+p_2,q_2)]\gamma_\nu(q_1+q_2,p_1;p_2)\\
&\quad+\epsilon^{acf}\epsilon^{bfd}\Delta((p_1+q_1)^2)[1+R(q_1,p_1,p_2+q_2)]C_\nu(p_2,p_1+q_1,q_2)+iS_\nu^{cdab}(q_1,q_2;p_1,p_2)\}\\
&\quad+p_2^\nu\Delta^{-1}(q_2)\{c\leftrightarrow d, q_1\leftrightarrow q_2\},
\end{aligned}\tag{7.20}$$

where the expression in the curly bracket on the left-hand side is identical to that on the left-hand side of Eq. (7.3). Equation (7.20) is the generalization of Eq. (7.3) and it allows us to determine the values of the vertices $C_{\mu\nu}^{abcd}$ and C_μ^{abc} at some subtraction points, in terms of that of the vertex $\Gamma_{\mu\nu\lambda}^{abc}$. It is so, because the subtraction constant for γ_ν is known from Eq. (6.9); the subtraction constant for R is known in terms of the value of C_μ at the subtraction point through Eq. (7.17); and S_μ is superficially convergent, so that the right-hand side of the Eq. (7.20) contains the value we seek and no other unknowns.

APPENDIX A: DERIVATIONS OF EQS. (3.5) AND (3.9)

We recall the definition

$$\Delta_L[i\delta/\delta\vec{J}_\mu]=\exp\left[\mathrm{Tr}\ln\left(1-ig\vec{t}\cdot\frac{\delta}{\delta\vec{J}^\mu}\partial^\mu\frac{1}{\partial^2}\right)\right]. \tag{A1}$$

In evaluating

$$\Delta_L[i\delta/\delta\vec{J}_\mu]J_\lambda^a(x)\Delta_L^{-1}[i\delta/\delta\vec{J}_\mu]$$

it is convenient to make the following mapping:

$$\delta/\delta\vec{J}_\mu\to-\vec{\xi}_\mu,\quad \vec{J}_\mu\to\delta/\delta\vec{\xi}_\mu, \tag{A2}$$

which is canonical. We see that

$$\begin{aligned}
&\exp\left[\mathrm{Tr}\ln\left(1+ig\vec{t}\cdot\vec{\xi}^\mu\partial_\mu\frac{1}{\partial^2}\right)\right]\frac{\delta}{\delta\xi_\lambda^a(x)}\exp\left[-\mathrm{Tr}\ln\left(1+ig\vec{t}\cdot\vec{\xi}^\mu\partial_\mu\frac{1}{\partial^2}\right)\right]\\
&\quad=\frac{\delta}{\delta\xi_\lambda^a(x)}-ig\int d^4y d^4z\sum_{b,c}\left\langle y,b\left|\frac{1}{1+ig\vec{t}\cdot\vec{\xi}^\mu\partial_\mu 1/\partial^2}\right|z,c\right\rangle(t^a)^{cb}\delta^4(x-z)\partial_z^\lambda\left\langle z\left|\frac{1}{\partial^2}\right|y\right\rangle\\
&\quad=\frac{\delta}{\delta\xi_\lambda^a(x)}-ig\,\mathrm{tr}\,t^a[\partial_x^\lambda H(x,y;-i\vec{\xi})]_{y=x},
\end{aligned}\tag{A3}$$

which is Eq. (3.5).

As for Eq. (3.9), we begin by noting that

$$\begin{aligned}
&igD_x^\lambda[\vec{A}]^{ab}t^{cbd}[\partial_\lambda H^{dc}(x,y;\vec{A})]_{y=x}\\
&\quad=igt^{cad}\frac{\partial^2}{\partial x^\mu\partial x_\mu}H^{dc}(x,y;\vec{A})\bigg|_{y=x}+igt^{cad}\frac{\partial^2}{\partial x^\mu\partial y_\mu}H^{dc}(x,y;\vec{A})\bigg|_{x=y}-ig^2t^{aeb}t^{cbd}\frac{\partial}{\partial x_\mu}H^{dc}(x,y;\vec{A})\bigg|_{x=y}A_\mu^e(x).
\end{aligned}\tag{A4}$$

Since

$$\frac{\partial^2}{\partial x_\mu\partial x^\mu}H^{dc}(x,y;\vec{A})=\delta^{dc}\delta^4(x-y)+gt^{def}A_\mu^e(x)\frac{\partial}{\partial x_\mu}H^{fc}(x,y;\vec{A}), \tag{A5}$$

we can write the right-hand side of Eq. (A4) as

$$ig^2(t^{cad}t^{def}-t^{aed}t^{cdf})\partial H^{fc}(x,y;\vec{A})/\partial x_\mu|_{x=y}A^e_\mu(x)+ig\,t^{cad}\partial^2 H^{dc}(x,y;\vec{A})/\partial x^\mu\partial y_\mu|_{y=x}$$

$$=ig\,\mathrm{Tr}\,t^a\left[\frac{\partial^2}{\partial x^\mu\partial y_\mu}H(x,y;\vec{A})+\frac{\partial}{\partial x^\mu}H(x,y;\vec{A})g\,\vec{t}\cdot\vec{A}^\mu(x)\right] \tag{A6}$$

We have used the Jacobi identity

$$t^{cad}t^{def}+t^{aed}t^{dcf}=t^{adf}t^{ecd}.$$

Equation (A6) is precisely the right-hand side of Eq. (3.9).

APPENDIX B: GENERATING FUNCTIONAL OF PROPER VERTICES

In the usual manner, the generating functional $\Gamma[\vec{B}_\mu]$ may be obtained by a Legendre transformation from $Z[\vec{J}_\mu]$.

We define

$$-\vec{B}_\mu(x)=\delta Z[\vec{J}_\mu]/\delta\vec{J}_\mu(x) \tag{B1}$$

and

$$\Gamma[\vec{B}_\mu]=Z[\vec{J}_\mu]+\int d^4x\,\vec{J}^\mu(x)\cdot\vec{B}_\mu(x). \tag{B2}$$

It follows that

$$\vec{J}_\mu(x)=\delta\Gamma[\vec{B}_\mu]/\delta\vec{B}_\mu. \tag{B3}$$

The expansion coefficients of $\Gamma[\vec{B}_\mu]$ in terms of $\vec{B}_\mu$ are the proper vertices. The proof may be found in Jona-Lasinio.[27] It is possible to construct $\Gamma[\vec{B}_\mu]$ perturbatively by the functional integration technique. First consider

$$\exp iZ[\vec{J}_\mu]=\int[d\vec{A}]\exp\left\{iS_\alpha[\vec{A}_\mu]-i\int d^4x\,\vec{A}_\mu(x)\cdot\vec{J}^\mu(x)+\mathrm{Tr}\ln\left(1-g\vec{A}^\mu\cdot\vec{t}\partial_\mu\frac{1}{\partial^2}\right)\right\},$$

where $S_\alpha[\vec{A}_\mu]$ is the gauge-dependent action

$$S_\alpha[\vec{A}_\mu]=\int d^4x\left[\mathcal{L}(x)-\frac{1}{2\alpha}(\partial^\mu\vec{A}_\mu(x))^2\right].$$

We can perform the functional integration by the steepest-descent method. Let $\vec{A}^0_\mu$ be defined by

$$\delta S_\alpha[\vec{A}^0_\mu]/\delta\vec{A}^0_\mu(x)=\vec{J}_\mu(x),$$

i.e., $\vec{A}^0_\mu$ is the solution of the classical equation of motion in the presence of the external source, so that

$$\vec{A}^0_\mu=\vec{A}^0_\mu(x,\vec{J}).$$

Now we may write, after DeWitt,[20]

$$\exp iZ[\vec{J}_\mu]=\exp i\left\{S_\alpha[\vec{A}^0_\mu]-\int d^4x\,\vec{A}^0_\mu(x)\cdot\delta S_\alpha[\vec{A}^0_\mu]/\delta\vec{A}^0_\mu(x)\right\}$$

$$\times\int[d\vec{A}]\Delta_L[\vec{A}_\mu]\exp i\left\{S_\alpha[\vec{A}_\mu]-S_\alpha[\vec{A}^0_\mu]-\int d^4x\frac{\delta S_\alpha[\vec{A}^0_\mu]}{\delta\vec{A}^0_\mu(x)}[\vec{A}_\mu(x)-\vec{A}^0_\mu(x)]\right\}.$$

The tree approximation consists in approximating $Z[\vec{J}_\mu]$ by

$$Z^{\text{tree}}[\vec{J}_\mu]=S_\alpha[\vec{A}^0_\mu]-\int d^4x\,\delta S_\alpha[\vec{A}^0_\mu]/\delta\vec{A}^0_\mu(x)\cdot\vec{A}^0_\mu(x),$$

$$[\vec{B}_\mu(x)]^{\text{tree}}\equiv-\delta Z^{\text{tree}}[\vec{J}_\mu]/\delta\vec{J}_\mu(x)=\vec{A}^0_\mu(x).$$

Therefore, in the tree approximation we have

$$\Gamma^{\text{tree}}[\vec{B}_\mu]=S_\alpha[\vec{B}_\mu].$$

The one-loop approximation consists in evaluating the functional integral by the steepest descent approximation. We then have

$$Z[\vec{J}_\mu] \simeq Z^{\text{tree}}[\vec{J}_\mu] + \tfrac{1}{2} i \operatorname{Tr} \ln \frac{\delta^2 S_\alpha[\vec{A}^0_\mu]}{\delta \vec{A}_\mu \delta \vec{A}_\nu} - i \operatorname{Tr} \ln \left(1 - g\vec{t}\cdot \vec{A}^0_\mu \partial^\mu \frac{1}{\partial^2}\right)$$

and

$$\Gamma[\vec{B}_\mu] \simeq S_\alpha[\vec{A}^0_\mu] + \int d^4x \frac{\delta S_\alpha[\vec{A}^0_\mu]}{\delta \vec{A}^0_\mu(x)} [\vec{B}_\mu(x) - \vec{A}^0_\mu(x)] + \tfrac{1}{2} i \operatorname{Tr}\ln\{\delta^2 S_\alpha[\vec{B}_\mu]/\delta\vec{B}_\mu \delta\vec{B}_\nu\} - i \operatorname{Tr}\ln(1 - g\vec{t}\cdot\vec{B}^\mu \partial_\mu/\partial^2)$$

$$= S_\alpha[\vec{B}_\mu] + \tfrac{1}{2} i \operatorname{Tr}\ln\{\delta^2 S_\alpha[\vec{B}_\mu]/\delta\vec{B}_\mu \delta\vec{B}_\nu\} - i \operatorname{Tr}\ln(1 - g\vec{t}\cdot\vec{B}_\mu \partial^\mu/\partial^2).$$

APPENDIX C: WARD-TAKAHASHI IDENTITY FOR THE GENERATING FUNCTIONAL OF PROPER VERTICES

We begin by rewriting Eq. (3.8) as

$$D^\lambda_x[i\delta/\delta\vec{J}]^{ab}\left[\frac{i}{\alpha}\partial_\lambda\partial_\mu \frac{\delta}{\delta J^b_\mu(x)} - J^b_\lambda(x)\right]W - ig\int d^4y \operatorname{tr} t^a[\partial_\mu H(x,y;i\delta/\delta\vec{J})]D^\mu_y[i\delta/\delta\vec{J}]\,\delta^4(x-y)W = 0. \tag{C1}$$

The last term on the left-hand side may be written as

$$-igt^{bac}[\partial_\mu H^{cd}(x,y;i\delta/\delta\vec{J})]_{x=y}(-ig)t^{deb}\frac{\delta W}{\delta J^e_\mu(x)} + igt^{bac}[\partial^2 H^{cb}(x,y;i\delta/\delta\vec{J})/\partial x_\mu \partial y^\mu]_{x=y} W. \tag{C2}$$

Noting that

$$(\partial^2 - g\partial^\nu \vec{A}_\mu\cdot\vec{t})^{ab}_y H^{cb}(x,y;\vec{A}) = \delta^{ac}\delta^4(x-y)$$

or

$$H^{cb}(x,y;\vec{A}) = -\delta^{cb}\tilde{D}_F(x-y) - gt^{bad}\int d^4z\, \partial^\nu_y \tilde{D}_F(y-z)A^a_\nu(z)H^{cd}(x,y;\vec{A}), \tag{C3}$$

we can cast the second term of Eq. (C2) into

$$igt^{bac}[\partial^2 H^{cb}(x,y;i\delta/\delta\vec{J})/\partial x_\mu \partial y^\mu]_{x=y} W = -(ig)^2 t^{bac}t^{bed}\int d^4z\, \partial^\mu_y \partial^\nu_y \tilde{D}_F(y-z)\partial_\mu H^{cd}(x,y;\vec{A})\Big|_{x=y}\frac{\delta W}{\delta J^e_\nu(z)}.$$

Thus the last term on the left-hand side of Eq. (C1) can be written as

$$-g^2 t^{bac}t^{deb}\int d^4z\, \partial^\mu_x H^{cd}(x,y;i\delta/\delta\vec{J})g^{\text{tr}}_{\mu\nu}(x-z)\frac{\delta W}{\delta J^e_\nu(z)}\Big|_{x=y},$$

where

$$g^{\text{tr}}_{\mu\nu}(x-y) = g_{\mu\nu}\delta^4(x-y) + \partial_\mu\partial_\nu \tilde{D}_F(x-y).$$

So finally, we obtain

$$\frac{i}{\alpha}\partial^2\partial_\mu \frac{\delta W}{\delta J^a_\mu(x)} + \frac{1}{\alpha}t^{abc}\frac{\delta}{\delta J^b_\lambda(x)}\partial_\lambda\partial_\mu\frac{\delta W}{\delta J^c_\mu(x)} - \partial^\lambda J^a_\lambda(x)W - it^{abc}J^b_\lambda(x)\delta W/\delta J^c_\lambda(x)$$
$$+ g^2 t^{abc}t^{bde}\int d^4z\, \partial^\mu_x H^{cd}(x,y;i\delta/\delta\vec{J})g^{\text{tr}}_{\mu\nu}(x-z)\frac{\delta W}{\delta J^e_\nu(z)}\Big|_{x=y} = 0. \tag{C4}$$

Equation (C4) may be translated into an expression involving the generating functional of the proper vertices. We recall that

$$W[\vec{J}_\mu] = \exp iZ[\vec{J}_\mu] \tag{B1}$$

and

$$-\vec{B}_\mu = \frac{\delta Z[\vec{J}_\mu]}{\delta\vec{J}_\mu(x)};\quad \vec{J}_\mu = \frac{\delta\Gamma[\vec{B}_\mu]}{\delta\vec{B}_\mu}. \tag{B3}$$

Thus

$$\frac{1}{\alpha}\partial^2\partial^\mu B^a_\mu(x) - \frac{1}{\alpha}t^{abc}B^b_\lambda(x)\partial^\lambda\partial^\mu B^c_\mu(x) - \partial^\lambda \frac{\delta\Gamma[B_\mu]}{\delta B^a_\lambda(x)} + t^{abc}B^b_\lambda(x)\frac{\delta\Gamma[B_\mu]}{\delta B^c_\lambda(x)}$$
$$= \frac{1}{\alpha}t^{abc}\frac{\delta}{\delta J^b_\lambda(x)}\partial_\lambda\partial^\mu B^c_\mu(x) + ig^2 t^{abc}t^{bde}\int d^4z\,\partial^\mu_x H^{cd}(x,y;\vec{\mathrm{B}} - i\delta/\delta\vec{\mathrm{J}})g^{\mathrm{tr}}_{\mu\nu}(x-z)B^{e\,\nu}(z)\bigg|_{x=y} \equiv -G^0. \tag{C5}$$

This equation is somewhat simplified if we define Γ^0 by

$$\Gamma^0[\vec{\mathrm{B}}_\mu] = \Gamma[\vec{\mathrm{B}}_\mu] + \frac{1}{2\alpha}\int d^4x[\partial^\mu\vec{\mathrm{B}}_\mu(x)]^2, \tag{C6}$$

which satisfies

$$[\partial_\lambda\delta^{ac} - t^{abc}B^b_\lambda]\frac{\delta\Gamma^0[\vec{\mathrm{B}}_\mu]}{\delta B^c_\lambda(x)} = G^a(x). \tag{C7}$$

If $G^a(x)$ were identically zero, then Eq. (C7) would imply that $\Gamma^0[\vec{\mathrm{B}}_\mu]$ is invariant under the local gauge transformation

$$\vec{\mathrm{B}}_\mu \to \vec{\mathrm{B}}_\mu + D_\mu[\vec{\mathrm{B}}_\mu]\vec{w}.$$

By an explicit computation of $\delta^2 G^a(x)/\delta B^b_\mu(y)\delta B^c_\nu(z)$, we have verified that $G^a(x)$ cannot be identically zero, however.

APPENDIX D: POWER COUNTING

Let $N_{n,m}$ be the number of vertices of the form $\partial^n\Phi^m$, i.e., m-point vertices with n derivatives, and let I and E be, respectively, the number of internal and external lines in a proper diagram. By L we denote the number of loops in the diagram.

We have two topological relations:

$$E + 2I = \sum_{n,m} mN_{n,m}, \tag{D1}$$

$$L = I + 1 - \sum_{n,m} N_{n,m}. \tag{D2}$$

The superficial degree of divergence D of the diagram is given by

$$D = \sum_{n,m} nN_{n,m} + 4L - 2rI, \tag{D3}$$

where the propagator is assumed to have the asymptotic behavior $(p^2)^{-r}$. Eliminating L and I in favor of E by the use of Eqs. (D1) and (D2), we write Eq. (D3) as

$$D = E(n-2) + 4 + \sum N_{n,m}[n + (2-r)m - 4]. \tag{D4}$$

For the gauge-invariant terms discussed in Sec. V, we have generally

$$n + m = 2 + 2r. \tag{D5}$$

We, therefore, have

$$D = E(r-2) + 4 + (1-r)\sum_{n,m} N_{n,m}(m-2). \tag{D6}$$

The two statements in Sec. V, for which we referred the reader to this Appendix, can be justified on the basis of Eq. (D6).

*On leave of absence from SPT, CEN, Saclay, B.P. 2, 91 Gif-sur-Yvette, France.

[1]J. Schwinger, Ann. Phys. (N.Y.) 2, 407 (1957).
[2]S. L. Glashow, Nucl. Phys. 22, 579 (1961).
[3]A. Salam and J. Ward, Phys. Letters 13, 168 (1964).
[4]S. Weinberg, Phys. Rev. Letters 13, 1264 (1967).
[5]S. Salam, in *Elementary Particle Theory*, edited by N. Svartholm (Almquist and Forlag A. B., Stockholm, 1968).
[6]J. Schechter and Y. Ueda, Phys. Rev. D 2, 736 (1970).
[7]G. 't Hooft, Nucl. Phys. B35, 167 (1971).
[8]B. W. Lee, Phys. Rev. D 5, 823 (1972).
[9]S. Weinberg, Phys. Rev. Letters 27, 1688 (1971).
[10]P. Higgs, Physics 12, 132 (1966).

[11]T. W. B. Kibble, Phys. Rev. 155, 1554 (1967).
[12]Further references can be found in Ref. 8.
[13]B. W. Lee, Nucl. Phys. B9, 649 (1969); J.-L. Gervais and B. W. Lee, *ibid.* B12, 627 (1969); B. W. Lee, Chiral Dynamics (Gordon and Breach, New York, to be published).
[14]N. N. Bogoliubov and O. S. Parasiuk, Acta. Math. 97, 227 (1957).
[15]N. N. Bogoliubov and D. B. Shirkov, *Introduction to the Theory of Quantized Fields* (Interscience, New York, 1959).
[16]K. Hepp, Commun. Math. Phys. 1, 95 (1965); *Théories de la Renormalisation* (Springer, Berlin, 1969).
[17]W. Zimmermann, in *Brandeis University Summer Institute in Theoretical Physics*, edited by M. Chretien *et al.* (MIT Press, Cambridge, Mass., 1970).
[18]A. Slavnov, Kiev Report No. ITP-71-83E (unpublished).
[19]R. P. Feynman, Acta. Phys. Polon. 26, 697 (1963); (unpublished).
[20]B. DeWitt, Phys. Rev., 162, 1195 (1967); 162, 1239 (1967).
[21]L. D. Faddeev and V. N. Popov, Phys. Letters 25B, 29 (1967); Kiev Report No. ITP-67-36 (unpublished).
[22]G. 't Hooft, Nucl. Phys. B33, 173 (1971).
[23]A. Slavnov, Nucl. Phys. B31, 301 (1971).
[23a]R. Jackiw and L. D. Faddeev (private communication).
[24]A. Slavnov, Kiev Report No. ITP-71-131E (unpublished).
[25]K. Johnson (private communication).
[26]K. Symanzik, Lett. Nuovo Cimento 2, 10 (1969); Commun. Math. Phys. 16, 48 (1970).
[27]G. Jona-Lasinio, Nuovo Cimento 34, 1790 (1964).

Spontaneously Broken Gauge Symmetries. II. Perturbation Theory and Renormalization

Benjamin W. Lee
National Accelerator Laboratory, P. O. Box 500, Batavia, Illinois 60510
and Institute for Theoretical Physics, State University of New York, Stony Brook, New York 11790

and

Jean Zinn-Justin*
Institute for Theoretical Physics, State University of New York, Stony Brook, New York 11790
(Received 10 March 1972)

The second paper in this series is devoted to the formulation of a renormalizable perturbation theory of Higgs phenomena (spontaneously broken gauge theories). In Sec. II, we reformulate the renormalization prescription for massless Yang-Mills theories in terms of gauge-invariant renormalization counterterms in the action. Section III gives a group-theoretic discussion of Higgs phenomena. We discuss the possibility that an asymmetric vacuum is stable, and show how the symmetry of the physical vacuum determines the mass spectrum of the gauge bosons. We show further that in a special gauge (U gauge), all unphysical fields can be eliminated. Section IV discusses the quantization of a spontaneously broken gauge theory in the R gauge, where, as we show in Sec. V, Green's functions are made finite by the renormalization counterterms of the symmetric theory (in which the gauge invariance is not spontaneously broken). The R-gauge formulation makes use of redundant fields for the sake of renormalizability. Section VI is a discussion of the low-energy limits of propagators in the R-gauge formulation. In Sec. VII we show that the particles associated with redundant fields peculiar to the R-gauge formulation are unphysical, i.e., they do not contribute to the sum over intermediate states.

I. INTRODUCTION

In this paper we give a renormalization method and a proof of finiteness of renormalized Green's functions of spontaneously broken gauge theories. For definiteness we consider a very simple model in which SU(2) gauge bosons are coupled to a triplet of scalar mesons. There is an extra complication when chiral fermions are included in the model, as pointed out by Veltman,[1] and more recently by Gross and Jackiw.[2] This difficulty can be circumvented in a realistic model of electromagnetic and weak interactions. We shall not discuss this problem further in this paper, but postpone the discussion until we deal with the renormalizability of a realistic theory in a sequel to this paper.

We give in this paper a method of renormalization which is based on the observation that, in a spontaneously broken symmetry theory, divergences in Feynman integrals can be classified according to, and identified with, those of a comparison theory in which the symmetry is not broken.[3] This method is successfully used for the σ model and we borrow many concepts and techniques from that study.

Let us summarize the contents of this paper. In Sec. II, we give a brief recapitulation of the results of the first paper on the renormalization of a massless Yang-Mills theory. We write down explicitly the effective action in terms of renormalized fields and gauge-invariant counterterms. The renormalized version of the Ward-Takahashi identity for the generating functional of renormalized Green's functions is recorded. The reader who is not particularly interested in the details may be able to gather enough background for the subsequent discussions by studying Secs. II and V of the previous paper concurrently with this section.

Section III is a discussion of group theory of Higgs phenomena. To a large extent, this section is a review of Kibble's work.[4] The discussion here is carried out in the context of classical field theory. We show how the instability of the symmetric vacuum arises, and how the symmetry of the physical vacuum determines the mass spectrum of gauge bosons. The study culminates in a theorem which shows which gauge bosons become massive in a spontaneously broken gauge theory. The theorem is an analog of that due to Bludman and Klein,[5] which shows in what quantum channels Goldstone bosons appear in a spontaneously broken symmetry theory.

There exists a special choice of gauge in which Goldstone boson fields combine with gauge boson fields to form massive vector fields with three degrees of polarizations. This is the gauge used by Kibble[4] in his discussion of Higgs phenomena. In this gauge, there are no redundant fields and the physical interpretation of the theory is straightforward. We shall call this gauge the U gauge (unitary gauge). Unfortunately the renormalizability of the U-gauge formulation is not obvious, even though indications are that the T matrix in this formulation is renormalizable.[6,7]

In Sec. IV, we quantize the simple model mentioned at the beginning in a class of gauges, which includes, in quantum electrodynamics, the transverse Landau gauge and the Feynman gauge. We shall call these gauges collectively as R gauge (renormalizable gauge). The R-gauge formulation contains redundant field components so that the unitarity of the S matrix is not manifest. As we show in Sec. V, Green's functions in the R-gauge formulation are rendered finite by the renormalization counterterms of the corresponding symmetric theory. Here lies the advantage of this formulation. Since the renormalization counterterms which make the theory finite are gauge invariant, the renormalized Green's functions of a spontaneously broken gauge theory satisfy appropriate Ward-Takahashi identities.

In Sec. VI we discuss the low-energy limits of propagators in the R-gauge formulation.

In Sec. VII we show that renormalized T-matrix elements are independent of the parameter which characterizes a particular R gauge chosen, and the redundant massless scalar fields peculiar to the R-gauge formulation are unphysical, i.e., they do not contribute to the sum over intermediate states when one computes the absorptive part of T-matrix elements by the Landau-Cutkosky rule.[8,9] These discussions are based on the Ward-Takahashi identities. For the proof of unitarity of the R-gauge formulation, we have identified the set of relations that are needed. The proof is worked out in detail for intermediate states containing one, two, and three such unphysical quanta.

In the sequel we wish to consider the renormalizability aspect of Weinberg's theory of weak and electromagnetic interactions in detail and the equivalence of the S matrix in the U and R gauges.

II. GAUGE-INVARIANT COUNTERTERMS

As Bogoliubov and Shirkov[10] have shown, the R operation can be formally implemented by the inclusion of counterterms in the Lagrangian. The discussion in the previous paper implies that these counterterms are themselves gauge-invariant. We can in fact reexpress the effective action (I2. 10) (where the prefix I refers to the equations of paper I) in terms of the renormalized field $\vec{A}_r^\mu$ and the renormalized coupling constant g_r,

$$\vec{A}^\mu = Z_3^{1/2}\vec{A}_r^\mu ,$$

$$g = g_r Z_1/Z_3^{3/2}$$

and making explicit the renormalization counterterms. With

$$\alpha = \alpha_r Z_3 ,$$

we write

$$\int d^4x\{-\tfrac{1}{4}(\partial^\mu\vec{A}_r^\nu-\partial^\nu\vec{A}_r^\mu-g_r\vec{A}_r^\mu\times\vec{A}_r^\nu)^2-(1/2\alpha_r)(\partial_\mu\vec{A}_r^\mu)^2-\tfrac{1}{4}(Z_3-1)(\partial^\mu\vec{A}_r^\nu-\partial^\nu\vec{A}_r^\mu)^2$$
$$+\tfrac{1}{2}g_r(Z_1-1)\vec{A}_{r\mu}\times\vec{A}_{r\nu}\cdot(\partial^\mu\vec{A}_r^\nu-\partial^\nu\vec{A}_r^\mu)-\tfrac{1}{4}g_r^2(Z_1^2/Z_3-1)(\vec{A}_r^\mu\times\vec{A}_r^\nu)^2\}$$
$$-i\,\mathrm{Tr}\ln(1-g_r\vec{t}\cdot\vec{A}_r^\mu\partial_\mu/\partial^2)-i\,\mathrm{Tr}\ln\left\{1+\frac{1}{\partial^2-g_r\vec{t}\cdot\vec{A}_r^\mu\partial_\mu}[(\tilde{Z}_3-1)\partial^2-g_r(\tilde{Z}_1-1)\vec{t}\cdot\vec{A}^\mu\partial_\mu]\right\}, \tag{2.1}$$

with

$$Z_1/Z_3=\tilde{Z}_1/\tilde{Z}_3 \tag{2.2}$$

which is a restatement of Eq. (I 6.4). We may choose Z_3 and $\tilde{Z}_3$ such that

$$\lim_{k^2\to -a^2}[\Delta_{\mu\nu}(k)]_r=\left(g_{\mu\nu}+\frac{k_\mu k_\nu}{a^2}\right)\left(\frac{1}{-a^2}\right)+\text{gauge-dependent terms}, \tag{2.3}$$

$$[\mathcal{G}(-a^2)]_r=-1/a^2 \tag{2.4}$$

and Z_1, so that

$$\lim_{p^2=q^2=r^2=a^2} i\Gamma^{abc}_{\lambda\mu\nu}(p,q,r)=\epsilon^{abc}\{[(p-q)_\nu g_{\lambda\mu}+(q-r)_\lambda g_{\mu\nu}+(r-p)_\mu g_{\nu\lambda}]+\cdots\}, \tag{2.5}$$

as we described in Eq. (I 6.7).

Clearly the construction of Eq. (2.1) can be extended when there are matter fields present in the Lagrangian. The part that has to do with the gauge invariance, for the triplet of scalar fields discussed in paper I, is

$$\tfrac{1}{2}(\partial_\mu\vec{\phi}_r-g_r\vec{A}_{r\mu}\times\vec{\phi}_r)^2+\tfrac{1}{2}(Z_2-1)(\partial_\mu\vec{\phi}_r)^2+g_r\left[Z_1\left(\frac{Z_2}{Z_3}\right)-1\right]\vec{A}_r^\mu\cdot(\vec{\phi}_r\times\partial_\mu\vec{\phi}_r)+\tfrac{1}{2}g_r^2\left[\frac{Z_1^2}{Z_3}\left(\frac{Z_2}{Z_3}\right)-1\right](\vec{A}_r^\mu\times\vec{\phi}_r)^2, \tag{2.6}$$

where Z_2 may be chosen to ensure the normalization condition for the scalar propagator, Eq. (I 7.15),

$$\lim_{k^2\to -a^2}[\Delta^{-1}(k^2)]_r\sim k^2+a^2-M^2. \tag{2.7}$$

It is perhaps useful to rephrase the Bogoliubov-Parasiuk-Hepp (BPH) renormalization procedure in terms of the Lagrangian of (2.1) and (2.6). First we include the regulator term (I 5.7) and other regulator terms in the Lagrangian. Feynman integrals are now finite and we can choose the renormalization constants, Z's, which depend on the cutoff Λ^2, in such a way that the renormalization conditions (2.2)–(2.5) and (2.7) are satisfied. As $\Lambda^2\to\infty$, the renormalized Feynman amplitudes are well defined and finite.

If we make the scale change

$$\vec{J}^\mu=Z_3^{-1/2}\vec{J}_r^\mu,$$
$$\vec{K}=Z_2^{-1/2}\vec{K}_r$$

in the definition of the generating functional of Green's functions, then functional derivatives of Z with respect to the renormalized sources are the renormalized Green's functions. The Ward-Takahashi identity (3.13) may be written in terms of renormalized quantities:

$$\frac{i}{\alpha_r}\partial_\mu\frac{\epsilon W}{\delta J^a_{r\mu}(x)}+\partial^\mu J^a_{r\mu}(x)W-ig_r\tilde{Z}_1\int d^4y\,d^4z\left[J_r^{c\mu}(y)g^{\mathrm{tr}}_{\mu\nu}(y-z)t^{cbd}\frac{\delta}{\delta J^b_{r\nu}(z)}\right]G_r^{da}(z,x;i\delta/\delta\vec{J}_r)W=0, \tag{2.8}$$

where

$$g^{\mathrm{tr}}_{\mu\nu}(x-y)=g_{\mu\nu}\delta^4(x-y)+\partial_\mu\partial_\nu\tilde{D}_F(x-y)$$

and

$$G(x,y;i\delta/\delta\vec{J})=\tilde{Z}_3G_r(x,y;i\delta/\delta\vec{J}_r),$$
$$G_r(x,y;i\delta/\delta\vec{J}_r)=\left\langle x\left|\left[\partial^2-ig_r\vec{t}\cdot\partial_\mu\frac{\delta}{\delta\vec{J}_{r\mu}}+\left((\tilde{Z}_3-1)\partial^2-ig_r(\tilde{Z}_1-1)\vec{t}\cdot\partial_\mu\frac{\delta}{\delta\vec{J}_{r\mu}}\right)\right]^{-1}\right|y\right\rangle. \tag{2.9}$$

III. GROUP THEORY OF HIGGS PHENOMENA

We will describe here the Higgs phenomenon[11,4] in the context of classical (nonquantized) field theory. Alternatively, one may interpret the following discussion as applying to the tree approximation to quantum field theory. The following discussion is essentially a review of Kibble's work.[4] We include it here, mainly to make this paper self-contained and to establish notations, terminology and concepts. For simplicity, we shall consider the system of gauge bosons interacting with scalar mesons.

Let G be the local gauge symmetry (compact, but not necessarily semisimple) of the Lagrangian. We denote by $\{L_a\}$ the set of generators of the group G. The Yang-Mills gauge bosons $\{A_a^\mu\}$ belong to the adjoint representation of the group G, so they can be put in one-to-one correspondence with the generators $\{L_a\}$. We assume that there are scalar multiplets $\phi^{(\alpha)}$ of dimensionalities n_α,

$$\phi^{(\alpha)}=\begin{pmatrix}\phi_1^{(\alpha)}\\ \cdot\\ \cdot\\ \cdot\\ \phi_{n_\alpha}^{(\alpha)}\end{pmatrix}. \tag{3.1}$$

The multiplet $\phi^{(\alpha)}$ transforms like an irreducible representation of the group G. We denote by $\{L^{(\alpha)}\}$ the matrix representation of the generators. The renormalizable Lagrangian in which the gauge bosons are coupled in the minimal way is of the form

$$\mathcal{L}=\sum_\alpha (D_\mu\phi^{(\alpha)})^\dagger\cdot(D^\mu\phi^{(\alpha)}) - \tfrac{1}{4}\sum_a(\partial^\mu A_a^\nu-\partial^\nu A_a^\mu-gf_{abc}A_b^\mu A_c^\nu)^2-V(\phi), \tag{3.2}$$

where D_μ is the covariant derivative

$$D_\mu=\partial_\mu+ig\vec{L}^{(\alpha)}\cdot\vec{A}_\mu, \tag{3.3}$$

f_{abc} is the structure constant,

$$[L_a, L_b]=if_{abc}L_c, \tag{3.4}$$

and $V(\phi)$ is an invariant polynomial in the $\phi^{(\alpha)}$, which is at most quartic in the scalar fields. The Lagrangian (3.2) is invariant under the local gauge transformation

$$\begin{aligned}\phi^{(\alpha)}&\to e^{i\vec{L}^{(\alpha)}\cdot\vec{\omega}}\phi^{(\alpha)},\\ \vec{A}_\mu\cdot\vec{L}&\to e^{i\vec{L}\cdot\vec{\omega}}\vec{A}_\mu\cdot\vec{L}e^{-i\vec{L}\cdot\vec{\omega}}\\ &\quad-\frac{i}{g}(\partial_\mu e^{i\vec{L}\cdot\vec{\omega}})e^{-i\vec{L}\cdot\vec{\omega}},\end{aligned} \tag{3.5}$$

where ω_a is a function of space-time.

The vacuum expectation values of the scalar fields $\phi^{(\alpha)}\equiv v^{(\alpha)}$ are determined by the conditions

$$\left.\frac{\delta V(\phi)}{\delta\phi_i^{(\alpha)}}\right|_{\phi=v}=0, \tag{3.6}$$

$$\left.\frac{\delta^2 V(\phi)}{\delta\phi_i^{(\alpha)}\delta\phi_j^{(\beta)}}\right|_{\phi=v}\geqslant 0. \tag{3.7}$$

The second condition (3.8) is necessary in order that the physical masses of the scalar particles be non-negative. The solutions of Eqs. (3.6) and (3.7),

$$\phi^{(\alpha)}=v^{(\alpha)}, \tag{3.8}$$

may be null vectors, in which case the vacuum is invariant under G. It may be that the minimum of V occurs at some finite $v^{(\alpha)}$. Let $\{l\}$ be the subset of $\{L\}$ which map all of $v^{(\alpha)}$'s to null vectors:

$$l_i^{(\alpha)}v^{(\alpha)}=0. \tag{3.9}$$

Then the set $\{l\}$ generates a subgroup S of G. We call S the little group of the vacuum.

The nature of the little group S depends on the polynomial $V(\phi)$. We give some examples below.

Example 1. Let

$$V(\phi)=\tfrac{1}{2}\mu^2\phi^2+\tfrac{1}{4}\lambda(\phi^2)^2,$$

where ϕ is an n-dimensional real vector. The group G of invariance is $O(n)$. The parameter λ has to be $\geqslant 0$ in order that $|\phi|$ is bounded, or the Hamiltonian is positive definite. If $\mu^2\geqslant 0$, the minimum of $V(\phi)$ occurs at $\phi=0$ and the little group S is equal to $O(n)$. If $\mu^2<0$, the minimum lies in the orbit $|\phi|^2=-\mu^2/\lambda$. Because of the invariance of $V(\phi)$ under $O(n)$ we can always put v in the standard form

$$v=\begin{pmatrix}(-\mu^2/\lambda)^{1/2}\\ 0\\ \cdot\\ \cdot\\ 0\end{pmatrix}.$$

The little group of the vacuum is $O(n-1)$.

Example 2. Let

$$M_\alpha^\beta=\sum_{i=0}^{8}(\lambda_i)_{\alpha\beta}(s^i+ip^i),$$

where λ_i, $i=0,\ldots,8$ are Gell-Mann's 3×3 matrices with $\lambda_0=(2/3)^{1/2}\,\underline{1}$ and $\alpha,\beta=1,2,3$. s and p are nonets of scalar and pseudoscalar fields. We consider

$$V(s,p)=\alpha\,\mathrm{Tr}(MM^\dagger)^2+\beta[\mathrm{Tr}(MM^\dagger)]^2+\gamma(\det M+\det M^\dagger)+\delta\,\mathrm{Tr}(MM^\dagger),$$

which is SU(3)×SU(3) invariant. Let us concentrate on the case in which parity is conserved, so that the minimum V lies on the hyperplane $p_i = 0$, $i = 0, \ldots, 8$. Let us assume that the minimum occurs at

$$M = M^\dagger = v\,,$$

where v is a 3×3 Hermitian matrix. We can diagonalize v by an SU(3) transformation so v takes the form

$$v = \begin{pmatrix} a & & \\ & b & \\ & & c \end{pmatrix}.$$

Equation (3.6) then demands that

$$4\alpha a^3 + 4\beta a(a^2 + b^2 + c^2) + 2\gamma bc + 2\delta a = 0$$

and two more equations obtained from the above by cyclic permutations of a, b, and c. The three equations imply that the three eigenvalues a, b, c cannot be all unequal. Therefore, the little group S cannot be smaller than SU(2).

When $\phi^{(\alpha)}$'s have nonvanishing vacuum expectation values we can perform nonlinear canonical transformations on $\phi^{(\alpha)}$'s and eliminate a certain number of field components from $V(\phi)$. Let the dimensionalities of G and S be N and M, respectively. There are, then, $m = N - M$ generators, $\{t\}$ of G, which span the cosets $S^{-1}G$:

$$\{l\} + \{t\} = \{L\}\,. \tag{3.10}$$

We may choose the generators to be orthonormal with respect to the Cartan inner product. Let us write

$$\phi^{(\alpha)} = D^{(\alpha)}[e^{i\vec{\xi}\cdot\vec{t}}](v^{(\alpha)} + \rho^{(\alpha)})\,, \tag{3.11}$$

where $\vec{\xi}$ has m components and choose $\rho^{(\alpha)}$'s, such that the mapping

$$\phi^{(\alpha)} \to (\vec{\xi}, \rho^{(\alpha)})$$

is canonical. [A nonlinear mapping $\phi_i \to \rho_j(\{\phi_i\})$ is called canonical if $(\delta\phi_i/\delta\rho_j)|_{\rho=0}$ is a nonsingular matrix.] Both ξ and $\rho^{(\alpha)}$'s have null vacuum expectation values. The collection of $\rho^{(\alpha)}$'s will have $[(\sum_\alpha n_\alpha) - m]$ components. Clearly, $V(\phi)$ is independent of the fields $\vec{\xi}$ since the invariance of V under G implies $V(\phi) = V(v + \rho)$. If there were no gauge bosons, the Lagrangian would depend on $\vec{\xi}$ only through $\partial_\mu\vec{\xi}$, arising from the terms $(\partial_\mu\phi^{(\alpha)})^\dagger \times (\partial^\mu\phi^{(\alpha)})$ in the Lagrangian. Consequently, the fields $\vec{\xi}$ would represent massless scalar particles, coupled to other particles gradiently. They would be the Goldstone fields.

When the theory is invariant under local gauge transformations, the $\vec{\xi}$ fields can be eliminated from the Lagrangian completely. We define the vector fields B_μ^a by

$$\vec{L}\cdot\vec{A}_\mu = e^{i\vec{\xi}\cdot\vec{t}}\,\vec{L}\cdot\vec{B}_\mu e^{-i\vec{\xi}\cdot\vec{t}} - \frac{i}{g}(\partial_\mu e^{i\vec{\xi}\cdot\vec{t}})e^{-i\vec{\xi}\cdot\vec{t}}\,. \tag{3.12}$$

The mapping $(A_\mu, \phi^{(\alpha)}) \to (B_\mu, \rho^{(\alpha)})$ expressed in Eqs. (3.11) and (3.12) is a gauge transformation (3.5) which leaves the Lagrangian (3.2) invariant. We have

$$\mathcal{L} = \sum_\alpha [\Delta_\mu(v^{(\alpha)} + \rho^{(\alpha)})]^\dagger \cdot [\Delta^\mu(v^{(\alpha)} + \rho^{(\alpha)})] - \tfrac{1}{4}\sum_a (\partial^\mu B_a^\nu - \partial^\nu B_a^\mu - g f_{abc} B_b^\mu B_c^\nu)^2 - V(v + \rho)\,. \tag{3.13}$$

Here Δ_μ stands for

$$\Delta_\mu = \partial_\mu + gi\vec{L}^{(\alpha)}\cdot\vec{B}_\mu\,. \tag{3.14}$$

Some of the gauge bosons are no longer massless. As the vector-meson mass term, we have

$$g^2 \sum_\alpha (v^{(\alpha)}, L_a^\dagger L_b v^{(\alpha)}) B_\mu^a B_\nu^b g^{\mu\nu}$$

so that the vector-meson mass matrix is given by

$$(M^2)_{ab} = 2g^2 \sum_\alpha (v^{(\alpha)}, L_a^\dagger L_b v^{(\alpha)})\,. \tag{3.15}$$

It is convenient to adopt the following convention: We order L_a's so that L_a, $a = 1, 2, \ldots, M$ form the set $\{l\}$. We see from Eq. (3.15) that M^2 is block diagonal, the upper $M\times M$ diagonal matrix being zero. The lower $m \times m$ matrix is positive definite (the lower matrix cannot have a null eigenvalue, for if it did, the little group would have a dimension larger than M).

Let us summarize the result of this section in a theorem [Kibble's theorem[4]]: Let G be the gauge symmetry of the Lagrangian and S, $G \supset S$, be the little group of the vacuum. The generators $\{L\}$ of G can be divided into two sets, the generators $\{l\}$ of S and the rest $\{t\}$. The gauge bosons corresponding to $\{l\}$ are massless. The gauge bosons corresponding to $\{t\}$ are massive. This theorem is an analog of that of Bludman and Klein[5] to spontaneously broken gauge theories.

If the symmetry is not spontaneously broken, i.e., $G = S$, the gauge bosons are endowed with the two transverse polarizations. If the symmetry is broken, some gauge bosons become massive and have three polarizations. How do the longitudinal components of massive vector bosons come about? We see from Eq. (3.12) that

$$\vec{L}\cdot\vec{B}_\mu = \vec{L}\cdot\vec{A}_\mu - \frac{1}{g}\vec{t}\cdot\partial_\mu\vec{\xi} + O(\vec{\xi}^{\,2}),$$

i.e., the would-be Goldstone fields serve as the longitudinal components of the massive vector bosons.

The discussions given above can be generalized to quantum field theory, if we use the generating functional of proper vertices instead of $\mathcal{L}$ in Eqs. (3.2), (3.6), and (3.7). This was done for the σ model in the last paper of Ref. 3.

IV. QUANTIZATION OF HIGGS PHENOMENA

In the preceding section we disposed of the general group-theoretical problem associated with the Higgs mechanism in the context of classical field theory. We shall now proceed to the quantization problem. To be specific we consider a simple model: SU(2) gauge bosons coupled to an isotriplet of scalar fields. The inclusion of fermions will be discussed in a sequel, when we discuss a more realistic model.

The Lagrangian of this model is, with $\mu^2 < 0$,

$$\mathcal{L} = -\tfrac{1}{4}\vec{F}_{\mu\nu}\cdot\vec{F}^{\mu\nu} + \tfrac{1}{2}(D_\mu\vec{\phi})^2 - \tfrac{1}{2}\mu^2\vec{\phi}^2 - \tfrac{1}{4}\lambda(\vec{\phi}^2)^2 - \tfrac{1}{2}\delta\mu^2\vec{\phi}^2, \tag{4.1}$$

where

$$\begin{aligned} D_\mu &= \partial_\mu - g\vec{A}_\mu\times, \\ \vec{F}_{\mu\nu} &= \partial_\mu\vec{A}_\nu - \partial_\nu\vec{A}_\mu - g\vec{A}_\mu\times\vec{A}_\nu, \end{aligned} \tag{4.2}$$

and $\delta\mu^2$ is the scalar mass counterterm. If μ^2 is positive, we can quantize the theory in the manner described in Sec. II, and choose, for example, $M^2 - a^2 = \mu^2$, where M^2 and a^2 are defined in Eqs. (2.6) and (2.7).

Irrespective of the sign of μ^2, we can write the generating functional of the Green's functions as

$$W = \exp\{iZ[\vec{J}_\mu, \vec{K}]\} = \int [d\vec{A}][d\vec{\phi}]\exp\left[i\left(S_\alpha[\vec{J}_\mu, \vec{K}] + \int d^4x\,[\vec{K}\cdot\vec{\phi} - \vec{J}_\mu\cdot\vec{A}^\mu](x)\right)\right], \tag{4.3}$$

where S_α is the effective action:

$$S_\alpha = \int d^4x\left[\mathcal{L}(x) - \frac{1}{2\alpha}[\partial_\mu\vec{A}^\mu(x)]^2\right] - i\,\mathrm{Tr}\ln\left(1 - g\vec{t}\cdot\vec{A}_\mu\partial^\mu\frac{1}{\partial^2}\right). \tag{4.4}$$

The important fact one should bear in mind is that Eq. (4.3) applies equally well to the broken-symmetry case as it does to the symmetric case, and therefore *the same functional Ward-Takahashi identity* (I 7.1),

$$\frac{i}{\alpha}\partial_\mu\frac{\delta W}{\delta J^a_\mu(x)} - \int d^4y\,J^c_\lambda(y)D^\lambda_y[i\delta/\delta\vec{J}]^{cb}G^{ba}(y,x;i\delta/\delta\vec{J})W + ig\int d^4y\,K^c(y)t^{cdb}\frac{\delta}{\delta K^b(y)}G^{da}(y,x;i\delta/\delta\vec{J})W = 0, \tag{4.5}$$

holds in the broken-symmetry case also.

If we were to write down the Feynman rules for the Lagrangian (4.1) as in the symmetric case, then we would get imaginary masses for scalar bosons. The correct way of generating the perturbation expansion for the generating functional (4.3) is to expand the $V(\phi)$ about its minimum

$$V(\phi) = \tfrac{1}{4}\lambda(\vec{\phi}^2)^2 + \tfrac{1}{2}\mu^2\vec{\phi}^2 \tag{4.6}$$

and define the free Lagrangian as the quadratic part in the new expansion parameters of the Lagrangian. Let

$$\left.\frac{\partial V}{\partial\vec{\phi}}\right|_{\vec{\phi}=\vec{v}} = 0, \quad \left.\frac{\partial^2 V}{\partial\phi_i\partial\phi_j}\right|_{\vec{\phi}=\vec{v}} \geq 0. \tag{4.7}$$

We shall write

$$\vec{v} = v\vec{\eta}, \tag{4.8}$$

where

$$v^2 = -\mu^2/\lambda \tag{4.9}$$

and $\vec{\eta}$ is a unit vector in the isospin space, pointing in the 3 axis, say. We shall denote the components of an isovector transverse to $\vec{\eta}$ by the subscript t: thus

$$(\phi_t)_i = (\delta_{ij} - \eta_i\eta_j)\phi_j\,. \tag{4.10}$$

We shall further define

$$\vec{\eta}\cdot\vec{\phi} = v + \psi, \tag{4.11}$$

$$\vec{\eta}\cdot\vec{A}^\mu = A^\mu . \tag{4.12}$$

We shall insist that v is the vacuum expectation value of the field $\vec{\eta}\cdot\vec{\phi} = \phi_3$, so that

$$\left.\frac{\delta Z}{\delta K_3(x)}\right|_{\vec{J}_\mu = \vec{K} = 0} = v . \tag{4.13}$$

Equation (4.9) should really be thought as defining the part μ^2 of $\mu_0{}^2 = \mu^2 + \delta\mu^2$.

The generating functional (4.3) may be written as

$$W = \int [d\vec{A}_t][dA][d\vec{\phi}_t][d\psi] \exp i \left\{ S^0_\alpha[\vec{A}^\mu_t, A^\mu, \vec{\phi}_t, \psi] + S^I[\vec{A}^\mu_t, A^\mu, \vec{\phi}_t, \psi] + \int d^4x [\vec{K}_t\cdot\vec{\phi}_t + K(v+\psi) - \vec{J}^\mu_t\cdot\vec{A}_{t\mu} - J^\mu A_\mu](x) \right\}, \tag{4.14}$$

where $K = K_3$ and $J^\mu = J^\mu_3$. In Eq. (4.14), S^0_α and S^I are respectively

$$S^0_\alpha = \int d^4x \{\tfrac{1}{2}(\partial_\mu\vec{\phi}_t)^2 + \tfrac{1}{2}(\partial_\mu\psi)^2 - \tfrac{1}{2}(2\lambda v^2)\psi^2 - \tfrac{1}{4}(\partial_\mu\vec{A}_\nu - \partial_\nu\vec{A}_\mu)^2 + \tfrac{1}{2}(gv)^2(\vec{\eta}\times\vec{A}_\mu)^2 + gv\vec{\eta}\cdot(\vec{A}^\mu\times\partial_\mu\vec{\phi}) - (1/2\alpha)(\partial_\mu\vec{A}^\mu)^2\} \tag{4.15}$$

and

$$S^I = \int d^4x \{\tfrac{1}{2}g\vec{A}_\mu\times\vec{A}_\nu\cdot(\partial^\mu\vec{A}^\nu - \partial^\nu\vec{A}^\mu) - \tfrac{1}{4}g^2(\vec{A}_\mu\times\vec{A}_\nu)^2 + g\vec{A}_\mu\cdot(\vec{\phi}\times\partial^\mu\vec{\phi}) + \tfrac{1}{2}g^2(\vec{\phi}\times\vec{A}_\mu)^2 - \tfrac{1}{4}\lambda(\vec{\phi}_t{}^2 + \psi^2)^2 - \lambda v\psi(\psi^2 + \vec{\phi}_t{}^2) - \tfrac{1}{2}\delta\mu^2(\vec{\phi}_t{}^2 + \vec{\psi}^2) - v\delta\mu^2\psi\} - i\,\mathrm{Tr}\ln(1 - g\vec{t}\cdot\vec{A}_\mu\partial^\mu/\partial^2). \tag{4.16}$$

The perturbation expansion for the generating functional (4.14) is obtained from the formula

$$W[\vec{J}_\mu, \vec{K}] = \left[\exp iv \int d^4x K(x)\right] \left\{\exp i S^I\left[\frac{i\delta}{\delta\vec{J}_{t\mu}}, \frac{i\delta}{\delta J_\mu}, -\frac{i\delta}{\delta\vec{K}_t}, -\frac{i\delta}{\delta K}\right]\right\} W_0[\vec{J}_\mu, \vec{K}], \tag{4.17}$$

where

$$W_0[\vec{J}_\mu, \vec{K}] = \int [d\vec{A}][d\vec{\phi}_t][d\psi] \exp i \left\{ S^0_\alpha + \int d^4x [\vec{K}_t\cdot\vec{\phi}_t + K\psi + \vec{J}_\mu\cdot A^\mu](x) \right\}. \tag{4.18}$$

The right-hand side of Eq. (4.18) may be evaluated by the elementary method[12] and yields

$$W_0[\vec{J}_\mu, \vec{K}] = \exp \frac{i}{2} \int d^4x \int d^4y \{\vec{K}_t(z)\cdot\bar{D}_F(x-y)\vec{K}_t(y) + K(x)\bar{\Delta}(x-y; 2\lambda v^2)K(y) - J^\mu(x)\bar{D}^{(\alpha)}_{\mu\nu}(x-y)J^\nu(y) - \vec{J}^\mu_t(x)\cdot\bar{\Delta}^{(\alpha)}_{\mu\nu}(x-y; g^2v^2)\vec{J}^\nu_t(y) + 2\vec{\eta}\cdot\vec{J}^\mu_t\times\bar{\Delta}^{(\alpha)}_\mu(x-y)\vec{K}_t(y)\}, \tag{4.19}$$

where

$$\begin{Bmatrix} \bar{D}_F(x-y) \\ \bar{\Delta}(x-y;\mu^2) \\ \bar{D}^{(\alpha)}_{\mu\nu}(x-y) \\ \bar{\Delta}^{(\alpha)}_{\mu\nu}(x-y;m^2) \\ \bar{\Delta}^{(\alpha)}_\mu(x-y) \end{Bmatrix} = \int \frac{d^4k}{(2\pi)^4} e^{ik\cdot(x-y)} \begin{Bmatrix} \dfrac{1}{k^2+i\epsilon} \\ \dfrac{1}{k^2-\mu^2+i\epsilon} \\ \left[\dfrac{1}{k^2+i\epsilon}\left(g_{\mu\nu} - \dfrac{k_\mu k_\nu}{k^2}\right) + \alpha\dfrac{k_\mu k_\nu}{(k^2)^2}\right] \\ \left[\dfrac{1}{k^2-m^2+i\epsilon}\left(g_{\mu\nu} - \dfrac{k_\mu k_\nu}{k^2}\right) + \alpha\dfrac{k_\mu k_\nu}{(k^2)^2}\right] \\ gvk_\mu(k^2+i\epsilon)^{-2} \end{Bmatrix}. \tag{4.20}$$

Equation (4.17), together with Eq. (4.19), gives the Feynman rules and the Dyson-Wick expansion theorem.

It is convenient to expand the Green's functions in powers of g with g^2v^2 and λv^2 fixed (this implies $\lambda \sim g^2$).

It was shown[3] that such an expansion coincides with the expansion in the number of loops in Feynman diagrams.

The interaction Lagrangian in Eq. (4.16) contains the term linear in ψ

$$-v\delta\mu^2\psi . \tag{4.21}$$

Since ψ is supposed to have no vacuum expectation value,

$$\begin{aligned}\frac{\delta Z}{\delta K_3}\bigg|_{\vec{J}_\mu=\vec{K}=0} - v &= 0 \\ &= \int [d\vec{A}][d\vec{\phi}_t][d\psi]\psi(x)e^{i[S\mathcal{G}+SI]} ,\end{aligned} \tag{4.22}$$

the role of the term (4.21) is to cancel the ψ-to-vacuum diagrams with one or more loops (the so-called tadpole diagrams). Let $ivS(v,\lambda)$ be the sum of the contributions from such diagrams. Then

$$v[S(v,\lambda)-\delta\mu^2]=0 , \tag{4.23}$$

which determines $\delta\mu^2\equiv\mu_0^2+\lambda v^2$. As we shall see, we can express Eq. (4.23) more elegantly:

$$v\Delta_{\phi_t}{}^{-1}(0)=0 , \tag{4.24}$$

where $\Delta_{\phi_t}(k^2)$ is the full propagator for the $\vec{\phi}_t$ field. Equation (4.24) is the mathematical expression for the Goldstone theorem. A detailed consideration shows that $\Delta_{\phi_t}(0)$ does not suffer from infrared divergence. Contributions from intermediate states of two massless particles to the self energy of $\vec{\phi}_t$ are explicitly proportional to k^2 to within logarithm in the Landau gauge for example, so that $\Delta_{\phi_t}(0)$ is finite (see Appendix D).

In the next section, we shall show that Green's functions are finite if we choose $\delta\mu^2$ to satisfy Eq. (4.23) or (4.24), and renormalize fields and sources according to

$$\begin{aligned}(v,\psi,\vec{\phi}_t) &= Z_2{}^{1/2}(v,\psi,\vec{\phi}_t)_r ,\\ A_\mu &= Z_3{}^{1/2}(A_\mu)_r ,\\ \vec{J}_\mu &= Z_3{}^{-1/2}(\vec{J}_\mu)_r ,\\ \vec{K} &= Z_2{}^{-1/2}(\vec{K})_r ,\end{aligned} \tag{4.25}$$

and coupling constants according to

$$\begin{aligned}g &= g_r Z_1/Z_3{}^{3/2} = g_r \tilde{Z}_1/\tilde{Z}_3 Z_3{}^{1/2} ,\\ \lambda &= \lambda_r Z_4/Z_2{}^2 ,\end{aligned} \tag{4.26}$$

where Z_1, Z_2, Z_3, Z_4, $\tilde{Z}_1$, and $\tilde{Z}_3$ are to be chosen to make the symmetric theory (the theory with the same λ and g but with $\mu^2>0$) finite. These renormalizations can be implemented in Eq. (4.14) if we write S_α^0 and S^I in terms of renormalized quantities and add to S^I counterterms. We shall omit the subscript r. The expression S_α^0 remains the same as Eq. (4.15) and

$$\begin{aligned}S^I = {} & \text{right-hand side of Eq. (4.16)}\\ & + \int d^4x\Big\{(Z_2-1)[\tfrac{1}{2}(\partial_\mu\vec{\phi}_t)^2+\tfrac{1}{2}(\partial_\mu\psi)^2] + (Z_1-1)\lambda v^2\psi^2 - (Z_3-1)\tfrac{1}{4}(\partial_\mu\vec{A}_\nu-\partial_\nu\vec{A}_\mu)^2\\ &\quad + \tfrac{1}{2}(gv)^2\Big(\frac{Z_1^2}{Z_3}\frac{Z_2}{Z_3}-1\Big)(\vec{\eta}\times\vec{A}_\mu)^2 + gv\Big(Z_1\frac{Z_2}{Z_3}-1\Big)\vec{\eta}\cdot\vec{A}^\mu\times\partial_\mu\vec{\phi}\\ &\quad + \tfrac{1}{2}g(Z_1-1)\vec{A}_\mu\times\vec{A}_\nu\cdot(\partial^\mu\vec{A}^\nu-\partial^\nu\vec{A}^\mu) - \tfrac{1}{4}g^2\Big(\frac{Z_1^2}{Z_3}-1\Big)(\vec{A}_\mu\times\vec{A}_\nu)^2 + g\Big(Z_1\frac{Z_2}{Z_3}-1\Big)\vec{A}_\mu\cdot(\vec{\phi}\times\partial^\mu\vec{\phi})\\ &\quad + \tfrac{1}{2}g^2\Big(\frac{Z_1^2}{Z_3}\frac{Z_2}{Z_3}-1\Big)(\vec{\phi}\times\vec{A}_\mu)^2 - \tfrac{1}{4}\lambda(Z_4-1)(\vec{\phi}_t{}^2+\psi^2) - \lambda v(Z_4-1)\psi(\vec{\phi}_t{}^2+\psi^2)\\ &\quad - \tfrac{1}{2}\delta\mu^2(Z_2-1)(\vec{\phi}_t{}^2+\psi^2) - v\delta\mu^2(Z_2-1)\psi\Big\}\\ & - i\,\mathrm{Tr}\ln\Big\{1+\frac{1}{\partial^2-g\vec{t}\cdot\vec{A}^\mu\partial_\mu}[(\tilde{Z}_3-1)\partial^2-g(\tilde{Z}_1-1)\vec{t}\cdot\vec{A}^\mu\partial_\mu/\partial^2]\Big\} .\end{aligned} \tag{4.27}$$

V. PROOF OF FINITENESS

The discussion in the previous section may suggest to the alert reader that all we have to do to renormalize the Green's functions of the spontaneously broken gauge theory is to construct the generating functional (4.3) of the *renormalized* Green's functions for $\mu^2>0$, and then continue the resulting functional analytically to $\mu^2<0$. Unfortunately, the Green's functions are not analytic in μ^2 at $\mu^2=0$,[3] so that we need a little bit of machinery to implement the above idea.

Let us set up this machinery. We consider the generating functional of Eq. (4.3) for $\mu^2>0$ and expand the generating functional about $\vec{J}_\mu=0$ and $\vec{K}=\vec{\gamma}$, where $\vec{\gamma}$ is a constant vector in the isospin space. The expansion coefficients are the Green's functions of the theory whose formal action is given by

$$S_\alpha(\vec{\gamma})=\int d^4x\{\mathcal{L}(x)-(1/2\alpha)[\partial^\mu\vec{A}_\mu(x)]^2+\vec{\gamma}\cdot\vec{\phi}(x)\}-i\,\mathrm{Tr}\ln(1-g\vec{t}\cdot\vec{A}_\mu\partial^\mu/\partial^2). \tag{5.1}$$

Of course, the action of Eq. (5.1) does not follow from any local Lagrangian which makes sense. The action (5.1) is just our device of connecting the $\mu^2>0$ and $\mu^2<0$ cases, as we shall see.

The term $\int d^4x\gamma\cdot\phi(x)$ induces a vacuum expectation value of $\phi(x)$. Let

$$v_i(\vec{\gamma})=\delta Z/\delta K_i(x)|_{\vec{J}_\mu=0,\vec{K}=\vec{\gamma}}\ . \tag{5.2}$$

As we shall see $\vec{v}$ and $\vec{\gamma}$ are necessarily parallel, and we write

$$\vec{\gamma}=c\vec{\eta}\ , \tag{5.3}$$

$$\vec{v}(\vec{\gamma})=v_c\vec{\eta}\ . \tag{5.4}$$

We shall now decompose the fields ϕ and A_μ as

$$\vec{\phi}=\vec{\phi}_t+\vec{\eta}(v_c+\psi)\ , \tag{5.5}$$

$$\vec{A}^\mu=\vec{A}_t^\mu+\vec{\eta}A^\mu\ , \tag{5.6}$$

with $\vec{\eta}\cdot\vec{\phi}_t=\vec{\eta}\cdot\vec{A}_t^\mu=0$. The action (5.1) can be written as

$$S_\alpha(\vec{\gamma})=S_\alpha^0(\vec{\gamma})+S^I(\vec{\gamma})\ , \tag{5.7}$$

where

$$S_\alpha^0(\vec{\gamma})=\int d^4x\{\tfrac{1}{2}(\partial_\mu\vec{\phi}_t)^2-\tfrac{1}{2}m^2\vec{\phi}_t{}^2+\tfrac{1}{2}(\partial_\mu\psi)^2-\tfrac{1}{2}(m^2+2\lambda v_c{}^2)\psi^2-\tfrac{1}{4}(\partial_\mu\vec{A}_\nu-\partial_\nu\vec{A}_\mu)^2+\tfrac{1}{2}g^2v_c{}^2(\vec{A}_t^\mu)^2$$
$$+gv_c\vec{\eta}\cdot(\vec{A}^\mu\times\partial_\mu\vec{\phi})-(1/2\alpha)(\partial^\mu\vec{A}_\mu)^2\}\ , \tag{5.7'}$$

with

$$m^2=\mu^2+\lambda v_c{}^2 \tag{5.8}$$

and S^I is given by Eq. (4.16) with $v=v_c$ except that the linear term in ψ should now be written as

$$[c-v_c(m^2+\delta\mu^2)]\psi\ . \tag{5.9}$$

Note that Eqs. (4.15) and (4.16) are recovered from Eqs. (5.7)–(5.9) as we let $c=0$ and $m^2=0$. Again, the role of the term (5.9) is to cancel the ψ-to-vacuum diagrams with one or more loops. Let iv_cS_c be the sum of the contributions from such diagrams. Then

$$v_c(m^2+\delta\mu^2-S_c)=c. \tag{5.10}$$

We will now give a brief summary of the ensuing argument. We will first show that the Green's functions for the action (5.1) are renormalized by the counterterms of the symmetric theory ($\mu^2>0$, $c=0$). We shall then show that the renormalized Green's functions of the spontaneously broken gauge theory ($\mu^2<0$, $c=0$) are obtained from those of the action (5.1) in the limit $c=0$, $m^2=0$. We shall make precise the meaning of this limit in due course. In the course of our discussion, it is important to note whether μ^2 or m^2 is kept fixed.

Following Jona-Lasinio, we will introduce the generating functional Γ of the proper (i.e., single-particle irreducible) vertices. First define

$$\Phi_i(x)=\delta Z/\delta K_i(x) \tag{5.11}$$

and

$$-g^{\mu\nu}\mathfrak{A}^a_\nu(x) = \delta Z/\delta J^a_\mu(x). \tag{5.12}$$

The generating functional Γ is obtained from Z by a Legendre transformation:

$$\Gamma[\vec{\mathfrak{A}}_\mu, \vec{\Phi}] = Z[\vec{J}_\mu, \vec{K}] - \int d^4x[\vec{K}\cdot\vec{\Phi} - \vec{J}^\mu\cdot\vec{\mathfrak{A}}_\mu](x). \tag{5.13}$$

We have the Maxwell equations dual to Eqs. (5.11) and (5.12):

$$-K_i(x) = \delta\Gamma/\delta\Phi_i(x) \tag{5.14}$$

and

$$g^{\mu\nu}J^a_\nu(x) = \delta\Gamma/\delta\mathfrak{A}^a_\mu(x). \tag{5.15}$$

In particular, we obtain

$$\vec{\Phi}(x)|_{\vec{J}_\mu=0,\ \vec{K}=\vec{\gamma}} = \vec{v}(\vec{\gamma}) = v_c\vec{\eta} \tag{5.16}$$

and its dual

$$-\gamma_i = \delta\Gamma/\delta\Phi_i|_{\vec{\mathfrak{A}}_\mu=0,\ \vec{\Phi}=\vec{v}(\vec{\gamma})}. \tag{5.17}$$

According to the analysis of Jona-Lasinio, the expansion coefficients of Γ about $\mathfrak{A}_\mu = 0$, $\vec{\Phi} = \vec{v}(\vec{\gamma})$ are the proper vertices for the action (5.1):

$$\tilde{\Pi}(x_1\cdots x_n; y_1\cdots y_m; z_1\cdots z_l|v) = \frac{\delta^{n+m+l}\Gamma[\vec{\mathfrak{A}}_\mu, \vec{\Phi}]}{\delta\mathfrak{A}(x_1)\cdots\delta\Phi_t(y_1)\cdots\delta\Phi_l(z_1)\cdots}\Bigg|_{\mathfrak{A}_\mu=0,\ \vec{\Phi}=\vec{v}}, \tag{5.18}$$

where we have written

$$\vec{\Phi} = \vec{\Phi}_t + \vec{\eta}\Phi_l$$

and suppressed all isospin and tensor indices. We define the Fourier transform Π by

$$(2\pi)^4\delta(\Sigma p+\Sigma q+\Sigma r)\Pi(p_1\cdots p_n; q_1\cdots q_m; r_1\cdots r_l|v)$$

$$= \int\prod_{i=1}^n d^4x_i e^{ip_i\cdot x_i}\prod_{j=1}^m d^4y_j e^{iq_j\cdot y_j}\prod_{k=1}^l d^4z_k e^{ir_k\cdot z_k}\tilde{\Pi}(x_1\cdots x_n; y_1\cdots y_m; z_1\cdots z_l|v). \tag{5.19}$$

The expansion coefficients of Γ about $\vec{\mathfrak{A}}_\mu = 0$, $\vec{\Phi} = 0$ are the proper vertices $\Pi(\cdots|v=0)$ of the symmetric theory. Therefore, we have for $\mu^2 > 0$, and μ^2 held fixed,

$$\Pi(p_1\cdots p_n; q_1\cdots q_m; r_1\cdots r_l|v, g, \lambda) = \sum_{s=0}^\infty \frac{(v)^s}{s!}\Pi(p_1\cdots p_n, q_1\cdots q_m, r_1\cdots r_l 00\cdots 0|0, g, \lambda). \tag{5.20}$$

Where $00\cdots 0$ consists of s factors. Equation (5.20) expresses a proper vertex for the action (5.1) in terms of those of the symmetric theory which we know how to renormalize. The proper vertices appearing in the right-hand side contain $(l+s)\phi_3$ lines of which s lines disappear into the vacuum. We recall from Sec. II that the renormalized vertex $\Pi_r(\cdots|0, g_r, \lambda_r)$,

$$\Pi_r(p_1\cdots p_n; q_1\cdots q_m; r_1\cdots r_l\cdots r_{l+s}|0, g_r, \lambda_r)$$

$$\equiv (Z_3)^{n/2}(Z_2)^{(m+l+s)/2}\Pi(p_1\cdots p_n; q_1\cdots q_m; r_1\cdots r_{l+s}|0, g_r Z_1 Z_3^{-3/2}, \lambda_r Z_4 Z_2^{-2}), \tag{5.21}$$

is finite with an appropriate choice of Z_1, Z_2, Z_3, Z_4, $\tilde{Z}_1$, and $\tilde{Z}_3$ with $Z_1Z_3^{-1} = \tilde{Z}_1\tilde{Z}_3^{-1}$. We define the renormalizations of the left-hand side of Eq. (5.17) by

$$\Pi_r(p_1\cdots p_n; q_1\cdots q_m; r_1\cdots r_l|v_r, g_r, \lambda_r)$$

$$\equiv (Z_3)^{n/2}(Z_2)^{(m+l)/2}\Pi(p_1\cdots p_n; q_1\cdots q_m; r_1\cdots r_l|Z_2^{1/2}v_r, g_r Z_1 Z_3^{-3/2}, \lambda_r Z_4 Z_2^{-2}). \tag{5.22}$$

Then we see that

$$\Pi_r(\cdots|v_r, g_r, \lambda_r) = \sum_{s=0}^\infty \frac{(v_r)^s}{s!}\Pi_r(\cdots 00\cdots 0|0, g_r, \lambda_r) \tag{5.23}$$

where $00\cdots0$ consists of s factors. It shows that, if we renormalize the wave functions and coupling constants, and choose the mass counterterm $\delta\mu^2$ as in the symmetric theory, then the proper vertex $\Pi_r(\cdots|v_r)$ is finite if $v_r = Z_2^{-1/2}v$ is. [*Note added in proof.* The expansion of Eq. (5.17) entails in part developing the vector-boson propagators in powers of v^2:

$$\frac{1}{k^2-(gv)^2}=\frac{1}{k^2}+\sum_{n=1}^{\infty}\frac{1}{k^2}\left(g^2v^2\frac{1}{k^2}\right)^n .$$

The terms with $n \geq 1$ will cause infrared divergence of the integral. This is a reflection of the fact that Π is not analytic in v near $v=0$. To circumvent this difficulty we may replace the nth term ($n \geq 1$) by

$$\frac{1}{k^2+i\epsilon}\left(g^2v^2\frac{1}{k^2+\lambda^2+i\epsilon}\right)^n .$$

Since primitively divergent parts which include this term are at most logarithmically divergent, renormalization constants of the symmetric theory render them finite. *After* the summation indicated on the right-hand side of Eq. (5.20) is carried out, λ^2 may be let go to zero. This process gives the desired $\Pi_r(\cdots|v_r, g_r, \lambda_r)$, which is independent of λ^2 and finite. Alternatively, we may define a modification of the series in Eq. (5.17). We substitute for massive vector-boson propagators the expression

$$\frac{1}{k^2+i\epsilon}+\frac{1}{k^2+i\epsilon}(gv)^2\frac{1}{k^2+\lambda^2+i\epsilon}$$

and for scalar-boson propagators similar expressions. A Feynman integral for Π becomes a sum of terms. In these terms, subdiagrams consisting entirely of the propagators of the symmetric theory are made finite by the counterterms of the symmetric theory. Subdiagrams in which one of the propagators is replaced by the second term above are at most logarithmically divergent. Divergence in such a subdiagram is also removed by a symmetric counterterm. See reference for a similar discussion for the σ model.] In the symmetric theory, $\delta\mu^2$ and Z_2 may be chosen to satisfy Eq. (2.7) with $\mu^2 = M^2 - a^2$, for example. For the purpose of making $\Pi_r(\cdots|0)$ finite, however, we need not choose the finite parts of $\delta\mu^2$ and Z_2 in this manner. For our purpose, it is more convenient to choose $\delta\mu^2$ so that $\Delta_{\phi_t}(0)$ has the value

$$\Delta_{\phi_t}(0) = -m^{-2}, \tag{5.24}$$

where m^2 is the quantity appearing in Eq. (5.8) (see Appendix D). Obviously the vertices $\Pi_r(\cdots|v_r)$ may be regarded as functions of m^2 rather than of μ^2. Henceforth we shall treat m^2 defined by Eq. (5.24) as an independent variable.

How does one determine v_r in Eq. (5.23)? It must be determined from Eq. (5.10) which is the condition that ψ have a null vacuum expectation value. To determine the structure of S_c, we turn to our sheep, the Ward-Takahashi identity (4.5). We show in Appendix A that Eq. (4.5) implies

$$\epsilon^{abc}\int d^4x\left[J_\mu^b(x)\frac{\delta}{\delta J_\mu^c(x)}+K^b(x)\frac{\delta}{\delta K^c(x)}\right]W=0. \tag{5.25}$$

Differentiating Eq. (5.25) with respect to K and taking the limit $\vec{J}_\mu = 0$ and $\vec{K} = \vec{\gamma}$, we obtain

$$\vec{\gamma}\Delta_{\phi_t}(0) = -\vec{v}(\vec{\gamma}), \tag{5.26}$$

which shows $\vec{\gamma}$ and $\vec{v}$ are parallel [see Eqs. (5.3) and (5.4)] and

$$c_r = v_r m^2, \tag{5.27}$$

where

$$c_r = cZ_2^{1/2} . \tag{5.28}$$

Equation (5.27) is the renormalized version of Eq. (5.10). Thus if c_r is finite, so is v_r.

Let us summarize the results so far. We have shown that the Green's functions for the action (5.1) become finite if we renormalize the fields, sources and coupling constants according to

$$\begin{aligned}
(\vec{\phi},\vec{v}) &= Z_2^{1/2}(\vec{\phi},\vec{v})_r ,\\
(\vec{K},\vec{\gamma}) &= Z_2^{-1/2}(\vec{K},\vec{\gamma})_r ,\\
\vec{A}_\mu &= Z_3^{1/2}(\vec{A}_\mu)_r ,\\
\vec{J}_\mu &= Z_3^{-1/2}(\vec{J}_\mu)_r ,\\
g &= g_r(Z_1/Z_3^{3/2}) = g_r(\tilde{Z}_1/\tilde{Z}_3 Z_3^{1/2}),\\
\lambda &= \lambda_r(Z_4/Z_2^2),
\end{aligned} \tag{5.29}$$

and choose $\delta\mu^2$ and Z_2 to satisfy Eq. (5.24), and other Z's to be those of the symmetric theory. By the regularization method developed in paper I, the renormalizations implied in Eqs. (5.21) and (5.23) are made unambiguous and to preserve gauge invariance. That is to say, the Green's functions constructed from $\Pi_r(\cdots|v)$ of Eq. (5.23) satisfy the Ward-Takahashi identities generated from Eq. (4.5) by expanding Z about $(\vec{J}_\mu)_r = 0$ and $\vec{K}_r = \vec{\gamma}_r$.

Equation (5.27) is the Goldstone theorem. In the spontaneously broken case, $c_r = c = 0$, so that $m^2 = 0$, which is Eq. (4.24). In this case, the renormalization conditions given in this section reduce to those of the last section. The finiteness proof of Eqs. (5.23) and (5.27) applies to the spontaneous breaking case (i.e., $c_r = 0$, v_r finite) as well.

VI. LOW-ENERGY BEHAVIORS OF PROPAGATORS

In this and the following sections we will deal exclusively with renormalized quantities. We shall therefore drop the subscripts r consistently.

From Eq. (4.5) we learn that the longitudinal part of the vector meson propagator is unrenormalized. The derivation of this fact is completely analogous to that given in Sec. IV of the previous paper. Therefore, the full vector propagator has the form

$$\Delta_{\mu\nu}(k)=(g_{\mu\nu}-k_\mu k_\nu/k^2)f(k^2)+\alpha k_\mu k_\nu/(k^2)^2. \quad (6.1)$$

For the $a=1$ and 2 components of the vector propagator, Eq. (6.1) leads to useful relations. Let $\Gamma_{\mu\nu}$, Γ_μ, and Γ be defined by

$$\tilde{\Gamma}^{\mu\nu}(x-y)=\delta^2\Gamma[\vec{\mathfrak{a}}_\mu,\vec{\Phi}]/\delta\mathfrak{a}^1_\mu(x)\delta\mathfrak{a}^1_\nu(y)|_{\vec{a}_\mu=0,\vec{\Phi}=\vec{v}},$$

$$\tilde{\Gamma}^{\mu}(x-y)=\delta^2\Gamma[\vec{\mathfrak{a}}_\mu,\vec{\Phi}]/\delta\mathfrak{a}^1_\mu(x)\delta\Phi^2(y)|_{\vec{a}_\mu=0,\vec{\Phi}=\vec{v}}, \quad (6.2)$$

$$\Gamma(x-y)=\delta^2\Gamma[\vec{\mathfrak{a}}_\mu,\vec{\Phi}]/\delta\Phi^2(x)\delta\Phi^2(y)|_{\vec{a}_\mu=0,\vec{\Phi}=\vec{v}},$$

and

$$\begin{bmatrix}\tilde{\Gamma}^{\mu\nu}\\ \tilde{\Gamma}^{\mu}\\ \tilde{\Gamma}\end{bmatrix}(x-y)=\int\frac{d^4k}{(2\pi)^4}e^{ik\cdot(x-y)}\begin{bmatrix}\Gamma^{\mu\nu}\\ \Gamma^{\mu}\\ \Gamma\end{bmatrix}(k). \quad (6.3)$$

In Eq. (6.2) $\vec{v}$ is chosen to be along the third axis of the isospin space, and is determined by the condition

$$\left.\frac{\delta\Gamma}{\delta\Phi_i}\right|_{\vec{a}_\mu=0,\vec{\Phi}=\vec{v}}=0. \quad (6.4)$$

The propagators defined as

$$\delta^2 Z/\delta K^2(x)\delta K^2(y)|_{\vec{J}_\mu=\vec{K}=0}=-\int\frac{d^4k}{(2\pi)^4}e^{ik\cdot(x-y)}\Delta(k^2),$$

$$\delta^2 Z/\delta J^1_\mu(x)\delta K^2(y)|_{\vec{J}_\mu=\vec{K}=0}=\int\frac{d^4k}{(2\pi)^4}e^{ik\cdot(x-y)}\Delta^\mu(k), \quad (6.5)$$

$$\delta^2 Z/\delta J^1_\mu(x)\delta J^1_\nu(y)|_{\vec{J}_\mu=\vec{K}=0}=\int\frac{d^4k}{(2\pi)^4}e^{ik\cdot(x-y)}\Delta^{\mu\nu}(k),$$

and the proper vertices of Eq. (6.3) are the inverses of each other, in the sense that

$$\begin{pmatrix}\Gamma(k^2) & \Gamma_\lambda(-k)\\ \Gamma_\mu(k) & \Gamma_{\mu\lambda}(k)\end{pmatrix}\begin{pmatrix}1 & \\ & -g^{\lambda\rho}\end{pmatrix}\begin{pmatrix}\Delta(k^2) & \Delta_\nu(-k)\\ \Delta_\rho(k) & -\Delta_{\rho\nu}(k)\end{pmatrix}=\begin{pmatrix}1 & \\ & -g_{\mu\nu}\end{pmatrix}. \quad (6.6)$$

We shall parametrize the proper vertices of Eq. (6.3) as

$$\Gamma_{\mu\nu}(k)=-g_{\mu\nu}A(k^2)+k_\mu k_\nu B(k^2),$$
$$\Gamma_\mu(k)=ik_\mu C(k^2), \quad (6.7)$$
$$\Gamma(k^2)=k^2D(k^2).$$

When the above expressions are substituted into Eq. (6.6), we obtain for $\Delta_{\mu\nu}$

$$\Delta_{\mu\nu}(k)=(g_{\mu\nu}-k_\mu k_\nu/k^2)A^{-1}+\frac{k_\mu k_\nu}{k^2}\frac{D}{D(A-k^2B)+C}. \quad (6.8)$$

Comparing the longitudinal parts of Eqs. (6.1) and (6.8), we obtain

$$\alpha(AD-BDk^2+C^2)=k^2D, \quad (6.9)$$

which is the desired relation.

The propagators in Eq. (6.5) can be written as

$$\Delta_{\mu\nu}(k)=\left(g_{\mu\nu}-\frac{k_\mu k_\nu}{k^2}\right)A^{-1}+\alpha\frac{k_\mu k_\nu}{(k^2)^2},$$

$$\Delta_\mu(k)=i\alpha\frac{k_\mu}{(k^2)^2}\left(\frac{C}{D}\right), \quad (6.10)$$

and

$$\Delta(k^2)=\frac{1}{k^2D(k^2)}-\frac{\alpha}{(k^2)^2}\left(\frac{C}{D}\right)^2.$$

Let us consider the low-energy limits of propagators in Eq. (6.10). Now taking the limit $k^2\to 0$ in Eq. (6.9), we find that

$$\lim_{k^2\to 0}(AD+C^2)=0. \quad (6.11)$$

It is instructive to see what happens in the $a=3$ channel. The invariances of the Lagrangian and the vacuum expectation value under $\phi_{1,3}\to+\phi_{1,3}$, $\phi_2\to-\phi_2$ and $A_\mu{}^2\to+A_\mu{}^2$, $A_\mu^{1,3}\to-A_\mu^{1,3}$ imply that $C(k^2)=0$ in Eq. (6.7). Equation (6.9) becomes

$$\alpha(A-k^2B)=k^2.$$

Writing $A=k^2J$ we see that

$$\Gamma_{\mu\nu}(k)=-(k^2g_{\mu\nu}-k_\mu k_\nu)J-(1/\alpha)k_\mu k_\nu. \quad (6.12)$$

VII. GAUGE INDEPENDENCE AND THE UNITARITY OF THE S MATRIX

By using Eq. (2.8) repeatedly, we obtain, for $k \leq l$,

$$\left(\frac{i}{\alpha}\right)^k \frac{\partial}{\partial x_1^{\mu_1}}\cdots\frac{\partial}{\partial x_k^{\mu_k}} \frac{\delta^{k+l} iZ}{\delta J_{\mu_1}(x_1)\cdots\delta J_{\mu_k}(x_k)\delta J_{\nu_1}(y_1)\cdots\delta J_{\nu_l}(y_l)}\bigg|_{J_\mu=K=0}$$

$$=\sum_{\text{part}}\sum_{\text{perm}(k)} W^{-1}\left\{\prod_{i=1}^{k}\int d^4z_i\, g^{\text{tr}}_{\nu_{ji},\lambda_i}(y_{j_i}-z_i)\left[-ig\tilde{Z}_1\frac{\delta}{\delta J_{\lambda_i}(z_i)}\right]G(z_i, x_i; i\delta/\delta J)\right\}\left[\prod_{m=k+1}^{l}\frac{\delta}{\delta J_{\nu_{jm}}(y_{j_m})}\right]W\bigg|_{J_\mu=K=0}, \tag{7.1}$$

where $\sum_{\text{part}}$ is the summation over all possible partitions of $(1, 2, \ldots, l)$ into two subsets, $\{j_i\}$, $i=1,\ldots,k$ and $\{j_m\}$, $m=k+1,\ldots,l$, and $\sum_{\text{perm}(k)}$ is the summation over all permutations of k elements of $\{j_i\}$. We have suppressed all references to the isospin which is not crucial in our discussion. We used the symbol $g^{\text{tr}}_{\mu\nu}$ for

$$g^{\text{tr}}_{\mu\nu}(x-y)=g_{\mu\nu}\delta^4(x-y)+\partial_\mu\partial_\nu\bar{D}_F(x-y)\,.$$

For $k=l+1$, we have

left-hand side of (7.1)

$$=\sum_{\text{perm}(l)} W^{-1}\left\{\prod_{i=1}^{l}\int d^4z_i\, g^{\text{tr}}_{\nu_{ji},\lambda_i}(y_{j_i}-z_i)\left[-ig\tilde{Z}_1\frac{\delta}{\delta J_{\lambda_i}(z_i)}\right]G(z_i, x_i; i\delta/\delta J)\right\}\frac{i}{\alpha}\frac{\partial}{\partial x_k^{\mu_k}}\frac{\delta}{\delta J_{\mu_k}(x_k)}W\bigg|_{J_\mu=K=0}. \tag{7.2}$$

There are $(k-1)$ more equations of this kind in which the privileged role of (x_k, μ_k) on the right-hand side is taken up by $(x_1, \mu_1), \ldots, (x_{k-1}, \mu_{k-1})$. For $k > l+1$, we have simply

$$\text{left-hand side of (7.1)}=0. \tag{7.3}$$

The three equations above are the bases of our discussion on the gauge independence and the unitarity of the S matrix. By the gauge independence of the S matrix, we mean that the on-shell S matrix is independent of α in the gauge defining term in the action (2.1). The proof given in Ref. 13 can be carried over to our case. First note that

$$\left(\prod_{i=1}^{m}\frac{i}{\alpha}\frac{\partial}{\partial x_i^{\mu_i}}\right)\frac{\delta^m Z}{\delta J_{\mu_1}(x_1)\cdots\delta J_{\mu_m}(x_m)}\bigg|_{J_\mu=K=0}=0, \tag{7.4}$$

which is a special case of (7.3). Equation (7.4) corresponds exactly to Eq. (6.11) of Ref. 13, and by the argument given there we conclude that the T matrix is independent of the parameter α.

We wish, next, to show that the massless scalar particles we encounter in the construction of Green's functions are unphysical, i.e., do not contribute to the sum over intermediate states when we compute the absorptive part of a physical (i.e., on-shell) amplitude by the Landau-Cutkosky rule.[8,9] Recall that there are in general three different massless scalars: the negative metric scalar excitation (the first kind) associated with the transverse vector propagator

$$\sim\frac{k_\mu k_\nu}{k^2}\frac{1}{A}$$

the Goldstone boson (the second kind), with the propagator

$$1/k^2D$$

and the fermion scalars associated with the gauge field quantization (the third kind).

Let us begin with the simplest example. Let $T_\mu(k\cdots)$ be the amputated Green's function with one vector boson off the mass shell and all other lines on the mass shell. We have shown explicitly the momentum k and the tensor index μ for the vector boson, but suppressed all other variables. Let $T(k\cdots)$ be the amputated Green's function with one Goldstone boson off shell (with momentum k) and all other external lines on shell. Consider now the combination

$$T^{(2)}_\mu\left(\frac{k^\mu k^\nu}{k^2+i\epsilon}\frac{1}{A}\right)T^{(1)}_\nu+T^{(2)}\frac{1}{(k^2+i\epsilon)D}T^{(1)} \tag{7.5}$$

and compute the absorptive part of this amplitude arising form the two kinds of scalars being on the mass shell. By the Cutkosky rule it is given by

$$- T_1^{(2)*} T_1^{(1)} + T_2^{(2)*} T_2^{(1)}, \tag{7.6}$$

where

$$T_2 = \lim_{k^2 \to 0} T(k^2)/[D(k^2)]^{1/2}$$

is the amplitude for the Goldstone boson (massless particle of the second kind), and

$$T_1 = i \lim_{k^2 \to 0} k_\mu T^\mu(k)/[A(k^2)]^{1/2}$$

is the normalized amplitude for the massless scalar particle of the first kind. (The infrared divergences in D and A *always* cancel the similar ones in the vertices to which the propagators are attached, so that T_1 and T_2 are free of divergences as $k^2 \to 0$). Since

$$\frac{i}{\alpha} \partial_\mu \frac{\delta Z}{\delta J_\mu(x)}\bigg|_{J_\mu = K = 0} = 0$$

we have the relation

$$\frac{1}{\alpha} k_\mu \left\{ \left[\left(g^{\mu\nu} - \frac{k^\mu k^\nu}{k^2} \right) A^{-1} + \alpha k^\mu k^\nu (k^2)^{-2} \right] T_\nu + i\alpha \frac{k^\mu}{(k^2)^2} \frac{C}{D} T \right\} = -i \frac{\sqrt{A}}{k^2} \left(\frac{ik^\mu}{\sqrt{A}} T_\mu - \frac{1}{\sqrt{D}} T \right) = 0,$$

which gives, in the limit $k^2 = 0$,

$$T_1 = T_2 . \tag{7.7}$$

Therefore the expression (7.6) is identically zero, and neither of the scalars contributes to the sum over states.

To proceed further, it is necessary to extract more information from Eqs. (7.1)–(7.3). Let

$$T_{i_1 i_2 \cdots i_s}(1, 2, \ldots, s)$$

be the amplitude for s massless scalar excitations of the first and second kinds, the subscripts $i_1, \ldots, i_s$, which take the value 1 or 2 indicating which kinds are involved. We suppress, as before, all references to other particles which are on their mass shells. Let

$$G_{i_1 \cdots i_s}(1, 2, \ldots, s \,|\, (j_1, k_1), (j_2, k_2), \ldots, (j_t, k_t))$$

be the amplitude for $s + 2t$ massless scalar excitations, s being either of the first or second kind, and $2t$ being of the third kind. The ghost "particles" of the third kind appear in pairs, and their pairings are unambiguous, because the ghost lines are continuous. In the pair (j_n, k_n), the ordering is important, because the ghost line is orientable (say, from the dotted end j_n to the undotted end k_n). Equation (7.1) tells us that for $k \leq l$

$$\sum_{i_1=1}^{2} \cdots \sum_{i_k=1}^{2} T_{i_1 \cdots i_k 11 \cdots 1}(1, \ldots, k, k+1, \ldots, k+l) = \sum_{\text{part}} \sum_{\text{perm}(k)} G_{\{i_{jm}\}}(\{jm\} \,|\, (j_1, 1), \ldots, (j_k, k)), \tag{7.8}$$

where $11 \cdots 1$ consists of l factors and, as before, $\sum_{\text{part}}$ means the summation over all possible partitions of $(k+1, k+2, \ldots, k+l)$ into two subsets, $\{j_i\}$ $i = 1, 2, \ldots, k$ and $\{j_m\}$ $m = k+1, \ldots, k+l$, and $\sum_{\text{perm}(k)}$ means the summation over all permutations of $\{j_i\}$, $i = 1, 2, \ldots, k$. For $k = l+1$, Eq. (7.2) tells us that

$$\text{left-hand side of Eq. (7.8)} = \sum_{i_k=1}^{2} \sum_{\text{perm}(l)} G_{i_k}(k \,|\, (j_1, 1), (j_2, 2), \ldots, (j_l, l)) \tag{7.9}$$

For $k > l+1$, we have from Eq. (7.3)

$$\text{left-hand side of Eq. (7.8)} = 0 . \tag{7.10}$$

We claim that Eqs. (7.8)–(7.10) are sufficient to prove that the contributions from three kinds of zero-mass excitations always cancel in the sum over intermediate states, no matter how many massless excitations there are in a given intermediate state. To see how it works, let us consider two cases in detail.

Suppose there are two massless scalars in the intermediate states. The unitarity sum is

$$U = \sum_{i_1, i_2} e^{i\pi(i_1+i_2)} T_{i_1 i_2}^{(?)*}(1,2) T_{i_1 i_2}^{(1)}(1,2) - G^{(2)*}(\,|(1,2)) G^{(1)}(\,|(2,1)) - G^{(2)*}(\,|(2,1)) G^{(1)}(\,|(1,2)) . \tag{7.11}$$

The last two terms have negative signs because the scalars of the third kind are fermions. Equation (7.8) gives

$$\sum_{i_1} T_{i_1 1}(1,2) = G(\,|(21)) \equiv G(21)\,,$$
$$\sum_{i_2} T_{1 i_2}(1,2) = G(\,|(12)) \equiv G(12)\,, \tag{7.12}$$

and Eq. (7.10) gives

$$\sum_{i_1, i_2} T_{i_1 i_2}(1,2) = 0\,. \tag{7.13}$$

Equations (7.12) and (7.13) allow us to express T_{11}, T_{12}, and T_{21} in terms of others:

$$T_{11} = T_{22} + G(12) + G(21)\,,$$
$$T_{21} = -\,T_{22} - G(21)\,,$$
$$T_{12} = -\,T_{22} - G(12)\,.$$

When the above expressions are substituted in Eq. (7.11), we find

$$U = 0\,.$$

Now consider the case of three massless scalars. We will use the abbreviations $T_{i_1 i_2 i_3} = T_{i_1 i_2 i_3}(1,2,3)$, $G_{i_1}(1\,|23) = G_{i_1}(1\,|(2,3))$. The unitarity sum is

$$U = \sum_{i_1 i_2 i_3} e^{i\pi(i_1+i_2+i_3)} T^{(2)*}_{i_1 i_2 i_3} T^{(1)}_{i_1 i_2 i_3}$$
$$-\sum_{k=1}^{2} e^{i\pi k}[G_k^{(2)*}(1\,|23)G_k^{(1)}(1\,|32) + G_k^{(2)*}(1\,|32)G_k^{(1)}(1\,|23) + G_k^{(2)*}(2\,|31)G_k^{(1)}(2\,|13)$$
$$+ G_k^{(2)*}(2\,|13)G_k^{(1)}(2\,|31) + G_k^{(2)*}(3\,|12)G_k^{(1)}(3\,|21) + G_k^{(2)*}(3\,|21)G_k^{(1)}(3\,|12)]\,. \tag{7.14}$$

The relations among various amplitudes we can get from Eqs. (7.8)–(7.10) are

$$\sum_i T_{i11} = G_1(3\,|12) + G_1(2\,|13)\,,$$
$$\sum_i T_{1i1} = G_1(3\,|21) + G_1(1\,|23)\,,$$
$$\sum_i T_{11i} = G_1(2\,|31) + G_1(1\,|32)\,,$$
$$\sum_{i,j} T_{ij1} = \sum_j G_j(2\,|13) = \sum_i G_i(1\,|23)\,, \tag{7.15}$$
$$\sum_{i,j} T_{i1j} = \sum_j G_j(3\,|12) = \sum_i G_i(1\,|32)\,,$$
$$\sum_{i,j} T_{1ij} = \sum_j G_j(3\,|21) = \sum_i G_i(2\,|31)\,,$$
$$\sum_{i,j,k} T_{ijk} = 0\,.$$

The relations (7.15) are enough to show that U of (7.14) is identically zero.

This process can be pushed *ad infinitum*. We have not found a sufficiently convenient and compact notation to carry out the calculation efficiently for N massless particles. In verifying the cancellation for $N=4$, for example, it is important to bear in mind the fermion nature of the particles of the third kind, so that in the unitarity sum we have

$$+G^{(2)*}(\,|(12),(34))\,G^{(1)}(\,|(21),(43)) - G^{(2)*}(\,|(12),(34))\,G^{(1)}(\,|(41),(23))\,.$$

Note the relative signs.

APPENDIX A: THE σ-MODEL-LIKE IDENTITY

The simplest way of deriving Eq. (5.25) is to consider a constant gauge transformation on the variables of integration in the functional integral (4.3). We give here an alternative derivation of Eq. (5.25) from Eq. (4.5).

From Eq. (5.25) we obtain

$$\left(\partial^2 - i\vec{t}\cdot\frac{\delta}{\delta\vec{J}_\mu}\partial_\mu\right)^{ab}\frac{i}{\alpha}\partial_\lambda\frac{\delta W}{\delta J^b_\lambda} + \partial^\mu J^a_\mu W + ig\left(J_\mu\vec{t}\cdot\frac{\delta}{\delta\vec{J}_\mu} + K\vec{t}\cdot\frac{\delta}{\delta\vec{K}}\right)^a W$$

$$- i\int d^4y\,\delta(y-x)t^{acd}D^\lambda_y[i\delta/\delta J]^{cb}\frac{\partial}{\partial x^\lambda}G^{bd}(y,x;i\delta/\delta\vec{J})W = 0\,. \tag{A1}$$

Since

$$\epsilon^{abc}\partial_\mu\left[\frac{\delta}{\delta J^b_\mu}\partial_\lambda\frac{\delta W}{\delta J^c_\nu}\right] = \epsilon^{abc}\frac{\delta}{\delta J^b_\mu}\partial_\mu\partial_\lambda\frac{\delta W}{\delta J^c_\nu}$$

and

$$\epsilon^{ccd}\left\{D^\lambda_y[i\delta/\delta\vec{J}]^{cb}\frac{\partial}{\partial x^\lambda}G^{bd}(y,x;i\delta/\delta\vec{J})\right\}_{x=y} = \epsilon^{acd}\frac{\partial}{\partial x^\lambda}\{D^\lambda_y[i\delta/\delta\vec{J}]^{cb}G^{bd}(y,x;i\delta/\delta\vec{J})\}_{x=y},$$

we can write all but the third term on the left-hand side of Eq. (A1) as divergences of vectors. Equation (5.25) follows upon integration over x.

APPENDIX B: CONSTRUCTION OF RENORMALIZABLE MASSIVE VECTOR-MESON THEORIES

In this appendix, we pose and discuss the following problem: How does one construct a theory in which all of the gauge bosons associated with the gauge group G become massive while the vacuum is invariant under the little group S, which is not a local gauge group? The construction here may be of interest in providing models of strong interactions.

We shall now consider the following set of groups:

$$G^{(L)}\times S^{(R)}\supset S^{(L)}\times S^{(R)}\supset S^{(D)}\,.$$

$S^{(L)}, S^{(R)}, S^{(D)}$ are isomorphic to S, and $S^{(D)}$ is the diagonal subgroup of $S^{(L)}\times S^{(R)}$.

We construct a theory with the following properties:

(1) The Lagrangian is invariant under local gauge transformations of the group $G^{(L)}$ and constant gauge transformations of the group $S^{(R)}$. A^a_μ are the gauge fields associated with the group $G^{(L)}$.

(2) $\phi^{(\alpha)}$ is a set of scalar fields, with nonzero vacuum expectation value $v^{(\alpha)}$. The little group of the vacuum is $S^{(D)}$.

(3) All other fields present in the Lagrangian are invariant under transformations of the group $S^{(R)}$.

In the notation of Sec. III, $\{L\}$ are the generators of $G^{(L)}\times S^{(R)}$, $\{l\}$ are the generators of $S^{(D)}$. The generators of $G^{(L)}$ complete the set of generators $\{L\}$. $\{t\}$ will be this set:

$$\{t\}+\{l\}=\{L\}\,.$$

Now one can choose fields $\rho^{(\alpha)}$ and $\vec{\xi}$ such that

$$\phi^{(\alpha)} = D[e^{i\vec{\xi}\cdot\vec{t}}](v^{(\alpha)}+\rho^{(\alpha)})\,.$$

Using the local gauge invariance, one can eliminate the fields $\vec{\xi}$ and all the gauge fields A^a_μ become massive. In this way one has constructed a theory in which there appears a set of *massive* Yang-Mills fields associated with a given spontaneously broken symmetry G, the theory remaining symmetric under a subgroup S of G. (S is not a local gauge group).

In order to illustrate this mechanism, we will give some examples:

(1) Let G and S be isomorphic to SU(2). ϕ belongs to the $(\frac{1}{2},\frac{1}{2})$ representation of SU(2)×SU(2). In this model massive Yang-Mills are associated with an exact SU(2) symmetry. This is one of the models proposed by 't Hooft.[14]

(2) G is SU(2)×SU(2), S is isomorphic to SU(2). We let $\phi^{(L)}$ belong to a $(\frac{1}{2},0,\frac{1}{2})$ representation; $\phi^{(R)}$ to a $(0,\frac{1}{2},\frac{1}{2})$; $(\sigma,\vec{\pi})$ to a $(\frac{1}{2},\frac{1}{2},0)$. In this way one can construct a model in which a set of massive Yang-Mills fields is associated with the broken chiral symmetry SU(2)×SU(2).

(3) G is isomorphic to SU(3), S is isomorphic to SU(3) or SU(2). ϕ belongs to the $(3,\bar{3})+(\bar{3},3)$ representation of SU(3)×SU(3) or the $(3,\frac{1}{2})+(\bar{3},\frac{1}{2})$ representation of SU(3)×SU(2).

(4) SU(3)×SU(3) can be treated by a combination of the two preceding methods.

APPENDIX C: MASSIVE YANG-MILLS THEORY AS A LIMIT OF SPONTANEOUSLY BROKEN GAUGE THEORIES

In all the models discussed previously, it is easy to see that the masses corresponding to the fields having nonzero vacuum expectation values are free parameters. When these masses become infinite, one finds as a limit ordinary massive Yang-Mills field models. This can be most easily seen in the U gauge, in which the would-be Goldstone bosons have been eliminated.

In this sense these theories, when the masses are finite, can be considered as regularization of the ordinary massive Yang-Mills theories, in the same way as the linear σ model can be understood as a regularization of the nonlinear σ model.[15] It is possible that this limit, as in the case of the σ model, is less singular than the direct power counting of the limiting theory suggests.

We shall study a particular model here, but all the arguments will be completely general.

1. The Lagrangian. We set

$$\mathcal{L}=-\tfrac{1}{4}\mathrm{Tr}\{\partial_\mu V_\nu-\partial_\nu V_\mu+ig[V_\mu,V_\nu]\}^2+\tfrac{1}{2}\mathrm{Tr}\{\partial_\mu M+ig[V_\mu,M]\}\{\partial^\mu M^\dagger-ig[M^\dagger,V^\mu]\}$$
$$+V(M)+\text{other matter fields}\,.$$

The gauge group is SU(n), V_μ^b is a Hermitian traceless $n\times n$ matrix and M is an $n\times n$ complex matrix. $V(M)$ is a polynomial in M and $M^\dagger$ which can be chosen such that the vacuum expectation value of M is of the form

$$\langle M\rangle_0=F\,,$$

where F is a real diagonal matrix. Furthermore, when the masses of the M fields become infinite $V(M)$ gives in the limit in the Feynman path integral a δ function of the form $\delta(MM^\dagger-F^2)$. In order to quantize the theory we add to the Lagrangian:

$$\delta\mathcal{L}=-\frac{1}{2\alpha}\mathrm{Tr}[\partial_\mu V^\mu-i\lambda(FM^\dagger-MF)]^2+\mathrm{Tr}\{\bar{c}\partial^2c+ig\bar{c}\partial_\mu[V^\mu,c]\}+\lambda g(\bar{c}FM^\dagger c+cMF\bar{c})\,,$$

where c and $\bar{c}$ are $n\times n$ matrices representing the usual scalar fermions ghosts, and λ can be chosen such that the term $-i(\lambda/\alpha)\mathrm{Tr}\,\partial^\mu V_\mu(FM^\dagger-MF)$ cancels the corresponding term in the Lagrangian which is obtained when one replaces M by $M'+F$. This is the gauge introduced by 't Hooft.[14]

Now in the limit of the infinite mass of scalar fields M the generating functional becomes

$$\exp iZ=\int[dV_\mu][dM]\cdots\prod_x\delta(MM^\dagger-F)\exp i\int d^4x[\mathcal{L}+\delta\mathcal{L}+\text{source terms}]\,.$$

We can make the following change of variable:

$$M=(e^{iH})\Omega\,,$$

where $H=H^\dagger$ and $\Omega=\Omega^\dagger$. The generating functional can now be written

$$\exp iZ=\int[dV_\mu][dH]J(H)d\cdots\exp i\int d^4x[\mathcal{L}+\delta\mathcal{L}+\text{source term}]\,,$$

where we have used the δ function, and $J(H)$ is the Jacobian.

2. Power counting. It is well known that in the unitary gauge, the most divergent graphs have a superficial degree of divergence δ of the form:

$$\delta=6L\,,$$

where L is the number of loops. But it has been shown[16] that on the mass shell, cancellations occur which reduce the degree of divergence.

We will give here a new derivation of this result, using another gauge. We shall use the following identities in order to calculate the superficial degree of divergence δ of a graph:

$$\delta=4-E_B-\tfrac{3}{2}E_F+\sum n_i(\delta_i-4)\,,$$
$$\delta_i=m_i+v_i^B+\tfrac{3}{2}v_i^F\,,$$
$$E_B+E_F+2I=\sum n_i(v_i^B+v_i^F)\,,$$
$$L=I+1-\sum n_i\,,$$

where E_B and E_F are the number of external bosons (or ghosts) and fermions, respectively, n_i is the number of vertices of type i, m_i is the number of derivatives at the vertex, v_i^B is the number of bosons, and v_i^F is the number of fermions at the vertex i. I is the number of internal lines. A straightforward calculation gives

$$\delta = 2L + 2 - \tfrac{1}{2}E_F + \sum n_i(m_i + \tfrac{1}{2}v_i^F - 2).$$

Not returning to the Lagrangian one sees that either

$$v_i^F = 2 \text{ and } m_i = 0$$

or

$$v_i^F = 0 \text{ and } m_i \leq 2.$$

The most divergent contributions are given by $\mathrm{Tr}\,\partial_\mu M \partial_\mu M^\dagger$ with $m_i = 2$. So,

$$\delta \leq 2L + 2.$$

A closer examination, using the fact that we are not interested in Green functions with external particles associated to the fields H, c and $\bar{c}$, shows actually

$$\delta \leq 2L.$$

3. *One-loop approximation.* When the current is conserved, we have

$$F = f\underline{1}.$$

In the one-loop approximation the Lagrangian can be replaced by the following effective Lagrangian:

$$\mathcal{L} = -\tfrac{1}{4}\mathrm{Tr}\{\partial_\mu V_\nu - \partial_\nu V_\mu + ig[V_\mu, V_\nu]\}^2 - (1/2\alpha)\mathrm{Tr}(\partial^\mu V_\mu)^2 + \tfrac{1}{2}(fg)^2\,\mathrm{Tr}\,V_\mu{}^2 + \tfrac{1}{2}f^2\mathrm{Tr}\{(\partial_\mu H)^2 - 2g^2 f^2 H^2 + igV^\mu[H, \partial_\mu H]\}$$
$$+ \mathrm{Tr}\{\bar{c}\partial^2 c + ig\bar{c}\partial_\mu[V^\mu, c]\}.$$

With this effective Lagrangian it is clear that the massless Yang-Mills theory is not the limit of the massive Yang-Mills theory. The massless Yang-Mills theory is obtained for $f = 0$. If f is different from zero, one can integrate over H, c, and $\bar{c}$, and obtains

$$\exp iZ = \int [dV^\mu] \exp i[S + \text{source terms}]\, \Delta_1(V^\mu)\Delta_2(V^\mu),$$

$$\Delta_1(V^\mu) = \exp[\mathrm{Tr}\ln(\partial^2 + ig\overleftarrow{\partial}^\mu[V^\mu, \,])],$$

$$\Delta_2(V^\mu) = \exp\{-\tfrac{1}{2}\mathrm{Tr}\ln(\partial^2 + \tfrac{1}{2}ig(\overleftarrow{\partial}_\mu - \overrightarrow{\partial}_\mu)[V^\mu, \,])\}.$$

In the Landau gauge ($\alpha = 0$) we have the relation

$$\Delta_1\Delta_2 = (\Delta_1)^{1/2}$$

because the two expressions inside Tr ln differ only by a term proportional to $\partial_\mu V^\mu$.

We, therefore, see in this way the origin of the difference of a factor 2 in front of the ghost loops, between the massless and the massive Yang-Mills cases.

APPENDIX D: INFRARED PROBLEM

Let us consider the contributions of intermediate states of two or more massless particles to the inverse of the propagator $\Delta_{\phi_t}(k^2)$. Since the phase space for N mass particles, ρ_N, goes as $(k^2)^{N-2}$, the integral

$$\int_0^\infty \frac{ds'}{s' - k^2}\rho_N(s')|T_N(s')|^2$$

is not infrared divergent for $N \geq 3$.

It suffices, therefore, to consider only the intermediate states of two massless particles. There are two such states: $\phi_1(p) \to \phi_2(p-q) + A_\mu{}^3(q)$ and $A_\mu{}^3(q)$ + the massless scalar associated with the longitudinal part of the A_μ' propagator. Since in the Landau gauge the $A_\mu{}^3$ propagator is purely transverse,

$$(g^{\mu\nu} - q^\mu q^\nu/q^2)[q^2 J(q^2)]^{-1},$$

and since any vector to be contracted with μ or ν of the above propagator may be expressed as a linear combination of $p_\mu(p_\nu)$ and $q_\mu(q_\nu)$ we see that the contributions of two massless particles to the self-energy are necessarily of order $p_\mu p_\nu$, disregarding logarithmic factors.

---

*On leave of absence from SPT, CEN Saclay, B.P. 2, 91 Gif-sur-Yvette, France.

[1]M. Veltman (private communication).

[2]D. Gross and R. Jackiw, Phys. D (to be published).

[3]B. W. Lee, Nucl. Phys. B9, 649 (1969); J.-L. Gervais and B. W. Lee, *ibid.* B12, 627 (1969); B. W. Lee, Chiral Dynamics (Gordon and Breach, New York, to be published).

[4]T. W. B. Kibble, Phys. Rev. 155, 1554 (1967).

[5]S. Bludman and A. Klein, Phys. Rev. 131, 2363 (1963).

[6]S. Weinberg, Phys. Rev. Letters 27, 1688 (1971).

[7]T. Appelquist and H. Quinn (unpublished).

[8]L. D. Landau, Nucl. Phys. 13, 181 (1959); J. D. Bjorken, doctoral dissertation, Stanford University, 1959 (unpublished).

[9]R. E. Cutkosky, J. Math. Phys. 1, 429 (1960).

[10]N. N. Bogoliubov and D. B. Shirkov, *Introduction to the Theory of Quantized Fields* (Interscience, New York, 1959).

[11]P. Higgs, Phys. Letters 12, 132 (1966).

[12]R. P. Feynman and A. R. Hibbs, *Quantum Mechanics and Path Integrals* (McGraw-Hill, New York, 1965); J. Schwinger *Particles, Sources and Fields* (Addison-Wesley, Reading, Mass., 1970).

[13]B. W. Lee, Phys. Rev. D 5, 823 (1972).

[14]G. 't Hooft, Nucl. Phys. B35, 167 (1971).

[15]D. Bessis and J. Zinn-Justin, Phys. Rev. D 5, 1313 (1972).

[16]S. L. Glashow and J. Iliopoulos, Phys. Rev. D 3, 1043 (1971).

Spontaneously Broken Gauge Symmetries. III. Equivalence

Benjamin W. Lee

National Accelerator Laboratory, P. O. Box 500, Batavia, Illinois 60510
and Insitute for Theoretical Physics, State University of New York, Stony Brook, New York 11790

and

Jean Zinn-Justin*

Institute for Theoretical Physics, State University of New York, Stony Brook, New York 11790

(Received 10 March 1972)

We discuss the equivalence of the *S* matrix in the *R*- and *U*-gauge formulations of spontaneously broken gauge theories. We give definitions of the *U*-gauge Green's functions in terms of the *R*-gauge ones, for both Abelian and non-Abelian cases. Based on the equivalence theorem, we give a renormalization prescription of the *U*-gauge formulation.

I. INTRODUCTION

In this paper, we wish to demonstrate the equivalence of the *S* matrix in the *R*- and *U*-gauge formulations of spontaneously broken gauge theories. We have discussed the advantages and disadvantages of the two formulations in a previous paper (paper II).

We shall carry out this demonstration by expressing Green's functions in the *U* gauge in terms of those in the *R* gauge. What we shall show in this paper is a concrete realization of the remarks made previously by Weinberg[1] and by Salam and Strathdee[2] about the equivalence of the two formulations. But more importantly, the present work gives *definitions* of the *U*-gauge Green's functions in terms of the well-defined *R*-gauge ones.

This paper is organized as follows. In Sec. II we consider the equivalence of the two formulations for the Abelian model considered previously. In Sec. III, we give some illustrations of the equivalence and formulate the renormalization prescription in the *U* gauge. In Sec. IV, we deal with the generalization to non-Abelian cases.

It is empirically known that the *T* matrix for the Abelian case computed in the *U* gauge is finite.[1,3]

This is a corroboration of our general arguments in this paper.

II. ABELIAN CASE

We recall the model discussed in Ref. 4. It consists of the gauge boson A_μ, coupled to a complex scalar field ϕ:

$$\mathcal{L} = -\tfrac{1}{4}(\partial_\mu A_\nu - \partial_\nu A_\mu)^2 + (\partial_\mu + ieA_\mu)\phi^*(\partial^\mu - ieA^\mu)\phi - \mu^2(\phi^*\phi) - \lambda(\phi^*\phi)^2 - \delta\mu^2(\phi^*\phi), \tag{2.1}$$

with $\mu^2 < 0$. The symmetric vacuum is unstable, and an asymmetric vacuum becomes stable. Let v be the vacuum expectation of ϕ. We can adjust the phase of ϕ so that v is real.

The R-gauge formulation is suggested by the parametrization

$$\phi = (1/\sqrt{2})(v + \psi + i\chi), \tag{2.2}$$

where ψ and χ are real fields with null vacuum expectation value, and requires the subsidiary condition

$$\partial^\mu A_\mu(x) = 0. \tag{2.3}$$

The U-gauge formulation, on the other hand, is based on the choice of fields

$$\begin{aligned} \phi &= (1/\sqrt{2})(u + \rho)e^{i\zeta}, \\ B_\mu &= A_\mu - (1/e)\partial_\mu \zeta, \end{aligned} \tag{2.4}$$

where ρ and ζ are real. Under the gauge transformation of the second kind, we have

$$\begin{aligned} &B_\mu \to B_\mu, \\ &\rho \to \rho, \\ &\zeta \to \zeta^\theta = \zeta + \theta. \end{aligned} \tag{2.5}$$

The constant u is to be so adjusted that $\langle\rho\rangle_0 = \langle\zeta\rangle_0 = 0$. (Only in the tree approximation does one have $u = v$). The gauge condition is chosen to be

$$\zeta = c, \tag{2.6}$$

where c is a constant, which we shall choose to be zero.

Since

$$\int [d\theta] \prod_x \delta[\zeta^\theta(x)] = 1$$

we may write the generating functional of the U-gauge Green's functions as

$$\exp i Z_U[J_\mu, K] = \int [dB_\mu] d\rho [d\zeta] \prod_x \delta[\zeta(x)] J[\rho] \times \exp i \left\{ S_U[B_\mu, \rho] + \int d^4x [-J_\mu B^\mu + K\rho] \right\}, \tag{2.7}$$

where S_U is the action expressed in terms of the U-gauge variables, and $J[\rho]$ is the Jacobian of the functional transformation $(\psi, \chi) \to (\rho, \zeta)$:

$$J[\rho] \sim \exp\left\{ \delta^4(0) \int d^4x \ln\left[1 + \frac{1}{u}\rho(x)\right]\right\}. \tag{2.8}$$

We may restrict the source J_μ to be transverse in Eq. (2.7):

$$\partial^\mu J_\mu = 0. \tag{2.9}$$

Since

$$\int [d\theta] \prod_x \delta[\partial^\mu A_\mu^\theta(x)] = \text{constant}, \tag{2.10}$$

independent of A_μ, we may insert the factor (2.10) on the right-hand side of Eq. (2.7), and revert to the R-gauge variables. We obtain thereby

$$\exp i Z_U[J_\mu, K] = \int [dA_\mu][d\psi][d\chi] \prod_x \delta(\partial^\mu A_\mu(x)) \exp i\left(S_R[A_\mu, \psi, \chi] + \int d^4x \left\{ -J_\mu A^\mu + K([(v+\psi)^2 + \chi^2]^{1/2} - u)\right\}\right). \tag{2.11}$$

Equation (2.11) allows us to evaluate the U-gauge Green's functions by the Feynman rules of the R gauge.

It follows immediately from Eq. (2.11) that the transverse parts of the vector meson propagators are the same in both gauges. In particular, if we write

$$\lim_{k^2 \to m^2} \Delta_{\mu\nu}(k; U) \sim \left(g_{\mu\nu} - \frac{k_\mu k_\nu}{m^2}\right) \frac{X_3}{k^2 - m^2} + k_\mu k_\nu \text{ term},$$

$$\lim_{k^2 \to m^2} \Delta_{\mu\nu}(k; R) \sim \left(g_{\mu\nu} - \frac{k_\mu k_\nu}{m^2}\right) \frac{Z_3}{k^2 - m^2} + k_\mu k_\nu \text{ term},$$

we have

$$X_3 = Z_3. \tag{2.12}$$

Let us examine the scalar source term in Eq. (2.11). It is

$$K([(v+\psi)^2 + \chi^2]^{1/2} - u) = K\left[(v - u) + \psi + \frac{1}{2v}\chi^2 - \frac{1}{2v^2}\psi\chi^2 + \cdots\right]. \tag{2.13}$$

The scalar propagators behave near $k^2 = \mu^2$ like

$$\lim_{k^2\to\mu^2} \Delta(k^2; U) \sim \frac{X_2}{k^2-\mu^2},$$
$$\lim_{k^2\to\mu^2} \Delta(k^2; R) \sim \frac{Z_2}{k^2-\mu^2}. \qquad (2.14)$$

The ratio $(X_2/Z_2)^{1/2}$ is not equal to unity, due to the possibility of exciting the physical scalar meson by the nonlinear terms in Eq. (2.13). Since the series on the right-hand side of Eq. (2.13) is infinite, X_2 contains high-order divergences. The mass shifts of the scalar boson in both gauges are the same, however.

From Eq. (2.11) we can compute a Green's function in the U gauge. After the amputation of external lines, and after letting external momenta on the mass shell, we obtain

$$X_2^{E_s/2}\,T(U) = Z_2^{E_s/2}\,T(R), \qquad (2.15)$$

where E_s is the number of external scalar lines and $T(U)$ is the T matrix in the U gauge. Note that Eq. (2.15) holds between the T matrices. There are no such simple relations between proper vertices in the two formulations.

III. RENORMALIZATION

The Lagrangian in the U gauge is

$$\begin{aligned}\mathcal{L} = &-\tfrac{1}{4}(\partial_\mu B_\nu - \partial_\nu B_\mu)^2 + \tfrac{1}{2}(eu)^2 B_\mu^2 \\ &+\tfrac{1}{2}(\partial_\mu\rho)^2 - \tfrac{1}{2}(2\lambda u^2)\rho^2 - \tfrac{1}{4}\rho^4 - \lambda u\rho^3 \\ &+\tfrac{1}{2}e^2 B_\mu^2(2u\rho+\rho^2) - \tfrac{1}{2}\delta\mu^2\rho^2 - u\delta\mu^2\rho. \qquad (3.1)\end{aligned}$$

The discussion in Sec. II implies that the T matrix becomes finite if we choose $\delta\mu^2$ so that the vacuum expectation value of ρ is zero, and renormalize fields and constants according to

$$\begin{aligned} B_\mu &= B_\mu^r X_3^{1/2}, \\ \rho &= \rho^r X_2^{1/2}, \\ u &= v^r (X_2')^{1/2}, \qquad (3.2)\\ e &= e_r X_1 X_2^{-1} X_3^{-1/2}, \\ \lambda &= \lambda_r X_4 X_2^{-2}. \end{aligned}$$

We will now discuss how X_1, X_4 and X_2' may be chosen.

Let $e^2 u\,\Gamma_{1,2}(U)$ be the on-shell T-matrix element with one scalar and two vector particles. We will define X_1 by

$$\Gamma_{1,2} = \frac{X_2}{X_1}\left(\frac{X_2}{X_2'}\right)^{1/2}. \qquad (3.3)$$

Then the renormalized T matrix,

$$e^2 u \Gamma_{1,2}(U) X_3 X_2^{1/2} = e_r^2 v_r, \qquad (3.4)$$

is finite. Similarly, we write $\lambda u \Gamma_{3,0}(U)$ for the on-shell T-matrix for three scalar particles, and define X_4 by

$$\Gamma_{3,0}(U) = \frac{1}{X_4}\left(\frac{X_2}{X_2'}\right)^{1/2}. \qquad (3.5)$$

We may choose X_2' so that the renormalized vector propagator has the low-energy limit

$$\lim_{k_\mu\to 0}\frac{1}{X_3}\Delta_{\mu\nu}(k) = -g_{\mu\nu}\frac{1}{(e_r v_r)^2}. \qquad (3.6)$$

The physical masses m^2 and μ^2 are finite functions of of e_r, λ_r, and v_r.

The ratios $(X_2/Z_2)^{1/2}$ and $(X_2'/Z_2')^{1/2}$ may be computed perturbatively from the structure of Eq. (2.13), where Z_2 is defined by Eq. (2.13) and Z_2' by the relation $v = v_r(Z_2')^{1/2}$. [The ratio (Z_2/Z_2') is finite.] We have

$$\left(\frac{X_2}{Z_2}\right)^{1/2} = 1 + i\lambda\left[\int\frac{d^4k}{(2\pi)^4}\frac{1}{k^2}\frac{1}{(p-k)^2}\right]_{p^2=\mu^2} - i\frac{1}{2v^2}\int\frac{d^4k}{(2\pi)^4}\frac{1}{k^2}, \qquad (3.7)$$

in the one-loop approximation. The requirement that $\langle\rho\rangle_0 = 0$ translates into

$$\delta Z_v/\delta K = 0. \qquad (3.8)$$

Combining Eqs. (3.8), (2.7), and (2.13), we see that

$$u = v + \frac{i}{2v}\int\frac{d^4k}{(2\pi)^4}\frac{1}{k^2}$$

or

$$(X_2'/Z_2')^{1/2} = 1 + \frac{i}{2v^2}\int\frac{d^4k}{(2\pi)^4}\frac{1}{k^2}, \qquad (3.9)$$

in the one-loop approximation.

There is an interesting check on Eq. (3.9). In the one-loop approximation, the transverse self-energy of the vector boson in the R gauge is given by, with $m^2=(ev)^2$, $\mu^2=\lambda v^2$,

$$\begin{aligned}\Sigma_{\mu\nu}(p;R) &= i\,4e^2(ev)^2\int\frac{d^4k}{(2\pi)^4}\frac{1}{(p-k)^2-\mu^2}\frac{1}{k^2-m^2}\left(g_{\mu\nu}-\frac{k_\mu k_\nu}{k^2}\right) \\ &\quad - ie^2\int\frac{d^4k}{(2\pi)^4}\frac{1}{(p-k)^2-\mu^2}\frac{1}{k^2}4k_\mu k_\nu + ie^2 g_{\mu\nu}\int\frac{d^4k}{(2\pi)^4}\left[\frac{1}{k^2-\mu^2}+\frac{1}{k^2}\right] \\ &= \Sigma_{\mu\nu}(p;U) + ie^2 g_{\mu\nu}\int\frac{d^4k}{(2\pi)^4}\frac{1}{k^2}.\end{aligned}$$

This difference between the self-energies in the two gauges must be accountable by the difference in the first-order expressions. From Eq. (3.9) we see that

$$e^2u^2 = e^2v^2 + ie^2\int \frac{d^4k}{(2\pi)^4}\frac{1}{k^2}.$$

IV. NON-ABELIAN CASE

In this section we will derive the generalization of Eq. (2.11) to the non-Abelian case. We shall use the concepts and notations developed in Sec. III of paper II. Corresponding to the decomposition of generators (3.11), we can write a finite transformation as

$$g = e^{i\vec{\alpha}\cdot\vec{L}} = e^{i\vec{\beta}\cdot\vec{T}}e^{i\vec{\gamma}\cdot\vec{t}}, \quad g \in G \tag{4.1}$$

where $\{\alpha\}$ and $\{\beta,\gamma\}$ are parametrizations of the group manifold. Under this transformation, $\vec{\xi}$ and ρ defined by Eq. (II3.11),

$$\phi = D[e^{i\vec{\xi}\cdot\vec{t}}](v+\rho),$$

transform nonlinearly[5] as

$$\vec{\xi} \to \vec{\xi}'(\vec{\xi}, g) \tag{4.2}$$

where $\xi'(\xi,g)$ is defined by

$$g e^{i\vec{\xi}\cdot\vec{t}} = e^{i\vec{t}\cdot\vec{\xi}'(\vec{\xi},g)}e^{i\vec{u}(\vec{\xi},g)\cdot\vec{T}} \tag{4.3}$$

and

$$\rho \to \rho'(g) = D[e^{i\vec{T}\cdot\vec{u}(\vec{\xi},g)}]\rho. \tag{4.4}$$

The vector fields $\vec{B}_\mu$ and $\vec{C}_\mu$ defined by

$$\{\vec{B}_\mu\cdot\vec{t} + \vec{C}_\mu\cdot\vec{T}\} = e^{-i\vec{\xi}\cdot\vec{t}}\vec{A}_\mu\cdot\vec{L}e^{i\vec{\xi}\cdot\vec{t}} + (i/g)e^{-i\vec{\xi}\cdot\vec{t}}\partial_\mu e^{i\vec{\xi}\cdot\vec{t}} \tag{4.5}$$

transform like

$$\vec{B}_\mu\cdot\vec{t} + \vec{C}_\mu\cdot\vec{T} \to \vec{B}'_\mu(g)\cdot\vec{t} + \vec{C}'_\mu(g)\cdot\vec{T} = e^{i\vec{u}\cdot\vec{T}}[\vec{B}_\mu\cdot\vec{t} + \vec{C}_\mu\cdot\vec{T}]e^{-i\vec{u}\cdot\vec{T}} - (i/g)(\partial_\mu e^{i\vec{u}\cdot\vec{T}})e^{-i\vec{u}\cdot\vec{T}}, \quad \vec{u} = \vec{u}(\vec{\xi}, g) \tag{4.6}$$

We define the U-gauge Jacobian Δ_U by

$$\Delta_U[\vec{C}_\mu, \vec{\xi}]\int[dg]\prod_x[\delta(\partial^\mu\vec{C}'_\mu(x,g))\delta(\vec{\xi}'(x,g))] = 1, \tag{4.7}$$

where dg is the invariant Hurwitz measure over the group manifold. The generating functional of the U-gauge Green's functions is

$$\exp iZ_U[\vec{I}_\mu, \vec{J}_\mu, K] = \int[d\vec{C}_\mu][d\vec{B}_\mu][d\vec{\xi}][d\rho]\Delta_U[\vec{C}_\mu, \vec{\xi}]J_\phi[\rho, \vec{\xi}]$$
$$\times \prod_x \delta(\partial^\mu\vec{C}_\mu(x))\delta(\vec{\xi}(x))\exp i\left\{S_U[\vec{B}_\mu, \vec{C}_\mu, \rho] + \int d^4x[K\cdot\rho - \vec{I}_\mu\cdot\vec{C}^\mu - \vec{J}_\mu\cdot\vec{B}^\mu]\right\}, \tag{4.8}$$

where $J_\phi[\rho, \xi]$ is the Jacobian of the functional transformation $\phi \to (\rho, \xi)$. Because of the δ functions in the integrand of Eq. (4.8) we may replace Δ_U by 1, and J_ϕ by $J_\phi[\rho, 0]$.

We may insert

$$\Delta_L[\vec{A}_\mu]\int[dg]\prod_x\delta[\partial^\mu\vec{A}'_\mu(x;g)] = 1 \tag{4.9}$$

in the integrand of Eq. (4.8) and revert to the original variable $\vec{A}_\mu$ and ϕ (which we shall assume to be real). We obtain, making use of the invariance of Δ_U, Δ_L, and S under the transformation,

$$\exp iZ_U = \int[d\vec{A}_\mu][d\phi]\Delta_L[\vec{A}_\mu]\prod_x\delta(\partial^\mu A_\mu(x))e^{iS_R[\vec{A}_\mu,\phi]}\Delta_U[\vec{C}_\mu, \vec{\xi}]$$
$$\times\int[dg]\prod_x\left\{\delta(\partial^\mu\vec{C}'_\mu(g))\delta(\vec{\xi}(g))\exp i\int d^4x[K\cdot\rho'(g) - \vec{I}^\mu\cdot\vec{C}'_\mu(g) - \vec{J}^\mu\cdot\vec{B}'_\mu(g)\right\}, \tag{4.10}$$

where ρ, ξ, B_μ, and C_μ on the right-hand side are to be regarded as nonlinear functionals of A_μ and ϕ.

Let $g_0 = g(\beta_0, \gamma_0)$ be such that

$$\vec{\xi}'(x; \vec{\xi}, g_0) = 0 \tag{4.11}$$

and

$$\partial^\mu \vec{C}'_\mu(x; g_0) = 0 . \tag{4.12}$$

Then Eq. (4.10) becomes

$$\exp i Z_U = \int [d\vec{A}_\mu][d\phi]\Delta_L[\vec{A}_\mu]\prod_x \delta(\partial^\mu \vec{A}_\mu(x)) \exp i\Big\{S_R[\vec{A}_\mu, \phi] + \int d^4x[K\cdot\rho'(g) - \vec{I}^\mu \cdot \vec{C}'_\mu(g) - \vec{J}^\mu \cdot B'_\mu(g)]\Big\}, \tag{4.13}$$

where $\rho'(g_0)$, $C'_\mu(g_0)$, and $B'_\mu(g_0)$ need still be expressed in terms of A_μ and ϕ. Equation (4.11) is satisfied if we choose

$$\vec{\gamma}_0 = -\vec{\xi}, \tag{4.14}$$

in which case we have also $\vec{u}(\vec{\xi}, g_0) = \vec{\beta}_0$. The parameters $\vec{\beta}_0$ are then determined by the requirement that $\vec{C}'_\mu$,

$$\vec{T}\cdot\vec{C}'_\mu(g_0) = e^{i\vec{\beta}_0\cdot\vec{T}}\,\vec{T}\cdot\vec{C}_\mu\, e^{-i\vec{\beta}_0\cdot\vec{T}} - (i/g)(\partial_\mu e^{i\vec{\beta}_0\cdot\vec{T}})e^{-i\vec{\beta}_0\cdot\vec{T}}, \tag{4.15}$$

be divergenceless.

As an illustration, we shall derive the expressions for $\rho'(g_0)$, $C'_\mu(g_0)$, and $B'_\mu(g_0)$ for the model discussed in Sec. IV of paper II; SU(2) gauge bosons interesting with an isotriplet of scalar mesons $\vec{\phi}$, with $\langle\varphi_3\rangle_0 = v$. First, we have

$$\begin{aligned}(\xi_1, \xi_2) &= (\phi_2, -\phi_1)\frac{1}{(\phi_1^2+\phi_2^2)^{1/2}}\arcsin\frac{(\phi_1^2+\phi_2^2)^{1/2}}{u+\rho}\\ &= -(\gamma_1, \gamma_2)_0\end{aligned} \tag{4.16}$$

and

$$\rho = (\phi_1^2+\phi_2^2+\phi_3^2)^{1/2} - u . \tag{4.17}$$

The parameter β_0 is determined from

$$\partial^\mu C'_\mu(g_0) = \partial^\mu[C_\mu + (1/g)\partial_\mu\beta_0] = 0 ,$$

so that

$$\beta_0 = -g(1/\partial^2)\partial^\mu C_\mu . \tag{4.18}$$

The fields C_μ and $\vec{B}_\mu$ may be computed from Eq. (4.5):

$$\begin{aligned}C_\mu &= \tfrac{1}{2}\operatorname{Tr}\tau^3[e^{-i\vec{\xi}\cdot\vec{\tau}/2}\vec{A}_\mu\cdot\vec{\tau}\,e^{i\vec{\xi}\cdot\vec{\tau}/2} + (2i/g)\partial_\mu(e^{-i\vec{\xi}\cdot\vec{\tau}/2})e^{i\vec{\xi}\cdot\vec{\tau}/2}]\\ &= A_\mu^3 + (\vec{\xi}\times\vec{A}_\mu)^3 - (1/2g)(\vec{\xi}\times\partial_\mu\vec{\xi})^3 + \cdots\end{aligned} \tag{4.19}$$

and

$$\begin{aligned}B_\mu^\perp &= \tfrac{1}{2}\operatorname{Tr}\tau^\perp[e^{-i\vec{\xi}\cdot\vec{\tau}/2}\vec{A}_\mu\cdot\vec{\tau}e^{i\vec{\xi}\cdot\vec{\tau}/2} + (2i/g)(\partial_\mu e^{-i\vec{\xi}\cdot\vec{\tau}/2})e^{i\vec{\xi}\cdot\vec{\tau}/2}]\\ &= A_\mu^\perp - \tfrac{1}{2}\partial_\mu\xi^\perp + (\vec{\xi}\times\vec{A}_\mu)^\perp + \cdots .\end{aligned} \tag{4.20}$$

So, finally, we have

$$\begin{aligned}\rho'(g_0) &= \rho = (\phi_1^2+\phi_2^2+\phi_3^2)^{1/2} - u ,\\ C'_\mu(g_0) &= (g_{\mu\nu} - \partial_\mu\partial_\nu/\partial^2)C^\nu ,\end{aligned} \tag{4.21}$$

and

$$[B'_\mu(g_0)]^\perp = \tfrac{1}{2}\operatorname{Tr}\tau^\perp[e^{i\beta_0\tau_3/2}\vec{B}_\mu\cdot\vec{\tau}\,e^{-i\beta_0\tau_3/2}],$$

where β_0, C_μ, and $\vec{B}_\mu$ are to be expressed in terms of $\vec{A}_\mu$ and ϕ by the use of Eqs. (4.16)–(4.20).

Incidentally,

$$\beta_0 = -g\frac{1}{\partial^2}[(\partial^\mu \vec{\xi} \times \vec{A}_\mu)^3 - \frac{1}{2g}(\vec{\xi} \times \partial^2 \vec{\xi})^3] + \cdots .$$

In the non-Abelian case $X_3 \neq Z_3$ in general, and the equivalence of the T matrix in two gauges is expressed as

$$X_2^{E_s/2} X_3^{E_v/2}\, T(U) = Z_2^{E_s/2} Z_3^{E_v/2}\, T(R), \tag{4.22}$$

where E_v is the number of external vector lines. Equation (4.22) shows that the T matrix in the U gauge is finite after renormalization, and the T matrix in the R gauge is unitary and devoid of infrared divergences.

---

*On leave of absence from SPT, CEN Saclay, B.P. 2, 91 Gif-sur-Yvette, France.

[1]S. Weinberg, Phys. Rev. Letters 27, 1688 (1972).
[2]A. Salam and J. Strathdee (unpublished).
[3]T. Appelquist and H. Quinn (unpublished).
[4]B. W. Lee, Phys. Rev. D 5, 823 (1972).
[5]S. Coleman, J. Wess, and B. Zumino, Phys. Rev. 177, 2239 (1969).

Spontaneously Broken Gauge Symmetries. IV. General Gauge Formulation

Benjamin W. Lee*
Institute for Theoretical Physics, State University of New York at Stony Brook, Stony Brook, New York 11790

and

Jean Zinn-Justin
Service de Physique Théorique, Centre d'Etudes Nucléaires de Saclay, B.P.2, 91 Gif-sur-Yvette, France
(Received 30 October 1972)

The advent of the dimensional-regularization procedure allows the study of renormalizability of spontaneously broken gauge theories formulated in a wide class of gauges. We derive and study the Ward-Takahashi identities appropriate to such gauges. A consequence of the Ward-Takahashi identities is that the physical S matrix is invariant under a variation of the gauge condition. As remarked before, since the variation of a parameter in the R_ξ-gauge formulation shifts the masses of unphysical excitations, the above result, the ξ independence of the physical S matrix, implies that the unphysical excitations cannot contribute to the sum over intermediate states, establishing the unitarity of the S matrix. We also give the renormalization procedure of a model formulated in the R_ξ gauge.

I. INTRODUCTION

The advent of a very powerful regularization procedure for Feynman integrals – the so-called dimensional regularization[1] – permits us to discuss intelligently the renormalizability question of spontaneously broken gauge theories formulated in a fairly general class of gauges.[2] The present paper is dedicated to the derivation of the Ward-Takahashi (WT) identities in such gauges, which can be used to prove the renormalizability and unitarity of the theory in question. It has been observed[3] that in the so-called R_ξ-gauge formulation invariance of the physical S matrix under the variation of a gauge-specifying parameter (i.e., ξ) implies the unitarity of the S matrix, that is to say, that unphysical excitations do not contribute to sums over intermediate states. Thus, the ability to formulate quantum theory of spontaneously broken gauge symmetry in a general class of gauge conditions, in a way that reflects the gauge invariance of the action as expressed through the WT identities, is extremely useful in showing that the theory is in fact unitary and renormalizable. [Let us recall that the proof (in paper II) of the unitarity in the R gauge was extremely complicated.]

In our discussion we shall assume that all expressions are dimensionally regulated, so that formal manipulations of the kinds necessary in showing the WT identities for Feynman amplitudes are justified.[1] In the next section, we consider a general class of gauge conditions and derive the WT identities appropriate to the gauges being considered. An important lemma we use in the derivation, which is a generalization of the results of Fradkin and Tyutin, and Slavnov[4] for the Landau gauge formulation, is proved in the Appendix. In Sec. III we give a demonstration that the physical S matrix is invariant under an infinitesimal change in the gauge condition. This, together with the remark in Ref. 3, establishes the unitary of the physical S matrix. In Sec. IV we discuss renormalization conditions of a model formulated in the R_ξ gauge. Here we illustrate how renormalization constants may be chosen in accordance with the WT identities. Finally the Lagrangian of the model is written down in terms of renormalized fields and parameters.

II. WT IDENTITIES

We shall discuss a theory consisting of Bose fields. Recently Bardeen[5] has given a discussion of renormalizability of gauge theories with fermions. He has shown that the problem associated with the anomalies[6-8] in fermion loops can be isolated from the general problem of renormalizability and gauge invariance, and if fermion loop anomalies are absent, or canceled among themselves *in lowest order*, the presence of fermions does not hinder the WT identities from being valid.

Let ϕ_i be the set of all fields including the gauge fields transforming as a linear (in general reducible) representation of a compact Lie group. The infinitesimal transformation is given by

$$\phi_i^g = \phi_i + (\Gamma_{ij}^\alpha \phi_j + \Lambda_i^\alpha) g_\alpha , \tag{1}$$

where our notation is such that the indices i and α stand for the space-time variables as well as the internal symmetry indices, and summation *and integration* over repeated indices will always be understood; g_α is the space-time-dependent parameter of the Lie group and Γ_{ij}^α is a reducible representation of the generator labeled by α, and Λ_i^α may involve space-time derivatives. The Lagrangian $\mathcal{L}$ is invariant under the transformation (1).

We choose as the gauge condition

$$F_\alpha(\phi) = a_\alpha , \tag{2}$$

and consider the integral

$$\Delta_F^{-1}(\phi) = \int [dg] \prod_\alpha \delta(F_\alpha(\phi^g) - a_\alpha) , \tag{3}$$

where $[dg] = \prod_x dg$ is the product of Hurwitz integrals at every space-time point and a_α is independent of ϕ_i and g. We need only compute, for our purpose, Δ_F with the restriction to the manifold $F_\alpha(\phi) = a_\alpha$. It is given by

$$\Delta_F(\phi) = \det M_{\alpha\beta} \tag{4}$$

for ϕ such that $F_\alpha(\phi) = a_\alpha$, where

$$M_{\alpha\beta} = \frac{\delta F_\alpha}{\delta \phi_i} (\Gamma_{ij}^\beta \phi_j + \Lambda_i^\beta) . \tag{5}$$

Popov and Faddeev[9] have shown that the vacuum-to-vacuum amplitude for a gauge theory should be written as

$$W(a_\alpha) = \int [d\phi] e^{iS[\phi]} \det M \prod \delta(F_\alpha(\phi) - a_\alpha) , \tag{6}$$

where $S[\phi]$ is the gauge-invariant action, $S[\phi] = \int d^4x \mathcal{L}(\phi)$. Consider the transformation (1) with g_α however restricted by the condition

$$g_\alpha = [M^{-1}(\phi)]_{\alpha\beta} \lambda_\beta , \tag{7}$$

where λ_α is an arbitrary infinitesimal number independent of ϕ. We show in the Appendix that $(\det M)[d\phi]$ is a measure invariant under the transformation (1) with the restriction (7). From Eq. (6), it follows then that W is invariant under an infinitesimal change of a_α, $\delta a_\alpha = \lambda_\alpha$,

$$\frac{dW(a_\alpha)}{da_\alpha} = 0 , \quad \text{for all } a_\alpha , \tag{8}$$

so W is independent of a_α. One can therefore integrate over a_α the right-hand side of Eq. (6) without changing the result,[10] up to some irrelevant normalization:

$$\begin{aligned} W &= \int [da] \prod_\alpha H(a_\alpha) \int [d\phi]\, e^{iS[\phi]} \\ &\quad \times \det M \prod_\alpha \delta(F_\alpha(\phi) - a_\alpha) \\ &= \int [d\phi]\, e^{iS[\phi]} \det M \prod_\alpha H(F_\alpha(\phi)) . \end{aligned} \tag{9}$$

We shall specialize $H(a_\alpha)$ to a Gaussian function in the following:

$$\prod_\beta H(a_\beta) = \exp(-\tfrac{1}{2} i a_\beta^2) . \tag{10}$$

Equation (9) can be written as[10]

$$W = \int [d\phi] \exp(iS[\phi] - \tfrac{1}{2} i F_\alpha^2) \det M . \tag{11}$$

Furthermore $\det M$ can be described as loops generated by a set of complex scalar fields c_α and $\bar{c}_\alpha$ obeying Fermi-Dirac statistics.[9,11] Equation (11) can thus be written as[9,10]

$$W=\int[d\phi][dc][d\bar{c}]\exp(iS_{\rm eff}[\phi,c,\bar{c}])\,, \tag{12}$$

where

$$S_{\rm eff}[\phi,c,\bar{c}]=S[\phi]-\tfrac{1}{2}F_\alpha{}^2+\bar{c}_\alpha M_{\alpha\beta}c_\beta\,. \tag{13}$$

So far we have been silent about permissible choices of the gauge condition $F_\alpha=a_\alpha$. The gauge condition must be so chosen that the operator $M_{\alpha\beta}$ is nonsingular, i.e.,

$$\det M\neq 0\,, \tag{14}$$

and in order that the Green's functions of the theory be renormalizable the effective action, Eq. (13), must not contain terms of dimension higher than four. This requires in particular that the dimension of the function F not exceed two.

We will now introduce source terms J_i for the fields ϕ_i in order to generate Green's functions:

$$W[J]=\int[d\phi]\det M\exp i(S[\phi]-\tfrac{1}{2}F_\alpha{}^2+\phi_iJ_i)\,. \tag{15}$$

Performing the nonlinear gauge transformations previously defined by Eq. (7) and remembering that a change of variables does not change the value of an integral, we obtain the identity

$$0=\int[d\phi]\det M\exp\{i(S[\phi]-\tfrac{1}{2}F_\alpha{}^2+J_i\phi_i)\}\frac{\delta}{\delta\lambda_\beta}(S[\phi]-\tfrac{1}{2}F_\alpha{}^2+J_i\phi_i)\,,$$

or

$$\int[d\phi]\det M\exp\{i(S[\phi]-\tfrac{1}{2}F_\gamma{}^2+J_i\phi_i)\}[-F_\alpha+J_i(\Gamma^\beta_{ij}\phi_j+\Lambda^\beta_i)(M^{-1})_{\beta\alpha}]=0\,. \tag{16}$$

Equation (16) can be translated into an equation for $W[J]$:

$$\left\{-F_\alpha\left(\frac{1}{i}\frac{\delta}{\delta J}\right)+J_i\left(\Gamma^\beta_{ij}\frac{1}{i}\frac{\delta}{\delta J}+\Lambda^\beta_i\right)\left[M^{-1}\left(\frac{1}{i}\frac{\delta}{\delta J}\right)\right]_{\beta\alpha}\right\}W[J]=0\,. \tag{17}$$

This is the WT identity for generating functional $W[J]$ in the gauge defined by $F_\alpha(\phi)=a_\alpha$.

In the special case where F_α is a linear function of fields ϕ_i,

$$F_\alpha(\phi)=F_{\alpha i}\phi_i\,, \tag{18}$$

a set of solutions of Eq. (17) can be obtained in the following way. Let us compute $W[J]$ with $J_i=K_\alpha F_{\alpha i}$,

$$W[K_\alpha F_{\alpha i}]=\int[d\phi]\det M\times\exp(i\{S[\phi]-\tfrac{1}{2}(F_\alpha-K_\alpha)^2+\tfrac{1}{2}K_\alpha{}^2\})\,. \tag{19}$$

Using the fact that one can add to F_α an arbitrary function of space-time by a succession of gauge transformations which satisfy Eq. (7) and leave the metric $\det M[d\phi]$ and the action $S[\phi]$ invariant, we can perform the integral and obtain

$$W[K_\alpha F_{\alpha i}]=W[0]\,e^{iK_\alpha{}^2/2}$$

or

$$Z[K_\alpha F_{\alpha i}]\equiv -i\ln W[K_\alpha F_{\alpha i}]=\tfrac{1}{2}K_\alpha{}^2+\text{const}\,. \tag{20}$$

Equation (17) requires knowing the quantity

$$\left[M^{-1}\left(\frac{1}{i}\frac{\delta}{\delta J}\right)\right]_{\beta\alpha}W[J]\,.$$

To this end we consider

$$W_{\alpha\beta}[J]=\int[d\phi][dc][d\bar{c}]c_\alpha\bar{c}_\beta\times\exp\{i(S[\phi]-\tfrac{1}{2}F_\alpha{}^2+\bar{c}_\alpha M_{\alpha\beta}c_\beta+J_i\phi_i)\}\,. \tag{21}$$

The functional $W_{\alpha\beta}[J]$, the Green's function for c in the presence of external sources, satisfies the equation

$$M_{\alpha\beta}\left(\frac{1}{i}\frac{\delta}{\delta J}\right)W_{\beta\gamma}[J]=\delta_{\alpha\gamma}W[J]$$

or

$$\left[M^{-1}\left(\frac{1}{i}\frac{\delta}{\delta J}\right)\right]_{\alpha\beta}W[J]=W_{\beta\gamma}[J]\,. \tag{22}$$

We believe that the discussion here of the WT identities for a general class of gauges parallels the combinatoric discussion of 't Hooft and Veltman[12] on the same subject.

III. SOME CONSEQUENCES OF WT IDENTITIES

The WT identities implied by Eq. (17) are satisfied by the dimensionally regulated Green's functions of the theory for general n, the number of space-time dimensions.[1] If the effective action $S_{\rm eff}$ in Eq. (12) is that of a renormalizable theory, then the singularities of the Green's functions at

$n=4$ are removed by the renormalization of fields and parameters of the Lagrangian and $F_\alpha(\phi)$. The renormalized Green's functions then satisfy the renormalized WT identities obtained by rescaling parameters and sources J_i in Eq. (17), as we have shown for the R-gauge formulation in paper II.

In this section we shall demonstrate that, for general n, a small variation in the unrenormalized expression for F_α leaves the physical S matrix invariant. It follows from the discussion in the last paragraph that when the theory is renormalized as we discussed, the physical S matrix is invariant under a variation of renormalized parameters appearing in the gauge condition.

As alluded to in the Introduction, a variation of the parameter ξ in the R_ξ-gauge formulation[3] causes the masses of unphysical particles to vary; in fact, in the limit $\xi \to 0$, these become infinite. Thus it is seen that the ξ-independence of the S matrix implies the unitarity of the S matrix; that is to say, the unphysical particles decouple from physical ones on mass shell.

We will give a small variation ΔF_α to F_α and compute the variation ΔW of $W[J]$:

$$\Delta W[J] = \int [d\phi] \det M \exp\{i(S[\phi] - \tfrac{1}{2}F_\alpha{}^2 + J_i\phi_i)\}\left(-iF_\alpha \Delta F_\alpha + \frac{\delta \Delta F_\alpha}{\delta\phi_i}(\Gamma^\beta_{ij}\phi_j + \Lambda^\beta_i)(M^{-1})_{\beta\alpha}\right). \tag{23}$$

We will now use Eq. (16). Applying $i\Delta F_\alpha(\delta/i\delta J)$ on it, we obtain

$$0 = \int \left[-iF_\alpha\Delta F_\alpha + \left(J_i \Delta F_\alpha + \frac{\delta\Delta F_\alpha}{\delta\phi_i}\right)(\Gamma^\beta_{ij}\phi_j + \Lambda^\beta_i)(M^{-1})_{\beta\alpha}\right]\exp\{i(S[\phi] - \tfrac{1}{2}F_\alpha{}^2 + J_i\phi_i)\}\det M[d\phi]. \tag{24}$$

Combining Eqs. (23) and (24), we finally obtain

$$\Delta W[J] = \int [d\phi]\exp\{i(S[\phi] - \tfrac{1}{2}F_\alpha{}^2 + J_i\phi_i)\}\det M J_i \Delta F_\alpha(\phi)(\Gamma^\beta_{ij}\phi_j + \Lambda^\beta_i)(M^{-1})_{\beta\alpha}. \tag{25}$$

The S-matrix element, not necessarily connected, is obtained from W by differentiating with respect to J's around $J=0$, truncating external lines, and setting every external momentum on its mass shell. Let S_F be the quantity so obtained in the gauge specified by F_α. Then the structure of ΔW in Eq. (25) implies

$$S_F + \Delta S_F = \left[1 + \tfrac{1}{2}\sum_e \left(\frac{\delta Z_F}{Z_F}\right)_e\right] S_F$$

or (26)

$$S_{F+\Delta F} = S_F \prod_e \left(\frac{Z_{F+\Delta F}{}^{1/2}}{Z_F{}^{1/2}}\right)_e,$$

where Z_F is the wave-function renormalization constant in the gauge specified by F_α, and the summation $\sum_e$ and the product $\prod_e$ run over all external lines e. The change in the renormalization constant, δZ_F, is obtained from the change in the propagator:

$$i\frac{\delta \Delta W[J]}{\delta J_i \delta J_k}\bigg|_{J=0} = -\int [d\phi]\det M \exp\{i(S[\phi] - \tfrac{1}{2}F_\alpha{}^2)\}[\phi_i \Delta F_\alpha(\Gamma^\beta_{kj}\phi_j + \Lambda^\beta_k)(M^{-1})_{\beta\alpha} + (i \leftrightarrow k)]. \tag{27}$$

The quantity $(\delta Z_F)_e$ is the residue of the pole of the Fourier transform of the above expression when i and k take the internal quantum numbers of e. Equation (26) shows that the renormalized, physical S-matrix element,

$$S = S_F \prod_e (Z_F{}^{-1/2})_e, \tag{28}$$

is independent of the gauge chosen, F_α, to compute it.

IV. RENORMALIZATION

In this section we apply the considerations of previous sections to a model, partly to illustrate the general arguments presented abstractly there, and partly to illustrate the renormalization procedure based on the WT identities is the R_ξ gauge.

We choose as our model the system of a quartet of scalar mesons. They form a representation of $[SU(2)]_G \times SU(2)$ where the first factor is a gauge symmetry. The $SU(2)\times SU(2)$ symmetry is spontaneously broken, leaving the diagonal SU(2) as invariance of the vacuum. In this model the triplet of gauge bosons all become massive.[13] In a gauge theory of weak and electromagnetic interactions, one of the gauge bosons remains massless and causes infrared divergences. In this model we avoid the infrared problem completely. In any case the infrared problem in the former is no worse than in quantum electrodynamics when formulated in the R_ξ gauge and can be resolved in the usual way.

The Lagrangian of the model is written as

$$\mathcal{L} = -\tfrac{1}{4}(\partial_\mu \vec{A}_\nu - \partial_\nu \vec{A}_\mu + g\vec{A}_\mu \times \vec{A}_\nu)^2 + \tfrac{1}{2}[(\partial_\mu \psi)^2 + (\partial_\mu \vec{\chi})^2] + \tfrac{1}{2} g \vec{A}^\mu \cdot (\psi \partial_\mu \vec{\chi} - \chi\, \partial_\mu \psi + \vec{\chi} \times \partial_\mu \vec{\chi})$$
$$+ \tfrac{1}{8} g^2 \vec{A}_\mu{}^2(\psi^2 + \vec{\chi}^2) - \tfrac{1}{2}\beta(\psi^2 + \vec{\chi}^2) - \tfrac{1}{4}\alpha(\psi^2 + \vec{\chi}^2)^2 . \qquad (29)$$

The infinitesimal form of the local $[\mathrm{SU}(2)]_G$ transformations is

$$\delta\phi_i = (\Gamma^\alpha_{ij}\phi_j + \Lambda^\alpha_i)\omega_\alpha ,$$
$$\delta\vec{A}_\mu = -\vec{\omega} \times \vec{A}_\mu + \frac{1}{g}\partial_\mu \vec{\omega} , \qquad (30)$$
$$\delta\chi = -\vec{\omega}\cdot\vec{\psi} ,$$
$$\delta\vec{\chi} = \vec{\omega}\psi - \vec{\omega} \times \vec{\chi} .$$

We assume that the parameters α and β are so arranged that the field ψ develops a vacuum expectation value,

$$\langle\psi\rangle = v . \qquad (31)$$

We choose the gauge condition

$$F_\alpha(\phi) = 0 , \qquad (32)$$
$$\sqrt{\xi}\left(\partial_\mu \vec{A}^\mu + \frac{\lambda}{\xi}\vec{\chi}\right) = 0 ,$$

where λ and ξ are parameters to be specified. The operator M is given by

$$M_{\alpha\beta} = \frac{\delta F_\alpha}{\delta\phi_i}(\Gamma^\beta_{ij}\phi_j + \Lambda^\beta_i) , \qquad (33)$$
$$\frac{g}{\sqrt{\xi}} M = \partial^2 + g\,\partial_\mu \vec{A} \times + \frac{\lambda g}{2\xi}(\psi + \vec{\chi}\times) .$$

We shall write the generating functional of Green's functions as

$$W[\vec{\eta}_\mu, \vec{J}, K] = \int [d\phi] \det M \exp\{i(S[\phi] - \tfrac{1}{2}F_\alpha{}^2)\} \exp\left(i\int d^4x(-\vec{\eta}_\mu\cdot\vec{A}^\mu + \vec{J}\cdot\vec{\chi} + K\psi)\right). \qquad (34)$$

The identity (20) translates into

$$-i \ln W[\partial_\mu \vec{\Lambda}, (\lambda/\xi)\vec{\Lambda}, 0] = \frac{1}{2\xi}\vec{\Lambda}^2 + \text{const} , \qquad (35)$$

which yields

$$\left(i\partial_\mu \frac{\delta}{\delta\eta^a_\mu(x)} + \frac{\lambda}{\xi}\frac{1}{i}\frac{\delta}{\delta J^a(x)}\right)\left(i\partial_\nu \frac{\delta}{\delta\eta^b_\nu(y)} + \frac{\lambda}{\xi}\frac{1}{i}\frac{\delta}{\delta J^b(y)}\right) Z = -\frac{1}{\xi}\delta^4(x-y)\delta^{ab} , \qquad (36)$$

where $Z = -i \ln W$, and a and b are isospin indices. Equation (36) yields a set of relations useful to renormalization of propagators and the parameters ξ and λ.

It is convenient to parametrize the $\vec{A}_\mu$ and $\vec{\chi}$ propagators as

$$-i\langle (A^a_\mu(x) A^b_\nu(y))_+\rangle_0 \sim \left(g_{\mu\nu}\frac{1}{a} + p_\mu p_\nu \frac{c^2 - bd}{a[(a+bp^2)d - c^2p^2]}\right)\delta^{ab} ,$$
$$-i\langle (A^a_\mu(x)\chi^b(y))_+\rangle_0 \sim ip\delta^{ab}\frac{c}{(a+bp^2)d - p^2c^2} , \qquad (37)$$
$$-i\langle (\chi^a(x)\chi^b(y))_+\rangle_0 \sim \delta^{ab}\frac{a+bp^2}{(a+bp^2)d - p^2c^2} ,$$

where $a(p^2)$, $b(p^2)$, $c(p^2)$, and $d(p^2)$ are free of poles, at least in perturbation theory. Equation (36) gives

$$p^2\left(d + 2\frac{\lambda}{\xi}c\right) + \left(\frac{\lambda}{\xi}\right)^2(a + bp^2) = -\frac{1}{\xi}[(a+bp^2)d - c^2p^2] . \qquad (38)$$

We shall adjust the value of λ so that

$$c(0)=0\,. \tag{39}$$

Then we must have

$$d(0)=-\frac{\lambda^2}{\xi} \tag{40}$$

and

$$a(0)d'(0)=\lambda^2\,. \tag{41}$$

We can renormalize fields and parameters so that

$$a(p^2)\underset{p^2\to 0}{\sim} -Z_3^{-1}(p^2-m^2)\,, \qquad d(p^2)\underset{p^2\to 0}{\sim} Z_\chi^{-1}(p^2-\mu^2)\,. \tag{42}$$

Then from Eqs. (40) and (41) we find that

$$\lambda=m(Z_3Z_\chi)^{-1/2} \tag{43}$$

and

$$\mu^2=\frac{1}{\xi_r}m^2\,, \tag{44}$$

where

$$\xi_r=\xi Z_3\,. \tag{45}$$

How is the constant m related to the fundamental parameters of the theory? In lowest order, we have $m=\frac{1}{2}(gv)$. The vacuum expectation value v may be used as a fundamental parameter of theory instead of β in Eq. (29). What is more convenient, we may use m as a fundamental parameter instead of v. Thus the renormalized theory can be specified completely in terms of m and ξ_r, in addition to g_r and α_r to be defined shortly.

Certain useful information is obtained from the WT identity which follows from differentiating Eq. (17) with respect to J_i and then putting all external sources equal to zero:

$$\left\{-F_\alpha\left(\frac{1}{i}\frac{\delta}{\delta J}\right)\frac{\delta}{\delta J_i}+\left(\Gamma^{\beta}_{ij}\frac{1}{i}\frac{\delta}{\delta J_i}+\Lambda^{\beta}_i\right)\left[M^{-1}\left(\frac{1}{i}\frac{\delta}{\delta J_i}\right)\right]_{\beta\alpha}\right\}W[J]\bigg|_{J=0}=0\,. \tag{46}$$

Specializing to $J_i=J^a(y)$, we learn that the inverse ghost propagator $G^{-1}(p^2)$,

$$-i\langle(c^a(x)\bar{c}^b(y))_+\rangle\sim G(p^2)\delta^{ab}\,, \tag{47}$$

is proportional to

$$G^{-1}(p^2)\sim[(a+p^2b)d-c^2p^2]\,, \tag{48}$$

so that $G(p^2)$ has a pole where the χ propagator does. Moreover, the low-energy limit of $G^{-1}(p^2)$ is given by

$$g^{-1}(p^2)\underset{p^2\to 0}{\sim} -\tilde{Z}_3^{-1}\left(p^2-\frac{1}{\xi_r}m^2\right), \tag{49}$$

where $\tilde{Z}_3$ is cutoff-dependent.

Differentiating Eq. (17) with respect to J_k and J_l and letting all external sources vanish, we obtain

$$\left\{-F_\alpha\left(\frac{1}{i}\frac{\delta}{\delta J}\right)\frac{\delta}{\delta J_k}\frac{\delta}{\delta J_l}+\left[\left(\Gamma^{\beta}_{kj}\frac{1}{i}\frac{\delta}{\delta J_j}+\Lambda^{\beta}_k\right)\right]\left[M^{-1}\left(\frac{1}{i}\frac{\delta}{\delta J}\right)\right]_{\beta\alpha}\frac{\delta}{\delta J_l}+(k\leftrightarrow l)\right\}W[J]\bigg|_{J=0}=0\,, \tag{50}$$

specializing to the case $J_i=\eta^b_\mu$, $J_l=\eta^c_\nu$, where η^b_μ and η^c_ν are transverse, $\partial^\mu\eta^b_\mu=0$, we obtain a relation between the $(A_\mu)^3$ and $A_\mu\bar{c}c$ vertices. Denoting the former by $i\Gamma^{abc}_{\lambda\mu\nu}(p,q,r)$ and the latter by $i\gamma^{abc}_\lambda(p,q;r)$ with $p+q+r=0$ [see Fig. 1 of paper I (Ref.14)], and expressing the low-energy limits of these quantities by

$$\begin{aligned} i\Gamma^{abc}_{\lambda\mu\nu}(p,q,r)&\underset{p,q,r\to 0}{\sim}\epsilon^{abc}\left(g\frac{1}{Z_1}[(p-q)_\nu g_{\lambda\mu}+(q-r)_\lambda g_{\mu\nu}+(r-p)_\mu g_{\nu\lambda}]+\text{cutoff-independent terms}\right),\\ i\gamma^{abc}_\lambda(p,q;r)&\underset{p,q,r\to 0}{\sim}\epsilon^{abc}p_\lambda\left(g\frac{1}{\tilde{Z}_1}+\text{cutoff-independent terms}\right), \end{aligned} \tag{51}$$

we find from Eq. (50) that[15]

$$Z_1/Z_3=\tilde{Z}_1/\tilde{Z}_3\,. \tag{52}$$

The renormalized coupling constant g_r is defined as

$$g=g_r\frac{Z_1}{Z_3^{3/2}}=g_r\frac{\tilde{Z}_1}{\tilde{Z}_3 Z_3^{1/2}}\,. \tag{53}$$

[Caution: Actually the quantity appearing in Eq. (50) is not γ_λ^{abc}, but a related quantity whose cutoff-dependent part is the same as γ_λ^{abc}.]

We may also define α_r to be the value of the renormalized $(\chi)^4$ coupling when all external momenta vanish. By the use of Eq. (17) it is possible to show that all remaining renormalization parts can be expressed finitely in terms of g_r, α_r, and m. We have done this. However, the process is much too arduous to reproduce here. We shall be content to give a renormalization procedure based on this study by specifying renormalization constants.

The effective Lagrangian of the theory is

$$\begin{aligned}\mathcal{L}_{\rm eff}=&-\tfrac{1}{4}(\partial_\mu\vec{A}-\partial_\nu\vec{A}_\mu+g\vec{A}_\mu\times\vec{A}_\nu)^2+\tfrac{1}{2}[(\partial_\nu S)^2+(\partial_\mu\vec{\chi})^2]+\tfrac{1}{2}g\vec{A}^\mu\cdot(S\partial_\mu\vec{\chi}-\vec{\chi}\partial_\mu S+\vec{\chi}\times\partial_\mu\vec{\chi})+\tfrac{1}{2}gv\vec{A}^\mu\cdot\partial_\mu\vec{\chi}\\&+\tfrac{1}{8}g^2v^2\vec{A}_\mu^2+\tfrac{1}{4}g^2v\vec{A}_\mu^2S+\tfrac{1}{8}g^2\vec{A}_\mu^2(S^2+\vec{\chi}^2)-\tfrac{1}{4}\alpha(S^2+\chi^2)-\alpha vS(S^2+\vec{\chi}^2)-\alpha v^2S^2-\delta\mu^2(S^2+\vec{\chi}^2)-v\delta\mu^2S\\&+\vec{\bar{c}}\left(\partial^2+\frac{\lambda gv}{2\xi}\right)\vec{c}+g\vec{\bar{c}}\cdot\partial^\mu[\vec{A}_\mu\times c]+\frac{\lambda g}{2\xi}(\vec{\bar{c}}\cdot\vec{c}S+\vec{\bar{c}}\cdot\vec{\chi}\times\vec{c})-\tfrac{1}{2}\xi\left(\partial_\mu A^\mu+\frac{\lambda}{\xi}\chi\right)^2,\end{aligned} \tag{54}$$

where $\delta\mu^2=\alpha v^2+\beta$, and $\psi=v+S$. We shall renormalize fields and parameters according to

$$\begin{aligned}&\vec{A}_\mu=Z_3^{1/2}(\vec{A}_\mu)_r\,,\\&\vec{\chi}=Z_\chi^{1/2}\vec{\chi}_r\,,\\&(S,v)=Z_\psi^{1/2}(S,v)_r\,,\\&\vec{c}=Z_3^{1/2}\vec{c}_r\,,\\&g=g_rZ_1/Z_3^{3/2}\,,\\&\alpha=\alpha_rZ_4/Z_\chi^2\,,\end{aligned} \tag{55}$$

and define

$$m=\tfrac{1}{2}g_rv_r\,. \tag{56}$$

We recall

$$\xi=\frac{1}{Z_3}\xi_r\,, \tag{57}$$

$$\lambda=m(Z_3Z_\chi)^{-1/2}\,. \tag{58}$$

We determine Z_3, Z_χ, and $\tilde{Z}_3$ from Eqs. (42) and (49). We choose Z_ψ in such a way that m^2 determined by Eq. (42) is the same as $\frac{1}{4}(g_rv_r)^2$ by rescaling v_r and S_r. [This does not make the S field normalized to the unit amplitude asymptotically, but no matter. The renormalized S propagator is finite.] The constants Z_1 and Z_4 are determined by requiring that the renormalized coupling constants associated with the $(A_3)^3$ and $(\chi)^4$ vertices at zero external momenta are g_r and α_r, respectively. The counterterm $\delta\mu^2$ is determined by the condition that the S field does not have a vacuum expectation value (see paper II).

Once they are renormalized according to Eqs. (55)–(58), Green's functions of the theory are finite in terms of m, g_r, α_r, and ξ_r. The S matrix is independent of ξ_r. Since the poles of the χ propagator, the longitudinal part of the A_μ propagator, and the ghost propagator $G(p^2)$ at

$$p^2=\frac{1}{\xi_r}m^2+\text{a finite correction}$$

arising from the zero of $[(a+bp^2)d-c^2p^2]$ vary as ξ_r is varied, while the S matrix is independent of ξ_r, we see that the contributions from these poles must cancel in all S-matrix elements. In fact, in the limit $\xi_r\to 0$, we find that the poles in question recede to infinity and the corresponding particles do not contribute to sum over intermediate states at any finite energy. The limit $\xi_r\to 0$ defines the U-gauge formulation discussed in paper III (see also Ref. 16) as an analytic limit of an infinite set of renormalizable gauge formulations.

APPENDIX

We wish to show the invariance of the metric

$$dV=[d\phi]\Delta_F,\quad \Delta_F=\det M\,, \tag{A1}$$

under the gauge transformation (1) constrained by Eq. (7). We shall first compute the Jacobian J of the transformation (1) to first order in g,

$$\begin{aligned}&J=\exp[\Gamma_{ii}^\alpha g_\alpha+(\Gamma_{ij}^\alpha\phi_j+\Lambda_i^\alpha)\delta g_\alpha/\delta\phi_i]\,,\\&\delta g_\alpha/\delta\phi_i=-(M^{-1})_{\alpha\beta}(\delta M_{\beta\gamma}/\delta\phi_i)g_\gamma\,.\end{aligned} \tag{A2}$$

Since the measure $[d\phi]$ is invariant under the group of linear transformations, $\exp(\Gamma_{ii}^\alpha g_\alpha)=1$, and we have

$$J=\exp[-(\Gamma^{\alpha}_{ij}\phi_j+\Lambda^{\alpha}_i)(M^{-1})_{\alpha\beta}(\delta M_{\beta\gamma}/\delta\phi^i)g_\gamma]\,. \tag{A3}$$

We shall now compute the change in Δ_F to this order,

$$\Delta'_F=\det(M+\delta M)=\Delta_F\exp\operatorname{Tr}M^{-1}\delta M\,, \tag{A4}$$

where

$$\delta M_{\beta\gamma}=\frac{\delta M_{\beta\gamma}}{\delta\phi_i}(\Gamma^{\delta}_{ij}\phi_j+\Lambda^{\delta}_i)g_\delta\,. \tag{A5}$$

So we have

$$\delta[\ln\Delta_F]=\frac{\delta M_{\beta\gamma}}{\delta\phi_i}(M^{-1})_{\gamma\beta}(\Gamma^{\delta}_{ij}\phi_j+\Lambda^{\delta}_i)g_\delta\,. \tag{A6}$$

The effect of the transformation on the measure dV is then

$$\delta[\ln\Delta_F]+\ln J=\left(\frac{\delta M_{\beta\gamma}}{\delta\phi_i}(\Gamma^{\delta}_{ij}\phi_j+\Gamma^{\delta}_i)(M^{-1})_{\gamma\beta}-(\Gamma^{\gamma}_{ij}\phi_j+\Lambda^{\gamma}_i)(M^{-1})_{\gamma\beta}\frac{\delta M_{\beta\delta}}{\delta\phi_i}\right)g_\delta\,. \tag{A7}$$

The expression (A7) vanishes for every g if we have

$$\left(\frac{\delta M_{\beta\gamma}}{\delta\phi_i}(\Gamma^{\delta}_{ij}\phi_j+\Lambda^{\delta}_i)-\frac{\delta M_{\beta\delta}}{\delta\phi_i}(\Gamma^{\gamma}_{ij}\phi_j+\Lambda^{\gamma}_i)\right)(M^{-1})_{\gamma\beta}=0\,. \tag{A8}$$

We must now compute $\delta M/\delta\phi$:

$$\frac{\delta M_{\beta\gamma}}{\delta\phi_i}=\frac{\delta^2F^\beta}{\delta\phi_i\delta\phi_j}(\Gamma^{\gamma}_{jk}\phi_k+\Lambda^{\gamma}_k)+\frac{\delta F^\beta}{\delta\phi_j}\Gamma^{\gamma}_{ji}\,. \tag{A9}$$

When Eq. (A9) is substituted in Eq. (A8), the term $\delta^2F/\delta\phi_i\delta\phi_j$ drops out. So Eq. (A8) becomes

$$\frac{\delta F^\beta}{\delta\phi_k}[\Gamma^{\gamma}_{ki}(\Gamma^{\delta}_{ij}\phi_j+\Lambda^{\delta}_i)-\Gamma^{\delta}_{ki}(\Gamma^{\delta}_{ij}\phi_j+\Lambda^{\gamma}_i)](M^{-1})_{\gamma\beta}=0\,. \tag{A10}$$

Let $f_{\alpha\beta\gamma}$ be the structure constant of the gauge group. Then

$$\Gamma^{\gamma}_{ki}(\Gamma^{\delta}_{ij}\phi_j+\Lambda^{\delta}_j)-\Gamma^{\delta}_{ki}(\Gamma^{\gamma}_{ij}\phi_j+\Lambda^{\gamma}_i)=f_{\gamma\delta\alpha}(\Gamma^{\alpha}_{kj}\phi_j+\Lambda^{\alpha}_k)\,. \tag{A11}$$

Therefore Eq. (A10) can be written as

$$f_{\gamma\delta\alpha}\frac{\delta F^\beta}{\delta\phi_k}(\Gamma^{\alpha}_{kj}\phi_j+\Lambda^{\alpha}_k)(M^{-1})_{\gamma\beta}=0\,,$$

or (A12)

$$f_{\gamma\delta\alpha}(M^{-1})_{\gamma\beta}M_{\beta\alpha}=0\,.$$

But Eq. (A12) is true because $f_{\alpha\delta\alpha}=0$ for any compact Lie group.

---

*Work supported in part by the National Science Foundation Grant No. GP-32998X.

[1]G. 't Hooft and M. T. Veltman, Nucl. Phys. B44, 189 (1972); C. G. Bollini, J. J. Giambiagi, and A. Gonzales-Dominguez, Nuovo Cimento 31, 550 (1964); G. Cicuta and Montaldi, Lett. Nuovo Cimento 4, 329 (1972).

[2]B. W. Lee and J. Zinn-Justin, Phys. Rev. D 5, 3121 (1972); 5, 3137 (1972); 5, 3155 (1972). These papers will be referred to as papers I, II, and III, respectively.

[3]K. Fujikawa, B. W. Lee, and A. I. Sanda, Phys. Rev. D 6, 2923 (1972). See also Y. P. Yao, Phys. Rev. D 6, 3129 (1972).

[4]E. S. Fradkin and I. V. Tyutin, Phys. Rev. D 2, 2841 (1970); A. Slavnov, Kiev Report No. ITP-71-83E (unpublished).

[5]W. A. Bardeen, in Proceedings of the Sixteenth International Conference on High Energy Physics, National Accelerator Laboratory, Batavia, Ill., 1972 (unpublished).

[6]J. S. Bell and R. Jackiw, Nuovo Cimento 51, 47 (1969).

[7]S. L. Adler, Phys. Rev. 177, 2426 (1969).

[8]In the context of spontaneously broken gauge theories, see C. Bouchiat, J. Iliopoulos, and Ph. Meyer, Phys. Letters 38B, 519 (1972); D. Gross and R. Jackiw, Phys. Rev. D 6, 477 (1972); H. Georgi and S. L. Glashow, *ibid.* 6, 429 (1972).

[9]V. Popov and L. D. Faddeev, Kiev Report No. ITP-67-36 (unpublished) [English translation available as NAL Report No. NAL-ThY-57 (unpublished)]; L. D. Faddeev and V. Popov, Phys. Letters 25B, 29 (1967).

[10]G. 't Hooft, Nucl. Phys. B33, 173 (1971).

[11]R. P. Feynman, Acta Phys. Polon. 26, 697 (1963).

[12]G. 't Hooft and M. T. Veltman, Utrecht report, 1972 (unpublished).

[13]For constructions of this kind, see Ref. 3, Appendix B; K. Bardakci and M. Halpern, Phys. Rev. D 6, 696 (1972).

[14]There is an error in Fig. 1 of paper I: The bare expression for $G(p^2)$ should be $(-p^2)^{-1}$.

[15]J. C. Taylor, Nucl. Phys. B33, 436 (1971); Slavnov, Ref. 4.

[16]S. Weinberg, Phys. Rev. Letters 27, 1688 (1971).

RENORMALIZATION OF A UNIFIED THEORY OF WEAK AND ELECTROMAGNETIC INTERACTIONS

D.A. ROSS and J.C. TAYLOR
Department of Theoretical Physics, Oxford University, Oxford, England

Received 25 July 1972

Abstract: We study the renormalization of the unified theory of weak and electromagnetic interactions due to Salam and Weinberg. The infinities can be absorbed by defining bare and renormalized Lagrangians, L_B and L_R, which have the same structure and the same number of independent parameters. This is demonstrated partly by appeal to Ward identities, partly by consideration of the underlying theory without spontaneous symmetry breaking, and partly by explicit calculation to second order. Because of complications with Ward identities for non-Abelian gauge symmetries, L_B and L_R are in general invariant under gauge groups containing different parameters. Since the theory accounts for many more masses and couplings than it has independent constants, the parameters appearing in L_R are not all directly related to simply measurable physical couplings. The connection between masses and coupling constants in L_R is the same as in L_B.

1. Introduction

The theory of weak and electromagnetic interactions due to Weinberg [1] and Salam [2], and based on earlier work of Higgs [3], Kibble [4] and others [5], has been shown by 't Hooft [6, 7] to be renormalizable, in the sense that the divergent integrals have been identified and are no worse than in quantum electrodynamics. Actually to implement the renormalization requires more discussion. Abelian models have been studied by Lee [8] and by Fradkin and Tyutin [9]. We consider here extra complications that arise in non-Abelian cases.

To do the renormalization, one must exhibit the bare and renormalized Lagrangians, L_B and L_R, and one must define a subtraction procedure. The difficulty with the present model is that a large number of physical masses and couplings derive from relatively few independent parameters in the Lagrangian. If the nature of the theory is to be preserved in renormalization, L_B and L_R should have similar structures. This means that the many infinities in the theory must satisfy a number of constraints in order to be cancelled by the counter terms in $L_B - L_R$.

The main purpose of this paper is to verify that these constraints are satisfied. To do this, we employ three kinds of argument: –

(a) Because of the local gauge group in the theory, we expect Ward identities to give constraints analogous to $Z_1 = Z_2$ in quantum electrodynamics. In non-Abelian gauge theories the corresponding relations are in general more complicated [10, 11]. They relate the renormalizations of the couplings of physical particles to those of "fictitious" [12] ones. In fact, one finds that two of the parameters, g and F, occuring in the gauge transformations are themselves renormalized; so that L_B and L_R are invariant under different transformations.

(b) The Higgs mechanism [3] starts with a Lagrangian containing massless vector fields and having an unbroken symmetry group. Most, but not all, of the infinities are logarithmic with dimensionless coefficients; and these are the same in the actual theory as in the unbroken-symmetry starting point. These logarithmic infinities are therefore constrained by the symmetry of the starting theory. However, this type of constraint does not apply to the *finite* part of any subtraction.

(c) There remain some constraints which should be satisfied but which we have not been able to derive from the general type of argument in (a) and (b). An example is Weinberg's relation,

$$M_Z^2/M_W^2 = (g^2 + g'^2)/g^2 \quad , \tag{1}$$

connecting masses with coupling constants. The mass renormalization cannot, in any obvious way, be related to the unbroken symmetry, since the vector fields are massless at the start. To confirm that relation (1) is preserved by renormalization, we have therefore had to resort to explicit calculation of the relevant infinite parts, and we have done this to second order only.

All three kinds of argument above refer to the *infinite* parts of the renormalization constants. We know of no general subtraction prescription, for the disparate processes involved, that would guarantee all the finite parts to conform to the constraints. Therefore we are in general driven to sacrifice the usual prescriptions that renormalized vertex parts vanish on mass-shell, etc. An exception is mass renormalization itself. We do define the masses in L_R to be the physical ones. We have also found it possible, and probably convenient, to maintain the usual definition of electric charge. But the coupling constants in L_R are not simply defined in terms of any physical process, and the renormalized fields are in general not defined to have unit matrix element between vacuum and one-particle states. The price of this is that any calculation with the renormalized Lagrangian will have to include finite multiplicative renormalization factors for the external lines.

It would be possible, of course, to carry through the renormalization so as to get rid of these finite corrections; but then the renormalized Lagrangian would contain many constants which would have to be *calculated* in terms of the fundamental parameters.

The lay out of the paper is as follows. Sect. 2 states the form of the bare Lagrangian, as related to the renormalized one. In order to avoid repeating long expressions, we also use this opportunity to display the renormalized Lagrangian, together with the "Lagrangians" for the fictitious particles, and the source terms which are used in the derivation of Ward identities. The renormalization of tadpole graphs, which is a special feature of the Higgs mechnism, is discussed in sect. 3. Self-energy part renormalization is treated in sect. 4. Sect. 5 deals with coupling constant renormalisation and the relation (1). Finally, the coupling of the scalar fields is mentioned in sect. 6.

Throughout the paper, the model considered is that of ref. [1]. However, other models [13] of the same family seem to have similar complications: non-Abelian groups, masses related to couplings, and mixing of neutral particle states.

2. Bare and renormalized Lagrangians and Ward identities

We begin by displaying a bare Lagrangian L_{B1} for Weinberg's [1] model *. The suffix 1 is because we are going to add other bits to it later on. To save space, we will subsequently state how to obtain the renormalized Lagrangian L_{R1} from L_{B1}.

$$L_{B1} = -\tfrac{1}{4}Z_W(\partial_\mu W_\nu - \partial_\nu W_\mu + g_0 W_\mu \wedge W_\nu)^2 - \tfrac{1}{4}Z_B(\partial_\mu B_\nu - \partial_\nu B_\mu)^2$$
$$+ Z_\phi |(\partial_\mu - \tfrac{1}{2}ig_0 \tau\cdot W_\mu + \tfrac{1}{2}ig_0' B_\mu)\Phi'|^2 - \tfrac{1}{2}\lambda_0(\Phi'^+\Phi' - \tfrac{1}{2}E_0^2)^2$$
$$+ Z_L \bar{L}(i\partial_\mu + \tfrac{1}{2}g_0 \tau\cdot W_\mu + \tfrac{1}{2}g_0' B_\mu)\gamma^\mu L$$
$$+ Z_R \bar{R}(i\partial_\mu + g_0' B_\mu)\gamma^\mu R - G_0(\bar{L}\Phi' R + \bar{R}\Phi'^+ L) \ , \qquad (2)$$

where

$$\Phi' = \begin{pmatrix} 0 \\ 2^{-\frac{1}{2}} F_0 \end{pmatrix} + \Phi \ . \qquad (3)$$

The renormalized Lagrangian is got from (2) and (3) by the substitutions

$$Z_W, Z_B, Z_\Phi, Z_L, Z_R \to 1 \ ,$$
$$g_0 \to g \ , \qquad g_0' \to g' \ , \qquad G_0 \to G \ ,$$
$$\lambda_0 \to \lambda \ , \qquad E_0 \to E \ , \qquad F_0 \to F \ . \qquad (4)$$

In addition, the condition

$$E = F \qquad (5)$$

is imposed, but note that

$$E_0 \neq F_0 \ . \qquad (6)$$

For simplicity we have included only, say, electrons and electron neutrinos. Muons

* We use $g_{00} = +1$, $g_{11} = -1$ etc., $\gamma_0^+ = \gamma_0$, $\gamma_i^+ = -\gamma_i$. Otherwise the notation is similar to that of ref. [1].

and their neutrinos can be included in the same way, with different constants, G, G_0, Z_R and Z_L.

Our choice of L_{B1} comes from the requirement that it should, as far as possible, have the same structure as L_{R1}. We find that one requires just one extra parameter in L_{B1} as shown by eqs. (5) and (6). (The Z-factors are not really extra parameters since they could be absorbed by re-scaling the fields. We have found it convenient, however, to write L_{B1} in terms of the renormalized, not the bare, fields.) The main purpose of this paper is to verify that the counter terms $(L_{B1} - L_{R1})$ are sufficient to absorb the infinities.

We now discuss how the Lagrangian L_1 must be supplemented to obtain the Feynman rules [12, 7]. The bare Lagrangian (2) is invariant under infinitesimal gauge transformations

$$\begin{aligned}
\Delta_0 W_\mu &= \partial_\mu \eta_0 - g_0 \eta_0 \wedge W_\mu \;, \\
\Delta_0 B_\mu &= \partial_\mu \eta_{40} \;, \\
\Delta_0 \Phi &= \tfrac{1}{2} i (g_0 \tau \cdot \eta_0 - g_0' \eta_{40}) \left[\Phi + \begin{pmatrix} 0 \\ 2^{-\frac{1}{2}} F_0 \end{pmatrix} \right] \;, \\
\Delta_0 L &= \tfrac{1}{2} i (g_0 \tau \cdot \eta_0 + g_0' \eta_{40}) L \;, \\
\Delta_0 R &= i g_0' \eta_{40} R
\end{aligned} \tag{7}$$

On the other hand, L_{R1} is invariant under a different set of gauge transformations obtained from (7) by the substitutions

$$\begin{aligned}
&g_0 \to g \;, \qquad g_0' \to g' \;, \qquad F_0 \to F \;, \\
&\Delta_0 \to \Delta \;, \qquad \eta_0 \to \eta \;, \qquad \eta_{40} \to \eta_4 \;,
\end{aligned} \tag{8}$$

where

$$\eta_0 = Y_W \eta \;, \qquad \eta_{40} = Y_B \eta_4 \;, \tag{9}$$

the Y factors being renormalization constants somewhat akin to the Z factors.

Following 't Hooft [7] we select a gauge by adding L_2 to L_1, where

$$\begin{aligned}
L_{B2}(\alpha) = L_{R2}(\alpha) = &- (2\alpha)^{-1} (\partial \cdot W + \alpha M \phi - J_W)^2 \\
&- (2\alpha)^{-1} (\partial \cdot B + \alpha M' \phi_3 - J_B)^2 \;.
\end{aligned} \tag{10}$$

Here we have used the notation

$$M = \tfrac{1}{2} g F \;, \qquad M' = \tfrac{1}{2} g' F \;, \tag{11}$$

$$\Phi = 2^{-\frac{1}{2}} (\chi + i \tau \cdot \phi) \begin{pmatrix} 0 \\ 1 \end{pmatrix} \;. \tag{12}$$

L_2 is designed to cancel the mixing terms

$$M\phi\cdot\partial^\mu W_\mu + M'\phi_3\,\partial^\mu B_\mu \tag{13}$$

in L_{2R}. The parameter α specifies the gauge of the renormalized theory. Any finite value of α gives a manifestly renormalizable gauge. The Feynman gauge corresponds to $\alpha = 1$, the Landau gauge to $\alpha = 0$ and the manifestly unitary Weinberg gauge to $\alpha = \infty$. For most of this paper we shall set $\alpha = 1$, because this gauge seems to be computationally the simplest (though the Landau gauge avoids some of the complications in renormalization). The external currents J are devices for generating Ward identities, and may be disregarded for the moment.

The important point to note is that L_2 is *not* renormalized. This will be confirmed in sect. 3. As a consequence, the mixing terms analogous to (13) in L_{B1} are not cancelled, nor does α determine the gauge of the bare theory in any simply way. Thus, when (10) is chosen, it is with the purpose of obtaining a convenient gauge for the renormalized theory not the bare theory.

It is correct to add L_2 to the Lagrangian only if fictitious particle closed loops are taken into account [12, 6, 7]. These can be described by an additional effective "Lagrangian" L_3. In the bare case this may be written

$$L_{B3} = -\xi\cdot\Delta_0(\partial\cdot W + \alpha M\phi) - \xi_4\,\Delta_0(\partial\cdot B + \alpha M'\phi_3)$$

$$+\,\xi\cdot J_W + \xi_4 J_B\ , \tag{14}$$

where Δ_0 is defined by eq. (7). The renormalized form, L_{R3}, is got from (13) by the substitution

$$\Delta_0 \to \Delta$$

which is defined in (8). In using L_3, we can think of ξ and η_i (or η_0) as ordinary fields. The kinetic part of (13) is

$$L_{R3} = -\xi\cdot(\Box + M^2)\,\eta - \xi_4(\Box + M'^2)\,\eta_4 + \ldots\ , \tag{15}$$

and so the propagator converts ξ into η and vice versa. However, vertices do the same, and so closed loops can be generated *. A minus sign must accompany each closed loop, which is why L_3 is only described as a "Lagrangian" in quotation marks.

The J terms in (14) could begin an open fictitious line; but since there is at present no term linear in η, there is nowhere for such a line to end. Thus the sources J may again for the present be disregarded.

In the path-integral formalism, L_3 contributes, when the ξ-functional integrals are done,

$$\delta\{\Delta_0(\partial\cdot W + \alpha M\phi - J_W)\}\ \delta\{\Delta_0(\partial\cdot B + \alpha M'\phi_3 - J_B)\} \tag{16}$$

* 't Hooft [6] takes ξ to be the Hermitean conjugate of η. We find the origin of L_3 clearer if this is not done. It is also sometimes convenient to use complex ξ and η to represent charged fictitious particles.

The η-functional integrals then give the inverse of the required functional Jacobian factor [12].

In order to generate Green's function by path-integral methods, it is convenient to include a source term, $j(x)$, for each field transforming homogeneously under gauge transformations. The Feynman path-integral then gives a functional $S(J, j)$. Partial differentiation with respect to the j gives Green's functions. Differentiation with respect to the J will, as we shall see, give Ward identities.

Before introducing the sources j, we define fields to diagonalize the kinetic terms in L_{R1}:

$$Z^\mu = \cos\theta\, W_3^\mu + \sin\theta\, B^\mu \quad ,$$

$$A^\mu = -\sin\theta\, W_3^\mu + \cos\theta\, B^\mu \quad , \tag{17}$$

where

$$\tan\theta = g'/g \tag{18}$$

We also use the notation

$$W^\mu = 2^{-\frac{1}{2}}(W_1^\mu + i\, W_2^\mu) \quad , \qquad \phi = 2^{-\frac{1}{2}}(\phi_1 + i\,\phi_2) \quad ,$$

$$\psi = \phi_3 \quad , \qquad \bar{g} = (g^2 + g'^2)^{\frac{1}{2}} \quad , \qquad \bar{M} = \tfrac{1}{2} F \bar{g} \quad . \tag{19}$$

Then the source terms are

$$L_{B4} = 2\, j_W^\mu (W_\mu^* - M_0^{-1}\partial_\mu \phi^*) + j_Z^\mu (Z_\mu - \bar{M}_0^{-1}\partial_\mu \psi)$$

$$+ j_A^\mu A_\mu + 2\bar{L} j_L + 2\bar{R} j_R + j_\chi \chi \quad , \tag{20}$$

where each j is the source of the field with which it is labelled. Here and subsequently we will adopt the convention that the real part is to be understood for any non-Hermitean term in a Lagrangian. Thus $2\bar{L}j_L$ is to be understood as $(\bar{L}j_L + \bar{j}_L L)$. The combinations $(W_\mu - M_0^{-1}\partial_\mu\phi)$ etc. are chosen because they transform homogeneously (under infinitesimal gauge transformations). Without loss of generality (because of the presence of the J) one may take $\partial \cdot j_A = 0$. To get L_{R4} from (20) one substitutes

$$M_0 \to M \quad , \qquad \bar{M}_0 \to M_0 \quad . \tag{21}$$

Before the addition of L_4, the theory was independent of J, and Ward identities would follow [6] by differentiating $S(J, j = 0)$ with respect to J. The addition of L_4 to L_1 spoils the gauge invariance and so we lose the Ward identities. For *infinitesimal* J, however, we may recover them as follows. An infinitesimal J may be removed from L_2 in eq. (10) by an infinitesimal gauge transformation η_0 (or η in the renormalized case). The correct η_0 to do this is given by (16) and therefore by (14) (this is the reason that we put J into L_3). Thus expression (14) plays two roles. It generates closed fictitious loops, and it determines the η_0 which correspond to J.

The complete Lagrangian, including L_{B4}, will therefore be independent of J if we transform L_{B4} by the same infinitesimal gauge transformation η_0 that appears in (14). Thus we are led to add yet another piece to the Lagrangian,

$$L_{B5} = \Delta_0 L_{B4} \quad \text{or} \quad L_{R5} = \Delta L_{R4} \tag{22}$$

where Δ_0 is defined in (7).

The operator Δ acting on the fields in L_4 gives terms linear in η. There will then be *open* fictitious lines, beginning at a J in (14) and ending at a j in (22). These open lines do not contribute to S-matrix elements since, unlike the contributions from the j in L_4, they do not have mass-shell poles.

Let us summarize the properties of the complete functional $S(J, j)$ derived from $L_1 + L_2 + L_3 + L_4 + L_5$.

(i) $S(J = 0, j)$ generates Green functions.

(ii) For $j = 0$ or for S-matrix elements it is independent of J (even for finite J), and differentiation with respect to J yields on-shell Ward identities [6].

(iii) $S(J, j)$ is independent of J for infinitesimal J, and off-shell Ward identities [10, 11] come from a single differentiation with respect to J. (These are required in order to derive the analogue of $Z_1 = Z_2$).

To close this section, we write out the pieces of the Lagrangian in detail. We use the following notation in addition to (12), (11), (17), (18) and (19):

$$m = 2^{\frac{1}{2}} G F \ , \qquad \mu^2 = \lambda F^2 \ ,$$

$$L = \begin{pmatrix} \nu \\ L_e \end{pmatrix} \ , \qquad \delta\theta = \theta_0 - \theta \ , \tag{23}$$

$$Z_A = \cos^2\theta\, Z_B + \sin^2\theta\, Z_W \ ,$$

$$Z_Z = \sin^2\theta\, Z_B + \cos^2\theta\, Z_W \ ,$$

$$Z_{Az} = \cos\theta \sin\theta\, (Z_B - Z_W) \ . \tag{24}$$

Then eqs. (2) and (10) may be written

$$L_{B1} + L_2 = -\tfrac{1}{2} Z_W |\partial_\mu W_\nu - \partial_\nu W_\mu|^2 - \tfrac{1}{4} Z_Z (\partial_\mu Z_\nu - \partial_\nu Z_\mu)^2$$

$$- \tfrac{1}{4} Z_A (\partial_\mu A_\nu - \partial_\nu A_\mu)^2 - Z_{AZ} (\partial_\mu Z_\nu - \partial_\nu Z_\mu)\, \partial^\mu A^\nu$$

$$- \alpha^{-1} |\partial \cdot W|^2 - (2\alpha)^{-1} (\partial \cdot Z)^2 - (2\alpha)^{-1} (\partial \cdot A)^2$$

$$+ Z_\Phi M_0^2 |W_\mu|^2 + \tfrac{1}{2} Z_\Phi \overline{M}_0^2 (\cos\delta\theta\, Z_\mu - \sin\delta\theta\, A_\mu)^2$$

$$+ \tfrac{1}{2} Z_\Phi \{(\partial_\mu \chi)^2 + (\partial_\mu \psi)^2 + 2|\partial_\mu \phi|^2\} - \alpha M^2 |\phi|^2 - \tfrac{1}{2} \alpha \overline{M}^2 \psi^2$$

$$
\begin{aligned}
&- (Z_\Phi M_0 - M)\, \phi^* \partial\cdot W - (Z_\Phi \overline{M}_0 \cos\delta\theta - M)\, \psi\, \partial\cdot Z - Z_\Phi \overline{M}_0 \sin\delta\theta\; \psi\, \partial\cdot A \\
&- \tfrac{1}{2}\mu_0^2 \chi^2 - (\mu_0^2 g_0/4 M_0)\, \chi(2|\phi|^2 + \psi^2 + \chi^2) + (\mu_0^2 g_0^2/16 M_0^2)(2|\phi|^2 + \psi^2 + \chi^2)^2 \\
&- \tfrac{1}{4}\mu_0^2 \{1 - (E_0/F_0)^2\}\{2|\phi|^2 + \psi^2 + \chi^2 + (4 M_0/g_0)\,\chi\} \\
&+ i Z_{\mathrm{L}} \overline{L}\gamma\cdot\partial L + i Z_{\mathrm{R}} \overline{R}\gamma\cdot\partial R - m_0 \bar{e} e + 2^{\frac{1}{2}} Z_{\mathrm{L}} g_0 \bar{\nu}\gamma\cdot W^* L_{\mathrm{e}} \\
&+ \tfrac{1}{2} Z_{\mathrm{L}} \bar{g}_0 \overline{L}_{\mathrm{e}} \gamma_\mu L_{\mathrm{e}} \{\sin(\theta+\theta_0) A^\mu - \cos(\theta+\theta_0) Z^\mu\} \\
&+ Z_{\mathrm{R}} g_0' \overline{R}\gamma_\mu R (\cos\theta\, A^\mu + \sin\theta\, Z^\mu) + \tfrac{1}{2} Z_{\mathrm{L}} \bar{g}_0 \bar{\nu}\gamma_\mu \nu (\cos\delta\theta\, Z^\mu + \sin\delta\theta\, A^\mu) \\
&- 2^{-\frac{1}{2}} G_0 \bar{e}(\chi - i\gamma_5\psi) e - 2i G_0 \bar{\nu} R \phi^* \\
&- i Z_\Phi \bar{g}_0 \phi^* \partial_\mu \phi \{\cos(\theta+\theta_0) Z^\mu - \sin(\theta+\theta_0) A^\mu\} \\
&- \tfrac{1}{4} Z_\Phi \bar{g}_0^2 |\phi|^2 \{\cos(\theta+\theta_0) Z^\mu - \sin(\theta+\theta_0) A^\mu\}^2 \\
&- \tfrac{1}{2} Z_\Phi \bar{g}_0 (\psi\partial_\mu\chi - \chi\partial_\mu\psi)(\cos\delta\theta\, Z^\mu + \sin\delta\theta\, A^\mu) \\
&+ \tfrac{1}{8} Z_\Phi \bar{g}_0^2 (\psi^2 + \chi^2)(\cos\delta\theta\, Z^\mu + \sin\delta\theta\, A^\mu)^2 \\
&+ Z_\Phi g_0 \{\phi\,\partial^\mu(\psi - i\chi) - (\psi - i\chi)\partial^\mu\phi\} W_\mu^* \\
&+ \tfrac{1}{4} Z_\Phi g_0^2 (\chi^2 + \psi^2 + 2|\phi|^2)|W_\mu|^2 + Z_\Phi g_0 M_0 \chi |W_\mu|^2 \\
&+ Z_\Phi g_0' \{g_0(\psi + i\chi) + 2i M_0\} \phi^* W_\mu (\cos\theta\, A^\mu + \sin\theta\, Z^\mu) \\
&+ \tfrac{1}{2} Z_\Phi \bar{g}_0 \overline{M}_0 \chi (\cos\delta\theta\, Z^\mu + \sin\delta\theta\, A^\mu)^2 \\
&- 2 i g_0 Z_{\mathrm{W}} \{W_\nu^*(\partial_\mu W^\nu - \partial^\nu W_\mu) - W_\mu \partial\cdot W^*\}(\cos\theta\, Z^\mu - \sin\theta\, A^\mu) \\
&- \tfrac{1}{2} g_0^2 Z_{\mathrm{W}} |W_\mu^* W^\mu|^2 + \tfrac{1}{2} g_0^2 Z_{\mathrm{W}} W_\mu^* W^{\mu *} W^\nu W_\nu \\
&- g_0^2 Z_{\mathrm{W}} (W_\lambda^* W^\lambda g_{\mu\nu} - W_\mu^* W_\nu)(\cos\theta\, Z^\mu - \sin\theta\, A^\mu)(\cos\theta\, Z^\nu - \sin\theta\, A^\nu)
\end{aligned}
\tag{25}
$$

This is the bare form. $L_{\mathrm{R1}} + L_2$ is obtained from (25) by removing the suffix zero everywhere it occurs and replacing all the factors Z_{W} etc. by 1.

With the further notation

$$\xi = 2^{-\frac{1}{2}}(\xi_1 + i\xi_2)\ , \qquad \eta = 2^{-\frac{1}{2}}(\eta_1 + i\eta_2)\ ;$$

$$\xi_Z = \cos\theta\,\xi_3 + \sin\theta\,\xi_4\ , \qquad \xi_A = -\sin\theta\,\xi_3 + \cos\theta\,\xi_4\ ,$$

$$\eta_Z = \cos\theta\,\eta_3 + \sin\theta\,\eta_4\ , \qquad \eta_A = -\sin\theta\,\eta_3 + \cos\theta\,\eta_4\ ,$$

$$Y_Z = \cos^2\theta\, Y_W + \sin^2\theta\, Y_B\ , \qquad Y_A = \sin^2\theta\, Y_W + \cos^2\theta\, Y_B\ ,$$

$$Y_{AZ} = \cos\theta\sin\theta\,(Y_B - Y_W)\ , \tag{26}$$

we may write

$$\begin{aligned}
L_{B3} + L_{B4} = &- 2Y_W\,\xi^*\,\Box\,\eta - Y_Z\,\xi_Z\,\Box\,\eta_Z - Y_A\,\xi_A\,\Box\,\eta_A \\
&- Y_{AZ}\,(\xi_Z\,\Box\,\eta_A + \xi_A\,Z\,\eta_Z) - 2\alpha M M_0\, Y_W\,\xi^*\,\eta \\
&- \alpha \overline{M}\,\overline{M}_0\,\xi_Z\,(Y_W\cos\theta_0\,\eta_3 + Y_B\sin\theta_0\,\eta_4) \\
&+ 2J_W^*\,\xi + J_A\,\xi_A + J_Z\,\xi_Z \\
&+ 2ig_0\,Y_W\,W_\mu\,\eta^*\,\{\cos\theta\,(j_Z^\mu + \partial^\mu\,\xi_Z) - \sin\theta\,(j_A^\mu + \partial^\mu\,\xi_A)\} \\
&- 2ig_0\,Y_W(j_W^\mu + \partial^\mu\,\xi)(W_{3\mu}\,\eta^* - W_\mu^*\,\eta_3) \\
&- 2g_0\,Y_W\,(\alpha M\,\xi - M_0^{-1}\,\partial\cdot j_W)\,\eta^*\,(\chi - i\psi) \\
&- 2i\,(\alpha M\,\xi - M_0^{-1}\,\partial\cdot j_W)(Y_W\,g_0\,\eta_3 - Y_B\,g_0'\,\eta_4)\,\phi^* \\
&+ 2ig_0\,Y_W\,(\alpha\overline{M}\,\xi_Z - \overline{M}_0^{-1}\,\partial\cdot j_Z)\,\phi^*\,\eta \\
&- \tfrac{1}{2}\,(\alpha\overline{M}\,\xi_Z - \overline{M}_0^{-1}\,\partial\cdot j_Z)(Y_W\,g_0\,\eta_3 - Y_B\,g_0'\,\eta_4)\,\chi \\
&- 2g_0\,Y_W\,j_\chi\,\phi^*\,\eta - \tfrac{1}{2}j_\chi\,\psi\,(Y_W\,g_0\,\eta_3 - Y_B\,g_0'\,\eta_4) \\
&+ i\,2^{\frac{1}{2}}\,g_0\,Y_W\,\bar{j}_e\,\nu\,\eta^* + i\,\bar{j}_e\,L_e\,(Y_W\,g_0\,\eta_3 + Y_B\,g_0'\,\eta_4) \\
&+ 2ig_0'\,Y_B\,\bar{j}_e\,R\,\eta_3 + i\,2^{\frac{1}{2}}\,g_0\,Y_W\,\bar{j}_\nu\,L_e\,\eta \\
&- i\,\bar{j}_\nu\,\nu\,(g_0\,Y_W\,\eta_3 - g_0'\,Y_B\,\eta_4)
\end{aligned}$$

In writing (27) we have for brevity sometimes used a mixed notation. In using (27), η_3 and η_4 should be eliminated with the aid of (26).

Once again, $L_{3R} + L_{5R}$ is got from (27) by dropping all suffixes zero and replacing all Y-factors by 1.

The ensuing sections are concerned to verify that the counter terms in $L_B - L_R$ defined by (25) and (27), are capable of absorbing all infinities.

3. Tadpole graphs

The renormalized Lagrangian contains one fewer parameter than does the bare one, as is seen in eqs. (5) and (6). This is because the vacuum expectation value of Φ is required to be zero; and tadpoles are to be omitted in calculating with the renormalized Lagrangian, since they are cancelled by counter-terms.

To second order, the relevant counter terms obtained from (25) are

$$\lambda(\delta F^2 - \delta E^2)(F\chi + \phi^* \phi + \tfrac{1}{2}\chi^2 + \tfrac{1}{2}\psi^2) \ , \tag{28}$$

where we write

$$\delta F^2 = F_0^2 - F^2, \text{etc.} \tag{29}$$

The χ-tadpole graph is quadratically divergent, and explicit second order calculation (using the gauge-invariant regulators of ref. [6]) gives

$$-\lambda F(\delta F^2 - \delta E^2) = (g^2/32\pi^2)\Big[\tfrac{3}{2}\Lambda^2\Big(2 + \frac{\mu^2}{M^2} + \frac{\overline{M}^2}{M^2} + \frac{2m^2}{3M^2}\Big)$$

$$+ \frac{3}{4}\frac{\mu^4}{M^2}\{\ln(\Lambda^2/\mu^2) + 1\} + (\tfrac{1}{2}\mu^2 + 3M^2)\{\ln(\Lambda^2/M^2) + 1\}$$

$$+ \Big(\frac{1}{4}\frac{\mu^2\overline{M}^2}{M^2} + \frac{3}{2}\frac{\overline{M}^4}{M^2}\Big)\{\ln(\Lambda^2/\overline{M}^2) + 1\} + \frac{m^4}{M^2}\{\ln(\eta^2/m) + 1\}\Big] \ , \tag{30}$$

where Λ is the cut-off momentum.

In calculating self-energy graphs in the next section, the contribution from the bilinear terms in (28) must be included. When this is taken into account, all remaining divergences in the theory are found to be only logarithmic.

4. Ward identities for self-energy parts

In this section, we show that the Ward identities for the self-energy parts of the W_μ, Z_μ, A_μ, ϕ and ψ fields guarantee that their infinities may be absorbed by the counterterms given in sect. 2.

We begin by treating the W_μ and ϕ self-energy parts. The two relevant Ward identities derived from sect. 2 are

$$\delta^2 S/\delta J_W^2 = 0 \ , \tag{31}$$

$$\delta^2 S/\delta J_W \, \delta j_W^\lambda = 0 \ , \tag{32}$$

where it will always be understood that all source currents are put to zero after the differentiation. In momentum space, call the $(W_\mu W_\nu^+)$, $(W_\mu \phi^+)$ and $(\phi\phi^+)$ self-energy parts, in the bare theory, respectively

$$a_0(k^2)(g_{\mu\nu} - k_\mu k_\nu/k^2) + b_0(k^2)\, k_\mu k_\nu/k^2$$

$$i\, c_0(k^2)\, k_\mu \ ,$$

$$d_0(k^2)\, k^2 \ , \tag{33}$$

where

$$a_0(0) + b_0(0) = 0 \ , \tag{34}$$

and the removal of the factor k^2 from d_0 will be justified immediately. Computing the exact propagators involves 2 X 2 matrices, because of the $W_\mu - \phi^+$ mixing. When this is carried out, eq. (31) gives

$$[d_0(k^2) - Z_\phi][b_0(k^2) - Z_\Phi M_0^2] = [c_0(k^2) + Z_\Phi M_0]^2 \tag{35}$$

and eq. (32) gives

$$\alpha M\, b_0(k^2) + (k^2 + \alpha M M_0)\, c_0(k^2) + M_0\, k^2\, d_0(k^2)$$

$$= Z_\Phi M_0[\{1 - Z_\Phi^{-1}\, d_0(k^2)\}\, k^2 - \alpha M\{M_0 + Z_\Phi^{-1}\, c_0(k^2)\}]\, \breve{e}_0(k^2) \ , \tag{36}$$

in which $\breve{e}_0(k^2)$ is the contribution from the open fictitious line. Eq. (35) justifies the separation of the factor k^2 from d_0 in (33). The two Ward identities (35) and (36) imply the simpler relations

$$c_0(k^2) + \{1 + \breve{e}_0(k^2)\}\, M_0\, d_0(k^2) = Z_\Phi M_0\, \breve{e}_0(k^2) \ , \tag{37}$$

$$b_0(k^2) + \{1 + \breve{e}_0(k^2)\}\, M_0\, c_0(k^2) = -\, Z_\Phi M_0^2\, \breve{e}_0(k^2) \tag{38}$$

Note that these are independent of the gauge parameter α. In the renormalized theory there are corresponding equations to (35) to (38) obtained by removal of the suffix zero everywhere and replacement of Z_Φ by 1.

The novel features of these Ward identities are that they are non-linear and that they are different for the bare and renormalized functions.

We now show that the counter-terms given by eq. (25) contributing to the self-energy parts (35) are consistent with the Ward identities. These counter-terms are

$$\delta a \equiv a_0 - a = -\, k^2(Z_W - 1) + Z_{\Phi} M_0^2 - M^2 \ ,$$

$$\delta b \equiv b_0 - b = Z_\Phi M_0^2 - M^2 \ ,$$

$$\delta c \equiv c_0 - c = -\, Z_\Phi M_0 + M \ ,$$

$$\delta d \equiv d_0 - d = Z_\Phi - 1 \ . \tag{39}$$

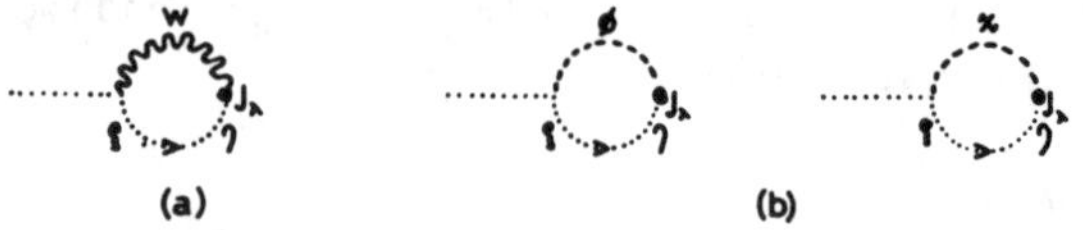

Fig. 1. Typical graphs contributing to e_0. Wavy lines represent vector fields, dashed lines scalar fields and dotted lines fictitious fields.

It is clear that these equations are consistent with (35) and the corresponding renormalized Ward identity. The other Ward identity contains e which is not a function appearing in the S-matrix elements. It is, however, related to the self-energy of the fictitious particle, as we shall now show. For simplicity we shall now restrict ourselves to the second order, so that the non-linear terms in (35) . . . (38) can be neglected.

The contributions to $\breve{e}_0$ in (36) are of two sorts, typified by graphs (a) and (b) of fig. 1.

The contributions from graphs like (a) have the structure

$$e_{0\lambda}^{(a)} = k_\lambda \, e_0^{(a)}(k^2) \ , \tag{40}$$

and from graphs like (b)

$$M_0^{-1} \, k_\lambda \, e_0^{(b)} (k^2) \ . \tag{41}$$

They come from the terms in (27) containing respectively

$$j_\lambda + \partial_\lambda \xi \tag{42}$$

$$M_0^{-1} \, \partial \cdot j - \alpha M \xi \tag{43}$$

The total contribution to e_0 is

$$\breve{e}_0 = e_0^{(a)} + M_0^{-1} \, e_0^{(G)} \tag{44}$$

Let $e_0(k^2)$ be the $(\xi \, \eta^*)$ self-energy function. Typical graphs are shown in fig. 2. where (a) and (b) come from the terms (42) and (43). Thus the same functions $e_0^{(a)}$ and $e_0^{(b)}$ occur here also, and we have

$$e_0(k^2) = k^2 \, e_0^{(a)}(k^2) + \alpha \, M \, e_0^{(G)}(k^2) \tag{45}$$

Fig. 2. Typical graphs contributing to e_0. Arrows and fictitious lines run from ξ to η.

Comparing (44) and (45),

$$\alpha MM_0 \breve{e}_0 (\alpha MM_0) = e_0 (\alpha MM_0) \tag{46}$$

This is the required relation. In the normalized case, the suffices zero are removed.

The counter terms in (27) contributing to e are

$$\delta e \equiv e_0 - e = (Y_W - 1)(k^2 - \alpha MM_0) - \alpha M(M_0 - M) \tag{47}$$

At the particular value $k^2 = \alpha MM_0$, one may now readily verify the consistency of eqs. (39) and (47) with the identities (37) and (38) and their renormalized counterparts, using (46).

It is necessary to check that the counter-terms (39) and (47) are sufficient to absorb the infinities in the self-energy functions. The tadpole renormalization is done before the self-energy renormalization, in the manner of the previous section. After the inclusion of the self-energy counter terms in (28), it is found that all self-energy parts are only *logarithmically* divergent. Therefore one further subtraction is sufficient in each case, except that $e_0(k^2)$ requires two because of the structure of (45).

Finally we discuss how to fix the subtraction points defining the renormalized functions $a(k^2)$ etc. in (39) and (47). For simplicity, we now specialize to the Feynman gauge, $\alpha = 1$, and restrict ourselves to second order. As usual, we shall impose

$$a(M^2) = 0 \ , \qquad e(M^2) = 0 \ , \tag{48}$$

whence, from the renormalized counterpart of (46),

$$\breve{e}(M^2) = 0 \ .$$

Also, if follows from (35) that (to second order) the combination

$$F(k^2) \equiv M^2\, b(k^2) + 2\, M\, k^2 c(k^2) + k^4\, d\,(k^2) \tag{49}$$

vanishes at $k^2 = M^2$. But (49) is the function that appears in

$$\delta^2 S / \delta\, j^\mu_W\, \delta\, j^\nu_W \ ,$$

and is therefore the physically relevant combination, occuring for example when a W is exchanged between leptons. It follows that, although b, c and d do not individually vanish at $k^2 = M^2$, the poles of the S-matrix are in the correct places.

In conventional renormalization theory, one would also insist that $a(k^2)$ and $e(k^2)$ should each have a second order zero at $k^2 = M^2$. As we shall see, because of the constraints (24), we are not able to do this. Nevertheless we will, in passing, note what would be the consequences of this condition. It would fix the four renormalization constants M_0, Z_W, Z_ϕ and Y_W appearing in (39) and (47), and therefore it would fix the remaining renormalized quantities $b(k^2)$, $c(k^2)$ and $d(k^2)$. Further, (37) and (38) imply that the physical combination $F(k^2)$ in (49) satisfies

$$F(k^2) = (k^2 - M^2)^2\, d(k^2) - 2\, M^2 (M^2 - k^2)\, \breve{e}(k^2) \tag{50}$$

Thus F also have a double zero at $k^2 = M^2$, and the poles of the S-matrix would have the required residue.

Next we treat the self-energy functions of the neutral mesons. The method is essentially the same as for the charged ones, but there are complications from mixing. We restrict ourselves to the Feynman gauge and to second order. We call the analogous functions to (33) a^A, a^Z, a^{AZ}, $c^{A\psi}$ etc., corresponding to the fields A and Z and to the mixed cases.

The Z self-energy is, to the second order, entirely analogous to the W one, with M replaced by $\overline{M}$ and Z_W replaced by Z_Z.

The A self-energy is also similar with M replaced by zero, except that the Ward identities (32) and (36) are inapplicable (because $\partial \cdot j_A = 0$). The counter-terms are like (39) and (47), with all masses zero. (To second order, the term $\overline{M}_0^2 \sin^2 \delta\theta$ is negligible). Conventional renormalization would require

$$a^A(0) = 0 \;, \qquad e^A(0) = 0 \;, \tag{51}$$

$$a^{A\prime}(0) = 0 \tag{52}$$

but we will discuss these conditions below.

The mixing terms obey (to second order in the renormalized version with the Feynman gauge) the Ward identities

$$b^{ZA}(k^2) + \overline{M} c^{A\psi}(k^2) = 0 \;, \tag{53}$$

$$\overline{M}\, b^{ZA}(k^2) + k^2\, c^{A\psi}(k^2) = \overline{M}(k^2 - \overline{M}^2)\, e^{ZA}(k^2) \tag{54}$$

The counter-term contributions are (to second order)

$$\delta a^{ZA} = -\,k^2(Z_{AZ} - 1) + \overline{M}^2\, \delta\theta \;,$$

$$\delta b^{ZA} = -\,\overline{M}^2\, \delta\theta \;,$$

$$\delta c^{Z\psi} = \overline{M}\, \delta\theta \;,$$

$$\delta e^{ZA} = Y_{ZA}(k^2 - \overline{M}^2) - \overline{M}^2\, \delta\theta \;,$$

where $\delta\theta = \theta_0 - \theta$. These are consistent with (53) and (54).

Conventional renormalization procedure would require a^{ZA} to vanish at both $k^2 = 0$ and $k^2 = \overline{M}^2$. We do not impose either of these conditions. To second order, this does not affect the positions of the poles; they are still at $k^2 = 0$ and $k^2 = \overline{M}^2$. It does, however, mean that the physical states are created by finite mixtures of the A and Z fields. We shall see this in more detail later.

Finally in this section, we must mention the constraints on the renormalization constants. The total number of independent constants is 7: M_0^2, $\overline{M}_0^2$, Z_Φ, Z_W, Z_B, Y_W, Y_B. We will not be concerned with the last two. On the remaining 5, eqs. (48) and the two analogous equations for the Z selfenergy impose 4 conditions. In conventional theory there would be 5 further conditions: the second order zeros in a^{WW}, a^{ZZ}, a^{AA} and the two zeros in a^{ZA}. Clearly we cannot satisfy all these conditions. We can satisfy one of them, and we choose to impose eq. (52). This is a rather arbitrary decision, but it is motivated by a desire to treat the photon in a normal manner. As we shall see in the next section, photon vertices can also be treated conventionally. With this choice, the five constants, Z_W, Z_A, Z_Φ, M_0^2, $\overline{M}_0^2$ are determined.

Since we have not been able to do the wave-function renormalization in the usual way, any calculation of physical quantities will have to include *finite* multiplicative correction terms. For example, a process in which a single Z-meson is emitted will be described (to second order) by an amplitude of the form

$$[1 - \tfrac{1}{2} a^{Z\prime} (\overline{M}^2)] f_Z - \tfrac{1}{2} \overline{M}^{-2} [a^{AZ}(\overline{M}^2) - a^{AZ}(0)] f_A \qquad (56)$$

in which f_Z and f_A are the amplitudes calculated for the Z and A fields *. Similarly, the Z-coupling constant is not measured by the residue at a Z-exchange poles, but requires a correction factor of

$$[1 - a^{Z\prime}(\overline{M}^2)] \qquad (57)$$

To close this section, we report some results of explicit second-order calculations. The Ward identities have been verified. Eqs. (24) imply that

$$(Z_A - Z_Z) \cos\theta \sin\theta = Z_{AZ}(\cos^2\theta - \sin^2\theta) \quad ,$$

$$(Z_A - Z_W) \sin^2\theta = (Z_Z - Z_W) \cos^2\theta \qquad (58)$$

The *infinite* parts of the constants must satisfy these constraints, because of the symmetry of the underlying massless theory. We have also checked this explicitly. The infinite parts are found to be

$$Z_B - 1 \sim \tfrac{5}{6} g'^2 L \quad ,$$

$$Z_W - 1 \sim \tfrac{21}{6} g^2 L \quad ,$$

$$Z_{\overline{\Phi}} - 1 \sim [1 + \tfrac{1}{2} M^{-2}(\overline{M}^2 + m^2)] g^2 L \quad , \qquad (59)$$

where

$$L = (16\pi^2)^{-1} \ln \Lambda^2 \quad ,$$

and Λ is a cut-off. We have included only one lepton, of mass m.

The complete self-masses, calculated from the fictitious particle self-energy functions, are

* A simpler example occurs with the lepton. The left handed electron and the neutrino have in (25) a single wave function renormalization constant Z_L. If we arbitrarily choose to renormalize the neutrino in the usual manner, then each external left-handed electron will require a factor $[1 - \tfrac{1}{2} \Sigma_L'(m)]$.

$$\delta M^2 = (g^2 M^2/16\,\pi^2)\,[(\tfrac{1}{4}\tan^2\theta - \tfrac{5}{4})\ln(\Lambda^2/M^2) - 2$$

$$+\int_0^1 \mathrm{d}x \left\{(\tfrac{3}{2} - \sin^2\theta - \tfrac{1}{4}\tan^2\theta)\ln\left(\frac{\bar{M}^2}{M^2}(1-x)+x^2\right) - \tfrac{1}{4}\ln\left(\frac{\mu^2}{M^2}(1-x)+x^2\right)\right\}\Bigg]$$

$$\delta \bar{M}^2 = (g^2 \bar{M}^2/16\,\pi^2)\left[(\tfrac{1}{4}\tan^2\theta - \tfrac{5}{4} + 2\sin^2\theta)\ln(\Lambda^2/M^2)\right.$$

$$\left. +\int_0^1 \mathrm{d}x \left\{(\tfrac{3}{2} - 2\sin^2\theta)\ln\left(1-\frac{\bar{M}^2}{M^2}x(1-x)\right) - \tfrac{1}{4}\sec^2\theta\ln\left(\frac{\mu^2}{M^2}(1-x)+\frac{\bar{M}^2}{M^2}x^2\right)\right\}\right] \qquad (60)$$

Eqs. (39) and (48) and their analogues for Z-mesons imply that

$$\bar{M}^{-2}\,[a_0^Z(\bar{M}^2) - \delta\bar{M}^2] - M^{-2}\,[a_0^W(M^2) - \delta M^2] = Z_W - Z_Z \qquad (61)$$

This equation, together with (52) and (58) determine Z_W and Z_Z.

5. Vertex parts and leptons

We begin this section by discussing the coupling of W-mesons to leptons, and the renormalization of the coupling constant g. The Ward identity which controls the renormalization of g is

$$\delta^3 S/\delta j_L\,\delta \bar{j}_L\,\delta J_W = 0 \qquad (62)$$

The corresponding identity for the massless Yang-Mills field has been discussed in ref. [10], where it was shown to imply that

$$Z_1/Z_L = Y_1/Y_W\ , \qquad (63)$$

where Z_1 is the W-lepton vertex-part renormalization constant, and Y is a constant defined in terms of open fictitious lines and may be said to renormalize the $\bar{j}_L\,\nu\eta$ coupling in (27). From the relevant terms in (25) and (27), one sees that (63) ensures that g is renormalized consistently in L_1 and L_5.

Since the quantities in (63) are each logarithmically divergent, the divergent parts are the same in the Weinberg theory as they are in the massless limit. Therefore (63) applies to the *infinite* parts in the Weinberg theory. The exact Ward identity (62) involves vertex parts for the coupling of the ϕ-field to leptons (because of the coupling of J in (9)), and we have not found that it gives useful information about the W-couplings. So, for the moment, we use (63), giving information in the Weinberg theory about infinite parts only. It then relates the infinite parts of the renormalization of g in the lepton couplings in (25) and in the lepton-current coupling in (27).

Similar arguments apply to the renormalization of g in the $W^+\phi\psi$, W^+WA interactions: the infinite parts are each related to the corresponding fictitious particle

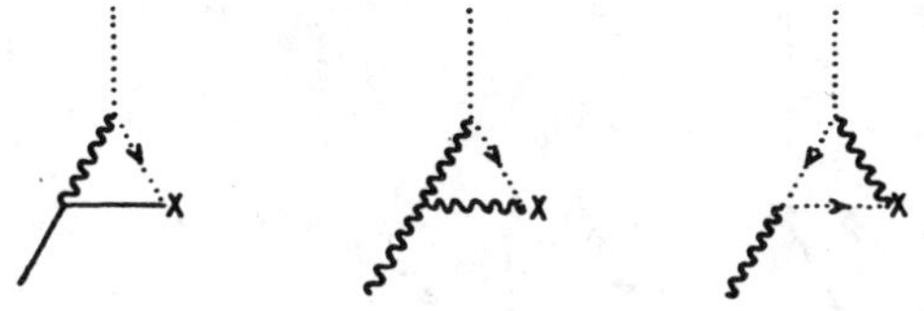

Fig. 3. Graphs contributing to the Y_1 renormalization constant. Continuous lines are leptons, wavy lines vector mesons, dotted lines fictitious particles. The cross indicates the action of the source current j.

couplings. It does not follow from these Ward identities that the different W coupling in (25) are renormalized in the same way. However, as we shall now show, the Ward identities may be used as the basis of an argument to prove the required universality of the infinite parts.

Working to second order, we compare the coupling of W-mesons to leptons with their coupling to themselves. Since we are only concerned with dimensionless logarithmically divergent quantities, we may work with L, in its original form (2), and neglect all masses. We compare the quantities Y_1 in (63) for the two sorts of interaction. For leptons there is only one graph, (a) in fig. 3; for W-mesons there are two graphs, (b) and (c). In the graphs the left-handed external lines represent on-shell particles. Since we are only concerned with divergent parts, all external momenta may be neglected. Therefore the graphs of fig. 3 may be replaced by the graphs of fig. 4 (with a minus sign). Then, using the basic Ward identities (eq. (4.4a) of ref. [6]), we obtain fig. 5.

Since external momenta can be neglected, graphs (b′) and (c′) of fig. 5 cancel one another out. The remaining two graphs, (a) and (b), contribute identical factors to the Y_1 constants for the two different processes. Hence, using (63), we conclude that the infinite part of g renormalization is universal.

On the other hand, for the g′ couplings, one finds that $\delta g' = g'_0 - g'$ is *finite.* The reason is that g' comes in the Abelian U(1) group in (7), and so the right-hand side

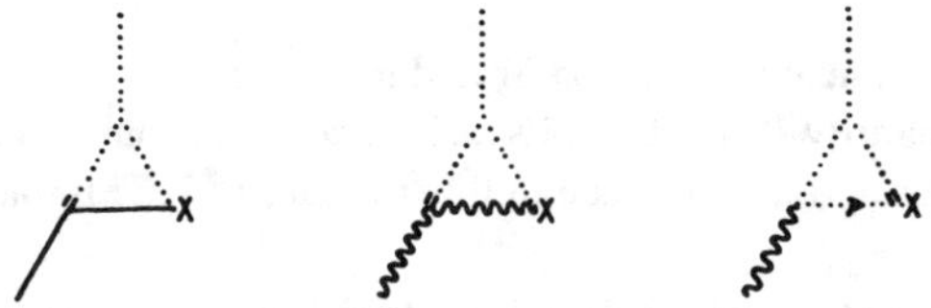

Fig. 4. Graphs obtained from infinite parts of graphs in fig. 3. The arrow on a fictitious (dotted) line indicates that its momentum is to be contracted into the vertex from which it points. The = sign indicates the momentum is contracted into that vertex (following ref. [6]).

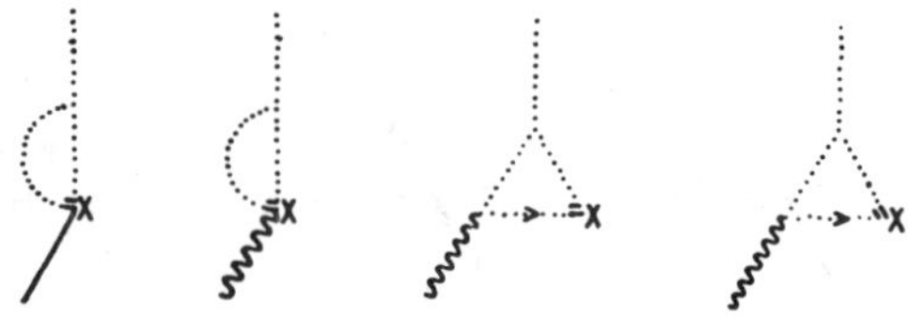

Fig. 5. Graphs obtained from fig. 4.

of an equation like (63) is unity in this case, just as it is in quantum electrodynamics. Therefore, the infinite part, which is common to the actual theory and to the massless limit, is zero.

We have verified the foregoing statements by explicit second-order calculation. We find the infinite part of δg to be

$$\delta g \sim 2 g^2 L \tag{64}$$

where L is defined after eq. (59).

Thus far in this section, we have considered the infinite parts only of renormalization constants. We have not discussed the exact definitions, nor where the subtractions are to be made. Once again, there are far too many physical processes (involving particles with a number of different masses) for it to be possible to define all the vertex-part renomalizations by subtraction on-shell, for there are only two relevant parameters at our disposal, δg and $\delta g'$. It is possible, however, and does seem to be convenient, to renormalize photon vertices in the usual manner, subtracting at zero photon 4-momentum with the charged particle on mass-shell. In Weinberg's model the electric charge is

$$e = gg'/\bar{g} \quad . \tag{65}$$

Therefore, from eq. (25) *,

$$\delta e \equiv e_0 - e = \frac{g_0 g_0'}{\bar{g}_0} - \frac{gg'}{\bar{g}} \quad , \tag{66}$$

and this provides one constraint on δg and $\delta g'$.

The exact quantity δe (not just its infinite parts) is common to all the photon couplings, if subtractions are made in the usual way **. This may be shown by a

* In finding e_0 from (25), contributions from the AZ mixing term must be included.

** This fact does not follow immediately by the usual arguments of quantum electrodynamics, because the Lagrangian L_2 in (9) is not invariant under photon gauge transformations (if it were, δe would vanish!). It is possible to modify L_2 to make it gauge invariant [7], but then the $SU_2 \times U_1$ symmetry in the massless limit is not obvious any more.

slight extension of the previous arguments involving figs. 3, 4 and 5. Firstly, the neglect of the external momentum at the top vertices of fig. 3 is *exact* for photons, since δe is defined by subtracting a zero photon momentum. Secondly, the sum of graphs (b′) and (c) in fig. 5 gives a term proportional to p_μ, where p is the momentum emerging at the right-hand vertex and μ is the Lorentz index of the current there. But the $W^+_\mu j^\mu_{\mathrm{W}} \eta_{\mathrm{A}}$ vertex (see eq. (27)) does not have this tensor structure; so graphs (b′) and (c) cannot contribute to its renormalization.

Explicit calculation of δe from the electron-photon vertex gives

$$\delta e = e^2 L + \mathrm{O}\left(e^2 \frac{m^2}{M^2}\right) \tag{67}$$

(L is defined after eq. (59)).

Finally in this section, we discuss an important point. That is the consistency of the mass renormalisation in the previous section with the coupling constant renormalization in this, bearing in mind the constraints

$$\delta M^2 = \tfrac{1}{4}\,\delta(g^2 F^2)\ , \qquad \delta \overline{M}^2 = \tfrac{1}{4}\,\delta(\overline{g}^2 F^2) \tag{68}$$

coming from eq. (11). As far as the finite parts go, eq. (68) taken together with eq (66) may be used to fix δg, $\delta g'$ and δF. But we must also check that the infinite parts thus found are consistent with the known infinite part of δg and $\delta g'$ (the former given by (64), the latter zero). Eq. (68) gives

$$\frac{\delta \overline{M}^2}{\overline{M}^2} - \frac{\delta M^2}{M^2} = \frac{\delta g^2}{g^2} - \frac{\delta \overline{g}^2}{\overline{g}^2} = \sin^2\theta\left(\frac{\delta g^2}{g^2} - \frac{\delta g'^2}{g'^2}\right)\ , \tag{69}$$

and the infinite parts given by (60) and (64) do satisfy this equation.

Eq. (69), linking coupling constant and mass renormalization, is an interesting feature of Weinberg's model. We have only been able to verify it by explicit calculation to second order; but there is probably a general proof that it is satisfied.

6. Lepton masses

In this section, we discuss the renormalization of the couplings of the scalar fields ϕ, ψ, χ to leptons, and their relation to lepton masses.

The ϕ and ψ couplings are controlled by Ward identities like (62) specialized to have each lepton on mass-shell, so that there are no contributions from open fictitious lines. Let $F_\mu(k)$ and $H(k)$ be the vertex functions for the coupling a W and a ϕ respectively to an electron of momentum p and a neutrino of momentum q. Then (62) implies that

$$\bar{u}_{\mathrm{e}}\,[k^\mu F_\mu - M H + g\,\Sigma_{\mathrm{e}}(p) - g\,\Sigma_\nu(q)]\,u_\nu = 0 \tag{70}$$

where Σ_{e} and Σ_ν are the respective self-energy functions. Let us define the renormalized Σ's to vanish on shell. This fixes the lepton mass renormalization $\delta m = \delta(GF)$.

Since δF is already known from the previous section, δG is determined. Eq. (70) then fixes the subtraction to be made in $H(k)$, since the renormalization of $F_\mu(k)$ is settled in sect. 5. One may verify that the counter-terms in (25) are consistent with (70). Similar considerations apply to the Z and ψ couplings.

Since δG is fixed, the renormalization of the χ couplings to leptons is also determined. The symmetry of the theory in the zero-mass limit guarantees that *infinite* parts of χ-vertices will have the correct values to be cancelled by the counter-terms. However, neither the wave-function renormalization nor the vertex-part renormalization for the χ is defined at any particular value of the external momenta.

The χ mass-renormalization is done in the conventional manner, using the counter-term $(\mu_0^2 - \mu^2)$ coming from eq. (25).

Note added in proof

A recent paper by G. 't Hooft and M. Veltman (Combinatorics of Gauge Fields, Nucl. Phys. B50 (1972) 318) shows generally how gauge theories are renormalized.

References

[1] S. Weinberg, Phys. Rev. Letters 19 (1967) 1264.
[2] A. Salam, Proc. of the 8th Nobel Symposium (Almqvist and Wicksel, Stockholm, 1968).
[3] P.W. Higgs, Phys. Rev. Letters 13 (1964) 508; Phys. Letters 12 (1964) 133; Phys. Rev. 145 (1966) 1156.
[4] T.W.B. Kibble, Phys. Rev. 155 (1967) 1554.
[5] G.S. Guralnik, G.R. Hagan and T.W.B. Kibble, Phys. Rev. Letters 13 (1964) 585.
[6] G. 't Hooft, Nucl. Phys. B33 (1971) 173.
[7] G. 't Hooft, Nucl. Phys. B35 (1971) 167.
[8] B.W. Lee, Phys. Rev. D5 (1972) 823;
B.W. Lee and J. Zinn-Justin, Phys. Rev. D5 (1972) 3121, 3137, 3155.
[9] E.S. Fradkin and E.V. Tyutin, Theory of neutral gauge fields with spontaneous symmetry breaking, P.N. Lebedev Physical Institute preprint N55 (1972).
[10] J.C. Taylor, Nucl. Phys. B33 (1971) 436.
[11] A.A. Slavnov, Generalized ward identities and gauge fields, Kiev preprint (1971),
[12] L.D. Faddeev and V.N. Popov, Phys. Letters 25B (1967) 29;
[13] S. Glashow and H. Georgi, Phys. Rev. Letters 28 (1972) 1494;
B.W. Lee, A model of weak and electromagnetic interactions, NAL THY–51 (1972);
J. Prentki and B. Zumino, CERN preprint th1504 (1972).

ERRATUM

D.A. Ross and J.C. Taylor, Renormalization of a unified theory of weak and electromagnetic interactions, Nucl. Phys. B51 (1973) 125.

In eqs. (36), (37) and (38), $\widetilde{e}_0(k^2)$ should everywhere be replaced by

$$[1 + (k^2 - \alpha MM_0)^{-1} (\alpha MM_0 \widetilde{e}_0 - e_0)]^{-1} \widetilde{e}_0(k^2).$$

Using eqs. (44) and (45), this quantity may be written

$$[1 - e_0{}^{(a)}(k^2)]^{-1} \widetilde{e}_0(k^2).$$

The appearance of e_0 above comes from self-energy insertions in the open fictitious line contribution to eq. (32). The resulting Ward identity is simplified with the use of eq. (35) to get it into the form (36). The above correction does not affect the Ward identities to lowest order.

In eqs. (44) and (45), read $e_0{}^{(b)}$ for $e_0{}^{(G)}$.

Renormalization of Gauge Theories

C. BECCHI,* A. ROUET,† AND R. STORA‡

Centre de Physique Théorique, CNRS 31, chemin Joseph Aiguier, F-13274 Marseille Cedex 2, France

Received December 8, 1975

Gauge theories are characterized by the Slavnov identities which express their invariance under a family of transformations of the supergauge type which involve the Faddeev Popov ghosts. These identities are proved to all orders of renormalized perturbation theory, within the BPHZ framework, when the underlying Lie algebra is semi-simple and the gauge function is chosen to be linear in the fields in such a way that all fields are massive. An example, the SU2 Higgs Kibble model is analyzed in detail: the asymptotic theory is formulated in the perturbative sense, and shown to be reasonable, namely, the physical S operator is unitary and independent from the parameters which define the gauge function.

1. INTRODUCTION

In a previous article [1] devoted to the renormalization of the Abelian Higgs Kibble model, we have developed a number of technical tools which will be applied here to the renormalization of nonabelian gauge models.

Our analysis relies on the Bogoliubov Paraziuk Hepp Zimmermann [2] version of renormalization theory. The extensive use of the renormalized quantum action principle of Lowenstein [3] and Lam [4] allows to push the analysis of the algebraic structure of gauge field models.

Very few properties of the perturbation series are actually used here, namely, nothing more than the general consequences of locality [5], sharpened by the theory of power counting [5], which, through the fundamental theorem of renormalization theory [5] insure the existence of a basis [2, 6] of local operators of given dimension and of linear relationships [2] between local operators of different dimensions. More precisely, we shall never use the information contained in the detailed structure of the coefficients involved in such relations, thus forbidding ourselves to envisage no renormalization type theorems [7].

The algebraic structure of classical gauge theories [8] is then deformed by

* University of Genova.

† Present address: Max Planck Institut für Physik und Astrophysik, München.

‡ Centre de Physique Théorique, CNRS, Marseille.

quantum corrections in a way which can be completely analyzed with the above mentioned economical means, through a systematic use of the implicit function theorem for formal power series [1], provided that it is rigid enough: in technical terms, if some of the cohomology groups [9] of the finite Lie algebra which characterizes the gauge theory, vanish. The analysis can thus be successfully carried out when this Lie algebra is semisimple—if the obstruction provided by the Adler Bardeen [10] anomaly is absent. We shall thus carry out most of our analysis in the semisimple case, and only make a few remarks concerning some of the phenomena which occur when abelian components are involved. For instance, we have noted that the fulfillment of discrete symmetries allows favourable simplifications which in particular lead to complete analyses of massive electrodynamics [11] in the Stueckelberg gauge, and of the abelian Higgs Kibble model in charge conjugation odd gauges [1]. A complete analysis of abelian cases would require proving nonrenormalization theorems [7], which would take us far beyond the technical level of the present work.

Another conservative limitation, due to the present status of renormalization theory, is to discard the study of models in which massless fields are involved, which does not pose new algebraic problems [12] but would force us to rely on work in progress [13] and to go into analytic details which would obscure the main line of reasoning. In the same spirit, we have not included here the analysis needed to deal with nonlinear renormalizable gauge functions [1], which would have made this article considerably longer without adding much to our understanding. Similarly, although the main subject of this paper has to do with algebraic properties, we have decided to include one important application which we have illustrated on a specific example.

The heart of the matter is to prove that one can fulfill to all orders of renormalized perturbation theory a set of identities, the so-called Slavnov identities [1, 14], which express the invariance of the Faddeev Popov ($\Phi\Pi$) [15] Lagrangian under a set of nonlinear field transformations [1] which explicitly involve the Faddeev Popov Fermi scalar ghost fields. Furthermore the theory should be interpreted whenever possible as an operator theory within a Fock space with indefinite metric involving the $\Phi\Pi$ ghost fields. When this interpretation is possible, the Slavnov identities allow to define a "physical" subspace of Fock space within which the norm is positive definite. The restriction of the S operator to this subspace is then both independent from the parameters which label the gauge function [11], and unitary in the perturbative sense.

This article is divided into two main sections, and a number of appendices devoted to some technical details.

Section 2 covers the algebraic discussion of the Slavnov identities.

Section 3 deals with a specific model (the SU2 Higgs Kibble model [16]) for which the operator theory is discussed.

Appendix A summarizes a number of well-known definitions and facts about the cohomology of Lie algebras [17].

Appendix B is devoted to the resolution of some trivial cohomologies encountered in Section 2.

2. Slavnov Identities

A. *The tree approximation*

Let G be a compact Lie group, $\mathfrak{h}$ its Lie algebra. Let $\{\varphi_a\}$ be a matter field multiplet corresponding to a fully reduced unitary representation D of G, with the infinitesimal version:

$$\{\omega_\alpha\} \to t^\alpha \omega_\alpha\,, \qquad \{\omega_\alpha\} \in \mathfrak{h} \tag{1}$$

Let $\{a_{\alpha\mu}\}$ be a gauge field associated with $\mathfrak{h}$, $\{C_\alpha\}$, $\{\bar{C}_\alpha\}$, the corresponding Faddeev Popov ghost fields.

We start with a classical Lagrangian of the form:

$$\begin{aligned} \mathscr{L} &\equiv \mathscr{L}(\{\varphi_a\}, \{a_{\alpha\mu}\}, \{C_\alpha\}, \{\bar{C}_\alpha\}) \\ &\equiv \mathscr{L}_{\text{inv.}}(\{\varphi_a\}, \{a_{\alpha\mu}\}) - (1/\kappa)[\tfrac{1}{2}\mathfrak{G}_\alpha \mathfrak{G}^\alpha - C_\alpha(M\bar{C})^\alpha]. \end{aligned} \tag{2}$$

$\mathscr{L}_{\text{inv.}}$ is the most general dimension four local polynomial invariant under the local gauge transformation:

$$\begin{aligned} \delta_\omega \varphi_a(x) &= \int [\delta_{\omega_\alpha(y)} \varphi_a(x)]\, \omega_\alpha(y)\, dy \\ \delta_\omega a_{\alpha\mu}(x) &= \int [\delta_{\omega_\beta(y)} a_{\mu\alpha}(x)]\, \omega_\beta(y)\, dy \end{aligned} \tag{3}$$

with

$$\begin{aligned} \delta_{\omega_\alpha(y)} \varphi_a(x) &= \delta(x-y)\, t^\alpha_{ab}[\varphi_b(x) + F_b] \\ \delta_{\omega_\beta(y)} a_{\alpha\mu}(x) &= \partial_\mu{}^x \delta(x-y)\, \delta_\alpha{}^\beta + g f_\alpha^{\beta\gamma} a_{\gamma\mu}(x)\, \delta(x-y) \end{aligned} \tag{4}$$

where $\{f_\alpha^{\beta\gamma}\}$ are the structure constants of $\mathfrak{h}$, $\{F_b\}$ some field translation parameters. The gauge function $\{\mathfrak{G}_\alpha\}$ will be chosen to be linear in the fields and their derivatives:

$$\mathfrak{G}_\alpha = g_\alpha{}^\beta \partial^\mu a_{\beta\mu} + g_\alpha{}^a \varphi_a\,. \tag{5}$$

Indices of $\mathfrak{h}$ are raised and lowered by means of an invariant nondegenerate symmetric tensor. The role of the gauge term is to remove the degeneracy of the quadratic part of $\mathscr{L}_{\text{inv.}}$, which is connected with gauge invariance. The field dependent differential operator M involved in the Faddeev Popov part of $\mathscr{L}$ is defined through the kernel

$$M_{xy}^{\alpha\beta} = \delta_{\omega\beta(y)} \mathfrak{G}^{\alpha}(x). \tag{6}$$

The essential property of this Lagrangian is its invariance under the following infinitesimal transformations which we shall call the Slavnov transformations:

$$\begin{aligned} \delta\varphi_a(x) &= \delta\lambda\, \delta_{\omega_\alpha(y)}\varphi_a(x)\, \bar{C}_\alpha(y) \equiv \delta\lambda\, s\varphi_a(x), \\ \delta a_{\alpha\mu}(x) &= \delta\lambda\, \delta_{\omega\beta(y)} a_{\alpha\mu}(x)\, \bar{C}_\beta(y) \equiv \delta\lambda\, s a_{\alpha\mu}(x), \\ \delta C_\alpha(x) &= \delta\lambda\, \mathfrak{G}_\alpha(x) \equiv \delta\lambda\, s C_\alpha(x), \\ \delta\bar{C}_\alpha(x) &= (\delta\lambda/2)\, f_\alpha^{\beta\gamma}(\bar{C}_\beta \bar{C}_\gamma)(x) \equiv \delta\lambda\, s\bar{C}_\alpha(x), \end{aligned} \tag{7}$$

where the summation over repeated indices and the integration over repeated spacetime variables are understood.

$\delta\lambda$ is a spacetime independent infinitesimal parameter which commutes with $\{\varphi_a\}$, $\{a_{\alpha\mu}\}$, but anticommutes with $\{C_\alpha\}$, $\{\bar{C}_\alpha\}$, and, for two transformations labeled by $\delta\lambda_1$, $\delta\lambda_2$, $\delta\lambda_1$, and $\delta\lambda_2$ anticommute.

This invariance can be checked immediately, by using the composition law for gauge transformations

$$[\delta_{\omega_\alpha(x)}, \delta_{\omega\beta(y)}] = f_\gamma^{\alpha\beta} \delta(x-y)\, \delta_{\omega_\gamma(y)} . \tag{8}$$

Conversely, it is interesting to know whether $\mathscr{L}$ is up to a divergence the most general Lagrangian leading to an action invariant under Slavnov transformations, and carrying no Faddeev Popov charge $Q^{\phi\pi}$:

$$\begin{aligned} Q^{\phi\pi} C &= C, \\ Q^{\phi\pi} \bar{C} &= -\bar{C}, \\ Q^{\phi\pi} \varphi_a &= Q^{\phi\pi} a_{\alpha\mu} = 0. \end{aligned} \tag{9}$$

Let

$$\underline{\Psi} = (\{\varphi_a\}, \{a_{\alpha\mu}\}, \{C_\alpha\}, \{\bar{C}_\alpha\}) \tag{10}$$

and given a functional $\mathscr{F}(\underline{\Psi})$, let us denote

$$\delta\lambda\, s\mathscr{F}(\underline{\Psi}) = \delta\underline{\Psi}(x)\, \delta_{\underline{\Psi}(x)} \mathscr{F}(\underline{\Psi}) \tag{11}$$

where $\delta\Psi = \delta\lambda s\Psi$ is the variation of Ψ under the Slavnov transformation of parameter $\delta\lambda$ (cf. Eq. (7)]. A remarkable property of s is:

$$s^2\mathscr{F}(\underline{\Psi}) = (M\bar{C})_\alpha\,(x)\, \delta_{C_\alpha(x)} \mathscr{F}(\underline{\Psi}). \tag{12}$$

This property actually summarizes the group law as follows. Let

$$\begin{aligned}
\delta\varphi_a &= \delta\lambda(\theta^\alpha_{ab}\varphi_b + q_a{}^\alpha)\,\bar{C}_\alpha \equiv \delta\lambda\, s\varphi_a\,,\\
\delta a_{\alpha\mu} &= \delta\lambda(\theta^{\beta\gamma}_\alpha a_{\gamma\mu}\bar{C}_\beta + q^{\ \beta}_\alpha\,\partial_\mu\bar{C}_\beta) \equiv \delta\lambda\, s a_{\alpha\mu}\,,\\
\delta\bar{C}_\alpha &= \delta\lambda\,\tfrac{1}{2}f^{\beta\gamma}_\alpha\bar{C}_\beta\bar{C}_\gamma \equiv \delta\lambda\, s\bar{C}_\alpha\,,\\
\delta C_\alpha &= \delta\lambda\,\mathfrak{G}_\alpha(\underline{\varphi}, \underline{a}) \equiv \delta\lambda\, sC_\alpha\,.
\end{aligned} \tag{13}$$

Equation (12) implies

$$s^2\varphi_a = s^2 a_{\alpha\mu} = s^2\bar{C}_\alpha = 0, \tag{14}$$

which in terms of Eq. (13) reads

$$\begin{aligned}
&\text{(a)}\quad f^{\beta\lambda}_\alpha f^{\gamma\delta}_\lambda + f^{\gamma\lambda}_\alpha f^{\delta\beta}_\lambda + f^{\delta\lambda}_\alpha f^{\beta\gamma}_\lambda = 0\\
&\text{(b)}\quad [\theta^\alpha, \theta^\beta]_{ab} - f^{\alpha\beta}_\gamma\theta^\gamma_{ab} = 0,\\
&\text{(c)}\quad [\theta^\alpha, \theta^\beta]_{\delta\eta} - f^{\alpha\beta}_\gamma\theta^\gamma_{\delta\eta} = 0,\\
&\text{(d)}\quad (\theta^\alpha q^\beta - \theta^\beta q^\alpha - f^{\alpha\beta}_\gamma q^\gamma)_a = 0,\\
&\text{(e)}\quad (\theta^\alpha q^\beta - f^{\alpha\beta}_\gamma q^\gamma)_\delta = 0.
\end{aligned} \tag{15}$$

Equation (15a) is the Jacobi identity. If we choose a solution corresponding to $\mathfrak{h}$, Eq. (15b) and Eq. (15c) assert that θ^α is a representation of $\mathfrak{h}$. If $\mathfrak{h}$ is semisimple, then all solutions (cf. Appendix A) of Eq. (15d) are of the form

$$q_a{}^\alpha = (\theta^\alpha q)_a \tag{16}$$

for some fixed $\{q_a\}$. Finally Eq. (15e) states that $q_\beta{}^\alpha$ intertwines $\theta^\alpha_{\beta\gamma}$ and $f^\alpha_{\beta\gamma}$ (the adjoint representation of $\mathfrak{h}$). In the semisimple case $\theta^\alpha_{\beta\lambda}$ is thus equivalent to the adjoint representation if $\{q_\beta{}^\alpha\}$ does not vanish identically, a requirement which belongs to the definition of $\{a\}$ as a gauge field. We may then choose in this case without any loss of generality Eq. (13) to be identical with Eqs. (3), (4), (7).

Let us now come back to the Lagrangian $\mathscr{L}$.

If $\mathscr{L}$ is Slavnov invariant namely such that

$$s\int \mathscr{L}(x)\,dx = 0, \tag{17}$$

a fortiori

$$s^2\int \mathscr{L}(x)\,dx = 0. \tag{18}$$

Now, $\mathscr{L}$ is of the form

$$\mathscr{L} = \mathscr{L}_{\text{inv.}}(\{\varphi_a\}, \{a_{\alpha\mu}\}) + C_\alpha(x)\, K_{xy}^{\alpha\beta} \bar{C}_\beta(y) + \Delta\mathscr{L}(\{\varphi_a\}, \{a_{\alpha\mu}\}) + L^{\alpha\beta\gamma\delta}(C_\alpha C_\beta \bar{C}_\gamma \bar{C}_\delta)(x) \tag{19}$$

where $\mathscr{L}_{\text{inv.}}$, invariant under gauge transformations, is by itself a solution of Eq. (17). By Eq. (12) it is obvious from Eq. (18) that

$$L^{\alpha\beta\gamma\delta} = 0 \tag{20}$$

and that

$$\int (M\bar{C})_{\alpha,x}\, (K\bar{C})_x^\alpha\, dx = 0. \tag{21}$$

Writing out the general form of $K_{xy}^{\alpha\beta}$ yields if $\mathfrak{h}$ is semisimple

$$C_\alpha(x)\, K_{xy}^{\alpha\beta} \bar{C}_\beta(y) = C_\alpha(x)\, \Gamma_{\alpha'}^{\alpha} M_{xy}^{\alpha'\beta} \bar{C}_\beta(y) \tag{22}$$

where $\Gamma_{\alpha'}^{\alpha}$ is a numerical symmetrical matrix. If $\mathfrak{h}$ has an abelian invariant part $\mathscr{A}$ this is not the general solution, the $\Phi\Pi$ mass term being left undetermined.

Going back to Eq. (17, 19) yields in the semisimple case:

$$\mathscr{L} = \mathscr{L}_{\text{inv.}}(\{\varphi_a\}, \{a_{\alpha\tau}\}) + C_\alpha \Gamma_{\alpha'}^{\alpha}(M^{\alpha'\beta} \bar{C}_\beta) - \tfrac{1}{2} \mathfrak{G}_\alpha \Gamma^{\alpha\alpha'} \mathfrak{G}_{\alpha'} \tag{23}$$

which is identical with Eq. (2) modulo a redefinition of $\mathfrak{G}_\alpha$ and C_α. In the abelian case, there may arise an ambiguity unless the gauge function $\mathfrak{G}_\alpha$, $(\alpha \in \mathscr{A})$ contains besides the $\partial^\mu a_\mu$ terms a part which is invariant under $\mathscr{A}$.

For instance in quantum electrodynamics in the Stueckelberg gauge [11], there may arise the Slavnov invariant photon-$\Phi\Pi$ mass term:

$$(a_\mu a^\mu/2) + C\bar{C}. \tag{24}$$

This is however not the case in the t'Hooft–Veltman [18] gauge which contains an $a_\mu a^\mu$ term.

A similar phenomenon, which occurs in the abelian Higgs Kibble model produces quite spectacular complications [1] if one makes a comparison with its SU2 analog (cf. Section 3)! This is but one pathology associated with the abelian parts of $\mathfrak{h}$ which make it unstable under deformations.

For the reasons explained in the Introduction we shall from now on assume that $\mathfrak{h}$ is semisimple (and compact!).

B. *Perturbation Theory: The Slavnov Identities*

The problem is to find a renormalizable effective Lagrangian

$$\mathscr{L}_{\text{eff}}(\{\varphi_a\}, \{a_{\alpha\mu}\}, \{C_\alpha\}, \{\bar{C}_\alpha\}) \tag{25}$$

whose lowest order term in $\hbar$ is given by Eq. (2), assuming that all the parameters in Eq. (2) have been chosen in such a way that all mass parameters are strictly positive, and which is furthermore invariant in the renormalized sense [1] under the Slavnov transformation

$$\begin{aligned} \delta\phi_i &= \delta\lambda\, N_2[(T^{\alpha}_{ij}\phi_j + Q_i^{\alpha})\,\overline{C}_{\alpha}] \equiv \delta\lambda\, P_i\,, \\ \delta\overline{C}_{\alpha} &= \delta\lambda\, \tfrac{1}{2}N_2[F^{\beta\gamma}_{\alpha}\overline{C}_{\beta}\overline{C}_{\gamma}] \equiv \delta\lambda\, P_{\alpha}\,, \\ \delta C_{\alpha} &= \delta\lambda\, \mathfrak{G}_{\alpha} \equiv \delta\lambda\, \mathfrak{G}_{\alpha}{}^{i}\phi_i\,, \end{aligned} \tag{26}$$

where

$$\begin{aligned} \{\phi_i\} &= \{\{\varphi_a\}, \{a_{\alpha\mu}\}\} \\ Q^{\alpha}_{\beta\mu} &= Q_{\beta}{}^{\alpha}\partial_{\mu}\,, \qquad \mathfrak{G}^{\beta\mu}_{\alpha} = \mathfrak{G}_{\alpha}{}^{\beta}\partial^{\mu} \end{aligned} \tag{27}$$

and $\mathfrak{G}_{\alpha}{}^{i}$, $F^{\beta\gamma}_{\alpha}$, T^{α}_{ij}, $Q_i{}^{\alpha}$ are to be found as formal power series in $\hbar$ [1] whose lowest order terms are the corresponding tree parameters, namely

$$\begin{aligned} P_i &= \mathring{P}_i + O(\hbar), \\ P_{\alpha} &= \mathring{P}_{\alpha} + O(\hbar), \\ \mathfrak{G}_{\alpha}{}^{i} &= \mathring{\mathfrak{G}}_{\alpha}{}^{i} + O(\hbar), \end{aligned} \tag{28}$$

with

$$\begin{aligned} \mathring{P}_i &= s\Phi_i\,, \\ \mathring{P}_{\alpha} &= s\overline{C}_{\alpha}\,, \\ \mathring{\mathfrak{G}}_{\alpha}{}^{i} &= g_{\alpha}{}^{i} \end{aligned} \tag{29}$$

(cf. Eqs. (4), (5), (7)].

In order to deal with radiative corrections, let us add to $\mathscr{L}_{\text{eff}}$ the external field and source terms [19].

$$\gamma^i P_i + \zeta^{\alpha} P_{\alpha} + J^i\Phi_i + \bar{\xi}^{\alpha} C_{\alpha} + \xi^{\alpha}\overline{C}_{\alpha} \tag{30}$$

where γ^i is assigned dimension two, $\Phi\Pi$ charge $+1$, Fermi statistics, ζ^{α} is assigned dimension two, $\Phi\Pi$ charge $+2$, Bose statistics; J^i, ξ^{α}, $\bar{\xi}^{\alpha}$ are sources of the basic fields: J^i of the Bose type, ξ^{α}, $\bar{\xi}^{\alpha}$ of the Fermi type.

Performing the Slavnov transformation Eq. (26), and using the quantum action principle yield in terms of the Green functional [1] $Z_c(\mathscr{J}, \eta)$:

$$\begin{aligned} \mathscr{S} Z_C(\mathscr{J}, \eta) &\equiv \int dx\, (J^i\delta_{\gamma^i} - \xi^{\alpha}\delta_{\zeta^{\alpha}} - \xi^{\alpha}\mathfrak{G}_{\alpha}{}^{i}\delta_{J_i})(x) \\ \ldots Z_C(\mathscr{J}, \eta) &= \int dx\, \Delta(x)\, Z_C(\mathscr{J}, \eta) \end{aligned} \tag{31}$$

where $\mathscr{J}$ stands for the sources and η for the external fields:

$$\mathscr{J} = \{J^i, \xi^\alpha, \bar{\xi}^\alpha\},$$
$$\eta = \{\gamma^i, \zeta^\alpha\}. \tag{32}$$

$\int \Delta(x)\, dx$ is the most general dimension five insertion carrying [1] $\phi\pi$ charge-1, which is of the form:

$$\int dx\, \Delta(x) = \int dx\, N_5[-s^P(\mathscr{L}_{\text{eff}} + \gamma^i P_i + \zeta^\alpha P_\alpha) + \hbar Q] \tag{33}$$

where s^P is the naive transformation Eq. (11) corresponding to Eq. (26), and $\hbar Q$ lumps together all radiative corrections.

Making explicit the external field dependence, we shall write

$$\Delta = \Delta_0 + \gamma^i \Delta_i + \zeta^\alpha \Delta_\alpha\,,$$
$$Q = Q_0 + \gamma^i Q_i + \zeta^\alpha Q_\alpha\,. \tag{34}$$

Introducing a new classical field β carrying $\Phi\Pi$ charge $+1$ and linearly coupled to Δ, the Lagrangian becomes:

$$\mathscr{L}_{\text{eff}}^{(\eta,\beta)} = \mathscr{L}_{\text{eff}} + \gamma^i P_i + \zeta^\alpha P_\alpha + \beta\Delta$$
$$= \mathscr{L}_{\text{eff}}^{(\eta)} + \beta\Delta. \tag{35}$$

Performing now the quantum variation of the fields:

$$\delta\phi_i(x) = \delta\lambda\, N_2 \left[\delta_{\gamma^i} \int \mathscr{L}_{\text{eff}}^{(\eta,\beta)}(y)\, dy\right](x),$$
$$\delta\bar{C}_\alpha(x) = \delta\lambda\, N \left[\delta_{\zeta^\alpha} \int \mathscr{L}_{\text{eff}}^{(\eta,\beta)}(y)\, dy\right](x), \tag{36}$$
$$\delta C_\alpha(x) = \delta\lambda(\mathfrak{G}\,{}^i\phi_i)(x),$$

the quantum action principle yields:

$$\mathscr{S} Z_C(\mathscr{J}, \eta, \beta) = \int dx\, \delta_{\beta(x)} Z_C(\mathscr{J}, \eta, \beta)$$
$$+ \int dx\, [\beta\{s^\Delta(\mathscr{L}_{\text{eff}} + \gamma^i \mathring{P}_i + \zeta^\alpha \mathring{P}_\alpha) + s\Delta\} + O(\hbar\Delta, \beta)](x)$$
$$\ldots Z_C(\mathscr{J}, \eta, \beta) \tag{37}$$

where s is the tree Slavnov transformation; s^Δ generates the variation corresponding to

$$\delta\phi_i = \delta\lambda\, N_2(-sP_i + \hbar Q_i),$$
$$\delta\bar{C}_\alpha = \delta\lambda\, N_2(sP_\alpha - \hbar Q_\alpha), \tag{38}$$
$$\delta C_\alpha = 0,$$

and $Z_c(\mathscr{J}, \eta, \beta)$ is the Green functional corresponding to $\mathscr{L}^{(\eta\beta)}_{\text{eff}}$. Thus, computing

$$\mathscr{S}^2 Z_C(\mathscr{J}, \eta) = -\int dx\, [\bar{\xi}^\alpha \mathfrak{G}_\alpha{}^i \delta_{\gamma^i}](x)\, Z_C(\mathscr{J}, \eta) \tag{39}$$

where we have used the abbreviation

$$Z_C(\mathscr{J}, \eta) = Z_C(\mathscr{J}, \eta, 0)$$

we get

$$\begin{aligned} -\int dx\, &[\bar{\xi}^\alpha \mathfrak{G}_\alpha{}^i \delta_{\gamma^i}](x)\, Z_C(\mathscr{J}, \eta) \\ &= \mathscr{S} \int dx\, \delta_{\beta(x)} Z_C(\mathscr{J}, \eta, \beta)\Big|_{\beta=0} \\ &= -\int dx\, [s^\Delta(\mathscr{L}_{\text{eff}} + \gamma^i \mathring{P}_i + \zeta^\alpha \mathring{P}_\alpha) + s\Delta + O(\hbar\Delta)](x)\, Z_C(\mathscr{J}, \eta) \end{aligned} \tag{40}$$

since

$$\int dx\, dy\, \delta_{\beta(x)} \delta_{\beta(y)} Z_C(\mathscr{J}, \eta, \beta) = 0. \tag{41}$$

In terms of the vertex functional $\Gamma(\Psi, \eta) = \Gamma(\phi_i, C_\alpha, \bar{C}_\alpha, \eta)$ which is the Legendre transform of $Z_C(\mathscr{J}, \eta)$ with respect to $\mathscr{J}$, Eq. (40) reads:

$$\begin{aligned} \int dx\, &[\delta_{C_\alpha} \Gamma \mathfrak{G}_\alpha{}^i \delta_{\gamma^i} \Gamma](x) \\ &= \int dx\, [s^\Delta(\mathscr{L}_{\text{eff}} + \gamma^i P_i + \zeta^\alpha P_\alpha) + s\Delta + O(\hbar\Delta)](x) \\ &= \int dx\, \Delta'(x) + O(\hbar\Delta). \end{aligned} \tag{42}$$

Equation (42) leads to a consistency condition for Δ. Indeed Eq. (42) is a perturbed version of the equation

$$\int dx\, [\delta_{C_\alpha} \Gamma \mathfrak{G}_\alpha{}^i \delta_{\gamma_i} \Gamma](x) = 0 \tag{43}$$

with a perturbation of order Δ'. Equation (43) is itself a quantum extension of:

$$\int dx \left[\delta_{C_\alpha} \left\{\int dy\, \mathscr{L}^{(\eta)}_{\text{eff}}(y)\right\} \mathfrak{G}_\alpha{}^i \delta_{\gamma_i} \left\{\int dz\, \mathscr{L}^{(\eta)}_{\text{eff}}(z)\right\}\right](x) = 0 \tag{44}$$

since for any field ω

$$\delta_{\omega(x)} \Gamma(\Psi, \eta) = \delta_{\omega(x)} \int dy\, \mathscr{L}^{(\eta)}_{\text{eff}}(y) + O(\hbar). \tag{45}$$

We have seen in the previous section that Eq. (44), as an equation for $\delta_{C_\alpha}\mathscr{L}_{\text{eff}}$ possesses the general solution

$$\delta_{C_\alpha}\int \mathscr{L}^{(\eta)}_{\text{eff}}(x)\,dx = \Gamma^{\alpha\alpha'}\mathfrak{G}_{\alpha'}^{\;i}\delta_{\gamma^i}\int \mathscr{L}^{(\eta)}_{\text{eff}}(x)\,dx \tag{46}$$

where $\Gamma^{\alpha\alpha'}$ is a numerical symmetrical matrix. This provides a solution to Eq. (43):

$$\delta_{C_\alpha}\Gamma = \Gamma^{\alpha\alpha'}\mathfrak{G}_i^{\alpha'}\delta_{\gamma_i}\Gamma. \tag{47}$$

This is the general solution of Eq. (43) because Eq. (43) is a quantum perturbation of Eq. (44). Hence the general solution of Eq. (42) is given by

$$\delta_{C_\alpha}\int \mathscr{L}^{(\eta)}_{\text{eff}}(x)\,dx = \Gamma^{\alpha\alpha'}\mathfrak{G}_{\alpha'}^{\;i}\delta_{\gamma^i}\int \mathscr{L}^{(\eta)}_{\text{eff}}(x)\,dx + O(\Delta') \tag{48}$$

which, taking into account the structure of the γ couplings can be written in the form

$$\delta_{C_\alpha}\int \mathscr{L}^{(\eta)}_{\text{eff}}(x)\,dx = \Gamma^{\alpha\alpha'}\mathfrak{G}_{\alpha'}^{\;i}\delta_{\gamma^i}\int \mathscr{L}^{(\eta)}_{\text{eff}}(x)\,dx + \delta_{C_\alpha}\int R^{(\Delta')}(x)\,dx \tag{49}$$

where $R^{(\Delta')}$ is of the order of Δ'. Substituting Eq. (49) into Eq. (42) leads to

$$\int dx\,\delta_{C_\alpha(x)}\int dy\,R^{(\Delta')}(y)\,\mathfrak{G}_\alpha^{\;i}\delta_{\gamma^i(x)}\int dz\,\mathscr{L}^{(\eta)}_{\text{eff}}(z) = \int dx\,\Delta'(x) + O(\hbar\Delta). \tag{50}$$

Recalling Eqs. (28), (29), one has

$$\mathfrak{G}_\alpha^{\;i}P_i = g_\alpha^{\;i}\mathring{P}_i + O(\hbar). \tag{51}$$

Hence, using Eq. (26),

$$\begin{aligned}\mathfrak{G}_\alpha^{\;i}\delta_{\gamma^i(x)}\int dy\,\mathscr{L}^{(\eta)}_{\text{eff}}(y) &= sg_\alpha^{\;i}\phi_i + O(\hbar)\\ &= s^2C_\alpha + O(\hbar).\end{aligned} \tag{52}$$

We have thus obtained the consistency condition:

$$\begin{aligned}&s^\Delta\int dx\,(\mathscr{L}_{\text{eff}} + \gamma^i\mathring{P}_i + \zeta^\alpha\mathring{P}_\alpha)(x) + s\int dx\,\Delta(x)\\ &= -s^2\int R^{(\Delta')}(x)\,dx + O(\hbar\Delta).\end{aligned}$$

Now, the most general form of Δ is

$$\begin{aligned}\Delta(x) = {}&(\Delta^\alpha\bar{C}_\alpha)(x) + (C_\alpha\Delta^{\alpha\beta\gamma}_{yz}\bar{C}_\beta(y)\,\bar{C}_\gamma(z))(x)\\ &+ (C_\alpha C_\beta\bar{C}_\gamma\bar{C}_\delta\bar{C}_\eta)(x)\,\Delta^{\alpha\beta,\gamma\delta\eta} + (\gamma^i\Delta_i + \zeta^\alpha\Delta_\alpha)(x)\end{aligned} \tag{54}$$

where dimensions and quantum numbers are taken into account by

$$\Delta_i(z) = \tfrac{1}{2}[\Theta_{ij}^{\alpha\beta}\phi_j\bar{C}_\alpha\bar{C}_\beta + \bar{C}_\alpha\Sigma_i^{\alpha\beta}\bar{C}_\beta](x) \tag{55}$$

with

$$\Sigma_{\gamma\mu}^{\alpha\beta} = \Sigma_\gamma^{\alpha,\beta}\partial_\mu \tag{56}$$

and

$$\Delta_\alpha(x) = -\tfrac{1}{6}\Gamma_\alpha^{\beta\gamma\delta}(\bar{C}_\beta\bar{C}_\gamma\bar{C}_\delta)(x) \tag{57}$$

for some c-number coefficients $\Theta_{ij}^{\alpha\beta}$, $\Sigma_\gamma^{\alpha,\beta}$, $\Gamma_\alpha^{\beta\gamma\delta}$. Since, due to power counting $R^{(\Delta)}$ cannot depend on the external fields (γ^i, ξ^α), the external field dependent part of the left hand side of Eq. (53), must be $O(\hbar\Delta)$, namely,

$$s^\Delta \int dx\,(\gamma^i\mathring{P}_i + \zeta^\alpha\mathring{P}_\alpha)(x) + s\int dx\,\Delta(x) = O(\hbar\Delta). \tag{58}$$

It is shown in Appendix B, by cohomological methods, that, as a result, the external field dependent part of Δ is of the form

$$\int (\gamma^i\Delta_i + \zeta^\alpha\Delta_\alpha)(x)\,dx = -s^{\mathring{P}+\Pi}\int [\gamma^i(\mathring{P}+\Pi)_i + \zeta^\alpha(\mathring{P}+\Pi)_\alpha](x)\,dx + O(\hbar\Delta) \tag{59}$$

where Π_i, Π_α can be parametrized in terms of numerical coefficients $\hat{\Theta}_{ij}$, $\hat{\Sigma}_i^{\alpha}$, $\hat{\Sigma}_\beta^{\alpha}$ according to

$$\begin{aligned}\Pi_i &= \hat{\Theta}_{ij}^{\alpha}\phi_j\bar{C}_\alpha + \hat{\Sigma}_i^{\alpha}\bar{C}_\alpha,\\ \Pi_\alpha &= \hat{\Gamma}_\alpha^{\beta\gamma}\bar{C}_\beta\bar{C}_\gamma,\end{aligned} \tag{60}$$

with

$$\hat{\Sigma}_{\beta,\mu}^{\alpha} = \hat{\Sigma}_\beta^{\alpha}\partial_\mu. \tag{61}$$

The coefficients $\hat{\Theta}$, $\hat{\Sigma}$, $\hat{\Gamma}$ can be chosen and will be chosen to be linear in Θ, Σ, Γ, as shown in Appendix B.

Now, we know from the quantum action principle that the external field dependent part of Δ is of the form (cf. Eqs. (33), 34):

$$\begin{aligned}&\int dx\,(\gamma^i\Delta_i + \zeta^\alpha\Delta_\alpha)(x)\\ &\quad = -s^P\int dx\,(\gamma^iP_i + \zeta^\alpha P_\alpha)(x) + \hbar\int(\gamma^iQ_i + \zeta^\alpha Q_\alpha)(x)\,dx\end{aligned} \tag{62}$$

where Q_i, Q_α are formal power series in $\hbar$ and in the coefficients of $\mathscr{L}_{\text{eff}}^{(n)\text{int}}$. From the previous observation that Π_i, Π_α are linear in Δ_i, Δ_α, it follows that Π_i, Π_α are formal power series of the same type as Q_i, Q_α.

Finally the system

$$\Pi_i = \Pi_\alpha = 0 \tag{63}$$

which insures that (cf. Eq. (59)

$$\int dx\, (\gamma^i \Delta_i + \zeta^\alpha \Delta_\alpha)(x) = O(\hbar\Delta) \tag{64}$$

is soluble for P_i, P_α as formal power series in $\hbar$: considering Π_i, Π_α as formal power series in $\hbar$, $P_i - \mathring{P}_i$, $P_\alpha - \mathring{P}_\alpha$, and comparing (59) with (62) we see that, to lowest non vanishing order in $\hbar$, $P_i - \mathring{P}_i$, $P_\alpha - \mathring{P}_\alpha$,

$$\begin{aligned} \Pi_i &\simeq P_i - \mathring{P}_i , \\ \Pi_\alpha &\simeq P_\alpha - \mathring{P}_\alpha . \end{aligned} \tag{65}$$

Hence system (63) assumes the form

$$\begin{aligned} P_i - \mathring{P}_i &= \Theta_i(\hbar, P - \mathring{P}), \\ P_\alpha - \mathring{P}_\alpha &= \Theta_\alpha(\hbar, P - \mathring{P}), \end{aligned} \tag{66}$$

where both Θ_i, Θ_α are $O(\hbar, (P - \mathring{P})^2)$. This system has a unique solution $O(\hbar)$ for $P_i - \mathring{P}_i$, $P_\alpha - \mathring{P}_\alpha$.

Let us now look at the external field independent terms in Δ, which is $O(\hbar\Delta)$ in the external fields. The consistency condition now reads

$$s^2 \int R^{(\Delta')}(x)\, dx = -s \int dx\, \Delta_0(x) + O(\hbar\Delta), \tag{67}$$

i.e.,

$$s \int dx\, (sR^{(\Delta')} + \Delta_0)(x) = O(\hbar\Delta) \tag{68}$$

where

$$\Delta = \Delta\, |_{\gamma=\zeta=0} \tag{69}$$

If we can prove that $\int \Delta_0(x)\, dx$ is of the form

$$\int dx\, \Delta_0(x) = s \int dx\, \hat{\Delta}_0(x) + O(\hbar\Delta), \tag{70}$$

comparing with Eq. (33), (34) which can be cast into the form

$$\Delta_0 = -s\mathscr{L}_{\text{eff}} + \hbar\tilde{Q}_0(\hbar, \mathscr{L}^{\text{int}}_{\text{eff}}) \tag{71}$$

where $\hbar\tilde{Q}_0 = \hbar Q_0 + (s - s^P)\mathscr{L}_{\text{eff}}$ is a formal power series in the indicated arguments, it follows that $\tilde{Q}_0$ is of the form

$$\tilde{Q}_0(\hbar, \mathscr{L}_{\text{eff}}^{\text{int}}) = s\hat{Q}_0(\hbar, \mathscr{L}_{\text{eff}}^{\text{int}}) + O(\hbar\Delta). \tag{72}$$

Since the equation

$$\mathscr{L}_{\text{eff}} - \mathscr{L} - \hbar\hat{Q}_0(\hbar, \mathscr{L}_{\text{eff}}^{\text{int}}) = 0 \tag{73}$$

is soluble for

$$\mathscr{L}_{\text{eff}} - \mathscr{L} = O(\hbar) \tag{74}$$

its solution leaves us with

$$\Delta_0 = O(\hbar\Delta). \tag{75}$$

Recalling that also (cf. Eq. (64))

$$\Delta_i = O(\hbar\Delta), \qquad \Delta_\alpha = O(\hbar\Delta) \tag{76}$$

we conclude that

$$\Delta = O(\hbar\Delta) \tag{77}$$

hence

$$\Delta = 0, \tag{78}$$

which we want to achieve.

Now the consistency condition shows that

$$\int dx\, \Delta_0(x) = -s \int R^{(\Delta')}(x)\, dx + \Delta_\natural + O(\hbar\Delta) \tag{79}$$

where

$$s\Delta_\natural = 0 \tag{80}$$

Thus, there remains to prove that any $\Delta_\natural$ can be put in the form

$$\Delta_\natural = s\hat{\Delta}_\natural \tag{81}$$

Using for $\Delta_\natural$ the expansion (cf. Eq. (54))

$$\begin{aligned}\Delta_\natural(x) = \Delta_\natural{}^\alpha(x)\, \bar{C}_\alpha(x) + C_\alpha(x)\, \Delta_{\natural xyz}^{\alpha\beta\gamma}\bar{C}_\beta(y)\, \bar{C}_\gamma(z)\\ + C_\alpha(x)\, C_\beta(x)\, \bar{C}_\gamma(x)\, \bar{C}_\delta(x)\, \bar{C}_\eta(x)\, \Delta_\natural^{\gamma\beta,\gamma\delta\eta}\end{aligned} \tag{82}$$

since (80) implies

$$s^2 \Delta_\natural = 0 \tag{83}$$

a discussion similar to that found at the end of Section A (cf. Eqs. (17), 18)) shows that

$$\begin{aligned} \Delta_\natural^{\alpha\beta,\gamma\delta\eta} &= 0 \\ \Delta^{\alpha\beta\gamma}_{\natural xyz} &= \Gamma^{\widehat{\alpha\delta}\gamma}\delta(x-z)\, M_\delta{}^\beta(x, y) \end{aligned} \tag{84}$$

where the numerical tensor $\Gamma^{\widehat{\alpha\delta}\gamma}$ is symmetric in α and δ. Using now

$$s\Delta_\natural = 0$$

yields

$$\Gamma^{\widehat{\alpha\delta}\gamma} = 0 \tag{85}$$

and

$$\delta_{\omega_\alpha(x)} \Delta_\natural{}^\beta(y) - \delta_{\omega_\beta(y)} \Delta_\natural{}^\alpha(x) - f^{\alpha\beta}_\gamma \delta(x-y)\, \Delta_\natural{}^\gamma(y) = 0. \tag{86}$$

One can show [20] that the general solution of Eq. (86), the first cohomology condition for the gauge Lie algebra, which is nothing else than the Wess Zumino [9] consistency condition, is

$$\Delta_\natural{}^\alpha(x) = \delta_{\omega_\alpha(x)} \hat{\Delta}_\natural + g_\alpha(x) \tag{87}$$

where $g_\alpha(x)$, the Bardeen [10] anomaly has the form

$$g^\alpha(x) = \partial^\mu \epsilon_{\mu\nu\rho\sigma} \left[D^{\alpha\beta\gamma} \partial^\nu a_\beta{}^\rho a_\gamma{}^\sigma + F^{\alpha\beta\gamma\delta} a_\beta{}^\nu a_\gamma{}^\rho a_\delta{}^\sigma\right] \tag{88}$$

where $D^{\alpha\beta\gamma}$ is a totally symmetric invariant tensor with indices in the adjoint representation of $\mathfrak{h}$ and

$$F^{\alpha\beta\gamma\delta} = (1/12)[D^{\alpha\beta\lambda} f^{\gamma\delta}_\lambda + D^{\alpha\gamma\lambda} f^{\delta\beta}_\lambda + D^{\alpha\delta\lambda} f^{\beta\gamma}_\lambda]. \tag{89}$$

Such an anomaly can only arise if the tree Lagrangian contains $\epsilon_{\mu\nu\rho\sigma}$ or γ_5 symbols and if there is a nontrivial D tensor. In the absence of such an anomaly, one has:

$$\bar{C}_\alpha(x)\, \Delta_\natural{}^\alpha(x) = \bar{C}_\alpha(x)\, \delta_{\omega_\alpha(x)} \hat{\Delta}_\natural = s\hat{\Delta}_\natural \tag{90}$$

which completes the proof of the Slavnov identity in such cases.

C. *The Faddeev Popov Ghost Equation of Motion*

It will be of interest in the following to write down the Faddeev Popov ghost equation of motion in terms of the Slavnov identity. It follows from Eq. (42) that once the Slavnov identity has been proved,

$$\int [\delta_{C_\alpha}\Gamma \mathfrak{G}_i{}^\alpha \delta_{\gamma_i}\Gamma](x)\,dx = 0 \tag{91}$$

and we know that the general solution of this equation is

$$\delta_{C_\alpha(x)}\Gamma = \Gamma^\alpha_{\alpha'} \mathfrak{G}_i^{\alpha'} \delta_{\gamma_i(x)}\Gamma \tag{92}$$

where $\Gamma^\alpha_{\alpha'}$ is a symmetrical matrix. Taking the Legendre transform of Eq. (92) yields

$$\Gamma^\alpha_{\alpha'} \mathfrak{G}_i^{\alpha'} \delta_{\gamma_i(x)} Z_C(\mathscr{J}, \eta) = \bar{\xi}^\alpha(x) \tag{93}$$

which is therefore the $\Phi\Pi$ equation of motion. One should also note that $\{\Gamma^\alpha_{\alpha'}\}$ is invertible in the tree approximation and therefore to all orders so that Eq. (93) may be written in the form

$$\mathfrak{G}_i{}^\alpha \delta_{\gamma_i(x)} Z_C(\mathscr{J}, \eta) = (\Gamma^{-1})^\alpha_{\alpha'}\, \bar{\xi}^{\alpha'}(x) \tag{94}$$

3. Physical Interpretation: An Example [16]

Given a gauge theory which has been renormalized in such a way that a Slavnov identity holds, there remains to show that one can interpret it in physical terms, which is not obvious since many ghost fields are involved (∂a, C, $\bar{C}$,...). First one should specify the connection between the parameters left arbitrary in the Lagrangian, and physical parameters, (masses, coupling constants) through the fulfillment of suitable normalization conditions. Then, once the theory has been set up within the framework of a fixed Fock space, one has to specify a physical subspace within which the theory is reasonable, e.g., the S operator is unitary and independent from unphysical parameters among which the $g_\alpha{}^i$ [cf. Eq. (26)].

In order to make this program explicit we shall treat in some details the SU2 Higgs Kibble model [16] in a way which parallels our treatment of the abelian Higgs Kibble model [1].

A. *The Classical Theory*

The basic fields are $\Psi = \{\sigma, \pi_\alpha, a_{\alpha\mu}, C_\alpha, \bar{C}_\alpha\}$, $\alpha = 1, 2, 3$. At the classical level, the Lagrangian is invariant under the Slavnov transformation:

$$\begin{aligned}
\delta\pi_\alpha &= \delta\lambda\,[-(e/2)\,\epsilon_\alpha^{\beta\gamma}\pi_\beta\bar{C}_\gamma + (e/2)(\sigma + F)\,\bar{C}_\alpha] \equiv \delta\lambda\,\mathfrak{T}_\alpha\,,\\
\delta\sigma &= \delta\lambda[-(e/2)\,\pi^\alpha\bar{C}_\alpha] \equiv \delta\lambda\,\mathfrak{T}_0\,,\\
\delta a_{\alpha\mu} &= \delta\lambda\,[\partial_\mu\bar{C}_\alpha - e\epsilon_\alpha^{\beta\gamma}a_{\beta\mu}\bar{C}_\gamma] \equiv \delta\lambda\,\mathfrak{T}_{\alpha\mu}\,,\\
\delta\bar{C}_\alpha &= \delta\lambda\,[(e/2)\,\epsilon_\alpha^{\beta\gamma}\bar{C}_\beta\bar{C}_\gamma] \equiv \delta\lambda\,\bar{\mathfrak{T}}_\alpha\,,\\
\delta C_\alpha &= \delta\lambda\,[\partial^\mu a_{\alpha\mu} + \rho\pi_\alpha] \equiv \delta\lambda\,\mathfrak{G}_\alpha\,,
\end{aligned} \tag{95}$$

where $\epsilon^{\alpha\beta\gamma}$ are the SU(2) structure constants, e plays the role of a coupling constant, and F is the σ field translation parameter. This particular choice of transformation laws implies the invariance under a global SU2 symmetry transforming $\boldsymbol{\pi}$, $\mathbf{a}_\mu$, C, $\bar{C}$ as vectors and leaving σ invariant. Even if this symmetry is to be preserved Eq. (95) is not the most general transformation law fulfilling the compatibility conditions Eqs. (14), (15), which depends on four parameters besides those which label the gauge function: e, F, and wavefunction renormalizations for the σ and $\bar{C}$ fields. Given these parameters and introducing external fields $\{\eta\} = \{\gamma^\alpha, \gamma^0, \gamma^{\alpha\mu}\xi^\alpha\}$ coupled to $\{\mathfrak{T}\} = \{\mathfrak{T}_\alpha, \mathfrak{T}_0, \mathfrak{T}_{\alpha\mu}, \mathrm{C}_\alpha\}$, and sources $\{\mathscr{J}\} = \{J^\alpha, J^0, J^{\alpha\mu}, \xi^\alpha, \bar{\xi}^\alpha\}$ coupled to $\{\Psi\} = \{\pi_\alpha, \sigma, a_{\alpha\mu}, \bar{C}_\alpha, C_\alpha\}$ the Slavnov identity reads:

$$\begin{aligned}
\mathscr{S}Z_C(\mathscr{J}, \eta) = \int dx\,[&J^\alpha(x)\,\delta_{\gamma^\alpha(x)} + J^0(x)\,\delta_{\gamma^0(x)} + J^{\alpha\mu}(x)\,\delta_{\gamma^{\alpha\mu}(x)}\\
&- \xi^\alpha(x)\,\delta_{\zeta^\alpha(x)} - \bar{\xi}^\alpha(x)(\partial^\mu\delta_{J^{\mu\alpha}(x)} + \rho\delta_{J^\alpha(x)})]\,Z_C(\mathscr{J}, \eta)\\
= 0.&
\end{aligned} \tag{96}$$

We have seen that the most general action compatible with Eq. (96) is of the form

$$\begin{aligned}
\int \mathscr{L}^{\mathrm{cl}}_{(x)}(x)\,dx = \int dx\,\Big\{&\mathscr{L}_{\mathrm{inv.}}(x) - \frac{1}{\kappa}\Big[\mathfrak{G}^\alpha(x)\,\mathfrak{G}_\alpha(x) - \int dy\,C_\alpha(x)\,M^{\alpha\beta}_{xy}\bar{C}_\beta(y)\Big]\\
&+ \gamma^\alpha(x)\,\mathfrak{T}_\alpha(x) + \gamma^0(x)\,\mathfrak{T}_0(x) + \gamma^{\alpha\mu}(x)\,\mathfrak{T}_{\alpha\mu}(x) + \zeta^\alpha(x)\,\bar{\mathfrak{T}}_\alpha(x)\Big\}
\end{aligned} \tag{97}$$

with

$$\begin{aligned}
\mathscr{L}_{\mathrm{inv.}} = &-\frac{Z_a}{4}\,G^\alpha_{\mu\nu}G^{\mu\nu}_\alpha + \frac{Z_1}{2}\,D_\mu\underline{\Phi}\,D^\mu\underline{\Phi}\\
&+ \frac{\mu^2}{2}\,\underline{\Phi}\cdot\underline{\Phi} - \frac{\lambda^2}{4!}\,(\underline{\Phi}\cdot\underline{\Phi})^2 - \Big[\frac{\mu^2}{2}\,F^2 - \frac{\lambda^2}{4!}\,F_J\Big]
\end{aligned} \tag{98}$$

where

$$G_{\alpha\mu\nu} = \partial_\mu a_{\alpha\nu} - \partial_\nu a_{\alpha\mu} - e\epsilon_\alpha^{\beta\gamma}a_{\beta\mu}a_{\gamma\nu}\,,$$

$$\underline{\Phi} = \begin{pmatrix}\{\pi_\alpha\}\\ \sigma + F\end{pmatrix} = \begin{pmatrix}\boldsymbol{\pi}\\ \sigma + F\end{pmatrix}; \qquad \underline{\Phi}\cdot\underline{\Phi} = \boldsymbol{\pi}^2 + (\sigma + F)^2,$$

$$D_\mu = \partial_\mu \mathbb{1} - \frac{e}{2}\left[\begin{array}{c|c} \boldsymbol{\epsilon}\cdot\mathbf{a}_\mu & \mathbf{a}_\mu \\ \hline -\mathbf{a}_\mu & 0 \end{array}\right],$$

$$(\boldsymbol{\epsilon}\cdot\mathbf{a}_\mu)_\alpha^\beta = -\epsilon_\alpha^{\beta\gamma} a_{\gamma\mu}, \tag{99}$$

$$M_{xy}^{\alpha\beta} = \partial^\mu(\partial_\mu{}^x \delta(x-y)\,\delta^{\alpha\beta} + e\epsilon^{\alpha\beta\gamma} a_{\gamma\mu}(x)\,\delta(x-y))$$
$$+ (\rho e/2)(\epsilon^{\alpha\beta\gamma}\pi_\gamma(x) + \delta^{\alpha\beta}(\sigma(x)+F))\,\delta(x-y)$$

and where the coefficients are so adjusted that the coefficient of the term linear in σ vanishes:

$$\mu^2 - F^2\lambda^2/3! = 0. \tag{100}$$

The theory thus depends on ten parameters, four specifying the transformation law, one related with the field vacuum expectation value, five specifying the external field independent part of the Lagrangian, constrained by conditions (14), (15). One can alternatively specify the following physical parameters: m_a, M, $m_{\phi\pi}$, which give the positions of the poles in the transverse photon, σ, C, $\bar{C}$ propagators respectively:

$$\begin{aligned} &\text{(a)}\quad \Gamma_{a^T a^T}(m_a{}^2) = 0,\\ &\text{(b)}\quad \Gamma_{\sigma\sigma}(M^2) = 0,\\ &\text{(c)}\quad \Gamma_{C\bar{C}}(m_{\phi\pi}^2) = 0, \end{aligned} \tag{101}$$

the residues of these poles:

$$\begin{aligned} &\text{(a)}\quad \Gamma'_{a^T a^T}(m_a{}^2) = Z_a\\ &\text{(b)}\quad \Gamma'_{\sigma\sigma}(M^2) = Z_1\\ &\text{(c)}\quad \Gamma'_{C\bar{C}}(m_{\phi\pi}^2) = 1/\kappa \end{aligned} \tag{102}$$

the value of the coupling constant

$$\Gamma_{a^T a^T \sigma}(m_a{}^2, m_a{}^2, M^2) = \epsilon \tag{103}$$

These normalization conditions together with Eq. (100) which is equivalent to

$$\langle\sigma\rangle = 0 \tag{104}$$

fix the values of Z_a, Z_1, μ^2, λ^2, κ, ρ, e, F, leaving free two parameters in the definition of the transformation laws. For simplicity we shall of course choose

$$Z_a = Z_1 = 1. \tag{105}$$

These normalization conditions together with the Slavnov symmetry actually imply that the masses associated with the $\partial \mathbf{a}$, $\boldsymbol{\pi}$ channel are pairwise degenerate with those of the $C\bar{C}$ channel. Thus, in view of the residual SU2 symmetry, all the ghost masses are degenerate.

Within the Fock space defined by the quadratic part of the Lagrangian, we shall define the bare physical subspace generated by application on the vacuum of the asymptotic fields $a_\mu^{T\,\mathrm{in}}$, σ^{in} which explains that C, $\bar{C}$, $\partial \mathbf{a}$, $\boldsymbol{\pi}$ are considered as "ghosts."

According to this definition, it is easy to see [21] that the matrix elements of the S operator between bare physical states do not depend on the gauge parameters κ, $m_{\phi\pi}$.

B. *Radiative Corrections*: *Slavnov Identities, Normalization Conditions*

Now, according to the analysis of Section 2, it is possible to find an effective Lagrangian such that the Slavnov identity (96) holds to all orders (where now ρ is to be determined as a formal power series in $\hbar$). We also know that the Faddeev Popov equation of motion is (cf. Eq. (94)):

$$[\partial^\mu \delta_{\gamma\alpha\mu(x)} + \rho \delta_{\gamma\alpha(x)}]\, Z_C(\mathscr{J}, \eta) = \bar{\kappa}\bar{\xi}_\alpha(x) \tag{106}$$

where $\bar{\kappa}$ is some formal power series in $\hbar$. The theory depends on ten formal power series whose lowest order terms specify the tree approximation Lagrangian. Eight of them can be fixed by imposing the normalization conditions (101), (102), (103), (104). These normalization conditions are enough to interpret the theory in the initial Fock space, because, as a consequence of the Slavnov identity, the ghost mass degeneracy still holds:

Expressing the Slavnov identity in terms of the vertex functional Γ yields:

$$\int dx\, \{\delta_{\pi_\alpha(x)}\Gamma\, \delta_{\gamma\alpha(x)}\Gamma + \delta_{\sigma(x)}\Gamma\, \delta_{\gamma^0(x)}\Gamma + \delta_{a_{\alpha\mu}(x)}\Gamma\, \delta_{\gamma\alpha\mu(x)}\Gamma$$

$$+ \delta_{\bar{C}_\alpha(x)}\Gamma\, \delta_{\zeta\alpha(x)}\Gamma + \delta_{C_\alpha(x)}\Gamma\, [\partial^\mu a_{\alpha\mu}(x) + \rho\pi_\alpha(x)]\} = 0. \tag{107}$$

In particular, one gets the following information on the two point functions:

$$\tilde{\Gamma}_{\pi^2}(p^2)\, \tilde{\Gamma}_\gamma(p^2) + \tilde{\Gamma}_{\pi a_L}(p^2)\, \tilde{\Gamma}_{\gamma L}(p^2) + \rho\tilde{\Gamma}_{\bar{C}C}(p^2) = 0$$

$$\tilde{\Gamma}_{\pi a_L}(p^2)\, \tilde{\Gamma}_\gamma(p^2) + \tilde{\Gamma}_{(a_L)^2}(p^2)\, \tilde{\Gamma}_{\gamma L}(p^2) - p^2\tilde{\Gamma}_{\bar{C}C}(p^2) = 0 \tag{108}$$

where

$$\tilde{\Gamma}_{\pi^2}(p^2) = \tilde{\Gamma}_{\pi_\alpha\pi^\alpha}(p^2),$$

$$\tilde{\Gamma}_{\pi a_L}(p^2) = ip_\mu\tilde{\Gamma}_{\pi^\alpha a_{\alpha\mu}}(p),$$

$$\tilde{\Gamma}_{(a_L)^2}(p^2) = -p_\mu p_\nu \tilde{\Gamma}_{a_{\alpha\mu} a_\nu{}^\alpha}(p), \tag{109}$$
$$\tilde{\Gamma}_\gamma(p^2) = \tilde{\Gamma}_{\bar{C}_\alpha \gamma^\alpha}(p),$$
$$\tilde{\Gamma}_{\gamma L}(p^2) = -(ip^\mu/p^2)\, \tilde{\Gamma}_{\bar{C}_\alpha \gamma^{\alpha\mu}}(p).$$

On the other hand the $\Phi\Pi$ equation of motion (106) yields:

$$\rho \tilde{\Gamma}_\gamma(p^2) - p^2 \tilde{\Gamma}_{\gamma L}(p^2) = \bar{\kappa} \tilde{\Gamma}_{C\bar{C}}(p^2). \tag{110}$$

It follows that:

$$\det \begin{pmatrix} \tilde{\Gamma}_{\pi^2} & \tilde{\Gamma}_{\pi a_L} \\ \tilde{\Gamma}_{\pi a_L} & \tilde{\Gamma}_{a_L{}^2} \end{pmatrix} (p) = [\tilde{\Gamma}_\gamma(p^2)\, \tilde{\Gamma}_{\gamma L}(p^2)]^{-1}\, [\bar{\kappa} \tilde{\Gamma}_{a_L \pi}(p^2) - \rho p^2]\, \tilde{\Gamma}^2_{C\bar{C}}(p^2)$$

which shows that this determinant has a double zero at

$$p^2 = m^2_{\Phi\pi}\,.$$

C. *The Physical S Operator* : *Gauge Invariance*

According to the LSZ asymptotic theory, the physical S operator is given in the perturbative sense in terms of the Green's functional $Z = \exp(i/\hbar) Z_C$, by:

$$S_{\text{phys}} = S_{\text{phys}}(\mathscr{J})|_{\mathscr{J}=0} \tag{112}$$

where

$$S_{\text{phys}}(\mathscr{J}) = :\exp \int dx\, \{\sigma^{\text{in}}(x)\, K^\sigma_{xy} \delta_{J^0(y)} + a^{T\text{in}}_{\alpha\mu}(x)\, K^{\alpha\mu\beta\nu}_{xy} \delta_{J\beta\nu(y)}\}: Z(\mathscr{J}, \eta)|_{\eta=0}$$
$$\stackrel{\text{def}}{=} \Sigma_{\text{phys}} Z(\mathscr{J}, \eta). \tag{113}$$

In the above formula, σ^{in}, K^σ; $a^{T\text{in}}_{\alpha\mu}$, $K^{\alpha\mu\beta\nu}$ are the canonically quantized asymptotic fields and differential operators involved in their equations of motion, derived from the asymptotic Lagrangian determined by Γ, which in the present case, can be read off from the tree approximation.

We want to prove:

$$\frac{\partial S_{\text{phys}}}{\partial m^2_{\Phi\pi}} = \frac{\partial S_{\text{phys}}}{\partial \kappa} = 0. \tag{114}$$

Owing to Lowenstein's action principle [3], one has

$$(\hbar/i)\, \partial_\lambda Z(\mathscr{J}, \eta) = \Delta_\lambda Z(\mathscr{J}, \eta) \tag{115}$$

where λ is one of the parameters κ, $m^2_{\Phi\pi}$ and Δ_λ is a dimension four insertion obtained by differenciating $\int dx\, \mathscr{L}_{\text{eff}}(x)$ with respect to λ. Using the Slavnov identity, we are going to show that Δ_λ can be written as:

$$\Delta_\lambda = \sum_{i=1}^{5} C_\lambda^{0,i} \Delta_i{}^0 + \sum_{i=1}^{5} C_\lambda^{s,i} \Delta_i{}^s \tag{116}$$

where the $\Delta_i{}^o$'s are such that

$$\Sigma_{\text{phys}} \Delta_i{}^0 Z(\mathscr{J}, \eta) = 0, \qquad i = 1,..., 5, \tag{117}$$

and therefore leave unaltered the physical normalization conditions (101a, b), (102a, b), (103). In the following these insertions will be called nonphysical. On the other hand the $\Delta_i{}^s$'s are symmetric insertions, namely

$$\mathscr{S} \Delta_i{}^s Z(\mathscr{J}, \eta) = 0, \qquad i = 1,..., 5. \tag{118}$$

Thus applying Eqs. (116), (117) to the physical normalization conditions (101a, b), (102a, b), (103), yields a linear homogeneous system of equations of the form

$$\sum_{i=1}^{5} C_\lambda^{s,i} \Delta_i^{s,j} = 0 \qquad j = 1,..., 5, \tag{119}$$

for which we are going to see that [11, 1]

$$\det \| \Delta_i^{s,j} \| \neq 0 \tag{120}$$

since this will prove to be true in the tree approximation. Hence it follows that

$$C_\lambda^{s,i} = 0 \qquad i = 1,..., 5, \tag{121}$$

and the gauge invariance of the S operator follows from Eqs. (112), (113), (115), (116), (117).

The decomposition of Δ_λ given in Eq. (116) follows first from the Slavnov identity:

$$0 = \partial_\lambda \mathscr{S} Z = (\partial_\lambda \mathscr{S})\, Z + (i/\hbar)\, \mathscr{S} \Delta_\lambda Z. \tag{122}$$

Hence

$$\begin{aligned} (\partial_\lambda \mathscr{S})\, Z &\equiv -\frac{\partial \rho}{\partial \lambda} \int \xi^\alpha(x)\, \delta_{J^\alpha(x)}\, dx\, Z \\ &= \frac{i}{\hbar} [\Delta_\lambda, \mathscr{S}]\, Z. \end{aligned} \tag{123}$$

Noticing that

$$\int dx\, \xi^\alpha(x)\, \delta_{J^\alpha(x)} = -\left[\int dx\, (J^\alpha(x)\, \delta_{J^\alpha(x)} + \gamma^\alpha(x)\, \delta_{\gamma^\alpha(x)}), \mathscr{S}\right], \tag{124}$$

we define $\Delta_1{}^0$ by

$$\Delta_1{}^0 Z = \frac{\hbar}{i} \int dx\, [J^\alpha(x)\, \delta_{J\alpha(x)} + \gamma^\alpha(x)\, \delta_{\gamma\alpha(x)}]\, Z \tag{125}$$

which is a dimension four insertion as follows from Lowenstein's action principle [3]. Hence

$$[\Delta_\lambda - (\partial\rho/\partial\lambda)\Delta_1{}^0, \mathscr{S}] = 0 \tag{126}$$

thus

$$\Delta_\lambda = (\partial\rho/\partial\lambda)\, \Delta_1{}^0 + \Delta_\lambda{}^s \tag{127}$$

where $\Delta_\lambda{}^s$ is a dimension four-symmetric insertion. We are thus left with finding a basis of symmetrical insertions. Since we know that, given the Slavnov identity $\mathscr{L}_{\text{eff}}$ depends on nine parameters, namely four to specify the couplings with the external fields and five to specify the remaining part of the Lagrangian, there are nine independent symmetrical insertions.

We shall first construct the four missing unphysical insertions. The method consists in constructing insertions which are relized by differential operators as a consequence of the action principle, and study their commutators with $\mathscr{S}$.

First consider [11], [1]

$$\begin{aligned} Q_\pi{}^\epsilon &= \frac{\hbar^2\rho}{\bar{\kappa}} \int dx\, \delta_{\xi\alpha(x+\epsilon)} \delta_{J_\alpha(x)} \\ Q_{a_L}^\epsilon &= \frac{\hbar^2}{\bar{\kappa}} \int dx\, \delta_{\xi\alpha(x+\epsilon)} \partial_\mu \delta_{J_{\alpha\mu}(x)} \end{aligned} \tag{128}$$

and define $\Delta_i{}^\epsilon$, $i = (\pi, a_L)$, by

$$\Delta_i{}^\epsilon Z = :\mathscr{S} Q_i{}^\epsilon:\, Z \tag{129}$$

where as usual the dots indicate the substraction of the disconnected part. One has:

$$\begin{aligned} [\Delta_i{}^\epsilon, \mathscr{S}]\, Z &= -\mathscr{S}^2 Q_i{}^\epsilon Z = -[\mathscr{S}^2, Q_i{}^\epsilon]\, Z \\ &= \begin{cases} \rho \int dx\, \xi^\alpha(x+\epsilon)\, \delta_{J\alpha(x)} Z, & i = \pi, \\ \int dx\, \xi^\alpha(x+\epsilon)\, \partial^\mu \delta_{J\alpha\mu(x)} Z, & i = a_L. \end{cases} \end{aligned} \tag{130}$$

These commutators have obviously finite limits as $\epsilon \to 0$. Consequently the infinite part of Δ_i^ϵ is symmetric. Substracting the infinite parts, we get in the limit $\epsilon \to 0$ some Δ_i's ($i = \pi, a_L$) such that

$$[\Delta_i, \mathscr{S}]\, Z = \begin{cases} \rho \int dx\, \xi^\alpha(x)\, \delta_{J\alpha(x)} Z, & i = \pi, \\ \int dx\, \xi^\alpha(x)\, \partial^\mu \delta_{J\alpha\mu(x)} Z, & i = a_L . \end{cases} \tag{131}$$

Furthermore

$$\begin{aligned} &\Sigma_{\text{phys}} \Delta_i^\epsilon Z \,|_{\mathscr{J}=0} \\ &\quad = \Sigma_{\text{phys}} :\mathscr{S} Q_i^\epsilon: Z \,|_{\mathscr{J}=0} \\ &\quad = \Sigma_{\text{phys}} \int dx\, [J^0(x)\, \delta_{\gamma^0(x)} + J^{\alpha\mu}(x)\, \delta_{\gamma^{\alpha\mu}(x)}]\, Q_i^\epsilon Z \,|_{\mathscr{J}=0} \\ &\quad = \Sigma_{\text{phys}} \int dp\, [\tilde{J}^0(p)\, \tilde{\Gamma}^\epsilon_{0i}(p)\, \delta_{\tilde{J}^0(p)} + \tilde{J}^{\alpha\mu}(p)\, \tilde{\Gamma}^\epsilon_{a,i}(p)\, \delta_{\tilde{J}\alpha\mu(p)}]\, Z \,|_{\mathscr{J}=0} \\ &\quad = \Sigma_{\text{phys}} \int dx\, [\tilde{\Gamma}^\epsilon_{0,i}(M^2)\, J^0(x)\, \delta_{J^0(x)} + \tilde{\Gamma}^\epsilon_{a,i}(m_a^2)\, J^{\alpha\mu}(x)\, \delta_{J\alpha\mu(x)}]\, Z \,|_{\mathscr{J}=0} \end{aligned} \tag{132}$$

where

$$\begin{aligned} \tilde{\Gamma}^\epsilon_{0,i}(p^2) &= \delta_{\tilde{\sigma}(p)} \delta_{\tilde{\gamma}^0(-p)} Q_i^\epsilon \Gamma \,|_{\Psi=\eta=0} \\ \tilde{\Gamma}^\epsilon_{a,i}(p^2) &= \tfrac{1}{3}(g^{\mu\nu} - p^\mu p^\nu / p^2)\, \delta_{\tilde{a}_\alpha{}^\mu(p)} \delta_{\tilde{\gamma}^{\alpha\nu}(-p)} Q_i^\epsilon \Gamma \,|_{\Psi=\eta=0} . \end{aligned} \tag{133}$$

Now let

$$\Delta_i^{0\epsilon} = \Delta_i^\epsilon - \tilde{\Gamma}^\epsilon_{0i}(M^2)\, \Delta_\sigma^\epsilon - \tilde{\Gamma}^\epsilon_{ai}(m_a^2)\, \Delta_a^s \tag{134}$$

where the symmetrical insertions

$$\begin{aligned} \Delta_\sigma^s &= \int dx\, [J^0(x)\, \delta_{J^0(x)} + \gamma^0(x)\, \delta_{\gamma^0(x)}], \\ \Delta_a^s &= \int dx\, [J^{\alpha\mu}(x)\, \delta_{J\alpha\mu(x)} + \gamma^{\alpha\mu}(x)\, \delta_{\gamma^{\alpha\mu}(x)} + \xi^\alpha(x)\, \delta_{\xi\alpha(x)} \\ &\qquad + J^\alpha(x)\, \delta_{J\alpha(x)} + \gamma^\alpha(x)\, \delta_{\gamma^\alpha(x)}], \end{aligned} \tag{135}$$

respectively coincide with $\int dx\, J^0(x)\, \delta_{J^0(x)}$ and $\int dx\, J^{\alpha\mu}_{(x)} \delta_{J^{\alpha\mu}_{(x)}}$ under application of Σ_{phys}, and restriction at $\mathscr{J} = 0$. After what was seen before the finite part Δ_i^0

of $\Delta_i^{0\epsilon}$ is nonphysical and satisfies Eq. (131). From these commutation relations one can check that

$$\begin{aligned} \Delta_1^{0,s} &= \int dx\, [\bar{\xi}^\alpha(x)\, \delta_{\xi^\alpha(x)}] - \Delta_\pi{}^0 - \Delta_a{}^0, \\ \Delta_2^{0,s} &= \int dx\, [\gamma^\alpha(x)\, \delta_{\gamma^\alpha(x)} + J^\alpha(x)\, \delta_{J^\alpha(x)}] + \Delta_\pi{}^0, \end{aligned} \tag{136}$$

are symmetrical, (and obviously nonphysical). In the same way one can check that

$$\Delta_3^{0s} = \int dx\, [\xi^\alpha(x)\, \delta_{\xi^\alpha(x)} + \zeta^\alpha(x)\, \delta_{\xi^\alpha(x)}] \tag{137}$$

is symmetrical (obviously nonphysical). Finally Δ_4^{0s} defined by

$$\Delta_4^{0s} Z = \int dx\, J^0(x)\, Z \tag{138}$$

which can be realized by performing a σ field tranlation, is certainly symmetrical and nonphysical. These four symmetrical nonphysical insertions are independent as can be checked in the tree approximation (Δ_4^{0s} has a σ^3 term which the others do not have; Δ_2^{0s} has a π^4 term, not contained in the others; $\Delta_1^{0,s}$ has a $\mathfrak{G}^\alpha \mathfrak{G}_\alpha$ term which is not in $\Delta_3^{0,s}$). To complete the basis of symmetrical insertions there remains to find five others which are independent on the physical normalization points. According to the general argument, i.e., the implicit function theorem for formal power series [1], we know that there are four of them whose tree approximations are the four terms of $\mathscr{L}_{\text{inv.}}$ (98). A fifth one is $\Delta_a{}^s$. Their independence on the normalization points, i.e., Eq. (120) can be decided at the tree level and can indeed be verified. This completes the proof.

D. *Radiative Corrections* : *Unitarity of the Physical S Operator*

Usually, the unitarity of the S operator follows from the relationship between time ordered and antitime ordered products, together with the hermiticity of the Lagrangian. Here we are investigating the unitarity of the physical S operator so that we have to show the cancellation of ghosts in intermediate states. Furthermore the Lagrangian is not hermitian. We shall however show that the unitarity of the S operator can be derived from the usual relation between time ordered and antitime ordered products, thanks to an additional symmetry property of the Lagrangian, and the Slavnov identity. Together with

$$Z(\mathscr{J}) = \left\langle T \exp \left\{ \frac{i}{\hbar} \int dx\, [\mathscr{L}_{\text{eff}}^{\text{int}}(x) + \mathscr{J}(x)\, \Psi(x)] \right\} \right\rangle \tag{139}$$

we introduce

$$\bar{Z}(\mathscr{J}) = \left\langle \bar{T} \exp \left\{ -\frac{i}{\hbar} \int dx\, [\mathscr{L}_{\text{eff}}^{\text{int}}(x) + \mathscr{J}(x)\, \Psi(x)] \right\} \right\rangle \tag{140}$$

where T and $\bar{T}$ respectively denote time ordered and antitime ordered products. We first define the S operator in the full Fock space (including the ghosts) through the LSZ formula:

$$S = \Sigma Z(\mathscr{J})|_{\mathscr{J}=0} \tag{141}$$

where $\Sigma = \exp \int dx\, dy\, \Psi^{\text{in}}_{i(z)} K^{ij}_{zy} \delta_{J^j_{(x)}}$: which is the straightforward extension of S_{phys}. One has:

$$\Sigma Z(\mathscr{J}) \cdot \Sigma \bar{Z}(\mathscr{J}) = 0 \tag{142}$$

and if the Lagrangian is hermitian,

$$\Sigma \bar{Z}(\mathscr{J}) = (\Sigma Z(\mathscr{J}))^{\dagger} \tag{143}$$

Let now E_0 be the projector on the bare physical subspace. By application of Wick's theorem one obtains:

$$\Sigma_{\text{phys}}\, Z(\mathscr{J}) \exp \mathscr{A} \Sigma_{\text{phys}}\, \bar{Z}(\mathscr{J}) = E_0 \tag{144}$$

where

$$\mathscr{A} = i\hbar \int dx\, [\overleftarrow{\delta}_{\mathscr{J}^g} \overleftarrow{K}^g S_+{}^g * K^g \delta_{\mathscr{J}^g}](x) \tag{145}$$

the index g indicating that the summation is restricted to the ghost fields, and $i\hbar S_+{}^g$ being the positive frequency part of the asymptotic ghost field commutators.

We are now first going to see that

$$[Z(\mathscr{J})]^{\dagger} = \tilde{\mathscr{C}} \bar{Z}(\mathscr{J}) \tag{146}$$

where $\mathscr{J}^{i\dagger}$ is the coefficient of $\Psi_i{}^{\dagger}$ in $\mathscr{J} \cdot \Psi$,

$$\begin{aligned} \tilde{\mathscr{C}} \bar{\xi} &= \xi, \\ \tilde{\mathscr{C}} \xi &= -\bar{\xi}, \\ \tilde{\mathscr{C}} \mathscr{J} &= \mathscr{J}^{\dagger} \text{ otherwise.} \end{aligned} \tag{147}$$

This is a consequence of the corresponding property at the Lagrangian level:

$$\mathscr{L}^{\dagger}_{\text{eff}} = \mathscr{C} \mathscr{L}_{\text{eff}} \tag{148}$$

where

$$\begin{aligned} \mathscr{C} C &= \bar{C}, \\ \mathscr{C} \bar{C} &= -C, \\ \mathscr{C} \eta &= \eta^{\dagger} \quad \text{(all classical fields)}, \\ \mathscr{C} \psi &= \psi^{\dagger} \quad \text{otherwise.} \end{aligned} \tag{149}$$

This property is due to the fact that the $\mathscr{C}$ operation and the hermitian conjugation transform the Slavnov identity in the same way, leave the normalization conditions unchanged (the $\mathscr{C}$ and † operations are defined in a natural way on Γ), and that the theory is uniquely defined by the normalization conditions and the Slavnov identity (101), (102), (103), (104), (105), (96).

Thus Σ_{phys} is hermitian since the asymptotic Lagrangian is hermitian, which follows from (146), hence (142) can be rewritten

$$\Sigma_{\text{phys}}\, Z(\mathscr{J})\, [\exp \mathscr{A}^{\mathscr{C}}]\, [\Sigma_{\text{phys}}\, Z(\mathscr{J})]^{\dagger} = E_0 \tag{150}$$

where

$$\mathscr{A}^{c} = i\hbar \int dx\, [\overleftarrow{\delta}_{\mathscr{J}^g} \overleftarrow{K}^g S_+{}^g * \overrightarrow{K}^g \delta_{\tilde{\mathscr{J}}^g}](x) \tag{151}$$

with

$$\tilde{\mathscr{J}}^g = \mathscr{C}\mathscr{J}^g \tag{152}$$

Let us consider

$$U(\lambda) = \exp \lambda \mathscr{A}^{\mathscr{C}} \tag{153}$$

Then

$$\partial_\lambda S_{\text{phys}}(\mathscr{J})\, U(\lambda)\, S^{\dagger}_{\text{phys}}(\mathscr{J})|_{\mathscr{J}=0} = S_{\text{phys}}(\mathscr{J})\, \mathscr{A}^{C} U(\lambda)\, S^{\dagger}_{\text{phys}}(\mathscr{J})|_{\mathscr{J}=0} \tag{154}$$

Introducing

$$\mathfrak{G}_\alpha = \frac{\rho \pi_a - \partial^\mu a_{\alpha\mu}}{2\rho \Gamma_\gamma(m^2_{\phi\pi})} \tag{155}$$

whose variation under a Slavnov transformation reduces on mass shell to $\bar{C}_\alpha$, and using the restricted 't Hooft gauge [22] defined by

$$\Gamma_{a_L \pi}(m^2_{\phi\pi}) = 0 \tag{156}$$

in which the ghost propagators have only simple poles [23], allows to rewrite

$$\begin{aligned}\mathscr{A}^{\mathscr{C}} = i\hbar \int dx\, [&\overleftarrow{\delta}_{\bar{\mathfrak{G}}}(\overleftarrow{K}S_+ * \overrightarrow{K})_{\bar{\mathfrak{G}}\mathfrak{G}}\, \overrightarrow{\delta}_{\mathfrak{G}} + \overleftarrow{\delta}_{\mathfrak{G}}(\overleftarrow{K}S_+ * \overrightarrow{K})_{\mathfrak{G}\bar{\mathfrak{G}}}\, \overrightarrow{\delta}_{\bar{\mathfrak{G}}} \\ &+ \overleftarrow{\delta}_{\mathfrak{G}}(\overleftarrow{K}S_+ * \overrightarrow{K})_{\mathfrak{G}\mathfrak{G}}\, \overrightarrow{\delta}_{\mathfrak{G}} + \overleftarrow{\delta}_{\bar{\xi}}(\overleftarrow{K}S_+ * \overrightarrow{K})_{C\bar{C}}\, \overrightarrow{\delta}_{\xi} + \overleftarrow{\delta}_{\xi}(\overleftarrow{K}S_+ * \overrightarrow{K})_{\bar{C}C}\, \overrightarrow{\delta}_{\bar{\xi}}](x).\end{aligned} \tag{157}$$

Now making use of the Slavnov identity on the ghost mass shell and of the vanishing source restriction allows [1] to reduce (154), (157) to

$$\begin{aligned}&\partial_\lambda S_{\text{phys}}(\mathscr{J})\, U(\lambda)\, S^{\dagger}_{\text{phys}}(\mathscr{J})|_{\mathscr{J}=0} \\ &\quad = S_{\text{phys}}(\mathscr{J}) \left[i\hbar \int dx\, \{\overleftarrow{\delta}_{\mathfrak{G}}(\overleftarrow{K}S_+ * \overrightarrow{K})_{\mathfrak{G}\mathfrak{G}}\, \overrightarrow{\delta}_{\mathfrak{G}}\}(x)\right] U(\lambda)\, S^{\dagger}_{\text{phys}}(\mathscr{J})|_{\mathscr{J}=0}\end{aligned} \tag{158}$$

which upon integration with respect λ leads to:

$$S_{\text{phys}}(\mathscr{J})\, U(1)\, S^{\dagger}_{\text{phys}}(\mathscr{J})|_{\mathscr{J}=0}$$
$$= S_{\text{phys}}(\mathscr{J}) \exp \left\{ i\hbar \int dx\, [\vec{\delta}_{\mathfrak{G}} (\vec{K} S_{+} * \vec{K})_{\mathfrak{G}\mathfrak{G}}\, \vec{\delta}_{\mathfrak{G}}](x) \right\} S^{\dagger}_{\text{phys}}(\mathscr{J})|_{\mathscr{J}=0} . \tag{159}$$

Since the expectation value between physical states of the time ordered product of an arbitrary number of gauge operators is disconnected [21]

$$S_{\text{phys}} S^{\dagger}_{\text{phys}} = E_0 , \tag{160}$$

i.e.,

$$E_0 S E_0\, S^{\dagger} E_0 = E_0 \tag{161}$$

Similarly, one can prove that:

$$E_0 S^{\dagger} E_0\, S E_0 = E_0 \tag{162}$$

which shows that the physical S operator $E_0 S E_0$ obeys perturbative unitarity.

Conclusion

Gauge theories can be characterized by the fulfillment of Slavnov identities when the underlying Lie algebra is semisimple, i.e., is sufficiently rigid against perturbations. Then, simple power counting arguments are sufficient to prove that indeed Slavnov identities can be fulfilled, in the absence of Adler Bardeen anomalies.

In particular, we have shown, on the SU2 Higgs Kibble model whose particle interpretation can be completely analyzed that the gauge independence and unitarity of the physical S-operator follow.

It is believed that both the lack of rigidity of the underlying Lie algebra and the possible occurrence of Adler Bardeen anomalies can only be mastered by more sophisticated tools based on a closer analysis of the consistency conditions involving the behaviour of the theory under dilatations [7, 20].

The analysis of gauge independent local operators, although not touched upon here [1] should also be tractable in terms of the methods used here.

APPENDIX A: Cohomology of Lie Algebras

This appendix is a brief summary of definitions and results needed here which are not easily found in classical text books [24].

DEFINITION. Let $\mathfrak{h}$ be a Lie algebra with structure constants $f^{\alpha\beta}_{\gamma}$, which is the sum of a semisimple algebra $\mathscr{S}$ and an abelian algebra $\mathscr{A}$. A cochain of order n with value in a representation space V on which $\mathfrak{h}$ acts through a completely reduced representation:

$$\mathfrak{h} \ni h^{\alpha} \to t^{\alpha}$$

is a totally antisymmetric tensor built on $\mathfrak{h}$ whose components

$$\Gamma^{\alpha_1 \cdots \alpha_n}$$

are elements of V. The set of such cochains is called $C^n(V)$. We define the coboundary operator d^n:

$$C^n(V) \xrightarrow{d^n} C^{n+1}(V) \tag{A1}$$

by:

$$(d^n \Gamma)^{\alpha_1 \cdots \alpha_{n+1}} = \sum_{k=1}^{n+1} (-)^{k-1}\, t^{\alpha_k} \Gamma^{\alpha_1 \cdots \hat{\alpha}_k \cdots \alpha_{n+1}} + \sum_{k<l=1}^{n+1} (-)^{k+l}\, f^{\alpha_k \alpha_l}_{\lambda} \Gamma^{\lambda \alpha_1 \cdots \hat{\alpha}_k \cdots \hat{\alpha}_l \cdots \alpha_{n+1}} \tag{A2}$$

in which capped indices are to be omitted. The fundamental property is:

$$d^{n+1} \circ d^n = 0 \tag{A3}$$

a consequence of the commutation relations

$$[t^{\alpha}, t^{\beta}] = f^{\alpha\beta}_{\gamma} t^{\gamma} \tag{A4}$$

and the Jacobi identities.

An element Γ of $C^n(V)$ is called a cocycle if

$$d^n \Gamma = 0. \tag{A5}$$

The set of cocycles is denoted $Z^n(V)$.

An element Γ of $C^n(V)$ is called a coboundary if

$$\Gamma = d^{n-1} \hat{\Gamma} \tag{A6}$$

for some $\hat{\Gamma} \in C^{n-1}(V)$.

Obviously every coboundary is a cocycle (cf. Eq. (A3)). The converse is not always true. However in the present case where the representation is fully reduced, the parametrization of all coboundaries can be found as follows:

We first split Γ into an invariant and a non invariant part:

$$\Gamma = \Gamma_{\natural} + \Gamma_{\flat} \tag{A7}$$

such that

$$t^2\Gamma_\natural = 0, \qquad t^2\Gamma_\flat \neq 0.$$

Here we have defined

$$t^2 = t^\alpha t_\alpha \tag{A8}$$

where indices are raised and lowered by means of a non degenerate invariant symmetrical tensor.

The restriction of Eq. (A5) to $\Gamma_\flat$ yields through multiplication by t_{α_1} and summation over α_1:

$$\Gamma_\flat = d^{n-1}\hat{\Gamma} \tag{A9}$$

where

$$(\hat{\Gamma})^{\alpha_1\cdots\alpha_{n-1}} = (t_\alpha/t^2)\, \Gamma_\flat^{\alpha\alpha_1\cdots\alpha_{n-1}}. \tag{A10}$$

(Use has been made of the commutation rules between the t^α's.)

Next we look at $\Gamma_\natural$. The cocycle condition reduces to:

$$\sum_{k<l=1}^{n+1} (-)^{k+l} f_\lambda^{\alpha_k\alpha_l}\Gamma_\natural^{\lambda\alpha_1\cdots\hat{\alpha}_k\cdots\hat{\alpha}_l\cdots\alpha_{n+1}} = 0. \tag{A11}$$

We define the two operations

$$\begin{gathered} {}^nI^\alpha\colon C^n(V)\to C^{n-1}(V)\colon \\ ({}^nI^\rho\Gamma)^{\alpha_1\cdots\alpha_{n-1}} = \Gamma^{\rho\alpha_1\cdots\alpha_{n-1}} \end{gathered} \tag{A12}$$

and

$$\begin{gathered} {}^n\theta^\rho\colon C^n(V)\to C^n(V) \\ ({}^n\theta^\rho\Gamma)^{\alpha_1\cdots\alpha_n} = t^\rho\Gamma^{\alpha_1\cdots\alpha_n} + \sum_{k=1}^{n} (-)^{k-1} f_\lambda^{\rho\alpha_k}\Gamma^{\lambda\alpha_1\cdots\hat{\alpha}_k\cdots\alpha_n} \end{gathered} \tag{A13}$$

(θ^ρ transforms Γ according to the sum of n adjoint representations of $\mathfrak{h}$ and t^ρ).

One can check that

$${}^{n+1}I^\rho \circ d^n + d^{n-1} \circ {}^nI^\rho = {}^n\theta^\rho \tag{A14}$$

and furthermore

$$d^n \circ {}^n\theta^\rho = {}^{n+1}\theta^\rho \circ d^n. \tag{A15}$$

Thus if $d^n\Gamma_\natural = 0$

$${}^n\theta^\rho\Gamma_\natural = d^{n-1} \circ {}^nI^\rho\Gamma_\natural \tag{A16}$$

Hence

$$({}^n\theta)^2 \Gamma_\natural = d^{n-1} \circ {}^{n-1}\theta_\rho \circ {}^n I^\rho \Gamma_\natural \, . \tag{A17}$$

Reducing the antisymmetrized product of n adjoint representations according to

$$\Gamma_\natural = \Gamma_\natural{}^\natural + \Gamma_\natural{}^\flat \tag{A18}$$

with

$$({}^n\theta)^2 \Gamma_\natural{}^\natural = 0, \qquad ({}^n\theta)^2 \Gamma_\natural{}^\flat \neq 0$$

we find that

$$\Gamma_\natural{}^\flat = [1/({}^{n-1}\theta)^2] \, d^{n-1} \circ {}^{n-1}\theta_\rho \circ {}^n I^\rho \Gamma_\natural{}^\flat$$

and $\Gamma_\natural{}^\natural$ is arbitrary.

To sum up, in the case of a completely reduced representation and a Lie algebra which is the sum of a semisimple and an abelian algebra, every cocycle is a coboundary up to totally invariant cochains. When $\mathfrak{h}$ is semisimple however there is no such cochain for $n = 1, 2$. (In this last case, the invariance which implies the cocycle condition and the non degeneracy of the Killing form yield the result.)

Appendix B

We want to show here that the equation:

$$\begin{aligned} & s^\Delta(\gamma^i(x)\, \mathring{P}_i(x) + \zeta^\alpha(x)\, \mathring{P}_\alpha(x)) + s(\gamma^i(x)\, \Delta_i(x) + \zeta^\alpha(x)\, \Delta_\alpha(x)) \\ & \quad \equiv \gamma^i(x)[(\Delta_j(y)\, \delta_{\Phi_j(y)} - \Delta_\alpha(y)\, \delta_{\bar{C}_\alpha(y)})\, \mathring{P}_i(x) \\ & \qquad - (\mathring{P}_j(y)\, \delta_{\Phi_j(y)} + \mathring{P}_\alpha(y)\, \delta_{\bar{C}_\alpha(y)})\, \Delta_i(x)] \\ & \qquad + \zeta^\alpha(x)[-\Delta_\beta(y)\, \delta_{\bar{C}_\beta(y)} \mathring{P}_\alpha(x) + \mathring{P}_\beta(y)\, \delta_{\bar{C}_\beta(y)} \Delta_\alpha(x)] = 0 \end{aligned} \tag{B1}$$

implies the structure:

$$\begin{aligned} & \gamma^i(x)\, \Delta_i(x) + \zeta^\alpha(x)\, \Delta_\alpha(x) \\ & \quad = -[s^\pi(\gamma^i(x)\, \mathring{P}_i(x) + \zeta^\alpha(x)\, \mathring{P}_\alpha(x)) + s(\gamma^i(x)\, \Pi_i(x) + \xi^\alpha(x)\, \Pi_\alpha(x))] \\ & \quad \equiv \gamma^i(x)[(\Pi_j(y)\, \delta_{\Phi_j(y)} + \Pi_\alpha(y)\, \delta_{\bar{C}_\alpha(y)})\, \mathring{P}_i(x) + (\mathring{P}_j(y)\, \delta_{\Phi_j(y)} + \mathring{P}_\alpha(y)\, \delta_{\bar{C}_\alpha(y)})\, \Pi_i(x)] \\ & \qquad - \zeta^\alpha(x)(\Pi_\beta(y)\, \delta_{\bar{C}_\beta(y)} \mathring{P}_\alpha(x) + \mathring{P}_\beta(y)\, \delta_{\bar{C}_\beta(y)} \Pi_\alpha(x)). \end{aligned} \tag{B2}$$

From this result it follows that Eq. (59) which is a perturbation of order $\hbar\Delta$ of Eq. (B2) is a consequence of the consistency condition Eq. (58) which is a perturbation of the same order of Eq. (B1).

Let us recall the notations

$$\mathring{P}_i(x) = t_{ij}^{\alpha}[\Phi_j \bar{C}_\alpha](x) + q_i{}^{\alpha}\bar{C}_\alpha(x) \tag{B3}$$

where:

$$q_a{}^{\alpha} = t_{ab}^{\alpha} F_b\,,$$

$$q_{\beta,\mu}^{\alpha} = q_\beta{}^{\alpha}\partial_\mu\,. \tag{B4}$$

Since $q_\beta{}^\alpha$ is an invariant tensor it can always be chosen of the form:

$$q_\beta{}^\alpha = q\delta_\beta{}^\alpha \tag{B5}$$

Also,

$$\mathring{P}_\alpha(x) = \tfrac{1}{2} f_{\alpha}^{\beta\gamma}(\bar{C}_\beta \bar{C}_\gamma)(x) \tag{B6}$$

Δ_i, Δ_α, Π_i, Π_α are given by

$$\Delta_i(x) = \tfrac{1}{2}[\Theta_{ij}^{\alpha\beta}\Phi_j \bar{C}_\alpha \bar{C}_\beta + \bar{C}_\alpha \Sigma_i^{\alpha\beta}\bar{C}_\beta](x),$$

$$\Delta_\alpha(x) = -\tfrac{1}{6}\Gamma_{\alpha}^{\beta\gamma\delta}(\bar{C}_\beta \bar{C}_\gamma \bar{C}_\delta)(x),$$

$$\Pi_i(x) = [\hat{\Theta}_{ij}^{\alpha}\Phi_j \bar{C}_\alpha + \hat{\Sigma}_i^{\alpha}\,\bar{C}_\alpha](x), \tag{B7}$$

$$\Pi_\alpha(x) = \tfrac{1}{2}\hat{\Gamma}_{\alpha}^{\beta\gamma}(\bar{C}_\beta \bar{C}_\gamma)(x),$$

with

$$\Sigma_{\gamma,\mu}^{\alpha\beta} = \Sigma_\gamma^{\alpha,\beta}\partial_\mu\,,$$

$$\Sigma_{\gamma,\mu}^{\alpha} = \Sigma_\gamma{}^{\alpha}\partial_\mu\,. \tag{B8}$$

Following these definitions, we shall reduce (B1) and (B2) to c numbers. First, substituting Eqs. (B3), (B6), and Eq. (B7) into Eq. (B1) yields:

$$\begin{aligned}
&\tfrac{1}{2}\gamma^i(x)[t_{ik}^{\alpha}\Theta_{kj}^{\beta\gamma} - \Theta_{ik}^{\alpha\beta}t_{kj}^{\gamma} + \tfrac{1}{3}\Gamma_{\lambda}^{\alpha\beta\gamma}t_{ij}^{\lambda} - f_{\lambda}^{\alpha\beta}\Theta_{ij}^{\lambda\gamma}](\phi_j \bar{C}_\alpha \bar{C}_\beta \bar{C}_\gamma)(x) \\
&\quad + \tfrac{1}{2}\gamma^a(x)[t_{ab}^{\alpha}\Sigma_b^{\beta\gamma} - \Theta_{ab}^{\alpha\beta}q_b{}^{\gamma} + \tfrac{1}{3}\Gamma_{\lambda}^{\alpha\beta\gamma}q_a{}^{\lambda} - f_{\lambda}^{\alpha\beta}\Sigma_a^{\lambda\gamma}](\bar{C}_\alpha \bar{C}_\beta \bar{C}_\gamma)(x) + \cdots \\
&\quad + \tfrac{1}{2}(\gamma^{\eta,\mu}\bar{C}_\alpha \bar{C}_\beta)(x)[t_{\eta\zeta}^{\alpha}\Sigma_\zeta^{\beta,\gamma} - \Theta_{\eta\zeta}^{\alpha\beta}q_\zeta{}^{\gamma} + \Gamma_{\lambda}^{\alpha\beta\gamma}\,q_\eta^{\lambda} - \tfrac{1}{2}f_{\lambda}^{\alpha\beta}\Sigma_\eta^{\lambda,\gamma} + \Sigma_\eta^{\lambda,\gamma}f_{\lambda}^{\beta\gamma}]\,\partial_\mu \bar{C}_\gamma(x) \\
&\quad + \zeta^\eta(x)[-\tfrac{1}{6}\Gamma_{\lambda}^{\alpha\beta\gamma}f_{\eta}^{\lambda\delta} + \tfrac{1}{4}f_{\lambda}^{\alpha\beta}\Gamma_{\eta}^{\lambda\gamma\delta}](\bar{C}_\alpha \bar{C}^\beta \bar{C}_\gamma \bar{C}_\delta)(x) = 0
\end{aligned} \tag{B9}$$

which in terms of the coefficients writes:

$$-\frac{1}{24}\left[\sum_{i=1}^{4}(-)^{i+1} f_{\eta}^{\alpha_i\lambda}\Gamma_{\lambda}^{\alpha_1\cdots\hat{\alpha}_i\cdots\alpha_4}+\sum_{l<k=1}^{4}(-)^{l+k} f_{\lambda}^{\alpha_l\alpha_k}\Gamma_{\eta}^{\lambda\alpha_1\cdots\hat{\alpha}_l\cdots\hat{\alpha}_k\cdots\alpha_4}\right]=0 \quad \text{(B10)}$$

$$\frac{1}{6}\left[\sum_{i=1}^{3}(-)^{i+1}[t^{\alpha_i},\Theta^{\alpha_1\cdots\hat{\alpha}_i\cdots\alpha_3}]_{ij}+\sum_{l<k=1}^{3}(-)^{l+k} f_{\lambda}^{\alpha_l\alpha_k}\Theta_{ij}^{\alpha_1\cdots\hat{\alpha}_l\cdots\hat{\alpha}_k\cdots\alpha_3}+\Gamma_{\lambda}^{\alpha_1\alpha_2\alpha_3}t_{ij}^{\lambda}\right]=0 \quad \text{(B11)}$$

$$\frac{1}{6}\left[\sum_{i=1}^{3}(-)^{i+1} t_{ab}^{\alpha_i}\Sigma_{b}^{\alpha_1\cdots\hat{\alpha}_i\cdots\alpha_3}+\sum_{l<k=1}^{3}(-)^{l+k} f_{\lambda}^{\alpha_l\alpha_k}\Sigma_{a}^{\lambda\alpha_1\cdots\hat{\alpha}_l\cdots\hat{\alpha}_k\cdots\alpha_3}\right.$$

$$\left.-\sum_{i=1}^{3}(-)^{i+1}\Theta_{ab}^{\alpha_1\cdots\hat{\alpha}_i\cdots\alpha_3}t_{bc}^{\alpha_i}F_c+\Gamma_{\lambda}^{\alpha_1\alpha_i\alpha_3}t_{ab}^{\lambda}F_b\right]=0 \quad \text{(B12)}$$

where, for convenience, we have replaced α, β, γ, δ of Eq. (B9) by α_1, α_2, α_3, α_0. Finally,

$$\tfrac{1}{4}[(\tau^{\alpha}\Sigma^{\beta})_{\eta}^{\gamma}-(\tau^{\beta}\Sigma^{\alpha})_{\eta}^{\gamma}-f_{\lambda}^{\alpha\beta}\Sigma_{\eta}^{\lambda,\gamma}-2\Theta_{\eta\zeta}^{\alpha\beta}q_{\zeta}{}^{\gamma}+2\Gamma_{\lambda}^{\alpha\beta\gamma}q_{\eta}{}^{\lambda}]=0 \quad \text{(B13)}$$

where τ^{α} is defined by

$$(\tau^{\alpha}\Sigma^{\beta})_{\eta}^{\gamma}=t_{\eta\zeta}^{\alpha}\Sigma_{\zeta}^{\beta,\gamma}+f_{\lambda}^{\gamma\alpha}\Sigma_{\eta}^{\beta,\lambda}. \quad \text{(B14)}$$

In Eqs. (B10), (B12) we have followed the conventions introduced in Appendix A. Thus, taking into account the definition of the cocycle condition given in Appendix A and putting

$$\begin{aligned}(t^{\rho}\Gamma^{\alpha\beta\gamma})_{\lambda}&=f_{\lambda}^{\rho\zeta}\Gamma_{\zeta}^{\alpha\beta\gamma},\\(t^{\rho}\Theta^{\alpha\beta})_{ij}&=[t^{\rho},\Theta^{\alpha\beta}]_{ij},\\(t^{\rho}\Sigma^{\alpha\beta})_{a}&=t_{ab}^{\rho}\Sigma_{b}^{\alpha\beta},\\(t^{\rho}\Sigma^{\alpha})_{\eta}^{\gamma}&=(\tau^{\rho}\Sigma^{\alpha})_{\eta}^{\gamma},\end{aligned} \quad \text{(B15)}$$

we see that Eq. (B10), (B11), (B12) can be written as:

$$(d^{3}\Gamma)_{\lambda}^{\alpha_1\cdots\alpha_4}=0 \quad \text{(B16)}$$

$$(d^{2}\Theta)_{ij}^{\alpha_1\alpha_2\alpha_3}+\Gamma_{\lambda}^{\alpha_1\alpha_2\alpha_3}t_{ij}^{\lambda}=0 \quad \text{(B17)}$$

$$(d^{2}\Sigma)_{a}^{\alpha_1\alpha_2\alpha_3}-\sum_{i=1}^{3}(-)^{i+1}\Theta_{ab}^{\alpha_1\cdots\hat{\alpha}_i\cdots\alpha_3}t_{bc}^{\alpha_i}F_c+\Gamma_{\lambda}^{\alpha_1\alpha_2\alpha_3}t_{ab}^{\lambda}F_b=0. \quad \text{(B18)}$$

Comparing the coboundary operators which operate on $\Theta_{ab}^{\alpha\beta}$ and $\Sigma_a^{\alpha\beta}$ we see that:

$$(d^2\Theta)_{ab}^{\alpha_1\alpha_2\alpha_3} F_b = [d^2(\Theta F)]^{\alpha_1\alpha_2\alpha_3} - \sum_{i=1}^{3} (-)^{i+1} \Theta_{ab}^{\alpha_1\cdots\hat{\alpha}_i\cdots\alpha_3} t_{bc}^{\alpha_i} F_c \tag{B19}$$

Hence, taking into account Eq. (B17), Eq. (B18) assumes the form:

$$[d^2(\Sigma - \Theta F)]_a^{\alpha_1\alpha_2\alpha_3} = 0. \tag{B20}$$

Finally Eq. (B13) writes:

$$(d^1\Sigma)_\eta^{\alpha\beta,\gamma} - 2q(\Theta_\eta^{\alpha\beta,\gamma} - \Gamma_\eta^{\alpha\beta\gamma}) = 0. \tag{B21}$$

In much the same way as for Eq. (B1) expressing Eq. (B2) in terms of its coefficients yields:

$$\Gamma_\eta^{\alpha_1\alpha_2\alpha_3} = -(d^2\hat{\Gamma})_\eta^{\alpha_1\alpha_2\alpha_3}, \tag{B22}$$

$$\Theta_{ij}^{\alpha_1\alpha_2} = -[(d^1\hat{\Theta})_{ij}^{\alpha_1\alpha_2} - \hat{\Gamma}_\lambda^{\alpha_1\alpha_2} t_{ij}^\lambda], \tag{B23}$$

$$\begin{aligned}\Sigma_a^{\alpha_1\alpha_i} &= -[(d^1\{\hat{\Sigma} - \hat{\Theta}F\})_a^{\alpha_1\alpha_i} + (d^1\Theta)_{ab}^{\alpha_1\alpha_i} F_b - \hat{\Gamma}_\lambda^{\alpha_1\alpha_2} t_{ab}^\lambda F_b], \\ &= -[(d^1\{\hat{\Sigma} - \hat{\Theta}F\})_a^{\alpha_1\alpha_i} - \Theta_{ab}^{\alpha_1\alpha_2} F_b],\end{aligned} \tag{B24}$$

$$\Sigma_\eta^{\alpha,\beta} = -2[(\tau^\alpha\hat{\Sigma})_\eta^\beta + q(\hat{\Theta}_\eta^{\alpha,\beta} - \hat{\Gamma}_\eta^{\alpha\beta})]. \tag{B25}$$

We are now going to show that Eqs. (B22), (B23), (B24), (B25) are consequences of the results of Appendix A, and that consequently $\hat{\Gamma}$, $\hat{\Theta}$, $\hat{\Sigma}$ can be chosen linear in Γ, Θ, Σ.

First, Eq. (B22) is indeed a consequence of Eq. (B16) since the cochain Γ takes values in the adjoint representation of $\mathfrak{h}$ which is semisimple so that $\Gamma_\xi = 0$. Then, owing to the relation

$$[d^2(\hat{\Gamma}_\lambda t^\lambda)]_{ij}^{\alpha_1\alpha_2\alpha_3} = (d^2\hat{\Gamma})_\lambda^{\alpha_1\alpha_i\alpha_3} t_{ij}^\lambda = -\Gamma_\lambda^{\alpha_1\alpha_2\alpha_3} t_{ij}^\lambda. \tag{B26}$$

Equation (B17) becomes

$$[d^2(\Theta - \hat{\Gamma}_\lambda t^\lambda)]_{ij}^{\alpha_1\alpha_i\alpha_3} = 0. \tag{B27}$$

Since no antisymmetric invariant tensor of rank two on $\mathfrak{h}$ exists if $\mathfrak{h}$ is semisimple we see that Eq. (B27) and Eq. (B20) imply Eq. (B23) and Eq. (B24) respectively. Finally substituting Eq. (B23) into Eq. (B21) we get

$$(d^1\Sigma)_\eta^{\alpha\beta,\gamma} + 2q(d^1\hat{\Theta})_\eta^{\alpha\beta,\gamma} + 2q(\Gamma_\eta^{\alpha\beta\gamma} - \hat{\Gamma}_\lambda^{\alpha\beta} f_\eta^{\lambda\gamma}) = 0. \tag{B28}$$

Now considering $\hat{\Gamma}_\eta^{\alpha\gamma}$ as a cochain of order one with values in the tensor product of the adjoint representation with itself and applying the coboundary operator d we get:

$$(d^1\hat{\Gamma})_\eta^{\alpha\beta,\gamma} = f_\eta^{\alpha\lambda}\hat{\Gamma}_\lambda^{\beta\gamma} - f_\eta^{\beta\lambda}\hat{\Gamma}_\lambda^{\alpha\gamma} + f_\lambda^{\gamma\alpha}\hat{\Gamma}_\eta^{\beta\lambda} - f_\lambda^{\gamma\beta}\hat{\Gamma}_\eta^{\alpha\lambda} - f_\lambda^{\alpha\beta}\hat{\Gamma}_\eta^{\lambda\gamma}. \tag{B29}$$

Comparing with Eq. (B22),

$$-\Gamma_\eta^{\alpha\beta\gamma} = f_\eta^{\alpha\lambda}\hat{\Gamma}_\eta^{\beta\gamma} + f_\eta^{\beta\lambda}\hat{\Gamma}_\lambda^{\gamma\alpha} + f_\eta^{\gamma\lambda}\hat{\Gamma}_\lambda^{\alpha\beta} - f_\lambda^{\alpha\beta}\hat{\Gamma}_\eta^{\lambda\gamma} - f_\lambda^{\beta\gamma}\hat{\Gamma}_\eta^{\lambda\alpha} - f_\lambda^{\gamma\alpha}\hat{\Gamma}_\eta^{\beta\gamma}, \tag{B30}$$

we see that Eq. (B28) has the form

$$[d^1(\Sigma + 2q\{\hat{\Theta} - \hat{\Gamma}\})]_\eta^{\alpha\beta,\gamma} = 0. \tag{B31}$$

Since no invariant tensor of rank one on $\mathfrak{h}$ exists if $\mathfrak{h}$ is semisimple, Eq. (B31) can be solved according to

$$\Sigma_\eta^{\alpha,\beta} = -2(\tau^\alpha\hat{\Sigma})_\eta^\beta - 2q(\hat{\Theta}_\eta^{\alpha,\beta} - \hat{\Gamma}_\eta^{\alpha\beta}). \tag{B32}$$

The possibility of choosing $\hat{\Gamma}$, $\hat{\Sigma}$, $\hat{\Theta}$ linear in Γ, Θ, Σ stems from the explicit construction given in Appendix A.

Acknowledgments

Two of us C. B., A. R., wish to thank CNRS for the kind hospitality extended to them at the Centre de Physique Théorique, Marseille, where this work was started. A. R. wishes to thank C. E. A. for financial support and Prof. W. Zimmermann for the kind hospitality extended to him at the Max Planck Institut für Physik und Astrophysik, in Munich. Two of us, A. R., R. S., wish to acknowledge the collaboration of E. Tirapegui, L. Quaranta at a very early stage of this work. We wish to thank the Service de Physique théorique, CEN Saclay where some results quoted here were obtained. One of us, R. S. wishes to thank G. 't Hooft, B. W. Lee, C. H. Llewellynn Smith, M. Veltman, J. Zinn Justin for informative discussions on gauge theories. We wish to thank J. H. Lowenstein and W. Zimmermann, M. Bergère and Y. M. P. Lam, F. Jegerlehner, and B. Schroer, B. Zuber and H. Stern for keeping us currently informed on their work, prior to publication, as well as H. Epstein, V. Glaser, K. Symanzik, and J. Iliopoulos for their kind interest in this work.

References

1. C. Becchi, A. Rouet, and R. Stora, *Phys. Lett.* **52B** (1974), 344; Renormalization of the Abelian Higgs Kibble Model, *Commun. Math. Phys.*, to be published.
2. W. Zimmermann, *Ann. Phys.* **77** (1973), **77** (1973), 570.
3. J. H. Lowenstein, *Commun. Math. Phys.* **24** (1971), 1.
4. Y. M. P. Lam, *Phys. Rev.* **D6** (1972), 2145; **D7** (1973), 2943. M. Bergere and Y. M. P. Lam, to be published.

5. H. Epstein and V. Glaser, Ann. Inst. Henri Poincaré, XIX, no 3, p. 211 (1973); in "Statistical Mechanics and Quantum Field Theory," (C. de Witt, R. Stora, Ed.) Les Houches Summer School of Theoretical Physics, 1970, Gordon and Breach New York 1971; Colloquium on Renormalization Theory, CNRS, Centre de Physique Théorique, Marseille June 1971, CERN TH 1344, June 1971.
6. M. Gomes and J. H. Lowenstein, *Phys. Rev.* **D7** (1973), 550.
7. J. H. Lowenstein and B. Schroer, *Phys. Rev.* **D7** (1973), 1929; C. Becchi, *Commun. Math. Phys.* (1973), 33–97.
8. A complete bibliography about the theory of gauge fields can be found for instance in: E. S. Abers and B. V. Lee, *Phys. Reports* **9C** (Nov. 1973), no. 1; M. Veltman, Invited talk presented at the International Symposium on Electron and Photon Interactions at High Energies, Bonn, 27–31 August 1973 (The algebraic structure referred to here is however along the lines of [1].); The results of the present paper have been announced by C. Becchi, A. Rouet, R. Stora, CNRS Centre de Physique Théorique, Marseille, Colloquium on Recent Progress in Lagrangian Field Theory, June 1974 and summarized in J. Iliopoulos, Progress in gauge theories, Rapporteur's talk, 17th International Conference on High Energy Physics, London 1974. The symmetry property described in [1] is applied to traditional renormalization procedures in J. Zinn Justin, Lectures given at the International Summer Institute for Theoretical Physics, Bonn 1974.
9. L. C. Biedenharn, in Colloquium on Group Theoretical Methods in Physics, CNRS Marseille June 5–9 (1972) (where however the implications of renormalizability thanks to which all algebraic problems reduce to essentially finite dimensional ones, have not been investigated); C. H. Llewellynn Smith, *Phys. Lett.* **46B** (1973), 233, and private discussions; J. Wess and B. Zumino, *Phys. Lett.* **37B** (1971), 95.
10. S. L. Adler, *Phys. Rev.* **177**, 2426 (1969); "Lectures on Elementary Particles and Quantum Field Theory" (S. Deser, M. Grisaru, H. Pendleton Ed.), 1970 Brandeis University Summer Institute of Theoretical Physics, M.I.T. Press, Cambridge, Mass., 1970; W. Bardeen, *Phys. Rev.* **184** (1969), 1848.
11. J. H. Lowenstein and B. Schroer, *Phys. Rev.* **D6** (1972), 1553.
12. J. H. Lowenstein, to be published; C. Becchi, to be published; T. Clark and A. Rouet, to be published.
13. J. H. Lowenstein and W. Zimmermann, to be published; Ph. Blanchard and R. Seneor, CERN/TH 1420 preprint, Nov. 1971; CNRS Centre de Physique Théorique Marseille, Colloquium on Renormalization Theory, June 1971, and to be published in Ann. Inst. Henri Poincaré.
14. A. A. Slavnov, T M Ø **10** (1972), 153.
15. L. D. Faddeev and V. N. Popov, *Phys. Lett.* **25B** (1967), 29.
16. G. 't Hooft, *Nuclear Phys.* **B35** (1971), 167; G. 't Hooft and M. Veltman, CNRS Marseille, Colloquium on Yang Mills Fields, June 1972, CERN/TH 1571.
17. Séminaire Sophus Lie, École Normale Supérieure Paris 1954: Théorie des Algèbres de Lie. In [9], the relevance of cohomology theory in the present context can be detected.
18. G. 't Hooft and M. Veltman, CERN Yellow Report TH/73/9 "Diagrammar" (1973); *Nuclear Phys.* **B50** (1972), 318.
19. The treatment of non linear gauges would require the introduction of two more multiplets of external fields, one coupled to the renormalized gauge function, one coupled to the renormalized Slavnov variation of the gauge function, with appropriate dimensions and quantum numbers. Cf. Ref [1].
20. C. Becchi, A. Rouet, and R. Stora, Lectures given at the International School of Elementary Particle Physics, Basko Polje, Yugoslavia 14-29 Sept. 1974, and to be published.

21. Using the LSZ formula, it is enough to compute $\partial Z_c(\mathcal{J})/\partial\lambda$, $(\lambda = \kappa, m^2_{\phi\pi})$, and, from the Legendre transform structure of $Z_c(\mathcal{J})$, to show that the matrix elements of the gauge function between physical states vanish, as a consequence of the Slavnov identity. More generally, one can check that physical matrix elements of time ordered products of gauge functions are disconnected.
22. G. 'T HOOFT, *Nuclear Phys.* **B35** (1971), 167.
23. The treatment of double poles given in [1] can easily be adapted modulo the modifications due to the non hermiticity of the Lagrangian described here.
24. The main reference is: "Séminaire Sophus Lie 1. 1954/1955, "Théorie des Algèbres de Lie," École Normale Supérieure, Secrétariat Mathématique, 11, rue Pierre Curie, Paris 5e. A summary can be found in N. Bourbaki, Groupes et Algèbres de Lie, Chap. I, Sect. 3, Problem no. 12, Hermann, Paris, 1960.